Exploring the Solar System

Exploring the Solar System

Second Edition

Peter Bond

FRAS (Fellow of the Royal Astronomical Society)
FBIS (Fellow of the British Interplanetary Society)

WILEY Blackwell

This second edition first published 2020
© 2020 John Wiley & Sons Ltd.

Edition History
Wiley-Blackwell (1e, 2012)

Registered Office(s)
John Wiley & Sons, Inc., 111 River Street, Hoboken, NJ 07030, USA
John Wiley & Sons Ltd, The Atrium, Southern Gate, Chichester, West Sussex, PO19 8SQ, UK

Editorial Office
9600 Garsington Road, Oxford, OX4 2DQ, UK

For details of our global editorial offices, customer services, and more information about Wiley products visit us at www.wiley.com.

Wiley also publishes its books in a variety of electronic formats and by print-on-demand. Some content that appears in standard print versions of this book may not be available in other formats.

Library of Congress Cataloging-in-Publication Data

Name: Bond, Peter, 1948- author.
Title: Exploring the solar system / Peter Bond, FRAS (Fellow of the Royal
   Astronomical Society), FBIS (Fellow of the British Interplanetary
   Society).
Description: Second edition. | Hoboken, NJ : Wiley-Blackwell, 2020. |
   Includes index.
Identifiers: LCCN 2019058265 (print) | LCCN 2019058266 (ebook) | ISBN
   9781119384908 (paperback) | ISBN 9781119384892 (adobe pdf) | ISBN
   9781119384915 (epub)
Subjects: LCSH: Solar system.
Classification: LCC QB501 .B69 2020 (print) | LCC QB501 (ebook) | DDC
   523.2–dc23
LC record available at https://lccn.loc.gov/2019058265
LC ebook record available at https://lccn.loc.gov/2019058266

Cover Design: Wiley
Cover Image: © NASA

Set in 9/11.5pt, MinionPro by SPi Global, Chennai, India.

10 9 8 7 6 5 4 3 2 1

Dedication

To my wife Edna, with whom I have travelled the rocky road of life. Also to the next generation of space travellers: Holly, Jack, Wren, Willow, Pearl, and Olive.

Conversion Table

The metric units used in this book can be converted to English units by using the approximate conversions given below:

Length

1 kilometer = 0.62 of a mile
1 meter = 39.37 inches
1 centimeter = 0.39 inches
1 millimeter = 0.039 inches

Area

1 sq. kilometer (km^2) = 0.04 sq. miles
1 sq. meter (m^2) = 1.2 sq. yards
1 sq. centimeter (cm^2) = 0.155 sq. inches

Temperature

To convert ° Celsius to ° Fahrenheit, multiply ° C by 1.8 and add 32

Speed

1000 km/h = 277.8 m/s = 621.37 mph

Volume

1 cubic cm (cm^3) = 0.061 cubic inches

Contents

Introduction to the First Edition ix

Introduction to the Second Edition xi

About the Companion Website xii

1 Beginnings 1

2 Sun 25

3 Earth 57

4 The Moon 92

5 Mercury 121

6 Venus 146

7 Mars 173

8 Jupiter 231

9 Saturn 269

10 Uranus 314

11 Neptune 336

12 Pluto and the Kuiper Belt 354

13 Comets, Asteroids, and Meteorites 384

14 Exoplanets 438

Appendices 475

Glossary 500

Further Reading 506

Index 521

Introduction to the First Edition

"We shall not cease from exploration and the end of all our exploring will be to arrive where we started... and know the place for the first time." – T.S. Elliot

This book is about a unique corner of the Universe, a small expanse of largely empty space that surrounds an ordinary star in the suburbs of the Milky Way galaxy. Known as the Solar System, this region is populated by the Sun, eight planets, dozens of satellites and dwarf planets, and a multitude of smaller objects.

Why is it important to explore and understand the Solar System? Because the third planet from the Sun is our home: Earth is the only place yet discovered where living organisms and intelligent life exist, or have ever existed. This unique "Goldilocks" world is the cradle of humankind, a fragile oasis in the vastness of space.

However, spaceship Earth is subject to many threats and stresses. Some are human-made, such as deforestation, atmospheric pollution, or emissions of ozone-destroying chemicals. Some are natural planetary processes, such as crustal movement and changing sea level. Others are external, including solar flares and marauding asteroids.

As news reports of natural disasters constantly remind us, Earth is an ever-changing world, subject to ice ages, hurricanes, earthquakes, volcanic eruptions, and devastating cosmic impacts. Since its birth some 4.5 billion years ago, the planet has endured all of these natural forces to evolve into the largely benign place we see today. If we can understand how this evolution occurred, then we will have a better chance of predicting how it will change in the future.

This is where studies of the Sun, planets, and other inhabitants of the Solar System come to the fore. Only by comparing and contrasting the evolution of these very different objects can we hope to understand the past, present, and future of our Earth.

This scientific endeavor has been made possible by the advent of the Space Age. During this great age of discovery, modern technology has enabled us to construct automated spacecraft and robots that can act as surrogate explorers, venturing forth into the vast, hostile ocean of space to seek out and study new worlds.

Over more than half a century, hundreds of robotic spacecraft have been sent from Earth to examine at close quarters all of the planets, and many other objects, in our Solar System. This book is based on the flood of data sent back by these probes, which has enabled scientists to assemble, piece by piece, a realistic picture of our Solar System. For the first time, human eyes have been able to see towering cliffs, dust devils, erupting volcanoes, dry river beds and ice formations on dozens of distant worlds, most of them totally alien to our experience here on Earth.

Many years ago, my imagination was captured by books that described the family of alien worlds that circle our Sun, although, at that time, most of the information available was pure speculation. I have been fascinated by the many and varied members of the Solar System ever since. It is my hope that readers of this book will be similarly fascinated and inspired.

Exploring the Solar System has been written as an introductory text book for undergraduate students with a modest background in science. However, it is also intended to inform and inspire anyone who looks up at the night sky and wishes to know more about the alien worlds that inhabit our corner of the Universe.

After an introductory chapter which provides an overview of the Solar System, the book sets out to systematically describe the main characteristics of each major planet and its retinue of satellites, as well as the smaller members of the Sun's retinue. The final chapter enables the reader to compare and contrast our Solar System with systems around distant stars, where huge numbers of strange and exotic exoplanets are now being discovered.

Questions at the end of each chapter have been added to help students to recognize and comprehend the main points of each chapter, and to compare each planetary system. Useful reference material is provided in the form of numerous appendices, an extensive reading list, and a comprehensive glossary.

This book would not have been possible without the support and encouragement of Ian Francis, Senior Commissioning Editor for Wiley-Blackwell, and Delia Sandford, the Managing Editor for this project. I am most grateful for their patience and forbearance as the book has edged towards completion.

My sincere thanks also go to Kelvin Matthews of Wiley-Blackwell, who has checked all of the illustrations, to the production team, especially Kathy Syplywczak, and to the various reviewers whose helpful comments and criticisms played such an important role in shaping the final text.

Much of the information in this book is based on original scientific papers, many of which are listed in the final pages. Numerous other sources – many now available on the Internet – were also used, including magazine articles, press releases, and other information provided by space agencies – particularly NASA - and universities. I am also very grateful to everyone who helped me to obtain, or provided me with, the spectacular images that illuminate this story of outreach and discovery.

Finally, I would like to thank my wife, Edna, who first encouraged me to describe and explain the wonders of our Solar System.

Peter Bond

Introduction to the Second Edition

Eight years after the first edition of this book was published, I am delighted to introduce a second edition. Although the structure of the book has not changed, the contents have been considerably revised and updated to reflect the flood of new information sent back by our robotic explorers.

The list of landmark events that have taken place since 2012 is impressive.

An entirely new book could be devoted to the discoveries Cassini made during its 13 years in orbit around beautiful Saturn. The completion of the Cassini mission in 2017 saw the first exploration of the gap between the inner ring and the planet. Other remarkable discoveries were made at cloud-shrouded Titan and on the icy geyser world of Enceladus, as well as the giant planet and its ever-changing rings.

The nuclear-powered New Horizons spacecraft revealed sheets of nitrogen ice, mountains, and deep valleys on distant Pluto and Charon, worlds that were previously believed to be inactive balls of ice. This success was followed by the first rendezvous with an even more remote Kuiper Belt object. Double-lobed 2014 MU69, a leftover remnant from the birth of the Solar System, seems to have been assembled during a low-speed collision.

More than 40 years after they left Earth, two more nuclear-powered craft, Voyagers 1 and 2, have left the Sun's realm and made the first crossings into interplanetary space.

The Juno orbiter is probing the invisible depths of Jupiter, providing new insights into the colorful cloud layers and deep interior of the gas giant.

Meanwhile, the MESSENGER spacecraft completed the first detailed reconnaissance of iron-hearted Mercury, whilst Japan's Akatsuki entered orbit around Venus and began imaging the super-rotating clouds.

Numerous robot explorers continue to study Mars from orbit and the surface, confirming the long-held beliefs that the Red Planet once supported rivers and large bodies of surface water – possible habitats for hardy, primitive organisms.

The smaller denizens of the Solar System have also attracted considerable attention. China achieved the first landing on the far side of the Moon, touching down on the unexplored South Pole-Aitken Basin.

Europe's Rosetta spacecraft made history when it flew alongside a comet for two years and released a lander onto its icy surface. Spacecraft from the U.S. and Japan have rendezvoused with small asteroids, revealing rocky rubble piles, and, following the success of Hayabusa 1, they are in the process of grabbing surface samples for analysis in labs back on Earth.

Following the release of huge amounts of new data from the armada of pioneering space missions, the scientific literature has expanded dramatically with the publication of new models and hypotheses – some contradictory, some revolutionary. Although planetary (and solar) science is in a continuous state of flux, I have tried to include many of these ground-breaking results and theories in this book, in an effort to showcase the latest research.

One of the most exciting research fields is the study of exoplanets, where space-based observatories, such as Kepler, and new ground-based instruments are opening new windows on an astonishing variety of alien worlds, many unlike anything that exists in our Solar System.

Only by studying distant worlds, whether in the Solar System or much further afield, can we hope to understand how our planetary system came about and how it may evolve in the future. There can be few more exciting areas of research, and I hope that the readers of this volume will be enthused by the evolving story of exploration described within these pages.

Peter Bond, September 2019

About the Companion Website

This book is accompanied by a companion website:

The URL is www.wiley.com/go/Bond-Solar-System2e

The website includes:

Figures and tables from the book

ONE
Beginnings

For millennia, people have studied the heavens and wondered about the nature and origins of the Sun, Moon, and planets. Indeed, Solar System studies dominated the field of astronomy until the introduction of powerful telescopes and advanced instruments in the 19th century. In the last 50 years, spacecraft have flown past or orbited all of the major planets and two dwarf planets, and landed on the Moon, Mars, Titan, a comet, and an asteroid. They have also brought back samples of Moon rock, comet and asteroid dust, as well as the solar wind. This era of robotic and human exploration has revolutionized scientists' knowledge of our corner of the Galaxy, and further astounding revelations are expected in the decades to come.

Wandering Stars

Since time immemorial, people have stared in wonder at the night sky. In previous millennia, when the darkness of the sky was not degraded by artificial lighting, it was easy to recognize how the stellar patterns drifted from horizon to horizon as the night progressed, and how they changed as the seasons passed.

However, in addition to the familiar, twinkling stars, observers noted seven objects that moved with varying speeds against the background of "fixed" stars.[1] In order of greatest apparent brightness, they were the Sun, Moon, Venus, Jupiter, Mars, Mercury, and Saturn. The ancient Greeks called them "planetes" ("wandering stars"), a designation we still use for all but the Sun and Moon.

For ancient astrologers and astronomers – the two disciplines were inextricably intertwined for many centuries – the most important of the wanderers were the Sun, which was responsible for daylight, and the Moon, which dominated the night. Both of these objects displayed visible disks and moved quite rapidly across the sky.

Careful study of their regular motions and apparitions enabled people to devise calendars and introduce convenient ways of measuring time. Thus, a year was the period before the Sun returned to the same place in the sky, while a month was the period that elapsed between each new or full Moon.

The other five planets were rather less noticeable, though each had its own peculiar characteristics. For example, Mercury and Venus never strayed far from the Sun in the twilight skies of morning or evening (Figure 1.1). The other three moved more slowly from constellation to constellation, sometimes describing loops in the sky as they appeared to temporarily reverse direction.

It was also evident that the seven planets often came together in the sky or even passed behind the Moon during occultations. They always remained within a narrow band on the sky, known as the zodiac (after the Greek word for "animal"). The Sun's annual path across the sky, called the ecliptic, ran along the center of this celestial highway. Clearly, the planes of the planets' orbits were closely aligned with each other.

The Earth-Centered Universe

Until the mid-16th century, it was accepted as an established fact by most civilizations that Earth lay at the center of the universe.[2] Like the axle of a wheel, everything else rotated around it.

---

[1] For a time, the ancient Greeks thought there were nine planets. Venus was named both as the Evening Star (Hesperus) and the Morning Star (Phosphorus). Similarly, Mercury was thought to be two different planets – Lucifer and Hermes.

[2] A Sun-centered (heliocentric) model of the universe was proposed by the Greek astronomer Aristarchus in the 3rd century BCE, but it was not widely accepted.

Exploring the Solar System, Second Edition. Peter Bond.
© 2020 John Wiley & Sons Ltd. Published 2020 by John Wiley & Sons Ltd.
Companion Website: www.wiley.com/go/Bond-Solar-System2e

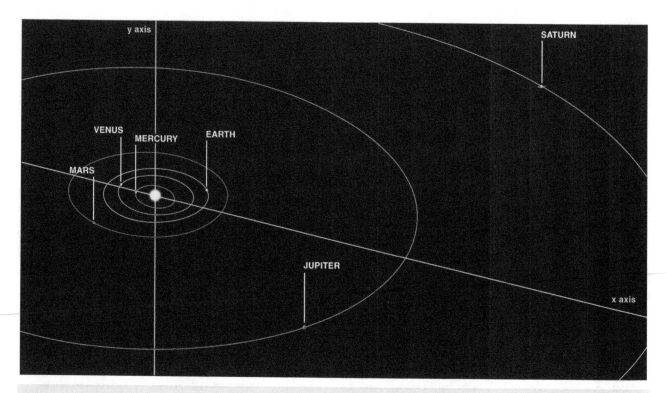

Figure 1.1 The relative sizes of the orbits of the "planets" visible to the naked eye and recognized by ancient astronomers. All the orbits are slightly elliptical and nearly in the same plane as Earth's orbit (the ecliptic). The diagram is from a view above the ecliptic plane and away from the perpendicular axis that goes through the Sun. (Lunar and Planetary Institute)

The reasons for this thinking seemed self evident. All the celestial objects, including the Sun, moved across the sky from east to west (with the occasional exception of a comet or shooting star). However, since no one experienced any of the sensations that would be expected if Earth was continually spinning, it seemed logical to believe that it was the heavens which were in motion around Earth.

According to this geocentric theory, the Sun, Moon, and planets were carried by invisible, crystalline spheres which were centered on the Earth. A much larger celestial sphere carried the fixed stars around the central Earth once every day.

Although early civilizations accepted the visual evidence that Earth is (more or less) flat, this idea was contradicted by several lines of evidence (see Chapter 3). For example, different star patterns or constellations are visible from different places. However, if Earth is flat, then the same constellations should be visible everywhere at a certain time.

One key piece of evidence was the curved outline of Earth's shadow as it drifted across the face of the full Moon during a total lunar eclipse. This was the case no matter where the observation was made or at what time it took place. Since only a spherical body can cast a round shadow in all orientations, it seemed clear that Earth was round.

Similarly, observations of a sailing ship disappearing over the horizon showed that, instead of simply becoming smaller and smaller, its hull disappeared from view before the sails and mast. This could only be explained on a curved ocean.

Measuring Distances and Sizes

One of the most fundamental problems facing early astronomers was the scale of the universe. How big were the Earth, Sun, and Moon, and how far away were they? It seemed evident that Earth was huge compared with every other object, and since it was the home of humanity, it was assumed that Earth was pre-eminent.

The question of the size of the spherical Earth was solved in the 3rd century BCE by Eratosthenes, who compared the length of shadows made at different locations at the time of the spring equinox (see Chapter 3). Some facts were also known about the relative sizes and distances of other objects.

Since its shadow easily covered the entire Moon during lunar eclipses, Earth had to be substantially larger than its satellite. During a solar eclipse, the Moon passed in front of the Sun, so the latter had to be further away. However, since their apparent sizes were identical, the Sun must be considerably larger than the Moon. Similarly, the Moon sometimes occulted or passed in front of stars and planets, so these, too, had to be much more remote.

Calculations by the Greek astronomers Aristarchus (c.310–c.230 BCE) and Hipparchus (c.190–120 BCE), based on the size of Earth's shadow, suggested that the Moon's diameter is about one third that of Earth and that its distance is nearly 59 times Earth's radius. This established the scale of the Earth–Moon system with a fair degree of accuracy. However, their simple geometric methods grossly underestimated the Sun's distance.

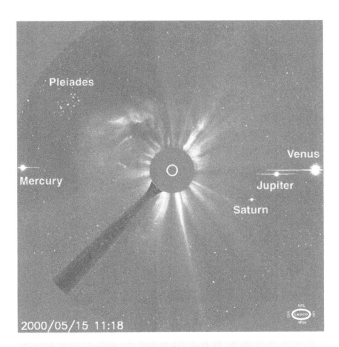

Figure 1.2 All the major planets follow orbits that lie within 8° of the Sun's path across the sky – the ecliptic. This narrow celestial belt is known as the zodiac. In this image from the SOHO spacecraft, four planets appear close to the Sun (whose light is blocked by an occulting disk). Also in view are some background "fixed" stars, including the Pleiades cluster. (ESA-NASA)

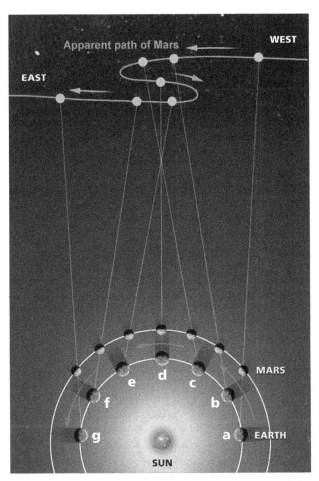

Figure 1.3 The apparent retrograde ("backward" or east–west) motions of Mars, Jupiter, and Saturn are now known to be caused by the relative orbital movement of the planets and Earth. Since Earth moves faster along its orbit than the more distant planets, it overtakes them on the inside track. As Earth approaches and passes Mars, the slower moving outer planet appears to move backward for a few months against the backcloth of "fixed" stars. (After NASA)

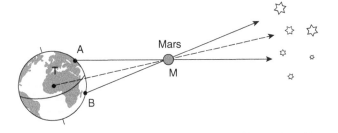

Figure 1.4 The distance of a planet such as Mars can be calculated by measuring its angle of sight – its location against the background of fixed stars – from two or more places on Earth. If the length of the baseline (e.g. the distance between two viewing sites, A-B) is known, the distance can be found by using simple trigonometry. (ESO)

Determination of the planetary distances remained problematic for a long time. It soon became clear to observers in the classical world that some planets move more slowly through the constellations of the night sky. Since a slow-moving planet such as Saturn was also fainter than the faster-moving objects, Mars and Jupiter, it seemed logical that Saturn was further away from Earth.

It was also clear that the Sun, Moon, and planets did not move at uniform speeds or follow simple curved paths across the sky. One of the most difficult observations to explain was an occasional "loop" in the motions of the more distant planets. This occurred when Mars, Jupiter, and Saturn were shining brightly around midnight (Figure 1.3 and Box 1.1). At such times, the planet's nightly eastward ("prograde") motion would gradually come to a stop. It would then reverse direction toward the west, becoming "retrograde," before resuming its general movement toward the east.

The explanation for this motion had to wait until astronomers realized that the Sun was at the center of the planetary system, and that Earth orbited the Sun (see The Central Sun). The loops could then be accounted for by Earth traveling along a smaller orbit so that it would catch up with, then overtake, the outer planets (see Figure 1.3) – like an athlete on an inside track.

Accurate calculations of planetary distances also had to wait until the 17th century, when observers were able to measure angular distances with reasonable accuracy. The basic geometrical method they used was called parallax (Figure 1.4).

This involved measurement of the apparent shift in position of an object when viewed from two different locations. To illustrate this, hold one finger upright in front of your nose and close first one eye and then the other. The finger seems to shift position against the background, although it is, of course, stationary. When the finger is moved closer, the shift appears larger, and vice versa.

Astronomers realized that, if a parallax shift in a planet's position could be measured from two widely separated locations, then its distance could be calculated. This method was first used by a French astronomer, Jean Richer, working in Cayenne (French Guiana), together with Giovanni Domenico Cassini and Jean Picard in Paris. They made simultaneous parallax observations of Mars during its closest approach in 1671, using the recently invented pendulum clocks to ensure that the measurements were made at precisely the same moment.[3]

Cassini's calculations led to a value of about 140 million km for the astronomical unit (AU) – the mean Sun–Earth distance. Now that this distance was known with reasonable accuracy, Kepler's third law (see Box 1.2) could be used to calculate the distances of the Sun and planets for the first time.

During the 18th century a great deal of time, money, and effort was spent in attempting to refine these figures. One method was to observe rare transits of Venus across the face of the Sun from many different locations. The most famous transit observations took place in 1761 and 1769 when the British explorer, Captain James Cook, sailed to the Pacific as part of an army of 150 observers scattered across the globe, but these gave very inaccurate results (see Chapter 6).

More successful was the worldwide effort to determine the parallax of the asteroid Eros when it passed close to Earth in 1931. Highly accurate measurements were possible since Eros has no atmosphere and appears as a mere point of light in even the largest telescopes. The value of the astronomical unit turned out to be 149.6 million km.

Since then, more sophisticated techniques have been introduced to refine the scale of the Solar System. One of the most successful is radar, when radio signals are reflected from the surfaces of distant objects (see Chapters 5, 6, and 13). Since the velocity of these microwaves is known and the time taken between emission and reception can be measured to a fraction of a second, the distance can be readily calculated. (Radar has also revealed the sizes and shapes of hundreds of asteroids.) A similar technique used to calculate changes in the Earth–Moon distance involves the use of laser pulses bounced off special reflectors left on the lunar surface.

Once an object's distance is accurately known, the diameter can be determined from its apparent angular size, as seen in a telescope. Unfortunately, this is very difficult for the smaller or more distant members of the Solar System, particularly if their albedo, or surface reflectivity, is uncertain.

In general, the larger an object, the more light its surface reflects. However, some objects are much better mirrors than others. A small, reflective object can have the same apparent brightness as a large, dark object. For example, observations of some Kuiper Belt objects, beyond the orbit of Pluto, indicate that their albedos are greater than previously believed. Since they are more reflective than anticipated, astronomers have revised their diameters downwards.

Another method, involving the occultation of a star by a planet or other object, is especially valuable in relation to objects which are normally difficult to observe. The object's diameter is calculated from the length of time during which it hides the star from view. This technique has been used to discover the rings of Uranus and Neptune, and to study Pluto's largest moon, Charon, for example. It is also invaluable for the detection and observation of exoplanets in orbit around distant stars (see Chapter 14). Unfortunately, if the object possesses a dense, cloudy atmosphere, the occultation only gives the diameter at the cloud tops.

The Central Sun

The difficult task of breaking with tradition and accepting the Sun as the center of the universe began with a Polish priest and astronomer named Nicolaus Copernicus (1473–1543). He decided that the only way to make sense of the planetary orbits was to relegate Earth to the status of a planet that orbited the Sun. The movement of the stars across the sky was then explained by the rotation of the spherical Earth, while the calendar of seasons and changing constellations in the heavens were accounted for by its year-long journey around the Sun.

Copernicus' most significant work, called *De Revolutionibus Orbium Celestium* (Concerning the Revolutions of the Celestial Spheres), was published shortly before his death. Curiously, this did not provoke a violent reaction by the establishment of the day, nor did it immediately lead to any major upheaval in scientific thought. Lacking enough evidence to swing the argument one way or the other, the great minds of the day were faced with an impasse.

Half a century passed before the interventions of two great scholars swung the argument in favor of Copernicus' heliocentric theory. The first breakthrough was made in 1609 by a young German named Johannes Kepler. By one of those strange twists of irony, Kepler was a pupil of Tycho Brahe, one of the leading opponents of the Copernican order. Given the unenviable task of finding an explanation for the retrograde motion of Mars (see Figure 1.3), Kepler was able to draw upon the excellent observational data recorded by his employer.

Brahe died in 1601, but Kepler continued to laboriously examine the problem before finally arriving at his eureka moment. The planetary orbits, he declared, were not circles but ellipses (regular oval shapes).[4] Within a short time, Kepler was able to draw up the first two laws of planetary motion (see Box 1.2). His third, and probably most important law, followed in 1619.

As a result, the relative distance of each planet from the Sun could be calculated accurately. Saturn, the most remote planet known at the time, turned out to be nearly 10 times further from the Sun than Earth. Since the actual distances remained unknown, the standard unit of measurement became the astronomical unit,

---

[3] A by-product of this experiment was the discovery that a pendulum swung more slowly at Cayenne than at Paris, showing that gravity is slightly weaker at the equator. Isaac Newton later used this result to show that Earth's diameter is greatest at the equator.

[4] Kepler's task was made slightly easier by the fact that, of the five known planets, only Mercury followed a more elliptical path than Mars.

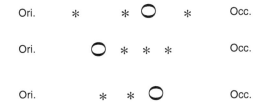

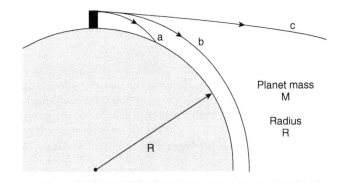

Planet mass
M

Radius
R

Figure 1.5 In January 1610, Galileo Galilei used his simple refracting telescope to discover three "stars" aligned on either side of Jupiter. Over a period of several weeks, a fourth "star" appeared. As they shifted positions, Galileo correctly deduced that these were satellites. *Occ.* is the Latin abbreviation for "west" and *Ori.* stands for "east." (NASA)

Figure 1.6 (a) If a spacecraft does not accelerate to orbital velocity, it will fall back to the planet's surface. (b) If it reaches orbital velocity, it will remain in a closed path (orbit) around the planet under free fall conditions. (c) If the spacecraft reaches escape velocity, it will be able to break free from the planet's gravitational pull and travel to another planet. The same rules apply to planets and spacecraft in orbit around the Sun. (NASA)

so Saturn's distance from the Sun was about 10 astronomical units or 10 AU.

In the same year that Kepler discovered elliptical orbits, an Italian scientist named Galileo Galilei made a simple refracting telescope, comprising two lenses at either end of a narrow tube, and began to study the heavens. Within a short time, despite the small magnification offered by his "optic tube," he had obtained visual evidence to support the theories of Copernicus and Kepler. Galileo became the first person in history to see the phases of Venus caused by its movement around the Sun. He also observed mountains and craters on the Moon, and saw the planets as disks, rather than points of light.

Most significant of all was his discovery of four star-like objects close to Jupiter (Figure 1.5). By watching their daily motions, he was able to calculate their orbital periods and show that they were Jovian moons (see Chapter 7). The discovery of the first planetary satellites (other than the Moon) supported theories that Earth was not at the center of the universe and confirmed that everything did not revolve around our world.

Galileo's discoveries caused a sensation, although the leaders of the Roman Catholic Church obstinately continued to support a geocentric universe. In 1633, Galileo was brought before the Inquisition and forced to recant under threat of torture.

Newton and Gravity

The next challenge was to find an explanation for Kepler's laws. Although Galileo conducted numerous experiments into the effects of gravity, he did not realize the full significance of his discoveries. This was left to an Englishman, Isaac Newton, who was born in 1642, the year that Galileo died.

One anecdote attributes Newton's discovery of universal gravitation to him observing an apple falling from a tree. Whatever the truth, by 1684 Newton was able to explain planetary motions. His law of gravitation stated that all objects attract each other, and that the strength of this gravitational attraction is proportional to their mass (see Chapter 8).

Clearly, since the Sun has nearly all the mass in the Solar System, it should pull all of the other bodies into it. Newton explained that this did not happen because their orbital velocities are just sufficient to counteract the Sun's gravity. The result is that the planets fall towards the Sun in such a way that the curve of their fall takes them completely around it (Figure 1.6). This is sometimes known as free fall. (This same explanation, of course, applies to artificial satellites.)

Newton's law also stated that the strength of gravitational attraction decreases with distance. For example, if planet A is twice as far from the Sun as planet B, then the gravitational force exerted by the Sun on planet A is one quarter that exerted on planet B.

In practical terms, this means that a satellite in low Earth orbit must travel at 8 km/s, whereas the Moon only has to circle the Earth at 1 km/s in order to avoid crashing into our planet. Similarly, planets further from the Sun are able to move more slowly around their orbits than those in the inner Solar System. Newton's law also explained why a planet's orbital speed increased as it approached perihelion (closest point to the Sun) and slowed near aphelion (furthest point from the Sun).

From this time on the orbital mechanics of the Solar System were very well understood. With the exception of Mercury, whose orbital motion refused to obey Newton's law (see Chapter 5), the only significant problems involved minor variations in orbits caused by gravitational interactions between the planets, particularly those involving massive Jupiter. Careful study of unexpected changes in the orbital velocity of Uranus may even have enabled the position of an unknown planet, Neptune, to be successfully calculated (see Chapter 11) – although there are those who consider the discovery to be pure chance.

What Is A Planet?

In the ancient world, astronomers counted eight planets. When the Sun, Earth, and Moon are removed from their list, the number of planets visible to the naked eye is reduced to five: Mercury, Venus, Mars, Jupiter, and Saturn.

Box 1.1 Orbits

The direction a spacecraft or other body travels in orbit can be prograde, when a satellite moves in the same direction as the planet (or star) rotates, or retrograde, when it goes in a direction opposite to the planet's (or star's) rotation. All of the planets in the Solar System orbit the Sun in a prograde direction – west to east or counterclockwise as observed from above the Sun's north pole. However, many comets and some satellites move in a retrograde (clockwise) direction.

Various technical terms are used to describe the characteristics of these orbits. The time an object takes to complete one orbit is known as the orbital period. The closest point of an orbit has the prefix "peri" – hence perigee for a satellite of the Earth and perihelion for an object orbiting the Sun. (Helios = Sun.) The furthest point in an orbit has the prefix "ap" – as in apogee and aphelion.

The plane of Earth's orbit around the Sun is called the ecliptic. The orbits of the other planets, comets, and asteroids are tilted to this plane. The angle of the tilt is the orbital inclination. The inclination of a satellite's orbit is measured with respect to the planet's equator. Hence, an orbit directly above the equator has an inclination of 0°, while one passing over a planet's poles has an inclination of 90°.

A planet, asteroid, or comet crosses the ecliptic twice during each orbit of the Sun. The points where an orbit crosses a plane are known as nodes. When an orbiting body crosses the ecliptic plane going north, the node is referred to as the ascending node. Going south, it is the descending node. The line that joins the ascending node and the descending node of an orbit is called the line of nodes.

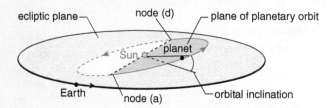

Figure 1.7 Some important characteristics of a planet's orbit. Here the planet is inferior, i.e. closer to the Sun than Earth. Its orbit is inclined to the ecliptic – the plane of Earth's orbit. The planet's orbit crosses the ecliptic at two nodes – the ascending node (a) and the descending node (d). (Peter Bond, after Open University)

One of the most important orbital, or Keplerian, elements, is the semi-major axis, the average distance of an object from its primary (planet or Sun). The shape of the orbit is described by its eccentricity, measured as a number between zero and 1. An eccentricity of zero indicates a circular orbit. A parabola has an eccentricity of 1.

With the invention of the telescope, the possibility arose of finding fainter, more remote planets. The first newcomer, Uranus, was discovered far beyond the orbit of Saturn by William Herschel in 1781. The list was further increased in 1801, when Giuseppe Piazzi found Ceres in the gap between the orbits of Jupiter and Mars. Pallas, Juno, and Vesta – objects in similar orbits to Ceres – were discovered between 1802 and 1807. Since they were clearly much smaller and less substantial than the other planets, they were soon downgraded to "minor planets" or "asteroids" (star-like objects).

Almost 40 years passed before the eighth planet, Neptune, was discovered by Johann Galle and Heinrich D'Arrest. However, neither Uranus nor Neptune seemed to be following its expected path, suggesting that an even more distant planet might be influencing the movements of its neighbors. The search for this world concluded in 1930 when Clyde Tombaugh observed the tiny image of Pluto on a photographic plate.

For many years, it was generally accepted that there were nine planets, despite growing concerns that Pluto seemed to be too small and lacking in mass to deserve this title. The crunch came in 2003, when Mike Brown discovered 2003 UB313 (now named Eris), an object that is comparable in size to Pluto. With the introduction of ever more sensitive detectors, it seemed likely that there would soon be dozens of Pluto-sized planets.

Aware that there was no generally accepted definition of the term "planet" and faced with a fierce debate over whether Pluto should be demoted, members of the International Astronomical Union gathered in Prague for the 2006 General Assembly.

After a lengthy discussion, they agreed to define a planet as a celestial body that: (a) is in orbit around the Sun, (b) has sufficient mass for its self-gravity to overcome rigid body forces so that it assumes a hydrostatic equilibrium (nearly round) shape, and (c) has cleared other objects from the neighborhood of its orbit.

Based on these criteria, the Solar System now consists of eight planets: Mercury, Venus, Earth, Mars, Jupiter, Saturn, Uranus, and Neptune. A new distinct class of objects called "dwarf planets" was also introduced (Figure 1.8). To be classified as a dwarf planet, an object must orbit the Sun and have a nearly round shape. The first dwarf planets to be announced were Ceres (the largest asteroid), Pluto, and Eris, followed by three more. Many others are expected to be discovered in the future.

Figure 1.8 In the "new" Solar System, as defined by the International Astronomical Union in 2006, there are eight planets: Mercury, Venus, Earth, Mars, Jupiter, Saturn, Uranus, and Neptune (shown in order of their distance from the Sun). A new, distinct class of objects called "dwarf planets" includes the largest asteroid, Ceres, and the two largest known Kuiper Belt objects, Pluto and Eris. The relative sizes of the planets and the Sun are shown. Jupiter's diameter is about 11 times that of Earth, and the Sun's diameter is about 10 times that of Jupiter. The distances of the planets are not shown to scale. (IAU)

This decision has not met with universal approval. One common criticism relates to what exactly is meant by a planet "clearing its neighborhood." For example, critics argue that Neptune is accepted as a planet, even though many Kuiper Belt objects (including Pluto) cross its orbit. Perhaps, they suggest, it would be more appropriate to use size as a criterion, particularly bearing in mind the diameters of objects that are large enough for gravity to dominate structural strength. There is also some discomfiture with defining Ceres – the largest of the asteroids – as a dwarf planet.

Another complication arises when the current definition is extended to extrasolar planets, i.e. planets orbiting other stars (see Chapter 14). Size is not a useful factor, since many of these planets are similar in size and mass to small, cool "failed stars" known as brown dwarfs.

Instead, astronomers attempt to distinguish between a giant extrasolar planet and a brown dwarf by determining how they were born. A star is formed during the gravitational collapse of a gaseous nebula, whereas a planet is the product of collisions and accretion (snowball-like growth) between particles in a disk of gas and dust around a central star. Even so, this method of differentiation is difficult to apply, especially in the case of planet-sized objects that have been flung into interstellar space and no longer orbit any star.

The Solar System

50 years ago, the population of the Solar System included one central star, nine planets, 31 satellites, and thousands of comets and asteroids. However, since the arrival of the Space Age and the development of ever more sensitive ground-based instruments, the inventory of objects has swollen remarkably.

Today, the astronomical community recognizes eight planets and five dwarf planets, the tally of planetary satellites has passed 150, and the number of identified small objects is climbing rapidly as increasingly sensitive searches discover thousands of Sun-grazing comets and icy Kuiper Belt objects that orbit beyond Neptune.

In terms of numbers, the Solar System is dominated by debris, in the form of comets, asteroids, meteorites, and dust. These are the leftovers from the formation of the planets, 4.5 billion years ago. The main asteroid belt, between Mars and Jupiter, is populated by millions of rocky objects that are shepherded by the powerful gravity of the nearby gas giant. They are thought to represent planetesimals – small planetary building blocks – that were unable to accrete due to the gravitational interference of Jupiter.

Beyond the orbit of Neptune are two more swarms of small objects, this time largely made of ice (Figure 1.9 and Figure 1.10). The inner region, known as the Kuiper Belt, is where short-period

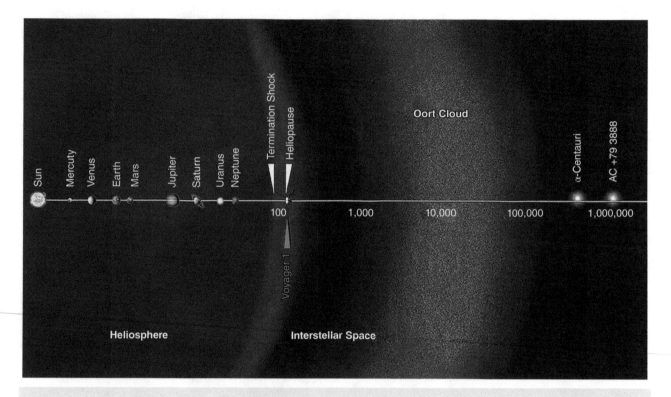

Figure 1.9   The size of the Solar System. The scale bar is in astronomical units, with each marked distance beyond 1 AU representing 10 times the previous distance. One AU is the distance from the Sun to the Earth, which is about 150 million km. The Kuiper Belt, which extends beyond Neptune from about 30 to 55 AU, is not shown. Two distant stars are also shown (right). (NASA/JPL-Caltech)

comets originate. Pluto and Eris are the largest known inhabitants. The orbital periods of Kuiper Belt objects range from 200–400 years for objects such as Pluto to 1,000 years or longer for those which follow very elliptical orbits that take them far from the Sun.

The Kuiper Belt poses a serious challenge for theories of planet formation, since it contains less than 1% of the mass of the proto-solar nebula. If the Kuiper Belt objects formed like the terrestrial planets, growing by accumulating smaller objects as they orbit the Sun, the shortage of local building material means it would take longer than the age of the Solar System to make one KBO!

Even further out – indeed, so far that none of the objects have ever been observed in situ – is the postulated Oort Cloud, the home of most long-period comets.

The basic characteristics of the Solar System are straightforward to describe. Close to the Sun, where temperatures are higher, there are four quite small, but dense, "terrestrial" planets that are composed largely of rock (Figure 1.11 and Table 1.1). Beyond Mars, where temperatures are always well below zero, is the realm of the gas giants, Jupiter and Saturn, and the ice giants, Uranus and Neptune.

As noted above, the orbits of the major planets are approximately circular, and close to the ecliptic plane. All of the planets and main belt asteroids circle the Sun in the same direction – counterclockwise as seen from above the Sun's north pole. This is also the direction of the Sun's rotation. However, the beautiful symmetry breaks down when it comes to the smaller members of the Solar System. Comets can arrive from any direction, and the orbits of the Kuiper Belt objects have no particular orientation, suggesting that there is a spherical swarm of these objects surrounding the Sun and major planets.

Of the four inner planets, Venus and Earth both possess dense atmospheres – though they are very different in nature – while Mercury is too lightweight to have retained a substantial gaseous envelope. Whereas the most common gas on both Venus and Mars is carbon dioxide, Earth is something of an oddball, with an atmosphere dominated by nitrogen and oxygen. This latter gas can be accounted for by the fact that Earth is – as far as we know – the only abode of life in our Solar System, and it is those life forms that pump oxygen into the air. Satellites are rare: Earth is orbited by the Moon, while Mars has two small companions that are generally considered to be captured asteroids.

As their name suggests, the gas and ice giants are characterized by their large size – tens to thousands of times bigger than Earth – and low bulk densities which can be accounted for by the dominance of hydrogen and helium in their interiors. All four of the giants have ring systems composed of dust, ice, and rocky debris, and their gravitational influence is such that they retain dozens of satellites – most of them captured billions of years ago.

Since they are relatively close to the Sun, all the terrestrial planets have high orbital velocities with periods of less than two Earth years (see Box 1.2: Kepler's Third Law). In contrast, their axial rotations are slow and their axial inclinations are very different.

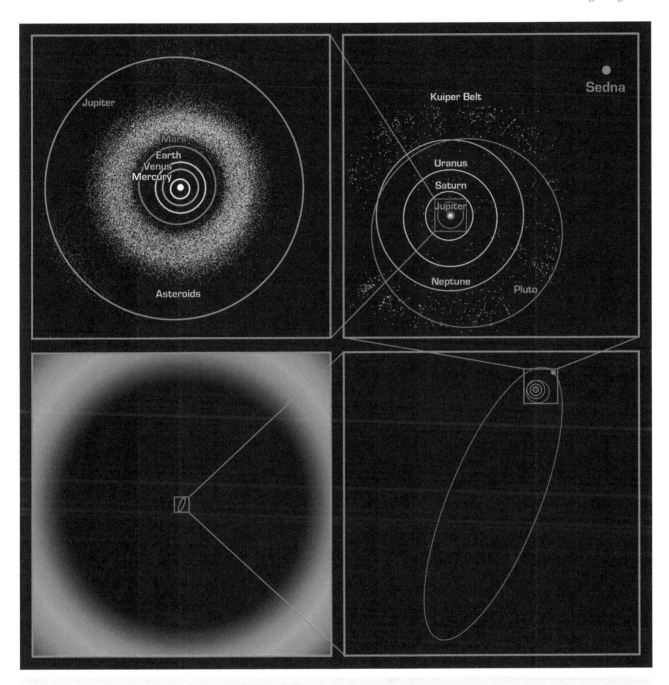

Figure 1.10   These four panels show the scale of the Solar System as we know it today. At top left are the orbits of the inner planets and the main asteroid belt. Top right shows the orbits of the outer planets and the Kuiper Belt. Lower right shows the orbit and current location of Sedna, one of the most distant known objects in the Solar System. Lower left shows that even Sedna's highly elliptical orbit, which takes it nearly 1,000 AU from the Sun, lies well inside the proposed Oort Cloud (shown in blue). This spherical cloud contains millions of icy bodies orbiting at the limits of the Sun's gravitational pull. (NASA/JPL/R. Hurt, SSC-Caltech)

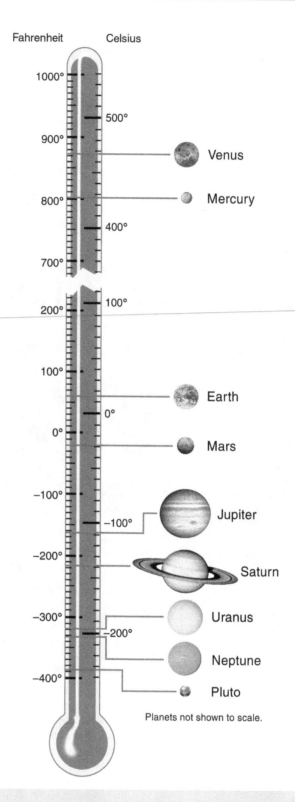

Figure 1.11   In general, a planet's surface temperature decreases with its distance from the Sun. Venus is the exception, since its dense carbon dioxide atmosphere traps infrared radiation. The runaway greenhouse effect raises its surface temperature to 467°C. Mercury's slow rotation and thin atmosphere result in the night-side temperature being more than 500°C colder than the dayside temperature shown above. Temperatures for Jupiter, Saturn, Uranus, and Neptune are shown for an altitude in the atmosphere where pressure is equal to that at sea level on Earth. Earth lies in the center of the "habitable zone," where water can exist as a liquid and conditions are favorable to life. (NASA / Lunar and Planetary Institute)

Table 1.1

The Planets: Relationship Between Solar Distance and Mean Density

Planet	Distance from Sun (AU)	Mean Density (g/cm^3)
Mercury	0.3871	5.43
Venus	0.7233	5.24
Earth	1.0	5.52
Mars	1.5237	3.91
Jupiter	5.2028	1.33
Saturn	9.5388	0.69
Uranus	19.1914	1.29
Neptune	30.0611	1.64

Mercury's axis is almost at right angles to its orbit. It takes 58 days to rotate once, or about two-thirds of the time it takes to orbit the Sun. Venus resembles a top that has been knocked completely upside down. As a result, it rotates in a retrograde direction that takes 243 Earth days, longer than its orbital period. Earth and Mars have very similar days and seasons – at least in the present epoch – since their sidereal periods of axial rotation are both around 24 hours and both axes are inclined about 24–25° to their orbits (Figure 1.12).

The motions of the outer planets are very different. Their large distances from the Sun require modest velocities to maintain their orbits. Orbital periods range from almost 12 years for Jupiter to about 165 years for Neptune. However, despite their swollen spheres, they all spin much faster on their axes than their terrestrial siblings, with sidereal periods in the range of 10–20 hours.[5] However, there is considerable variation in their axial tilts. Jupiter is almost upright, Saturn and Neptune are inclined more than Earth and Mars, while Uranus spins on its side so that the polar regions alternately point toward or away from the Sun.

The orbits and axial inclinations of the planets (and satellites) are not fixed, e.g. the axial tilt of Mars changes dramatically over millions of years.

The Birth of the Solar System

The Sun, which contains over 99% of the Solar System's mass, completes one rotation in about 24 days. In contrast, the largest planets, Jupiter and Saturn, rotate once in about 10 hours. When combined with their orbital motion, it turns out that Jupiter accounts for some 60% of the Solar System's angular momentum, with another 25% accounted for by Saturn. This compares with about 2% for the sluggardly Sun.

Any theory of cosmogony that attempts to account for the formation of the Solar System must take into account the angular momentum of the Solar System objects, as well as the facts that all of the planets travel in the same direction and more or less in the same plane. The obvious conclusion is that they all formed in the same manner and at about the same time.

Scientists have usually considered two main possibilities: the planets were either created by material derived from the Sun or a nearby companion star, or they formed from a cloud of diffuse matter that surrounded the Sun. However, theorists have struggled for centuries to match the hypotheses to the known facts, in order to choose between them.

One of the earliest, and most successful, attempts to explain how the Solar System came about was the nebular hypothesis – the idea that the Sun and planets formed from a vast, slowly rotating disk of gas and dust. A modified version of this hypothesis is the generally accepted explanation today.

Some of the key evidence comes from modern observations of distant star systems. Today, spaceborne telescopes can peer into the hearts of giant molecular clouds, such as the Orion Nebula, and search for young, Sun-like stars that replicate the conditions that prevailed in our Solar System some 4.6 billion years ago.

These observations show that so-called protoplanetary disks, or proplyds, exist around most very young stars – those less than 10 million years old (Figure 1.16). Many of the disks are larger than our Solar System. Observations of slightly older stars show how these disks evolve as time goes by, with the formation of swarms of rocky and icy debris and gaps in the clouds created by fledgling planets.

As currently envisaged, the Solar System began with the collapse of a cloud of interstellar gas. The trigger for this collapse may have been the passage of an externally generated shock wave from a supernova explosion, density waves passing through the galaxy, or a major reduction in the cloud's magnetic field or temperature.

The first of these explanations is the prime candidate, since many stars form in clusters within clouds containing thousands of solar masses of material. When the giant stars of the cluster run through their short life spans, they are likely to produce a series of supernovas, preceded by powerful stellar winds.

Evidence from meteorites and dynamical modeling of supernova shock wave propagation into giant molecular clouds indicate that a supernova explosion compressed part of a cloud, causing this region to collapse. The shock wave would also have injected material from the exploding star into the solar nebula. Scientists have detected evidence of this material in the form of the decay products from radioactive isotopes, particularly iron-60. These are found in primitive meteorites and can only form in the giant stars that end their lives as supernovas.

Over millions of years, the original cloud may be broken up into smaller fragments, each mixed with heavier elements from the dying stars, as well as the ubiquitous hydrogen and helium gas. Once a fragment reaches a critical density, it is able to overcome the forces associated with gas pressure and begins to collapse under its own gravity.

The contracting cloud begins to rotate, slowly at first, then faster and faster – rather like when an ice skater pulls in his arms. Since material falling from above and below the plane of rotation collides at the mid-plane of the collapsing cloud, its motion is cancelled out. The cloud begins to flatten into a disk, with a bulge at the center where the protostar is forming. The disk was probably thicker

[5]The sidereal rotation period is the time a planet takes to spin once on its axis, with respect to a particular background star.

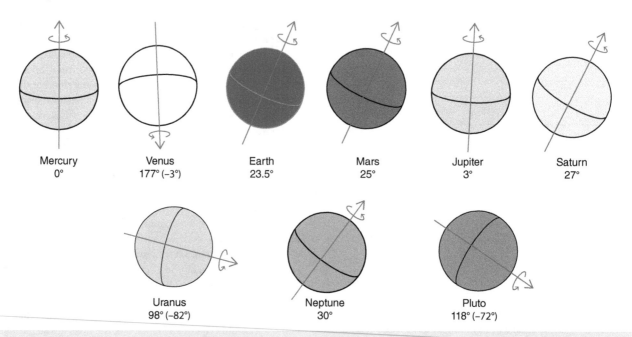

Figure 1.12 The axial inclinations (obliquities) of the planets and Pluto compared to their orbital planes. Most of the planets have axial tilts of less than 90°, so they rotate in a prograde direction, from west to east. Venus, Uranus, and Pluto have obliquities greater than 90°, so they are said to rotate in a retrograde (backwards) direction. (Peter Bond)

at a greater distance from the protostar, where gas pressure was lower.

Such a nebula would almost certainly rotate slowly in the early stages, but as it contracts, conservation of angular momentum causes the cloud to spin faster. If this process continues, the core forming at the center of the nebula will spin up so fast that it flies apart before it has a chance to form a star. Somehow, that angular momentum must be removed before a star can form.

Studies of other young stars and their surrounding disks provide evidence that, as the interstellar gas collapses, it also winds up the magnetic field which permeates the nebula. Gas which is rotating too fast to collapse is expelled and dispersed along the magnetic field.

This process naturally forms a spiral-shaped magnetic field that helps to generate polar jets and outflows associated with very young stars. At the same time, the jets remove angular momentum, allowing other material to accrete and collapse. Gravitational instability, turbulence, and tidal forces within the "lumpy" disk may also play a part, helping to transfer much of the angular momentum to the outer regions of the forming disk.

The protoplanetary disk is heated by the infall of material. The inner regions, where the cloud is most massive, become hot enough to vaporize dust and ionize gas. As contraction continues and the cloud becomes increasingly dense, the temperature at its core reaches the point where nuclear fusion commences. The emerging protostar begins to emit copious amounts of ultraviolet radiation. Radiation pressure drives away much of the nearby dust, causing the star to decouple from its nebula.

The young star may remain in this T Tauri stage for perhaps 10 million years, after which most of the residual nebula has evaporated or been driven into interstellar space. All that remains of the original cloud is a rarefied disk of dust grains, mainly silicates and ice crystals.

Meanwhile, the seeds of the planets have begun to appear. More refractory elements condense in the warm, inner regions of the nebula, while icy grains condense in the cold outer regions. Individual grains collide and stick together, growing into centimeter-sized particles. These swirl around at different rates within the flared disk, partly due to turbulence and partly as the result of differences in the drag exerted by the gas. After a few million years, these dusty or icy golf balls accrete into kilometer-sized planetesimals and gravity becomes the dominant force.

The Solar System now resembles a shooting gallery, with objects moving at high speed in chaotic fashion and enduring frequent collisions with each other. Some of these impacts are destructive, causing the objects to shatter and generate large amounts of dust or meteorite debris. Other collisions are constructive, resulting in a snowballing process. Over time, the energy loss resulting from collisions means that construction eventually dominates.

Eventually, the system contains a relatively small number of large bodies or protoplanets. Millions of years pass as they continue to mop up material from the remnants of the solar nebula and to collide with each other, finally resulting in a population of widely separated worlds occupying stable orbits and traveling in the same direction around the young central star.

It is likely that the largest planets in the Solar System, Jupiter and Saturn, formed first. They presumably accumulated their huge gaseous envelopes of hydrogen and helium before the solar nebula dispersed.

Box 1.2   Kepler's Laws of Planetary Motion

Johannes Kepler (1571–1630) was one of the most important characters in the story of unraveling how the Solar System works. The German-born mathematician was appointed assistant to Tycho Brahe (1546–1601), the most famous observer of the day. Granted access to Brahe's catalog of positional data, Kepler was given the task of explaining the orbit of Mars. After four years of calculations, Kepler finally realized in 1605 that the orbits of the planets were not perfect circles, but elongated circles known as ellipses.

Whereas a circle has one central point, an ellipse has two key interior points called foci (singular: focus). **The sum of the distances from the foci to any point on the ellipse is a constant.** For Solar System objects, the Sun always lies at one focus.

In order to draw an ellipse, place two drawing pins some distance apart and loop a piece of string around them. Place a pencil inside the string, draw the string tight and move the pencil around the pins. Now move one of the pins and repeat the process. Note how the shape of the ellipse has changed.

The amount of "stretching" or "flattening" of the ellipse is termed its eccentricity. All ellipses have eccentricities lying between zero and one. A circle may be regarded as an ellipse with zero eccentricity. As the ellipse becomes more stretched, its eccentricity approaches one.

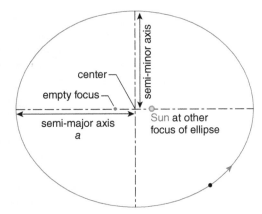

Figure 1.13   A circle has an eccentricity of zero. As the ellipse becomes more stretched (i.e. the foci move further apart) the eccentricity approaches one. Half of the major axis is termed the semi-major axis. The average distance of a planet from the Sun as it follows its elliptical orbit is equal to the length of the semi-major axis. The eccentricity is calculated by dividing the distance between the two foci by the length of the major axis. (Peter Bond)

In reality, most of the planets follow orbits that are only slightly elliptical. Their eccentricities are so small that they look circular at first glance. Pluto and Mercury are the main exceptions, with eccentricities exceeding 0.2.

Another key characteristic of an ellipse is its maximum width, known as the major axis. Half of the major axis is termed the semi-major axis. The average distance of a planet from the Sun as it goes around its elliptical orbit is equal to the length of the semi-major axis.

After intensive work on the implications of his discovery, Kepler eventually formulated his *Three Laws of Planetary Motion*.

- **Kepler's First Law: The orbits of the planets are ellipses, with the Sun at one focus of the ellipse.** (Generally, there is nothing at the other focus.)
- **Kepler's Second Law: The line joining the planet to the Sun sweeps out equal areas in equal times as the planet travels around the ellipse.** In order to do so, a planet must move faster along its orbit near the Sun and more slowly when it is far away. A planet's point of nearest approach to the Sun is termed perihelion; the furthest point from the Sun on its orbit is termed aphelion. Hence, a planet moves fastest when it is near perihelion and slowest when it is near aphelion.
- **Kepler's Third Law: The square of a planet's sidereal (orbital) period is proportional to the cube of its mean distance (semi-major axis) from the Sun.** This means that the period, or length of time a planet takes to complete one orbit around the Sun, increases rapidly with its distance from the Sun. Thus, Mercury, the innermost planet, takes only 88 days to orbit the Sun, whereas remote Pluto takes 248 years to do the same.

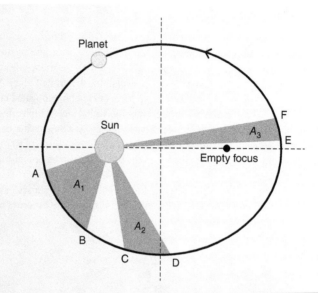

Figure 1.14   Kepler's first law states that the orbit of a planet about the Sun is an ellipse with the Sun at one focus. The other focus of the ellipse is empty. According to Kepler's second law, the line joining a planet to the Sun sweeps out equal areas in equal times. In this diagram, the three shaded sectors, A_1, A_2, and A_3, all have equal areas. A planet takes as long to travel from A to B as from C to D and E to F, because it moves most rapidly when it is nearest the Sun (at perihelion) and slowest when it is farthest from the Sun (at aphelion). (Peter Bond)

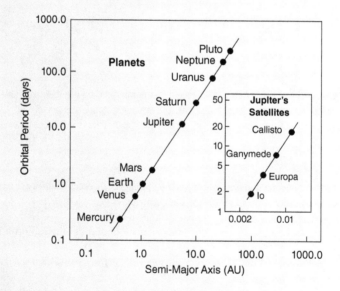

Figure 1.15   A graph showing the orbital periods of the planets plotted against their semi-major axes, using a logarithmic scale. The straight line that connects the planets has a slope of 3/2, verifying Kepler's third law which states that the squares of the orbital periods increase with the cubes of the planetary distances. This law applies to any bodies in elliptical orbits, including Jupiter's four largest satellites (inset). (Kenneth R. Lang, The Cambridge Guide to the Solar System)

This law can be used to make some useful, but fairly simple, calculations. For example, if the period is measured in Earth years and the distance is measured in astronomical units (AU), the law may be written in the simple form: $P(years)^2 = R(AU)^3$.

This equation may also be written as: $P(years) = R(AU)^{3/2}$. Thus, if we know that Pluto's average distance from the Sun (semi-major axis) is 39.44 AU, we can calculate that its orbital period $P = (39.44)^{3/2} = 247.69$ years. Similarly, if we know that Mars takes 1.88 Earth years to orbit the Sun, we can calculate that its semi-major axis $R = (1.88)^{2/3} = 1.52$ AU.

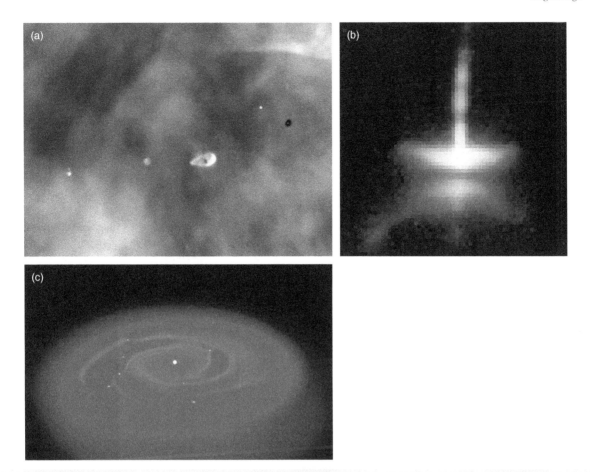

Figure 1.16 The early stages of star and planet formation. (a) A Hubble Space Telescope view of five young stars in the Orion Nebula. Four are surrounded by gas and dust trapped in orbit as the stars formed. These are possibly protoplanetary disks, or "proplyds," that might eventually produce planets. The bright proplyds are closest to the hottest stars of the parent star cluster, while the object farthest from the hottest stars appears dark. (C. R. O'Dell/Rice University; NASA) (b) This HST image shows Herbig-Haro 30, a young star surrounded by a thin, dark disk. The disk extends 64 billion km, dividing the nebula in two. The central star is hidden from direct view, but its light reflects off the upper and lower surfaces of the disk to produce the pair of reddish nebulas. Gaseous jets (green) remove material from above and below the disk and transfer angular momentum outwards. (Chris Burrows/STScI, the WFPC2 Science Team and NASA) (c) A computer simulation showing how a protoplanetary disk surrounding a young star begins to fragment and form gas giant planets with stable orbits. (Mayer, Quinn, Wadsley, Stadel, 2002, Science)

Observations of young star systems show that the gas disks that form planets usually have lifetimes of only 1 to 10 million years, which means the gas giant planets probably formed within this time frame. In contrast, Earth probably took at least 30 million years to form, and may have taken as long as 100 million years.

It is worth noting here that computer simulations of the early Solar System show that even the slightest differences in initial conditions can produce different planetary systems. Depending on exactly where each embryo started out, the orbital positions of new planets vary randomly from simulation to simulation. The total number of planets – and hence, their final masses – may also vary greatly. It seems that planet formation is a very chaotic process as evidenced by exoplanet systems which bear little resemblance to our Solar System (see Chapter 14).

Rocky Planets

Modeling suggests that collisions between planetesimals initially occur at low velocities, allowing them to merge and grow (Box 1.3). At the Earth's distance from the Sun, it takes only about 1,000 years for 1-km-sized objects to grow into 100-km objects. Another 10,000 years produces 1,000-km diameter protoplanets, which double in diameter over the next 10,000 years. Such models indicate that Moon-sized objects can form in a little over 20,000 years.

As planetesimals within the protosolar disk grow larger and more massive, their gravity increases, and once a few of the objects reach a size of 1,000 km, they begin to stir up the remaining smaller objects. Near encounters accelerate the smaller, asteroid-sized chunks of rock to higher and higher speeds.

Box 1.3   Key Steps in the Formation of Rocky Planets (after Kenyon and Bromley)

1. A molecular cloud made up of gas and dust begins to collapse.
2. A protostar begins to form at the core of the collapsing nebula.
3. A disk-shaped nebula of orbiting dust and gas develops in the protostar's equatorial plane.
4. Dust grains in the disk collide and merge.
5. Large (1 mm) dust grains fall into a thin, dusty sheet.
6. Collisions produce planetesimals 1 m to 1 km across.
7. More collisions between planetesimals produce planetary embryos.
8. Planetary embryos stir up the leftover planetesimals.
9. Planetesimals then collide and fragment.
10. A cascade of collisions reduces fragments to dust.
11. Planets sweep up some of the dust.
12. Radiation and a "wind" of charged particles from the central star remove the remaining gas and dust.

Eventually, they are traveling so quickly that when they collide, they pulverize each other instead of merging.

While the largest protoplanets continue to grow, the remaining rocky planetesimals grind each other into dust. Some of this dust is drawn in by the surviving planets, while much of the remainder is swept out of the Solar System when the Sun evolves into a hydrogen-burning star. (A cloud of micron-sized dust particles still exists in the ecliptic plane of the Solar System. Known as the zodiacal cloud, it is composed of silicate particles that are largely derived from collisions between main belt asteroids.)

One of the problems that must be solved by Solar System theorists is an explanation for the silicate and metal-rich nature of the terrestrial planets and the dominance of hydrogen and helium in the outer planets (Box 1.4). Clearly, the marked difference in composition between the inner and outer planets must be related to the materials that made up different regions of the disk.

The dense, rocky nature of the Earth and its neighbors suggests that they simply formed through the accretion of dust grains in the solar nebula. However, studies of primitive chondritic meteorites show the presence of millimeter-sized droplets (chondrules) that were once liquid.

It seems that, before they amalgamated to form the meteorites, these existed for a brief period as independent spheroids at temperatures above 1,500°C. Some chondrules seem to include other chondrules, indicative of being exposed to high temperatures on more than one occasion (see Chapter 13). The source of the heating is uncertain, although shock waves, solar heating, and collisions between planetesimals have been suggested.

Laboratory experiments indicate that these molten globules were cooled very rapidly, within 10 million years of the collapse of the molecular cloud. The cause of such sudden cooling events

remains unclear. What does seem certain is that the chondrules and dust began to stick together and grow in size, creating chunks of chondritic material. Drag from gas in the nebula encouraged the pebble-sized objects to creep inward, all the time gathering in more material.

Once a population of large planetesimals evolved, their destiny was determined largely by chance. A fast, head-on collision caused the objects to break apart. A slow, gentler encounter enabled the participants to merge into an even larger object. In this way, the terrestrial planets grew to more or less their current size over a period of some 10 million years.

The huge amounts of kinetic energy dumped in the planets by frequent, massive impacts caused partial or total melting and the creation of magma oceans. This led to internal differentiation, with the denser elements, such as iron, sinking to the core and the lighter ones rising to the surface to create silicate crusts.

Early atmospheres were generated by outgassing of volatile molecules such as water, methane, ammonia, hydrogen, nitrogen, and carbon dioxide. A final heavy bombardment, which ended about 3.8 billion years ago, is clearly marked in the crater record of the Moon, and this has been applied to other planets and satellites.

Occasionally a satellite was created as the by-product of a major impact. Such is thought to be the case with Earth and its Moon. Debris from an ancient collision between the young Earth and a Mars-sized planetesimal created a ring of debris that eventually came together to form the Moon. A similar explanation has been put forward for the satellites of Mars and the Pluto-Charon system (see Chapters 7 and 12).

Gas Giants and Ice Giants

In the outer reaches of the solar nebula, temperatures were low enough for ices to form. Indeed, it seems that ice particles were much more abundant than silicate dust particles. This being the case, any planetesimals born in the frigid outer zone would have resembled icy dirt balls, much like the comets we see today. However, the main constituents of Jupiter and Saturn are hydrogen and helium, rather than water. Since temperatures in the nebula would have been too warm for these gases to condense, accretion of hydrogen and helium snowflakes cannot have occurred. Another explanation must be found.

There seem to be two possibilities. Studies of gas giant interiors suggest that Jupiter and Saturn may possess rocky cores at least as large as the Earth. It may be, therefore, that the early stages of growth of these planets resembled the accretion taking place in the inner Solar System, with the growth of massive nuclei of ice and dust. Once these became sufficiently large, about five to 15 times the mass of Earth, they were able to attract and hold onto even the lightest gases in the surrounding solar nebula. As their mass and gravitational grasp grew, their spheres became ever more bloated.

Alternatively, they could simply have developed as the result of large-scale gravitational instabilities in the solar nebula. Since

Box 1.4   Mass and Density

Two of the basic properties of Solar System objects are mass and density. Mass is a measure of the amount of matter in a particle or object. The standard unit of mass in the International System (SI) is the kilogram (kg). This is usually determined by measuring the object's gravitational influence on other objects, e.g. natural or artificial satellites.

Once the volume of an object is known, its bulk density can be calculated. In this book, density is usually expressed in grams per cubic centimeter (g/cm^3). As a guide, the density of water is $1.0 \, g/cm^3$. Objects which have a density lower than water are able to float (assuming enough water is available!).

If a planet has a high density, it means that it is largely made of dense, rocky, or metallic materials. Objects often have low densities because they contain a lot of gases or ices, but few rocky materials. This is why all of the giant planets in the Solar System have low densities, despite their huge size.

The planet with the lowest density ($0.7 \, g/cm^3$) is Saturn. The reason that Saturn has such a low density is that it is mainly composed of gas, particularly hydrogen and helium. There is only a small rocky core at its center.

Other objects, including many small satellites and asteroids, have low densities because they are piles of loosely consolidated rubble or highly porous, i.e. they contain numerous empty spaces.

The densities of planets are also a reflection of their size and the layering of their interiors. Earth has the highest density of all the planets in the Solar System because it is made of dense, rocky materials. At the surface, crustal rocks have densities between 2.5 and $3.5 \, g/cm^3$. However, Earth's average density is much higher ($5.5 \, g/cm^3$).

This is partly because the denser elements, such as iron and nickel, have sunk to the center of the planet, while the less dense materials have risen to the surface. Many planets were internally differentiated in this way early in their lives.

The centers of the planets are also more compressed by the weight of the overlying material. In the case of Earth, for example, the normal, uncompressed density of its rocks is about $4.4 \, g/cm^3$, but the central core is compressed to greater than normal density by the overlying layers.

More massive planets should experience greater compression at their centers, and hence higher average densities, if they are made of the same rocky and metallic materials as Earth. The opposite should apply to smaller planets. However, the smallest of the rocky planets, Mercury, actually has an average density of $5.4 \, g/cm^3$, only slightly lower than Earth's.

Mercury's density rises to a remarkable $5.3 \, g/cm^3$ after it has been corrected for the effects of internal compression – much higher than Earth's. The only way to explain this is to assume that the little planet has a huge core of iron and nickel that takes up almost half of its interior (see Chapter 5).

the disk in the outer reaches contained both dust and condensed ices, there was plenty of raw material for large planets to develop and grow.

Any theory must also account for the fact that Jupiter and Saturn are huge hydrogen–helium planets, whereas Uranus and Neptune are notably smaller and contain sizeable amounts of elements that form ices: oxygen, carbon, and nitrogen. If the latter pair began as icy nuclei, they must have grown quite slowly in the more rarefied conditions of the presolar nebula beyond about 15 AU. By the time they were massive enough to draw in large amounts of gas, the nebula had dissipated, and the supply was cut off.

Migrating Planets

Our picture of the early Solar System is complicated by the likelihood that the giant planets migrated considerable distances before they ended up in their present positions. Such large-scale movement is supported by the discovery of numerous large, extrasolar planets that orbit within a fraction of an astronomical unit of a star.

In the case of our Solar System, this migration can be explained by the exchange of orbital momentum between giant planets and innumerable planetesimals. One current model (the Nice model – see Box 1.5 and Figures 1.17 and 1.18) envisages a chaotic early Solar System occupied by the major planets out to a distance of about 15 AU (closer than the present orbit of Uranus). Jupiter may have been born a little farther out in the Solar System than it is today, whereas the other giants were closer to the infant Sun than at present. Beyond the planets was a region swarming with leftover planetesimals.

Whereas Jupiter was massive enough to eject large numbers of planetesimals to the outer reaches of the Solar System or out of the system altogether, the three smaller giants were unable to do this. Instead, they flung similar numbers of planetesimals toward the Sun and away from it. Whenever Uranus or Neptune decelerated a nearby planetesimal, causing the object to move closer to the Sun, the planet gained a tiny amount of momentum. The resultant acceleration caused it to drift away from the Sun.

Over time, after billions of such gravitational interactions, Jupiter spiraled inward a modest distance, while Saturn drifted outward. When Jupiter reached a distance of 5.3 AU and Saturn arrived at 8.3 AU, the two planets were in a 2:1 orbital resonance, so that one orbit of Saturn lasted precisely two Jupiter orbits. The repeated gravitational pull of Jupiter caused Saturn's orbit to become much more elongated.

Box 1.5 The Nice Model and Jupiter's Grand Tack

In recent years, many thousands of exoplanets have been discovered orbiting distant stars (see Chapter 14). Most of these planetary systems are very different from our Solar System. Instead of all their planets traveling in near-circular orbits that lie in the same plane, they are made up of planets traveling around the central stars in unusual orbits that are difficult to explain with the traditionally accepted planetary formation process.

These remote systems include exoplanets whose paths lie in completely different planes, and worlds with extremely eccentric orbits that take millennia to complete.

At the other extreme are the "hot Jupiters," which orbit extremely close to their stars – much nearer than Mercury's distance from the Sun. At such close proximity to a star, temperatures would be too high for a massive planet to retain its gaseous envelope during formation.

If these worlds cannot have formed at their current locations, the obvious implication is that they formed further out and then their orbits were greatly modified. This would mean that the system evolved as the result of planetary migration.

The possibility of planetary migration has been considered for several decades, but the first detailed analysis of how this could occur came in 2005 with the introduction of the Nice Model (named after the French city).

Papers published in Nature by an international collaboration of scientists (Rodney Gomes, Hal Levison, Alessandro Morbidelli, and Kleomenis Tsiganis) suggested that planetary migration may have occurred as the result of an exchange of orbital momentum between the giant planets and innumerable planetesimals, over the course of a billion years.

The Nice Model envisaged an early Solar System in which the major planets were much more closely spaced and compact than at present, following near-circular orbits between ~5.5 and ~17 astronomical units (AU).

Jupiter may have been a little farther out in the Solar System than it is today, whereas the other giants were closer to the infant Sun than at present. Beyond the planets was a region swarming with leftover planetesimals (comets and Pluto-like objects).

The outermost planet, Neptune, began interacting with comets located at the inner edge of the young Kuiper Belt, so that some were scattered outward to interstellar space and others were sent inwards. Some of these entered the gravitational sphere of Uranus and were scattered again. This process of gravitational scattering was repeated with Saturn.

Whenever Saturn, Uranus, or Neptune decelerated a nearby planetesimal, causing the object to move closer to the Sun, the planet gained a tiny amount of momentum and accelerated. The overall result was the gradual outward migration of Neptune, Uranus, Saturn and the Kuiper Belt.

Unlike the three smaller giants, Jupiter was massive enough to eject large numbers of planetesimals to the outer reaches of the Solar System or out of the System altogether. Over time, after billions of such gravitational interactions, Jupiter spiraled inward a modest distance, at the same time as Saturn was drifting outward.

When Jupiter reached a distance of 5.3 AU and Saturn arrived at 8.3 AU, the two planets were in a 2:1 orbital resonance, so that one orbit of Saturn lasted precisely two Jupiter orbits. This would have occurred around 600–700 million years after they began to form.

The repeated gravitational pull of Jupiter caused Saturn's orbit to become much more elongated. This resulted in Saturn passing closer to Uranus and Neptune, so their orbits were also made more elliptical.

As the outer planets interacted chaotically with each other, it seems that Neptune and Uranus may have sometimes swapped places. One or both ice giants also plunged into the outer reservoir of planetesimals, scattering billions of them in all directions.

By the time the planets had cleared most of the intruders from their vicinities and the system had settled down again, Saturn had migrated out to about 9.5 AU. The effect on the outer planetary pair was even more extreme. Uranus had moved from about 13 to 19 AU, while Neptune had been catapulted from 15 to 30 AU.

Another consequence of this 500-million-year long planetary reshuffle was that the remaining planetesimals, perhaps 0.1% of the original population, were relocated beyond 30 AU, where they now reside as Kuiper Belt objects (KBOs).

The inward flux of planetesimals during the phase of dynamical instability also allows for chaotic capture of Jupiter's and Neptune's Trojan asteroid populations.

Furthermore, the asteroid belt was also strongly perturbed during Jupiter's migration, adding to the sudden, massive delivery of planetesimals to the inner Solar System. As their pockmarked surfaces show, the Moon and terrestrial planets appear to have suffered heavily during this Late Heavy Bombardment, around 3.9 billion years ago.

A modified version, the Nice 2 Model, suggests that the gradual scattering of planetesimals caused Jupiter and Saturn to fall into a 3:2 orbital resonance (not the originally proposed 2:1). This favours the development of a stable inner Solar System, where the rocky, terrestrial planets could form.

It also suggests that the mass of planetesimals hitting the regular satellites of Jupiter, Saturn etc. is smaller than assumed in studies based on the classic Nice Model by a factor of between 3 and 6. The impact rate is smaller in the Nice 2 Model because (at least in part) encounters with the planets cause the orbits of KBOs to become highly eccentric, resulting in less gravitational focusing by the planets.

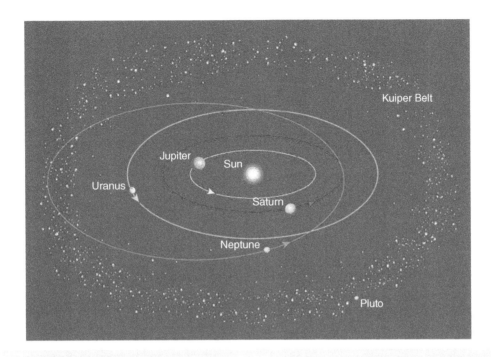

Figure 1.17 According to the Nice model, the outward migration of Saturn's orbit changed the orbits of Uranus and Neptune. In the scenario shown here, the orbits of Uranus and Neptune cross over. Meanwhile, Neptune plows into the cloud of icy planetesimals that make up the young Kuiper Belt. The gravitational interaction of Saturn, Uranus, and Neptune with the planetesimals sent billions of these objects inward, toward the Sun. As a result, these three planets migrated outward to their present orbits. This may also account for a possible period of heavy bombardment in the inner Solar System about 4 billion years ago. Some planetesimals, such as Pluto, were locked into orbital resonances with Neptune. (Nature)

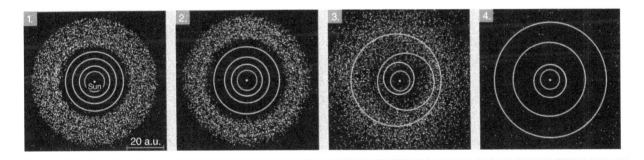

Figure 1.18 One version of the Nice Model. (1) The giant planets surrounded by a cloud of planetesimals. Neptune's orbit (blue) is closer to the Sun than that of Uranus (green). (2) As the giant planets scatter comets into deep space, Jupiter migrates inward, while the other three planets migrate outward. (3) After Jupiter and Saturn briefly enter a 2:1 orbital resonance, they change the orbits of Uranus and Neptune, causing them to scatter billions of planetesimals inward or outward. (4) The giant planets settle into their final orbits and the outer belt of planetesimals is heavily depleted. (Gomes, Levison, Morbidelli, and Tsiganis)

Suddenly, Saturn began to create havoc with the orbits of Uranus and Neptune, causing them to become more elliptical. They began to plow through the outer swarm of icy planetesimals, scattering billions of them in all directions. By the time the planets had cleared most of the intruders from their vicinities and the system had settled down again, Saturn had migrated out to about 9.5 AU. The effect on the outer planetary pair was even more extreme.

Uranus had moved from about 13 to 19 AU, while Neptune had been catapulted from 15 to 30 AU.

Another consequence of this 500-million-year long planetary reshuffle was that the remaining planetesimals, perhaps 0.1% of the original planet-building population, were relocated beyond 30 AU, where they now reside as Kuiper Belt objects.

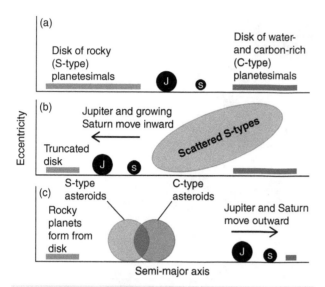

Figure 1.19 The main stages of the Grand Tack Model. Panel (a) shows the initial state, where Jupiter (black circle labeled J) and a not-complete Saturn (black circle labeled S) lie between an inner, warmish region of the Solar System populated by differentiated assorted kinds of asteroids (S-type), and an outer region of C-type asteroids containing water, carbon, etc. In panel (b), as Jupiter and proto-Saturn move toward the young Sun, S-type asteroids are scattered outward. Panel (c) shows the position of Jupiter and the fully-grown Saturn after they "tacked," or reversed course. The large circles in the center depict asteroids from the inner and outer Solar System orbiting mostly in separate regions of the asteroid belt. Eccentricity is a measure of how elliptical an orbit is; semi-major axis is the mean distance from the Sun. (Kevin Walsh)

Furthermore, the asteroid belt was also strongly perturbed during this burst of migration, adding to the sudden, massive delivery of planetesimals to the inner Solar System. As their pockmarked surfaces show, the Moon and terrestrial planets suffered heavily during this Late Heavy Bombardment, around 4 billion years ago.

According to a more recent theory, known as the Grand Tack Model, planetary migration was also a key factor during the first 5 million years of Solar System evolution (Figure 1.19). The model suggests that, directly after its formation out of the early solar nebula at about 3.5 AU, Jupiter migrated toward the Sun, as the dense gas in the nebula dragged it toward the Sun. Then, due to the growing gravitational influence of the newly formed Saturn, Jupiter halted its migration, then "tacked" (reversed direction) like a sailboat tacking around a buoy, when it reached about 1.5 AU. It then migrated outward toward its current position at 5.2 AU.

The migrating Jupiter depleted the mass concentrated at the current locations of Mars and the asteroid belt, thereby preventing Mars from growing bigger. The model also successfully explains the inclinations of asteroids and the transition from water-poor to water-rich asteroids in the middle of the main asteroid belt. The whole process took about 500,000 years.

Planetary Satellites

The Solar System contains well over 150 planetary satellites, but, as might be expected from the wide range of sizes and compositions, these seem to have arisen in several different ways (Figure 1.20).

As mentioned above, Earth's Moon is thought to have been born during a massive, grazing collision between the young Earth and a Mars-sized planetary embryo. The mixture of debris from both objects formed a ring around the scarred Earth, eventually accreting into a large satellite.

Other satellites may also have been created by sizeable impacts early in the Solar System's history. For example, the Pluto-Charon system may have originated during a collision between two large, icy planetesimals over 4 billion years ago. Simulations show that some of the debris from the collision would be blasted into orbit around the surviving protoplanet, eventually coalescing to form Charon and several smaller satellites.

Most of the major satellites seem to have followed a less traumatic path, gradually accreting from a protoplanetary disk, much like the planets. The most obvious example is the Jovian system, with its family of four Galilean moons. The inner pair, Io and Europa, are smaller but denser (with a higher proportion of rock) than the outer pair, ice-rich Ganymede and Callisto. All of them orbit Jupiter in the same direction and in more or less the same plane.

These characteristics can be explained if the moons were born from a spherical cloud of dust and gas being drawn inward from the solar nebula by a fledgling planet. As time went by, the cloud flattened into a disk around the protoplanetary core. This disk was hotter and denser near the center, allowing condensation and accretion of the less volatile materials. Further out, the icy volatiles could also condense and accrete to form Ganymede and Callisto.

Although the Saturnian family of satellites is dominated by planet-sized Titan, none of the members are particularly rocky, with many only slightly denser than water. Titan itself is similar in density to Ganymede and Callisto. If the proto-Saturn was surrounded by a collapsing cloud, it seems to have been only about one quarter the mass of Jupiter's. This suggests that the cloud contained less silicate (rocky) material and more ice than its counterpart in the warmer environs of Jupiter.

Certainly, there is a general increase in size and mass moving outward from Saturn toward Titan, with a marked decrease in both properties beyond Titan. This has led theorists to suggest that Titan grew sufficiently quickly to collect much of the solid material in the disk around Saturn, leaving only a modest amount for the medium-sized satellites to accumulate.

A modified version of the accretion scenario has been proposed by Robin Canup and William Ward of the Southwest Research Institute. They suggested that a growing satellite's gravity induces spiral waves in a surrounding disk of gas, primarily hydrogen.

Figure 1.20 The most significant satellites in our Solar System are shown beside the Earth, with their correct relative sizes and colors. Ganymede and Titan are larger than Mercury and eight satellites are larger than Pluto. Earth's Moon is the fifth largest, with a diameter of 3,476 km. Most of them are thought to have formed from a disk of gas and dust in orbit around their home planet. However, Triton and many of the smallest satellites are thought to be captured asteroids or Kuiper Belt objects that formed elsewhere in the Solar System. Earth's Moon, and possibly the moons of Mars, Uranus, and Pluto, are thought to have formed as the result of major impacts. (NASA)

Gravitational interactions between these waves and the satellite cause the moon's orbit to contract. This effect becomes stronger as a satellite grows, so that the bigger a satellite gets, the faster its orbit spirals inward toward the planet. They proposed that the balance between the inflow of material to the satellites and the loss of satellites through collision with the planet implies a maximum size for a satellite of a gas giant.

Numerical simulations and analytical estimates of the growth and loss of satellites showed that multiple generations of satellites were likely, with today's satellites being the last surviving generation to form as the planet's growth ceased and the gas disk dissipated.

The origin of the Uranian (and Neptunian) satellites is open to debate. Models suggest that the ice giants grew more slowly than their larger cousins. By the time they were large enough to gather a disk of material, most of the gas and dust in the protosolar disk had been dispersed, probably after the young Sun entered its active T Tauri phase. If the regular satellites of Uranus and Neptune could not have formed through large-scale accretion from a circumplanetary disk, how did they come about?

One idea is that the planets were larger and hotter during their accretion phase. As they subsequently cooled and contracted, they left behind a "spinout disk" from which small satellites could grow by accretion.

One complication is the fact that the Uranian satellites orbit in circles close to the planet's equator, even though it spins on its side. Neptune's rotation axis is also tilted quite markedly, aligned at about 30° to its orbital plane, while the orbits of its small satellites are circular and near-equatorial. This suggests that the planets were involved in major impacts early in their histories, and that the satellites were born during or after these collisions.

It may be that impacts with planet-sized objects blasted out clouds of hot material that formed orbiting disks around the ice giants. When the material cooled and condensed, the ice-rock ingredients were available for medium-sized satellites to form.

The major exception is Triton, the largest satellite of Neptune. One clue to its origin is that most of its bulk properties are very similar to those of Pluto, one of the largest known members of the Kuiper Belt. Furthermore, unlike the other Neptunian moons, it follows a retrograde path which is quite steeply inclined to the planet's equator. This unusual orbit has led to speculation that Triton was a Kuiper Belt object that ventured too close to Neptune and was somehow captured.

The Heliosphere

The motion of superhot plasma (electrified gas) inside the Sun generates a powerful magnetic field. The Sun's atmosphere extends into interplanetary space through the motion of the electrically charged particles (mainly electrons and protons) of the solar wind, which streams outward in all directions at typical speeds of between 400 and 7,500 km/s (see Chapter 2).

As the particles spiral around the Sun, they carve out an invisible bubble which extends outward for many billions of kilometers. Although electrically neutral atoms, cosmic rays, and dust particles from interstellar space can penetrate this bubble, virtually all of the atomic particles in the heliosphere originate in the Sun itself.

The region of space in which the Sun's magnetic field and the wind of charged particles (solar wind) dominate the interstellar medium is known as the heliosphere (Figure 1.21). The shape of the heliosphere and the distance of the heliopause are determined by three main factors: the motion of the Sun as it plows through the interstellar medium, the density of the interstellar plasma, and the pressure exerted on its surroundings by the solar wind.

From theoretical studies and spacecraft observations of planetary magnetospheres and the solar wind, it is known that the density of the solar wind decreases as the inverse square of its distance from the Sun. In other words, solar wind density at 4 AU is only one quarter its density at 2 AU. The strength of the Sun's magnetic field also weakens with distance, although at a slower rate. Eventually, the density and magnetic influence of the solar wind decrease so much that its outward motion is impeded by the sparse plasma of the interstellar medium.

The heliosphere acts like an island in a stream, causing interstellar plasma to be diverted around it. At first it was thought that the heliosphere really was spherical, but the two Voyager spacecraft, which are currently heading out of the Solar System on different paths, observed what seemed to be a "squashed" heliosheath.

In this new model, the heliosphere resembled a huge windsock or tadpole – much like a comet's elongated tail – that is shaped by the motion of the Sun as it plows through a hot, tenuous cloud of interstellar gas and dust. Studies of the motion of nearby stars show that the Sun is traversing the cloud at a velocity of 25.5 km/s.

The interstellar medium forces the solar wind to turn back and confines it within the heliosphere.

This picture had to be revised again in 2009, when data from the IBEX spacecraft and the Cassini spacecraft in orbit around Saturn showed that the heliosphere is roughly spherical – perhaps like an elongated balloon – after all. Instruments on the spacecraft were used to map the intensity of the energetic neutral atoms ejected from the heliosheath as the solar wind interacts with the interstellar medium. The data showed a belt of hot, high-pressure particles where the interstellar wind flows by the heliosphere. Their distribution indicates that the heliosphere resembles a huge bubble which expands and contracts under the influence of the local interstellar magnetic field as it sweeps past.

The interaction of the heliosphere with the interstellar medium takes place in several stages. For a spacecraft traveling out of the Solar System, the first boundary to be reached is the termination shock. This is a standing shock wave where the supersonic solar wind slows dramatically from more than 100 km/s to about half that speed.

Beyond the termination shock is a region known as the heliosheath, where particles of the solar wind and interstellar gas mix. We are able to learn about conditions in this remote region by studying data from NASA's two Voyager spacecraft, which are heading out of the Solar System in different directions.

Voyager 1 crossed the termination shock on December 17, 2004, becoming the first spacecraft to enter the heliosheath. Voyager 2 crossed the termination shock on August 30, 2007, 30 years after it was launched from Florida. The Voyager 2 crossing took place almost 1.6 billion km closer to the Sun than Voyager 1's, confirming that the outer boundary of the Solar System is curved.

Observations by the Magnetospheric Imaging Instrument (MIMI) on board Cassini showed that the heliosheath is about 40 to 50 AU (6 billion to 7.5 billion km) thick. Further out is the heliopause, the boundary between the interstellar medium and the heliosphere.

The Voyager 1 spacecraft made history once more in August 2012 when it crossed the heliopause and entered interstellar space, leaving the solar wind behind. The crossing took place about 19 billion km from the Sun. The spacecraft entered a region where the density is 40 times greater because it is generated by material from other stars and stellar explosions.

On November 5, 2018, Voyager 2 became the second spacecraft to record crossing the heliopause, at a distance of more than 18 billion km from the Sun. The crossing was marked by a steep decline in the speed of the solar wind particles, followed by an absence of solar wind flow around the intrepid spacecraft.

Beyond the heliopause, the interstellar ions flow around the heliosphere, modifying its size and shape. Still further out, there is probably a bow shock, another shock surface where the supersonic flow of the interstellar medium is suddenly slowed as it approaches the heliosphere. All of these boundaries are thought to be moving back and forth at speeds of up to 100 km/s as the heliosphere is squeezed and released due to gusts in the solar wind and variations in the interstellar magnetic field.

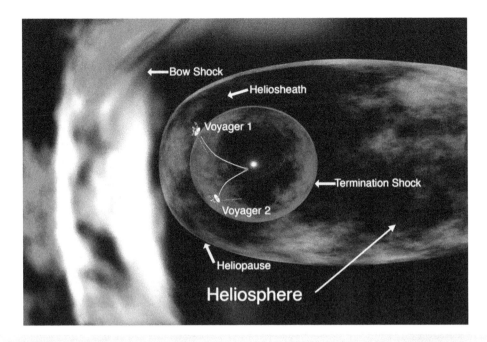

Figure 1.21 The heliosphere is a bubble in space, filled with the particles and magnetic fields carried in the solar wind. The speed of the solar wind drops abruptly at the termination shock, as it begins to feel the effects of the interstellar wind. The heliosheath is the outer region of the heliosphere, where the solar wind piles up as it presses outward against the approaching wind in interstellar space. The boundary between solar wind and interstellar wind is the heliopause, where the pressure of the two winds is in balance. This causes the solar wind to turn back and flow down the tail of the heliosphere. As the heliosphere plows through interstellar space, a bow shock forms ahead of it. Also shown are the two Voyager spacecraft which have now crossed the heliopause. (NASA/Goddard/Walt Feimer)

The Future

The Solar System is continually evolving and changing. The collision of comet Shoemaker-Levy 9 with Jupiter in July 1994 illustrated that impacts and planetary evolution are continuing today. More significantly, the Sun is also evolving, as nuclear fusion continues to create helium from hydrogen in its core.

Since its birth, 4.54 billion years ago, the Sun has grown 30% brighter and this change will continue (see Chapter 2). Over the next 1.2 billion years, its surface temperature will increase by about 150°C and its luminosity will increase by another 10%. By this time, the oceans will have boiled away. Over the next 2 billion years, even the water vapor will be lost, turning Earth into an arid planet comparable to Venus today.

Models suggest that, about 7 billion years into the future, the Sun will swell into a red giant with a diameter perhaps 200 times larger than today's value – large enough to reach almost to Earth's present orbit. However, an increase in the solar wind will cause up to 25% of the Sun's mass to be blown away.

This drop in mass will cause the orbits of the planets to expand outwards, so that Venus may recede to Earth's current orbit, while Earth may lie near the present orbit of Mars. However, this outward retreat will probably be partially balanced by solar tidal drag, which will cause our planet to spiral slowly inward. Earth's fate will hang in the balance.

Further out, Mars will briefly become warm enough to melt its icy volatiles, leading to a temporary spell of warmth with a dense atmosphere. However, the planet's gravity is not strong enough to maintain the situation for very long.

Jupiter's ice-rich Galilean moons will also develop thick atmospheres of water vapor, but again, these wet greenhouse conditions will be fairly short-lived. On Saturn's giant moon Titan, an ocean of liquid ammonia may survive for several hundred million years, perhaps providing a brief interlude when primitive life may evolve.

With its hydrogen now exhausted, the Sun will shrink and become 100 times less luminous as it switches to helium for its energy source. However, the fusion process that converts helium to carbon will only prolong its active life for a few hundred million years. As the helium becomes exhausted, the Sun will expand once more into a red giant. Riven by sudden pulsations in size, it may well consume Earth – if it still exists.

100 million years after the second red giant phase, the Sun will eject its outer layers, forming a beautiful (from the outside!) planetary nebula. All that will be left is a tiny, extremely hot, superdense core known as a white dwarf.

The final layout of the Solar System is hard to predict, but it may be that the scorched remnants of Earth and Mars, along with the outer giants, will continue to orbit the fading dwarf star, largely undisturbed, for hundreds of billions of years.

Meanwhile, our galactic environment will also have changed dramatically. About 4.5 billion years from now, the Andromeda galaxy and our Milky Way will collide, combining to form a single, football-shaped elliptical galaxy. By then, the Sun will be an aging star nearing the red giant phase and the end of its life.

Models suggest that the Solar System likely will reside 100,000 light years from the center of the new galaxy – four times further than the current distance. Any human descendants observing the future sky will experience a very different view. The band of the Milky Way will be gone, replaced by a huge bulge of billions of stars.

Questions

- What did the word "planet" originally mean? (b) Why were they given this name?
- How many planets were recognized before the invention of the telescope? (b) How many planets are recognized today in the Solar System? (c) What is the current definition of a planet?
- List six characteristics of the present Solar System that any theory of its formation must explain.
- Explain the importance of: (a) Johannes Kepler, (b) Galileo Galilei, and (c) Isaac Newton in improving our understanding of the Solar System.
- Explain the main processes by which: (a) rocky planets and (b) gaseous planets are believed to form.
- What are the main similarities and differences between gas giants and ice giants?
- Describe the main features of planetary migration, as hypothesized for the early Solar System. What relevance may this have had to the current Kuiper Belt?
- Explain three possible origins for planetary satellites. Give likely examples of each type.
- Describe the main features of the heliosphere.
- What is the likely fate of the Solar System beyond 1 billion years into the future?

TWO
Sun

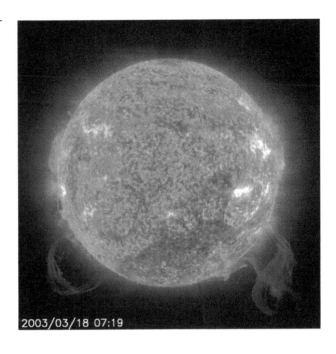

2003/03/18 07:19

The Sun is our nearest star, some 270,000 times closer than the next closest stars of the Alpha Centauri system. When compared with other stars, there is nothing out of the ordinary about the Sun. However, since it accounts for more than 99% of the mass of the Solar System, the Sun is undoubtedly the dominant influence over the local galactic neighborhood. It is the Sun's gravitational influence which holds the planets in their orbits and controls a region of space that extends in all directions for several light years.

The Sun influences every aspect of our lives, most obviously through its tremendous output of light and heat. Plants grow upward toward the Sun and soon die if the supply of light is cut off. Humans require sunlight to produce vitamin D. If sunlight is blocked by thick clouds or disappears at night, the temperature drops noticeably. In winter, when the Sun is above the horizon for shorter periods, the lower temperatures often result in frost and prolonged snow cover. If the Sun was extinguished tomorrow, Earth would soon turn into a frozen, lifeless planet, blanketed in ice.

In recent decades, space observatories have examined our nearest star in incredible detail and at many different wavelengths. Nevertheless, there are still many mysteries that remain to be unraveled, including the causes of solar flares, the mechanism that causes such extreme heating of the corona, and the origin of its 11-year periodic cycle of surface activity.

The Birth of the Sun

The structure and evolution of Sun-like stars are quite well understood (see Chapter 1). The Sun was formed some 5 billion years ago in a giant cloud of dust and gas, known as a molecular cloud, located in one of the spiral arms of the Milky Way galaxy. The cloud began to contract, possibly triggered by a shock wave from a nearby supernova explosion. Over tens of millions of years, the cloud continued to collapse under its own gravitation while pulling in material from its neighborhood.

As the gas cloud shrank to a fraction of its original size, and as the central density and pressure increased, a protostar began to form, surrounded by a rotating disk, or nebula. Jets and radiation from the growing star cleared away most of the gas in the surrounding cloud. As it shrank in size, it also began to rotate much faster. If this spin up process had continued for long enough, the fledgling star would have flown apart, but it was able to shed much of its angular momentum by accumulating (accreting) slow-spinning material from the surrounding disk of gas and dust, and by ejecting fast-spinning material through two powerful bipolar jets that formed perpendicular to the disk and were shaped by the local magnetic field.

When the core of the protostar reached a temperature of 10 million degrees Celsius, hydrogen fusion caused the Sun to begin to emit copious amounts of radiation. The remains of the accretion disk around the star eventually clumped together to form a planetary system. Today our star is approximately halfway through its life.

Exploring the Solar System, Second Edition. Peter Bond.
© 2020 John Wiley & Sons Ltd. Published 2020 by John Wiley & Sons Ltd.
Companion Website: www.wiley.com/go/Bond-Solar-System2e

The Sun as a Star

The Sun is one of more than 100 billion stars in the barred spiral galaxy that we call the Galaxy, or, more commonly, the Milky Way (Figure 2.1). By mapping the distribution of stars in the sky, astronomers have found that it is located in the Orion Arm, a small, partial arm located between the Sagittarius and Perseus arms of our Galaxy.

The Sun lies in the main disk of the Galaxy, about 28,000 light years from its center. This is quite a desirable location for a planetary system that supports life, far from the excessive gravitational disturbances and intense radiation that exist nearer the densely populated center. As the Galaxy rotates, it takes the Sun about 225 million years to complete one circuit, traveling at a velocity of 220 km/s. This amounts to about 20 trips over the course of its 4.5 billion year lifetime.

The Sun is the central star of our Solar System and by far the largest single object in the System. With a diameter of approximately 1.4 million km, it is more than 100 times wider than Earth. 1.3 million Earths would fit inside the Sun. Indeed, all of the planets combined could easily be swallowed by its enormous sphere. However, its size is modest compared with giant stars such as Betelgeuse in Orion, whose vast bulk would extend all the way to the orbit of Jupiter if it swapped places with the Sun.

The Sun's rotation was first detected by observing the motion of sunspots in the photosphere (Figure 2.2). Evidence for its gaseous composition is provided by different rotation rates at different latitudes – on solid bodies, such as rocky planets, all parts of the surface have the same period of rotation. At the equator, the Sun

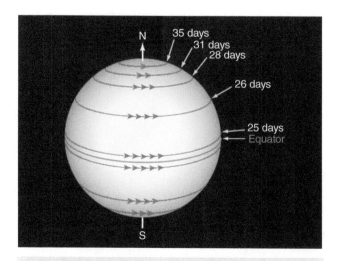

Figure 2.2 Since the Sun is a ball of gas/plasma, it does not rotate rigidly like the rocky planets. The rate of rotation generally decreases toward the poles. The Sun's equatorial regions rotate once every 25 days, compared with more than 30 days in the polar regions. This differential rotation was first detected by observing the motions of sunspots. The Sun's rotation axis is tilted by about 7.25 degrees from the axis of the Earth's orbit, so we see more of the Sun's north pole each September and more of its south pole in March. (NASA)

Table 2.1
The Sun

Diameter	1,392,000 km = 109 Earth diameters
Mass	$1.988,500 \times 10^{24}$ kg = 333,000 Earth masses
Volume	1.412×10^{33} cm^3 = 1.3 million Earths
Density	1.408 g/cm^3 (water = 1)
Temperature at the surface	5,772 K
Temperature in the corona	> 2,000,000 K
Temperature at the center	15,500,000 K
Luminosity (energy output)	$= 3.846 \times 10^{33}$ erg/s
Absolute magnitude	4.79
Apparent magnitude	–26.78
Rotation time at equator	25 days
Rotation time at 60° latitude	29 days
Photosphere composition	hydrogen (90.965%), helium (8.889%)

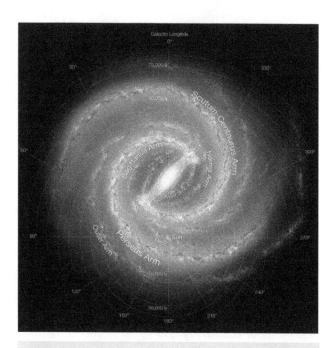

Figure 2.1 The Milky Way is a spiral galaxy with a bar-shaped mass of stars at the center. The Sun lies near a small, partial arm called the Orion Arm, or Orion Spur, about 28,000 light years from the center – a little over halfway towards the outer edge. (ESO)

spins on its axis once every 25 days, but this increases to more than 30 days at the poles. The cause of this differential rotation is an important area of current research.

The Sun's rotation axis is tilted about 7.25° to the pole of the ecliptic, so we see more of the solar north pole each September

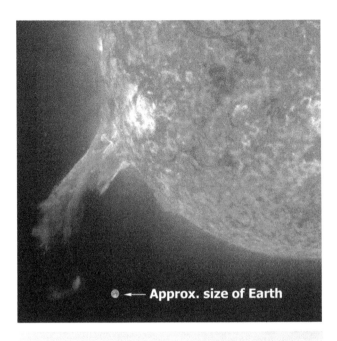

Figure 2.3 A SOHO image of a huge prominence rising from the solar disk, with Earth shown to scale. 1.3 million Earths would fit inside the Sun. (ESA/NASA)

Figure 2.4 On July 14, 2000, high-energy particles from a solar flare left their mark on the detectors of the SOHO spacecraft for several hours. SOHO is located about 1.5 million km from Earth, so that it can provide advance warning of solar storms heading for the planet. (ESA/NASA)

and more of its south pole in March. Its gaseous composition and low density result in a mass "only" 333,000 times that of Earth.

The temperature of the Sun's visible surface (the photosphere) is about 5,500°C, but about 2,000 km above this region is the much hotter corona – the Sun's outer atmosphere – which is usually only visible during a total solar eclipse. The temperature of the gas in the corona exceeds 2 million degrees Celsius. However, even this figure is dwarfed by solar flares, sudden releases of energy whose temperature typically reaches 10 or 20 million degrees Celsius, occasionally soaring to 100 million degrees.

The Solar Spectrum

The Sun is yellow in color, which correlates with a surface temperature of about 6,000°C, so it is classified as a yellow dwarf of spectral type G2. The visible white light of the solar spectrum can be split into many wavelengths (colors) by passing it through a raindrop or a prism. The red end of the spectrum is at a slightly longer wavelength (lower frequency) than the violet end.

However, most of the Sun's output of electromagnetic radiation is invisible to our eyes. The first to demonstrate the existence of this invisible light was William Herschel. 19 years after he discovered the planet Uranus (see Chapter 10), Herschel made another notable breakthrough when he used a prism to split sunlight into its colors and then placed a thermometer just beyond the red end

of the spectrum. Noting a higher temperature at this location, he concluded that it must be caused by the existence of "calorific rays" – now known as infrared (literally "below red") radiation.

Subsequent studies have shown that the Sun also emits invisible electromagnetic radiation at much longer and shorter wavelengths. In 1801, Johann Ritter proved the existence of ultraviolet radiation by using a prism to create a solar spectrum and then noting how paper soaked in silver chloride became much darker beyond the violet end of the visible spectrum.

In 1942, James Hey was assigned to find a solution to what seemed to be severe jamming of anti-aircraft radars by the Germans. Instead, he found that the direction of maximum interference seemed to follow the Sun. On checking with the Royal Observatory, he learned that a very active sunspot was traversing the solar disc. Hey had discovered the first radio emissions from the Sun.

Seven years later, solar X-rays were discovered using instruments on board U.S. suborbital rockets. The earliest detectors were simply pieces of photographic film shielded from visible and ultraviolet light by foil made of beryllium or aluminum. Later flights carried more sophisticated Geiger counters. By the 1950s it was possible to confirm that the solar corona consisted of million-degree plasma and that active regions on the Sun's surface were also a source of X-rays.[1]

---

[1] Plasma – often described as the fourth state of matter (after solids, liquids, and gases) – is an almost completely ionized gas, in which atoms or molecules are converted into ions by the addition or removal of one or more electrons. The hot mixture comprises negatively charged electrons and positively charged protons derived from high-speed collisions between ions. The electrons and protons neutralize each other, so there is no net charge.

Box 2.1   Solar Eclipses

Observation of the Sun usually requires specialized equipment and proper eye protection. However, there are rare occasions when anyone can stare at the Sun (or at least its outer regions) for several minutes without any protection. These spectacular apparitions are total solar eclipses.

Solar eclipses occur when the Moon passes directly in front of the Sun, covering its disk. This is possible because of a mathematical coincidence: the diameter of the Sun is some 400 times greater than the Moon's diameter, but, at this particular epoch of time, the Moon is approximately 400 times closer to Earth than the Sun.

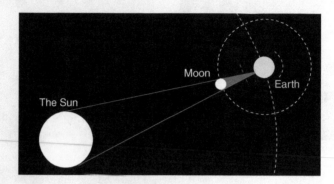

Figure 2.5   Once every four weeks, the Moon moves between the Sun and Earth. This is the lunar phase known as New Moon. Occasionally, the Moon passes precisely in front of the Sun, so that its shadow (the umbra) just reaches Earth. This results in a total solar eclipse for anyone located within the umbra. As the shadow travels across the surface of the rotating Earth, it traces a narrow path of totality, where the Sun's disk is hidden from view for up to 7½ minutes. (ESA)

Solar eclipses can only happen at the time of New Moon. If the Moon travels directly between the Sun and Earth, it is clear that our planet will pass through the lunar shadow and an eclipse will occur. However, such alignments do not take place every 4 weeks, partly because the plane of the Moon's orbit is inclined around 5° to the plane of Earth's orbit (see Chapter 4). Moreover, although the Moon crosses the plane of Earth's orbit twice every four weeks, one of these intersection points (nodes) must intersect a line joining the Sun and Earth for a solar eclipse to occur.

This means that, although the Moon passes between the Sun and Earth during each New Moon phase, the objects are usually not precisely aligned. As a result, there are only two to five total eclipses per year, when the Moon completely covers the Sun's disk.

At such times, the Moon's cone-shaped shadow (the umbra) stretches across space and grazes the Earth. For an observer situated inside the umbra, the eclipse is total. For someone located in the outer part of the shadow (the penumbra), only part of the Sun is masked, so the eclipse is partial.

As the planet rotates, the tip of the umbra travels over the surface at about 1,600 km/h, tracing a path no more than 270 km wide. Within the path of totality the sky goes dark, enabling the stars and planets to appear for up to 7½ minutes.[2]

Just before the Moon completely covers the photosphere, the Sun's light shines through gaps in lunar mountains to produce a glittering "diamond ring" effect. As totality sets in, the normally invisible solar atmosphere – the corona – appears as a pearl-white ring around the black Moon. (The corona is not usually visible in daylight because its luminosity is only about one millionth that of the photosphere.) A number of reddish, flame-like prominences may rise above the lunar limb, held in place by powerful magnetic fields.

The visible structure of the corona is related to the density of electrons in the solar atmosphere that are available to reflect light from the photosphere. Its appearance varies considerably. Near times of solar maximum, when sunspots are most numerous, the corona displays numerous bright "helmet" streamers that emanate all around the solar disk. When the Sun is less active, these streamers are fewer in number, and often missing altogether from the polar regions.

Sometimes, when the Moon is near its apogee, the lunar disk does not completely cover the Sun. During such an annular eclipse, the Moon's shadow does not reach Earth, so the dark circle of the Moon is surrounded by a bright ring of the Sun's surface. (Annular comes from annulus, the Latin word for ring.) Annular eclipses are slightly more frequent than total eclipses.

Partial eclipses are seen over a much larger area than total eclipses, so they are much more frequent in any given location on Earth. At these times, only part of the Sun is covered by the Moon, resembling a bite taken out of its disk.

Figure 2.6 The total solar eclipse of August 21, 2017, was seen across the United States. This image shows the Sun's modest activity as it neared solar minimum. In addition to the bright coronal streamers, three red prominences are visible (right). (David H. Hathaway)

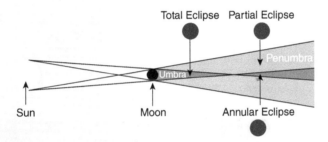

Figure 2.7 The Moon's shadow consists of two cone-shaped areas, known as the umbra and the penumbra. For an observer standing within the umbra, the eclipse is total. If Earth is slightly further away from the Moon, the outer part of the solar disk is not covered and the eclipse is annular (ring-like). For an observer standing in the penumbra, only part of the Sun is masked: the eclipse is partial. (ESA)

Occasionally, major outbursts on the Sun, involving the interaction of high-energy electrons, protons, and atomic nuclei, have been found to produce gamma rays – a form of electromagnetic radiation with extremely low wavelengths (high frequencies) and energies up to one million electron volts.[3]

Since most of the high-frequency radiation is screened out by Earth's atmosphere (fortunately for our health), studies of the high energy end of the spectrum generally require instruments to be lofted above the blanket of air, on balloons, sounding rockets, or satellites.

Considerable information on the Sun's composition and temperature can be obtained by spreading out the visible spectrum in a spectrograph (Figure 2.9). By passing light through a series of fine slits or a diffraction grating, it is possible to see hundreds of black lines in the spectrum. These are called absorption lines because they are created when different ionized atoms in the solar atmosphere absorb light at certain wavelengths.[4] The wavelength of each line indicates a specific ionized element in the Sun, while the darkness of each line shows its relative abundance.

This spectral fingerprinting makes it possible to unravel the chemical composition of the solar atmosphere. It turns out that the Sun is mainly composed of the two lightest elements, hydrogen and helium – both of which are rare on Earth.

71% of the Sun (by mass) consists of hydrogen, although this element actually accounts for 92% of the atoms in the Sun. Helium, the second-most abundant element in the Sun, is so rare on Earth that it was first discovered by studying the solar spectrum during the eclipse of 1868. Helium accounts for 27% of the Sun's mass and 7.8% of its atoms. This means that heavier elements make up only 0.1% of the atoms in the Sun.

Oxygen, carbon, and nitrogen are the Sun's three most abundant "metals," i.e. elements heavier than helium. It also has traces

[2] The longest total eclipse of the 21st century, which took place over Asia on July 22, 2009, lasted a maximum of 6 minutes 39 seconds. This duration will not be exceeded until 2132.

[3] An electron volt is a measure of energy. It is defined as the kinetic energy gained by an electron passing through a potential difference of one volt.

[4] The black lines are sometimes called Fraunhofer lines after Joseph von Fraunhofer who, in 1814, invented the diffraction grating. Many years passed before Gustav Kirchhoff discovered the link between the spectral lines and the chemical composition of the Sun.

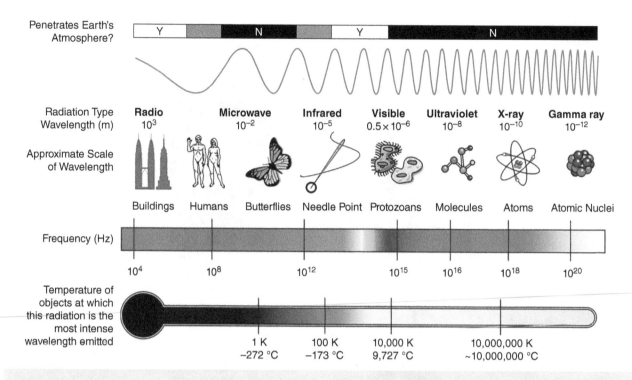

Penetrates Earth's Atmosphere?

	Radio	Microwave	Infrared	Visible	Ultraviolet	X-ray	Gamma ray
Radiation Type Wavelength (m)	10^3	10^{-2}	10^{-5}	0.5×10^{-6}	10^{-8}	10^{-10}	10^{-12}

Approximate Scale of Wavelength

Buildings　Humans　Butterflies　Needle Point　Protozoans　Molecules　Atoms　Atomic Nuclei

Frequency (Hz)

10^4　　10^8　　10^{12}　　10^{15}　10^{16}　10^{18}　10^{20}

Temperature of objects at which this radiation is the most intense wavelength emitted

1 K　　100 K　　10,000 K　　10,000,000 K
−272 °C　−173 °C　9,727 °C　~10,000,000 °C

Figure 2.8 The electromagnetic spectrum ranges from extremely short wavelength (high frequency) gamma rays to extremely long wavelength (low frequency) radio waves. The shorter wavelength radiation (gamma rays, X-rays, and ultraviolet) is associated with high temperatures and high energy processes. (NASA)

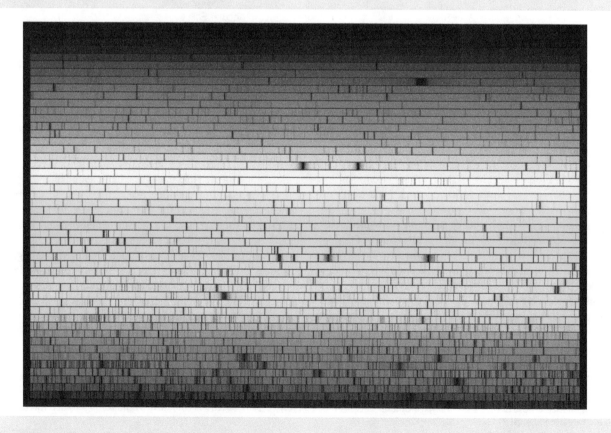

Figure 2.9 The solar spectrum in the visible, or white light, region. Spectrograms like this split the light up into different wavelengths (colors). Dark bands superimposed on the colors are absorption lines, created when atoms in the Sun's outer regions absorb light at certain wavelengths. These lines (sometimes called Fraunhofer lines) indicate specific ionized chemical elements in the Sun. (AURA/NOAO/NSF)

of neon, sodium, magnesium, aluminum, silicon, phosphorus, sulfur, potassium, and iron. Altogether some 67 elements have been detected in the solar spectrum.

Nuclear Fusion

Until the early 20th century, the source of the Sun's heat and light was unknown. Although the extreme conditions in the Sun's core were recognized by 1870, there was still no known physical process that could account for the solar furnace. For example, heat from the gravitational collapse of the Sun would only be sufficient to keep it shining for some 15 million years.

The explanation eventually came after a series of breakthroughs at the Cavendish Laboratory in Cambridge, England, and elsewhere. In 1905, Albert Einstein published his famous equation $E = mc^2$, which basically states that a small quantity of mass is equivalent to a very large amount of energy. The release of energy is always associated with a reduction in mass.

By 1911, Ernest Rutherford had shown that the atom comprised a tiny nucleus which was surrounded by particles called electrons. The nucleus itself was eventually found to be composed of other particles – protons and neutrons.

Nuclear reactions not only involved shifts in proton–neutron combinations, which released much greater amounts of energy than any chemical reactions, but they were also the cause of radioactivity – particles which are emitted from nuclei as a result of nuclear instability.

By the 1920s, Francis Aston had discovered isotopes, elements with the same chemical properties but different atomic masses. The isotopes resulted from the presence of different numbers of neutrons in an element's nucleus.

By now, it was known from laboratory experiments that some elements could be transformed into other elements. Arthur Eddington put together the pieces of the puzzle by proposing that stars such as the Sun were crucibles inside which hydrogen was transformed into helium. This transformation, he suggested, was the source of their energy.

The subsequent confirmation that hydrogen is the most abundant element in the Sun supported the idea that hydrogen nuclei (protons) are fused together inside the Sun's core, under conditions of extreme temperature and pressure. With the development of the atomic (fission) bomb in 1945, and then the hydrogen (fusion) bomb in 1952, there could be no doubt that such nuclear reactions produce vast amounts of radiation and energy.

According to the most generally accepted current theory, the Sun's energy output is the result of a series of nuclear fusion reactions, the most dominant of which is known as the proton–proton or p-p chain. These reactions take place in the central core (Figure 2.10).

In the first stage of the most common p-p chain reaction (known as P-P I), two positively charged protons collide with enough energy to overcome the repulsive electrical force between them, uniting to form a deuteron.[5]

However, since a deuteron comprises a single proton and a single neutron, one of the original protons must be transformed into a neutron by emitting a positively charged particle, known as

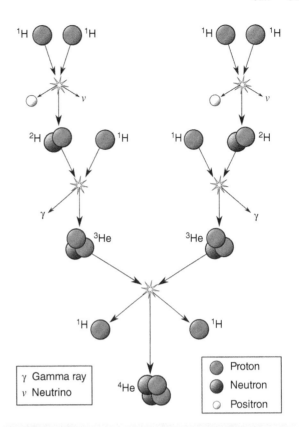

Figure 2.10 The P-P I proton-proton chain reaction has three stages. During this nuclear process, two pairs of hydrogen nuclei (protons) are united to create a helium nucleus. During this process, two of the protons are changed into neutrons, so the helium nucleus comprises two protons and two neutrons. This proton transformation is made possible by the release of a positively charged anti-particle, called a positron, and a neutrino (v), leaving behind a neutral (chargeless) neutron. Gamma rays (γ) are also released during this collision process. (Wikimedia)

a positron, along with a low-energy neutrino. Some of the energy liberated during this reaction is converted into radiation when the positron collides with an electron.

The second stage occurs less than a second later, when the deuteron collides with another proton to form a nucleus of light helium (helium-3), while releasing a gamma ray. Finally, two helium-3 nuclei meet and fuse to form a nucleus of normal helium. When this occurs, two protons are returned to the surrounding gas. This leaves a helium nucleus comprising two protons and two neutrons.

The entire process may take more than a million years, on average, but the sheer number of reactions that occur each second inside the Sun has a remarkable multiplier effect. Roughly 10 trillion trillion trillion helium nuclei are created every second, resulting in a reduction in the Sun's mass of about 4.3 million tonnes per second. This seems an incredible amount, but the loss is

---

[5] There are two other possible proton-proton reactions which are not described here.

insignificant compared with the Sun's total mass of two thousand trillion trillion tons. Since the helium nucleus is only 0.7% lighter than its original components, the Sun has actually lost only a few hundredths of one percent of its mass during its 5-billion-year history.

More serious is the rate of conversion of hydrogen into helium. About 600 million tons of hydrogen are transformed into helium each second in the Sun's core. Over its lifetime, approximately 37% of the hydrogen has already been converted into helium. After another 5 billion years of chain reactions, our star will run out of hydrogen and require a new source of fuel.

The Structure of the Sun

Although the Sun is a giant ball made almost entirely of plasma, it is not uniform throughout. Models of the solar interior, based on a combination of theoretical studies and observations, show a layered structure, with two main spherical shells surrounding a high density core (Figure 2.11 and Table 2.2).

Each of the regions has different physical characteristics. The core, the innermost 25%, is where the Sun's energy is generated. This energy diffuses outward in the form of radiation (mostly gamma rays and X-rays) through the radiative zone, then continues up toward the surface by convective fluid motion through the convection zone, the outermost 30%. A thin interface layer (the tachocline) between the radiative zone and the convection zone is where the Sun's magnetic field is thought to be generated.

Radiation takes over again in the thin, visible surface layer – the photosphere. Above the photosphere are the chromosphere and the corona. These two regions make up the Sun's sparse, but extremely hot atmosphere, which makes such spectacular viewing during total solar eclipses. This atmosphere is molded and modified by ever-changing magnetic fields and electrical currents. The moving currents and the rotation of the Sun result in a complex magnetic structure, whose strength varies and whose polarity changes in roughly 11-year cycles.

Relatively empty zones in the corona, known as coronal holes, often occur – particularly near the Sun's poles. Sometimes they spread towards the equator or even stretch all the way from pole to pole. They are particularly prominent at times of solar minimum. Since the magnetic fields in these holes are radially aligned to the surface, rather than curving back to the Sun, they offer an escape route for the charged particles in the high-speed solar wind (see Figure 2.34).

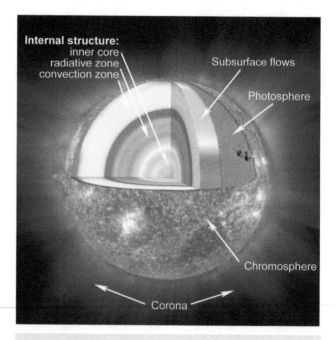

Figure 2.11 At the center of the Sun is an extremely dense core, where the temperature reaches 15 million degrees Celsius. The core accounts for 2% of the Sun's volume but 60% of its mass. Energy generated during nuclear reactions in the core travels to the surface through the radiative and convective zones. The radiative zone lies between about 25% and 70% of the distance from the center. The convective zone extends to just below the surface. In this region, the reduction in temperature toward the surface is so rapid that the gas becomes unstable, allowing convection currents of rising and falling plasma to dominate. (NASA)

The Core and the Radiative Zone

In the core, gas is squashed by the pressure of the overlying material, so that it occupies a much smaller volume. At the very center, the Sun's density is about 150 g/cm³, 10 times greater than that of solid lead, and the temperature soars to more than 15,000,000°C. As described above, nuclear reactions in the core consume hydrogen to form helium.

Both the temperature and the density decrease with distance from the center, so that nuclear fusion more or less ceases beyond

Table 2.2
Internal Zones of the Sun

Zone	Radial Distance from Center (%)	Temperature (K)	Density (g/cm³)	Energy Transport
Core	0–25	15 million–7 million	~150–20	Radiative
Radiative zone	25–70	7 million–2 million	20–0.2	Radiative
Convective zone	70–100	2 million–5,700	0.2–0.0000002	Convective

the outer edge of the core (about 25% of the distance to the surface or 175,000 km from the center). At that point the temperature is only half its central value and the density drops to about 20 g/cm^3.

The incredibly high temperature is a major factor in preventing the Sun from collapsing under its own gravity. Hydrogen nuclei near the center collide more frequently and at higher speeds than anywhere else in the Sun, so they exert a greater outward push against the overlying material. In addition to this gas pressure, the colliding particles also emit electromagnetic radiation which slowly escapes toward the surface, further helping to inflate the solar sphere. This latter force is called radiation pressure.

The main mechanism of energy transport in the inner 70% of the Sun is electromagnetic radiation in the form of photons which exhibit both wave-like and particle-like properties.

Although gamma ray photons are produced in the core, the photons that emerge from the Sun's surface are mainly in the form of visible light.

Each photon generated in the core travels only a very short distance before it encounters another particle. The photon is then scattered or absorbed, and then re-emitted. The photons produced by these encounters may emerge in any direction, so their zig-zag progress toward the surface via innumerable collisions may take hundreds of thousands of years.

As the photons diffuse outward, they move into regions where the plasma is cooler. As a result, they collide with electrons or ions that are generally moving at slower speeds, and the overall energy of the photon population is gradually reduced.

Above the core is the radiative zone, the Sun's largest internal "shell," which reaches out to 70% of the star's diameter. From the bottom to the top of the radiative zone, the density drops from 20 g/cm^3 (about the density of gold) to only 0.2 g/cm^3 (much less than the density of water). Meanwhile, the temperature falls from 7 million degrees Celsius to about 2 million degrees over the same distance.

The Interface Layer (Tachocline)

Between the radiative zone and the outer convective zone is a relatively thin transition region, known as the interface layer, or tachocline. There, the stable conditions of the radiative zone gradually disappear, to be replaced by the more turbulent fluid motions found in the convection zone.

Although it is extremely difficult to investigate conditions at such depth within the Sun, recent investigations indicate that the Sun's magnetic field is generated by a dynamo effect in this layer. Shear flows – changes in fluid flow velocities across the layer – stretch and increase the strength of magnetic field lines of force. Observations reveal that the speed of plasma in layers above and below the supposed dynamo region can change by 20% in six months. Furthermore, when the lower gas speeds up, the upper gas slows down, and vice versa.

There also seem to be sudden changes in chemical composition across the tachocline. Variations in temperature and pressure, originating in the tachocline and spreading into the convection zone through the transport of heat, may also be the key to differences in the rate of rotation of the Sun's surface.

The Convective Zone

Above the tachocline is the convective zone, which makes up the outer 30% of the solar sphere. The convective zone extends from a depth of about 200,000 km all the way to the visible surface, the photosphere. At the base of the convection zone the temperature is about 2 million degrees Celsius, but this decreases to about 6,000°C at the surface, where the density is only 0.0000002 gm/cm^3 (about 1/10,000th the density of air at sea level).

As its name suggests, energy is transported within this zone by convection – the turbulent motion resulting from hot gas rising toward the surface and cooler gas sinking back toward the center.

Although, to us, the temperature in the lower convective zone seems incredibly high, gas in the convective zone is actually cool enough for the heavier ions (such as iron, carbon, nitrogen, calcium, and oxygen) to hold on to some of their electrons. This means that the material is more opaque, making it harder for radiation to travel through.

Heat trapped in the lower part of the convective zone makes the plasma unstable, so that it starts to boil or convect. The hotter, less dense, material starts to bubble up from below, and continues to rise as long as its temperature exceeds that of its surroundings. These convective motions carry heat quite rapidly to the surface, although the gas expands and cools as it rises. On the surface the convection can be seen in the form of granulation and supergranulation, creating a carpet of gaseous cells that rise and fall like bubbles in boiling liquid (Figure 2.14).

Helioseismic observations by the MDI instrument on the SOHO spacecraft have also revealed two huge circulation cells within the convective zone, probably extending to a depth of at least 100,000 km (4% of the solar radius).

Plasma near the surface flows from the equator to the poles at a speed of 32–64 km/h. There is a return flow lower down in the convection zone. Since the plasma density is much higher at depth, this flow toward the equator is much slower, about 5 km/h. Plasma can take anywhere from 30 to 50 years to complete the full circuit (see Figure 2.32).

Figure 2.12 Rotation rates of the plasma near the bottom of the convection zone (white line) – the level of the suspected magnetic dynamo – can change markedly over six months. Faster/slower rates are shown in red/blue. Meanwhile, near the surface (shown on the left of each cutaway) bands of faster (red) and slower (green) rotation move towards the equator. (ESA/NASA)

Box 2.2   Helioseismology

Although direct study of its interior is impossible, insights into the conditions – temperature, composition, and motions of gas – within the Sun may be gained by observing oscillating waves, rhythmic inward and outward motions of its visible surface. The study of these solar oscillations is called helioseismology. In many ways, it resembles the study of seismic waves generated by earthquakes to learn about the Earth's interior.

The complex pattern of periodic throbbing motions appears on the surface due to acoustic (sound) waves that are trapped inside the Sun. Although they cannot be observed with the naked eye, the tiny motions can be detected as subtle shifts in the wavelength of the spectral absorption lines. Light from a region that is rising displays a shortening in wavelength which causes a small blue shift. Sinking columns are slightly redshifted. Images showing the overall pattern of these spectral shifts are known as Dopplergrams.

The movements are also visible as minuscule variations in the Sun's light output. Based on shifts in specific spectral lines, the helioseismic images reveal millions of vertical, gaseous motions generated by sound waves all over the photosphere. The most intense of these are low-frequency waves that oscillate on a time scale of about 5 minutes, coinciding with velocities of 0.5 km/s. However, the overall pattern is extremely complex – the result of millions of oscillations, both large and small, that simultaneously resonate with periods ranging from a few minutes to one hour. Motions as slow as a few millimeters per second have been detected, but they may also be remarkably long-lived, persisting for up to one year.

Although observations of the Sun from any single location on Earth (except the poles) are generally limited in duration due to the planet's rotation, a worldwide network of observatories known as the Global Oscillation Network Group (GONG) has been set up to enable 24-hour helioseismic studies. Meanwhile, spacecraft such as SOHO have an uninterrupted view of the Sun from orbit.

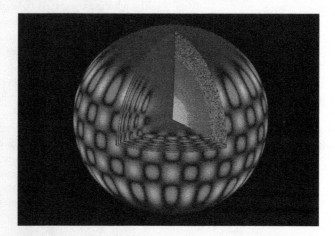

Figure 2.13   A computer representation of one of nearly ten million modes of sound wave oscillations of the Sun, showing receding regions in red and approaching regions in blue. By measuring the frequencies of many such modes and using theoretical models – a technique known as helioseismology – it is possible to infer a great deal about the Sun's internal structure and dynamics. Such oscillations are measured by the ground-based Global Oscillation Network Group (GONG), as well as orbiting observatories such as SOHO. (National Solar Observatory/AURA/NSF)

It turns out that the entire Sun is ringing like a bell, with global oscillations that may continue for weeks. Each of the 10 million sound waves reverberates around the interior before it reaches the surface. Waves of different frequencies descend to different depths. On their return journey, they are influenced by changes in temperature, density, and composition, just like seismic waves inside Earth.

The lower-pitched waves, with a frequency of about 3 MHz (a 5-minute period), have been used to probe the solar interior and even to make images of the far side of the Sun, when they give advance warning of flares and active regions before they appear around the limb and start to impact Earth.

One of the most significant results has been the recognition of a small change in sound speed at a radial distance from the center of 71.3%, marking the lower boundary of the convection zone. Measurements of the speed of the waves indicate a helium composition of 23–26% in the convection zone – consistent with other calculations of helium abundance.

The Photosphere

The photosphere is the Sun's visible surface, the region with which we are most familiar. This gaseous layer is only about 100 km deep – extremely thin compared to the 700,000 km radius of the Sun. It appears darker towards the limb, or edge of the visible disk. This limb darkening is the result of looking at the cooler and dimmer regions of the solar disk, whereas the observer looks straight down toward the hotter, brighter regions in the center of the disk.

With a temperature of only 5,700°C, the photosphere is cool enough for molecules to form, so it mostly consists of neutral (not ionized) gas. Most of the spectral Fraunhofer lines are formed here.

The most easily recognized features in the photosphere are dark sunspots, which occur singly or in groups. They grow and then fade over days or weeks. However, for reasons not yet understood, the sunspots increase in overall number, decrease, and then return to a peak every 11 years or so. This is called the sunspot cycle.

Convectional cells, found all over the photosphere, are marked by small, cell-like granules. About 1,000 km across, they show where gas is rising, cooling, and then sinking – similar to motions in a pan of boiling water. The granules are short-lived, lasting for only about 20 minutes before they are replaced by new, upwelling cells. The flow of gas within the granules can reach speeds exceeding 7 km/s, producing sonic "booms" and other noise that generates surface waves.

By measuring the Doppler effect in the solar spectrum, it is possible to measure the motion of material in the photosphere. The measurements show that much larger scale convectional motion occurs over the entire photosphere, creating supergranules about 35,000 km across. Individual supergranules survive for a day or two and have flow speeds of about 0.5 km/s.

Also visible are small, bright regions, known as faculae, which are often associated with sunspots. Faculae occur where strong magnetic fields greatly reduce the local density of the gas. The low density makes it nearly transparent, so the lower levels of granules are more easily visible. At these deeper layers, the gas is hotter and radiates more strongly, explaining the brightening.

The Chromosphere

The chromosphere (literally "color-sphere") is an irregular layer, about 5,000 km thick, which lies above the photosphere. It is named after the reddish prominences which can be seen arcing above the Sun during a solar eclipse. In this region, the temperature rises from 6,000°C to about 20,000°C, possibly due to turbulence and the action of shock waves.[6] At these higher temperatures, hydrogen emits reddish light known as H-alpha (Hα) emission.

When the Sun is viewed through a spectrograph or a filter that isolates the hydrogen-alpha light, many new features become visible. Huge prominences which tower above the limb are shaped

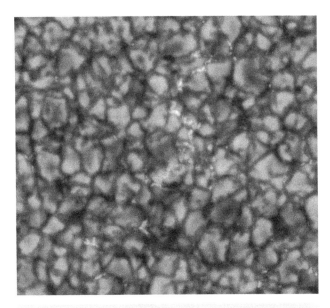

Figure 2.14 A detailed view of convection cells, or granulation, in the photosphere, taken by Japan's Hinode spacecraft. The lighter areas show where gas is rising from below, while the darker, intergranular lanes reveal where cooler gases are sinking. (JAXA/NASA/PPARC)

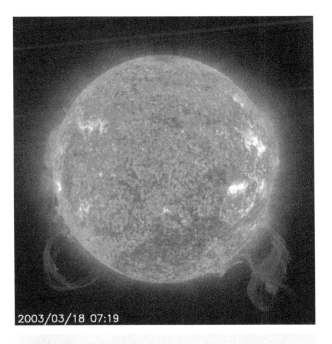

2003/03/18 07:19

Figure 2.15 Two giant prominences rising from the chromosphere on March 18, 2003. The shape of the plasma structures is molded by the local magnetic field lines. The loops rose to an altitude equal to 20 Earth diameters in a few hours, and may have been associated with a flare and coronal mass ejection. The SOHO image was taken in extreme ultraviolet light. (EIT consortium, ESA/NASA)

---

[6]The coolest part of the Sun is in the lower chromosphere, about 500 km above the photosphere, where the temperature is about 4,370°C.

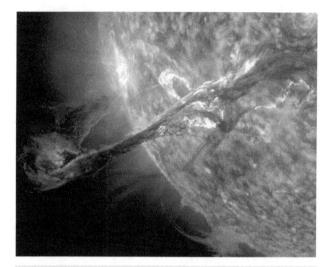

Figure 2.16   A huge solar filament erupted into space on August 31, 2012. The filament strand was stretched outward until it finally broke away. Such an erupting prominence occurs when its magnetic structure becomes unstable and bursts outward, releasing the trapped plasma. This extreme ultraviolet image was taken by NASA's Solar Dynamics Observatory. (NASA)

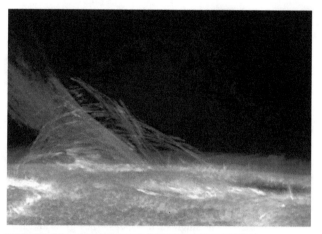

Figure 2.17   Taken by the Hinode satellite on January 12, 2007, this image reveals fine filaments of plasma in the chromosphere that extend outward from the top of the convection cells, or granulation, of the photosphere. Like a field of grass swaying in the wind, the structure results from the interaction of hot, ionized gas with constantly moving magnetic field lines. (Hinode JAXA/NASA/PPARC)

by local magnetic fields, often forming loops and arcs many times bigger than Earth.

When they occur above the main solar disk, these dense, gaseous tongues look very different. Instead of red eruptions, they appear as dark, thread-like features, so they are known as filaments. Both filaments and prominences can erupt to considerable heights over the course of a few minutes or hours.

Locally concentrated magnetic fields are also associated with plage (French for "beach"), bright patches surrounding sunspots that are best seen in H-alpha light. Plage also form part of the network of brighter regions that make up the chromospheric network. Once again, concentrated magnetic fields produced by fluid motions in the supergranules result in a patchwork that outlines the giant convectional cells. The network is best seen in H-alpha and the ultraviolet spectral line of calcium (Ca II K).

Spicules

Images taken by spacecraft such as Hinode and IRIS have revealed fine structure in the hot, ionized gas which extends up from the surface granulation toward the corona. This escaping gas can also be seen in a huge swarm of jets known as spicules.

A snapshot of the Sun taken at any moment shows 10 million spicules erupting from the star's surface. These small jets of plasma are propelled upward from the photosphere at speeds of up to 150 km/s. They are less than 600 km in diameter, but soar to heights of 10,000 km, extending to the chromosphere's outer boundary in only a few minutes.

Although each jet is short-lived, usually plunging back towards the surface, they often reform in the same spot every five minutes

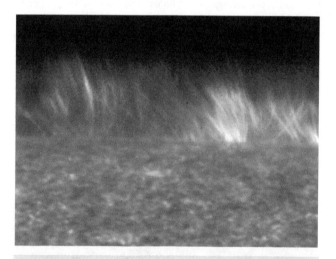

Figure 2.18   Like a field of waving grass, small jets of superheated gas, known as spicules, soar upward from the Sun's chromosphere toward the corona. Millions of these plasma jets are thought to contribute to the heating of the corona. This Hinode image was taken in extreme ultraviolet light. (JAXA/NASA)

or so. This five-minute periodicity is not a coincidence, since the surface of the Sun oscillates with the same period.

Spicules almost always occur very close to swarms of small magnetic flux tubes, where the magnetic field is highly concentrated. If the flux tube is inclined, rather than vertical, the dominant five-minute acoustic waves (see Box 2.2) can leak all the way up into the atmosphere. When they encounter lower densities of plasma, the waves grow exponentially with height and develop into shock waves. These shock waves then drive mass and energy upwards to create a supersonic jet of plasma: a spicule is formed.

The spicules may reach temperatures of more than 11,000°C and deliver huge amounts of energy and material into the corona, some of which is ejected into space in the form of the solar wind. They also generate Alfvén waves, a type of strong magnetic wave that scientists suspect is key to heating the Sun's atmosphere (see Coronal Heating).

The Transition Region

The transition region is a thin, very irregular layer that separates the chromosphere from the much hotter corona. In this region, the temperature increases rapidly from about 20,000°C to 1,000,000°C. At such high temperatures, hydrogen atoms are ionized (stripped of one or more electrons), making them difficult to see. Instead of hydrogen, the light emitted by this region is dominated by carbon, oxygen, and silicon ions. These charged particles emit ultraviolet light, so they can only be observed by spacecraft, such as Hinode and the Solar Dynamics Observatory.

The Corona

The corona is the Sun's outer atmosphere. It is extremely hot, with a temperature of about 2 million degrees Celsius, but the density of its plasma is quite low, so the corona is not very bright. Until the introduction of the coronagraph in 1930, it could only be seen during total eclipses of the Sun (Figure 2.6).[7]

The corona displays a variety of transient features, including streamers, plumes, and loops, as well as huge eruptions known as coronal mass ejections. The overall shape of the corona also changes during the 11-year sunspot cycle. Dark coronal holes are visible in UV and X-ray images of the polar regions, often extending toward the solar equator during times of solar minimum (see Figure 2.34). These holes are associated with "open" magnetic field lines which allow the high speed solar wind to escape into space. Long, thin streamers known as polar plumes also project outward along the open magnetic field lines at the poles.

Coronal loops are another common feature, sometimes rising over a million kilometers above the photosphere. These are flows of trapped plasma moving along "alleys" in the arch-shaped magnetic fields of the corona at speeds up to 320,000 km/h. Some loops are extremely hot, with temperatures of well over 1 million degrees Celsius.

Coronal loops are more common around solar maximum, when the Sun's magnetic field is highly disturbed and sunspots are numerous. Indeed, the loops are often found close to sunspots and active regions. The loops run between the north and south poles of a localised magnetic field, such as occurs in pairs of sunspots. Many coronal loops last for days or weeks, but most change quite rapidly.

One explanation is that the flows are caused by uneven heating at either extremity of a loop, with plasma racing from the hotter end to the cooler end. The reason for such heating remains uncertain (see Coronal Heating).

Figure 2.19 A series of coronal loops seen in ultraviolet light by the TRACE spacecraft. Shaped by magnetic field lines, the loops of extremely hot plasma extend 120,000 km above the photosphere. (NASA)

Observations of the corona's visible spectrum in the late 19th century revealed mysterious, bright emission lines at wavelengths that did not correspond to any known elements. The true nature of the corona was revealed in 1939, when it was shown that the lines were produced by highly ionized iron and calcium.

Such strong ionization could only happen in an environment where temperatures exceed 1 million degrees Celsius. At these temperatures, both hydrogen and helium (the two dominant elements) are completely stripped of their electrons. Even the less common solar constituents, such as carbon, oxygen, and nitrogen, are reduced to bare nuclei. Only the heavier elements are able to retain a few of their electrons.

The corona's high temperature means that it emits energy mainly at ultraviolet and X-ray wavelengths. Many active regions, flares, and other coronal features are clearly visible in X-ray images taken by space observatories such as SOHO, Hinode, IRIS, and the Solar Dynamics Observatory.

Coronal Heating

Why is the corona hundreds of times hotter than the photosphere? Since it is physically impossible to transfer thermal energy from the cooler surface to the much hotter corona, the mechanism for such intense heating has intrigued physicists for many years. However, there is now a general consensus that the Sun's localized, highly variable, and intense magnetic fields are the cause.

---

[7] **A coronagraph is an instrument that uses a disk to blot out the glare of the light from the Sun's disk.**

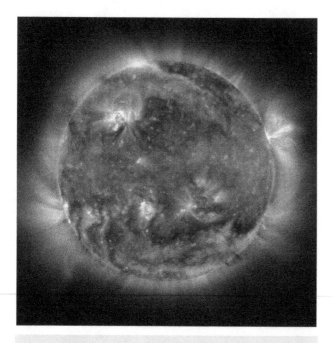

Figure 2.20   A false color image of the corona, made from three exposures taken by SOHO. Blue shows plasma at a temperature of 1–1.5 million degrees Celsius, green 1.5–2 million degrees, and red hotter than 2.5 million degrees. A dark coronal hole is visible at upper right. (ESA/NASA)

Observations show that coronal heating is more intense in regions with stronger magnetic fields. It is also generally accepted that the origin of coronal heating and activity probably lies in the photosphere and the underlying convective zone. Since convection causes rapid, turbulent motion of the gases in the photosphere, the magnetic field lines emerging from the surface are bent or mixed. This may result in the creation of waves along the magnetic field lines, or the formation of magnetic discontinuities and reconnections in the corona.

Several physical processes have been proposed to explain coronal heating. One is "wave-heating," involving waves in the gas that propagate along the magnetic field lines and then dissipate in the corona.[8]

Another is "microflare heating," when a great number of extremely small-scale flares dissipate magnetic energy in the corona through a process called magnetic reconnection. This occurs when magnetic field lines behave rather like rubber bands, snapping and then reconnecting with lines of opposite polarity.

A third candidate is spicules – the fountain-like jets of plasma that emanate from the chromosphere (see above). Some fast-moving spicules reach temperatures of more than 1 million degrees and cause a brightening of the corona. If even some of

that super-hot plasma stays aloft it would make a contribution to coronal heating. Spicules also carry electrical currents and generate magnetic waves.

High resolution X-ray imaging has revealed the presence of numerous, relatively small explosions, called microflares or nanoflares. These sudden bursts of energy occur within coronal loops – thin magnetic tubes filled with very hot plasma that arch high above the surface. Although they are so small that they cannot be studied individually, so many erupt at the same time that their combined effect is quite dramatic.

Nanoflares are associated with the build-up of considerable magnetic stress due to random shaking of the magnetic field lines that are rooted in the solar interior. In this scenario, energy is released through the reconnection of neighboring magnetic field lines with locally opposite polarity. This violent release of energy creates two jets of material at the reconnection site which are accelerated and repelled by the reconnected and highly curved magnetic field lines.

The Hinode spacecraft has measured plasma in active regions with temperatures so high that they can only be produced by impulsive energy bursts associated with storms of nanoflares.

Models based on the nanoflare theory suggest that the plasma strands within coronal loops are confined by magnetic field lines. When a nanoflare occurs, a low-temperature, low-density strand is rapidly heated to around 10 million degrees Celsius. Heat flows from the hot, upper part of the strand toward the base of the coronal loop, where conditions are cooler. Since the plasma at the base is denser, the heat input is only sufficient to raise its temperature to about 1 million degrees. This dense plasma then expands upward along the strand. Each coronal loop is, therefore, a collection of faint, 5–10 million degree strands, and bright, 1 million degree strands.

Although it seems that these processes play an important, and perhaps dominant, role in coronal heating, there is still some doubt over whether the amount of energy released by the nanoflares is sufficient to heat the corona.

Data from Hinode and NASA's Solar Dynamics Observatory indicate that a special kind of magnetic waves, known as Alfvén waves, may account for much of the coronal heating.[9] These waves travel at very high speeds along the magnetic field lines that extend from the photosphere and into the corona. However, observations have shown that short-lived spicules which shoot upward from the chromosphere wiggle sideways up to 1,000 km while they form. Advanced computer simulations suggest that these movements are caused by lateral motion of the magnetic field in the Alfvén waves.

Hinode images have also shown oscillations within solar prominences – large structures of relatively cool plasma that rise through the corona. These oscillations are widely thought to be caused by Alfvén waves propagating along the threads at speeds of about 20 km/s. Like waves crashing on a beach, they dump their energy in the corona, heating the plasma to millions of degrees.

---

[8]These are generally known as magnetohydrodynamic waves. The most important types involved in coronal heating are magneto-acoustic waves (sound waves that have been modified by the presence of a magnetic field) and Alfvén waves (similar to ULF radio waves that have been modified by interaction with plasma).

[9]The waves are named after Hannes Alfvén, who received a Nobel Prize in 1942 for his work in this field.

The Magnetic Sun

Magnetism is the key to understanding the Sun. This is because it is mostly composed of plasma, an electrically conducting gas in which the atomic nuclei have been almost entirely stripped of their electrons. The flow of electrically charged ions and electrons in the plasma is readily deflected by local magnetic fields. This is particularly noticeable in the corona, where the gases are extremely rarefied and thus easily shaped by magnetism, rather than gravity.

Observations of the photosphere's magnetic field are made by measuring the splitting of spectral absorption lines (known as the Zeeman effect) and the polarization of light. Various techniques are then used to determine both the strength and the direction of the magnetic field. These magnetic field observations can then be compared with the observed structures in the Sun's outer regions.

The Sun's magnetic field, like Earth's, resembles that of a bar magnet surrounded by a dipole field (i.e. it has two magnetic poles). The field lines flow out of the Sun at the north pole, and re-enter at the south pole. Usually, the magnetic axis roughly coincides with the rotation axis. During times of high activity, the Sun also features numerous dipoles with multiple bar magnets, each represented by an active region.

The complex "magnetic carpet" seen in the photosphere reaches the surface via the granular network of convection cells that are created as plasma wells up from below. Intense stirring causes magnetic dipoles to grow continually within the cells before being shed into the corona through magnetic reconnection.

The Sun's magnetic variability waxes and wanes over the 11-year sunspot cycle, changing in parallel with the number of active regions.[10] At solar maximum, the magnetic field is very complicated, featuring numerous, relatively small, structures in the form of active regions. The field is weaker and concentrated at the poles around solar minimum, when the lack of turbulent activity does not favor the formation of sunspots or active regions.

The active regions arise in unpredictable locations, emerging from the deep interior and breaking through the photosphere into the corona. Features frequently (but not always) associated with magnetically active regions include sunspots, coronal loops, flares, and coronal mass ejections.

Sunspots occur where very intense magnetic lines of force break through the surface. The sunspot cycle results from the recycling of magnetic fields by the flow of material in the interior. Prominences that rise above the surface are supported by, and threaded through, with magnetic field lines.

Streamers and loops seen in the corona, sometimes reaching heights of several hundred thousand kilometers, are also shaped by magnetic fields (see The Corona). The arches comprise bright strands of plasma that connect two areas in the photosphere with opposite magnetic polarities. Bright blobs of hot plasma race up and down coronal loops at tremendous speeds along threads that follow the magnetic field lines.

The polarity of the Sun's magnetic dipole switches at the end of each 11-year cycle, as the internal magnetic dynamo reorganizes itself (Figure 2.24). At solar minimum, the magnetic field

Figure 2.21 Based on SOHO data, this image shows irregular magnetic fields (the "magnetic carpet") on the photosphere. The carpet comprises a sprinkling of tens of thousands of magnetic concentrations which have both north and south magnetic poles. These are the bases of magnetic loops extending into the corona. Whiter areas represent more material at a temperature exceeding one million degrees Celsius, darker areas represent less. The black and white spots represent magnetic field concentrations with opposite polarities. Each spot is roughly 8,000 km across. (Stanford-Lockheed Institute for Space Research / NASA-GSFC)

---

[10] The 11-year sunspot cycle is actually half of a 22-year cycle of solar activity in which the Sun's magnetic polarity reverses roughly every 11 years, then flips back after 22 years.

Figure 2.22 Active regions on the Sun are made up of many relatively small magnetic structures emerging from the surface at adjacent locations. In this EUV image from the Solar Dynamics Observatory, the closely spaced, bright active regions are linked by a tangle of magnetic loops. The base of each arch has a different polarity. (NASA)

north pole. For nearly a month, the Sun had two north poles. The original south pole migrated north and, for a while, became a band of south magnetic flux smeared around the equator. By May 2000, it had returned to its usual location near the Sun's southern spin axis. Then, in 2001, the magnetic field completely flipped, so that the south and north magnetic poles swapped positions, which they retained throughout the rest of the cycle.

The Sun's magnetic field is carried out into the Solar System by the charged particles (electrons and protons) of the solar wind (see Solar Wind).

Sunspots

Sunspots are dark areas of irregular shape on the surface of the Sun. They are often large enough to be seen with the naked eye, and Chinese records of major sunspots go back to at least 28 BCE.[11]

Sunspots are much easier to observe than any other solar phenomena because of their dark appearance against much brighter surroundings. This is because they are typically about 1,500°C cooler than the rest of the photosphere – hence their darker appearance – although sunspots are actually fairly luminous. The spots vary in diameter from a mere 1,000 km to one million km. Many of them are larger than Earth, and easily visible to the naked eye under the right conditions.

Spectroscopic studies show that they are associated with strong, concentrated magnetic fields. A sunspot usually consists of a circular, dark core (the **umbra**), surrounded by a lighter region called **a penumbra** (Figure 2.25). The umbra is usually associated with a strong, vertical magnetic field, whereas the penumbra tends to have a weaker, horizontal field.

The penumbra displays light and dark linear features, known as filaments, which radiate away from the umbra. These filaments are the result of radial outward flows of gas guided by the nearly horizontal magnetic field.

In close proximity to sunspots, the strength of the magnetic field may increase more than 3,000 times. Field strengths are directly

resembles a simple dipole, with magnetic field lines running north–south. However, because the equator rotates much faster than the poles, the magnetic field becomes increasingly twisted as time goes by.

Over a period of 11 years, the twisted field lines wrap around the Sun, generating areas of intense magnetic fields that appear at the surface as sunspots. The process of tangling ends when the dynamo readjusts, recreating a dipole field, but with a reversal in polarity, i.e. the north magnetic pole switches to the south magnetic pole, and vice versa.

The change-over is not always smooth. In March 2000, for example, the south magnetic pole faded and was replaced by a

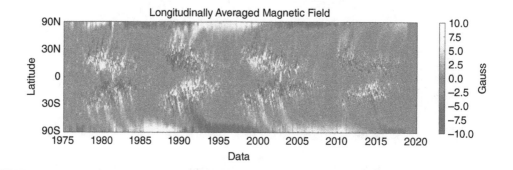

Figure 2.23 A magnetic butterfly diagram showing the distribution of the Sun's surface magnetic field (longitudinally averaged) over the last four solar cycles, i.e. since 1975. Sunspots appear in bands on either side of the equator. At the beginning of each cycle, the active regions emerge at latitudes of about 30 degrees. As the cycle progresses, the active regions emerge closer and closer to the equator. Cycles typically overlap by 2–3 years. The polarities reverse from one cycle to the next around the time of solar maximum. (David H. Hathaway)

---

[11] Direct, naked eye observation of the Sun is very dangerous and proper eye protection is essential.

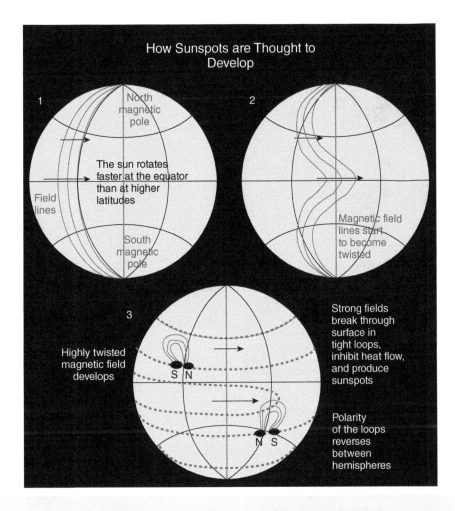

Figure 2.24 At the start of the 11-year sunspot cycle, the Sun's magnetic field resembles a large bar magnet, with two magnetic poles. However, because the equatorial region rotates much faster than the poles, the magnetic field gets progressively wrapped around the Sun, becoming stretched as it nears the equator. After many rotations the magnetic field becomes extremely complex. The twisted field lines wrap around the Sun, generating areas of intense magnetic field that are visible at the surface as sunspots. The tangling process ends when the dynamo readjusts, recreating a dipole field. (UCAR)

related to their physical size, ranging from 1,000 gauss in smaller examples to more than 6,000 gauss in giant sunspots. (This compares with an average 0.5 gauss for Earth's surface magnetic field.) These powerful fields partially block the flow of plasma and heat from below, causing the sunspot's temperature to drop more than a thousand degrees compared with its surroundings.

Sunspots are the most visible features associated with active regions, often seen alongside faculae and plage (see "The Chromosphere"). They almost always occur in groups or pairs of opposite polarity, where magnetic fields project through the photosphere from below: in the center of a sunspot the magnetic field lines are vertical. Isolated sunspot pairs tend to be aligned in an east–west direction.

Magnetograms show that the polarities of sunspot pairs located in the northern and southern hemispheres are reversed. In one hemisphere, the sunspot with a negative magnetic polarity almost always leads the sunspot which has a positive polarity (with respect to the westward apparent motion due to solar rotation). The behavior is similar in the other hemisphere, except for reversed magnetic polarities. This pattern is a direct result of the internal dynamo that generates the overall magnetic field.

Some sunspots survive only a matter of hours, but many persist for days, weeks, or even months in the case of the very biggest. Each spot resembles a whirlpool, where hot gas near the surface converges and then dives into the interior at speeds of up to 4,000 km/h. The sinking gas in the immediate vicinity of the sunspot reaches depths of only a few thousand kilometers. However, the descending flow removes the heat that accumulates beneath the spot and eventually brings it to the surface, far from the sunspot, where it is radiated into space.

The first evidence that sunspots are likely to form comes when sound waves deep in the convective zone begin to accelerate through a particular region. Within half a day, intense magnetic fields shoot upwards like a fountain, traveling at 4,500 km/h. A 20,000-km wide column of hot gas, confined

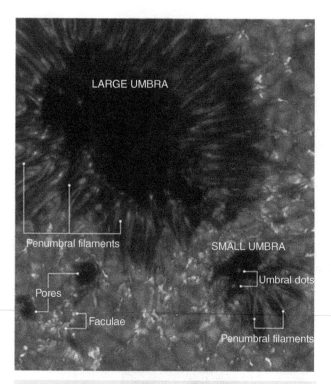

Figure 2.25 Sunspots have a darker, central region (the umbra) surrounded by a lighter outer section (the penumbra) with a linear filamentary structure. Around the sunspot are thousands of granules, the tops of convectional cells 1,000–2,000 km in diameter. Sunspots appear dark because they are cooler than their surroundings. A large sunspot might have a temperature of about 4,000°C compared with about 5,500°C for the nearby bright photosphere. This sunspot was observed with the Swedish 1-m Solar Telescope on La Palma. (Institute for Solar Physics, Royal Swedish Academy of Sciences)

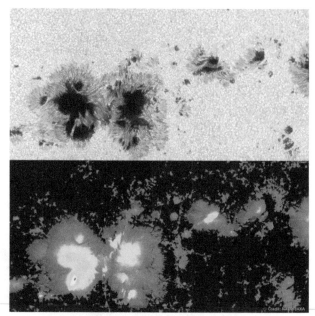

Figure 2.26 Sunspots observed by the Hinode spacecraft in February 2014. (Top) Visible light image. (Bottom) Magnetic field strength. Field strength varies from weak (purple, dark blue) to strong (green, yellow, red). Red indicates a strength of more than 6,000 gauss (600 mT), one of the highest figures ever recorded. Surprisingly, the strongest field was not in the umbra, as would be expected, but in a bright region between two umbrae. Horizontal gas flows from one umbra compressed the fields near another umbra, enhancing the field strength to over 6,000 gauss. (NAOJ/JAXA)

Sunspot Cycles

The rise in sunspot activity every 11 years or so is associated with greater magnetic activity, with an increase in the number of active regions, flares, and coronal mass ejections. This cycle was discovered in 1843 by German astronomer Heinrich Schwabe as a by-product of his search for a planet closer to the Sun than Mercury.

Astronomers give each 11-year solar cycle a number. For obscure historical reasons, solar cycle 1 was a fairly ordinary cycle which peaked in 1760. The most recent cycle, number 24, began in 2007 and ended in 2019. However, cycle 24 was extremely reluctant to start, with the Sun remaining largely blank well into the autumn of 2009, interrupted by an occasional sunspot with reversed magnetic polarity (indicative of the start of a new cycle). In addition to the deepest solar minimum in nearly a century, there was clear evidence of a decline in sunspot magnetism of about 50 gauss per year since 1992.

This was not the first time that sunspots largely disappeared. The solar minima of 1901 and 1913, for instance, were even longer than the 2008–2009 hiatus. In the 17th century, the Sun experienced a 70-year sunspot drought, known as the Maunder Minimum, that still baffles scientists. Between 1645 and 1715, the number of observed sunspots plummeted from thousands per year to a few dozen.

by an intense, rope-like magnetic field, reaches almost to the visible surface.

At a depth of 4,000 km, it may separate into strands that make their own way towards the surface, eventually forming smaller sunspots around the main spot complex. The intense magnetic fields prevent the normal upward flow of energy from the interior, leaving the sunspot much cooler than its surroundings. Immediately below the main spot is a cushion of cooler, less intensely magnetized gas. Each major sunspot complex is associated with a separate magnetic column in which the polarity is often aligned in the opposite direction.

Since more sunspots appear at solar maximum, the solar irradiance reaching Earth during that time might be expected to decrease. However, satellite radiometer measurements show that, while sunspots cause a decrease in the solar irradiance on time scales of days to weeks, the long-term solar irradiance actually increases by about 0.1% as sunspot (magnetic) activity increases. The source of this additional irradiance has been traced to the bright faculae near the limb of the Sun (see above).

The sunspot number is calculated by first counting the number of sunspot groups and then the number of individual sunspots. The final sunspot number represents the sum of the number of individual sunspots and ten times the number of groups. Since most sunspot groups contain, on average, about 10 spots, this formula has provided reliable numbers even when observing conditions were less than ideal and small spots were difficult to see.

Each cycle tends to start with a few small spots close to latitudes 40° north and south. As it progresses, the spots and associated active regions move closer to the equator, and larger, more long-lived spots appear. After approximately 11 years, the final spots of that particular cycle form at latitudes of about 5°, before disappearing altogether (Figure 2.31).

Why do the bands where sunspots form drift equatorward over time and then disappear? Data from space observatories, such as the Solar Dynamics Observatory, suggest that it may be caused by a giant circulation system, known as meridional plasma flow (Figure 2.32).

The meridional flow works something like a conveyor belt. The basic flow pattern shows two main areas of circulation, either side of the Sun's equator. Near the equator, the plasma rises and moves toward the poles, within 32,000 km of the Sun's surface. Flowing through the surface layers, where the plasma is less compressed, the material is able to move quite quickly, reaching 32–64 km/h.

Near the poles, the plasma sinks. Compressed plasma, 100,000 km below the surface, moves from the poles toward the equator at a speed of about 5 km/h – equivalent to a leisurely walking pace. The variable speed means that plasma can take anywhere from 30 to 50 years to complete the full circuit.

However, helioseismic data show that there is actually a double conveyor belt: the equatorward flow occurs in the middle of the convection layer, sandwiched between two streams of material moving toward the poles. This results in a double-cell system in which two elongated flow systems are stacked on top of each other.

Since the speed of this meridional circulation system changes slightly from one sunspot cycle to the next, it may act like an internal clock that sets the period of the sunspot cycle. The circulation is faster in cycles that are shorter than the average 11-year period and slower in longer-than-average cycles.

The absence of sunspots in 2008–2009 has been attributed to the fact that, from 1996 onwards, the deep, internal plasma flow associated with the next solar cycle was moving more slowly than usual, whilst the top of the conveyor belt was moving at record-high speed. This seems to have affected the internal dynamo, delaying the rearrangement of the solar magnetic field. This, in turn, prolonged the period of switchover between cycles, leading to an extended solar minimum.

Near-surface jet streams, which flow from the poles toward the equator at depths of 1,000 to 7,000 km, have also been associated with the sunspot cycle. By using helioseismology to keep track of gas moving below the surface, scientists from the National Solar Observatory, Arizona, found that the Sun generates new jet streams near its poles every 11 years. The streams migrate to the equator and are apparently associated with the production of sunspots once they reach a critical latitude of 22°.

Despite important progress in recent years, predictions of future sunspot cycles are not very accurate. Time will tell if the slow start to cycle 24 may mark the start of an extended quiet period of

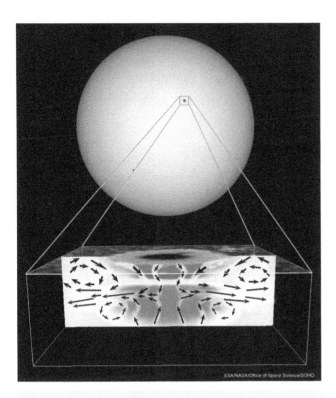

Figure 2.27 Plasma flows within and around a sunspot, derived from SOHO data. Red is hot gas, blue is cooler gas. Outflowing plasma at the surface is underlain by material rushing inward, like a giant whirlpool. This inflow is strong enough to pull the magnetic fields together and reduce the amount of heat that normally flows from the interior. The cooled material sinks to a depth of only a few thousand kilometers and then spreads out. (ESA/NASA)

Figure 2.28 An image of the chromosphere obtained by Hinode on November 20, 2006. Plasma aligned along the solar magnetic field lines is rising vertically from a sunspot (an area of strong magnetic field) toward the corona. (Hinode, JAXA/NASA/PPARC)

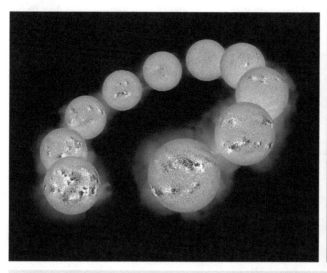
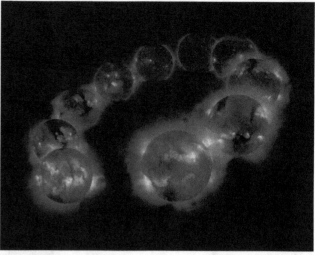

Figure 2.29 Snapshots of the changing solar magnetic field (left) and the soft X-ray corona (right) from 1991 to 2000 – almost an entire solar cycle. Obtained one year apart between one solar maximum (lower right) and the next, they show the evolution of coronal structure due to changes in the magnetic fields. Note the few magnetic features and lack of X-ray bright loops in the middle, at solar minimum. The strongest magnetic fields (shown in dark blue and white) occur in the active regions and coincide with the brightest coronal X-ray emissions. White shows an upward pointing magnetic field and dark blue a downward pointing field. Pale blue shows a weak field of "mixed" polarity. (Left: National Solar Observatory/NOAO/NSF. Right: Yohkoh/ISAS/Lockheed-Martin/NAOJ/NASA)

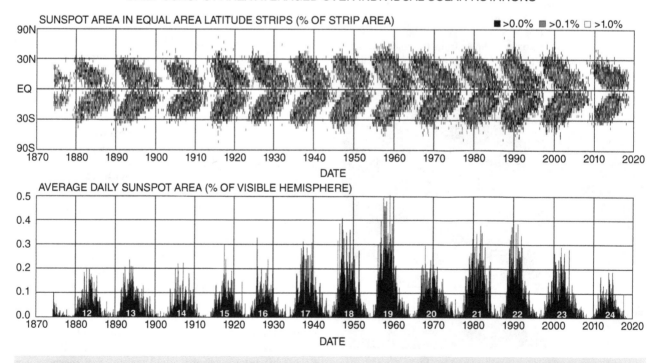

Figure 2.30 This butterfly diagram (top), named for its characteristic appearance, shows the average positions of sunspots for each rotation of the Sun, based on observations obtained by the Royal Greenwich Observatory, London, since May 1874. The bands first form at mid-latitudes, widen, and then move toward the equator as each cycle progresses. The smallest spots are shown in black, the largest ones in yellow. Below is a plot of the average area covered by sunspots over the same period. Note the decline during cycles 23 and 24. (David H. Hathaway)

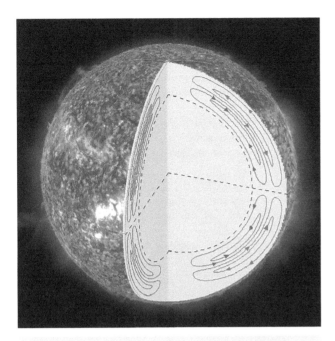

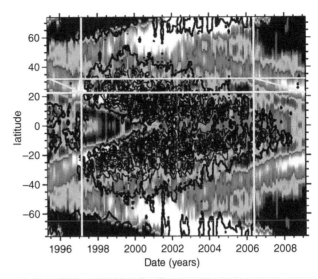

Figure 2.31 Plasma in the Sun's outer regions moves rather like a double conveyor belt. Near the surface, plasma flows slowly toward the poles and sweeping up knots of solar magnetism (decaying sunspots). It then sinks and returns toward the equator at a depth of 100,000 km, transporting magnetic flux along the way. This meridional plasma flow has two branches, north and south, each taking about 40 years to complete one circuit. Deeper still is another stream of material moving toward the poles. The structure and strength of this meridional flow is a major influence on the strength of the Sun's polar magnetic field, which then determines the duration of the sunspot cycle and number of sunspots. (NASA, Stanford University)

Figure 2.32 A diagram showing roughly east-to-west motion at a depth of about 7,000 km in the solar interior, as derived from helioseismology observations. The diagonal yellow bands are jet streams that are associated with the solar cycles. Flow cycle 23 is shown in yellow on the left side, while the flow for the next cycle (#24), which is notably slower, is shown in yellow on the right side. The jets move slowly toward the equator and when they reach 22° latitude, new sunspots begin to appear. (National Solar Observatory/AURA/NSF)

solar activity. However, exactly how the jet streams are generated, and the precise mechanism that enables them to trigger sunspot production, remains uncertain.

Solar Wind

The corona emits a continuous stream of charged particles, known as the solar wind. This solar wind flows radially outward in all directions at an average speed of about 400 km/s. Almost a million tonnes of solar material is ejected into interplanetary space each second. This outward flow of particles creates an invisible bubble in space, known as the heliosphere, and is responsible for the anti-sunward-trailing ion tails of comets and the shape of the magnetospheres around the planets.

The solar wind comprises an equal number of protons and electrons, plus a smaller number of heavier ionized atoms. These particles originate in the corona where the temperature is so high that not even the Sun's powerful gravity can hold on to them. As the particles stream toward the outer reaches of the Solar System, they carry their own magnetic fields.

Like winds on Earth, the solar wind is highly variable, changing with the state of the corona. During solar minimum, there is usually a fast, steady wind over the poles and a slow, variable wind at lower latitudes. At solar maximum, it is highly chaotic, with fast and slow streams of particles interrupted by more frequent coronal mass ejections at all solar latitudes.

The high-speed wind (750–800 km/s) is derived from plasma leaking through the corners of a magnetic honeycomb of plasma cells, mainly in large **coronal holes**. These regions are most commonly located near the poles – where the solar atmosphere is less dense and cooler than surrounding areas. Their open magnetic field lines allow a constant flow of high-density plasma to stream into space. This is particularly noticeable during solar minima, when coronal holes tend be larger and longer-lasting.

In contrast, the slow solar wind (300 km/s) originates over lower latitudes, typically over active regions and streamers, where closed magnetic field lines trap the electrically charged coronal gases, except at the edges of bright, wedge-shaped features called helmets. Many small mass ejections, driven by magnetic explosions, also contribute to the solar wind.

Massive compression and shock waves can result if a fast stream collides with a slow stream. As the Sun rotates, the various streams in the solar wind also rotate, producing a pattern similar to a rotating lawn sprinkler. If a slow stream is followed by a fast stream, the high speed material will catch up and plow into it.

When the high- and low-speed streams interact with one another, they create dense regions known as co-rotating interaction regions (CIRs) that trigger geomagnetic storms when they interact with Earth's atmosphere.

Box 2.3   SOHO

The Solar and Heliospheric Observatory (SOHO) is the most important solar observatory ever sent into space. The European-built spacecraft was launched on December 2, 1995, and transferred to the L1 Lagrangian point, approximately 1.5 million km from Earth on its sunward side. At this location, it is relatively easy to keep SOHO in a stable halo orbit from which its instruments can observe the Sun continuously, without being interrupted by eclipses.

SOHO carries 12 instruments that probe all aspects of the Sun. The spacecraft was designed to operate for two years, but it was still operating in 2019, despite some major technical problems which almost ended the mission.

SOHO has observed the Sun throughout the entire 11-year solar cycle, and its observations have led to numerous discoveries. These include:

- Helioseismic data from SOHO and the GONG network of ground stations detected currents of gas beneath the visible surface, giving new insights into the layers of the Sun's interior, the behavior of the magnetic field, and the change in sunspot numbers during the solar cycle.
- A 0.1% increase in the Sun's luminosity as the count of sunspots increased 1996–2000. Scientists estimate that high-energy ultraviolet rays from the Sun have become 3% stronger over the past 300 years.
- Until the launch of the STEREO mission, SOHO provided the only reliable way to identify coronal mass ejections that were heading towards Earth. This was done by linking expanding haloes around the Sun to shocks seen in the Earth-facing atmosphere. This gave 2–3 days' warning of these potentially damaging storms.
- Thousands of nanoflares occur every day, due to continual rearrangement of tangled magnetic fields. This helps to explain why the corona is far hotter than its visible surface.
- SOHO helped to locate the sources of the fast and slow solar winds.
- Charged atoms that feed the fast solar wind gain speed very rapidly, apparently driven by strong magnetic waves in the corona. Similar magnetic waves may accelerate the slow wind.
- SOHO found many elements in the solar wind, including the first detections of phosphorus, chlorine, potassium, titanium, chromium, and nickel. These give clues to conditions on the Sun and to the history of the Solar System.
- After a solar flare, SOHO observed waves spreading outward across the Sun's visible surface.
- SOHO discovered large tornadoes, where hot gas was spiraling outwards from the Sun's polar regions.
- A wind of particles from distant stars blows through the Solar System, partially counteracting the solar wind. SOHO fixed its direction (from the Ophiuchus constellation) and speed (21 km/s) more accurately.
- Two instruments, SWAN and MDI, detected sound waves reflected from far-side sunspots. This made it possible to "see" what was happening on the far side of the Sun, giving advance warning of active regions that had yet to appear on the Earth-facing hemisphere.
- More than 3,000 sungrazing comets have been discovered in LASCO images. This makes SOHO by far the most successful comet discoverer in history.

In addition to dramatic changes in speed, spacecraft have recorded the presence of magnetic clouds – clumps of solar particles with embedded magnetic fields – and variations in the composition of the particle population which reflect conditions in their coronal source regions. For example, instruments have shown a higher abundance of magnesium ions compared to oxygen ions in the slow solar wind than in high-speed streams.

Despite many years of observation, the precise mechanism of solar wind formation is still not fully understood. Although it is recognized that the fast solar wind originates from coronal holes, images of the outflowing material are still rare. Various source regions have been proposed for the slow solar wind. These include the boundary of coronal holes, helmet streamers located above closed loop structures in the corona, and the edges of active regions.

The suggestion that one of the sources of the slow solar wind is the boundary between coronal holes and active regions has been supported by high-resolution images from Hinode. These show a continuous outflow of plasma along apparently open field lines that rise from the edge of an active region adjacent to a coronal hole.

The mechanism that drives the solar wind is also poorly understood. However, recent observations have indicated that Alfvén waves in X-ray jets – fast-moving eruptions of hot plasma that occur near the poles – may accelerate the solar wind to hundreds of kilometers per second. These magnetic oscillations, possibly associated with short-lived spicules in the chromosphere (see Coronal Heating), appear to travel at very high speeds along open field lines that extend from the photosphere and into the corona. The wave energy that leaks into the corona may be sufficient to power the solar wind.

Flares

Flares are tremendous explosions on the surface of the Sun which can last from several seconds to a few hours. The largest flares are also the longest in duration, but they are usually quite rare, occurring only a few times a year until solar maximum

Figure 2.33 As the Sun rotates every 27 days, the solar wind becomes a complex spiral of high and low speeds, and high and low densities. When high speed solar wind overtakes slow speed wind, it creates a corotating interaction region. These interaction regions consist of solar wind with very high densities and strong magnetic fields. (NASA)

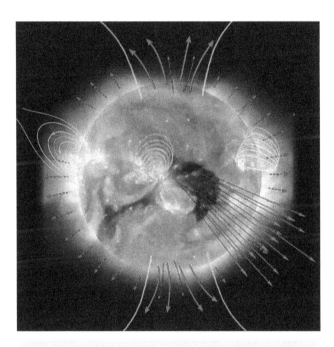

Figure 2.34 The corona is threaded with magnetic fields (yellow lines). Areas with closed magnetic fields give rise to a slow, dense solar wind (short, dashed red arrows), while coronal holes with open magnetic fields yield fast, less dense solar wind streams (longer, solid red arrows). In addition to the permanent coronal holes at the poles, coronal holes sometimes occur closer to the equator (right of center). The image was taken by SOHO's Extreme Ultraviolet Imaging Telescope. (ESA-NASA)

approaches. Many smaller flares occur down to the limits of detection of modern instruments at about 10^{27} ergs. These generally last for a short time, down to a few seconds, and they, too, are most frequent near solar maximum, when there may be several dozen flares per day.

As flares heat material to a temperature of up to 100 million degrees Celsius, they release energy equivalent to that released by millions of 100-megaton hydrogen bombs exploding simultaneously. The energy is released in many forms: electromagnetic radiation across virtually the entire spectrum, from radio waves to X-rays and gamma rays; energetic particles (protons and electrons); and mass flows.

The high-energy electromagnetic radiation travels at light speed across interplanetary space, sometimes heading towards Earth. Such increases in X-ray flux enhance the planet's ionosphere, causing it to decrease in altitude, but, fortunately, they are unable to penetrate the lower atmosphere.

Not far behind are the particles that are accelerated to near light speed by the flare. Streaming outward along magnetic field lines, they can reach the Earth–Moon system within 20 to 30 minutes, interfering with short wave radio communications, causing short circuits and computer reboots on satellites, and threatening the health of astronauts who are outside their spacecraft.

Flares are usually difficult to see in visible light against the bright background of the photosphere. Most are detected from the invisible radiation they emit. Radio and optical emissions can be observed with telescopes on the Earth, while orbiting observatories are needed to detect energetic emissions such as X-rays and gamma rays.

There are typically three stages to a solar flare, each lasting anything from a few seconds to more than an hour. First is the precursor stage, when the release of magnetic energy is triggered. The soft X-ray emission gradually increases but few, if any, hard X-rays or gamma rays are detected.

In the second, impulsive stage, protons and electrons are accelerated to energies exceeding 1 MeV, whilst hard X-rays and gamma rays are emitted, often rising in many short but intense "spikes," each lasting a few seconds to tens of seconds. The soft X-ray flux also rises more rapidly during this phase, often synchronizing with the hard X-ray profile.

In the third, decay stage, hard X-ray and gamma ray fluxes start to decay rapidly in a matter of minutes, whereas the soft X-ray flux continues to rise, reaching a peak before declining again, sometimes over a period of several hours.

Flares (and coronal mass ejections) occur near sunspots, usually along the neutral dividing line between areas of oppositely directed magnetic fields. A flare occurs when an enormous amount of magnetic energy that has built up in the corona is suddenly released as the result of magnetic reconnection – when magnetic field lines snap and reconfigure themselves.

When a coronal loop rises to great height, it may become stretched and distorted, so that the two sides of the loop move closer together. When magnetic field lines snap and then reconnect, the original loop splits in two, forming a smaller arch at the surface and a separate loop in the corona. The excess energy is released in an explosive flare, often coinciding with a coronal mass ejection which blasts huge amounts of material into space.

The flare generated in the small arch accelerates electrons down the magnetic field lines, causing them to slam into the denser plasma below, producing X-rays, microwaves, and a shock wave that heats the surface. The result is a major seismic wave in the Sun's interior, which can be observed as a series of outward moving

Box 2.4   Ulysses

Nearly all scientific spacecraft that observe the Sun or are dispatched across the Solar System follow paths that lie close to the ecliptic. However, the European Space Agency's Ulysses spacecraft followed a highly elliptical orbit that carried it above and below the solar poles, regions that are usually extremely difficult to observe.

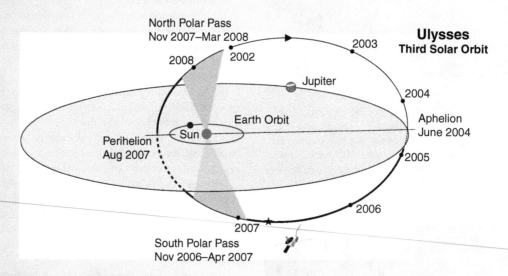

Figure 2.35   The Ulysses spacecraft was the first to leave the ecliptic and explore the Sun's polar regions. Launched in 1990, it was sent to Jupiter for a gravity assist that bent its orbit southward so that it could fly over the star's poles. Ulysses completed almost three orbits before it was shut down in 2009. The final orbit is shown here. (ESA)

Launched on the Space Shuttle Discovery in October 1990, Ulysses received a gravity assist from Jupiter in 1992. This bent its path southward, so that it would fly over the solar poles. Before its mission ended in 2009, the nuclear-powered spacecraft made nearly three complete orbits of the Sun, enabling it to observe the cyclical changes in solar activity.

Prior to Ulysses, it had only been possible to measure the solar wind near the ecliptic, giving the impression that it typically swept across the Solar System at about 400 km/s, with occasional faster gusts. Ulysses showed that this perception was incorrect. For much of the sunspot cycle, the dominant component is a fast wind from the cooler regions close to the poles that fans out to fill two-thirds of the heliosphere. Blowing at a fairly uniform speed of 750 km/s, this far outstrips the slower wind that emerges from the equatorial zone. Rather than being typical, the slow wind is a relatively minor player.

When Ulysses first flew over the polar regions in 1994 and 1995, solar activity was close to minimum, providing a view of the three-dimensional heliosphere at its simplest. The fast solar wind escaping from high latitudes flowed uniformly to fill a large fraction of the heliosphere, and solar wind variability was confined to a narrow region around the equator.

When Ulysses returned to high latitudes in 2000 and 2001, during solar maximum, the Sun displayed many active regions and solar storms were common. Solar wind flows from the poles appeared indistinguishable from flows at lower latitudes.

In 2007, Ulysses made its third polar passage. Compared with observations from the previous solar minimum, the strength of the solar wind pressure had decreased by 20%, and the field strength had decreased by 36%. Although the wind speed was almost the same, the density and pressure were significantly lower.

Of particular interest were Ulysses' observations of the Sun's magnetic field reversal at the change-over between solar cycles. Despite the apparent chaos and complexity of the magnetic field at the surface, the Sun's magnetic equator was quite well defined and stable, clearly separating the negative and positive magnetic hemispheres. However, during the solar minimum of 1994–1995, the magnetic equator was pushed 10° southward with respect to the Sun's rotational equator. The reason for this offset is still not fully understood.

Ulysses' measurements also showed that energetic particles originating in storms close to the Sun's equator are much more mobile than previously thought, and are even able to reach the polar regions. Since these particles tend to move along the magnetic field lines in the solar wind (much like beads on a wire), this indicated that the structure of the magnetic field is more complex than previously thought.

In addition to studying the Sun, instruments on Ulysses detected small dust particles, hydrogen ions, cosmic rays, and neutral helium atoms that entered the heliosphere from interstellar space. The dust flowing into our Solar System was 30 times more abundant than predicted.

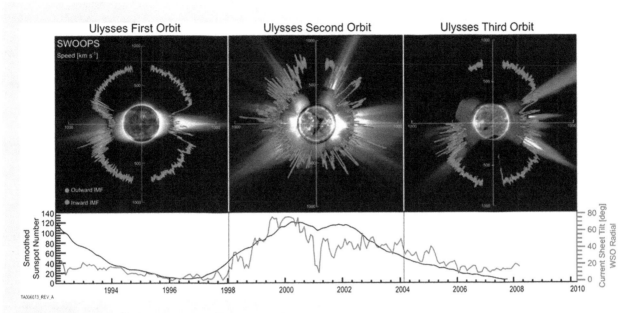

Figure 2.36 Polar plots of solar wind speed over each of Ulysses' orbits. The first orbit occurred during solar minimum, when there was a slow wind over the equator and a fast wind over the poles. The second orbit showed fast and slow winds at all latitudes, consistent with solar maximum. Data from its third orbit, during another solar minimum, indicated that the solar wind speed was similar, even though the density and pressure were significantly lower than in the previous solar minimum. The plot below shows the number of sunspots. (ESA/NASA, SOHO and High-Altitude Observatory, Mauna Loa, from McComas et al., 2008)

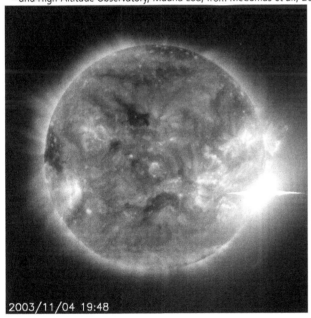

Figure 2.37 SOHO's Extreme Ultraviolet Imager took this image of the most powerful flare ever recorded by spacecraft, which erupted on November 4, 2003. Since the X-ray detector on board NOAA's GOES satellite was saturated by the radiation outburst, there is some uncertainty over its magnitude, with estimates ranging from X28 to X45. The record-breaking series of solar storms in October and November 2003 affected radio communications across the globe and caused temporary power outages in Europe due to fluctuations in Earth's magnetic field. (ESA/NASA)

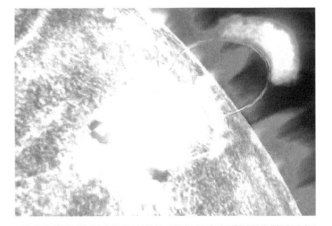

Figure 2.38 A flare is created when a magnetic loop in the corona rises to great height, becoming stretched and distorted. When the two sides of the loop get close enough, magnetic reconnection takes place. The loop splits in two, forming a smaller arch at the surface and a separate loop in the corona. The excess energy is released in explosive events like flares and coronal mass ejections (CMEs). The white flash represents a flare generated in the small arch as electrons are accelerated down its magnetic field and slam into the denser plasma near the solar surface, releasing high energy radiation. Plasma trapped in the coronal loop quickly rises and expands, propelling the CME plasma away from the Sun. (NASA)

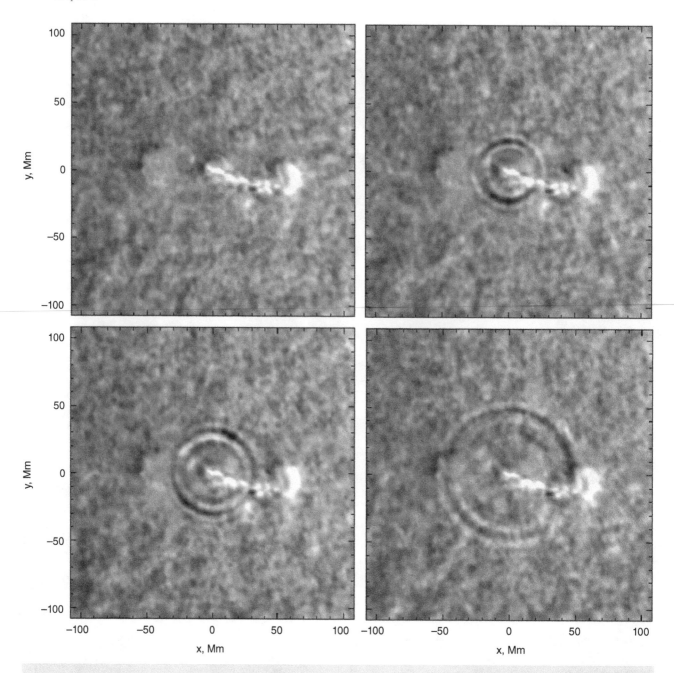

Figure 2.39 Seismic waves ripple away from the site of a moderate-sized solar flare. The image was taken by SOHO's Michelson Doppler Imager on July 9, 1996. Over the course of an hour, the waves traveled for a distance equal to 10 Earth diameters before fading into the photosphere. Unlike water ripples that travel outward at a constant velocity, the solar waves accelerated from an initial speed of 35,200 km/h to a maximum of 250,000 km/h before disappearing. (Alexander Kosovichev, Valentina Zharkova, ESA/NASA)

ripples – much like the ripples that spread from a rock dropped into a pool of water.

Over the course of an hour, the seismic waves may travel more than 100,000 km before they fade into the fiery background of the Sun's photosphere. The waves can accelerate from an initial speed of 35,200 km/h to a peak of 400,000 km/h before they disappear.

Flares may often erupt one after the other when a particularly active sunspot region appears. One of the most extraordinary

sequences of solar storms took place between October 18 and November 5, 2003, when more than 140 flares were observed, primarily associated with two large sunspot groups. Among them were 11 major X-class flares, including an X17 event on October 28, and an even bigger one on November 4.

The strength of the latter event was difficult to determine, since it saturated the spacecraft detectors, but, based on radio wave-based measurements of the X-rays' effects on Earth's upper

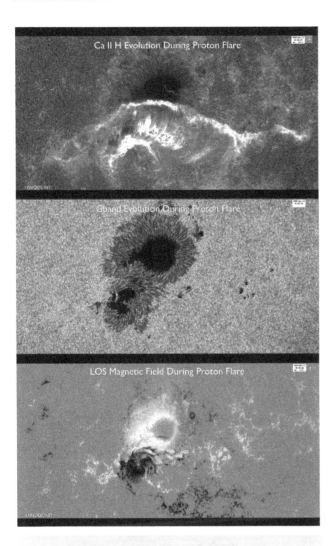

Figure 2.40 On December 13, 2006, Hinode's Solar Optical Telescope imaged at different wavelengths a new, developing sunspot colliding with an existing spot and then exploding into an X-class solar flare. The bright area in the upper image marks the footprint of a magnetic field that channeled suddenly released energy from the inner corona down to the surface. The lower image shows that the two sunspots had opposite magnetic polarities. The flare produced high-energy protons that reached Earth. (Hinode, JAXA-NASA)

atmosphere, it was later uprated to X45, making it by far the largest flare detected since the GOES satellites began their solar X-ray measurements in 1976.

Classification of Solar Flares

Scientists classify solar flares according to their brightness at X-ray wavelengths, i.e. their X-ray flux. There are five classes: A (the weakest), B, C, M, and X (the strongest). Each category has nine subdivisions, e.g., C1 to C9, M1 to M9, and X1 to X9. Very occasionally, extremely powerful flares are given a much higher designation. For example, one storm of November 4, 2003, has been variously classified as X28 or X45.

These scales are logarithmic, much like the seismic Richter scale, so an M flare is 10 times as strong as a C flare. The most important categories are:

- X-class flares – the most powerful. These major events can trigger radio blackouts around the world and long-lasting radiation storms in Earth's upper atmosphere which can damage or destroy satellites.
- M-class flares – medium-sized, releasing 10% of the energy of X-class flares. They generally cause brief radio blackouts that affect Earth's polar regions. Minor radiation storms sometimes follow an M-class flare.
- C-class flares – small, releasing 10% of the energy of M-class flares, with few noticeable consequences on Earth.

Coronal Mass Ejections

Flares are often associated with huge eruptions of ionized material, known as coronal mass ejections (CMEs), that can propel up to 10 billion tons of matter into space. These massive bubbles of plasma, threaded with magnetic field lines, are ejected from the Sun over the course of several hours.

We now know that CMEs are very common. During solar maximum, the Sun averages five of these outbursts per day, with about 100 of them heading Earthward per year. During solar minimum, the annual total may drop to about 180, with a strong CME every two days and about 10–15 per year directed at Earth.

Like all solar activity, CMEs are associated with releases of magnetic energy in active regions, particularly the concentrated magnetic fields close to sunspots. However, although they are often associated with flares and prominence eruptions, they can also occur without either of these events taking place.

Forecasting CMEs is not easy. One possible sign of something stirring is the emergence of an inverse-S shape (sigmoid structure) in the X-ray corona. This happens when a magnetic flux tube begins to twist, due to the rotation of one or both of the footprints of a cluster of loops located above a sunspot group.

In 2008, Hinode observed a CME in which one end of the tube spun clockwise, whilst the other rotated counterclockwise. This unfurling action caused the field lines to rupture and realign – a process called magnetic reconnection – producing a huge explosion that heated a huge cloud of material and propelled it away from the Sun in the form of a CME.

Although CMEs are so large, they are very sparse and spread out, containing only a few particles per cubic centimeter. Much of their ability to disrupt the flow of the solar wind and disturb planetary environs comes from their magnetic fields, particularly if their polarity is aligned in the opposite direction to a planet's field, leading to favorable conditions for magnetic reconnection to take place.

CMEs travel outward from the Sun at speeds ranging from less 250 km/s to around 3,000 km/s. The fastest CMEs can reach Earth in only 15–18 hours, though it usually takes two to four days to cross the 150-million-km gap.

During their journey, they expand in size as they move away from the Sun. Larger CMEs can reach a size comprising nearly a quarter of the space between Earth and the Sun by the time they reach our planet.

Box 2.5   Space-based Solar Observatories

Japan has been playing a leading role in solar studies for several decades. The first of its pioneering orbital observatories was Yohkoh, which was launched August 30, 1991. It carried four main instruments: a Hard X-ray Telescope, the first instrument ever to image high-energy X-ray flares; a U.S.–Japanese Soft X-ray Telescope with a field of view that covered the full solar disk but could also obtain a series of small-scale, high-resolution images of flares; a Wide-Band Spectrometer to observe solar radiation in soft X-rays, hard X-rays, and gamma rays, with a fourth detector monitoring Earth's radiation belts; and a UK-US-Japanese spectrometer to study specific spectral regions in soft X-rays.

Yohkoh was the first spacecraft to continuously observe the Sun in X-rays during an entire sunspot cycle. When it was launched, the Sun was near the peak of its 11-year cycle, so many active regions and flares were imaged. It then observed the subsequent decline and the start of sunspot cycle 23 in the late 1990s. The spacecraft sent back over six million X-ray images before it failed in late 2000.

Yohkoh provided important new data about the corona, including information about how and where this multimillion-degree layer is heated to temperatures hundreds of times greater than the solar surface. By tracking the evolution of the corona, it improved understanding of how the Sun's magnetic fields are deformed, twisted, broken, and reconnected during flares; and how the coronal plasma is heated to millions of degrees by flares. Various structures, known as sigmoids (S-shaped regions in the corona) and trans-equatorial interconnecting loops (TILs), were shown to be more likely to be the sites of solar eruptions.

Its successor, Hinode, was also a US-UK-Japan collaboration. It was launched on September 23, 2006, into a Sun-synchronous, near-polar orbit around Earth that allows it to remain in continuous sunlight for 9 months each year. It carries an optical telescope that includes a high-resolution imager, a magnetograph that makes rapid observations of the Sun's magnetic and velocity fields, and a spectropolarimeter that makes extremely precise observations of the solar magnetic field. Together these instruments are able, for the first time from space, to measure small changes in the strength and direction of the magnetic field, as well as how these changes coincide with events in the corona. One important product is vector magnetograms that illustrate variations in the strength of the Sun's magnetic field.

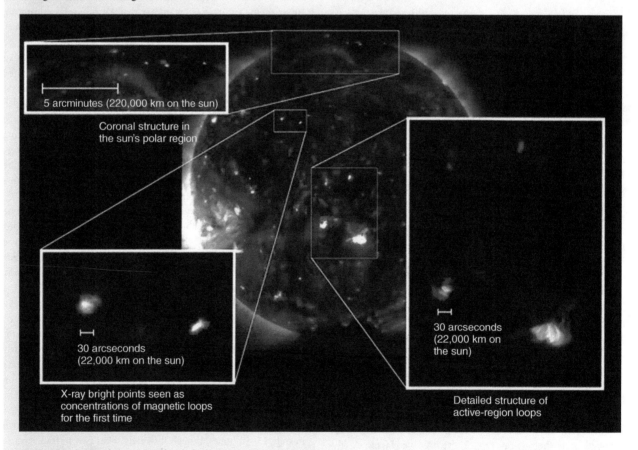

Figure 2.41   Hinode is equipped with the highest resolution solar X-ray telescope ever flown. This full disk image, taken early in the mission, shows features of the X-ray Sun with a spatial resolution of nearly 1 arc second. These include coronal activity within dark holes near the poles and coronal loops associated with active regions. (JAXA-NASA)

Also on board is an advanced version of the Soft X-ray Telescope flown on Yohkoh. The highest resolution solar X-ray telescope ever flown, it shows the structure and dynamics of the corona over a wide range of temperatures and a broad field of view. By combining optical and X-ray observations, it is possible to study how changes in the magnetic field trigger explosive solar events.

An Extreme Ultraviolet Imaging Spectrometer (EIS) provides a key link in the data by observing the chromosphere and transition region that separate the photosphere from the corona. The EIS measures the velocity of solar particles, and the temperature and density of solar plasma.

Its high-resolution images revealed gigantic arcing magnetic structures that dwarf the underlying sunspots. It found evidence for several mechanisms that may be contributing to the extraordinary heating of the corona, including twisted and tangled magnetic fields that snap and reconnect, extreme turbulence and acoustic waves in the lower atmosphere, magnetic Alfvén waves that propagate upwards at high speed, and X-ray jets and nanoflares that are continually exploding. Hinode also discovered extended structures at the edges of active regions where material is flowing rapidly outward, possibly contributing as much as a quarter of the total solar wind.

NASA has launched several spacecraft in recent years to investigate different aspects of solar activity. The twin Solar Terrestrial Relations Observatory (STEREO) spacecraft, launched in October 2006, were designed to provide the first simultaneous views of the Sun's Earth-facing hemisphere and the opposite hemisphere. With one flying ahead of Earth and the other behind it, scientists could produce 3-D images of Sun–Earth space. These made it possible to pinpoint the location and speed of a CME, and study how it interacted with its surroundings. In this way, a CME could be tracked all the way to Earth and its arrival predicted at least a day in advance. Communications with the STEREO-B craft were lost in October 2014.

The Solar Dynamics Observatory, which was launched in February 2010, was placed in a geosynchronous orbit around Earth, which enables it to observe the Sun without interruption. The observatory carries three instruments to determine how the Sun's magnetic field is generated, structured and converted into the solar wind, flares, and coronal mass ejections.

The most recent mission is the Parker Solar Probe, which was launched on August 12, 2018. The spacecraft is designed to fly through the Sun's outer atmosphere on numerous occasions to gather data on the processes that heat the corona and accelerate the solar wind. By the end of its seven-year mission, it will fly within 6 million km of the Sun's visible surface, deep inside the corona.

The Solar Probe carries four instruments. One measures the electric field around the spacecraft and uses three small magnetometers measure magnetic fields. The other instrument suites study energetic particles, and image the corona and solar wind. To withstand the intense temperatures, which will reach almost 1,400°C, the spacecraft and instruments are protected by a carbon-composite heat shield.

The European Space Agency is also developing the Solar Orbiter, which will provide complementary observations alongside those of other space observatories. Solar Orbiter was launched on February 9, 2020, the mission will provide close-up, high-latitude observations of the Sun from a highly elliptical orbit that will take it well inside the orbit of Mercury. By flying close to the Sun in an inclined orbit, the spacecraft will be able to observe the dynamic solar surface and its connection to the heliosphere for much longer periods than from near-Earth vantage points.

A 15–60-minute advance warning of an incoming CME is provided by spacecraft which are parked at the L1 Lagrange Point between the Sun and Earth, such as the Deep Space Climate Observatory (DSCOVR). They detect sudden increases in particle density, interplanetary magnetic field (IMF) strength and solar wind speed when the CME-associated interplanetary shock arrives ahead of the magnetic cloud.

Much of our knowledge of CMEs has come from the Large Angle and Spectrometric Coronagraph (LASCO) on SOHO. Advance warning of a CME that could be heading toward Earth was provided when LASCO imaged a "halo event," when the entire Sun appeared to be surrounded by the CME. However, it was still not possible to definitively say if a CME was coming Earthward. Another viewpoint was needed to provide the third dimension.

Between 2006 and 2014, two extra viewpoints were provided by NASA's twin Solar Terrestrial Relations Observatory (STEREO) spacecraft. With one flying ahead of Earth and the other behind it, scientists could produce 3D images of Sun–Earth space. These made it possible to pinpoint the location and speed of a CME, and study how it interacted with its surroundings. In this way, a CME could be tracked all the way to Earth and its arrival predicted at least a day in advance.

The STEREO data showed that almost all CMEs have a common shape, similar to a croissant. This shape is explained by the twisted magnetic flux tubes being wider in the middle and thinner at one end.

CMEs can create major disturbances in the interplanetary medium and in Earth's magnetic field. If they reach Earth, they result in beautiful polar auroras but may also cause large-scale power cuts, e.g. Quebec in 1989, and problems with spacecraft systems.

The most powerful CME to reach Earth in the last 160 years occurred on September 1, 1859, when solar science was in its infancy. British astronomer Richard Carrington saw a brilliant white flash on the Sun, the first ever observation of a solar flare. Only 17.6 hours later, a massive CME slammed into Earth's magnetic field – arriving much faster than most CMEs.

Campers in the Rocky Mountains woke up in the middle of the night, mistaking the glow of brilliant auroras for sunrise. Even as far south as Cuba, the red illumination of the Northern Lights was bright enough to enable people to read their morning paper.

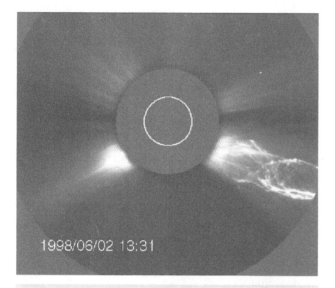

1998/06/02 13:31

Figure 2.42 A SOHO coronagraph image showing a spiral-shaped CME (lower right) erupting from the Sun on June 2, 1998. This CME was rather unusual since the width of the blast was fairly narrow and the strands of plasma were twisting. The LASCO instrument on SOHO blocks the Sun in order to observe coronal structures in visible light. The white circle represents the Sun. (ESA/NASA)

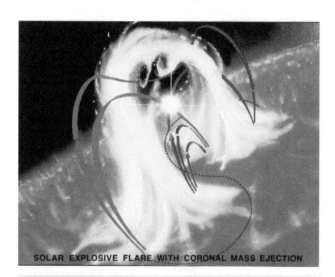

SOLAR EXPLOSIVE FLARE WITH CORONAL MASS EJECTION

Figure 2.43 Coronal mass ejections occur when solar magnetic field lines snake around each other, forming the letter "S." Usually, they go past each other, but if they connect, the mid-section breaks free and creates a mass ejection. In this artwork, a coronal mass ejection erupts as magnetic field lines reconnect. The yellow arch represents a magnetic flux rope filled with flare plasma. The red, blue, and green lines represent higher magnetic field lines connecting opposite polarities on the solar surface. The dotted line shows the sheared, low-lying magnetic field. The bright spot is the site of magnetic field reconnection. (NASA-MSFC)

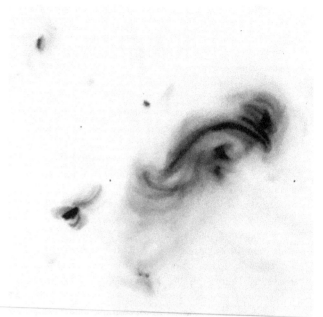

Figure 2.44 An S-shaped structure (sigmoid), in a solar active region is often a precursor to a CME. In this Hinode X-ray image, a bright sigmoid is observed (right) at the beginning of its eruption on February 12, 2007. The fine structure reveals that the sigmoid is really two opposing "J" shapes wrapping around each other. Sigmoid structures, defined by twisting magnetic fields, can often be observed for several days before a CME occurs. (Hinode, JAXA, NAOJ, David McKenzie-University of Montana)

Meanwhile, one of the largest recorded geomagnetic storms electrified telegraph lines, shocking technicians and setting their telegraph papers on fire. Magnetometers around the world recorded strong disturbances in the planetary magnetic field for more than a week.

A Carrington-class solar superstorm blasted off the Sun on July 23, 2012, but fortunately, it missed the Earth. However, the consequences for our modern society, which is largely dependent on electrical systems and satellites, could have been severe. If a solar storm of that magnitude did strike our planet, the cleanup might cost $2 trillion, according to a study by the National Academy of Sciences.

The Sun's Future

The Sun will not continue to shine for ever (see also Chapter 1). Like most stars, it is undergoing gradual evolution on the so-called main sequence. Calculations based on the amount of hydrogen in the Sun and the rate at which it is being consumed, allied with theories of stellar evolution based on observations of other Sun-like stars, indicate that it is currently almost halfway through its active life. Eventually, the Sun will leave the main sequence and swell into a **red giant** as it begins the final, dramatic phase of its existence.

Figure 2.45 The Cat's Eye nebula in the constellation Draco may resemble the remains of our Sun some 7 billion years from now. One of the first planetary nebulas discovered, the Cat's Eye is also one of the most complex examples of this kind of nebula, with 11 rings, or shells, of gas surrounding the dying star at the center. The star ejected a series of concentric shells in a series of pulses at 1,500-year intervals. (NASA, ESA, HEIC, and The Hubble Heritage Team STScI/AURA)

Some 7 billion years from now, the Sun will have grown to 2.3 times its present diameter and shine 2.7 times brighter than it does today. Within a few hundred million years, it will balloon outward, expanding to more than 100 times its current size, so that it engulfs Mercury, and possibly Venus, whilst swelling almost to the present orbit of Earth.[12] Meanwhile, the surface temperature will drop to around 4,000°C, making it appear red in color, and the bloated star will lose up to 25% of its mass in the form of a powerful solar wind.

The first red giant phase will end abruptly when there is no longer enough hydrogen available to continue the fusion process that leads to the creation of helium. As the core temperature soars to 100 million degrees Celsius, the intense heat and pressure initiate nuclear reactions that use helium as fuel. In this process, three helium atoms are fused to create one carbon atom.

The Sun will remain a fully fledged red giant for only a few hundred million years. During this time it will steadily burn its supply of helium, until this, too, begins to run out, leaving the star with a core made of nuclear fusion by-products – carbon and oxygen.

As the energy output begins to drop, the star balloons outward again, once more extending to at least the present orbit of Venus. During this "asymptotic branch" stage on the sequence of stellar evolution, it will experience enormous pulsations known as

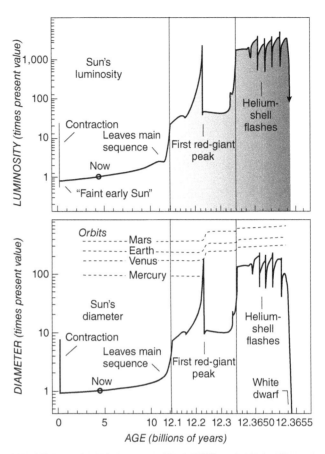

Figure 2.46 The evolution of the Sun will culminate in a red giant phase, beginning about 7 billion years in the future. The Sun will expand enormously in the first phase as hydrogen burning accelerates, possibly reaching the orbit of Venus. Once helium fusion takes over, it will suddenly shrink. A few hundred million years later, when most of the helium is converted to carbon and oxygen, it will expand once more, marked by periodic outbursts known as helium flashes. When the star runs out of helium it collapses to form a white dwarf. (I-J Sackman, Caltech)

helium flashes. However, these 10,000-year-long bursts of energy are almost the last throw of the dice.

The second red giant phase, which lasts for perhaps 100 million years, is marked by another huge mass loss as the remaining atmosphere is ejected into space, creating a beautiful, but short-lived, planetary nebula. Eventually, all nuclear fusion ceases. Without the outward push of radiation pressure, the stellar remnant collapses under its own gravity to form a white dwarf.

All that is left is an extremely hot (100,000°C) core no larger than Earth. This is composed of highly compressed matter which is so dense and hot that the atoms are stripped of almost all their electrons. One teaspoon of this degenerate matter would weigh around 5 tons.

---

[12] As it expands and loses mass, the Sun's gravitational pull will decrease, causing the planets to move away from the Sun. Earth and Mars will probably escape being swallowed by the swelling star.

Over hundreds of millions of years, the white dwarf will gradually cool and fade until it emits only a tiny amount of heat or light. The Sun will end its life as a black dwarf, a cold, dark object hidden amongst the surrounding stars and surrounded by what remains of its retinue of planets and satellites.

Questions

- In what ways does the Sun influence its planetary system?
- The Sun has been in existence for 4.5 billion years. How is such longevity possible?
- Name the internal layers of the Sun. Describe the main characteristics of each layer.

- Why was the extremely high temperature of the corona so unexpected? How might this temperature be accounted for?
- Briefly describe and explain: (a) solar flares; (b) coronal mass ejections; (c) sunspots; (d) spicules; (e) prominences; (f) solar eclipses.
- What is the solar cycle? Suggest reasons for the changes that occur during this cycle.
- Explain why most of the recent advances in solar science have been made by observatories in space, rather than ground-based observatories.
- Why will the Sun undergo major changes some 6–7 billion years from now? How will these changes affect Earth and the other planets?

THREE

Earth

Earth is our home, the most familiar planet in the Solar System. Of the four terrestrial planets, it is the largest and has the highest bulk density. It also has a strong magnetic field and thick atmosphere, which protect the planet from solar and cosmic radiation, and an ozone layer which acts as a barrier to solar ultraviolet light. Its unique nitrogen-oxygen atmosphere also contains greenhouse gases that maintain an equable temperature. Earth is also the only world in the Solar System where conditions are right for large bodies of liquid water to exist on the surface. It is the only planet that is known to support life.

Orbit and Rotation

Earth is the third planet from the Sun. It follows an almost circular orbit (eccentricity 0.017) at an average solar distance of 149,600,000 km. At perihelion, which occurs around January 3, the Sun–Earth distance is 147.1 million km. Aphelion, when Earth is farthest from the Sun, is reached around July 4. The planet is then 152.1 million km from the Sun. Moving at a mean orbital speed of about 30 km/s, Earth completes each orbit in one year or 365.25 days.

The planet rotates from west to east once every 24 hours, on average.[1] This period, from noon to noon, is called the solar day. As a result, Earth's equatorial region is moving eastward at a rate of 1,670 km/h, though this rate decreases to almost nothing near the poles.

The planet's rotation explains why the Sun, Moon, and stars appear to rise in the east and set in the west. However, the time of sunrise and sunset changes by up to 4 minutes each day because the planet is also moving around its orbit. This means that Earth has to rotate for approximately another 4 minutes before the Sun returns to the same place in the sky where it was the previous day.

On the other hand, Earth's sidereal day, measured relative to the position of the "fixed" background stars, is 23 hours 56 minutes 4 seconds (see Figure 3.1).

Earth's rotational axis is currently tilted 23.44° to the ecliptic plane. The axis is aligned in the same direction relative to the stars throughout the year, with the North Pole pointing approximately in the direction of the star Polaris.

However, the orbit and rotation rate are not stable and unchanging. The planet's motion is influenced by its gravitational interaction with the Sun, Moon, and other planets. There are four major variations over time:

- The shape of Earth's orbit slowly changes from almost circular to more elliptical on a regular basis. Astronomers say that its eccentricity changes from nearly zero to 0.06, and then back again, in a cycle that takes between 90,000 and 100,000 years. When the orbit is more elliptical, the amount of insolation (incoming solar radiation) received at perihelion is about 0.2% greater than at aphelion.
- The direction in which Earth's axis points changes slowly because the planet wobbles on its axis as it spins – an effect

[1] The braking action of the tides causes Earth's period of rotation to slow by between 1.5 and 2 milliseconds per century. In order to ensure that clocks remain linked to Earth's rotation, an extra second is sometimes added to the year. This is known as a leap second.

Exploring the Solar System, Second Edition. Peter Bond.
© 2020 John Wiley & Sons Ltd. Published 2020 by John Wiley & Sons Ltd.
Companion Website: www.wiley.com/go/Bond-Solar-System2e

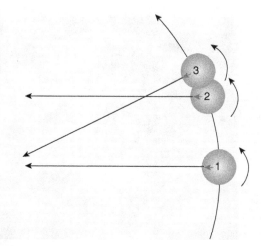

Figure 3.1 On a prograde planet like Earth, which rotates from west to east, the sidereal day is shorter than the solar day. At 1, the Sun and a certain distant star are both overhead. At 2, the planet has rotated 360° and the distant star is overhead again. The time period between 1 and 2 is one sidereal day. Only when Earth has rotated a little further is the Sun overhead again (3). The interval between 1 and 3 is one solar day. (Wikipedia)

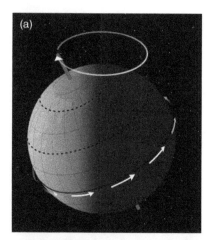

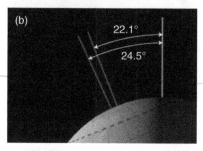

Figure 3.2 (a) Earth rotates (white arrows) once a day around its rotational axis (red). Like a spinning top, Earth's axis wobbles as it rotates. This effect, known as precession, causes the axis to describe a circle in space (white circle) once every 25,800 years. As time goes by, the planet's North Pole points toward different stars. (Robert Simmon, NASA-GSFC) (b) The change in the tilt of the Earth's axis (obliquity) affects the magnitude of seasonal change. At higher inclinations the seasons are more extreme, and at lower inclinations they are milder. The current axial tilt is 23.44°. (Robert Simmon, NASA-GSFC)

known as precession. The cause of the precession is the planet's equatorial bulge, caused by Earth's fairly rapid rotation. The pull of the Moon and Sun on the bulge makes the Earth precess. This wobbling motion affects the direction in which the rotational axis is inclined, but it does not affect the obliquity (tilt angle) of Earth. The planet completes a full wobble every 25,800 years. At present, the rotation axis is pointing almost exactly at Polaris. However, 13,000 years from now the axis will be aligned toward Vega, the brightest star in the constellation Lyra (see Figure 3.2a).

- Another type of precession is associated with the sideways rotation of Earth's elliptical orbit. Known as precession of the ellipse or orbital precession, this motion causes the positions of perihelion and aphelion to shift with regard to the stars.
- The obliquity or inclination of Earth's axis, i.e. its angle of tilt relative to the plane of its orbit, varies between 22.1° and 24.5° every 41,000 years. (It is currently decreasing). Such changes modify Earth's seasons. A higher inclination means more extreme seasons – warmer summers and colder winters. A lower inclination results in less severe seasons – cooler summers and milder winters (see Figure 3.2b).

All of these changes affect the timing and duration of the seasons. Today, northern summer is the longest season and northern winter the shortest; 10,000 years from now, the length of the seasons will be reversed. The gradual modifications of the orbit are also thought to explain Earth's periodic ice ages (see Ice Ages).

Seasons

Planets have seasons because their rotational axes are tilted, rather than perpendicular (upright) in relation to the planes of their orbits. Venus and Jupiter have negligible axial inclinations, so there is no difference between the amount of solar radiation arriving at their equator or poles throughout their year.[2] This is not the case with planets such as Earth, Mars, and Saturn, which have noticeable tilts. Even more unusual conditions occur on objects which are more or less spinning on their sides, such as Uranus and Pluto.

The northern hemisphere experiences summer when the North Pole is tilted toward the Sun. Six months later, when Earth has traveled halfway around its orbit, the northern hemisphere is tilted away from the Sun and experiences winter. (The seasons are reversed in the southern hemisphere.)

---

[2] Mercury is rather different because its orbit is much more elliptical than those of the other planets, so its surface is noticeably cooler when it nears aphelion.

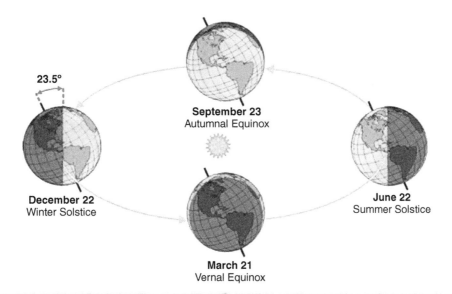

Figure 3.3 Since Earth's axis is inclined to its orbital plane, the amount of radiation any one place receives varies throughout the year. In June, the North Pole is tilted towards the Sun, resulting in longer days and warmer temperatures (i.e. summer). In December, the North Pole is tilted away from the Sun, causing longer nights and colder temperatures (winter). The seasons are reversed in the southern hemisphere. (NOAA)

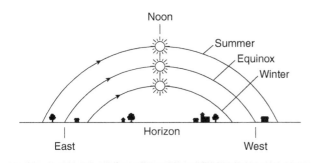

Figure 3.4 The Sun's motion across the sky, looking south. The maximum height of the Sun in the sky, and its rising and setting points on the horizon, change with the seasons. In the summer, the Sun rises in the north east, reaches its highest maximum height at noon, and stays up longest. The Sun rises in the south east and remains low in the winter when the days are shortest. The length of day and night are equal on the vernal, or spring, equinox (March 20) and on the autumnal equinox (September 23) when the Sun rises exactly east and sets exactly west. (NASA)

The dates on which Earth's axis is most directly tilted toward or away from the Sun are known as the solstices. They occur on or around June 21, when the Sun is overhead at midday at the Tropic of Cancer (23.44°N), and December 22, when the midday Sun is overhead at the Tropic of Capricorn (23.44°S).

Not only are days longer in summer, but the Sun moves noticeably higher above the horizon at that time of year, providing more

heat per square meter of surface area. The opposite is true in the winter. These two factors combined account for much of the difference in seasonal temperature.

When one of the poles is tilted towards the Sun, the surrounding regions at high latitude are bathed in permanent sunlight (hence the term, "the land of the midnight Sun"). The opposite polar region endures 24-hour darkness and extreme cold (Figure 3.3).

However, even in December or June, the noonday Sun is never far from the zenith at the equator, so the amount of insolation received shows little variation throughout the year (Figure 3.4). Hence, the equatorial regions are always hot.

Midway between the solstices, on or around March 21 and September 23, the Sun is directly overhead at noon at the equator. On those dates, Earth's axis is inclined away from the Sun. Day and night are equal in length across the globe – so they are known as the spring (or vernal) and autumnal equinoxes.

Earth's orbit is slightly elliptical. The planet reaches perihelion in early January, only about two weeks after the December solstice.[3] This means that the northern hemisphere winter and the southern hemisphere summer begin about the time that Earth is nearest the Sun. (Similarly, the southern hemisphere's winter and northern hemisphere's summer coincide with aphelion.)

However, the difference between the aphelion and perihelion distances is only about 5 million km or 0.3%. This difference results in a 6% increase in incoming solar radiation (insolation) from July to January – too small to cause any significant seasonal effects.

The dates of perihelion and the December solstice will not always be so close, since the date of perihelion is not fixed, but slowly regresses (becomes later in the year). The rate of orbital

[3] There are small year-to-year variations in the dates and times of solstice and perihelion due to the leap-year cycle in the calendar and the effect of the Moon on Earth's motion.

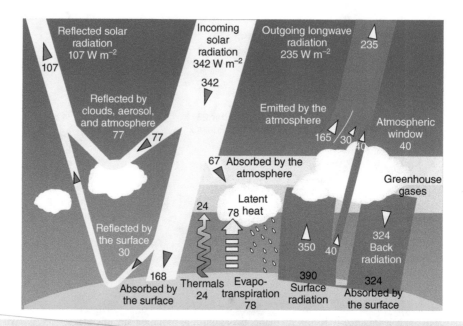

Figure 3.5 Earth's temperature is determined by its radiation budget – the amount of solar radiation it receives compared with the amount of heat it loses to space. This diagram shows what happens to incoming solar radiation and longer wavelength radiation emitted by the surface. Here, the overall amounts of incoming and outgoing radiation are balanced, so the temperature is in equilibrium. In reality, the global temperature has generally been increasing gradually since the 1970s. (NASA)

precession (see Orbit and Rotation) is about one full day every 58 years. There is some evidence that this long-term change in the date of perihelion influences Earth's climate.

Surface Temperature

Earth is located in the habitable zone of the Solar System, where temperatures are neither too hot not too cold for life (as we know it) to exist. Although extremes do occur, notably in hot deserts and at the poles, the planet's mean temperature is a hospitable 7°C. (However, without the greenhouse gases in the atmosphere, notably water vapor, carbon dioxide, and methane, Earth's temperature would be well below the freezing point of water.)

The global radiation budget is determined by the amount of incoming solar radiation and the amount of radiation absorbed or lost by the planet (Figure 3.5). When solar radiation passes through the atmosphere, some wavelengths are absorbed by gases, dust, and cloud droplets. Some visible light is scattered by the gases, giving the sky its blue appearance. About half of the total insolation is absorbed by the surface, compared with one-third which is reflected back into space from the clouds and the surface.

Of course, the surface does not retain all of its heat; some is re-radiated at infrared wavelengths. A fraction of this is absorbed by the greenhouse gases and clouds, then re-radiated in all directions.

The role of clouds and aerosols is crucial in this planetary balancing act. (Up to 70% of the world is covered with cloud at any one time.) They help to cool the planet by reflecting a substantial amount of solar radiation back into space. However, they can also raise its temperature by absorbing or re-radiating downwards much of the infrared radiation that would otherwise escape. Scientists are keen to know how the climate would react to a long-term increase or decrease in this cloud and aerosol blanket.

The pattern of surface temperatures is also influenced by a number of other factors. In winter, the amount of incoming solar radiation (**insolation**) is insufficient to compensate for the heat that escapes into space at night. The same effect, of course, on a much smaller scale, applies to day and night temperatures.

The daily and seasonal variation in overall insolation greatly influences surface temperatures, particularly over the continents. This is because land surfaces respond more rapidly than water to daytime heating and nighttime cooling. Oceans take a long time to warm, and an equally long time to cool. In effect, they act as storage heaters. (The lower "specific heat" of rock means that its temperature rises more quickly than that of water for the same input of energy.)

In Earth's southern hemisphere, which is largely covered in water, the slow response of the oceans to variations in insolation means that diurnal and seasonal temperature ranges are less extreme. (Places on the shores of large lakes experience similar moderating effects.) In the northern hemisphere, where land masses predominate, summer temperatures tend to be much higher, while thermometers plunge during the winter. As a result, places located in continental heartlands have a much greater variation in temperature than islands and coastal regions.

The nature of the surface also plays a part (Figure 3.6). Dark surfaces, such as those covered with vegetation and soil, have a low albedo (reflectivity). They heat up more quickly than lighter, more reflective surfaces, such as ice sheets. Ice and snow reflect

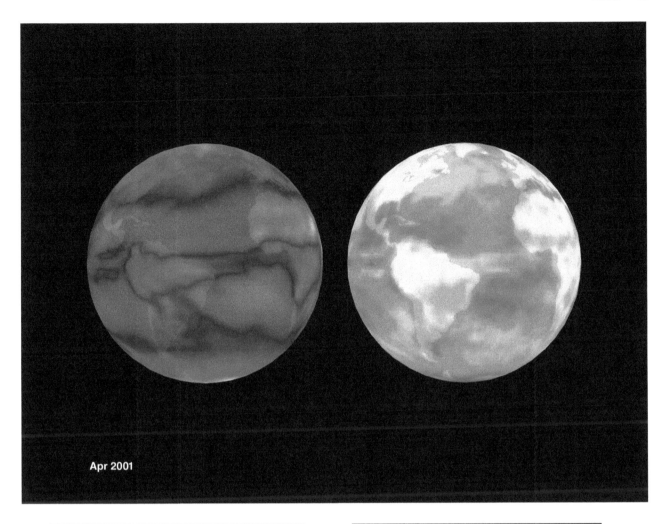

Apr 2001

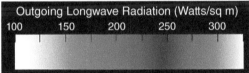

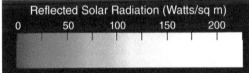

Figure 3.6 These images show the amount of long wavelength thermal radiation (heat) emitted to space from Earth's surface and atmosphere (left) and shorter wavelength solar radiation reflected by the ocean, land, aerosols, and clouds (right). The largest flux of long wave radiation is from cloud-free tropical regions. Clouds over the equator and at high latitudes reflect a lot of incoming solar radiation back into space. Such images help to determine the factors that influence Earth's radiation balance. (NASA)

some 80% of the solar energy they receive, compared with 20% for an area of grassland and 10% for a dry, black soil. This explains why scientists are concerned that the melting of polar ice sheets will contribute to global warming.

Atmosphere

Earth's atmosphere is unique in the Solar System, since nitrogen and oxygen together make up 99% by volume. Other gases which are naturally present in small, variable, amounts include carbon dioxide, methane, water vapor, and ozone. Minor

constituents include inert gases, particularly argon, along with human-made pollutants such as sulfur dioxide and nitrogen dioxide. Near-surface ozone and carbon dioxide are also generated by human activity.

Earth's original atmosphere was probably made up of hydrogen and helium from the solar nebula, but these light gases soon escaped. A secondary atmosphere was then created as the result of outgassing through volcanic activity. Its actual composition is uncertain, although its main constituents most likely included carbon dioxide, water vapor, ammonia (NH_3), and methane (CH_4). Large amounts of water vapor, organic (carbon-rich) compounds,

Box 3.1   Earth Observation Satellites

Since the early 1960s, numerous satellites have been launched to observe Earth. Some of them are fairly specialized, such as meteorology (weather) or radar satellites (see Box 3.2). Others, such as the various spacecraft in NASA's Earth Observing System or Europe's Sentinel satellites, carry a wide range of sensors for multiple uses, such as environmental monitoring, land use mapping, altimetry, etc.

Two main types of orbit are used. For large-scale or hemispheric observations, geostationary orbits are favored. From an altitude of 35,780 km above the equator, such satellites circle the planet once every 24 hours, so they appear to hover over the same spot, providing continuous monitoring of the same part of Earth's surface. These orbits are ideal for weather satellites such as the U.S. Geostationary Operational Environmental Satellite (GOES) series and Europe's Meteosats.

More detailed observations require much lower orbits. These are usually provided by satellites in near-polar orbits, inclined almost 90° to the equator. However, polar-orbiting spacecraft can only provide brief snapshots as they pass overhead. Such overpasses typically take place twice per day at any given spot on the surface. Global coverage is achieved by combining many swaths (strips) of data that are acquired as the planet rotates beneath the satellite. The other alternative is to orbit dozens or even hundreds of satellites that carry similar Earth observation instruments.

One of the most popular types of polar orbit is one in which illumination conditions on the surface remain constant. Such Sun-synchronous orbits are typically inclined about 98° to the equator, enabling the satellite to cross the equator at the same local time on each orbit. Perhaps the most famous example is the U.S. Landsat series, which has been providing continuous, medium-resolution imagery of Earth's land masses in different spectral wavelengths since 1972.

Multiple observations of the same area by different instruments on different satellites are also invaluable in providing information about changes over short and long periods of time (i.e. multi-temporal data). The most enterprising example is the Afternoon Constellation, often known as the "A-Train," which includes up to seven satellites from the U.S. and France that follow the same orbital path around Earth, one behind the other. The A-Train enables near-simultaneous coordinated measurements of the same regions, giving a more complete overview than would be possible from a single satellite.

Each satellite in the A-Train crosses the equator at around 1:30 p.m. local time, separated by intervals of seconds or minutes. Satellites such as Aqua and CloudSat observe many different surface and atmospheric phenomena, including hurricanes, clouds, and aerosols.

and other volatiles would also have arrived through the initial bombardment by comets and asteroids.

As the temperature dropped, condensation of water vapor caused the planet to be blanketed in cloud. The resultant global downpour led to the formation of the first oceans. Some carbon dioxide gas dissolved in the rain and oceans. At the same time, chemical reactions involving acidic rain and the first, primitive crust led to the formation of carbonate rocks and a further reduction in atmospheric carbon dioxide.

According to this scenario, nitrogen – a gas which is not very chemically active – continued to accumulate in the atmosphere as a result of outgassing and numerous impacts. Free oxygen was scarce, since it was soon removed through chemical reactions with rocks and other gases. The first single-celled organisms were probably also important in regulating the climate by generating substantial amounts of methane – a greenhouse gas.

The change to a more modern atmosphere began when bacteria and algae developed the ability to split water molecules by harnessing the energy of sunlight – a key part of photosynthesis. As a result, the amount of oxygen released into the atmosphere began to increase, although for a long time, most of the gas was removed by chemical reactions with rock minerals. To this day, the majority of Earth's oxygen produced over time is locked up in ancient, oxidized rock formations.

It was not until about one billion years ago that the rate of oxidation slowed sufficiently to enable free oxygen to stay in the air

and the levels of oxygen in the oceans to rise. Once oxygen gas was available, ultraviolet light began to split the molecules, producing a layer of ozone (O_3) high above the planet that acted as a shield against ultraviolet light. Only at this point did life move out of the oceans and respiration evolve.

Troposphere

Earth's atmosphere extends from the surface to an altitude of about 500 km, when it effectively fades into the vacuum of space (Figure 3.7). The lowest part of the atmosphere is called the troposphere. This region is where most weather phenomena take place, e.g. cloud formation. The troposphere contains about 80% of the atmosphere's mass, including almost all of its water.

Air temperature generally decreases at a steady rate (the "lapse rate") with altitude, until the tropopause, the upper boundary of the troposphere, is reached.[4] The height of this boundary extends up to 17 km near the equator, reducing to about 8 km above the polar regions.

Since the troposphere is characterized by a continuous decrease of temperature with increasing height, any parcel of air that is warmer than its surroundings is likely to become buoyant and unstable, sometimes rising to great heights. The result of this convection is towering clouds and heavy precipitation (rain, snow, hail, etc.).

[4]The average decrease in temperature – the normal lapse rate or environmental lapse rate – is 6.5°C per 1,000 m of altitude.

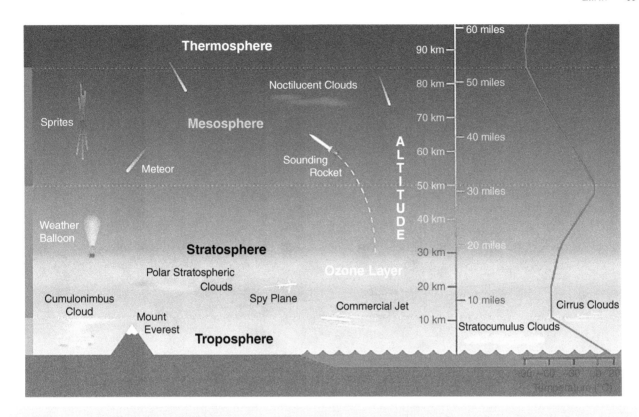

Figure 3.7 The vertical structure of Earth's atmosphere. Most of the atmosphere and weather is found in the lower region – the troposphere. Temperature decreases at a steady rate (the "lapse rate") until it reaches the top of the troposphere. It then increases in the stratosphere, where ozone gas absorbs solar radiation. In the ionosphere, about 100 km above the surface, positively charged particles (ions) reflect radio waves. (UCAR)

In the upper troposphere, the temperature drops to about −50°C. At this height, jet streams occur at the boundaries between air masses that have significantly different temperatures. The jet streams are relatively narrow bands of very high wind speeds (up to 360 km/h) which meander around the globe at about 60° latitude – above the polar – and at 30° latitude. During the winter, Arctic and tropical air masses create a greater temperature contrast, resulting in a stronger jet stream. However, during the summer, when surface temperature variations are less dramatic, the jet streams are weaker.

The Ozone Hole

The dust-free, almost cloudless stratosphere lies above the troposphere. It is characterized by an increase in temperature with altitude, due to the presence of ozone gas which absorbs solar radiation (and protects life on the planet's surface from harmful solar ultraviolet light). This temperature pattern means there is little convection and mixing in the stratosphere, so the layers of air there are quite stable.

The ozone layer is particularly marked above the poles, where the stratosphere becomes much warmer in summer. However, the lack of insolation in winter means that the upper air then becomes very cold, and this frigid air is trapped by polar-circling winds in the stratosphere.

In the late 1970s, British scientists discovered that the ozone layer over Antarctica was getting thinner every spring. By 1985, it was clear that an "ozone hole" was opening up above the continent. A similar, but smaller, feature was later discovered over the Arctic (Figure 3.8).

These holes last for several months before disappearing. The ozone hole that appears over Antarctica fluctuates in size, normally reaching its widest in the polar spring (September–October). The largest ozone hole, detected over Antarctica in 2006, covered 29.6 million sq. km, an area approximately three times the size of the United States. The deepest ozone hole (i.e. the hole containing least ozone) occurred in September 1994.

One major factor in the creation of these ozone holes was the injection of human-made pollutants, such as chlorofluorocarbons (CFCs), into the atmosphere. Due to the lack of vertical convection in the stratosphere, these chemicals can stay there for a long time.

When the Antarctic spring arrives, the combination of extremely cold temperatures in the stratosphere and the return of sunlight causes a complex series of chemical reactions involving ice clouds and the atmospheric pollutants. The resultant cocktail of compounds releases chlorine into the air, and this chlorine destroys the ozone gas.

After the introduction of the 1987 Montreal Protocol, which aims to phase out production of ozone-destroying chemicals, levels of CFCs have leveled off. However, the chemicals can stay in

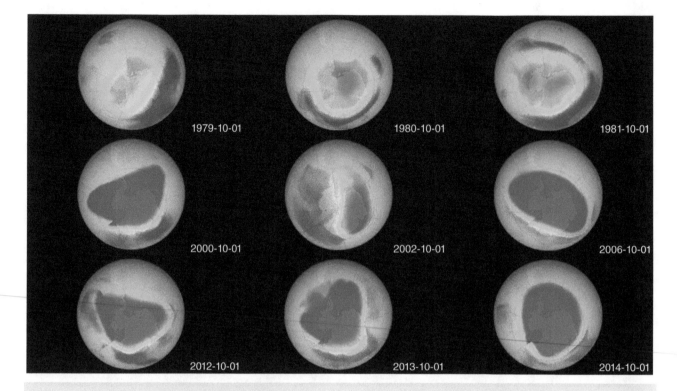

Figure 3.8 A series of images showing the changing size and location of the Antarctic ozone hole on October 1 for selected years between 1979 (top left) and 2014 (bottom right). Data from satellites and stratospheric balloons show that the ozone hole (dark blue and purple) grew rapidly from the mid-1980s. It is expected to take many years for the hole to disappear. (NASA-GSFC)

the atmosphere for decades, so it will be a long time before the stratospheric ozone recovers.

Upper Atmosphere

At the top of the stratosphere, the rise in temperature ceases. In the mesosphere (50–80 km) there is a continuous decrease of temperature, reaching a minimum of almost –100°C. The temperature then rises rapidly as the atmosphere becomes ever thinner in the thermosphere.

The upper part of the mesosphere and much of the thermosphere (80–500 km) is also known as the ionosphere. In this region, high-energy UV and X-ray solar radiation splits atoms and molecules – a process known as ionization. Radio stations take advantage of the fact that these free electrons and ions reflect radio waves.

The outermost region, above an altitude of about 500 km, is known as the exosphere. Its main constituents are the lightest gases, particularly hydrogen and helium, with some atomic oxygen. The exosphere is characterized by the escape of these gases into space (Figure 3.9). Air pressure is so low that it is effectively a vacuum. However, the orbits of low-flying satellites are gradually lowered by slight atmospheric friction, particularly during periods

of maximum solar activity when the upper atmosphere is heated more strongly and balloons outwards.

Atmospheric Circulation

On balance, regions above latitudes 35°N or S lose more energy than they receive. In contrast, the tropical regions receive more radiant energy than they lose. On a planet with no atmosphere or oceans, this would result in the poles becoming cooler and the tropics becoming hotter. Fortunately, Earth is very efficient at transferring heat towards the poles. One of the main mechanisms for doing this is through large-scale atmospheric circulation.

Surface temperature differences are largely responsible for Earth's winds. Warm air expands and rises. Colder air is dense and heavy, so it sinks. As a result, air is always in motion. This effect can be clearly seen in coastal areas where sea breezes blow. Air over the warm land rises, to be replaced at ground level by cooler air from offshore. This circulation is often reversed at night, when the land becomes cooler than the sea.

Similar atmospheric motions can also be seen on a global scale, in the form of six large cells of circulating air – three in each hemisphere. Nearest to the equator are the Hadley cells, named after the scientist who first proposed their existence. In temperate

Figure 3.9 Hydrogen is continuously escaping into space, creating a cloud of neutral atoms around the Earth that extends beyond the Moon's orbit. This halo of hydrogen is illuminated by reflected solar ultraviolet light, when it is known as the "geocorona." In this false color ultraviolet image, the geocorona is brighter on the sunlit side (top). The UV camera was operated by astronaut John Young on the Apollo 16 lunar mission. (NASA-JSC)

latitudes are the Ferrel cells, also named after their discoverer. Finally, there are the polar cells.

Each Hadley cell is driven by hot air rising through convection in equatorial regions, creating low pressure at the surface. This unstable air soars upward until it reaches the top of thetroposphere. The convective air currents result in towering thunderclouds and heavy downpours, particularly in the warmth of the afternoon.

The height of the top of the troposphere varies with latitude – it is lowest over the poles and highest at the equator. It also varies with the seasons – it is lower in winter and higher in summer. It can be as high as 20 km near the equator, and as low as 7 km over the poles in winter.

The air at high level then spreads poleward before sinking at about 30° latitude. This descent warms and dries the air, resulting in cloud-free skies – perfect conditions for hot deserts, such as the Sahara. The large areas of stable, sinking air have light winds – so they were often dreaded by crews of sailing ships.

At the surface the air then diverges. Some moves back toward the equator, completing the Hadley circulation. Some moves toward higher latitudes until it meets up with colder air spreading from the poles. The boundary between the warm, subtropical air and the cold, polar air is known as the **polar front**.

The polar front is like a battleground, where a cool air mass is always trying to undercut and displace the warmer, less dense, subtropical air. Air forced upward at the polar front results in the

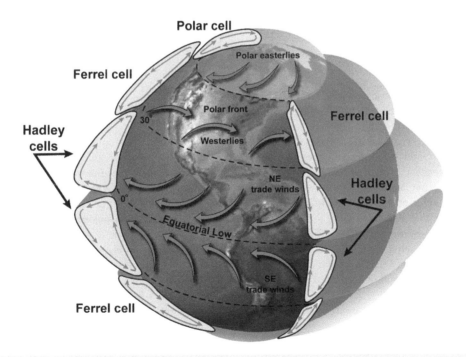

Figure 3.10 Earth's main zonal winds are associated with six large atmospheric circulation systems – three in each hemisphere. In the Hadley cells warm air rises near the equator, moves poleward, and then sinks in the subtropics. The rising section of the Hadley cell causes thick, deep, convective clouds and thunderstorms. Over land, the sinking section is associated with deserts. Over oceans, sinking air results in clear, dry air, and very light winds. Similar circulation systems, called the Ferrel cells and polar cells, occur nearer the poles. Where these cells meet, active uplift of air results in numerous low pressure systems (depressions) and rain belts. (After Barbara Summey, NASA Goddard VisAnalysis Lab)

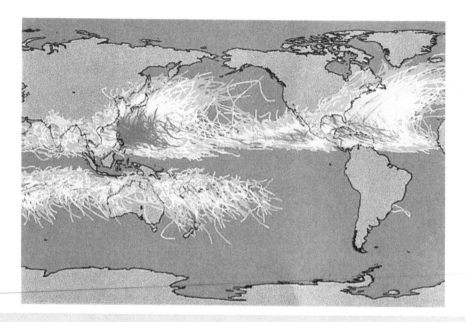

Figure 3.11   A map of tracks followed by tropical storms and cyclones that develop over tropical oceans. Red and orange show the most intense category 5 and 4 storms. As they move westward, driven by upper-level trade winds, they increase in intensity, then die out when they move over land or cooler water. (National Hurricane Center / Typhoon Warning Center/ NOAA)

formation of clouds and heavy rain. Low pressure areas known as depressions are driven eastward by the polar jet stream (see Troposphere).[5]

Zonal Winds

The simple north–south atmospheric circulation is modified by the Coriolis effect, caused by Earth's west to east rotation. Surface air near the equator moves eastward, in the same direction as Earth's rotation, with a speed of nearly 1,700 km/h. At higher latitudes, the surface and overlying air masses move eastward at slower speeds. Consequently, a parcel of air carried away from the equator by Hadley circulation moves eastward faster than the air or ground below it. An observer on the ground would see the air veer on a curved path to the east, as if moved by some force.

On the other hand, air near the poles starts off moving slowly to the east, since Earth's rotation speed is slower at high latitudes. As it moves towards the equator, a parcel of polar air travels over a faster-moving surface. As a result, it appears to move along a curved path to the west.

In effect, winds flowing north–south are diverted to the right in the northern hemisphere and to the left in the southern hemisphere. This explains, for example, why the tropical trade winds in the northern hemisphere blow towards the southwest, whereas the trade winds in the southern hemisphere flow toward the northwest (see Figure 3.10).

The Coriolis effect also explains why winds move around pressure systems in opposite directions in different hemispheres. There is a clockwise flow around a high-pressure area (anticyclone) in the northern hemisphere and a counterclockwise flow around an anticyclone in the southern hemisphere. Conversely, winds blow counterclockwise around low-pressure areas (cyclones or depressions) in the northern hemisphere and clockwise in the southern hemisphere.

Tropical Cyclones

The region where the trade winds of the two hemispheres meet is known as the Inter-Tropical Convergence Zone (ITCZ). There, warm, moist air is forced upward. When it cools, large amounts of water vapor condense, leading to massive cloud formation and high rainfall almost every day. Tropical cyclones, Earth's most powerful and destructive storms are also associated with the ITCZ (Figures 3.11 and 3.12).

Huge amounts of energy are transferred by these short-lived storms as they form over the tropical oceans, then move westward before heading towards higher latitudes and decaying. Most of them are born between June and October, when the surface temperatures of the oceans exceed 27°C. (They are rare close to the equator, since the Coriolis effect is weak and rapid rotation is less likely.) The number varies, but on average there are 40–50 tropical cyclones each year, with an equal number of less intense tropical storms.

These storms are given different names by the local inhabitants, but the causes are the same.[6] They begin as the result of rapid upward movement of air caused by divergence and an absence of

[5] It is this jet stream which causes flights from Europe to North America to take much longer than flights in the opposite direction.

[6] They are called hurricanes in the Atlantic, typhoons in the Asian Pacific, and cyclones in the Indian Ocean and Australia.

Figure 3.12 Katrina was the first Category 5 storm of the 2005 Atlantic hurricane season. One of the most powerful hurricanes ever recorded, it formed over the Bahamas on August 23, 2005, then crossed southern Florida before strengthening rapidly over the warm waters of the Gulf of Mexico (orange-red). A storm surge from Katrina caused catastrophic damage and flooding along the coastlines of Louisiana, Mississippi, and Alabama. (NASA)

strong winds at high altitude. This causes surface pressure to fall rapidly.

Warm, moist air flows inwards towards the developing center of low pressure. As the surface air pressure plummets, the wind speed increases dramatically. The strong updraft causes towering convectional clouds to form, causing heavy rain. The storms are also fed by an additional source of heat, which is released during the rapid condensation process (conversion of water vapor to water droplets). This is known as "latent heat."

A fully fledged storm can develop in less than a week. At its height, wind speeds around the central "eye" may exceed 120 km/h, occasionally producing sustained winds of 250 km/h in a Category 5 storm. Indeed, everything about tropical cyclones is extreme. The circulating system of strong wind and dense cloud may measure up to 1,000 km across. Spiral bands of convectional cloud tower 12 km or more above the ocean surface.

A well-developed tropical cyclone may pick up as much as two billion tonnes of water a day through evaporation and sea spray; 1,250 mm of rain may be dropped over any location along its track. As a result, flooding and mudslides are often more of a problem than damage caused by strong winds.

A central pressure as low as 880 mb also causes the ocean surface to bulge upwards. When combined with huge waves whipped up by the wind and the torrential rainfall, the outcome is a storm surge, which can inundate low-lying coastal areas.

An average tropical cyclone transfers a huge amount of energy. The daily energy release from a cyclone is equivalent to the output of the entire United States' electricity grid over six months.

The tracks followed by tropical cyclones are very hard to predict. However, the average storm moves east to west along a curved path, traveling at 15–50 km/h. They rapidly lose their potency when deprived of their supply of warm, humid air, so they usually die out when they move inland. Others lose intensity when they head away from the tropics and move over cooler water. However, downgraded hurricanes may still bring strong winds and heavy rainfall to temperate regions, particularly if they merge with other low-pressure areas or depressions.

Ocean Currents

The other major conveyor belts for heat from the tropics are the warm ocean currents (Figure 3.15). These fairly narrow ribbons of surface water display marked differences in temperature and salinity from the surrounding ocean.[7] They are also associated with differences in sea height, which can be measured by satellites. Warm currents are seen as rises and cold currents as valleys in the ocean surface.

Many of the currents are driven by surface winds blowing around the subtropical high pressure zones. This means that they

[7] Colder or saltier water is denser and tends to sink.

flow clockwise in the northern oceans and the opposite direction in the southern oceans, traveling at speeds of between 0.4 and 1.2 m/s (35–105 km per day).

Not only do the warm currents carry heat absorbed by sea water, but they also warm the air above them. (The oceans redistribute about half as much heat as the atmosphere.) The effects can be very noticeable at mid-latitudes. The most famous example is the North Atlantic Drift (commonly known as the Gulf Stream) which carries warm water from the Gulf of Mexico to the shores of Norway. As a result, winters in northwestern Europe are much milder than would be expected from their latitude (50–70°N).

Other places are cooled by the oceanic circulation. Cold currents flowing from high latitudes towards the equator tend to cool nearby coastal areas. Since cool air is relatively dense and stays near the surface, condensation of water vapor to form fog is quite common, but clouds and rain are rare. As a result, places such as southern California, northern Chile, and southwestern Africa experience desert conditions.

The foggiest places on Earth occur where warm and cold currents meet, e.g. off the coast of Newfoundland, where the interaction between warm, moist air and cooler air (also fairly moist) results in 150–200 days of sea fog per year.

Vertical motions in the oceans are also important in places such as Peru. Earth's rotation and strong winds push surface water away from some western coasts, so that cold water rises from the depths to replace it. This upwelling of nutrient-rich waters is a bonanza for marine life.

The cooling and sinking of cold water in the polar regions drive a much deeper, global circulation. This oceanic "conveyor belt" is set in motion when cold, dense water in the North Atlantic sinks and moves south (Figure 3.16). It circulates around Antarctica, and then moves into the Indian and Pacific basins, where it returns to the surface. Once at the surface it is carried back to the North Atlantic, and the cycle begins again. This circulation is extremely slow – water from the North Atlantic may take 1,000 years to find its way into the North Pacific.

El Niño and La Niña

Changes in oceanic circulation can lead to variations in heat transport and very different weather patterns. One of the most important of these periodic variations in the ocean current and atmospheric circulation is the El Niño – Southern Oscillation (ENSO) which occurs over the Pacific Ocean (Figure 3.17).

Under normal conditions, the trade winds blow towards the west across the tropical Pacific. These winds pile up warm water in the western Pacific, so that the sea surface is about 0.5 m higher in Indonesia than off the coast of Peru. As a result, the sea surface temperature is about 8°C higher in the western Pacific. Moist air rising over the warm water results in convectional clouds and heavy rainfall. In contrast, the upwelling of cold water from depth off the western coast of South America causes the overlying air to be stable, so rainfall is low.

This pattern changes every two to five years. During an El Niño[8], the trade winds weaken and the equatorial countercurrent strengthens, sending warm water eastward towards Ecuador and Peru. The storm pattern also shifts toward the east and the upwelling of cold water in the eastern Pacific is suppressed. Instead of drought conditions, the west coast of the Americas experiences heavy rainfall, but there is decreased rainfall in the western Pacific (e.g. Australia and Indonesia).

El Niño is often replaced by La Niña, which is characterized by unusually cold ocean temperatures in the equatorial Pacific and a tongue of cool water which extends farther westward than usual. The reasons for these major changes are not well understood.

Changes in the ENSO system of the Pacific Ocean also influence wind patterns, rainfall, and temperature in many other parts of the world.

Monsoons

One of the most important and predictable seasonal atmospheric cycles is associated with monsoons, particularly those of southeast Asia, where the surface winds change with the seasons over a vast area. Less-pronounced monsoons are found over eastern Africa, North America, northern Australia, China, and Japan.

The primary cause of the Asian summer monsoon is convection, when warm, rising air over the extremely hot Indian subcontinent creates an area of low atmospheric pressure. The rising air is replaced by extremely moist air flowing from the Indian Ocean, where pressure is relatively high. The result is a season of torrential downpours and floods, often made worse by tropical cyclones. Rainfall is particularly high over the southern slopes of the Himalayan mountains – the wettest place on Earth – where moist air is cooled as it is forced upward. The arrival of the summer monsoon usually occurs around the same date each year.

In winter, the air flow is reversed. The interior of the Asian land mass becomes very cold, creating an area of intense high pressure. The dense, sinking air flows away from the land and out to sea, resulting in dry conditions over most of continental Asia.

Ice Ages

As over-deepened, U-shaped valleys, gouged rock outcrops, and plains strewn with unsorted boulders or pulverized rocks show, large areas of Earth were once covered with sheets of moving ice. It is clear that, from time to time, extended periods marked by long, cold winters allowed a huge build-up of snow and ice. Under such ice age conditions, Earth's average temperature may drop by 5°C and polar ice sheets can advance over much of northern Eurasia and North America, covering up to 30% of the planet's land surface.

---

[8]In Spanish, El Niño means "the boy," a reference to the baby Jesus, since historically the phenomenon has been observed near Christmas. The reverse situation is called La Niña, "the little girl."

Box 3.2   Radar

Most remote sensing of planets and moons is by passive sensors, which simply detect and record natural radiation emitted by the atmosphere or surface (see Chapters 6, 9, and 13). However, in recent years, a number of spacecraft have been equipped to carry out active remote sensing with lidars (laser imaging detection and ranging systems) or radar.

One of the main uses of radar is altimetry. The distance between a spacecraft and the planet below can be calculated by precise measurement of the time it takes for a microwave signal to return to the spacecraft from the surface, irrespective of whether it is an ocean, ice, or land. Over time, such data can provide topographic maps of an entire planet.

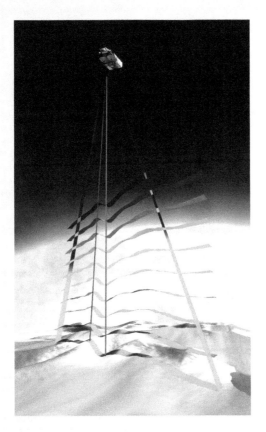

Figure 3.13   Spacecraft such as ESA's CryoSat use radar to map surface topography. The first radar echo comes from the nearest point to the satellite. CryoSat can measure the angle from which this echo originates, so that the source point can be located on the ground. This, in turn, allows the height of that point to be determined. (ESA)

The characteristics of the echoes also provide further information about the roughness of the surface, wave heights or wind speeds over the ocean. Such measurements can be made 24/7, regardless of cloud cover or night conditions.

One of the major uses of radar altimetry has been the investigation of variability in sea surface height and its impact on the general circulation of the oceans. Such measurements highlight the importance of eddies in shaping and controlling major ocean currents. They also reveal the growth and development of major climatic events, such as El Niño.

The accuracy of altimetry measurements depends on knowing the spacecraft's precise orbital position. Many Earth observation satellites carry a radio receiver and laser reflectors for precise orbit determination.

By combining two or more radar images of the same area, spaceborne systems also make it possible to produce maps showing surface change associated with both earthquakes and subsurface volcanic activity which are accurate to within a few millimeters.

Another area of interest is the determination of changes in the surface height and area of global ice cover, particularly in the Arctic Ocean, Greenland, and Antarctica. Measurements made over a number of years can reveal whether the ice sheets are losing or gaining mass and thickness – an important clue with regard to climate change.

Ground-based radar has been used for many years to reveal rain and snowfall from storm systems. Radar instruments installed in spacecraft are now being used to give a wider perspective on major storms. For example, the US-Japanese Tropical Rainfall Measuring Mission (TRMM) satellite carried the first radar flown in space to measure precipitation. The instrument worked by measuring the echoes backscattered from rain. Since the strength of the echo is roughly proportional to the square of the volume of falling water, the instrument produced very accurate estimates of rainfall.

Figure 3.14   A multi-temporal radar image of the Bay of Naples in Italy as seen on three separate occasions by ESA's ERS-2 satellite. With its numerous buildings, the city of Naples (top center) is very radar reflective and bright, as are mountain ridges facing the radar. To the east of Naples is Vesuvius, one of the most explosive volcanoes in Europe. To the west is a much older volcanic region, the Phlegrean Fields. The colored patches north of Naples are fields of crops. The colors offshore indicate sea surface roughness on each viewing date. (ESA)

Other spacecraft such as CloudSat carry a millimeter-wavelength, cloud-profiling radar which is over 1,000 times more sensitive than typical weather radar. It can not only take a vertical slice through clouds and storm systems, even in the polar winter, but also distinguish between cloud particles and precipitation.

Major glaciations are known to have taken place during the late Proterozoic (between about 800 and 600 million years ago), the Pennsylvanian and Permian (between about 350 and 250 million years ago), and the Quaternary (the last 4 million years).[9] There is evidence that at least two dozen warm-cold cycles have occurred during the past 1.6 million years, with the most recent glacial advance peaking about 20,000 years ago and ending about 10,000 years ago. Today, most of Earth's fresh water is locked up in the ice sheets that persist over Antarctica and Greenland.

Various theories to explain the ice ages have been proposed, including the emergence of supercontinents due to continental drift, reductions in solar activity, asteroid impacts, enormous volcanic eruptions, and cosmic dust clouds obscuring the Sun. However, the most widely accepted explanation for the recent ice advances is the astronomical theory popularized by the Serbian scientist, Milutin Milankovitch, between 1920 and 1941.

According to Milankovitch, Earth's current glacial-interglacial cycles are mainly the result of slow, but significant, changes in its orbit (see Orbit and Rotation). Three orbital parameters are especially important in causing the waxing and waning of ice sheets:

- Changes in the eccentricity of Earth's orbit over a period of 100,000 years;
- Changes in the tilt of Earth's axis over a period of 41,000 years;
- Precession of the orbit over a period of 22,000 years.

Milankovitch developed a mathematical model that was able to calculate how these variables would influence the amount of insolation reaching the planet's surface at different latitudes. He showed that Earth's changing orbital geometry can reduce the amount of solar radiation reaching high latitudes (around 65°N) by 10–15%. These periods of lower summer heating (and melting) coincide with advances of the polar ice sheets and mountain

[9]There is some evidence that, during at least two periods between 550 and 800 million years ago, ice sheets reached all the way from the poles to the equator, a situation known as "snowball Earth."

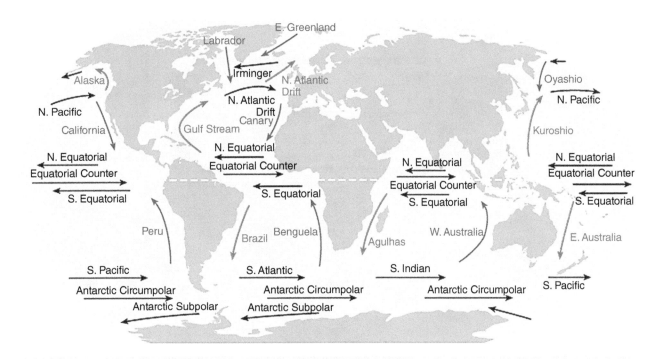

Figure 3.15 Earth's ocean currents are driven by the prevailing winds. Many of the currents circulate around the subtropical high pressure zones – so they flow clockwise in the northern hemisphere and counterclockwise in the southern hemisphere. Currents moving away from the equator are regarded as "warm"; those moving away from polar regions are "cold." Huge amounts of heat energy are transferred from the tropics to higher latitudes by warm currents. This helps to keep the northwestern coasts of Europe and North America warmer than would otherwise be the case. (NASA)

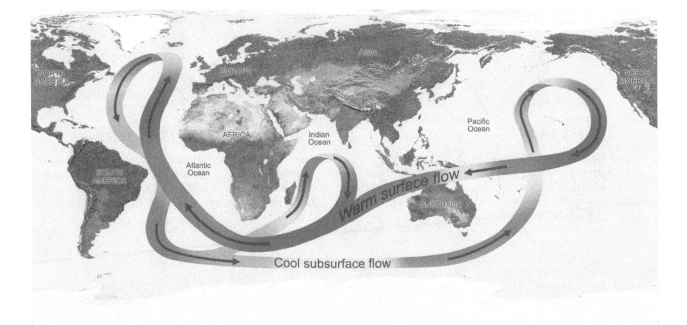

Figure 3.16 The oceanic conveyor belt. Warm water is carried by ocean currents to higher latitudes, where the stored heat is released into the atmosphere. As the water cools it becomes denser. In regions where the water is also very salty, such as the North Atlantic, the water becomes dense enough to sink to the bottom. This travels at depth around Antarctica and into the Indian and Pacific basins before returning to the surface. Carbon dioxide is also transported during this circulation. Cold water absorbs carbon dioxide from the atmosphere, and some is carried to great depth. When deep water returns to the surface in the tropics and is warmed, the carbon dioxide is released back into the atmosphere. (NASA)

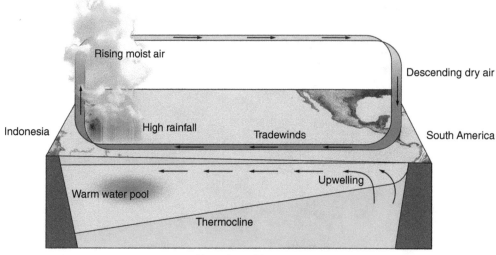

Rising moist air

Descending dry air

Indonesia

High rainfall

Tradewinds

South America

Upwelling

Warm water pool

Thermocline

Normal conditions

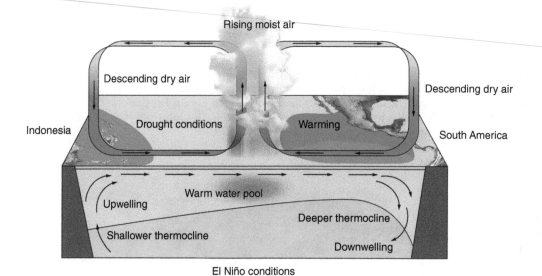

Rising moist air

Descending dry air

Descending dry air

Indonesia

Drought conditions

Warming

South America

Warm water pool

Upwelling

Deeper thermocline

Shallower thermocline

Downwelling

El Niño conditions

Figure 3.17   (Top) Normal conditions in the Pacific Ocean. The trade winds blow steadily toward the west, piling up warm ocean water. Moist, warm air rises, causing heavy convectional rain. In contrast, conditions are dry in the eastern Pacific, where cold water rises from depth and cools the air above. (Bottom) During El Niño, the trade winds weaken and warm water moves eastward, causing wet conditions in the Americas and dry conditions in the western Pacific. The red areas of warmer water correspond to higher sea surface. (ESA-ATG medialab)

glaciers. Nevertheless, the Milankovitch theory does not explain all of the fluctuations in ice cover during the Quaternary or in the distant past, and other factors almost certainly played a part.

What we do know is that, at the height of the last major advance, ice sheets about 2 km thick covered much of Europe and North America. Enough water was locked up in these ice sheets to cause sea levels to drop more than 100 m below present levels. Such dramatic changes in ice cover inevitably led to major shifts in surface drainage and vegetation cover. The subsequent melting of the massive ice sheets has caused the crust to rebound – a process called isostatic uplift which is still going on today.

Climate Change

By comparing historical records and modern measurements from weather stations and satellites, it has become clear that Earth's average temperature has generally been increasing since the 1970s. Studies suggest that the period from January 2000 to December 2009 was the warmest decade in the past 1,000 years. Since then, the planet's average temperature has remained very high by recent historical standards, and in 2016 Earth's surface temperature was the warmest since modern recordkeeping began in 1880 (Figure 3.18).

Furthermore, the planet's average surface temperature has risen about 1.1°C since the late 19th century and it is forecast to rise by at least another 1.4°C during the 21st century.

Such global warming is generally attributed to an increase in greenhouse gases, particularly carbon dioxide and methane, which absorb heat and prevent it from escaping into space. Most of these gases have been added to the atmosphere by human activity, notably the clearing of forests by slash and burn farming, changes in agricultural land use, increased use of fossil fuels, and rampant urbanization/industrialization. Since about 1860, levels of atmospheric carbon dioxide have increased by almost 40% and methane levels have more than doubled.

The role of greenhouse gases in influencing Earth's temperature seems to be confirmed by measurements of air bubbles trapped in ancient ice cores. These indicate that on the ice age Earth of 20,000 years ago there was 50% less carbon dioxide and an even greater reduction in methane compared with current levels.

As the orbital geometry initiated warmer conditions, it seems that the warming was accelerated by the spread of plant cover and the release of carbon dioxide and methane. The minor warming by these gases also triggered a larger increase in water vapor, the most common greenhouse gas.[10] In recent times, the process has been further accelerated by human activity.

Nevertheless, the warming trend since the end of the last glacial advance has been punctuated by some major fluctuations. Reconstructions of past climate show relatively warm conditions around the year 1000, which enabled the Vikings to settle in Greenland and even reach North America, followed by a relatively cold period, or "Little Ice Age," from roughly 1500 to 1850.

It is also worth noting that there are remarkably strong correlations between measures of past solar activity and global temperature. For example, the Little Ice Age, which began around 1300 and lasted until 1870, coincided with a period when few or no sunspots were observed. Moreover, the rise in temperature over the past 100 years has occurred when the Sun increased its output to its highest levels in the last millennium.

Although radiant heat from the more active Sun is not enough to explain the rise in 20th century temperatures, it has been suggested that a change in solar activity can have important side-effects. For example, an active Sun generates a more powerful magnetosphere. This provides a stronger shield from high-energy cosmic radiation. Scientists theorize that these cosmic rays affect our climate by ionizing particles and gases in our atmosphere. The ionized molecules act as nucleation points for water droplets and lead to the formation of clouds. These clouds reflect sunlight back into space but also trap heat radiated from the ground.

Scientific views on climate change will continue to be revised as new data become available, but the present consensus is that human-made CO_2 in the atmosphere is modifying the climate and is the main cause of global warming.

Size and Density

With an equatorial diameter of 12,756 km, Earth is the largest of the Solar System's four rocky planets. Although it is generally portrayed as a sphere, the planet is very slightly oblate, i.e. it has a slight outward bulge at the equator and flattening at the poles. This means that the polar diameter is 43 km less than its equatorial diameter. As is the case with other non-spherical planets, the reason for this difference is Earth's fairly rapid rotation, which causes it to bulge outward at the equator.

It is interesting to note that Earth's shape, defined by the planet's gravity field, or geoid, also changes a little over time. Satellite data have shown that the shape is influenced by major earthquakes and climatic events that cause changes in the mass of water stored in oceans, continents, and atmosphere.

Earth's bulk density is 5.52 g/cm³, higher than any of the other planets in the Solar System. Since the density of the rocks in Earth's crust is much lower than this, it follows that the material in the interior must compensate by having a much higher density.

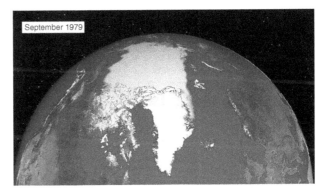

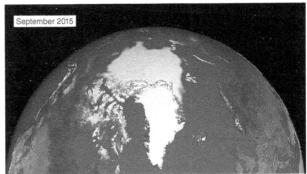

Figure 3.18 Satellite data show a significant decrease in the area of summer Arctic sea ice since the late 1970s. These maps show sea ice extent in September 1979 (top) and September 2015 (bottom). The September 2015 ice cover was 1.81 million sq. km lower than the 1981–2010 average. This decades-long decline, which affects how much solar radiation is reflected back into space, have been attributed to global warming. (NASA)

---

[10] It has been estimated that a 1% rise in water vapor could raise Earth's global average temperature by more than 4°C.

Interior

Seismology – the study of earthquake waves – provides a window into the planet's interior. Careful measurement of the way these waves are bent inside the Earth and the time they take to travel through the planet (and, hence, their speed) has enabled geophysicists to determine the density and physical properties of the interior. The data show that the interior is rather like an onion, with a number of concentric layers.

At the surface is a thin, rocky crust which is 20–60 km deep beneath the continents and 8–10 km deep on the ocean floor. The continental crust is less dense (average 2.7 g/cm³) than the oceanic crust (average 3.0 g/cm³). Whereas the continents have a very complex structure and variable composition, the oceanic crust has a simple, layered structure and a uniform composition of basaltic lava. The continental crust is generally much older, with parts of Canada and Australia dating back more than 3,500 million years. In contrast, the oceanic crust is geologically young, with a maximum age of about 200 million years.

At the base of the crust is the so-called Mohorovičić discontinuity (usually shortened to "Moho"). This boundary marks a sudden change to the mantle, which is mostly solid and composed of dense rock rich in the mineral olivine – a silicate containing magnesium and iron.

Geophysicists refer to the crust and the solid portion of the upper mantle as the lithosphere. The lithosphere is divided into many separate slabs, known as plates, that move independently.

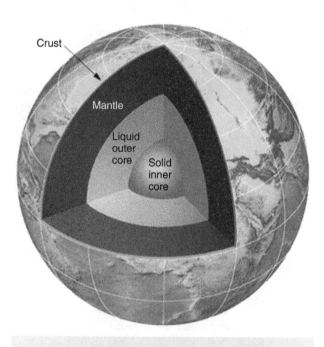

Figure 3.19 Earth's interior is composed of four main layers. At the center is a solid, iron-rich core, surrounded by a liquid outer core, which is the source of Earth's magnetic field. Above this is the mantle, where giant convection cells are thought to exist. The low density, solid crust "floats" on top of the "plastic" upper mantle. (Lawrence Livermore National Laboratory)

These lie on top of a mainly solid, "plastic" layer known as the asthenosphere, which extends to a depth of about 250 km.

The temperature of the rocky materials that make up the asthenosphere tends to be just below their melting point. This gives them a plastic-like quality comparable to glass. An increase in the temperature or pressure causes the material to deform and flow. If the pressure on the material is greatly reduced, so is its melting point, and the material may begin to melt quickly, providing a supply of magma for volcanoes.

The solid silicate mantle, which makes up 84% of Earth's total volume, extends to a depth of 2,900 km, where it meets the even denser, metallic core. Large convective cells in the mantle circulate heat and are probably partly responsible for the motions of the crustal plates.

The outer core, which is about 2,300 km thick, seems to be composed mainly of an iron–nickel alloy with trace amounts of lighter elements, such as sulfur. Swirls and eddies in the liquid outer core are responsible for generating Earth's magnetic field. The solid inner core, which is composed of iron with some nickel, makes up the final 1,200 km to the planet's center.

Surface Features

Earth's extremely varied surface ranges from high mountains to deep valleys. This varied topography is largely an expression of crustal processes associated with plate tectonics.

About 70% of the planet is covered by water, i.e. below sea level. Although the continental shelves bordering the major land masses are very shallow, they rapidly drop away to a deep abyssal plain. Much of the ocean floor lies at depths of about 5 km, but it is much deeper in narrow ocean trenches. The record holder is the Marianas Trench in the western Pacific, with a maximum depth of 10,911 m.

Although it is often blanketed by thick sediment, the ocean floor is not completely flat and low-lying. Running along the spine of the Atlantic Ocean is a ridge of submarine, volcanic mountains, 2–4 km in height and hundreds of kilometers wide. In places, the ridge rises above sea level, most notably in Iceland. Similar mountain ranges, where crustal plates are moving apart, are found beneath the other oceans.

High mountain ranges are also found on the continents, often near the coasts. These fold mountains, largely made up of highly compressed sediments, are usually associated with collision zones where two plates are inexorably pushing into each other. The highest of Earth's mountains are the Himalayas, between India and China, with Mt. Everest towering 8,848 m above sea level.

Large areas of the continents are fairly flat and close to sea level. The higher plains have been lowered and smoothed by weathering and erosion over millions or billions of years. Lowland plains are typically covered in thick layers of sediment. These may have been deposited by glaciers (either as unsorted debris or meltwater deposits), rivers, wind, or ocean waves.

Plate Tectonics

If you look at a map of the world, it is easy to see how the shape of Africa's west coast appears to mirror the eastern coast of South

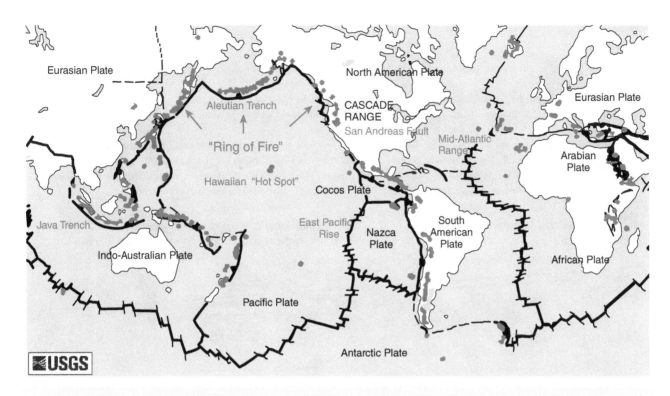

Figure 3.20 Earth's lithosphere is divided into seven large plates and a number of smaller ones which are in continual motion. Volcanoes (red dots) and fold mountains form along plate boundaries. Most of the world's earthquakes also occur where crustal plates meet. (USGS)

America. Studies of the geological structure and rock types on either side of the Atlantic also indicate that the two continents were once joined, e.g. the Appalachians and the Scottish Highlands.

Evidence such as this was used by a German meteorologist, Alfred Wegener (1880–1930), to develop a theory of continental drift. Although his vision of moving land masses was too revolutionary to be accepted at the time, it is now accepted that Earth's lithosphere is divided into a number of huge, jostling plates that are in perpetual motion (Figure 3.20). This is known as the theory of plate tectonics.

According to this theory, the solid lithosphere is "floating" on the denser mantle. The plates are slowly, but continuously jostling against each other. Some are moving apart (diverging), others are colliding (converging) or sliding laterally past each other. This movement, typically a few centimeters per year (comparable to the growth rate of fingernails), is responsible for most earthquakes and volcanoes.

It is also responsible for the gradual growth or shrinkage of the oceans. For example, Europe and North America are creeping further apart as new crust is created on the floor of the Atlantic Ocean. In contrast, the Pacific Ocean is gradually shrinking as oceanic crust is destroyed at the margins.

The lithospheric plates grow at the mid-ocean ridges, where two slabs of crust move apart and new crust is created. Hot material slowly rising beneath the ridge experiences a reduction in pressure which lowers its melting temperature. This enables rock from the asthenosphere or upper mantle to melt beneath the rift area,

forming basaltic lava flows which solidify on or beneath the ocean floor. Over millions of years, as the new oceanic crust piles up to form a huge mountain range, the ocean gradually becomes wider.

An example of a new ocean experiencing its birth pangs can be seen in East Africa, where the crust is pulling apart to form a huge rift valley. The northern end of the Great Rift Valley has already been flooded to form the Red Sea. Large lakes, such as Lake Tanganyika, also occupy parts of the valley.

Since we know that Earth is not increasing in size, it follows that sea-floor spreading must be balanced by the destruction of crustal material. This is now known to occur at ocean trenches, where a thick, relatively dense oceanic plate is melted as it dives beneath a neighboring plate which is less dense.

The descending slab contains a significant amount of water (Figure 3.23). As the plate descends deeper and deeper, encountering greater temperatures and pressures, this water is released into the overlying wedge of mantle. The addition of water has the effect of lowering the mantle's melting point. Magma produced in this way varies from basalt to andesite in composition.

As it rises to the surface, the magma feeds arcs of volcanic islands, such as Japan, Indonesia, and the Philippines, that lie parallel to the oceanic trench. If the plate sinks beneath a continent, the magma creates a similar belt of volcanoes near the coast, e.g. the Andes of South America, or the Cascades in North America.

These destructive plate margins, often known as subduction zones, are also associated with the deepest (and some of the most

Box 3.3   Flat Earth or Round Earth?

Before the Space Age, there were three main ways to prove that the Earth is round:

1. By watching ships sailing towards the horizon. A ship's hull follows the curve of the Earth and drops out of view first. Only later do the sails or masts disappear from sight. If Earth was flat, a ship should simply get smaller and smaller until it is no longer visible.
2. People traveling south see different constellations rise higher above the horizon. At the same time, the familiar northern constellations disappear below the horizon. If Earth was flat, travelers would always see the same star patterns.
3. During a lunar eclipse, the Moon passes through Earth's shadow. The edge of the shadow seen on the Moon is always curved, no matter how high the Moon is over the horizon. Only a sphere casts a circular shadow in every direction.

If Earth is spherical, how big is it? Once again, the ancient Greeks found an answer. It was clear from the size of its shadow during a lunar eclipse that our planet is bigger than the Moon. The actual size was first calculated by Eratosthenes (276–195 BCE).

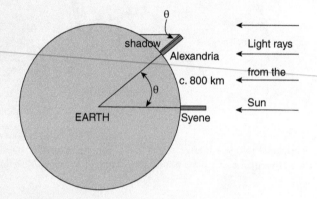

Figure 3.21   Simple geometry was used by Eratosthenes to calculate the Earth's size. The first step was to calculate the distance between Syene and Alexandria. He then measured the length of the shadows at these two places at midday on June 21, and calculated the angle of the Sun (*θ*) above the horizon at Alexandria (a little over 7°). Since there are 360° in a circle, he was then able to calculate Earth's circumference. (Peter Bond)

The first step was to place sticks in the ground at two places, 800 km apart. On the day of the summer solstice (June 21), he obtained measurements of the length of their shadows at noon. At one location, Syene (modern Aswan), the Sun was overhead and cast no shadow. At Alexandria, the Sun was about 7° away from the zenith, so there was a noticeable shadow. A simple calculation using the length of the shadows and the distance between the two sticks enabled him to work out Earth's circumference and diameter. (If a difference of a little over 7° is 800 km, then a full circle of 360° must be about 40,000 km. The diameter is equal to the circumference divided by 3.14, the value of *π* ["pi"]. This gives a diameter of about 12,700 km – very close to the actual figure.)

Later astronomers attempted to repeat his observations and obtained smaller figures. The belief in this "shrunken" Earth prevailed for centuries, and resulted in Christopher Columbus believing that the westward journey from Europe to Asia would be much shorter than it really is. His expedition was only saved by the unexpected discovery of the Americas.

severe) earthquakes on the planet, such as the one which caused the Indian Ocean tsunami disaster on December 26, 2004.

Not all plate boundaries involve the formation or destruction of crust. In some places, known as transform faults, two plates slide horizontally past each other. The most famous example is the San Andreas fault system in western California.

The San Andreas fault zone, which is about 1,300 km long and sometimes tens of kilometers wide, separates the North American Plate from the Pacific Plate. These have been grinding horizontally past each other for 10 million years, at an average rate of about 5 cm per year. Land on the western side of the fault zone (on the Pacific Plate) is moving in a northwesterly direction relative to

the land on the eastern side (on the North American Plate) – so Los Angeles and San Francisco are slowly moving away from each other. Sudden releases of built-up stress cause powerful earthquakes in this region.

By reversing current movements and studying the different magnetization of minerals in the rock, geologists can trace back the positions of the continents through time. On several occasions in the past, the present land masses have come together to form a supercontinent.

Approximately 550 million years ago, Africa, South America, Australia, Antarctica, and India were joined to form one gigantic land mass known as Gondwana. Over the next 300 million years,

Figure 3.22 Earth's surface exhibits a difference in height of about 20 km from the highest mountain to the deepest ocean trench. The highest land (apart from the ice sheets of Greenland and Antarctica) is found in ranges of fold mountains – shown in dark red. Most of the continental lowlands (green) are drained by large river systems and covered by sedimentary deposits. The ocean floors exhibit major ranges of volcanic mountains, isolated volcanic islands, and deep, narrow trenches. (NASA)

Gondwana was further enlarged by the addition of Europe and North America, so that all of the world's major land masses were eventually combined in a supercontinent known as Pangaea.

Pangaea began to break into two large masses, Gondwana and Laurasia, around 170 million years ago (Figure 3.25). Eventually, crustal rifting enabled the South Atlantic Ocean to open up as South America began to drift slowly westward away from Africa. Then India began to move northward, eventually colliding with Eurasia to form the Himalayan mountains and the Tibetan plateau. Other young fold mountains are growing around the Mediterranean Sea, where Africa is plowing into Europe, and along the west coast of the Americas, where oceanic plates are diving beneath the continents.

The driving forces behind the plate movements are not completely understood. It was formerly believed that the plates are like rafts, passively drifting on top of the moving asthenosphere. However, more recent studies suggest that they contribute to their own movement.

One hypothesis suggests that the weight of a cold slab of crust descending into the asthenosphere pulls the entire plate along. This is supported by normal faulting in the ocean crust, evidence that these plates experience tensional stresses. There is also a suggestion that gravity causes plates to slowly slide downhill, away from the high mid-ocean ridges. However, during initial rifting, plates begin to move without the help of such an elevated ridge.

It seems that convection cells also exist in the solid mantle. In this scenario, warm, rising rock from the mantle reaches the lithosphere and then spreads out laterally. The drag applied to the lithosphere causes it to move. The size of these convection cells is uncertain. It may be that they are confined to the asthenosphere and upper mantle, down to a depth of 700 km.

Alternatively, the convection cells could extend much deeper, reaching all the way down to the core boundary. There is considerable evidence for hotspots in the mantle beneath both the ocean floors and the continents. These are where thermal plumes, vertical columns of upwelling mantle 100–250 km in diameter, lift the overlying lithosphere and spread laterally at divergent plate margins, e.g. central Iceland.

Plates probably move as a result of a combination of these mechanisms. Plate separation may be initiated by mantle convection or a plume, but the subsequent formation of a topographically high spreading ridge may then drive the plates apart.

Earth is the only planet in the Solar System where such complex plate tectonics occur at the present time. On the Moon, Mercury, Venus, and Mars the rigid lithosphere forms a single layer, rather than separate slabs whose motions are driven by convection in the mantle.

Volcanoes

There are estimated to be more than 1,500 active or dormant volcanoes on land, while an unknown – but even larger – number of submarine volcanoes also exists. Some 600 of these have erupted in historical times, while about 50–70 volcanoes erupt each year. This makes Earth one of the most volcanically active places in the Solar System.

The vast majority of these volcanoes are located on islands or near to coastal regions (Figure 3.20). They are also generally located close to active plate boundaries, either along mid-ocean spreading ridges or alongside subduction zones at ocean trenches (see Plate Tectonics).

The main exceptions are chains of volcanic islands which occur above hotspots in the mantle, far from any plate boundary. The most famous example of hotspot volcanism is the island chain of Hawaii. Each of the islands has been created, one by one, as the Pacific Plate passed over a stationary plume of hot magma that has been in the same place for about 80 million years (Figure 3.26). The individual volcanoes erupt for a few million years before the movement of the plate carries them away from the rising plume.

This scenario is confirmed by the ages of the islands, which become progressively older with distance from today's center of volcanic activity on the main island of Hawaii. A bend in the seamount chain was probably caused by a shift in the direction of movement of the Pacific Plate, from a northward to a more northwesterly direction, about 47 million years ago.

Mauna Loa, which makes up half the area of the island of Hawaii, is the largest volcano on Earth. However, only a small part of the huge shield volcano is visible above sea level. With an estimated volume of about 75,000 cubic km, and a base that is more than 145 km in diameter, Mauna Loa is so massive that it deforms the oceanic plate on which it sits.

Although its overall height from base to summit is about 17 km, the average gradient of its slopes is quite gentle, a mere 4–5°. Since it began to form nearly one million years ago, the enormous structure has grown from accumulated layers of fluid, basaltic lava. At its summit is a large caldera, a crater-shaped basin created when subterranean magma drained away and caused massive subsidence.

The Hawaiian volcanoes are characterized by frequent eruptions of very hot, non-viscous, basaltic lava which sometimes produces fountains of liquid rock and small spatter cones. Since any gas the lava contains can easily escape, explosions are rare.

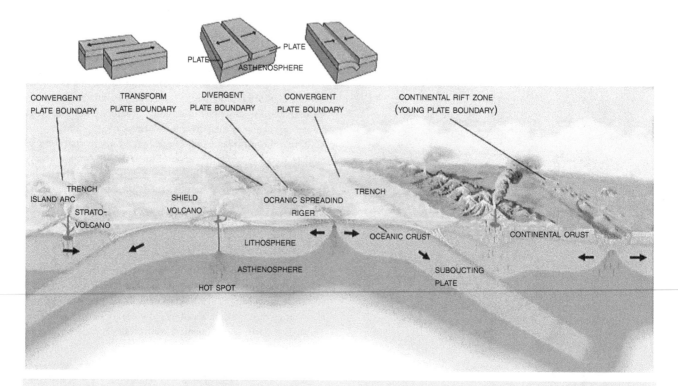

Figure 3.23 Oceanic plates grow at mid-ocean ridges, where magma rises from the mantle and creates new crust. Oceanic crust is destroyed at subduction zones, where a dense plate dives beneath a less dense plate and is slowly destroyed. Water from the melting plate helps create magma that rises to the surface and feeds nearby volcanic islands. The erratic descent of the subducted slab also sets off deep, sometimes catastrophic, earthquakes. Some volcanoes also form above hot spots in the mantle. (USGS)

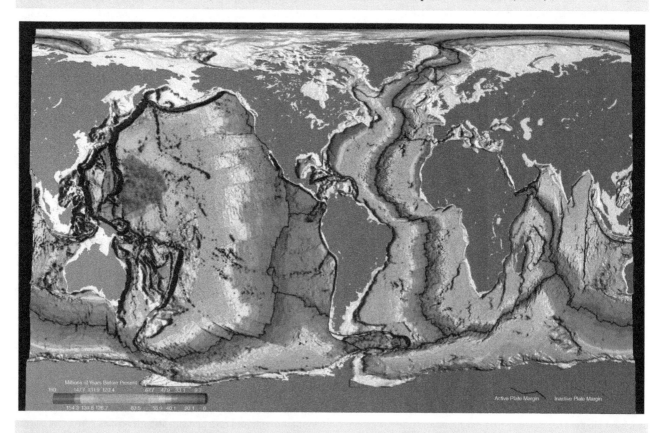

Figure 3.24 A map of ocean floor age shows the youngest areas (red) along the mid-ocean ridges, where new rock is forming from rising magma. The oldest oceanic crust is furthest from the ridge, often along the continental margins (blue). Oceanic crust is destroyed at ocean trenches. Nowhere is it more than about 180 million years old. This contrasts with the much older continents, parts of which date back at least 3,500 million years. (NOAA)

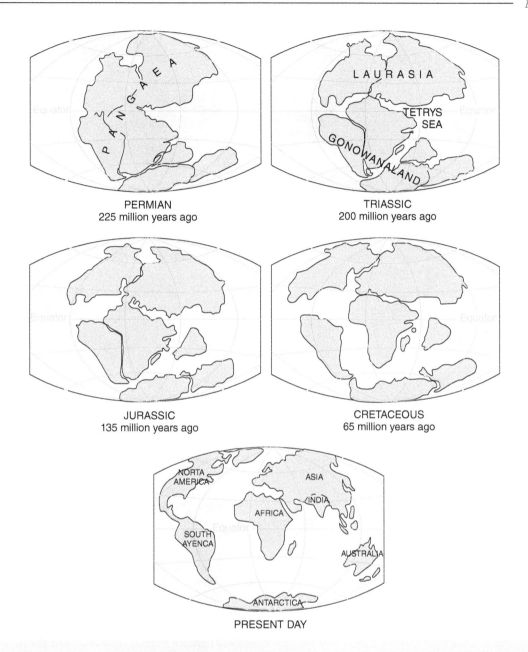

Figure 3.25 A series of maps showing the gradual break-up of the Pangaea supercontinent over the last 225 million years. Note how South America and Africa drifted apart with the opening up of the Atlantic Ocean, followed by Europe and North America. India moved north to collide with Eurasia, creating the Himalayan mountains and the Tibetan plateau, but Australia remained attached to Antarctica until quite recently. (USGS)

In other parts of the world, eruptions of shield volcanoes are associated with basaltic lava pouring quietly from long fissures, instead of central vents. The fluid lava repeatedly floods the surrounding land, creating multiple layers which build into broad plateaus. Lava plateaus of this type can be seen in Iceland, the northwestern USA, and the Deccan of India.

Eruptions are more violent where the lava is less fluid and gases can escape less easily. Pressure may build inside the volcano until the gases escape in a massive explosion, destroying part or all of the summit.

In the case of volcanoes such as Vesuvius in Italy and Mt. St. Helens in the USA, explosive activity is quite common, resulting in huge clouds of steam, gas, solid fragments, and ash which can rise several kilometers into the air and blanket the surrounding terrain. Glowing clouds of red-hot rock fragments, known as nuées ardentes, may flow downhill, incinerating everything in their path.

The combination of alternating layers of ash and lava results in composite volcanoes or stratovolcanoes. The classical shape of such a volcano is a cone, like Mt. Fuji in Japan, but the symmetry is often spoiled by smaller cones and lava flows on the flanks.

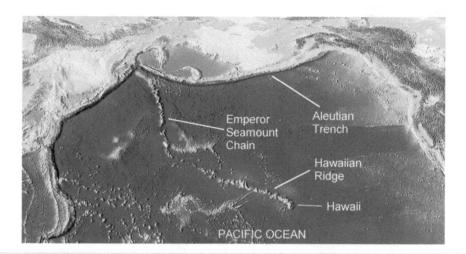

Figure 3.26   The Hawaiian Islands and Emperor Seamount Chain formed as the Pacific Plate drifted over a hot spot in the mantle. Today, only the Big Island of Hawaii – the youngest island – has active volcanoes. Mauna Loa, which makes up half of the Big Island, is the largest shield volcano on Earth. The age of each volcanic sea mount increases with distance from Hawaii. The bend in the chain may have been caused by a change in the Pacific Plate's direction of movement. (NGDC/USGS)

Many high volcanoes have summits covered with ice and snow, and heavy rainfall often accompanies large ash clouds. The plentiful supply of water may turn the ash deposits into mud flows (also known as lahars) which bury the surrounding landscape.

In extreme cases, the products of an explosive eruption may travel around the planet and even modify the climate (Figure 3.27). The most notable example of modern times was the 1991 eruption of Mt. Pinatubo in the Philippines, which ejected vast amounts of ash and gas as high as the stratosphere.

Pinatubo also injected about 15 million tonnes of sulfur dioxide into the stratosphere, where it reacted with water to form a hazy layer of aerosol particles composed primarily of sulfuric acid droplets. Over the course of the next two years, strong winds in the stratosphere spread these particles around the globe. The result was a drop of about 0.6°C in Earth's average temperature for a period of almost two years.

Mountains

Although volcanism accounts for the extensive mid-ocean mountain ridges and some large islands, such as those of the Philippines, Japan, and Indonesia, Earth also exhibits extensive mountain ranges which have been created by tectonic forces.

At the present time, the western side of the Americas, southern Europe, and much of southern Asia are dominated by relatively young, high fold mountains (Figure 3.22). Many of these have grown within the last 60 million years, and they are still increasing in height today.

The tallest mountains on the planet are the Himalayas, which are continuing to rise at an average rate of about 40 mm per year as the Indian tectonic plate slides north beneath the Eurasian Plate.

Like the Himalayas, all of the continental fold mountains have been created by compressional tectonic activity involving a collision between two lithospheric plates. There are three types of tectonic convergence: ocean-continent, arc-continent, or continent-continent.

When oceanic and continental plates collide, marine sediments are scraped off the descending oceanic plate, squeezed and accreted to the edge of the continent, forming a range of fold mountains parallel to the coast. Magma formed by melting of the descending plate feeds numerous volcanoes. This is the case with the Andes mountains of South America.

When a volcanic island arc collides with the edge of a continental plate, sediments are also scraped off the descending oceanic plate, but the volcanic islands collide with the continent and are also accreted to the margin of the continent. This type of collision may have been responsible for the creation of the Sierra Nevada range in California.

The most dramatic type of convergence occurs when an ocean basin closes and two continental plates collide. In this case, sediments that once lay on the old ocean floor and rocks in the continental crust are folded, fractured, and changed by heat and pressure.

Over time, the mountains are weathered and eroded. As material is removed by glaciers and rivers, the weight of the crust becomes progressively less. The continental crust makes an isostatic adjustment, causing it to rise slowly – rather like removing a weight from a floating raft. Such adjustment places considerable strain on the crust, causing it to fracture along fault lines or break up into separate blocks. A block of crust may rise to form a horst, or sink between two faults to form a graben or rift valley.

Over hundreds of millions of years, weathering and erosion lower the once-towering mountains and make them more rounded in shape. The modern-day Appalachians and Scottish Highlands are examples of this process, remnants of a period of mountain building that ended about 250 million years ago (see Plate Tectonics).

The Blue Planet

The presence of vast quantities of liquid water at the surface makes Earth unique among the bodies of the Solar System. 71% of the

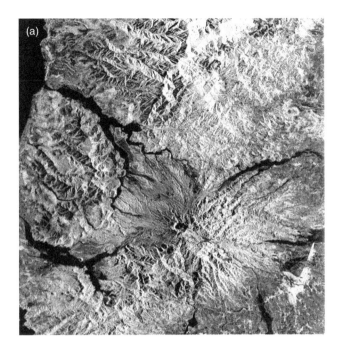

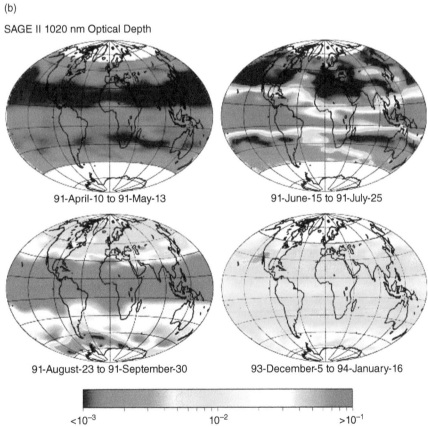

Figure 3.27 (a) A false color radar image, taken from the Space Shuttle on April 13, 1994, shows the after-effects of the 1991 eruption of Mt. Pinatubo in the Philippines. The red color shows deposits of ash on the higher slopes. The dark areas are mudflows which filled nearby river valleys after heavy rain. The area shown is approximately 45 × 68 km. (b) The eruption sent a cloud of fine particles and gases into the stratosphere and around the world, causing a noticeable cooling of the planet. A large amount of sulfur dioxide (green to red) was injected into the upper atmosphere. (NASA-JPL / NASA-GSFC)

Figure 3.28 The Himalayas are the highest mountains on Earth. They are still increasing in height as the Indian plate moves north beneath the Eurasian plate. Weathering and erosion by glaciers and rivers carve deep valleys and remove large amounts of material which is eventually carried to the sea as sediment. This false color image shows the Ganges river, which rises in the snow-covered Himalayas and flows 2,510 km to the Bay of Bengal. (ESA)

planet is covered by the saline oceans, equal to 97% of the world's free liquid water. In comparison, the volume of fresh water is very small. Most of this (2% of the total) is locked up in the ice sheets of Greenland and Antarctica, while the remaining 1% occurs in rivers and lakes.

Water plays a key role in the planet's evolution, since the temperature range on Earth enables it to exist as a gas (water vapor), a liquid, or a solid (ice). Even in the driest deserts there is likely to be some morning dew, night frost, or an occasional rainstorm.

Water plays a subtle, but key, role in breaking down Earth's rocks into small particles which can then be removed by wind or rain. This weathering process takes many forms, ranging from the expansion of ice in cracked rocks to chemical solution of rock minerals.

However, the impact of water is most noticeable through its ability to erode the land, i.e. wear away and remove surface material. In high mountains, frozen water in the form of moving glaciers slowly carves very wide and deep valleys. More commonly, innumerable streams and rivers flow from upland areas towards the ocean (or occasionally, large, landlocked lakes, such as the Caspian Sea).

The force of the flowing water removes loose material – anything from sand and silt to large boulders – from the stream's bed and banks. This in turn acts like sandpaper, further wearing away the surface. After millions of years, this process is capable of eroding mountains and forming broad valleys.

When the water slows down, it drops much of its load of sediment. This may happen during floods, or when the river enters a lake or the sea (Figure 3.34). Over thousands of years, the build-up of sediment at river mouths may lead to growth of large deltas, such as those of the Mississippi and the Nile.

Large-scale erosion and deposition also occur in coastal regions through the action of waves and currents. This may result in the growth of new land through the accumulation of mud, sand, and gravel, or the collapse of coastal cliffs which are undercut by wave action, especially during storms.

Deserts

Most places on Earth receive more than 25 cm of annual precipitation, with coastal mountains exposed to moist winds receiving at least 10 times that amount. However, there is considerable variation between different latitudes and between coastal or continental locations.

In temperate regions, where depressions are the major source, rainfall is usually spread throughout the year, with a modest increase in winter (when the low-pressure systems are deeper and more frequent).

At the equator, rainfall is also found throughout the year, though the maximum coincides with the time of strongest solar heating and convective uplift. This typically results in thunderstorm activity during the afternoon.

Between these two wet zones, the rainfall is more seasonal and unreliable. Many of the driest places occur at 20–30° latitude, where dry, sinking air results in a belt of hot deserts (see Zonal Winds). Other deserts occur in the center of Asia, in rain shadow regions which are cut off from moist, maritime air by high mountains.[11]

Deserts are characterized by cloudless skies, high daytime temperatures, low nighttime temperatures, and an annual precipitation of less than 25 cm. Some places may experience several years with no rain at all. This inevitably results in an absence of vegetation. When rain does fall, it usually comes in the form of short-lived, but torrential storms, which often lead to flash floods. However, any surface water soon sinks into the sand or is evaporated.

Although major rivers, such as the Nile and Colorado, carry sufficient water from their mountain sources to reach the sea, most desert river courses remain dry throughout the year, filling only occasionally when torrential downpours cause flash floods.

Under extreme day–night temperatures, rock minerals expand and contract. This process may also be aided by occasional wetting from rainfall or groundwater. As time goes by, the rock disintegrates into particles of sand or gravel. (90% of the Sahara is made up of rocky or stony areas, with only 10% covered by the more familiar sand – derived from the breakdown of sandstone rock.)

---

[11] The Arctic and Antarctic are also extremely dry. Only the slow rate of evaporation prevents these regions from being classed as deserts.

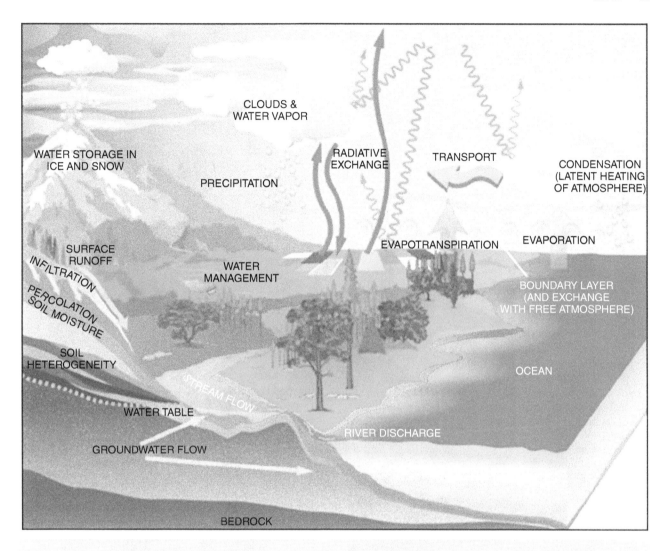

Figure 3.29 Earth's water cycle. Liquid water on the surface is warmed and evaporates, turning to water vapor. When this gas rises, it cools and condenses, turning back into cloud droplets of liquid water. These droplets collide, growing in size until they are heavy enough to fall towards the surface in the form of precipitation (rain, sleet, snow, etc.). Rain seeps into the ground or flows over the surface, feeding streams and rivers which flow downhill into lakes or the ocean. Melting snow and ice also feed the surface drainage systems. (NASA-GSFC)

With no surface moisture and no vegetation to slow the wind or bind the loose material together, wind becomes the main agent of erosion and transportation. The smallest dust particles (less than 0.2 mm) are carried in suspension in the air, while larger sand grains more commonly bounce across the surface, displacing other particles – a process known as saltation.

These grains, often traveling at high speed, act like sandpaper when they collide with bare rock surfaces. Wind erosion shapes a wide variety of strange rock formations, such as rock pedestals which have been eroded at their base to leave narrow necks, and yardangs – elongated ridges up to 8 m high, separated by troughs where less resistant material has been removed.

The loose material is always shifting location, though the finest dust travels the furthest. Sometimes the grains are removed by the wind, leaving deflation hollows. If these fill with temporary ponds or lakes, which then evaporate, layers of salt may be left behind. The most famous example is the Bonneville salt flats in Utah, USA, the remains of an ancient lake that formed during the last ice age and subsequently evaporated. The salts on the surface contain potassium, magnesium, lithium, and sodium chloride (common table salt).

Moving sand piles up when it meets an obstacle, forming dunes. These may take different shapes, depending on the variability of the wind. Barchans are crescent-shaped dunes with one steep side and one more gently sloping side. They form when the wind blows from one dominant direction. The sand moves up the gentler slope until it reaches the crest, then falls down the lee side. This process also drives the entire dune forward. Barchans can be up to 30 m high and 400 m long.

Longitudinal or linear dunes lie parallel to the prevailing wind (Figure 3.31). They form symmetrical ridges up to 100 m high,

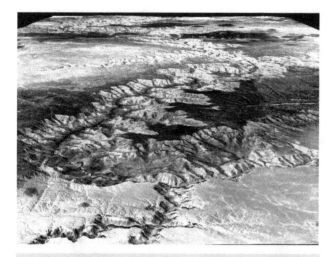

Figure 3.30   This false color Landsat image shows the Grand Canyon of northern Arizona which has been created by the Colorado river after millions of years of erosion. In this westward-looking image, the river cuts through the salmon-colored rock of the Colorado Plateau. The Little Colorado river enters from the east (bottom) across the Painted Desert. Nearby uplands are covered in green vegetation. The fairly rapid downward erosion by the river, combined with the lack of rainfall in this desert region, has preserved the canyon's remarkably steep sides. (NASA-GSFC)

Figure 3.32   A satellite image of a dust storm sweeping from the Algerian desert over the Canary Islands. Such storms may carry dust and aerosols all the way across the Atlantic Ocean to the Caribbean and Florida, blocking sunlight and affecting surface temperatures. (SeaWiFS Project, NASA-GSFC, and ORBIMAGE)

Figure 3.31   Linear dunes of the Namib Sand Sea imaged by an astronaut aboard the International Space Station. The highest dunes show smaller linear dunes along their crests. Linear dunes are generally aligned parallel to the formative wind – in this case, strong winds from the south. This simple pattern is disrupted near the Tsondab river valley (top left), which acts as a funnel for winds from the east. These less frequent but strong winter winds are channeled down the valley and usually carry large amounts of sand. These easterly winds significantly deflect all the linear dunes near the valley so that they point downwind (center). Further inland (right), the north-pointing and west-pointing patterns appear superimposed, making a rectangular pattern. (NASA-JSC)

600 m wide, and 80 km long. They generally form on the lee side of an obstacle where sand is abundant and the wind is constant and strong. At least some of these are derived from barchans which have been modified by cross winds. Transverse dunes lie at right angles to the prevailing wind direction. They have a gently sloping windward side and a steeply sloping leeward side.

Huge amounts of loose material can be removed from deserts or dry regions where there is little vegetation. Extensive deposits of fertile, windblown sediment are found in a broad zone from western Europe to China. This material, known as loess, was moved during the cold, dry periods of the recent ice age. On a smaller scale, dust from the Mongolian desert often blankets the Chinese city of Beijing, and Saharan dust frequently moves across Europe, with occasional excursions as far as North America (Figure 3.32).

Biosphere

Earth is unique in possessing a biosphere – an environment dominated by living organisms (Box 3.4). These organisms, which range from microscopic bacteria to plants, animals, and humans, have adapted so that at least some of them can survive in every type of environment on or near Earth's surface. Some microorganisms even exist within the crust, several kilometers underground.

The presence of these organisms greatly influences the planet's surface and atmosphere. Studies of newly formed, barren, volcanic landscapes show how plants and animals are able to colonize even

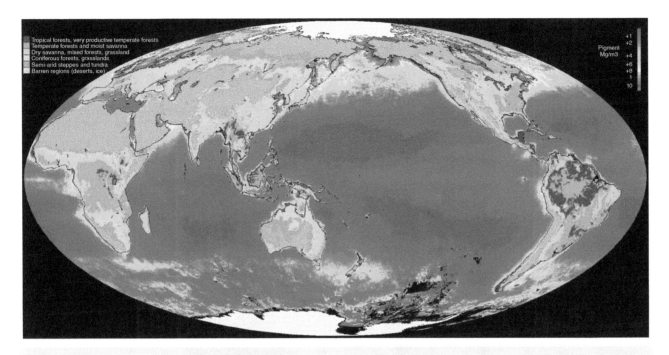

Figure 3.33 The global biosphere. Different plant ecosystems on the continents are shown in shades of green. Chlorophyll measurements (blue – purple) indicate the amount of phytoplankton in the ocean. Both vegetation and phytoplankton consume atmospheric carbon dioxide and release oxygen during photosynthesis. Note the large area of barren land (tan color), where there is little, if any, vegetation. (NASA-GSFC)

the harshest environments. First, mechanical and chemical weathering processes break down the rock into fragments of inorganic material. Later on, plant roots also intrude into every crack and pore in a rock, helping to break it apart.

Once a small amount of inorganic debris develops, plants, then animals, begin to invade the area. Over time, their waste and dead remains decompose, providing water, carbon dioxide, organic acids, and salts such as nitrates, which are then incorporated into the soil by fungi and bacteria. Earthworms and insects mix the soil as they consume and spread the dead organic material.

Certain bacteria, such as those which occur on the roots of leguminous plants like peas and beans, capture nitrogen from the air then incorporate it into proteins. Other bacteria then break these down into ammonia and nitrates, which can be absorbed by plants. This nitrogen cycle is only one method of circulating soluble nutrients around the biosphere. Micro-organisms also play key roles in the cycling of elements such as sulfur and phosphorus.

Vegetation also plays an important part in the planet's water cycle, since plants remove large quantities of moisture from the soil and release large amounts through their leaves into the atmosphere during the growing season. An increase in vegetation cover is generally associated with an increase in regional rainfall, and vice versa.

Plants also modify the local climate, for example by slowing the wind, thereby reducing erosion, and reducing the likelihood of frosts. High latitude forests reflect less solar radiation than grasslands or tundra blanketed by ice and snow.

Perhaps the most important process of all is the ability of many organisms – plants, algae, and some bacteria – to take up carbon

Figure 3.34 The lower reaches of the Betsiboka, the largest river in Madagascar, transport a huge amount of red sediment to the sea. The amount of sediment is increased by large-scale deforestation inland. The removal of forest cover accelerates soil erosion. Once the soil has been removed it cannot be replaced. Very little of the original forest now remains. (NASA)

dioxide from the atmosphere and produce oxygen through photosynthesis. Using sunlight as an energy source, they are able to fix (convert) carbon dioxide into carbohydrates, producing oxygen as a byproduct. It is this process which largely explains why Earth

Box 3.4   The Story of Life

No one knows how and when life began on Earth. Some believe it originated in a random manner, possibly in a chemical "soup" enriched by early atmospheric gases such as methane and carbon dioxide, with energy supplied by lightning or solar ultraviolet light. Others suggest the raw materials, or life itself, were delivered by comets and meteorites – a theory known as "panspermia."

Yet another possibility is that the first life forms lived deep beneath Earth's surface, where the temperature was far above 100°C, with sulfur in the rocks as its source of nutrition. Warm, chemical-rich, environments around deep sea volcanic (hydrothermal) vents, known as black and white smokers, are another alternative.

Figure 3.35   White hydrothermal smokers on the ocean floor near the Mariana Islands in the Pacific Ocean. Small white chimneys emit a cloudy white fluid, as well as columns of bubbles rising from the surrounding sediment. The vents release hot water, carbon dioxide, and minerals – one reason why the oceans are salty. Although they are often too deep to receive any sunlight, such smokers are often home to many organisms, such as bacteria, worms, and shrimp. They may have been sites where life first appeared on Earth. (NOAA)

Once the surface began to cool and life was able to establish a foothold on Earth, it was probably extremely precarious and short-lived. The massive bombardment by incoming asteroids and comets may well have wiped it out on a number of occasions until the rain of planetary debris tailed off about 3.8 billion years ago.

Nevertheless, the fossil record indicates that life was flourishing 3.5 billion years ago. Microfossils thought to be the remains of blue-green algae have been found in rocks of this age in parts of Australia. If this is the case, it shows that photosynthesis began very early in Earth's history.

Although micro-organisms seem to have spread quite rapidly across the oceans and the land surface, the oxygen produced did not build up immediately in the atmosphere, since it was involved in chemical reactions with rocks, particularly oxidation of iron. Even today, most of the oxygen produced over Earth's lifetime is locked up in ancient rock formations.

Levels of oxygen in the atmosphere seem to have remained quite low until the Great Oxidation Event, about 2.4 billion years ago. Only when the minerals in the crust became saturated and the rate of oxidation dropped dramatically did the oxygen content of the atmosphere increase – sporadically at first, and then irreversibly.

At the beginning of the Devonian period, about 400 million years ago, there was a wide variety of plants and animals in the oceans, including the first fish and sharks. Despite the steady evolution of life in the oceans, there seems to have been little change on the continents until the invasion of the land by algae, lichens, and mosses. Exactly when this occurred is uncertain, although fossils suggest land plants may have existed by 500 million years ago.

At that time, Earth's surface underwent a major change as North America collided with Europe, creating the Appalachian Mountains of the USA and the highlands now found in Norway, Scotland, and Greenland. Meanwhile, the huge land mass of Gondwanaland rested over the South Pole. This coincided with an explosion in the number and variety of land creatures.

Once large plants such as ferns moved onto the land, animals soon followed. The first were the arthropods, creatures such as crabs, spiders, and scorpions. Then came amphibians, fish-like animals that were able to breathe air and spend short times out of water. Insects began to appear about 325 million years ago, evolving from land-dwelling arthropods.

During the next 50 million years, there were major evolutionary changes in both plants and animals. By 280 million years ago, large areas were colonized by swampy forests with trees up to 30 m high and 1.8 m in diameter.

The final break of life's tie with water came with the arrival of the first reptiles, which laid hard-shelled eggs on the ground. These eventually developed into the most fearsome (and successful) creatures that ever roamed the Earth – the dinosaurs. Their reign came to an end 65 million years ago, after a massive asteroid impact coincided with huge volcanic eruptions, opening the way for the dominance of warm-blooded mammals.

is the only object in the Solar System to exhibit an abundance of atmospheric oxygen and an ozone layer in the stratosphere.

Vegetation cover protects and stabilizes the soil, reducing erosion by heavy rain and wind. When this ecological balance is upset, perhaps by logging or clearance for agriculture, the exposed soil may suffer large-scale erosion, causing rivers to fill with sediment. This results in landslides, increased flooding, and transport of soil to the sea.

Impacts

Like all of the terrestrial planets (and the icy objects in the outer Solar System), Earth has been subjected to innumerable impacts – large and small – by asteroids and comets. Studies of the ages of lunar craters indicate that Earth and the Moon underwent a very heavy bombardment prior to about 3.8 billion years ago. Indeed, the formation of the Moon is generally thought to be associated with a collision between Earth and a Mars-sized intruder (see Chapter 4). Such a violent episode would have been characterized by global melting of Earth's crust, perhaps delaying the establishment of life.

A glance at the heavily cratered Moon shows that there must have been many millions of impacts during Earth's 4.6 billion year history. Yet, at the present time, there are only 190 confirmed impact structures on Earth's surface. They range in size from the 2,000-km wide Vredefort feature in South Africa to a 15-m diameter hollow in Kansas.

Since about 70% of Earth is covered in water, it is logical to assume that most impact craters are hidden beneath the sea. However, this alone can not account for the extremely modest number of visible impact structures compared with the Moon.

The main factor is Earth's much younger crust. Plate tectonics, combined with active weathering, erosion, and deposition, mean that the ancient crust, which recorded impacts prior to 200 million years ago, has largely been destroyed or covered by sediments (see Surface Features). Only a few of the biggest impact structures, some dating back more than 2 billion years, can still be found on the continents.

Since the end of the heavy bombardment, the number and size of the impacts has decreased dramatically. Even so, Earth is still exposed to occasional collisions, with an estimated population of 1,000 near-Earth objects (NEOs) that are more than 1 km in diameter. However, a more likely threat comes from smaller objects, which are much more numerous. By February 2019, 1,947 NEOs had been classified as potentially hazardous (see Chapter 13).

An example of the potential threat was provided in 1908 when a 60-m wide object – probably a meteorite – exploded near Tunguska, Siberia. Although it disintegrated 6–10 km above the ground, the blast leveled 80 million trees over an area of 2,000 sq. km.

A similar, but smaller, event occurred on February 15, 2013, when a 10–20-m wide object exploded in the air above the Russian city of Chelyabinsk. The blast injured about 1,500 people and damaged more than 7,000 buildings, collapsing roofs and breaking thousands of windows.

Extinctions

Life on Earth has suffered periods of mass extinction, when many of the living organisms were wiped out within a short period of time – perhaps a few thousand years. Many of these undoubtedly occurred too long ago to be recognized in the fossil record, but between 5 and 20 such events are thought to have taken place in the last 540 million years. (The precise number depends on the definition of "mass extinction.")

Probably the most catastrophic of these events occurred about 250 million years ago. Sometimes known as the "Great Dying," the extinction marks the boundary between the Permian and Triassic periods in geological history. It is estimated that about 60% of all land species and 90% of all ocean species abruptly died out at that time.

The most famous extinction took place 65 million years ago and was responsible for ending the age of the dinosaurs. The most recent major extinction took place at the end of the Eocene period, about 35 million years ago.[12]

The causes of these mass extinctions are not well understood, but a major impact, or periodic increases in the rate of asteroid or comet impacts have been put forward as explanations for at least some of them. The most likely example of this is thought to be the event which coincided with the demise of the dinosaurs, as well as many other species of flora and fauna.

In 1980, a team of geologists, led by Luis and Walter Alvarez, discovered a relatively high concentration of iridium in a strip

Figure 3.36   Meteor crater (also known as Barringer crater) in Arizona is one of the most recent impact craters on Earth. It was created about 50,000 years ago, when an iron meteorite excavated a hollow some 1.2 km in diameter and 180 m deep. The crater has a simple basin shape with no central peak or rim terraces. Surrounding the basin is a wall of material 30–45 m high where the rocks have been uplifted and, in some cases, overturned. The 30-m wide meteorite probably weighed about 100,000 tonnes and struck the surface at a speed of around 12 km/s. The energy released was equivalent to about 2.5 megatonnes of TNT. (D. Roddy / USGS, Lunar and Planetary Institute)

---

[12] The disappearance of species from the Earth at the present time, resulting from human activity, is often regarded as another mass extinction.

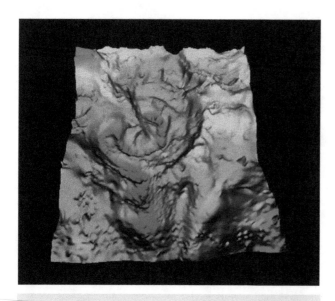

Figure 3.37   A computer-generated gravity map reveals the Chicxulub impact crater (center) buried beneath Mexico's Yucatan peninsula. The impact structure is widely thought to be the cause of a mass extinction 65 million years ago. Gravity measurements show a multi-ring basin with a fourth, outer ring about 300 km in diameter. It is one of the largest impact structures on Earth. (LPI/NASA)

of clay that runs through rocks around the world. Iridium is an element that is very rare on Earth but common in meteorites and asteroids. This clay-iridium layer marks the end of the Cretaceous period in geological history.

Searches for likely impact candidates revealed a crater up to 320 km across, buried beneath the Yucatan peninsula of Mexico. Geophysical surveys indicate that the multi-ringed Chicxulub crater was formed on the ocean floor. The explosive impact would have caused huge clouds of debris to blanket the planet and rain down from the sky, disrupting the climate and the food chain, while surrounding shorelines were devastated by tsunamis. This interpretation is supported by finds of shocked rocks, tektites (small pieces of natural glass), and widespread deposits of iridium.

Whether similar impacts are to blame for other extinctions remains uncertain. It has been suggested that most mass extinctions over the past 300 million years coincided with large-scale volcanic activity. For example, at the time of the Permian extinction, major eruptions were taking place in Siberia, where 1.5 million cubic kilometers of lava flowed from huge fissures in the crust. Other possible contenders include high-energy radiation from a nearby supernova and environmental changes brought about continental drift.

Even the impact theory for the Cretaceous event is disputed by some scientists, who point out that the extinction also coincides with the formation of the Deccan Traps in India, one of Earth's largest eruptions of fluid lava. Geologists have identified a 600-m thick lava sheet that may have piled up in as little as 30,000 years – potentially causing a deadly shift in atmospheric gases and global climate (see Volcanoes).

Magnetic Field

For centuries, explorers and navigators have relied on the presence of Earth's magnetic field – hence the value of the magnetic compass. In fact, this magnetic field is by far the strongest of all the terrestrial planets. Theory suggests that the field is generated by a dynamo effect created by currents circulating in the planet's liquid outer core.

The magnetic field resembles a dipole, with one North Pole and one South Pole, which is inclined about 10° to Earth's rotation axis. At these magnetic poles, which are fairly close to Earth's geographic poles, a compass needle will point straight down, or up, respectively.

However, the magnetic field is very dynamic. Studies of ancient magnetized rocks, both on the continents and the ocean floor, show that the magnetic poles have wandered across the surface throughout Earth's history. This motion is continuing today. During the 20th century, Earth's magnetic north pole moved over 1,100 km across the Canadian Arctic. In recent decades its rate of migration has increased significantly, and it is now moving across the Arctic Ocean towards Siberia.

Every so often, the magnetic field also reverses, so that the magnetic poles switch polarity, causing a compass needle to point towards the south instead of north.[13] This sequence of reversals is recorded in the once-molten rocks of the ocean floor, forming "stripes" that run parallel on either side of a mid-ocean ridge. As such, they provide key evidence supporting sea floor spreading, continental drift, and plate tectonics.

Although the last magnetic reversal occurred almost 800,000 years ago, the rate of reversals during the last 10 million years has averaged 4 or 5 every million years. However, research published in 2018 indicated that the field can reverse in less than two centuries. On the other hand, the interval has often been much longer. For example, during the Cretaceous era, 70 million years ago, the time between magnetic reversals was about one million years.

The magnetic field may also experience an "excursion" – a large, but temporary, decrease in its overall strength – rather than a reversal. Today, the field strength is steadily declining, leading to suggestions that a reversal may be on the way. Even so, the current strength of the magnetic field is still as high as it has been in the last 50,000 years.

There is no evidence to suggest that a reversal has any major impact on the planet's atmosphere or biosphere. Even in the absence of a magnetic field, the atmosphere is still able to filter out most cosmic radiation.

The area of space dominated by Earth's magnetic field – the magnetosphere – takes the shape of a tadpole or windsock, which is shaped by the flow and pressure of the incoming solar wind. This invisible magnetic bubble extends, on average, some 60,000 km toward the Sun and trails more than 300,000 km away on the planet's night side, forming a long magnetotail which stretches far beyond the Moon's orbit (Figure 3.39). The outer edge of the magnetosphere is called the magnetopause.

A curved bow shock – a type of shock wave – typically occurs about 90,000 km from Earth, though its size and location vary

[13] At present, Mercury is the only planet with the same polarity as Earth.

Box 3.5   Tides

Most of Earth's seas experience two high and two low tides each day. There is also a less noticeable tide in its solid crust, which causes a variation in height of about half a meter.

The tides are caused by the combined effects of the gravitational pull of the Moon and the Sun. Clearly, the pull of lunar gravity is greatest where Earth's surface is nearest the Moon. On the opposite side of Earth, the pull of lunar gravity is weakest. The overall result is that there are equal-sized oceanic bulges on opposite sides of the planet. The tidal bulges are fixed with respect to the direction of the Moon. However, the solid surface rotates beneath the oceans much more rapidly than the Moon orbits Earth. This gives rise to two high tides each day.

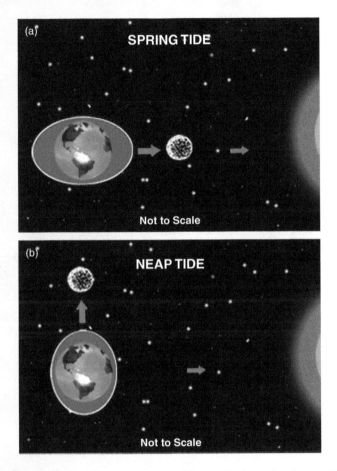

Figure 3.38   The highest tides, known as spring tides, occur when the Moon and Sun are aligned so that their combined gravitational pull increases the size of the tidal bulges in the oceans. When the Sun and Moon are 90° apart in the sky, their gravity pulls in different directions and the tides are least noticeable. These are the neap tides. (Windows to the Universe / Lisa Gardiner)

Precisely the same effects are created by the interaction of the Sun and Earth, although the tidal bulges are slightly less than half the size of those created by the Moon. About twice a month, the Sun and Moon line up and then their gravitational effects reinforce each other to produce the largest tides, the spring tides. When the Sun and Moon are 90° apart in the sky, the tidal bulges are at their minimum. These are known as the neap tides.

Although this simple pattern predominates in the deep ocean, the times of the tides can be delayed by several hours, even for places that are quite close together, in the complex waters of the continental shelf and river estuaries.

In reality, Earth's rotation carries the tidal bulges slightly ahead of the point directly beneath the Moon. This means that the force between the Earth and the Moon is not acting exactly along the line between their centers. This causes a net transfer of rotational energy from Earth to the Moon, slowing the planet's period of rotation by about 1.5 milliseconds per century. It also causes the Moon to raise its orbit by about 3.8 cm per year. Over a period of 100 million years, the Moon will recede from Earth by 3,800 km.

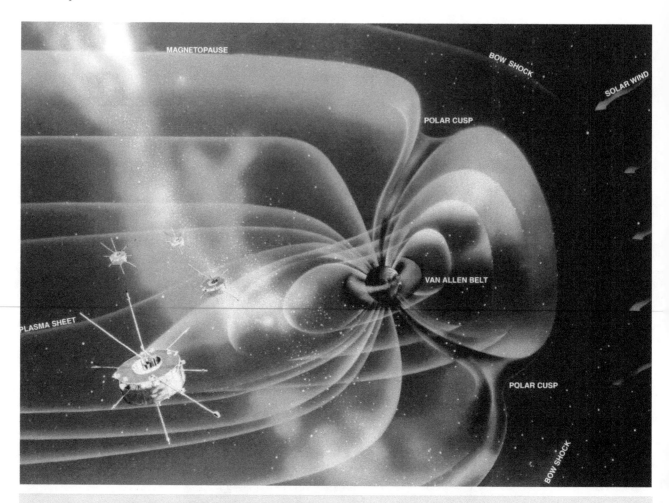

Figure 3.39   Earth's magnetic field creates a huge "bubble" – the magnetosphere – around the planet. This acts like a shield, diverting electrons and protons in the solar wind around the planet. The motion of the solar wind shapes the magnetosphere like a tadpole, with a blunt "head" and a very long "tail." Where the solar wind slows suddenly, upstream of the magnetosphere, a bow shock is formed. Some solar particles enter the magnetosphere at the polar cusps, above Earth's magnetic poles. These particles collide with the upper atmosphere and cause auroras. Other particles are trapped in the Van Allen radiation belts. Also shown are ESA's four Cluster spacecraft, which were launched to explore the magnetosphere and its interaction with the solar wind. (After ESA)

according to the density and pressure of the solar wind. The bow shock is created where the solar wind is slowed suddenly when it nears the outer magnetosphere – rather like water piling up in front of the bow of a ship.

Passing through the shock, which ranges in thickness from roughly 100 km to 2 Earth radii (approx. 12,700 km), the electrically charged particles of the solar wind are slowed, compressed, and heated. The region downstream of the bow shock, between the shock and the magnetopause, which is occupied by the shocked solar wind plasma, is known as the magnetosheath.

The magnetosphere acts as a protective shield against solar radiation and high-energy particles that flow from the Sun and other sources in the Galaxy. However, the magnetic shield is not impenetrable.

Its weakest points are the cusps above the magnetic poles, where some protons and electrons from the solar wind cross into the magnetosphere and spiral down the field lines into the upper atmosphere.

When the particles collide with atoms and molecules in the air, the gases, particularly oxygen and nitrogen, are excited and start to glow. Oxygen generates a green or brownish-red glow; nitrogen a blue or red glow. The result is oval-shaped auroras – commonly known as the northern and southern lights – that more or less continuously surround the magnetic poles (Figure 3.40).

Particles also build up in the magnetotail on Earth's night side, until gusts in the solar wind or coronal mass ejections from the Sun make the plasma unstable. This triggers a magnetic substorm, which causes plasma to flow away from the disturbance.

At such times, accelerated electrons spiral down towards the magnetic poles and bombard the thin upper atmosphere. Such magnetic storms often result in spectacular brightening of the auroral ovals. Meanwhile, electric currents in the ionosphere may cause major magnetic disturbances on the ground.

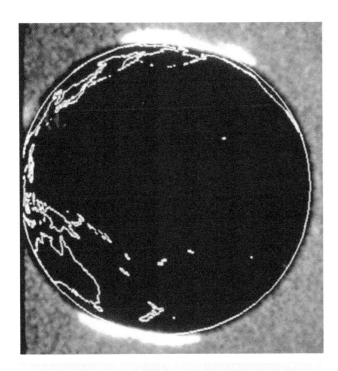

Figure 3.40 NASA's Dynamics Explorer spacecraft imaged both the aurora borealis ("northern lights") and the aurora australis ("southern lights"). These glowing ovals, centered above the magnetic poles, are caused by high energy particles striking the upper atmosphere. Each oval is about 500 km wide and 4,500 km in diameter. Green lines show outlines of land areas. Australia is at lower left. (NASA/University of Iowa)

Close to Earth are two donut-shaped regions, one inside the other, where high energy electrons and protons have been trapped by the magnetic field. They are known as the Van Allen radiation belts, after their discoverer.[14]

The inner belt was discovered by a Geiger counter carried on the first U.S. satellite, Explorer 1. It is located between 1,000 and 5,000 km above the equator and contains particles which have been captured from the solar wind or which originate from collisions between cosmic rays and atoms in the upper atmosphere.

Since Earth's magnetic field is offset from the axis of rotation, the inner belt dips down towards the surface over the South Atlantic Ocean, off the coast of Brazil. This South Atlantic Anomaly poses a threat to astronauts and satellites in low-Earth orbit.

The outer belt lies between 15,000 and 25,000 km above the equator, though it curves closer to the surface towards the magnetic poles. This region mainly contains particles from the solar wind.

The magnetosphere is constantly changing in response to levels of solar activity. Coronal mass ejections from the Sun and solar flares can prompt a sudden intensification of the magnetic field at ground level together with a rapid shift in orientation. Ground-level electrical currents induced by geomagnetic storms may affect power grids and electronics, causing blackouts and interference with radios and telephones.

Questions

- Summarize at least 6 major differences between Earth and the other terrestrial planets of the Solar System.
- Describe the processes that take place in Earth's water cycle.
- Outline the main characteristics of plate tectonics.
- What differences, if any, does the biosphere make to Earth's physical characteristics?
- Earth is sometimes described as a "Goldilocks" world, where conditions have been just right for life to evolve. With reference to examples, explain whether this is a reasonable description.
- Describe the main factors that influence: (a) Earth's surface temperature, (b) Earth's zonal winds, (c) Earth's atmospheric precipitation.
- What are the main features of Earth's magnetosphere? (b) Describe its interaction with the solar wind.
- How is human activity changing the physical characteristics of the planet?
- Describe some of the ways in which satellite technology being used to enhance our knowledge of planet Earth.

[14] A third, temporary, belt was discovered in September 2012, possibly caused by the arrival of a shock wave following a major eruption on the Sun a few days earlier.

FOUR

The Moon

Earth is the only terrestrial planet to be orbited by a large satellite, the Moon. As noted in Chapter 3, the Moon has had a considerable influence on Earth's history and evolution. Since the Earth–Moon system has been in existence for more than 4 billion years, both bodies have experienced similar rates of impact and solar activity. Whereas most of this history has been erased on Earth, it has been preserved on the Moon. Much of what we know about the early Solar System has been derived from studies of lunar samples brought back by the Apollo astronauts. Today, a second era of lunar robotic exploration is underway, providing new opportunities to increase our understanding of its origin and the dramatic changes that have taken place in the Earth–Moon system over the eons.

The most obvious characteristic of the Moon, apart from its close proximity, is that the same hemisphere is always visible. Until 1959, when the Luna 3 spacecraft sent back pictures from behind the Moon, the "far side" had never been seen.

We never see anything other than the familiar features of "the man in the Moon" because the orbital period of Earth's only natural satellite is 27 days 7 hours and 43 minutes – identical to its period of axial rotation.[1] This is a particular form of synchronous rotation, known as spin-orbit coupling (see also Chapter 2).

Rotating west to east, the same direction as its orbital motion, the Moon always keeps the same hemisphere towards Earth. However, a number of minor libration effects mean that some regions that are normally just out of view may occasionally become visible. Hence, 59% of the Moon's surface can be observed from Earth at some time or other.

The most significant libration effects result from the slight tilt of the Moon's axis and the elliptical shape of its orbit. Although the plane of the Moon's orbit is inclined about 5.1° to the ecliptic, its equator is inclined about 6.6°, resulting in a 1.5° inclination of the Moon's spin axis to its orbital plane around the Sun.

Libration of longitude – an apparent sideways "wobble" – occurs because the Moon moves faster around its orbit near perigee and more slowly near apogee, whereas its rotation rate remains the same. As a result, regions beyond the eastern or western limbs sometimes come into view (see Figure 4.2). Libration of latitude occurs because the Moon's equator is slightly inclined, allowing the poles to tilt alternately towards and away from Earth (see Figure 4.2).

A third, smaller effect, known as diurnal libration, enables observers to see a little more of the surface at the Moon's western limb when it is rising and more at the eastern limb when it is setting. There are also some slight irregularities in the Moon's motion caused by its non-spherical shape.

---

[1] On average, the Moon rises (and sets) about 50 minutes later each night because of its motion around the Earth. It moves eastward against the background stars by about 13° per day.

Exploring the Solar System, Second Edition. Peter Bond.
© 2020 John Wiley & Sons Ltd. Published 2020 by John Wiley & Sons Ltd.
Companion Website: www.wiley.com/go/Bond-Solar-System2e

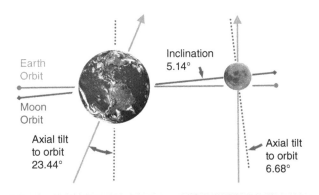

Figure 4.1 On average, the Moon's orbit is inclined 5.14° to the ecliptic plane, while its spin axis is tilted 1.54° to its orbit. This means that the Moon's axis of rotation is inclined 6.68° relative to Earth's north-south axis. As a result, the lunar north and south poles are alternately tilted slightly towards Earth. Not to scale. (Peter Bond)

Figure 4.2 Libration effects sometimes allow parts of the far side to become visible. (David Haworth)

Spin-orbit coupling is common in all major planetary satellites. It comes about because tidal interactions with the nearby planet modify a satellite's rotation (see also Chapter 5). In the distant past, the Moon rotated more quickly than it does today. However, the gravitational pull of the Earth on the slightly ellipsoidal Moon has slowed down the satellite's rotation over billions of years.

At present the Moon's distance from Earth (center to center) varies from 356,410 km at perigee to 406,700 km at apogee (see Figure 4.2). However, this separation has also varied over time. Once again, tidal effects are to blame.

The Sun and Moon create tidal bulges in Earth's oceans (see Chapter 3). However, frictional forces cause the bulges to be carried around the planet slightly forward of the line joining the centers of Earth and the Moon. In this position, the bulges exert a turning effect on the Moon that accelerates it slightly. This raises the Moon's orbit by about 4 cm per year. (It also slows the rotation of the Earth by about 1.5 milliseconds per century.) In a few hundred million years the Moon will be too far away to cause a total eclipse of the Sun. Only annular eclipses will be possible.

Figure 4.3 The apparent size of the full Moon varies considerably between apogee and perigee. (Sheridan Williams FRAS)

Phases

The Moon shines by reflecting sunlight. Only half of its surface can be illuminated at any one time, but the Moon's orbital motion results in a continuously changing position in relation to the Sun and Earth. As the days go by and the Moon moves around Earth, different proportions of its sunlit hemisphere are visible to us – it goes through phases.

When the Moon lies in the direction of the Sun, its unlit side is facing us. Since this night side is very dark and the Moon is very close to the Sun in the sky, it is sometimes impossible to see the Moon at all.[2] This is called the "New Moon."

A couple of days later, when the Moon has moved a short distance around its orbit, a narrow, sunlit crescent becomes visible on the its eastern side.

The Moon continues to "wax" or show more of its illuminated side. A week after New Moon, it has gone a quarter of the way round its orbit. At "First Quarter" we see half of the sunlit side, i.e. a quarter of the entire surface. Over the next few days the sunlit area grows larger than a semicircle and the phase is called "Gibbous."

Two weeks after New Moon, when the entire visible hemisphere is illuminated, a "Full Moon" occurs[3]. The Moon has now moved halfway around its orbit and is on the opposite side of the Earth to the Sun. In the northern hemisphere, this means that the Full Moon is due south in the sky at midnight.

Over the following two weeks the pattern of phases is reversed as the Moon appears to shrink or "wane." It becomes Gibbous again

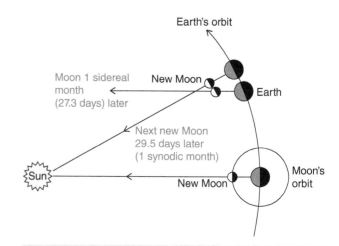

Figure 4.4 The Moon completes one orbit around Earth every 27.3 days. This is called the sidereal month. However, the motion of Earth (and Moon) around the Sun means that the period between each New Moon or Full Moon is 29.5 days. This is the synodic month – sometimes called the "lunation." (Peter Bond)

for a few days, then passes through "Last Quarter" to become a shrinking crescent, visible in the eastern sky just before sunrise. Finally, it vanishes in the morning twilight to become a New Moon again.

The complete cycle of phases, from New Moon to New Moon (known as the synodic month) lasts 29.5 days – more than two days longer than the Moon's orbital period (the sidereal month). The extra two days occur because the Earth is orbiting the Sun. As the Moon goes through its phases, the Earth moves approximately one-twelfth of the way around its orbit. This means that the Moon has to make slightly more than one full orbit of Earth before it is back in line with the Sun.[4]

Lunar Eclipses

Earth casts a cone-shaped shadow across space that measures about 1.4 million km in length. From time to time, the Moon passes through this shadow and becomes darker (Figure 4.5). This can only happen at the time of Full Moon, when it is directly opposite the Sun. However, Earth's shadow is not very wide – about two-and-a-half times the Moon's diameter at the lunar distance – and the satellite's orbit is inclined to the ecliptic. Hence, a lunar eclipse does not take place at every Full Moon. There are typically two or three each year.

For a lunar eclipse to occur, the Moon must be at or near one of its nodes – the point in space where it crosses the ecliptic. During a

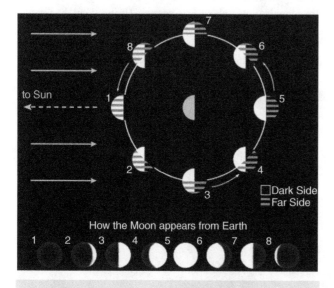

Figure 4.3 Lunar phases are caused by different amounts of the Moon's sunlit hemisphere being visible from Earth as the satellite moves around its orbit. (UCAR)

[2] This is not always the case, since the Moon's night side is often illuminated faintly by "Earthshine" – light reflected from the nearby, fully illuminated Earth.

[3] A Full Moon is bright enough to cast shadows, but this is deceptive. The Moon's surface is actually quite dark. Its albedo, the fraction of the total light that is reflected from it, is on average only 0.067, lower than all the planets except Mercury.

[4] The term "month" comes, of course, from the word Moon.

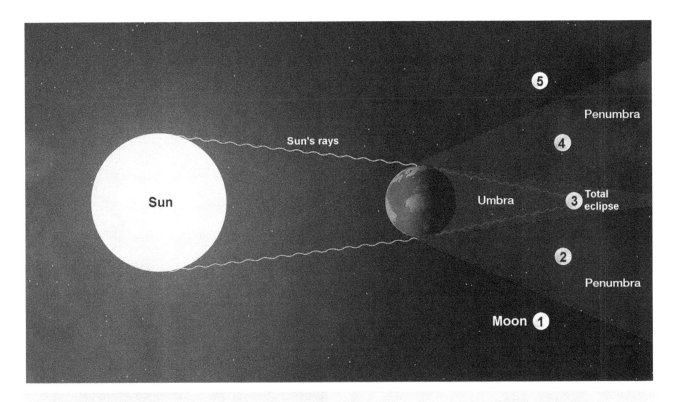

Figure 4.5 Geometry of the Sun, Earth, and Moon during a lunar eclipse. The inclined orbit of the Moon means that it does not enter Earth's shadow at each Full Moon. In some eclipses, the Moon enters only the penumbra, when its darkening is barely noticeable. In others, it is partly within the main shadow (umbra), resulting in a partial eclipse. A total eclipse occurs when it is completely in the umbra. Refraction of sunlight through Earth's atmosphere causes the Moon to appear a copper or reddish color (red lines). Sizes and distances are not to scale. (Peter Bond)

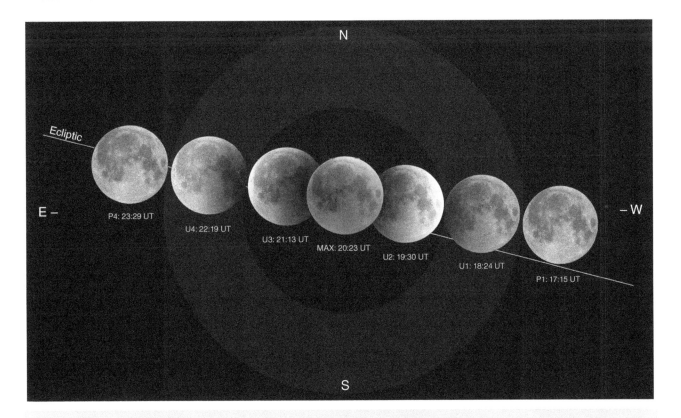

Figure 4.6 A sequence of images showing the lunar eclipse of July 27, 2018. Note the orange coloration of the fully eclipsed Moon, caused by sunlight passing through Earth's atmosphere. (Kosmas Gazeas)

total eclipse, the Moon passes first into the area of partial shadow, the "penumbra," then into the darker, full shadow, the "umbra." The duration of totality (when it is fully immersed in the umbra) may last up to 1 hour 44 minutes.

Lunar eclipses are visible on Earth wherever the Moon is above the horizon. Even during totality, the Moon does not completely disappear (Figure 4.6). Instead, it remains partially illuminated by sunlight scattered by Earth's atmosphere, sometimes taking on a blood red hue. Some eclipses, when the Moon only enters the penumbra, are barely noticeable with the naked eye.

Day and Night on the Moon

The slow rotation of the Moon means that most latitudes have two-week nights followed by two-week days. In the absence of any oceans or atmosphere, there is no mechanism to distribute heat from the equator to the poles, or from the day side to the night side. There are also no clouds or aerosols to reflect or scatter incoming solar radiation. As a result, the bare, arid terrain is strongly heated by the Sun during the lunar day, although heat radiates rapidly into space during the long night. Under these conditions, surface temperatures at the equator may soar to 123°C in the day, but plummet to –170°C at night.

However, in the polar regions, the Sun always appears close to the horizon. Solar energy spreads out over a larger surface area and daylight temperatures are a little less extreme. Some deep craters at the poles harbor areas that lie in permanent shadow. In these dark, cold traps, a thermometer would consistently register –233°C, allowing any water ice to remain deep frozen for millions of years (see Lunar Water).

Physical Characteristics

The Moon is the fifth-largest satellite in the Solar System. With a diameter of 3,476 km (over a quarter of Earth's diameter), it would just about cover Australia. Its average density, 3.34 times that of water, is also considerably lower than Earth's, so its overall mass is one-eightieth that of our planet.

These figures may not sound too impressive, but, with the exceptions of Pluto's satellite Charon and some Kuiper Belt objects, the Moon is the largest satellite in relation to its primary in the Solar System. This has led to the two neighbors being considered as a "double planet," with implications for how the Earth–Moon system originated (see The Birth of the Moon).

As a consequence of its lower density and smaller size, surface gravity on the Moon is approximately one-sixth of that on Earth. (This proved to be both a bonus and a problem for the Apollo astronauts. Lifting large loads was relatively easy, but they had to learn new techniques of locomotion, such as "kangaroo hops.")

Escape velocity is also low (2.37 km/s), which means that it is unable to hold onto any appreciable atmosphere. On the other hand, it is much easier to launch a rocket from the Moon to Earth than the other way around. A number of lunar meteorites have also been discovered on the Earth, having been blasted into space by large impacts (see Chapter 13).

Figure 4.7   Earth and Moon shown to scale. Earth's diameter is almost 4 times greater than the Moon's, so 50 Moons would fit inside Earth. Note the much lower albedo (surface reflectivity) of the Moon. (NASA)

Near Side, Far Side

Until the arrival of the Space Age, no one had ever seen the lunar far side, whereas the near side had been carefully sketched and photographed for some 350 years. The Earth-facing hemisphere was characterized by relatively flat, dark regions (called "maria" – meaning "seas") and lighter, heavily cratered uplands (Figure 4.9). As the Moon went through its phases, it was possible to observe changing shadows along the terminator and determine that the rugged lunar surface has many mountains and valleys, as well as impact craters.

Mankind's first sight of the hidden hemisphere came in 1959, when the Luna 3 spacecraft sent back a series of fuzzy pictures. The far side proved to be very different from the familiar near side, with few mare regions among the expansive, cratered highlands.

The brighter highlands cover nearly 85% of the lunar surface as a whole, including almost the entire far side and 65% of the near side. The rest of the surface is covered by smoother, less cratered mare material, divided into 22 separate "seas."

The reason for this dichotomy is related to the thickness of the lunar crust. Apollo seismic results and gravity data from lunar orbiters (e.g. GRAIL) indicate that the crust is about 35–45 km thick on the Earth-facing side and up to 60 km thick on the far side.

In some of the impact basins, the crust may be nonexistent, meaning that mantle material is exposed at the surface.

Figure 4.9 The Earth-facing side of the Moon, showing the contrast between the dark maria and the lighter highlands of the near side. Several bright ray craters are also obvious, particularly Tycho in the far south. The image was taken by the Galileo spacecraft. (NASA)

These basins include Mare Crisium on the near side and Moscoviense on the far, as well as a couple of small craters within the South Pole–Aitken basin on the far side. Japan's Kaguya orbiter detected olivine, a mantle mineral, exposed at the surface near Moscoviense, Crisium, and some other impact features (see Figure 4.11).

Early in the Moon's history, large asteroid impacts on the near side excavated giant basins, fracturing the crust and enabling magma to flood the basins from below. On the far side, the thicker crust acted as a shield, largely preventing magma from reaching the surface.

The asymmetry of the Moon also shows up in topographic maps, which show that the overall range of topography is about 19.8 km (see Figure 4.10). The highest regions occur on the north-central far side, perhaps due to impact ejecta. Although the near-side maria are below the general surface level, the deepest depression occurs on the far side, near the lunar south pole. The highest point (+10.75 km) is on the southern rim of the Dirichlet–Jackson Basin, located on the lunar far side, while the lowest point (−9.06 km) is inside Antoniadi crater in the South Pole–Aitken basin.

Although it is 2,500 km in diameter, making it the largest impact feature known anywhere in the Solar System, the South Pole–Aitken basin is not deep enough to contain much mare material.

The Moon has a much "bumpier" gravitational field than Earth, with small anomalies due to mass concentrations ("mascons") (Figure 4.12) on the surface, and a large overall asymmetry

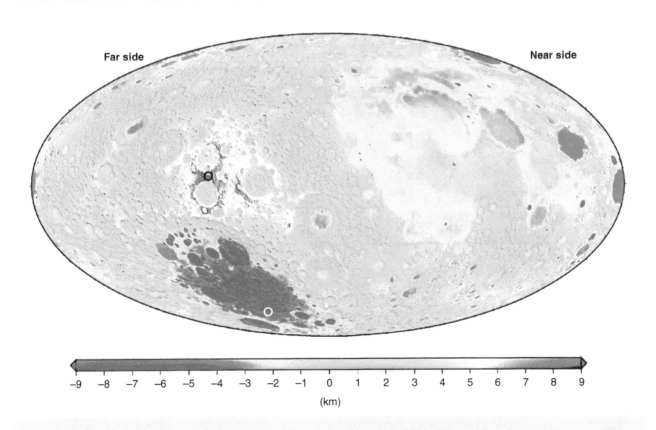

Figure 4.10 Lunar topography, based on data from Japan's Kaguya orbiter. The mare areas of the near side are revealed as lowlands, contrasting with the much higher, ancient terrain of the far side. However, the deepest and largest basin of all, the South Pole–Aitken basin, dominates much of the far side. (JAXA)

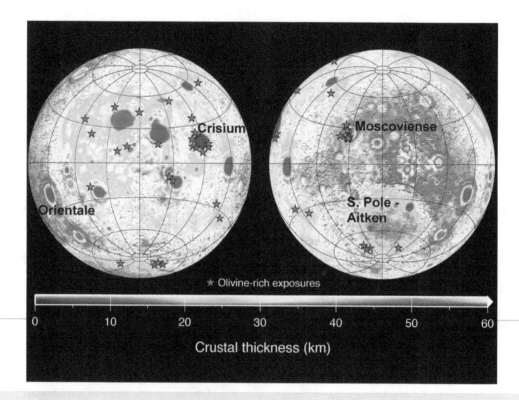

Figure 4.11   A map of lunar crustal thickness derived from analysis of GRAIL gravity data. The crust is thickest (c. 60 km) beneath the highlands of the far side and thinnest beneath some of the large multi-ring impact basins. The mantle may be near or at the surface in some of these basins, e.g. Crisium. Purple stars denote locations where the Kaguya orbiter detected olivine, a mantle mineral, exposed at the surface. (NASA-JPL/GSFC/MIT/IPGP)

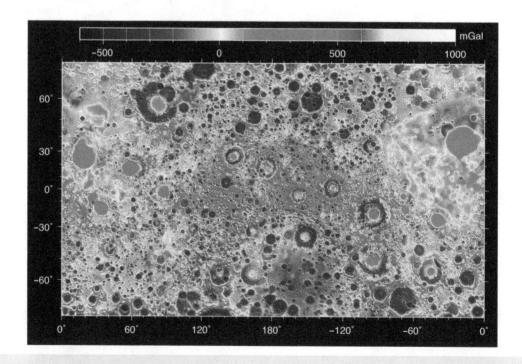

Figure 4.12   A map of the Moon's gravity field, as measured by NASA's GRAIL mission. The Mercator projection shows the far side of the Moon in the center, with the Earth-facing side to left and right. Reds correspond to mass excesses, areas of higher local gravity often associated with the mascons beneath the basalt-filled maria, particularly on the near side. Blues correspond to mass deficits which create areas of lower local gravity. Units are milliGalileos where 1 Galileo is 1 cm per second squared. (NASA/JPL-Caltech/GSFC/MIT)

Box 4.1   Lunar Exploration

Almost 80 lunar missions have been launched by the USA, Russia (formerly the USSR), and other countries since 1959. Activity was particularly frenetic during the 1960s, when the race to send men to the Moon resulted in both superpowers sending armadas of robotic craft to map the Moon and learn more about its surface.

The first detailed images came from three American Ranger probes which impacted the Moon 1964–1965. They were followed by various orbiters and soft-landers. Five U.S. Surveyor craft (1966–1968) sent back panoramic and close-up images, dug trenches, and conducted studies of the magnetic environment and elemental chemistry. NASA also launched five Lunar Orbiters (1966–1967) to image most of the lunar surface at high resolution.

The effort culminated with the American Apollo program, in which nine missions and 27 astronauts were sent to the Moon between 1968 and 1972. After two dress rehearsals in lunar orbit, there were six landings involving 12 crew. For safety reasons they were directed to sites fairly close to the lunar equator.

The last three Apollo missions carried battery-powered rovers which enabled the astronauts to travel much further from their lunar module and to transport more rock samples. These missions, primarily focused on science, were sent to Hadley Rille, Cayley Plain at Descartes, and the Taurus-Littrow valley. A total of 381.7 kg of rock samples were brought back, three-quarters of which are still preserved for future researchers.

Although the Soviets had lost the race, they continued to show interest in the Moon. During the early 1970s, three unmanned spacecraft (Luna 16, 20, and 24) recovered 320 g of regolith from the near side, while two automated Lunokhod rovers spent many months exploring the surface.

The next major advances came when two small American spacecraft, Clementine and Lunar Prospector, went into orbit around the Moon in 1994 and 1998. Clementine produced multispectral maps that showed the composition of the surface rock, and carried a laser altimeter to map global topography. Lunar Prospector also produced detailed maps of surface composition and gravity. Both spacecraft found circumstantial evidence for the existence of water ice in permanently shaded craters at both poles.

Since then, there has been a renewal of interest in lunar exploration. The European Space Agency's SMART-1 spacecraft used an ion drive to spiral out to lunar orbit in November 2004. On board were seven miniaturized instruments, which studied the lunar environment until the spacecraft was deliberately crashed into the Moon in September 2006.

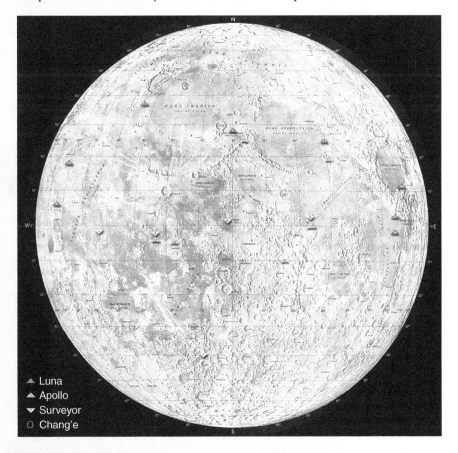

Figure 4.13   The locations of all the successful landings on the near side of the Moon. Apollo missions are green, Lunas are red, Surveyors are yellow, and China's Chang'e 3 is orange. Note the concentration near the lunar equator. (NSSDC)

Figure 4.14   Massive, broken boulders were visited by the Apollo 17 astronauts during their exploration of the Taurus-Littrow valley on December 13, 1972. The original boulder rolled downslope before it broke into five pieces. Dating indicates that it formed after a huge impact almost 4 billion years ago. It is made of breccias created by the shock of the impact. The lunar rover is parked to the right of the boulder. The ancient, rounded mountains in the distance were created during the major impact. South Massif, 8 km distant, forms the skyline on the right; East Massif is on the left-side horizon. Astronaut Harrison Schmitt is beside the boulder. (NASA)

The United States, China, Japan, and India have also initiated in-depth lunar surveys. NASA's Lunar Reconnaissance Orbiter has been sampling the radiation environment, searching for water ice, making topographic maps, and imaging the entire surface. The LCROSS mission crashed a small spacecraft and its upper stage booster into the Moon in a successful attempt to detect water in the debris cloud.

India's Chandrayaan-1 was launched in 2008 on a year-long mission to map the Moon and deliver a small impactor to the surface. On board was a U.S. radar instrument, which found evidence of water ice in polar craters, and a number of European instruments similar to those on SMART-1. A follow-on mission in 2019 will deliver another orbiter and a lander/rover.

Having already sent two small spacecraft to the vicinity of the Moon in 1980, Japan launched its Kaguya spacecraft in 2007. Kaguya was the largest spacecraft sent to the Moon since the Apollo era. The main orbiter carried a high resolution camera, various spectrometers, a laser altimeter, and a radar sounder. Three-dimensional tracking of two small subsatellites enabled the first direct gravity measurements of the lunar far side.

Figure 4.15   This panorama shows the landing site of the Chang'e 4 spacecraft on the lunar far side. The touchdown was in the southern portion of an ancient impact feature named von Karman crater. The image shows numerous small craters but very few rocks on top of the smooth regolith. Change'4 was the first spacecraft to land on the lunar far side. Von Karman lies within the giant South Pole–Aitken basin. (CNSA)

China also has a long-term plan to explore the Moon. Already, several Chang'e lunar orbiters have flown, a small lander/rover has explored northern Mare Imbrium, a data relay satellite has been placed at the L2 Earth-Moon Lagrange Point, and the first-ever touchdown on the lunar far side, inside the South Pole–Aitken Basin, was achieved by the Chang'e 4 lander and rover on January 3, 2019. Several automated sample return missions are planned. The ultimate objective is the delivery of Chinese astronauts to the Moon.

due to the fact that the crust is thicker on the far side of the Moon (Figure 4.11). Surprisingly, the mascons – areas of increased gravity – are particularly associated with the large, lava-filled impact basins in both hemispheres, rather than tall mountains or highlands, where the crust is thicker.

Mascons typically resemble a target with a gravity surplus at the central bullseye surrounded by a ring of gravity deficit and an outer ring of gravity surplus. This pattern arises as a natural consequence of crater excavation, collapse, and cooling following an impact.

The higher density and gravitational pull at the bullseye was caused by lunar material that melted from the heat of the asteroid impact. The melting causes the material to become more concentrated, stronger and denser, and pulls in additional material from the surrounding areas.

Cool lunar crust slid into the impact basin, forming a rigid, curved edge that held down the material beneath it and prevented it from fully rebounding to its original surface height. This causes a ring with less gravitational pull.

Major Impact Basins

Impact features more than 300 km across are termed impact basins. More than 40 such basins have been identified on the Moon, many of them filled with the dark, basaltic lavas that form the maria. In addition to these basalts, which are mainly confined to the central topographic depressions, the impact basins are associated with widespread faulting and other forms of deformation.

The maria are broad plains, pockmarked with a few large craters and many smaller impact hollows. They are concentrated on the near side of the Moon, where they cover 35% of the surface. In contrast, they cover only 5% of the hidden side, where even the largest, Mare Moscoviense, pales into insignificance compared with its near-side cousins.

Material ejected from these basins (ejecta) is distributed over wide regions, serving as a useful marker in trying to determine the broad chronology of events that shaped the Moon's surface. For example, if a crater lies on top of the ejecta, then it must be younger than the impact basin from which the ejecta originated. On the other hand, if a feature is partially buried by an ejecta blanket, it must be older than the corresponding impact basin.

Although the original floors of many of the basins are now hidden by lava infill, various features are visible in relatively young impact structures such as Mare Imbrium and Mare Orientale. Both of these display several concentric rings of mountains that formed when the basin floors bounced back upward and shock waves rippled outward immediately after the monumental impact events.[5]

The outermost ring in the Imbrium Basin has a diameter of 1,300 km and rises roughly 7 km above the mare surface. It is divided into several different ranges – the Carpathians to the south, Apennines to the southwest, and the Caucasus to the east – though it is less marked to the north and west. The middle ring is marked by the Alps in the north, and some isolated uplands near the large craters Archimedes and Plato. Least conspicuous is

Figure 4.16 The Orientale Basin has three concentric mountain rings – the Cordillera on the outside and two ranges of the Rook Mountains – around a fairly small area of basaltic mare material in the center. The outermost ring is 930 km in diameter. The image shows the basin's gravitational signature (red indicates excess mass – a mascon – and blue indicates mass deficits). (NASA/GSFC Scientific Visualization Studio)

the 600-km wide inner ring, which is now largely buried by lava and marked only by low hills that protrude through the plains and a series of curved mare ridges.

Extending some 800 km from the Imbrium Basin is a huge blanket of ejecta, known as the Fra Mauro formation, which was visited by the crew of Apollo 14. Also encircling the basin is a pattern of radial grooves – deep, linear furrows that were gouged by large projectiles that were blasted from the basin and skimmed across the surface.

The force of the colossal impact literally shattered the crust, creating a global pattern of faults, both radial and concentric to the basin. Directly opposite the basin, on the far side of the Moon, is a region of chaotic terrain associated with the Van de Graaff crater. This is thought to have been formed by focusing of seismic waves from the impact that traveled through the Moon's interior.

Orientale is the youngest and best-preserved example of a multi-ring basin anywhere in the Solar System. Studies based on GRAIL data indicate that the Orientale Basin was created after an impactor about 65 km across hit the surface while travelling at

[5] All of the mountains on the Moon are associated with impacts and ejecta blankets. With heights up to 7,800 m, they are comparable to the highest mountains on Earth, though generally not as steep.

about 14 km per second. A huge amount of material – equivalent to over 150 times the combined volume of the Great Lakes – was blasted outward, but the initial crater was quickly transformed by subsequent events.

As the crust rebounded following the impact, warm and ductile rocks in the subsurface flowed inward toward the impact point. That inward flow caused the crust above to crack and slip, forming the cliffs, several kilometers high, that compose the outer two rings.

The innermost ring was formed by a different process. The central peak that formed due to the rebound of the crust was too large to be stable. The peak's material flowed back outward, eventually mounding in a circular fashion, forming the inner ring. The cliffs and the central ring all formed within minutes of the initial impact.

Ejecta from the basin extend up to 500 km beyond the outer ring. Linear patterns in the rough, hummocky material radiate away from the basin, evidence of gouging by large chunks of surface-skimming ejecta.

However, the most surprising impact feature is the South Pole–Aitken Basin on the Moon's far side (see Figure 4.15). Little was known about it until 1994, when topographic maps made from Clementine orbiter data revealed an ancient basin about 2,600 km across and 12 km deep, making it one of the largest impact structures in the Solar System. Since its creation, the basin has been much modified by smaller impacts, hence the difficulty in recognizing it. Although it contains only modest pools of dark mare material, its floor is markedly darker than the surrounding highlands and shows an enrichment of iron and titanium.

The reason for this unusual composition is uncertain. It may be due to a mixture of material from the lower crust and rock dug up from the lunar mantle. However, the fact that mantle material seems to be unexpectedly scarce suggests that it was not a typical, high-velocity impact. One hypothesis suggests that it was formed by a low-velocity projectile that hit at a low angle (about 30° or less) and so did not plough very deeply into the Moon. If so, much of the debris would have been blasted into space, possibly providing a source of projectiles that gouged out other lunar basins between 3.85 and 3.95 billion years ago.

Maria

The floors of the maria, which vary from flat to gently undulating, are covered by a thin layer of powdered rock (regolith) that darkens them and accounts for the Moon's low albedo. The maria consist largely of basalt and were created between 3.9 and 3 billion years ago, when fluid lavas originating in the mantle periodically filled the major, pre-existing impact basins. The oldest maria are irregular in shape and found in broad lowlands (e.g. Mare Tranquillitatis and Mare Fecundidatis).

The largest of all is Oceanus Procellarum, on the west side of the Moon, which stretches 2,500 km from north to south and covers roughly 4 million sq. km. Unlike the other lunar maria, Procellarum is not contained within a single, well-defined impact basin. Around its edges are many smaller "bays" and "seas," including Mare Nubium and Mare Humorum to the south.

Although the enormous feature has generally been attributed to a massive impact, scientists studying GRAIL data believe they

have found evidence that the irregular outline of this roughly rectangular region is actually the result of the formation of ancient rift valleys.

The rectangular outline of Oceanus Procellarum, with its angular corners and straight sides, contradicts the theory that the feature is an ancient impact basin, since such an impact would create a circular landform. Instead, the GRAIL data suggest that processes beneath the Moon's surface dominated the evolution of this region.

According to this interpretation, the gravity anomalies discovered by GRAIL reveal the rifts and dykes that fed lava to the surface during ancient volcanic eruptions. The rifts are now buried beneath the dark volcanic plains and can be detected only in the gravity data. The lava-flooded rift valleys are unlike anything found anywhere else on the Moon and may once have resembled rift zones on Earth, Mars, and Venus.

Over time, the region would cool and contract, pulling away from its surroundings and creating fractures similar to the cracks that form in mud as it dries out, but on a much larger scale.

Another theory arising from recent data analysis suggests this region formed as a result of churning deep in the interior of the Moon that led to a high concentration of heat-producing radioactive elements in the crust and mantle of this region.

Figure 4.17 A topographic image of the Oceanus Procellarum region, based on data collected by a laser altimeter on NASA's Lunar Reconnaissance Orbiter. Linear gravity anomalies bordering the Procellarum region are superimposed in red or orange. The anomalies are derived from GRAIL data. They are interpreted as ancient lava-flooded rift zones buried beneath the volcanic plains (maria). (NASA/Colorado School of Mines/MIT/GSFC/Scientific Visualization Studio)

Figure 4.18   A view westward, across southwestern Mare Tranquillitatis, from the Apollo 11 orbiter. Tranquility Base (TB), where Armstrong and Aldrin made the first manned landing, lies on the terminator. As in most maria, the smooth surface of solidified lava is interrupted by craters, numerous ridges (R), straight rilles or grabens (G), sinuous rilles (S), and islands of unburied crustal material / terrae (T). Craters such as Maskelyne (M) are surrounded by extensive aprons of ejecta. Maskelyne is 210 km from the landing site. (NASA/LPI)

The more recent maria are roughly circular (such as Crisium, Serenitatis, Nectaris, Imbrium, and Orientale). They lie at successively lower levels to the east, with Mare Smythii – the easternmost of the mare basins on the near side – lying almost 5 km below the average lunar radius.

Various distinctive features favor a volcanic origin for the mare material, e.g. many broad, low domes with summit craters that closely resemble terrestrial shield volcanoes. In some places, irregular, steep-sided volcanic mountains can be seen. Also visible are broad, lobate flow fronts that mark the furthest limits of individual lava flows. These flow fronts may be several hundred kilometers long and up to 100 m high.

The maria are also crossed by numerous sinuous rilles, which resemble dry river valleys and were carved by flowing lava, and wrinkle ridges – tectonic features created when the basaltic lava cooled and contracted. Many of these rilles originate in craters near the higher margins of the mare basins and flow into the lowlands. Samples collected from the margins of Hadley Rille by the crew of Apollo 15 confirmed that these landforms were carved by basaltic lava flows.

Although the mare material is only a few hundred meters thick near the basin rims, the central regions may contain 2–4 km depth of ancient lava flows, e.g. Mare Humorum. This mass of dense rock has deformed the underlying crust, creating fractures and fault-like depressions (graben) around their outer edges and compressed wrinkle ridges, where the mare surface has buckled. The most spectacular fault is the 110-km long Straight Wall (Rupes Recta), which lies on the edge of Mare Nubium. Land on one side of the fracture is a few hundred meters higher than on the other side.

Mare Rocks

Mare and highland rocks differ in both appearance and chemical content, e.g. most of the mare basalts are richer in iron and titanium, but poorer in aluminum than highland rocks. However, analysis of the Apollo samples shows that the chemical composition of mare materials varies from place to place. By correlating these differences with the subtle color changes in spectra obtained from orbit, it is possible to map surface composition.

Figure 4.19 A mare basalt, showing numerous hollows (vesicles) formed by bubbles of dissolved gas escaping from the molten lava. Rock 15016 was collected at the Apollo 15 landing site near Hadley Rille. Radiometric dating shows it is 3.3 billion years old, younger than many other lunar basalts. (NASA)

The shapes of the mineral grains and the way they are intergrown in mare basalts confirm that they formed in lava flows. As on Earth, some of these were only one meter thick, others much thicker (perhaps up to 30 m). Many of the basalts also contain holes, called vesicles, that were formed by gas bubbles trapped when the lava solidified. The composition of these gases was probably a mixture of carbon dioxide and carbon monoxide, with some sulfur-rich gases. The absence of water or water-bearing minerals shows that little water vapor was present.

When the gases escaped in explosive fashion, the liquid rock formed tall lava fountains. Fragmented rock cascaded to the ground, forming deposits of pyroclastics. One such deposit was sampled by the crew of Apollo 17 and found to consist of numerous, small, orange glass beads (Figure 4.20). The molten rock cooled so rapidly that there was not enough time for crystals to form and grow, so it simply turned to glass.

Experiments conducted on mare basalts and pyroclastic glasses show that they formed when the interior of the Moon partially melted, typically at a temperature of 1,000–1,200°C. The melting, caused by heat from decaying radioactive isotopes, took place at depths of 100–500 km. The basalts were largely composed of the silicate minerals plagioclase (a type of feldspar) and pyroxene (containing iron, calcium, and magnesium), with varying amounts of olivine (another magnesium-iron silicate) and ilmenite (an iron-titanium oxide).

Highlands

The lunar highlands (terrae) are brighter and more rugged than the maria. They occupy two-thirds of the Earth-facing hemisphere and about 85% of the entire lunar surface. The greater age of the highlands is indicated by a much higher density of large craters and confirmed by radiometric dating of rock samples.

Most of the light-colored highlands are largely composed of anorthosite – a type of igneous rock that is rich in aluminum and

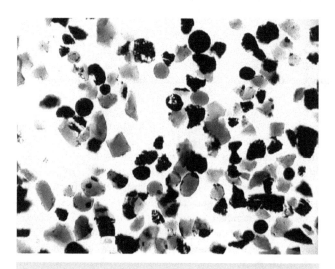

Figure 4.20 Beads of orange, volcanic glass found at Shorty Crater in the Taurus-Littrow valley. The glass particles, intermixed with black crystals of ilmenite, are 20–45 microns across – about the same size as silt particles on Earth. The orange color is caused by a titanium oxide (TiO_2) content of 8%. Other constituents included oxides of silicon (39%), iron (22%), magnesium (14%), and calcium (8%). The glass formed 3.64 billion years ago from material that melted about 400 km below the surface and then erupted onto the surface in a lava "fire fountain." (NASA)

Figure 4.21 When the Moon first formed its surface was probably composed mostly of feldspar-rich igneous rocks called anorthosite. Today, this rock type makes up the lunar highlands. This 4.4-billion-year-old sample was collected by the Apollo 16 crew. (NASA)

calcium, and relatively poor in iron, magnesium, and titanium, compared with the mare rocks. They appear to have formed when feldspar crystallized and floated to the top of a global magma ocean that covered the Moon soon after it formed (see Geological History).

One peculiar type of highland rock has been dubbed "KREEP," because it is rich in potassium (K), rare earth elements (REE), and phosphorus (P). It is also rich in the radioactive elements thorium and uranium, making it easy to map from orbit. These incompatible elements cannot fit into the crystal structures of the common lunar rock-forming minerals of plagioclase feldspar, pyroxene, and olivine. As a result, when lunar magma slowly crystallizes, the KREEP elements get left behind in the dwindling reservoir of melt.

The unique elemental signature suggests that the rock was formed during the final stages of the formation of the primitive lunar crust, when the cooling of a global magma ocean was nearing completion. The last stage in the crystallization of the magma produced rocks that are strongly enriched in KREEP material.

This ancient highland material was later spread over the mare regions by major impacts, hence the ability of the Apollo 12 astronauts to pick up samples from Oceanus Procellarum. The distribution mapped by Lunar Prospector supports the theory that the impact which formed the Imbrium Basin excavated KREEP-rich rocks and spread them over the Moon. The huge impact that created the South Pole–Aitken Basin also exposed KREEP-rich material.

Since their formation, some 4.4 billion years ago, the ancient highland rocks have been drastically modified by countless impact events. The cumulative effect of these collisions has transformed

Figure 4.22 An Apollo 8 image showing the contrast between a dark mare region and surrounding highlands. Tsiolkovsky crater, on the lunar far side, is 198 km in diameter. Its crater is partially filled by lava and the internal peaks are offset from the center. Note the extensive landslips around the rim. (NASA)

the nature of the surface, with the result that few samples recognizable as original crustal material were recovered by the Apollo astronauts.

Most of the highland rocks are classified as breccias – conglomerates formed from anorthosite that was shattered by an incoming meteorite and then welded together by the heat and pressure generated by the impact. These are often studded with small, glassy spheres that probably formed when the spray of molten rock from the impact solidified in mid-flight. In addition to metamorphism of the rock minerals by sudden impact shocks, huge amounts of ejecta were spread over great distances.

Other processes that have modified the highlands are tectonism and mass wasting. Tectonism is visible in numerous linear faults or fractures, sometimes in the form of grabens (rift valleys). Some of the largest linear structures, radiating from the edge of the Imbrium Basin, are obviously related to its formation (see Major Impact Basins).

Smaller linear or curvilinear fractures are widespread. These lobate scarps have steep slopes and are often segmented. They are similar to, but much smaller than, the lobate scarps found on Mercury and Mars, with a maximum relief of less than 100 m and lengths of only a few tens of kilometers. Their global distribution indicates that they may have resulted from the cooling and shrinking of the Moon, perhaps within the last 1 billion years.

Mass wasting has subdued the rugged highland landscape by moving loose regolith downslope. In the absence of running water the rate of movement is often very slow, but it may be hastened by moonquakes or shock waves from impacts. One photo taken by the Apollo astronauts clearly showed a track left by a boulder that had rolled down a steep slope.

Volcanism in the highlands is hard to identify, though a few areas, notably the Gruithuisen domes west of Mare Imbrium, have been suggested as evidence of volcanic activity.

Lunar Craters

On Earth, craters are associated with both volcanic cones and meteorite impacts. For many years, scientists could not agree on the origin of the lunar craters, until close-up images and other data from spacecraft closed the debate.

Today, we know that volcanic craters on the Moon are rare and comparatively small, whereas the landscape is covered by impact features of all sizes, ranging from pits less than a meter across to huge hollows more than 200 km across. It is estimated that there are at least 300,000 craters with a diameter larger than 1 km on the near side of the Moon alone. Most of these were created more than 3.8 billion years ago, during a heavy bombardment by large planetesimals left over from the formation of the Solar System.

The shape of lunar craters varies with their size and age. The smallest hollows, typically less than about 15 km in diameter, have a simple, bowl-like form. Larger examples display more complex forms, including shallow, flat floors made of solidified lava, one or more central peaks, and terraces on the inner-rim walls (see Box 4.2).

The impact structures become more complex as they increase in size. When they reach between 180 and 300 km in diameter, they change from craters with central peaks or groups of peaks to circular basins surrounded by two or more rings.

Figure 4.23   Copernicus, one of the youngest major impact craters on the Moon, displays several central peaks up to 800 m high, terraced walls, and a raised rim. Boulders from the central peak litter the crater floor. The crater is 93 km wide and more than 3 km deep. The image mosaic was taken by the Lunar Reconnaissance Orbiter. (NASA)

Large scale crustal faulting has produced small, linear rilles that slice through many craters. Volcanic activity has also filled many of the largest craters with lava, completely covering the original floors and, sometimes, their central peaks. Craters within or on the fringes of the maria have often been breached and flooded with lava, producing bays and "ghost rings."

Since larger impacts excavate larger basins and eject more debris, the major craters are surrounded by extensive debris aprons and numerous secondary craters created by falling ejecta. Relatively young craters, such as Copernicus and Tycho, also display bright rays of fragmented material that radiate outward for hundreds or thousands of kilometers.

One explanation for the brightness of the rays is that they are composed of light-colored highland material that is conspicuous against the darker mare material. The powdery nature of some rays also makes them more reflective. However, as time goes by, the constant bombardment by the solar wind and innumerable micro-impacts causes them to fade.

Copernicus, 93 km in diameter, is one of the most prominent features on the Moon's near side and displays most of the features associated with a large lunar crater. The well-defined crater is believed to have formed less than 1 billion years ago, by which time the crust had thickened too much to allow its interior to be flooded with lava. In the center are several peaks made of material that rebounded from deep in the crust as a result of the impact. Much of the crater floor is hummocky or uneven, while the inner part of the crater walls is broken up into a series of terraces marked by debris flows. Just beyond the crater rim is a continuous ejecta blanket, breaking up into discontinuous clumps as the distance increases. Chains of secondary impact craters are also prominent on the surrounding plains.

Like Copernicus, most of the craters on the Moon are circular, but there are a few exceptions, notably the neighboring Messier and Messier A in Mare Fecunditatis. The former feature is elongated, measuring about 11 km long and 8 km wide. About 20 km to the west is the double crater, Messier A, where one crater is superimposed on another. Two bright rays stretch 120 km towards the west. These craters and the rays were probably created by a very shallow (1–5°) grazing impact.

Two crater chains on the Moon are thought to have been formed by the impact of asteroids or comets that broke apart before they arrived. The first is the Davy chain, a 47-km long sequence of 23 craters, 1–3 km in size. More prominent is the 250-km long Abulfeda chain, which comprises 24 craters with diameters of 5–13 km. Neither of these features seems to be associated with debris ejected from large, nearby impact craters. (See also Chapters 8 and 13.)

Although large impacts are extremely rare on the Moon (and Earth) at the present time, NASA's Lunar Reconnaissance Orbiter has shown that the surface is being continually peppered by incoming meteoroids. In the seven years after it entered lunar orbit, LRO images enabled scientists to identify over 200 newly created impact craters that ranged in size from about 3 to 43 m in diameter.

Over 47,000 small surface changes, called "splotches," were also identified. They are most likely caused by small impacts. The presence of dense clusters of these splotches around new impact sites suggests that many of them may be secondary surface changes caused by material thrown out from the primary impact event.

Small flashes caused by meteoroids hitting the lunar surface are also observed on a daily basis.

Box 4.2   Impact Craters

Impact craters are found on the surfaces of all the terrestrial planets and satellites – with the exception of Io, which is being resurfaced continuously by volcanic activity. The craters are created by collisions involving an incoming projectile – an asteroid, comet, or meteorite. In the case of the Moon, these objects have widely ranging speeds, but average about 20 km/s.

On Earth, impact craters are not easily recognized because of weathering and erosion, but on worlds with no atmosphere, such as the Moon, they provide an impact record going back more than 4 billion years.

The shape and size of craters depends on many factors – the size, density, speed, and angle of arrival of the impactor, together with the physical and chemical properties of the target surface. One key factor is the amount of kinetic energy possessed by the impacting object.

Kinetic energy (energy associated with motion) is described mathematically as: $KE = \frac{1}{2}(mv^2)$ where m = mass and v = velocity. During impact, the kinetic energy of an asteroid is transferred to the target's surface, shattering rock and moving debris around. Generally speaking, the greater the kinetic energy, the larger the crater.

When a solid object strikes a planetary surface at high speed, an extremely powerful shock wave – millions of times Earth's normal atmospheric pressure – spreads out from the impact site. This compresses the rock and makes it deform so that it flows like a fluid.

As the shock wave radiates outward, decompressed material behind it is thrown upwards and outwards, excavating a volume of material much larger than that of the impactor. Rock strata that were once horizontal may even be bent back or turned upside down, so that the older rocks are now on top of younger rocks. Beneath the crater floor is a lens-shaped body of breccia – rock that has been broken and pulverized by the shock wave. The incoming object itself is melted or partly vaporized.

As a general rule, small (100 m) craters tend to be about 20 times the diameter of their impactor, whereas medium-sized craters (1–10 km) are about 10 times wider, and large (100 km) craters are about 5 times wider.

The critical diameters at which one type of crater changes to another depend inversely on the surface gravity. For example, the transition between simple and complex craters occurs at about 20 km diameter on the Moon but at only 2–4 km on Earth. The transition from a central peak structure to a ringed structure occurs at about 150–200 km on the Moon, but at only 20–25 km on Earth.

Small craters are bowl shaped, with smooth walls, a raised rim and a surrounding apron of hummocky ejecta (see Figure 4.25). Their depth is between one-fifth and one-tenth of the diameter. Immediately after its excavation, a simple crater is filled, to perhaps half its original depth, by a mixture of redeposited (fallback) ejecta and debris that slumps inward from the walls and rim.

Lunar craters with diameters of more than 20 km are shallower and more saucer-shaped. They develop one or more central peaks, caused by rebound of the surface and reflection of shock waves from depth. Steep internal slopes are marked by terraces where slumping has occurred. The rim may tower more than one kilometer above the surroundings, and an ejecta blanket spreads far from the crater in all directions (see Figure 4.24).

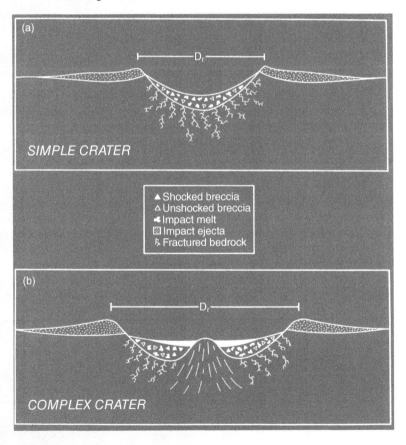

Figure 4.24 Schematic cross sections of (a) a simple impact crater and (b) a complex crater. Simple craters are typically bowl-shaped and small in size (less than 20 km across on the Moon or 4 km on Earth). Larger craters are more complex, with central peaks, relatively shallow floors, and terraced or slumped rims. In both cases there is a raised rim, a surrounding blanket of ejecta, and a scattering of smaller, secondary impact craters. (LPI)

Impacts by large pieces of debris excavate numerous secondary craters. Finer, more powdery ejecta travel hundreds or thousands of kilometers from the impact site, forming bright, radial ray deposits.

The largest craters, with diameters over 300 km, are known as basins. The central peak is replaced by a cluster of peaks (a peak ring). The largest craters contain a series of concentric mountain rings, which may be caused by a combination of slumping and shock waves traveling across the surface.

Figure 4.25   Lying amid a scattering of small, bowl-shaped craters is 18 km wide Dawes crater, imaged from Apollo 17. Located between Mare Serenitatis and Mare Tranquilitatis, Dawes is an example of a simple, slightly oval-shaped, lunar crater. Note the deposits of slumped or fallback material on its floor, the steep walls, and sharp rim. (NASA / USGS)

Volcanic Domes and Pyroclastics

The lunar maria were flooded by huge volumes of magma, but not all volcanic landforms are on such a grand scale. Many of the maria exhibit small, dome-shaped, volcanoes that were formed by quiet, repetitive outpourings of more viscous (sticky) lava than normal mare basalts. This makes their flanking slopes quite gentle, no more than 1–8°.

The domes are often less than 10 km across and a few hundred meters high. At their summits are pits around 1.5 km wide. Their modest dimensions make them quite hard to observe unless they are near the day-night terminator.

Most of the domes are found in clusters, often close to the edges of the maria, where the lava flows are relatively thin. This implies that their magma sources were near the surface, where pressures were low.

These domes probably formed during the later stages of volcanism on the Moon, which is characterized by a decreasing rate of lava extrusion and comparably low-temperature eruptions.

At the opposite extreme are pyroclastics, deposits of ash and cinders created during explosive volcanic eruptions. When gases trapped within viscous (sticky) magma suddenly escape, they expand rapidly, shredding the magma and producing clouds of fine material that are deposited around the source vent in a circular patch.

Lunar pyroclastics are not easy to detect, but they can be identified by their relatively dark color compared with typical mare lavas, e.g. on the floor of Alphonsus crater. The dark coloration is caused by tiny black beads of glass and crystals that formed from cooling of the basalt fragments as they were blasted from the volcanic vent.

The glass resulted from rapid cooling of liquid droplets, while the crystals condensed more slowly.

Some 75 pyroclastic deposits have been recorded on the Moon. Most of them are small, covering an area of less than 100 sq. km, but a few cover enormous areas. The largest of all, which occurs on and around the Aristarchus plateau, blankets an area of 50,000 sq. km, while the Imbrium ejecta south of Sinus Aestuum rille has a 10,000-sq. km coating of pyroclastic material. In this and several other cases, no obvious vents are visible, but sinuous rilles link the deposits to volcanic activity (see Sinuous Rilles).

Orbiter images of probable pyroclastic deposits prompted scientists to support the landing of Apollo 17 in the Taurus-Littrow valley, and their forecasts were confirmed when the astronauts brought back samples of black and orange glassy beads (Figure 4.20).

Sinuous Rilles

Snakelike valleys on the Moon (called sinuous rimae or rilles) have been the focus of much debate. One possibility is that they were cut by flowing water, but the absence of water in lunar samples makes this very unlikely. They were probably formed either by faulting and subsidence of the crust, or represent lava channels or collapsed lava tubes. The wide variety in shapes suggests that all of these processes have been involved. The valleys may have originated as fault troughs that were later modified and obscured by lava flows, impact ejecta, or landslides on the walls.

Those rilles that are broken into separate sections are almost certainly collapsed lava tubes. Molten lava flows often develop a solid crust, forming tunnels beneath the surface. When the lava

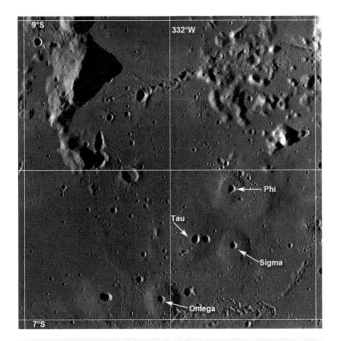

Figure 4.26 Four volcanic domes (Phi, Tau, Sigma, and Omega) in the Hortensius region. The domes display very little relief, making them visible only at low Sun angles. Summit craters of all the Hortensius Domes show no raised rims and are not circular, so they are probably volcanic calderas rather than impact craters. (NASA-GSFC/Arizona State University)

later drains from the tunnels, a long, hollow, cylindrical tube remains. All or part of the roof may subsequently collapse, leaving steep-sided troughs or chains of near-circular pits.

Some sinuous rilles contain a small channel within a larger valley, indicating at least two episodes of valley erosion. Other meandering channels are not associated with any obvious craters – they occur on lava plains that are so flat it is difficult to tell the direction of flow. Still other channels have formed on mountainsides covered by hummocky ejecta thrown outward from major impact basins, suggesting that they may have been formed by some other process.

Perhaps the best known example is Hadley Rille, a V-shaped gorge that starts in a volcanic crater on the flank of the Apennine Mountains and flows along the eastern edge of Mare Imbrium, eventually merging with a second rille about 100 km to the north. Its average width is about 1.5 km and depth is about 400 m.

The Apollo 15 astronauts explored Hadley Rille and found that layered basaltic lava flows are exposed in the valley walls (Figure 4.28). The walls themselves are quite steep, with an average slope of about 25°. The floor of the channel is very irregular, suggesting that eruptions of lava at different places along the channel may have added to the material flowing from the source crater. Its meandering shape indicates a great deal of erosion and modification by flowing lava of what may originally have been a fault trough.

Linear Rilles

There are a few dozen large trenches or linear rilles on the Moon. Unlike their sinuous cousins, these valleys have straight, steep

Figure 4.27 A view from the Apollo 15 orbiter, looking southeast across Vallis Schröteri (Schröter's Valley), the largest rille on the Moon, which originates on the Aristarchus Plateau. It comprises three main features: the rounded Cobra Head, the primary rille (155 km long), and the narrower inner rille (204 km long). Such rilles are believed to have formed as large volumes of very fluid magma erupted and flowed rapidly from a vent. (NASA/JSC/Arizona State University)

Figure 4.28   Apollo 15 astronaut James Irwin with the lunar rover at the edge of Hadley Rille. The view is looking northwest from the flank of St. George Crater. Loose blocks are visible on the sides and the floor, evidence that the rille has been widened and made shallower over time by wasting along its walls. (NASA)

walls and flat floors – features characteristic of grabens, valleys created when crust drops down between two faults. These grabens form in regions of horizontal extension, where the crust is forced apart.

How could such extension take place on a world where there are no crustal plate movements or fold mountains? One important clue is that at least one-third of the linear rilles are associated with nearby volcanic features, such as domes, cones, and pyroclastic deposits. One form of extension related to volcanic activity is the intrusion of magma, where sheets of liquid rock flow into near-vertical cracks in the crust and solidify. Magma rising towards the surface in such dikes pushes the crust apart.

If the magma reaches the surface, an eruption occurs and the dike is buried beneath its own lava flows. Where the eruption of magma onto the surface is limited, small cones and pyroclastics form at a few locations along the dike. If the dike does not reach the surface, no surface volcanism is visible.

Some linear rilles are very large and easily visible in ground-based telescopes. Ariadaeus Rille, for example, is a graben that measures 220 km long and 4 km wide, with near-vertical, parallel walls that drop a few hundred meters to a flat floor. Nearby are magma collapse pits and a volcanic dome.

Other linear valleys, such as the Sirsalis Rille, are associated with the formation of major basins and maria. They are roughly radial to the curved rim of the Oceanus Procellarum, and probably helped to feed the vast volumes of magma that filled the huge basin.

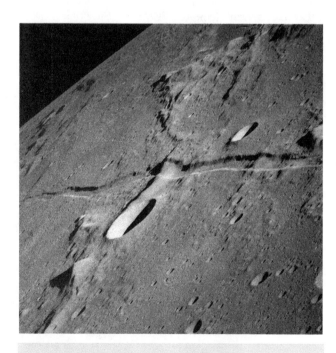

Figure 4.29   The linear Ariadaeus Rille is a tectonic graben feature where crustal faults cut across the bright plains to the west of Mare Tranquillitatis. The younger rille crosses a pre-existing ridge (center). The large crater at left is Silberschlag. (NASA)

Box 4.3   Dating the Moon

Crater densities are used for relative dating of planetary surfaces. Assuming that impactors strike all regions of a planet at approximately the same rate, the crater density – the number of craters in a given area – should always be the same. However, in many cases, another, more recent, process has covered or removed some of the oldest craters. Hence, regions with a higher crater density are generally older than regions with a lower crater density.

A common way for large numbers of craters to be removed is through volcanism. Early in the Moon's history, for example, lava flows flooded large areas, completely burying many craters. These areas were essentially wiped clean of craters about 3.5 billion years ago and thus their crater density only tells what happened to them after the lava flows ceased. The lunar highlands, in contrast, were rarely flooded by lava. Their crater density dates back to when they were formed, more than 4 billion years ago, and thus is much higher.

Crater size distributions can also be used to estimate the absolute age of a surface. The cratering rate was very high after the planets finished forming about 4.5 billion years ago and there was much leftover interplanetary debris to cause impacts. However, there is considerable evidence that nearly all of the major lunar impact basins were excavated by 3.8 billion years ago. Since then, the frequency (and size) of impacts has been steadily decreasing as the amount of interplanetary debris has declined due to continuous collisions and impacts (see Geological History).

By measuring the crater density on different areas of the Moon, and using radiometric dating techniques to determine the ages of rocks returned from different regions by Apollo astronauts, scientists can calibrate the cratering rate to actual surface ages. For example, rocks from the Fra Mauro region, ejected by the impact which created the Imbrium Basin, have been dated at 3.75–3.85 billion years. (The most common radiometric method compares the amounts of argon isotopes in lunar rocks. This ratio changes at a fixed rate over time.)

This relationship of cratering rate v time can then be extrapolated to other places in the Solar System, although other modifying factors also have to be taken into account, e.g. distance from the asteroid belt (Mars, located near the asteroid belt, may have a rate of crater formation roughly twice that of the Moon) and proximity to the Sun or large planets, such as Jupiter, whose gravitational field attracts impactors.

Present-day Volcanism?

The Moon is generally regarded as a geologically dead world. However, hundreds of observations of minor changes, usually flashes or glows, have been reported by amateur astronomers over the years. The most famous of these "transient lunar phenomena" took place on November 3, 1958, when Soviet scientist Nikolai Kozyrev obtained a spectrum of Alphonsus crater which he interpreted as proof of a gaseous emission from its central peak. Other red glows were reported near the crater Aristarchus by astronomers at Lowell Observatory.

Although no confirmation of any of these apparent glows or gaseous outbursts has been obtained, it has been shown that the phenomena are most frequently reported around the time of lunar perigee, when tidal forces are at their greatest and moonquakes are common. Orbital observations also revealed peaks in radon gas over Aristarchus, Grimaldi, and the fringes of the maria – presumably associated with gas seeping through surface fractures in these regions.

Although most volcanism on the Moon is thought to have ceased around 1 billion to 1.5 billion years ago, NASA's Lunar Reconnaissance Orbiter (LRO) has imaged sites which have seen activity much more recently. Scores of distinctive rock deposits observed by LRO on the Moon's near side are estimated to be less than 100 million years old. This time period corresponds to Earth's Cretaceous period, when dinosaurs ruled Earth. Some areas may be less than 50 million years old.

The deposits are scattered across the Moon's dark volcanic plains and are characterized by a mixture of smooth, rounded, shallow mounds next to patches of rough, blocky terrain. This combination of textures has caused researchers to refer to these unusual areas as irregular mare patches. The features are too small to be seen from Earth, averaging less than 500 m across at their widest.

Regolith and Moon Dust

Without protection from an atmosphere or a magnetic field, the lunar surface is bombarded continuously by micrometeorites, cosmic rays – mostly very-high-speed protons (hydrogen nuclei), helium nuclei (alpha particles), and carbon nuclei – and the charged particles of the solar wind. Brief flashes caused by small objects striking the surface have often been observed during meteor storms.

Over the Moon's lifetime, these tiny impacts have broken up the rocks and minerals on the surface, producing a layer of fine-grained regolith. Mixing of the regolith by the constant, small-scale bombardment is sometimes called "lunar gardening."

Over the ancient highlands the regolith may be more than 15 m thick, compared with 2–8 m on the maria. However, the material is very fine-grained, mostly in the form of dust particles less than 0.1 mm across. During the Apollo landings, rocket exhaust from the lunar module kicked up dust clouds that often became visible

Figure 4.30   Plumes of dust spurt from the wheels of the Lunar Roving Vehicle driven by Apollo 16 commander John Young. Note the numerous small craters and the rounded shape of the nearby uplands. (NASA)

at altitudes of 30–50 m, largely obscuring the surface during the final 10–20 m of descent.

The famous photo of Armstrong's bootprint in the lunar surface showed that the regolith easily supported the weight of the astronauts and their equipment. Even the lunar module footpads sank only 2–20 cm into the soil. However, the astronauts found the compacted regolith extremely difficult to penetrate, both when erecting the Stars and Stripes or drilling to obtain core samples. For sampling at greater depths, they used a battery-powered drill which could penetrate up to 3 m.

The all-pervasive, fine, gritty dust seeped into pressure suits and equipment. Some of the particles became electrostatically charged so that they clung to any objects with which they came into contact. The dark grains also absorbed sunlight, with the result that dust-coated equipment sometimes became extremely hot.

When inhaled, the astronauts likened its smell to that of gunpowder. Laboratory studies suggest that lunar dust could pose a health threat. Unlike particles on Earth, lunar dust is coarse and jagged. If inhaled, the grains could embed themselves in the lungs, eventually causing a long-term respiratory ailment similar to silicosis.

The lunar bootprints may not last as long as once expected. Before the launch of the LRO, it was thought that churning of the lunar regolith by meteoroid impacts took millions of years to overturn the surface down to a depth of 2 cm. However, images from LRO have revealed small surface changes that are transforming the surface much faster than previously thought. It seems the Apollo astronauts' tracks will be gone some tens of thousands of years from now.

Future visitors hoping to set up an astronomical observatory on the Moon may have to cope with "streamers" of suspended dust reported by the crew of Apollo 17. Scientists surmise that the carpet of dust must be electrostatically charged by periods of high

ultraviolet solar radiation and by the solar wind. As a result, the dust may rise 100 m or so in a phenomenon dubbed "fountaining." The particles then discharge, sink back to the surface, and the cycle starts again. At night, the particles in the solar wind that curve around behind the Moon and hit the surface drive a similar process.

Internal Structure

There are a number of factors that must be accounted for when modeling the lunar interior. The most fundamental of these is the bulk density (3.34 g/cm^3), which is very similar to the density of the surface rocks (3.3 g/cm^3). This implies that – in contrast with the Earth – the Moon does not possess a large, iron-rich core.

Nevertheless, studies of lunar rocks, surface features, and seismicity (moonquakes) make it clear that the Moon has a layered structure, and that the interior has a different composition from the surface (Figure 4.31). Furthermore, heat flow experiments set up by the crews of Apollo 15 and 17 show that the interior has a high temperature, presumably due to the presence of radioactive elements such as potassium, uranium, and thorium.[6]

Some of the most compelling evidence comes from seismometers that were left on the Moon. The instruments measured less than 3,000 events per year on average, none of them exceeding a value of 3 on the Richter scale – very weak by terrestrial standards. They recorded three different types of moonquake events: meteorite impacts, internal movements, and impacts by abandoned rocket stages or lunar modules.

The most surprising seismic records came from the artificial impacts, which set the Moon "ringing like a bell" for more than an hour – implying that most of the lunar mantle is solid. Other information came from an impact on July 17, 1972, which was thought to involve a meteorite weighing about 1,000 kg. The seismic waves indicated that the lunar rocks are hot enough to be molten below a depth of 1,000–1,200 km.

Some natural moonquakes originated deep inside the satellite and are thought to have been generated by tidal stresses created by the Sun and Earth. Shallow moonquakes are more likely to be caused by expansion and contraction of the surface rocks as a result of the large diurnal temperature range.

The velocities of the seismic waves provided useful information about the lunar interior. The upper few hundred meters are rubble generated by eons of meteorite bombardment. There is a change from regolith to firmer, more compacted material at a depth of 1 km.

The speed of seismic waves continues to increase as far down as 20 km, indicating fractured basaltic material whose density increases with depth. Below this, to a depth of 60 km, there may be a rock layer rich in anorthosite, the most common material in the lunar highlands. Deeper still, the sparse data suggest that there is a transition from the crust to the mantle.

The mantle rocks, which are relatively dense compared with the crust, are believed to be rich in olivine, pyroxene, iron, and magnesium. The rigid mantle – which extends to a depth of about

---

[6]Specifically radioisotope potassium-40. (Most potassium on Earth is stable and non-radioactive.)

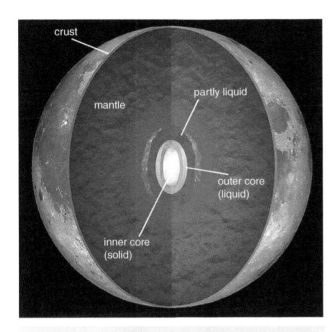

Figure 4.31 The likely internal structure of the Moon. Seismic studies indicate the presence of a solid inner core surrounded by a liquid outer core. There may be a semi-fluid layer at the core-mantle boundary. A thin, rocky crust lies on top of the rigid mantle. (NAOJ)

1,000 km – and the crust make up the lunar lithosphere. This thick layer makes it almost impossible at the present time for molten rock to reach the surface and for plate tectonics to take place.

The question of whether the Moon has a dense core, and whether this is still liquid, remains uncertain. Studies of Apollo data suggest that seismic waves slow down when they pass through the deep interior, indicate that a partially molten layer may exist below a depth of 1,000 km. One possible cause of this semi-liquid layer is internal heating caused by a continual change in the Moon's shape due to its tidal interaction with Earth.

Although this layer may extend all the way to the center, most geophysicists believe that it forms a thin shell surrounding a solid metallic core. Remanant magnetism in surface rocks indicates that this may be composed of iron or iron sulfide at a temperature of about 830°C. The radius of the core has been estimated as 340 km ± 90 km. For an iron-rich composition, this represents a mere 1 to 3% of the Moon's total mass.

Magnetic Fields

Although the Moon probably has a small, iron-rich core, it does not possess a global magnetic field. However, there is evidence that fluid motions in the core once created a dynamo that generated a weak, global field.

Isolated magnetic fields exist, both on a local and regional scale.[7] The magnetic field strength measured at four Apollo landing sites varied from a low of 6 nanoteslas for Apollo 15 to a high of 313 nanoteslas for Apollo 16. (Earth's field is about 30,000 nanoteslas.) In places, the field is strong enough to deflect solar wind particles. Older rocks are generally more magnetized, suggesting that the field was stronger in the distant past, gradually fading as the core cooled and solidified.

The presence of strong, isolated fields – producing mini-magnetospheres only 100 km in diameter – supports the theory that shock waves melted the crust, causing iron particles in the rocks to be magnetized by the global field that still existed about 3.6 to 3.8 billion years ago. In addition, high-speed impacts would have vaporized rock, producing hot gas that was partly ionized into electrons and positive ions. The ionized gases may also have amplified the magnetic field.

Box 4.4   Lunar X-rays

It has been known for many years that the Moon's sunlit side emits X-rays. These X-rays are generated by energetic solar electrons that strike the surface and cause the atoms in the rocks to fluoresce. The emissions can be used to determine the composition of the lunar surface.

By measuring the precise energy of each X-ray, the element that is emitting it can be identified. For example, the Chandra X-Ray Observatory has been used to detect oxygen, magnesium, aluminum, and silicon. ESA's SMART-1 spacecraft provided more detailed X-ray maps of the surface and made the first detection from orbit of the element calcium. By studying the abundance and distribution of such elements, scientists can gain a better understanding of how the Moon was formed.

Surface Resources

Future astronauts visiting the Moon for long periods will be encouraged to "live off the land" as much as possible. The most obvious local resource is the regolith, a fine-grained, cohesive mixture made up of glass spheres, dust, and coarse breccia. This could be used to protect any future habitation against ionizing radiation, extreme temperatures, and meteorites simply by piling it onto the roof. A straightforward sintering (heating and compressing) process could produce bricks for construction.

The composition of the regolith is generally similar to the Earth's crust, but with very different element ratios. In particular, the lunar soil is rich in refractory elements (i.e. ones with high boiling points) and low in volatile elements (i.e. ones with low boiling points).

As on Earth, oxygen (in the form of oxides) is the most abundant element on the Moon's surface. However, unlike most soils on Earth, the lunar regolith has high concentrations of sulfur, iron, magnesium, manganese, calcium, and nickel (again, often in the form of oxides).

[7] The unfortunate result is that it is not possible to navigate around the Moon using a compass.

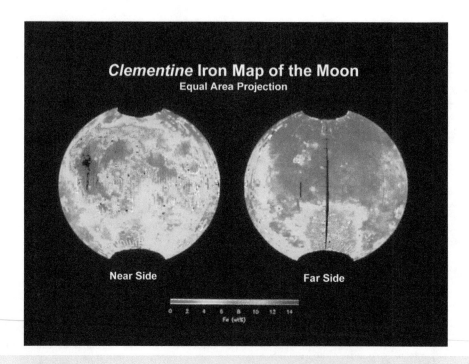

Figure 4.32   These Clementine maps show that iron-rich rocks are concentrated in the mare basalts of the near side, with very low amounts in the ancient, anorthositic rocks of the far side (with the exception of the South Pole–Aitken Basin). (NASA/LPI)

Elements such as magnesium, iron, and titanium are enriched in the surface of the maria, often in the form of the mineral ilmenite ($FeTiO_3$). The highland rocks consist mainly of the mineral feldspar, which is enriched with calcium and aluminum.

There are also traces of sulfur, phosphorus, carbon, hydrogen, nitrogen, helium, and neon in the soil. The carbon, hydrogen, helium, and nitrogen found in the regolith are almost entirely due to implantation resulting from constant exposure to the solar wind.[8]

An enhanced level of hydrogen found at the poles has been interpreted as evidence of trapped deposits of water ice (see Lunar Water). The water ice could be used either for production of oxygen, water supply, and growing crops, or for manufacturing hydrogen rocket fuel.

Solar energy will be a valuable resource, particularly on crater rims in the polar regions, which are almost permanently illuminated by the Sun. One "peak of eternal light" has been identified close to the lunar north pole, between three large impact craters called Peary, Hermite, and Rozhdestvensky. Another is on the rim of Shackleton crater near the south pole.

Lunar Water

The lunar surface is extremely dry. This was confirmed when rocks brought back from the Moon contained no moisture and revealed no evidence of alteration by aqueous processes. On the other hand, there is no doubt that comets and asteroids have delivered vast quantities of water to the Moon over billions of years. Since Earth's oceans may have been largely derived from water delivered by large planetesimals some 4 billion years ago, it would be illogical to assume that Earth's neighbor escaped this hydration process.

So what happened to the Moon's reservoir of water? Most water molecules are split by sunlight into their constituent atoms of hydrogen and oxygen, then lost to space. However, there is another possibility.

Some water released from impacting comets and water-rich asteroids may migrate by "hopping" to places where it could be stable and survive over long periods. This hopping process would involve sublimation and migration in the daytime, followed by condensation on the surface at night. Eventually, a small fraction of the incoming water could be deposited in craters filled by permanent, deep shadows – the only places to escape direct sunlight and soaring daytime temperatures.

The first direct evidence that such icy deposits may exist came in 1995, when the Clementine spacecraft confirmed that substantial areas of permanent shade exist at the bottom of deep craters near both lunar poles. Their temperatures never rise above −170°C and sometimes plummet to −249°C, the coldest temperatures yet measured in the Solar System. Under such conditions, water ice could exist for billions of years.

Data from Clementine's radar experiment also suggested that the radio signals were being reflected by water ice. Similar results have been returned by other spacecraft, e.g. India's Chandrayaaan-1 (Figure 4.33). Radar echoes from 40 craters near the north pole displayed the signatures of water ice, with an estimated content of at least 600 million tons.

---

[8]The helium-3 isotope, rare on Earth, has been proposed as a future fuel for nuclear fusion reactors.

Figure 4.33 Maps showing distribution of surface ice at the Moon's south pole (left) and north pole (right), based on data from NASA's Moon Mineralogy Mapper instrument. Blue shows the ice locations, whilst the gray scale corresponds to surface temperature (darker areas are colder and lighter areas are warmer). The ice is concentrated at the darkest and coldest locations, in the shadows of craters. (NASA)

Other supporting data came from spacecraft such as Lunar Prospector, which found concentrations of hydrogen close to the poles. The favored explanation was that the hydrogen was derived from water ice, which appeared to be mixed in with the lunar regolith at low concentrations, perhaps 0.3–1%.

Analysis indicated the possible presence of discrete, near-pure water ice deposits buried beneath 40 cm to 1 m of dry regolith. The hydrogen signature was stronger at the Moon's north pole, where the ice was thought to be spread over 10,000 to 50,000 square km, compared with 5,000–20,000 square km around the south pole.

Confirmation of the presence of small amounts of water near the south pole was provided in October 2009 by the LCROSS spacecraft and its Centaur rocket stage, which both impacted a crater called Cabeus. Spectral measurements led to estimates of about 100 kg of water – the equivalent of a dozen large buckets – in the ejecta cloud from the newly formed crater.

NASA's Lunar Reconnaissance Orbiter subsequently identified bright areas in craters near the Moon's south pole that are cold enough to have frost or ice present on the surface. The data showed that the coldest places near the south pole are also the brightest places – brighter than would be expected from soil alone – and that might indicate the presence of surface frost. The icy deposits appear to be patchy and thin, and it is possible that they are mixed in with the soil, dust, and small rocks of the regolith.

The clinching evidence came in 2018, when analysis of data from a U.S. sensor on board India's Chandrayaan-1 orbiter revealed a distinctive near-infrared spectral signature that is associated with water ice (Figure 4.33). Although almost all of the detections were in so-called cold traps marked by permanent shadow, ice was only found in 3.5% of the shadowed area. At the southern pole, most of the ice is concentrated in dark lunar craters, while the northern pole's ice is more widely, but sparsely, spread.

This suggests the ice is mixed with the lunar regolith and formed by slow condensation, either due to impact or water migration through the lunar exosphere. It seems that that the ice accumulation processes are different from those on Mercury or Ceres, where water ice is purer and more abundant than on the Moon.

Meanwhile, a new analysis of data from two lunar orbiters, published in 2018, found evidence that water is widely distributed across the entire surface and not confined to a particular region or type of terrain. The water appears to be present day and night, although it is likely created during the long lunar day. (Despite this finding, the Moon is still much drier than any desert on Earth.)

The finding of widespread and relatively immobile water suggests that it may be present primarily as OH (hydroxyl), a relative of H_2O which is more chemically reactive with minerals in rocks. The water is likely to be absorbed or trapped in glass and minerals at the lunar surface.

The results point toward OH and/or H_2O being created by the solar wind when hydrogen ions from the Sun strike the lunar surface, though it could also be released slowly from deep inside minerals where it has been locked since the Moon was formed.

Atmosphere

The Moon does have a sparse atmosphere – more correctly referred to as an exosphere. The entire atmosphere, altered to match Earth's surface temperature and pressure, would fit into a 63.5 m cube. As mentioned above, the escape velocity is so low that gases are easily lost to space.

Atoms of hydrogen, helium, argon, neon, sodium, and potassium have been identified as present in the atmosphere. Of these, the hydrogen and neon come from the solar wind, as does 90% of the helium, while most of the argon originates from radioactive decay of rocks in the lunar interior. The presence of sodium

Table 4.1
Lunar Geological Periods

System	Approx. age (billion years ago)	Main characteristics
Pre-Nectarian	4.5–3.92	Basins and craters formed before the Nectaris Basin. 30 multi-ring basins identified.
Nectarian	3.92–3.85	Period after formation of Nectaris basin and before formation of Imbrium Basin. Includes 12 large, multi-ring basins and some buried maria.
Imbrian	3.85–3.2	Period after formation of Imbrium Basin and before formation of Eratosthenes crater. Includes the youngest known mare lavas, the Orientale Basin, most visible maria, and many large-impact craters, e.g. Plato, Archimedes.
Eratosthenian	3.2–1.1	Period after formation of Eratosthenes crater and before formation of Copernicus crater. Includes mare lavas, e.g. in Oceanus Procellarum and Mare Imbrium, and craters without visible rays.
Copernican	1.1 to present	Period since the formation of Copernicus crater. Includes other fresh ray craters, e.g. Tycho.

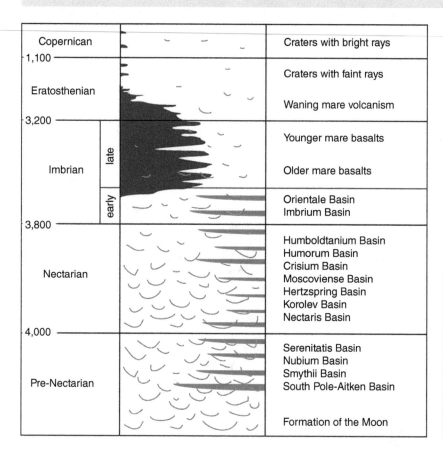

Figure 4.34 The geological history of the Moon, showing the principal events that have shaped its surface. Red lines denote formation of major impact basins. Brown "U"s denote smaller impact craters. The brown infill denotes the ebb and flow of mare volcanism. Numbers denote approximate age in millions of years. (Emily Lakdawalla after Tanaka & Hartmann 2012)

and potassium shows certain similarities with the atmospheres of Mercury and Io.

Geological History

Geologists recognize a number of major geological periods in lunar history (see Table 4.1 and Figure 4.34). This chronology is based on radiometric dating of returned samples, although there are currently few samples from younger terrains with which to check the dates.

- The oldest rocks on the Moon, found in the heavily cratered highlands, are 4.4 billion years old, so they belong to the Pre-Nectarian period. Since the Solar System formed 4.56 billion years ago, this implies that the accretion of the Moon took no more than 200 million years;
- The rocks found in the ejecta near ancient craters are up to 4.2 billion years old;
- The rocks close to the lunar maria are 3.9 billion years old;
- The basaltic rock that comprises the lunar maria is typically 3.7 billion years old;

Box 4.5   Lunar Meteorites

Samples of lunar rocks are available without actually visiting the Moon. Fragments ejected from the Moon by impacts are sometimes found on Earth as meteorites. It is possible to tell that they have come from the Moon by comparing their *chemical compositions, isotope ratios, mineral content, and textures* with the samples brought back by the Apollo astronauts and automated Lunas. They are an important source of information because they come from all over the surface, including the far side. However, it is very difficult to pin down their precise place of origin. (One meteorite found in Oman has been traced back to a small crater on the Mare Imbrium.)

Any rock on the lunar surface that is accelerated by the impact of a meteoroid to lunar escape velocity (2.38 km/s) or greater will leave the Moon's gravitational influence. Calculations suggest that impacts excavating craters as small as 450 m in diameter can launch lunar meteorites. Some of this ejected material is captured by Earth's gravitational field and lands on our planet within a few hundred thousand years (sometimes much less).

Other ejected material enters an orbit around the Sun, returning to strike Earth much later. Their time in space can be determined by studying radioactive isotopes (nuclides) created by exposure to high energy cosmic rays. Lunar meteorites Yamato 82192/82193/86032 and Dhofar 025 remained in space for perhaps 10 million and 20 million years respectively before finally landing on Earth.

The first lunar meteorite to be identified was found in the Allan Hills of Antarctica in 1981. Today, more than 330 named examples are known, although many of these are fragments of larger rocks, so the number of actual meteoroids is probably about 130. The largest single example is Kalahari 009, with a weight of 13.5 kg, but others weigh only a few grams.

After the surface of the Moon solidified, the interior heated up because of energy released by the radioactive decay of elements such as uranium and thorium. Differentiation of the interior into a mantle and a core took place, and molten lava welled up to fill the low-lying regions.

At first, during the intense bombardment by incoming asteroids and comets, many impacts pierced and fractured the thin crust. As the crust thickened, it became heavily cratered and marked by numerous multi-ringed basins. The bombardment shattered, melted, and mixed the highland crust, creating widespread breccias. Volcanism during this period produced lava flows of KREEP basalt, representing the final residue of the magma ocean. However, the extensive lava flows were destroyed and reprocessed or blanketed by ejecta from the incessant impacts.

About 3.85 billion years ago, a huge impact formed the Imbrium Basin, blanketing much of the near side with ejecta to create the Fra Mauro formation.[9] Shortly after, the bombardment eased sufficiently for the extensive lava flows that flooded the major impact basins to retain their dark color and their record of cratering.

The eruptions of basaltic lava that formed the maria continued on a large scale until about 3 billion years ago, after which they died down. Major volcanic activity ceased around 1 billion years ago. Since then, the surface has changed very little, apart from small, localised volcanic eruptions, continual "gardening" of the surface by small meteoroid impacts, and occasional large impacts that created "fresh" ray craters, such as Copernicus, Kepler, and Tycho.

The Birth of the Moon

Any explanation of the Moon's origin must take into account a number of unusual characteristics: the high angular momentum of the Earth–Moon system, the Moon's strange orbit inclined at 5.1° to the ecliptic, its large size and mass relative to the Earth, and a bulk density which is considerably lower than that of the four terrestrial planets.

One early theory favored the co-accretion of the Moon and Earth, in which both objects formed at the same time and in the same vicinity of the solar nebula, similar to the way the planets formed around the Sun. However, this double planet theory does not take into account the very different density and composition of both bodies. It was also difficult to account for the high angular momentum of the system and the Moon's orbital inclination.

Another possibility was that the Moon formed elsewhere and was captured by Earth during a close approach. Unfortunately, dynamic studies have shown that the capture of such a large passing body is extremely unlikely.

The most widely accepted explanation is the single impact hypothesis (Figure 4.35). This envisages a collision about 4.5 billion years ago between a Mars-sized planetary embryo and the young Earth. There is overwhelming evidence that such large embryos existed in the early Solar System during the final stages of planet formation.

Both the intruder and Earth are assumed to have already differentiated into a silicate mantle and an iron-rich core. The grazing

● No rocks younger than 2.9 billion years have been found, evidence that the Moon has been geologically dead for most of its history.

The sequence of events began with the birth of the Moon when a large planetary embryo collided with the Earth about 4.5 billion years ago (see The Birth of the Moon). Material from both bodies recondensed to form the Moon. The Moon was probably molten to a depth of at least 500 km during this accretion phase. As the accretion slowed, the magma ocean began to cool, allowing minerals to crystallize and rocks to form.

Since it was less dense than the magma, the cooled anorthosite rock began to float to the top, eventually accumulating to produce the first lunar crust. By about 4.4 billion years ago, the highland crust was in place.

[9]Some studies of argon isotopes in Apollo 14 rock samples suggest an age of 3.75 billion years.

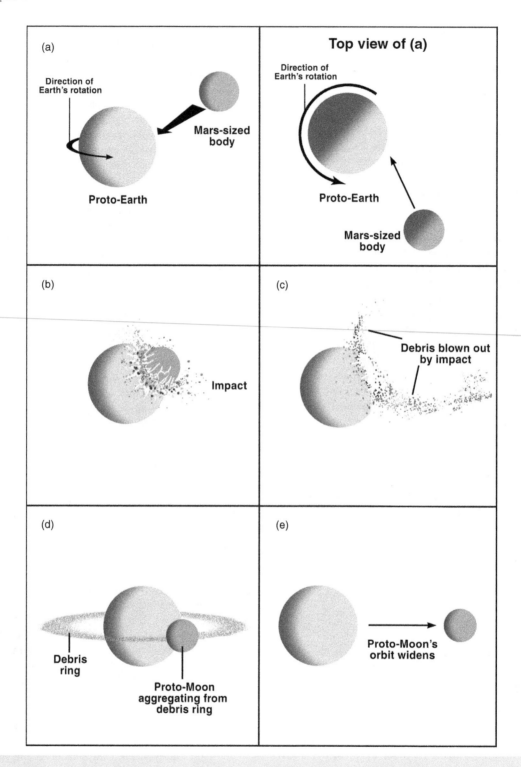

Figure 4.35   A simplified version of how the Moon might have been created by a grazing collision between a Mars-sized planetesimal and Earth. The vaporized material derived from the impactor and the Earth's mantle was ejected into orbit, but most of the iron fell back to Earth. The ejected rocky matter formed a disk that coalesces to form the Moon. (LPI)

Box 4.6   Apollo Experiments

38 scientific experiments were conducted during the Apollo missions, 16 from the orbiting Command-Service Module and the remainder on the surface. The Apollo 11 crew deployed a solar wind collector and the Early Apollo Scientific Experiments Package (EASEP), which included a passive seismic experiment and laser retroreflector. Subsequent missions, which stayed longer on the surface, deployed the more advanced Apollo Lunar Surface Experiments Package (ALSEP). A central station equipped with a radioisotope thermoelectric generator (RTG) supplied power for the ALSEP instruments and communications so that data could be relayed to Earth.

One of the most significant data sets came from active and passive seismic experiments. The Apollo 11 seismometer returned data for only three weeks, but more advanced seismometers deployed at the Apollo 12, 14, 15, and 16 landing sites transmitted data to Earth until September 1977. These were used to detect vibrations from natural moonquakes and impacts involving rocket stages or lunar modules, or small explosions set off in the vicinity of the sensors. The data provided key information about near-surface layering and the structure of the deep interior.

More information on the interior was provided by heat flow experiments on Apollo 15 and 17. The experiment involved drilling two holes about 2 m into the regolith and measuring the temperature at several depths. The results indicated a heat flow 18–24% of Earth's average heat flux of 87 milliwatts per sq. m.

Various instruments were used to measure the composition of the tenuous lunar atmosphere and the solar wind. A lunar dust detector measured the effects of dust accumulation, while a portable magnetometer measured the local magnetic field. (The build-up of dust proved to be much lower than expected.)

Among the more unusual experiments conducted on Apollo 17 were a surface electrical properties experiment, in which electrical signals were transmitted through the regolith and recorded on the lunar rover, and a portable gravimeter experiment.

One passive experiment is still in use today – the laser ranging retroreflectors deployed on Apollo 11, 14, and 15. These consisted of a series of corner-cube reflectors that return an incoming light beam in the direction it came from. (Similar reflectors were included on the Soviet Lunokhod rovers.) Decades later, scientists still calculate the distance to the Moon by sending laser pulses and measuring the precise time taken for them to return. Results show that the Moon is moving away from the Earth by 38 mm per year and that it also wobbles a little.

Figure 4.36   The Apollo Lunar Surface Experiments Package deployed by the crew of Apollo 17 at Taurus-Littrow valley. In the foreground is the Lunar Ejecta and Meteorites experiment. In the background are (L–R) the Lunar Surface Gravimeter, the Central Station, and the RTG. Note the rubbish pile at left. North Massif mountain rises in the distance (right). (NASA)

The Command-Service Modules of Apollo 15, 16, and 17 also carried some remote sensing experiments. These included panoramic and mapping cameras, laser altimeters, X-ray, alpha-ray and gamma-ray spectrometers, and mass spectrometers. Apollo 15 and 16 also carried a small subsatellite which was released in lunar orbit to measure gravitational anomalies, magnetic fields and the interaction of solar plasmas (charged particles) with the Moon.

blow blasted into orbit a mixture of largely vaporized material from both the impactor and Earth. Temperatures exceeded 10,000°C in some parts of the cloud.

The impactor's core was accreted by Earth within a matter of hours, as was the majority of the material from Earth's mantle. The remainder, a metal-poor mass of silicate largely derived from the impactor, rapidly collapsed to form an orbiting disk, then coalesced into a molten Moon within only a few years.

This hypothesis accounts for the geochemical evidence that at least half of the Moon was molten shortly after accretion. Since most of the material in the disk would have been white-hot vapor, the proto-Moon would have been depleted in volatiles (including water) and enriched in refractory elements (those that remain solid at high temperatures). The absence of metals in the accretion disk would explain the Moon's relatively low density. The hypothesis also accounts for Earth's relatively high spin rate and the Moon's inclined orbit.

The hypothesis assumes that the impactor came from the same broad region of the solar nebula as Earth. It also requires the Moon to condense from a debris disk orbiting quite close to Earth. Support for this assumption comes from laser measurements showing that the Moon is currently drifting away from the Earth at a rate of 3 cm per year.

Questions

- The Apollo program is said to have revolutionized scientific knowledge of the Earth–Moon system and the Solar System as a whole. Explain why you agree or disagree with this statement.
- The Moon is generally regarded as an arid world, with a minuscule amount of water. (a) Where is water ice thought to exist on the Moon? (b) Explain why this water is likely to be present.
- Discuss the main theories of lunar formation and the arguments for and against each of them.
- Describe and explain the major differences between the lunar highlands and the maria.
- Describe and explain the origins of the various impact structures that are found on the Moon.
- (a) What evidence is there for volcanism on the Moon? (b) How important has volcanism been during the lifetime of the Moon?
- (a) How have scientists been able to learn about the lunar interior? (b) What evidence is there for an iron-rich core?
- Many countries currently have programs to explore the Moon with automated orbiters, landers, rovers, and sample return missions. Explain why there is this renewal of scientific interest, despite five decades of intensive research.
- Describe the problems that will face human explorers and settlers on the Moon and explain how they may be overcome.

FIVE
Mercury

At first glance, the innermost planet in the Solar System, Mercury, looks like a close cousin of the Moon. Both are small, rocky worlds that are covered in impact craters. With no atmosphere worth mentioning, surface temperatures soar in the daytime, but plummet at night. However, first glances can be deceptive, particularly where planetary studies are concerned.

Of the five planets visible to the naked eye, Mercury is the most elusive since it is always low above the horizon and close to the Sun in the sky. Its existence has been known since at least the 3rd millennium BCE, when the Sumerians and Babylonians kept careful records of its periods of visibility and motion across the sky. Yet, although this little world has probably been recognized since prehistoric times, it remained one of the most mysterious inhabitants of our Solar System until recent years.

Apparitions of Mercury

Mercury is one of the four terrestrial planets, regularly approaching to within 80 million km of Earth – closer than any planet other than Venus and Mars. At such times it reaches magnitude −1.3 and rivals Sirius, the brightest star in the sky.

To an observer on Earth, the planet becomes visible for about two months as a "morning star," appearing just before dawn, to the west of the rising Sun. It then disappears from view, only to reappear a few weeks later as an "evening star," which lingers for another two months. The cycle then begins again. During 2018, for example, Mercury could be observed during four different apparitions in the morning, separated by three evening apparitions.

Even at its greatest elongation – its angular distance from the Sun in the sky – Mercury never moves more than 28° from the Sun, a little more than the width of fully spread hand held at arm's length (Figures 5.1 and 5.2). At such times, it appears above the horizon for about two hours after sunset or before sunrise – although for much of this short window of opportunity it is invisible in the solar glare. However, Mercury's highly elliptical orbit means that some elongations are less favorable, with the planet straying only 18° from the Sun.

Orbit

Mercury is termed an "inferior planet," because it orbits nearer to the Sun than Earth. Indeed, its twilight apparitions and rapid motion across the sky show that it is closer to the Sun than any of the other planets. Its mean distance from the Sun is only 57.9 million km, or about 0.39 AU. However, its orbit is much more eccentric than those of the other planets.

As a result, it approaches to within 46 million km of the Sun at perihelion, while its furthest distance (aphelion) is about 69.8 million km – half as far again. This orbit means that an observer on Mercury would see the Sun's apparent diameter grow from twice to three times its size, as seen from Earth, over a period of just six weeks.

Not only is the orbit far from circular, but it is inclined by 7° to the ecliptic, so that it dips far above and below the paths of all the other planets. The reason for Mercury's unusual orbital inclination and eccentricity is linked to periodic, long-term oscillations in its orbital characteristics, caused by perturbations from the other planets.

In order to avoid being pulled into the Sun, Mercury must also travel around its orbit faster than the other planets – in accordance with Kepler's third law of planetary motion (see Chapter 1). By

Exploring the Solar System, Second Edition. Peter Bond.
© 2020 John Wiley & Sons Ltd. Published 2020 by John Wiley & Sons Ltd.
Companion Website: www.wiley.com/go/Bond-Solar-System2e

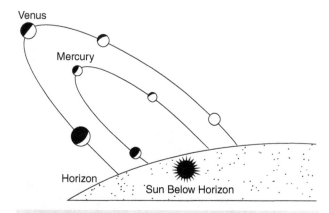

Figure 5.1 The changing positions and phases of Mercury and Venus in the evening sky. (NASA)

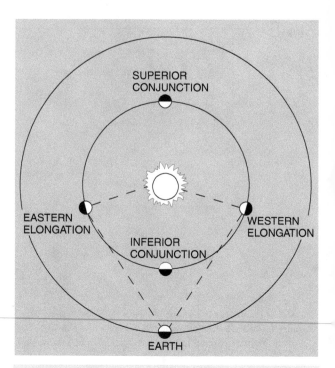

Figure 5.2 Key orbital positions of Mercury, observed from Earth. When Mercury (or Venus) is between the Sun and Earth, it is at inferior conjunction. Unless it is transiting the Sun's disk, it is then invisible. On the opposite side of the Sun from the Earth, Mercury (or Venus) is at superior conjunction. The planet is most easily observed at greatest elongation, when it is furthest from the Sun in the sky. (NASA)

traveling at an average speed of 48 km/s, it is able to overtake the more distant planets on the inside track, completing one journey around the Sun in only 88 Earth days. Mercury finishes four orbits before Earth completes one. However, the elliptical orbit means that its orbital velocity varies greatly, ranging from a relatively modest 39 km/s at aphelion to 56 km/s around perihelion.

Another orbital peculiarity is the substantial forward drift of the position of its perihelion (Figure 5.3). All of the planets undergo some shift of perihelion as a result of the gravitational influence of their neighbors. However, the advance of Mercury's perihelion, known as precession, is by far the largest of any planet.

In fact, the motion is still very small – less than 2° per century. At this rate it would take a quarter of a million years for Mercury's perihelion position to complete a full circle around the Sun. However, even this rate of precession is too large to be explained by Newtonian physics. When the effects of Earth's precession (the changing orientation of its axis in space) and the gravitational pull of the other planets are taken into account, 43 arc seconds of Mercury's perihelion movement remain unexplained.[1]

One possibility, put forward in 1859 by Urbain Le Verrier, was that an unknown planet was pulling Mercury off its path. (Neptune had been discovered only 13 years earlier as the result of a similar calculation.) However, we now know that such a planet does not exist.

The explanation had to wait until Albert Einstein published his General Theory of Relativity in 1915. One of its predictions was that massive objects such as the Sun significantly warp the space around them. In the case of Mercury, the planet is traveling in space that is much more curved at perihelion than at aphelion. The predicted advance of Mercury's perihelion by moving through this warped space is 0.1 seconds of arc per orbit (43 arcsec per century), exactly in agreement with the observed value. (One second of arc = 1/3,600 degrees). The success of this prediction has been cited as one proof of the validity of Einstein's groundbreaking theory.

Phases

When seen through a telescope, Mercury, like Venus and our Moon, appears to go through phases. At superior conjunction, on the far side of the Sun from Earth, it appears as a full disk, though it can never be properly viewed because of the intervening glare of the Sun (see Figure 5.2).

As it steadily approaches our world, Mercury displays a smaller illuminated area but grows in apparent size. At greatest elongation, the planet resembles a quarter Moon, then, as it comes ever nearer to the Earth, it shrinks to a slim crescent. By the time it reaches inferior conjunction, no part of the visible surface is bathed in sunlight and the planet cannot be seen – unless it happens to transit the Sun.

Transits

Although it is the closest planet to the Sun, transits of Mercury occur only a dozen times or so per century. This is because of the 7° inclination of the planet's orbit to the ecliptic, which means that

---

[1] All of the planetary orbits precess. Newton's theory of gravity predicts these effects, which result from the planets pulling on one another. With the exception of Mercury, this precession can be understood using Newton's equations.

Figure 5.3   Mercury's elliptical path around the Sun shifts slightly with each orbit, so that its closest point to the Sun (perihelion) moves forward (precesses). Newton's theory of gravitation predicted an advance only half as large as the one actually observed. According to Einstein's theory of relativity, the precession of Mercury's orbit is caused by the massive Sun warping the space around it. NASA's MESSENGER spacecraft, shown here, was used to detect tiny changes in Mercury's position. (NASA-GSFC)

Earth, Mercury and the Sun are rarely aligned precisely when the innermost planet is at inferior conjunction.

Transits of Mercury can only occur close to 8–9 May or 10–11 November, when its orbit crosses the ecliptic – either at the ascending or descending node (see Chapter 1). The May events take place when Mercury is near aphelion, while the autumn transits occur near perihelion. Although the latter events are twice as frequent as those in May, they last for a shorter time. The interval between successive crossings varies from 3 to 13 years, most recently with transits in May 2003, November 2006, November 2016 and November 2019. The next transit will take place on November 13, 2032.

The sharp outline of Mercury's disk during transits – in contrast to the blurry outline of Venus – provided important evidence that the planet does not possess much of an atmosphere. Transits also made it possible to determine the planet's orbit and diameter (Figure 5.4).

Spin-Orbit Resonance

Until the early 1960s, it was thought that Mercury was locked into a synchronous rotation which caused the planet to keep its same face toward the Sun – just as the Moon does when it orbits Earth (see Chapter 4). Under these circumstances, its rotation period would last 88 days, the same length as its year. One side of Mercury would always be exposed to the heat of the Sun, with the other hemisphere in perpetual darkness. Drawings and maps made during favorable viewing opportunities suggested that the same surface was always visible.

The first doubts arose in 1962, when studies of radio waves from Mercury suggested that the night hemisphere was warmer than expected. Then, in 1965, Gordon Pettengill and Rolf Buchanan Dyce used the 330 m Arecibo dish in Puerto Rico to bounce radio waves off the planet's surface and detect the echoes. These Doppler radar observations showed that Mercury's rotational velocity was

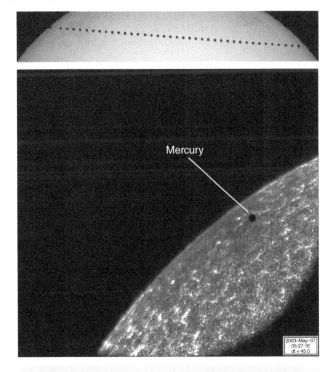

Figure 5.4   Two views of the transit of Mercury on May 7, 2003. The time sequence (top) was obtained by the SOHO spacecraft, while the image of Mercury against the Sun's photosphere (bottom) was obtained by TRACE. The entire transit lasted about 5½ hours. (NASA/ESA)

consistent with a rotation period of about 59 days, a figure subsequently refined to 58.65 days.

The error of early observers was soon explained by Giuseppe "Bepi" Colombo, a specialist in celestial mechanics, who noted

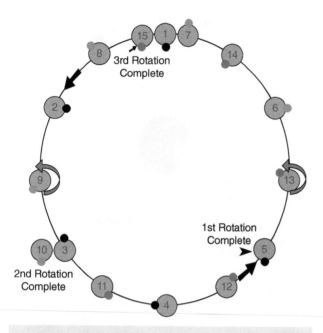

Figure 5.5   The spin-orbit coupling of Mercury means that the planet rotates three times during every two orbits of the Sun. (NOAA-National Weather Service)

that the 59-day rotation period was two-thirds of the Mercurian year (Figure 5.5). In other words, Mercury rotates three times during two orbits of the Sun – a type of spin-orbit resonance found nowhere else in our Solar System.

This 3:2 ratio explains how the telescopic observers came to the wrong conclusion. After three complete orbits, the same face of the planet is visible during the same phase, so observations made at the most favorable viewing times always revealed the same surface features. Presented with familiar, dusky markings at many apparitions, astronomers tended to ignore any occasional variations and jumped to the wrong conclusion.

In order to explain how Mercury entered this unusual, but stable, resonance, most studies assume that it originally rotated much faster than it does today. A rapid rotation rate – perhaps as short as eight hours – would cause the planet to bulge outwards at the equator. A combination of tidal friction – when the Sun raises tides in the solid body of the planet – and differential motions in the liquid core relative to the mantle would enable the Sun's gravity to slow the rotation.

Recent studies have also shown that the orbital eccentricities and inclinations of the inner planets are very unpredictable on timescales of millions of years. This makes it likely that Mercury's orbit was even more eccentric in the past, varying from near zero to 0.45 or more.

Research by Correia and Laskar shows that, when the eccentricity exceeds a value of 0.325, solar tidal effects lead to corresponding changes in the planet's spin rate. The result is that some resonant states – including the 3:2 spin-orbit resonance – are passed through many times, so that the probability of eventual capture into that resonance is greatly increased. Indeed, models suggest

that, over a 4 billion year period, the 3:2 spin-orbit resonance becomes the most likely outcome.

Two-Year Days and Double Sunsets

The spin-orbit resonance means that Mercury's solar day (from sunrise to sunrise) is equal to two Mercurian years, or 176 Earth days. As a result, it is the only planet to have its day longer than its year. The unusual spin-orbit coupling, combined with the high orbital eccentricity, would produce some very strange effects for anyone standing on Mercury's surface.

At the 0° and 180° meridians, an observer would see the Sun rise above the eastern horizon and then gradually increase in apparent size as it slowly moved toward the zenith over the next 1 ½ months. During perihelion, at local noon, the Sun would slow to a complete halt in its westward motion across the sky and then move a little in a retrograde direction (eastward) for eight days.

The reason for this reversal is that the planet's orbital angular velocity exceeds its spin angular velocity near perihelion. In other words, Mercury's orbital motion is so rapid around perihelion that its slow rotational motion is unable, in effect, to keep up.

The Sun's disk – three times larger than on Earth – would dominate the sky, remaining almost overhead for several weeks before resuming its path toward the horizon and decreasing in apparent size. The period of continuous daylight would last 88 days.

Observers elsewhere on Mercury would see different, but equally bizarre motions during a single perihelion passage. Someone standing at the 90° longitude would see a double sunrise, while another at 270° (on the opposite side of the planet) would witness a double sunset. At the next perihelion, the situations would be reversed.

The first sunrise occurs, as usual, when the planet's prograde spin brings the Sun above the horizon. However, near perihelion, the planet's orbital speed exceeds its spin rate, rotating the observer back into darkness as the Sun sets. Once the planet is past perihelion, the spin rate becomes the dominant factor once more, causing a second sunrise.

Hot Spots

Mercury's spin axis is almost perpendicular to the plane of its orbit (unlike most of the other planets), so there are no seasonal variations with latitude. Every part of the surface, apart from some deep, permanently shaded, craters at the poles, suffers from lengthy exposure to daytime heat and extreme cold at night.

Its proximity to the Sun means that Mercury receives up to 11 times more solar radiation than Earth, so its day side is much hotter than a domestic oven set at maximum. The slow rotation period and the lack of an atmosphere mean that Mercury has the harshest thermal environment of any planet in the Solar System, with a day-night variation of up to 610°C. The noon temperature on the sunlit side at perihelion soars to 430°C, hot enough to melt zinc, tin, and lead. With no atmosphere to transfer warmth from the day side, heat radiates into space during the long night and the temperature plummets to −183°C just before dawn.

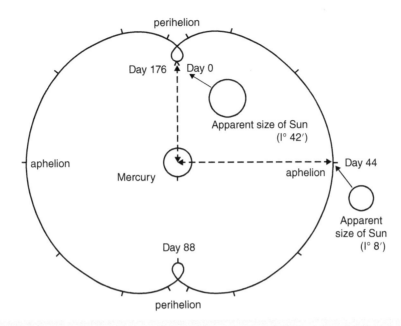

Figure 5.6 The Sun as seen by an observer on Mercury during two Mercurian years. The Sun's disk appears 66% larger at perihelion, so the planet's surface is much hotter at such times, compared with aphelion. The Sun's motion across the sky appears to reverse direction for a time around perihelion. (Peter Bond)

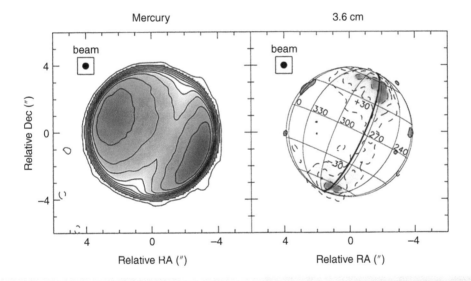

Figure 5.7 Radio emissions at a wavelength of 3.6 cm show the two hottest regions on Mercury (dark red in left image). The hot spots appear in both the day (left) and night (right) hemispheres, and coincide with regions which receive maximum solar radiation around perihelion. The Caloris Basin is located near the spot on the right – hence its name. The spot at left lies in a region of "peculiar" hilly terrain. The right hand image shows cooler (blue) areas at the poles and along the morning terminator (bold line, Sun is to the right). They are caused by shadows cast by hills and craters. (David Mitchell and Imke de Pater/NRAO)

However, the unusual spin-orbit coupling means that some parts of the planet, centered on 0° and 180° longitude, suffer more extreme thermal conditions than others. Since the planet rotates 1½ times during each orbit, these longitudes face the Sun at alternate perihelion passages. Hence, the hemisphere centered on 0° longitude is subjected to prolonged, maximum heating by the noon Sun at one perihelion. This

experience is then repeated at 180° longitude during the following perihelion.

The places where the Sun's zenith coincides with its closest approach are sometimes known as "hot poles" or "hot spots." These regions receive 2½ times more radiation overall than places at 90° and 270° longitude, which suffer the noonday Sun when the planet is near aphelion. As a result, maximum temperatures are noticeably lower, so the 90° and 270° sub-solar points are often called the "warm poles."

Images returned by spacecraft show that the 180° hot spot coincides with the location of a gigantic impact basin, appropriately known as Caloris ("Hot"), while its antipodal equivalent is marked by a "peculiar terrain" of hills and linear features.

Surface Observations

Until the visits of Mariner 10 in 1974–1975, very little was known about Mercury's surface. Apart from the difficulties caused by its phases and twilight apparitions (see Phases), observation were also complicated by its modest size.

Mercury is the smallest planet in the Solar System. With a diameter of 4,879 km, it is dwarfed by Jupiter's moon Ganymede and Saturn's moon, Titan. Even at its closest to Earth, Mercury's apparent diameter is only about 11 arc seconds, smaller than all but Uranus and Neptune.

As a result, features less than about 300 km across cannot be resolved in even the largest ground-based telescopes. Not surprisingly, the best maps of Mercury until 1974 merely showed vague variations in shading, though irregular markings along the terminator – the boundary between night and day – suggested a mountainous or cratered surface.

Evidence that Mercury's surface might resemble that of the Moon was supported by similarities in the way it reflected sunlight and radar pulses, and in its emission of infrared and radio

Figure 5.8   A global mosaic of Mercury captured through red, green, and blue filters by the MESSENGER spacecraft. Mercury's color variations are very subtle. Note the rays radiating from Hokusai, a small impact crater located near Mercury's north pole. (NASA/JHU-APL/CIW/color mosaic by Jason Perry)

waves. These suggested a similar blanket of fragmented material generated by eons of meteorite bombardment.

With a mass only 5% that of the Earth and an escape velocity of 4.25 km/s it seemed clear that Mercury could not hold on to a substantial atmosphere, even if it originally possessed one. Under these circumstances, the sky would appear black and full of stars even in the middle of the day, and there would be no natural protection from the Sun's ultraviolet radiation.

Size, Mass, and Density

Prior to the Space Age, the absence of a satellite made it difficult for ground observers to determine Mercury's mass or calculate its density. One of the most precise estimates was obtained by studying the way the planet perturbed the large asteroid Eros. The problem was finally solved by Mariner 10, which confirmed that its diameter is only 4,879 km, about one-third the diameter of Earth. Approximately 18 Mercurys would fit inside our world (Figure 5.10).

Measurements based upon the way it modified Mariner 10's trajectory also showed that Mercury is very massive for its size. Although it is 30% smaller than Mars, this is compensated by its large mass, which results in the two planets having similar surface gravity and escape velocities. Similarly, although its diameter is only 40% greater than that of the Moon, its surface gravity is almost twice that of Earth's satellite.

Calculations of Mercury's density – obtained by dividing its mass by the volume – were even more surprising. The planet's average density (5.4 g/cm^3) is second only to Earth in the Solar System. However, materials are much more compressed at the center of the larger Earth, thus increasing their density. Mercury would have a density of 5.3 g/cm^3 in an uncompressed state, compared with only 4.4 g/cm^3 for Earth (see Chapter 1).

The Iron Planet

How can Mercury have the greatest uncompressed density of any planet? The only possible explanation is that it contains a lot of dense, metallic elements. Since iron is the most abundant heavy element in the Solar System, and seismic data indicate that it is the dominant constituent of Earth's core, it seems reasonable to suppose that Mercury is also largely composed of iron (Figure 5.11). This theory is supported by the presence of a magnetic field and by studies of the probable composition of the inner regions of the early solar nebula (see Chapter 1).

Mercury's core is different from any other planetary core in the Solar System. Earth has a metallic, liquid outer core sitting above a solid inner core. Mercury appears to have a solid silicate crust and mantle overlying a solid, iron sulfide outer core layer, a deeper liquid core layer, and possibly a solid inner core of iron and sulfur.

Gravity field measurements from the MESSENGER spacecraft and Earth-based radar observations indicate that Mercury's iron core is about 2,000 km across.

Iron probably accounts for over 60% of Mercury's mass, with rocky, silicate-rich material accounting for the remainder. This is twice as much iron per unit volume as any other planet or satellite. Assuming that most of this metal is concentrated in the planet's

Box 5.1   Mariner 10

Until the MESSENGER spacecraft began to study Mercury in 2008, the source of most of our knowledge about Mercury was Mariner 10, which flew past the planet on three separate occasions 1974–1975.

Launched on November 3, 1973, Mariner 10 swept past Venus on February 5, 1974. The extremely accurate flyby modified the spacecraft's trajectory and speed, putting it on course for an encounter with Mercury.

Its first images of half-lit Mercury were acquired on March 24, 1974, from a distance of 5.4 million km. As the range closed, it became clear that the planet did, indeed, have an intensely cratered surface that closely resembled the ancient lunar highlands. To everyone's surprise, the plasma and particle data indicated the presence of a substantial magnetic field. Closest approach, at a distance of 703 km, took place on the planet's night side. Further mosaics of a different hemisphere were returned on the outward leg.

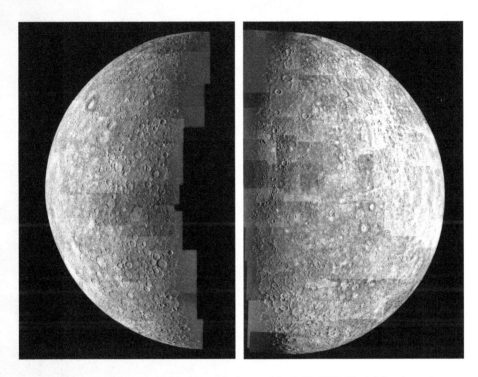

Figure 5.9   Mariner 10's first close encounter with Mercury took place on March 29, 1974. The left-hand mosaic was assembled from 18 photographs taken at a distance of about 200,000 km, 6 hours before closest approach. The outgoing view (right) is composed of 18 images taken 6 hours after closest approach at a distance of 210,000 km. The most notable feature, on the sunrise terminator, is the giant Caloris impact basin. (NASA-JPL)

The spacecraft's orbit was such that, with only a few minor adjustments, it could make repeated flybys of Mercury. The only drawback was that these took place at the same point in Mercury's orbit, when the same hemisphere was illuminated. (Mariner's orbital period was almost exactly twice Mercury's period – see Spin-Orbit Resonance.)

The second encounter was dedicated to imaging the southern hemisphere, while the third flyby primarily concentrated on magnetic field and particle measurements. In order to image the entire southern hemisphere, terrain that would link the two regions seen during the first encounter, the flyby on September 21, 1974, took place on the sunlit side at an altitude of 48,000 km. Although their spatial resolution was a modest 1–3 km, the 360 images extended the coverage of the illuminated hemisphere from about 50% to 75%.

The third and final encounter occurred on March 16, 1975. Closing to only 327 km, the spacecraft was expected to send back high-resolution images of some intriguing geological features. Unfortunately, a problem with the ground station meant that only one quarter of each picture could be received. Nevertheless, the measurements of the magnetic field confirmed that Mercury generates an internal field that is weaker, but similar to, that of Earth.

Altogether, more than 2,300 black and white images of Mercury were taken, many at moderate resolution (3–20 km/pixel) but a limited number with a resolution of only 140 m. Altogether, 45% of the planet was imaged.

Figure 5.10   Relative sizes of the terrestrial (rocky) planets – left to right, Mercury, Venus, Earth, and Mars. Mercury is much smaller than its neighbors, but has a surprisingly high bulk density. (LPI)

Table 5.1

Comparison of Mercury and Earth

	Mercury	Earth
Equatorial diameter (km)	4,879	12,756
Rotation period	58.65 days	23 h 56 m
Density (water = 1)	5.43	5.51
Mass (Earth = 1)	0.055	1
Surface gravity (Earth = 1)	0.38	1
Inclination of axis to orbit	0.034°	23.44°
Orbital period (days)	87.97	365.24
Average distance from Sun (million km)	57.91	149.6
Average temperature (°C)	179	7
Satellites	0	1

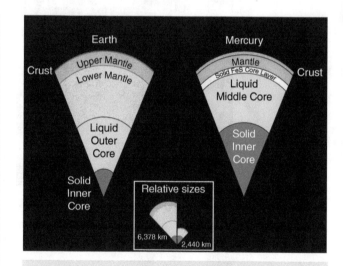

Figure 5.11   Comparison of the internal structures of Earth and Mercury. Mercury's interior has a much larger metallic core and less silicate rock material than Earth. The core seems to be partly liquid with a solid center. Mercury also appears to have a solid layer of iron sulfide that lies at the top of the core. The presence of this solid layer may influence the generation of the planet's magnetic field. The inset shows a comparison of the relative radial sizes of the Earth and Mercury. (NASA/Johns Hopkins University Applied Physics Laboratory/Carnegie Institution of Washington)

center, the iron core fills about 80–85% of Mercury's diameter and accounts for more than half of its volume.

The core is overlain by a silicate (rocky) mantle and crust. This rigid outer shell, or lithosphere, is approximately 400 km thick and is mainly composed of igneous rock that is made of calcium-rich feldspar, one of many common silicate minerals found on Earth and the Moon (see Chapter 4). The outer crust may be around 35 km thick.

Origin

How did an iron planet come about? It is generally accepted that the planets accreted from a solar nebula of gas and dust, and that

the composition of this nebula varied with temperature. However, the fact that non-volatile, rocky elements would be expected to condense in the hotter, innermost part of the nebula is not sufficient to account for Mercury's huge iron core.

Three main theories have been put forward to explain why Mercury is so different from Venus, Earth, and Mars. Each theory predicts a different composition for the rocks on Mercury's surface.

In the selective accretion model, drag by nebular gas favored accretion of dense particles as the planet formed from the cloud. These particles (such as metallic iron) condensed and were preferentially retained in the innermost regions near the Sun, whereas the lighter particles were swept away. As a result, Mercury became enriched in iron and other metals, but there was no comparable enrichment in silicate minerals. In this case, surface rock composition would be similar to that of the other terrestrial planets.

Another theory suggests that intense radiation from the young Sun (during its T Tauri stage) vaporized much of the silicates in proto-Mercury, while allowing the planet to hold on to its metal-rich core. This idea predicts a surface enriched in aluminum, calcium, and magnesium, but poor in easily vaporized elements like sodium and potassium. It also predicts rocks low in ferrous iron (FeO).

The third theory suggests that a giant impact took place between the young Mercury and a large planetesimal. This proto-Mercury may have consisted of a small iron core and larger rocky (silicate) mantle, with at least twice the mass of Mercury today. A high-speed collision would strip away most of the primordial crust and upper mantle, leaving a much smaller planet that is dominated by its iron core. Under these conditions, the present-day surface would be expected to be made of rocks highly depleted in those elements (silicon, aluminum, and oxygen) that were concentrated in the primordial crust and mantle.

X-ray fluorescence spectra obtained by the MESSENGER spacecraft indicate that the planet's surface differs in composition from those of other terrestrial planets. The data show low abundances of iron and titanium, important constituents on the Moon and other silicate bodies, ruling out a lunar-like feldspar-rich crust.

Despite the lack of surface iron, Mercury is darker than the Moon. MESSENGER revealed that the darkening agent for parts of the surface is an unusually high carbon content, likely in the form of graphite (the material found in "lead" pencils). This is surprising because carbon is found at typical concentrations of only a few hundred parts per million on the Moon, Earth, and Mars. The carbon-rich material appears as dark deposits excavated from depth by impact cratering.

Scientists believe that, like Earth's Moon and the other inner planets, Mercury likely had a global magma ocean when it was young and the surface was very hot. As this magma ocean cooled, the minerals that began to crystallize and solidify would sink – with the exception of graphite, which was buoyant and accumulated as the original crust of Mercury. Remnants of this ancient material probably persist beneath the present upper crust.

MESSENGER also detected higher-than-expected amounts of potassium and sulfur on the surface, suggesting that the giant impact hypothesis and vaporization of the crust and mantle did not occur. These volatile elements would have been vaporized and lost by the extreme heat generated by both these events. Their continuing presence on Mercury means that such a heat-intense episode could never have happened.

Magnetic Mercury

Mariner 10 was expected to find a geophysically dead world. A planet without a liquid core should not be able to support the fluid, internal dynamo that generates a magnetic field, and Mercury's small size and slow rotation indicated a Moon-like world with a metallic core that had cooled and solidified long ago. (Small planets have a high surface area compared with their volume. Hence, other factors being equal, smaller bodies radiate their internal heat to space more rapidly. This is why a pebble cools down faster than a large rock.)

However, Mariner 10 detected a weak magnetic field (about 300nT) at the surface, comparable to about 1% of Earth's magnetic field strength (Figure 5.12). With the exception of periods of high solar activity, this is sufficient to deflect the solar wind and prevent it from reaching the planet's surface. However, major, rapid fluctuations are caused by gusts in the solar wind, solar flares, and coronal mass ejections. The solar wind can also impact the surface in polar regions.

Observations of low-energy solar wind electrons reveal that the magnetosphere resembles a scaled-down version of Earth's magnetic bubble. Although the velocity of the solar wind is not much faster at Mercury than at Earth, the Sun is much closer, so its density is 8–10 times greater, depending on Mercury's distance from the Sun. When combined with a weaker planetary magnetic field, this compresses the dimensions of the magnetosphere to about 5% that of Earth.

Data from NASA's MESSENGER mission confirm that Mercury's field is dipolar and aligned with the planet's spin axis.[2] This means that on a global scale the planet behaves as if there is a giant bar magnet at its core. The polarity is the same as Earth's, with the south magnetic pole in the northern hemisphere.

However, MESSENGER also discovered that the field's source is offset 480 km (about 20% of Mercury's radius) towards the north pole, so it is stronger in the northern hemisphere than in the southern hemisphere. The axial symmetry and latitudinal offset of Mercury's magnetic field make it unique among planets in the Solar System.

A marked bow shock occurs above the sunward side of the planet, where the solar wind is slowed as it comes into contact with the planet's magnetosphere. The shock occurs only a few thousand kilometers above the planet, although the altitude varies as solar weather varies.

The solar wind compresses the field so much on Mercury's sunward side that there is no possibility of radiation belts like those of Earth or Jupiter. Solar wind particles are guided around the planet and sent down the magnetic tail rather than trapped. Like Earth, the magnetotail is divided by a magnetically neutral sheet.

At the magnetic field cusps, near the poles, charged particles from the solar wind travel down field lines to slam directly into Mercury's surface. However, the northward offset of the magnetic

---

[2] Most planets in the Solar System have magnetic fields that are inclined to their rotation axes. Mercury and Saturn are the exceptions.

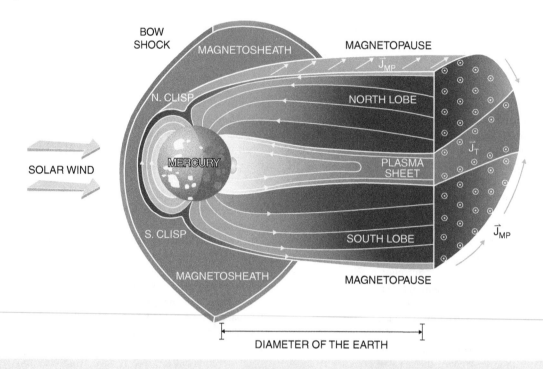

Figure 5.12   Despite its slow rotation, Mercury has a magnetic field which is stronger than those of Venus, Mars, or the Moon. Its strength at the equator is about 300 nanoteslas, or about 1% of Earth's surface magnetic field. Although the density of the solar wind is 10 times higher at Mercury than Earth, the modest magnetic field is usually strong enough to deflect most of the particles and create a teardrop-shaped magnetosphere which is 5% the size of the magnetic bubble surrounding the Earth. The axis of the magnetic field is aligned with the rotational axis. (James Slavin/NASA-GSFC)

dipole means that the south polar region is much more exposed than the north to charged particles in the solar wind.

The impact of those charged particles knocks atoms from Mercury's surface and kicks them into space in a process called sputtering, contributes to the generation of the planet's tenuous atmosphere and to the "space weathering" of surface materials.

The process also causes X-ray emission, which an X-ray imager can see as a brightening on Mercury's night side where the cusp of the magnetic field intersects the surface. This is the same process that makes auroras on Earth and other planets with atmospheres; at Mercury, which lacks an atmosphere, the aurora occurs at the surface.

MESSENGER's low altitude measurements revealed evidence of magnetized rocks in a part of Mercury's crust that, based on the presence of many craters from cosmic impacts, appears to be ancient. This leads scientists to believe that the magnetic field has existed for at least 3.7 to 3.9 billion years, most of Mercury' existence. It may once have been comparable in strength to Earth's field at the present time.

It has been suggested that the key to understanding the unusual magnetic field may be the type(s) and distributions of light elements in the liquid core. The distributions of these light elements may control the sources of buoyancy that drive convectional motion in the core.

The strength and other properties of the magnetic field are challenging to reconcile with current dynamo models. The offset suggests that the dynamo originates near the core-mantle boundary, rather than in the core itself.

The magnetic fields of terrestrial planets are thought to be produced by an electrically conducting, fluid outer core surrounding a solid inner core. Hence, if Mercury's field is generated by an internal dynamo, it is likely that part of the iron-rich core – probably an outer shell – is molten.

This raises the question of how the core has managed to remain molten for more than 4 billion years. There are several possibilities:

- The core is enriched in the radioactive elements thorium and uranium, which provide additional heat;
- The mantle prevents heat easily escaping to the surface;
- The presence of an element such as silicon or sulfur in the core lowers the melting point of iron.

The third of these is considered to be the most likely explanation. Calculations indicate that, with a sulfur abundance of 0.2% or less, the entire core should now be solid. At the other extreme, an abundance of 7% sulfur would enable the entire core to remain fluid at the present time. The most likely situation is somewhere between the two extremes, with a pure iron inner core and an outer core of iron and sulfur, with a temperature of around 1,000°C.

One laboratory study suggests that the convection currents in the outer core may be the result of differentiation of the molten, iron-sulfur mixture. As the liquid slowly cools, iron atoms condense into "flakes" that fall toward the planet's center. When the iron sinks and the lighter, sulfur-rich fluid rises, convection currents are created that power the dynamo and produce the magnetic field.

A Varied Surface

Until the MESSENGER mission, little was known about the surface of Mercury. Most of our knowledge was inferred from ground-based optical, thermal infrared, and radio observations. Such techniques relied on comparisons with the known properties of lunar and terrestrial rocks and soils.

Earth-based radar provided the first indications that Mercury is covered in craters and closely resembles the Moon. However, the MESSENGER mission has shown that Mercury has a very varied surface with a wide range of topography and landforms.

Each type of physical feature has its own system of nomenclature:

- impact craters are usually named for men and women who have made contributions to the arts and humanities;
- scarps (rupes) are named for ships used in exploration on Earth;
- ridges (dorsa) and valleys (valles) are named for astronomers and radio observatories, respectively;
- plains (planitia) are often named for the planet or god Mercury in various languages;
- mountains, usually the rims of large impact craters, are named for the nearest plains.

Together these features form the different physiographic provinces that reflect the geologic evolution of Mercury. Three principal types of terrain are found globally: heavily cratered terrain, intercrater plains, and smooth plains. Some distinct, regional landscapes are also found, notably those associated with the enormous impact feature called the Caloris Basin and an area of highly disrupted hills and depressions on the opposite side of the planet to Caloris.

The highest place on Mercury is 4,480 m above Mercury's average elevation (Figure 5.14). This high point is located just south of the equator in some of Mercury's oldest terrain. The lowest elevation, 5,380 m below Mercury's average, is found on the floor of Rachmaninoff basin, a double-ring impact basin that may host some of the most recent volcanic deposits on the planet.

The presence of numerous impact craters of all sizes implies an ancient surface that has been bombarded by asteroids and comets over billions of years. As on the Moon, Mercury's younger craters commonly display bright ray systems created by ejecta, and the larger craters contain central peaks. The most impressive display of linear rays radiates up to 1,000 km from the Hokusai impact crater, which is located at latitude 58°N.

Two general types of plains units are visible on Mercury: intercrater plains and smooth plains. The relative abundance of craters on the intercrater plains indicates that they are older.

The intercrater plains are widespread and, as their name suggests, they are found between the areas of heavily cratered terrain. Although they have relatively smooth or low, undulating relief, they are generally covered with many small secondary craters (mostly less than 10 km in diameter), which were created by ejecta from large impacts. However, the plains material partially infills some of the large craters and covers ejecta blankets. Many craters

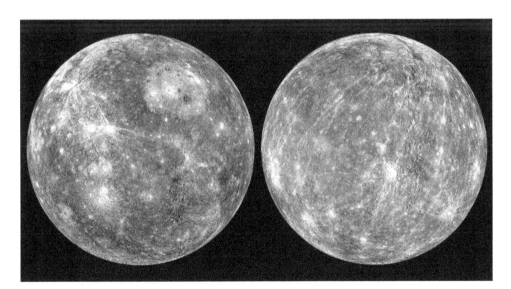

Figure 5.13 The two hemispheres of Mercury are shown in false color to reveal variations in composition and duration of exposure of materials on the surface. Linear rays of ejecta radiating from fresh impact craters appear light blue or white. Medium- and dark-blue areas are "low-reflectance material." Tan areas are plains formed by eruption of highly fluid lavas. The large circular area in the left image (top center) is the Caloris impact basin, whose interior is filled with smooth volcanic plains. Small orange spots are materials deposited by explosive volcanic eruptions. (NASA/Johns Hopkins University Applied Physics Laboratory/Carnegie Institution of Washington)

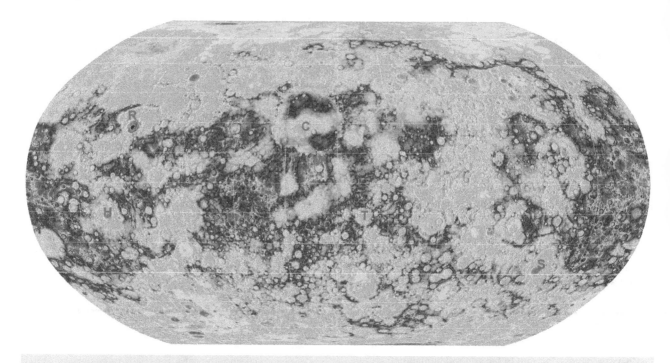

Figure 5.14   This topographic map of Mercury includes craters, volcanoes, and tectonic landforms. Higher elevations are brown, yellow, and red, while lower regions appear blue and purple. Selected features labelled are Caloris Planitia (C), Rachmaninoff (R), Holst (H), Mozart (M), Tolstoj (T), and Schubert (S). (USGS)

less than 50 km in diameter have been buried and destroyed by the intercrater plains formation.

Dating based on crater counts suggests that the intercrater plains formed between 4.0 and 4.2 billion years ago, during the period of heavy bombardment. As time went by, the amount of new intercrater plains material being created seems to have diminished.

About 30% of Mercury's surface is covered by smooth plains – comparable to the area on the Moon covered by mare basalts. (Mercury's smooth plains are very likely also basalt.) In particular, the north polar region, now called Borealis Planitia, is dominated by smooth plains, created by huge amounts of volcanic lava flooding the surface. The relative absence of craters indicates that these plains are younger than Mercury's rougher surfaces, but they are still thought to be about 3.6 billion years old.

Multispectral imaging by MESSENGER has confirmed that the surface displays three major color units: relatively high-reflectance smooth plains, average-reflectance cratered terrain, and low-reflectance material.

The widespread dark material is found in the giant Caloris and Tolstoj basins, as well as in southern, heavily cratered regions. In some places, impacts have excavated the material from depth and then thrown it beyond the newly formed crater as an ejecta blanket.

On a much smaller scale, the reflection of sunlight as viewed from different angles suggests a surface layer composed of fine, dark granular rock. Thermal infrared observations from Mariner 10 also showed that the surface is a good insulator, probably covered by a layer of fine silicate dust and sand at least tens of centimeters thick.

As is the case with the Moon, this mantle of loose material, or regolith, is thought to have been formed by the repeated pulverizing of surface rocks by the high-speed impacts of micrometeorites. However, projectiles, including microscopic dust, strike Mercury much faster than the Moon, presumably resulting in a surface that contains more glass and material condensed from impact-produced vapor.

Not everywhere is the same. Infrared observations also show areas of compacted soil, boulders or bedrock, indicating places where the regolith is much thinner or where fresh, unweathered surfaces are exposed.

Surface History

The history of Mercury seems to be very similar to that of the Moon (see Chapter 4). After the early period of planetesimal accretion, a thin crust formed over the molten planet. This initial crust may have been removed by a giant impact or vaporization in the hot solar nebula, but a thicker crust soon formed as the planet gradually cooled.

During the subsequent heavy bombardment, the planet experienced internal differentiation. Elements of comparatively low density, such as aluminum, rose into the upper crust. At the same time, iron and other heavy elements sank toward the center and formed a massive core.

The impact craters and basins which can still be seen today reflect the final phase of the late heavy bombardment, which may have ended about 3.8 billion years ago. These features were then

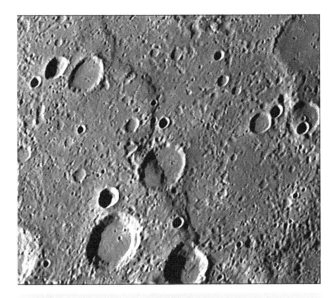

Figure 5.15 This high resolution view of Mercury's intercrater plains is centered at 3°N, 20°W. Most of the numerous small craters are secondary impact features created by ejecta from young, primary impact craters. The prominent Santa Maria Rupes scarp cuts both intercrater plains and older craters. The large crater (top right) has been flooded by intercrater plains material. (NASA-JPL)

Figure 5.16 Impact structures of different ages and sizes are visible in this Mariner 10 view of Mercury's ancient cratered terrain. A relatively young crater lies in the center of an older, larger crater. The newer crater is about 12 km across. The photo covers an area of 130 x 170 km. (NASA-JPL)

partly covered by eruptions of magmas of varying composition over an extended period of time.

The record of gigantic impact events is clearly visible today. One of the earliest examples is the 500-km diameter Tolstoj Basin in the southern hemisphere. This was followed by the most dramatic impact of all, which formed the Caloris Basin. Shortly after the Caloris impact, massive extrusions of flood lavas formed broad, smooth plains.

Thereafter, the number of impacts decreased rapidly. During the last 3 billion years, Mercury's surface has continued to be modified by volcanic eruptions and occasional impacts. The current low rate of cratering continues to produce a regolith of dust grains of different sizes that covers all surface features. Meanwhile, a slight shrinkage of the planet caused by its gradual cooling has resulted in the formation of lobate scarps and wrinkle ridges.

Impact Craters

As is the case with most solid bodies in the Solar System, impact craters are seen almost everywhere on Mercury – evidence of a bombardment by meteoroids and comets over billions of years.

The preservation of this impact record is not surprising. Mercury has no plate tectonics or atmosphere to erase the ancient landscapes. It has also been subjected to a more severe pounding than Earth or Mars. This is because the Sun's gravity accelerates incoming projectiles, causing them to impact the surface at higher velocities than on more remote planets.

On average, asteroids strike Mercury at a velocity of about 34 km/s, compared with 22 km/s on the Moon and 19 km/s on Mars. Comets on parabolic paths arrive at even faster velocities – typically 87 km/s, compared with 52 km/s for the Earth–Moon and 42 km/s on Mars.

Collisions with comets arriving from the outer Solar System should also be more frequent, partly because they often reach perihelion close to Mercury's orbit, and partly because Mercury's distance from the asteroid belt reduces the likelihood of an asteroid strike. Hence, about 40% of the craters on Mercury are likely to be created by comet impacts, compared with about 10% for the Earth–Moon and less than 3% for Mars.

Higher velocity impacts release more energy, so incoming objects of similar size will excavate larger craters on Mercury than on the Moon. The morphology of fresh craters and their ejecta blankets is also influenced by gravity. Mercury's surface gravity is 2.3 times greater than that of the Moon, and this permits ejecta to travel only 65% of the distance it would reach on the Moon.

Studies of similarly sized craters on the two worlds show that the amount of ejected material is roughly the same, but Mercury's stronger gravity causes a higher concentration of secondary craters and a thicker ejecta blanket close to the main craters. These factors also reduce the amount of modification of the surrounding terrain by ejecta that cover or degrade adjacent craters.

Crater counts on different terrains show remarkable similarities to lunar cratering. Although it is not certain that the same number of craters was formed at the same rate on both Mercury and the Moon, their histories have probably not been dramatically different.

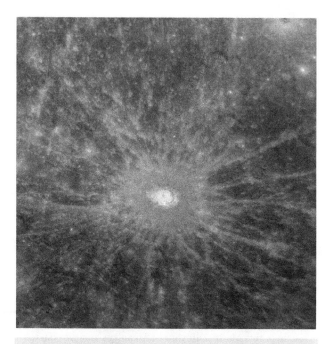

Figure 5.17   85 km diameter Debussy is a relatively young impact crater with a central peak and rays that extend for hundreds of kilometers. It was first identified by ground-based radar. (NASA/Johns Hopkins University Applied Physics Laboratory/Carnegie Institution of Washington)

Studies of lunar cratering suggest that the heavily cratered terrain on Mercury retains the imprint of the late heavy bombardment which ended about 3.8 billion years ago. This terrain is so crowded with closely grouped and overlapping craters more than 30 km in diameter that ejecta blankets and individual fields of secondary craters are difficult to detect.

Many of the craters appear degraded, with smoother walls, more rounded rims, and small craters on their floors. Their appearance probably reflects their greater age, the effectiveness of erosion, and burial by ejecta from adjacent craters. In some cases, lava infilling also seems to have occurred, with breached walls and smooth, interior plains. The youngest craters, such as Kuiper, display a sharp rim and prominent ejecta deposits, sometimes with bright rays that spread large distances from the impact site (see Chapter 4).

Some craters (e.g. Berkel and Basho) contain dark material in the center and in a ring immediately surrounding them, as well as rays of ejecta. Other craters, such as Debussy, have the rays but lack dark halos.

Mercury, the Moon, and Mars all have similar crater distributions, the main exception being the number of craters less than 50 km in diameter. The marked deficit of such craters on Mercury is due to burial by intercrater plains material during the period of heavy bombardment or flooding by lava.

Below a diameter of about 20 km, the abundance of craters increases sharply. The crater population superimposed on the smooth plains within and surrounding the Caloris Basin shows the same size/frequency distribution as the lunar highlands and

the ridged plains of Mars over the same diameter range, but the crater density is much lower. On the other hand, the crater density on Mercury's smooth plains is much greater than on the lunar maria.

Impact Basins

Mercury's ancient surface shows evidence of enormous collisions with asteroids and comets in the distant past. The largest impact feature is the Caloris Basin, about 1,525 km in diameter (see Caloris Basin), followed by Rembrandt (715 km), Beethoven (625 km), and Tolstoj (510 km) (Figure 5.18). Many other heavily degraded and buried basins have been proposed from inferred remnants of basin rings.

The data from MESSENGER indicate that volcanic activity and surface tectonics have modified many impact basins on the planet. One of the most unusual examples is Rembrandt (33°S, 88°E), which may have formed about 4 billion years ago, making it one of the youngest large impact basins on Mercury – comparable in age to Caloris.

Smooth plains in the center and near the northern rim are evidence of partial flooding by lava. It also displays a unique wheel-and-spoke-like pattern of radial and concentric wrinkle ridges and graben. The wrinkle ridges are crosscut by the graben, suggesting that deformation due to contraction occurred before

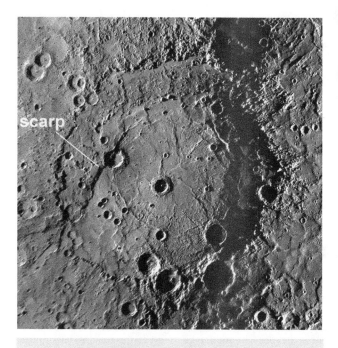

Figure 5.18   The multi-ringed Rembrandt basin has a diameter of about 715 km and may be the youngest large impact basin on Mercury. High cliffs and small craters are visible inside the basin. A 1,000-km long lobate scarp cuts through one of the large internal craters (left). (NASA/Johns Hopkins University Applied Physics Laboratory/Carnegie Institution of Washington)

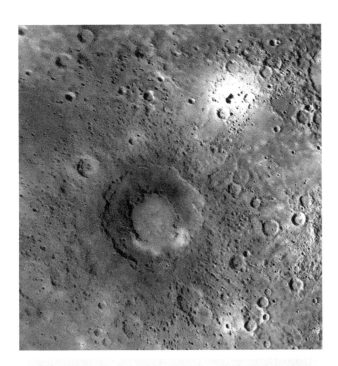

Figure 5.19 An enhanced color MESSENGER view of the 290-km wide Rachmaninoff double-ring impact basin. The bright yellow area (top right) is centered on a rimless depression that may have been an explosive volcanic vent. The basin's smooth interior may be the result of flooding by lava. Smooth plains, thought to be a result of earlier episodes of volcanic activity, also cover much of the surrounding area. (NASA/Johns Hopkins University Applied Physics Laboratory/Carnegie Institution of Washington)

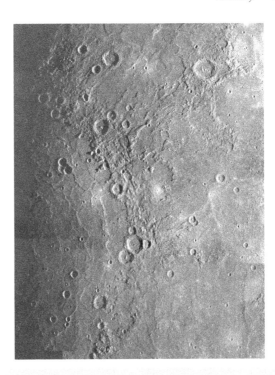

Figure 5.20 Although the Caloris Basin (left) was partly hidden by shadow during the Mariner 10 flybys, the spacecraft's cameras revealed an outer ring of smooth, rounded block mountains rising 1–2 km above the surrounding terrain, and an interior crossed by ridges and fractures. Smaller craters were also visible in the basin. (NASA-JPL)

the smooth plains were stretched and fractured due to crustal extensional deformation. The youngest deformation of the basin involved the formation of lobate scarps during the global cooling and contraction of Mercury.

Some parts of the crater floor seem to contain unusually high amounts of iron and titanium. These elements indicate that some exposed materials have not been covered by more recent lava flows, and so might date from a very early epoch in Mercury's formation.

Not all large, ancient basins have flat floors. Beethoven (20°S, 124°W) shows a 1-km deep trough near the subdued rim, then a broad topographic rise towards the interior and a depression in the centre. One interpretation is that the basin was flooded by lava, then underwent differential subsidence in the centre and near the margins.

Dozens of multiple ring impact basins larger than 200 km across (e.g. Bach) have also been recognized – far more than on the Moon. Once again, they are often associated with evidence of various types of volcanic activity.

One of the youngest examples is Rachmaninoff, a remarkably well-preserved double-ring basin (Figure 5.19). The low numbers of superposed impact craters and marked differences in color across the basin suggest that the smooth area within the innermost ring may be the site of some of the most recent extrusive volcanism

on Mercury. There is also evidence of an explosive volcanic vent beyond the outer ring (see Figure 5.19).

Caloris Basin

The best preserved impact basin on Mercury, and one of the youngest, is the Caloris Basin (30.5°N, 170°E). With a diameter of 1,525 km, it is also one of the largest impact basins in the Solar System.

Its rim is marked by a ring of smoothly rounded mountain blocks, up to 2 km high and 50 km square, bounded on the interior by a relatively steep escarpment (Figure 5.20). The ring of mountains is almost continuous, with the exception of a 300-km wide gap. Smooth plains appear to intrude into this gap from outside the basin.

Some valleys – possibly fault-bounded depressions (graben) – within the mountain chain also seem to be filled with plains material. A second, much smaller and less continuous escarpment occurs about 100 to 150 km farther out from the centre of the basin. This may be evidence of block faulting associated with basin deformation.

The interior of Caloris is occupied by smooth plains that show prominent wrinkle ridges and fractures, evidence of forces that deformed Mercury's crust long ago. The wrinkle ridges are arranged in a roughly polygonal pattern and probably formed as

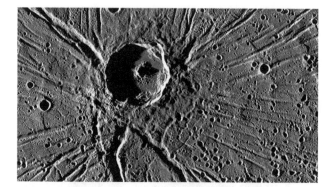

Figure 5.21 The 40-km wide Apollodorus crater near the center of Caloris Basin is surrounded by some 230 radial fractures and rift features. The crater, which is not in the exact center of the radial pattern, may have formed later and is probably not linked with the formation of the faults. (NASA/Johns Hopkins University Applied Physics Laboratory/Carnegie Institution of Washington)

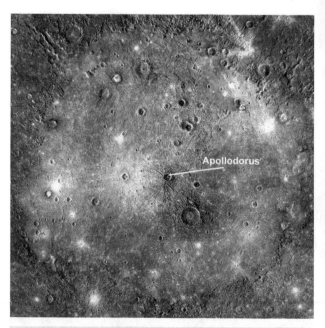

Figure 5.22 This enhanced color mosaic shows that Caloris Basin has been flooded by lavas (orange). Post-flooding craters have excavated material from beneath the surface. The larger craters have exposed low-reflectance material (blue) from beneath the surface lavas, likely exposing some of the original basin floor material. Analysis of these craters suggests the thickness of the volcanic layer is 2.5–3.5 km. (NASA/Johns Hopkins University Applied Physics Laboratory/Carnegie Institution of Washington)

the result of horizontal compression when the basin floor subsided under the weight of inflowing lava. The largest examples are a few hundred kilometers long, 1.5–12 km wide, and 500–700 m high. Although they bear some similarity to ridges found on the lunar maria, their radial alignment and large number makes them quite distinct.

The fractures in the center of the basin cross the wrinkle ridges at various angles, clear evidence that they formed at a later date, presumably when uplift created tensional forces in the crust. This uplift may have been the result of magma migrating towards the surface. However, they must be ancient features, since they do not deform the impact craters which they cross.

A unique series of some 230 radial fractures in the central basin is known as Pantheon Fossae. The largest of them are flat-floored grabens, about 700 m deep and up to 8 km across. The fault structures tend to increase in width towards the centre of Pantheon Fossae. Near the center of this spider-like feature is the 40-km diameter crater Apollodorus.

Pantheon Fossae is a rare example of crustal extension, rather than compression, on Mercury. It seems to have been created by fairly local upward doming of the crust, either caused by dykes radiating from subsurface magma chambers or upwelling of the underlying mantle and crust after the creation of Caloris.

Other fractures that occur near the outer edge of the basin are roughly concentric. This rare combination of fractures is only seen in large impact basins on Mercury. They probably formed when the floor of the basin was uplifted, causing horizontal stretching and breaking apart of the material that filled the basin. Also visible are many smaller craters that have been excavated more recently.

Besides lava flows, Caloris contains spots with diffuse boundaries which appear to be volcanic in origin. Sometimes the spots are centered on rimless depressions, possibly indicating that they were caused by explosive, pyroclastic eruptions.

Beyond the basin rim is so-called lineated terrain, characterized by long, hilly ridges separated by radial grooves that cut through pre-Caloris craters. This type of terrain can be traced for about 1,000 km from the rim. It bears a close resemblance to the sculptured terrain around the lunar Imbrium basin, and may represent fault troughs or chains of coalescing secondary craters formed by ejecta produced during the huge impact that created the basin.

Also visible beyond the main escarpment are several areas of smooth and hummocky plains. The hummocky terrain comprises low hills, some close together, others scattered, and less than 1 km across. They may represent basin ejecta composed of solid fragments and impact melt. The smooth plains are probably ancient lava flows (see Volcanism and Hollows).

The enormous impact that created Caloris seems to have modified the surface on the antipodal (opposite) side of the planet, with the result that the rims of all the old craters in this region have been broken up into a jumbled landscape of hills and depressions.

Termed hilly and lineated terrain, it is similar to formations antipodal to the Imbrium and Orientale basins on the Moon. These have been attributed to the transport and deposition of material ejected by impacts on the opposite side of the satellite. Ejecta from an impact travels outward in all directions, on roughly ballistic trajectories. If the impact has enough energy, the material may fly so far around the planet that the ballistic trajectories converge again, leaving a combined deposit.

Furthermore, studies suggest that the seismic waves from the huge impact were focused as they passed through the planet's

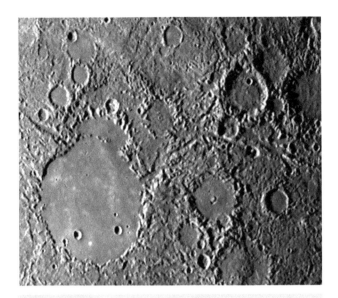

Figure 5.23 This Mariner 10 image shows part of the hilly and lineated terrain that was created on the opposite side of Mercury in the aftermath of the Caloris impact. Hills, depressions and valleys disrupt pre-existing landforms. The floor of the crater (left) was flooded by lava after the impact. (NASA-JPL)

large core, resulting in a sudden deformation of the rocks exactly opposite the Caloris Basin. The result was a 500-km wide landscape marked by roughly polygonal hills 5–10 km across and up to 1.5 km high. The localized nature of these features and the lack of similar damage to younger craters indicate a one-off, catastrophic event, in which the ground underwent vertical motions of more than 1 km in matter of minutes.

Volcanism and Hollows

Relatively flat or smoothly undulating plains cover large parts of the surfaces of the terrestrial planets. In most cases, they are the result of resurfacing, the covering or destruction of an earlier, rough topography.

This resurfacing may take place in at least three different ways:

- An increase in temperature may reduce the strength of the crust and its ability to retain high relief. Over millions of years, the mountains sink and the craters rise.
- The movement of material down slopes and along valleys leads to infilling of low areas. This may be through volcanic lava flows, or erosion and deposition of fine water-borne sediments.
- Fragmented material is deposited on the surface from above, eventually covering the rough surface, e.g. blanketing by impact crater ejecta or volcanic ash.

All three of these processes may have modified Mercury at different times, particularly during its early history.

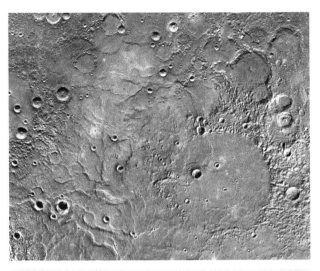

Figure 5.24 Mercury's northern volcanic plains, shown in enhanced color to emphasize different types of rocks. The 291-km diameter Mendelssohn impact basin (bottom right) was nearly filled with lava. Toward the bottom left, large wrinkle ridges, formed during lava cooling, are visible. Also in this region, the circular rims of "ghost" impact craters buried by the lava can be seen. Near the top of the image, the bright orange region shows the location of a volcanic vent, the source of one of the largest pyroclastic deposits on the planet. The north pole is toward the bottom left corner. (NASA/Johns Hopkins University Applied Physics Laboratory/Carnegie Institution of Washington)

Volcanic flooding seems to have been significant over a wide area. This resurfacing by volcanic activity implies that large sources of magma once existed in Mercury's upper mantle.

The lava flows were deep enough to bury craters, leaving only traces of their rims visible. These are called ghost craters. Wrinkle ridges and small troughs within the ghost craters formed as a result of the cooling of the lava. Some flooded impact features host tectonic structures – graben, ridges, and scarps – that formed during or after volcanic infilling.

The smooth plains cover about 27% of the surface. They are found all over the planet, but are more heavily concentrated in northern latitudes and in the hemisphere surrounding Caloris. They have a lower density of impact craters and typically fill low-lying areas such as impact craters and basins (Figures 5.24 and 5.25). Individual deposits range from hundreds of square kilometers to 1.7 million sq. km (in the interior of the Caloris Basin), rivaling the sizes of the largest flood basalt units on Earth or the Moon.

The majority of smooth plains (at least 65%) are probably volcanic in origin, although impact melt and basin ejecta may account for some of them. The relative absence of craters indicates that they are among the youngest features on Mercury.

In most respects, the smooth plains resemble the lunar maria, with abundant ridges and scarps. The main differences are that the Mercurian terrain has a similar albedo and color to its surroundings, and it displays more impact craters.

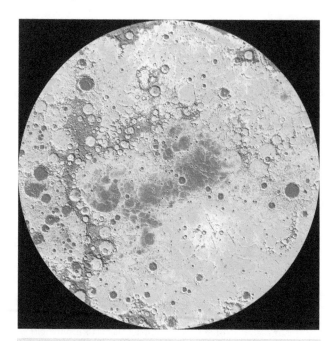

Figure 5.25 A topographic map of Mercury's northern hemisphere, showing the smooth volcanic plains. The lowest regions are shown in purple and blue, and the highest regions are red. The difference in elevation between the lowest and highest regions is roughly 10 km. The north pole is in the center. Low-lying craters near the pole host radar-bright materials, thought to be water ice. (NASA/Johns Hopkins University Applied Physics Laboratory/Carnegie Institution of Washington)

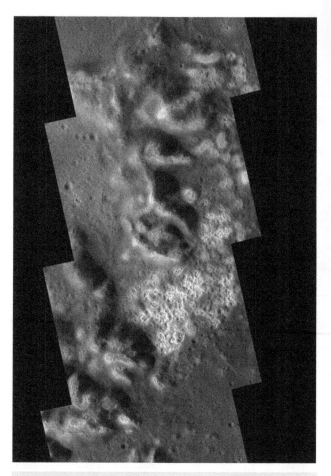

Figure 5.26 A false color view of dense clusters of hollows, shallow irregular depressions on the floor, and marginal mountains of Raditladi impact basin. These unique landforms may be created by the escape of trapped subsurface gases. (NASA/ Johns Hopkins University Applied Physics Laboratory/Carnegie Institution of Washington)

Volcanism seems to have occurred long after the major effusions of fluid lava ended. There are more than 150 vents surrounded by bright deposits of pyroclastic material – the result of explosive volcanism.

Such volcanism is driven by volatiles (substances that readily change from solid to vapor at low temperatures) in the magma. Like froth escaping from an opened champagne bottle, the released gases shred the magma, spraying out fragments that fall to the ground as pyroclastics. Some deposits can cover large areas, notable the pyroclastic region northeast of Rachmaninoff basin, which extends up to 130 km from its vent (see Figure 5.19).

MESSENGER's images also, unexpectedly, revealed numerous small, irregular depressions known as "hollows." Clusters of hollows, some hundreds of kilometers across, occur all over Mercury, but more commonly in impact craters. The flat-floored pits, about 50 m deep and ranging in diameter from tens of meters to several kilometers, suggest that the surface has somehow been eroded. Their fresh appearance, bright interiors and haloes, and sharp edges suggest that they may even be forming today.

The clusters are thought to be caused by the loss of volatile material from the planet's surface, rather than volcanic eruptions.

This seems to be confirmed by their distribution: hollows are most common in areas that receive more solar heating. This suggests that the volatiles are escaping from places where dark subsurface material (low-reflectance material) has been exposed and then strongly heated by the Sun.[3]

Lobate Scarps and Global Contraction

All of the major terrains on Mercury exhibit lobate scarps, long, sinuous cliffs with rounded fronts that are unique to the planet. They range in length from about 20 km to more than 500 km, often reaching heights of several kilometers (Figures 5.27 and 5.28).

Images of craters and other landforms sliced through by these scarps show surface lateral displacement of up to 10 km, clear evidence that they are faults or major fractures in the crust. The

---

[3] Similar depressions have been found in the carbon dioxide ice cap at the Martian south pole, giving it a "swiss cheese" appearance. However, the hollows on Mercury are found in rock and display bright floors and haloes.

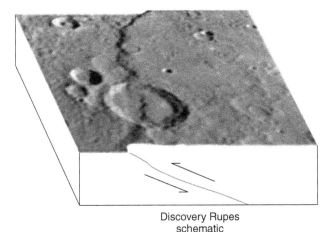

Discovery Rupes
schematic

Figure 5.27 With a length of about 550 km and a height of 1.6 km, Discovery Rupes is one of the largest lobate scarps on Mercury. This steep, sinuous feature was created by a thrust fault which disrupted an older impact crater. (PSRD, NASA)

appearance of landforms such as Discovery Rupes and Hero Rupes suggests that they are created by thrust faults, where one block of crust is pushed up over another (Figure 5.27).

The overthrust block, unable to support its weight, then collapses. The surface outline seems to meander because the fault plane slopes at a low angle and reaches the surface at different places according to elevation.

How did these huge fractures form? Their global distribution and more-or-less random orientation indicates that they are the result of a major episode of contraction and compression that encompassed the entire planet.

Studies of surface features indicate that the crust was fractured after the formation of the large, degraded craters of the highlands and the intercrater plains. Only the younger craters are seen to disrupt the scarps. It seems that the cliffs date from after the period of late heavy bombardment, some 3.8 billion years ago.

Calculations based on the overall amount of crustal shortening associated with these faults and wrinkle ridges indicate that the surface area of Mercury shrank by about 0.3%, as its diameter decreased by up to 7 km. The cause of such a unique shrinkage was probably thermal contraction due to cooling of the planet's interior, particularly its huge, liquid iron core.

There is also some evidence that the underlying crust may have been fractured by despinning of the planet. If Mercury once rotated much faster, then was subjected to tidal forces which slowed it until it was trapped in a 3:2 spin-orbit resonance (see Spin-Orbit Resonance), the planet's shape would have changed dramatically from an oblate spheroid (like a slightly squashed ball) to a sphere. As the equatorial diameter decreased and the polar regions became less flattened, stresses in the crust would have created a series of compressional and tensional stress fractures.

Curiously, such a global fracture pattern is not seen, with the possible exception of some linear ridges that trend northeast and northwest. It seems that any fractures caused by despinning occurred very early in Mercury's history, before the formation of the intercrater plains.

Figure 5.28 A color-coded image of the central part of Carnegie Rupes, where the giant lobate scarp cuts through Duccio crater. Red shows higher terrain, blue lower terrain. The scarp, which is clearly younger than the crater, is nearly 2 km high in places. (NASA/Johns Hopkins University Applied Physics Laboratory/Carnegie Institution of Washington)

MESSENGER discovered much smaller thrust faults which are only tens of meters high and a few kilometers in length. These are comparable in scale to small, young scarps on the Moon which have also survived bombardment by comets and meteoroids. Their small-scale, pristine appearance, crosscutting of impact craters, and association with small graben all indicate an age of less than 50 million years.

The young geological age of the small scarps, along with evidence for recent activity on large-scale scarps, imply that Mercury's interior continues to cool even today. Like Earth, Mercury may still be tectonically active.

Polar Ice?

As on the Moon, the low axial inclination of Mercury, combined with rugged polar relief, result in some large craters having permanently shaded floors (see Chapter 4). These near-polar craters have been associated with radar images obtained by ground-based radio telescopes that show small, highly reflective regions inside the craters.

Analysis showed that the echoes were very similar to those obtained from some icy satellites of the outer planets and from the residual water ice cap of Mars. If the strong reflection is due to water ice, it must be fairly uncontaminated and perhaps 2–20 m thick. However, the data indicate that it may be covered by a thin layer of dust or soil, or else does not completely cover the crater floors.

High-resolution (1.5 km) radar images of the north polar region have been able to pin down the highly reflective surfaces to localized areas that (with a few exceptions) coincide with fresh, deep craters, some less than 10 km across (see Figure 5.29). The donut shape of the deposits is explained by the presence of central

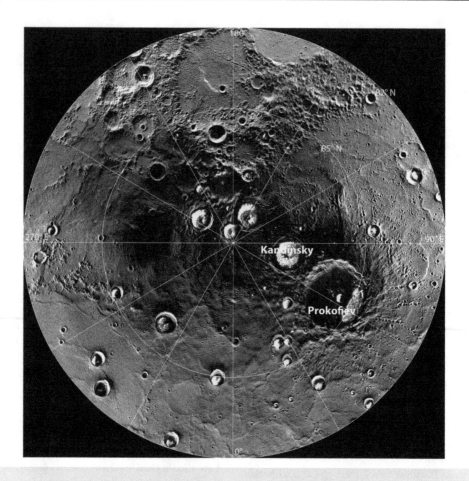

Figure 5.29   A high-resolution radar image of Mercury's north polar region, superimposed on MESSENGER images, shows radar-bright features associated with deep, shaded craters. The radar-bright materials likely contain water ice. All of the larger polar deposits are located on the floors or walls of impact craters. Deposits farther from the pole are concentrated on the north-facing sides of craters. Most of the craters linked with highly reflective material are located at latitudes above 85°N, but some are located down to 72°N. This suggests that a thin layer of regolith may cover and protect water ice deposits. (NASA/Johns Hopkins University Applied Physics Laboratory/Carnegie Institution of Washington/National Astronomy and Ionosphere Center, Arecibo Observatory)

peaks in the craters, while bright arcs coincide with their shaded rims.

The neutron spectrometer on the MESSENGER spacecraft provided further support for the water ice theory by identifying large concentrations of hydrogen at Mercury's north pole. These measurements indicate that ice is exposed at the surface in the coldest of those deposits. Where temperatures are slightly too warm for ice to be stable at the surface, most of the deposits are buried beneath a dark, insulating material – possibly a mix of complex organic compounds delivered to Mercury by the impacts of comets and volatile-rich asteroids.

How can the presence of ice be explained on a planet that is roasted by the nearby Sun? In fact, temperature is not a problem, since the interiors of large, permanently shaded craters remain perpetually cold, below –212°C.

Mercury's surface is extremely dry, but water could be supplied by an occasional comet or meteorite impact, or through outgassing from the interior. Some of this water vapor would be lost through a variety of processes, but a significant proportion would condense on the surface at night. Eventually, some of the water vapor might reach a dark haven in a deep polar crater. There, it could remain for billions of years (Figure 5.30).

Although the nature of the polarization of radar waves and the hydrogen signature strongly suggest the presence of water ice, other possibilities cannot be completely ruled out. Some of the apparent ice deposits are in craters too small and too far from the pole to be in permanent shadow. Thermal models suggest that 10-km diameter craters should not be capable of supporting even insulated ice. Another challenge involves the discovery of several "ice" features at latitudes between 71–75°N.

An alternative suggestion is that elemental sulfur could be the cause. Outgassing of sulfur and volatilization of sulfide minerals by meteorite impacts could release sulfur into the atmosphere. Over time, this could interact with the surface and migrate to permanently shaded polar regions. Estimates suggest that such a process could deposit some 35 m of sulfur over a period of one billion years. On the other hand, although sulfur is stable at a higher temperature than water ice, there are no high backscatter

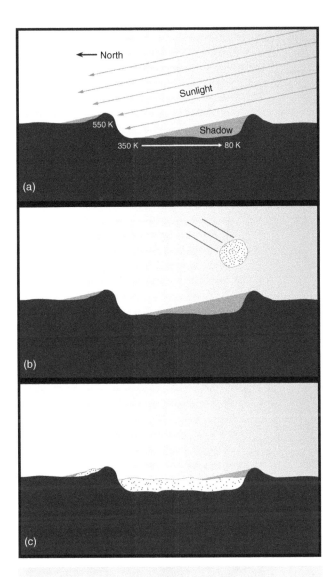

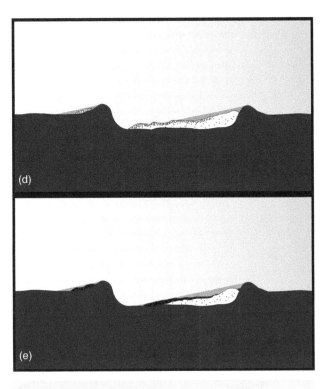

Figure 5.30 (*Continued*)

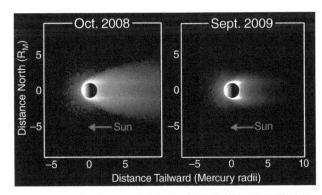

Figure 5.30 Diagrams illustrating how Mercury's polar ice deposits may form. (a) A high-latitude impact crater illuminated by the angled rays of the Sun creates a region of very warm temperatures on the illuminated rim, lower temperatures on the sunlit floor of the crater, and extremely cold temperatures in areas of permanent shadow. (b) A comet or water-rich asteroid that also contains organic compounds impacts Mercury. (c) The water and organic compounds are spread over a wide region, and a small fraction of both compounds migrates to the poles where they can become cold-trapped as ices. (d) Over time, the water ice in the warmer regions vaporizes, leaving behind the more stable organic impurities at the surface. (e) The ice retreats further to a stable, long-term configuration. In the coldest areas, water ice remains on the surface. In the warmer areas, the ice is covered by an ice-free surface layer that is rich in organic impurities that have been darkened by exposure to Mercury's space environment. (NASA/UCLA/Johns Hopkins University Applied Physics Laboratory/Carnegie Institution of Washington)

Figure 5.31 Comparisons of Mercury's neutral sodium tail, based on data obtained during flybys by MESSENGER on October 6, 2008, and September 29, 2009. Distinct north and south enhancements in the emission result from material being sputtered from the surface at high latitudes on the dayside. The sodium abundance in Mercury's exosphere and tail was much lower in September 2009, due to reduced solar radiation pressure. The tail extends anti-sunward from the planet. It is composed of sodium and other atoms sputtered off Mercury's surface.(NASA/Johns Hopkins University Applied Physics Laboratory/Carnegie Institution of Washington)

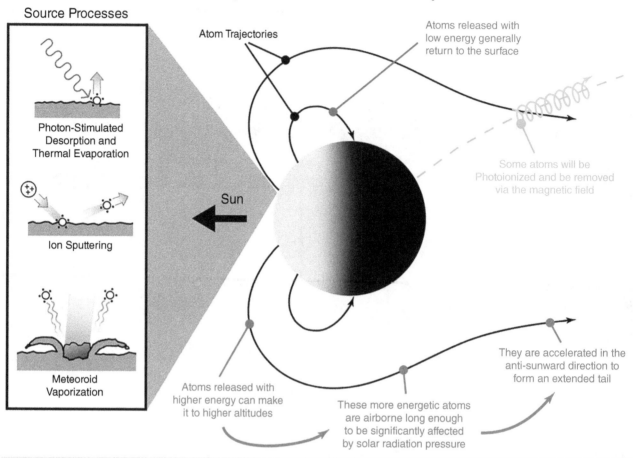

Mercury's Surface-Bounded Exosphere

Figure 5.32 The major processes that generate and maintain Mercury's exosphere. The panels (left) summarize the three primary sources. Photon-stimulated desorption occurs when sunlight excites atoms or molecules on the surface, releasing them to the exosphere. Solar heating causes atoms and molecules to evaporate. During both of these low-energy processes, most of the released material reaches only low altitudes and usually returns to the surface. Ion sputtering occurs when charged particles from the solar wind or Mercury's magnetosphere impact the surface, "knocking off" atoms and molecules. Vaporization occurs when incoming meteoroids, generally small dust particles, impact the surface at high speeds. Both ion sputtering and meteoroid vaporization are high-energy processes, and the released material can reach high altitudes. All material in the exosphere is accelerated away from the Sun by radiation pressure; atoms and molecules at high altitudes escape the planet's gravitational influence and enter Mercury's neutral tail. Some of the ions are returned to the surface by Mercury's magnetosphere. (NASA-JHU-APL)

radar signatures from other regions where temperatures are within the stability range of sulfur.

Atmosphere

Ground-based searches for a Mercurian atmosphere were unsuccessful during the 1960s and 1970s. This was not too surprising, since Mercury's weak gravity could not be expected to hold onto any atmosphere generated early in its history.

However, ultraviolet observations by Mariner 10 revealed that there is a very thin atmosphere. (The proper term is exosphere, since the few gaseous atoms on Mercury rarely collide, but bounce from place to place on the surface. In an atmosphere, the gaseous atoms are in constant collision.) The surface pressure is a mere 10^{-12} bar or ~1/1,000,000,000,000 of the sea-level pressure on Earth. However, the planet's highly elliptical orbit and proximity to the Sun lead to marked seasonal variations in exospheric density.

Gases detected include atomic hydrogen, helium, oxygen, sodium, magnesium, potassium, and calcium. All of the species were identified by searching for faint spectral features as sunlight reflected from the surface is scattered by the atoms. There may also be trace amounts of argon, carbon dioxide, water vapor, nitrogen, xenon, krypton, and neon.

High daytime surface temperatures cause atoms to travel so quickly that they exceed the escape velocity of 4.2 km/s and escape into space (Figures 5.31 and 5.32). Solar ultraviolet radiation also ionizes some of the atoms, enabling them to be carried away by the magnetic field in the solar wind.

Box 5.2   Missions to Mercury

Many of the mysteries involving Mercury should be answered in the next decade, after data from three orbital spacecraft are analyzed. The first of these, NASA's MESSENGER (MErcury Surface, Space ENvironment, GEochemistry and Ranging) spacecraft, entered a near-polar orbit on March 17, 2011, and completed its mission on April 30, 2015, with an impact on the planet's surface. Two more orbiters are planned in the European-Japanese BepiColombo mission, which was launched on October 20, 2018.

During a 7.9-billion km journey that included 15 circuits of the Sun, MESSENGER received gravity assists from Earth, Venus, and Mercury itself before entering orbit. Three Mercury flybys in January 2008, October 2008, and September 2009 enabled the spacecraft to match the planet's speed and location prior to the orbit insertion maneuver. During these encounters, the spacecraft mapped almost the entire planet and gathered data critical to planning the year-long orbit phase (equivalent to four Mercury years, or two Mercury solar days).

MESSENGER carried seven instruments to examine the planet's surface, atmosphere, and near-space environment. The multispectral imaging system has wide- and narrow-angle CCD imagers to map the surface. Nearly all of the planet was imaged in stereo to determine topographic variations and landforms.

Its X-ray, gamma-ray, and visible-infrared spectrometers studied the elemental and mineral composition of the rocks on the surface. Just as importantly, they also showed which elements and minerals are conspicuously absent.

The neutron spectrometer mapped variations in the neutrons that Mercury's surface emits when struck by cosmic rays. This was used to estimate the amount of hydrogen – possibly locked up in water molecules – and other elements.

A laser altimeter mapped Mercury's landforms and other surface characteristics using an infrared laser transmitter and a receiver that measures the round-trip time of individual laser pulses. The altimeter could view the planet from up to 1,000 km away with an accuracy of 30 cm. The data were also used to track the planet's slight forced libration – a slight wobble about its spin axis – that provides information about the size of its core and whether the core is partially liquid. Altimeter data, when compared with gravity field measurements gathered by tracking MESSENGER's orbital motion, helped to determine the thickness and structure of the crust and map the planet's gravitational field.

The magnetic field was studied by a magnetometer that was mounted on a 3.6 m boom to keep it away from the spacecraft's own magnetic field. The sensor determined the field's strength and how it varied with position, altitude, and time.

The composition of Mercury's tenuous atmosphere is measured by ultraviolet and energetic particle spectrometers. By comparing these data with X-ray and gamma-ray measurements of the surface rocks, scientists learn more about the origin of each element in the exosphere.

During the primary science mission, MESSENGER operated in a highly elliptical orbit around Mercury, 200 km above the surface at the closest point (periapse) and 15,193 km at the farthest (apoapse). The plane of the orbit was inclined 82.5° to Mercury's equator, and the low point in the orbit occurred at 60°N. MESSENGER orbited Mercury twice every 24 hours; for one-third of this period it was oriented to send data to Earth. The instruments were focused on global mapping during the first Mercurian solar day, with an emphasis on targeted science investigations during the second.

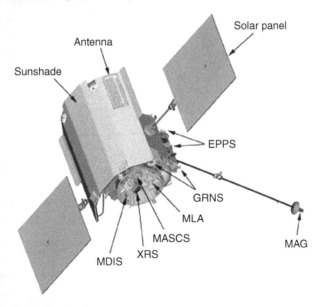

Figure 5.33   MESSENGER carried seven scientific instruments, including cameras that mapped the entire surface, a laser altimeter to gather topographic information, a magnetometer, and several spectrometers. In order to survive where solar radiation is up to 11 times more intense than on Earth, the spacecraft carried a sunshield. (NASA)

The mission was extended twice, allowing the spacecraft to make observations from extremely low altitude, capturing images and other data about the planet in unprecedented detail. During its final operational phase, it operated only 5 to 35 km above the surface.

BepiColombo is a joint European-Japanese mission. Each partner has provided a separate Mercury orbiter. A transfer module carries the two spacecraft to the planet, using a combination of solar power, electric propulsion, and nine gravity-assist flybys of Earth, Venus, and Mercury. When it arrives at Mercury in late 2025, the transfer module will be discarded and the two scientific satellites will make their separate ways to their operational orbits.

Figure 5.34   The BepiColombo mission comprises a European polar orbiter (right) to study Mercury's surface and interior, and Japan's spinning MIO polar orbiter to study the planet's magnetosphere. (ESA/ATG medialab; NASA/JHU-APL/CIW)

The European Space Agency and its member states have built the Mercury Planetary Orbiter, which carries a suite of 11 experiments and instruments to study the planet's surface and internal composition. It will be placed in a 480×1,500 km polar orbit around Mercury.

Japan's contribution is the Mercury Magnetospheric Orbiter, now named MIO. As its name suggests, the spinning spacecraft will concentrate on the planet's magnetic field and magnetosphere. To achieve this, it will fly in a higher, more elliptical polar orbit of 590×11,640 km.

In addition, Mercury's intrinsic magnetic field interacts with the interplanetary medium to modify the spatial distribution and density of solar wind plasma and high-energy charged particles that impact and sputter materials from the surface. This interaction leads to changes in exospheric densities that vary seasonally with Mercury's distance from the Sun, as well as on time scales as short as a few hours. Impacts also supply both meteoritic and volatilized surface materials to the exosphere.

This depletion of Mercury's atmosphere is counterbalanced by constant replenishment. The solar wind acts as a source of ions and atoms which are captured by the planet. Atomic hydrogen and helium come, at least partly, from the stream of hot, ionized gas emitted by the Sun. Some of the hydrogen and oxygen may also come from collisions with comets and meteorites.

Several different processes also liberate material from the surface and are responsible for supplying the exosphere: evaporation of elements due to strong ultraviolet light, meteoroid impacts, excitation of atoms on rock surfaces due to solar radiation, sputtering by impacts with solar wind ions, or emission of gases from the planet's interior.

Questions

- Why was little known about Mercury until the arrival of Mariner 10 in 1974?
- Mercury is sometimes called "the iron planet." Explain why this is so. How might this unusual composition have come about?
- Mercury and the Moon are often said to be near twins. To what extent is this true? (Make a list of the similarities and differences between the two objects.)
- Why does Mercury's surface have hot spots?
- What evidence is there to suggest that plate tectonics do not take place on Mercury?
- Mercury seems to have contracted slowly over billions of years. Why?
- Valleys are relatively rare on Mercury. Explain how those that have been imaged may have formed.
- Much of Mercury's surface is covered by volcanic material. Where are the volcanic regions? What is the evidence for such volcanism?
- What evidence is there that water ice may exist in polar craters? How might such deposits have been formed?
- Mercury does not have an atmosphere, only an exosphere. What does this mean? Why is this the case?
- Why were scientists surprised to find that Mercury has a significant magnetic field? How can such a magnetic field be accounted for in such a small planet?

SIX

Venus

Familiar to humans since prehistoric times, Venus is the brightest object in the night sky after the Moon. Unsurprisingly, the "morning star" or "evening star" played an important role in the astrological and religious beliefs of many ancient civilizations. It is named after the Roman goddess of beauty, love, laughter, and marriage. For centuries after the invention of the telescope, the cloud-shrouded planet was regarded as Earth's twin, a misconception which only ended after visiting spacecraft revealed its true, extremely inhospitable, nature.

Orbit and Size

Venus follows the most circular orbit of any planet around the Sun, at an average distance of only 108 million km – closer than all of the planets except Mercury. Since it is located inside Earth's orbit, it is never too far from the Sun in the sky. Whether it appears to the west of the rising Sun in the morning, or to the east of the setting Sun in the evening, Venus is a wonderful sight to behold. At its brightest, the planet reaches a dazzling magnitude –4.5, bright enough to cast a faint shadow on a clear moonless night. Venus is also observable during daylight under certain conditions.

Since there is little variation in its distance from the Sun, Venus' orbital velocity is also fairly constant. Traveling at an average speed of 35 km/s, it completes one circuit of the Sun every 225 Earth days. It comes closest to Earth every 584 days during inferior conjunction. (Since five intervals of 584 days are almost exactly equal to eight years of 365/366 days, apparitions of Venus in the sky repeat every eight years.)

By comparing measurements of its angular size with its distance, astronomers were able to calculate that Venus has a diameter of 12,104 km, only slightly smaller than Earth.[1] Determination of its mass and density was complicated by the absence of any moons, but it was generally assumed that Venus is a rocky planet, with an overall composition and internal structure very similar to that of our world. Subsequent measurements of its gravitational influence on visiting spacecraft have confirmed this view.

Phases

When seen through a telescope or binoculars, Venus resembles Mercury and the Moon by displaying phases (see Chapters 4 and 5). This means that the planet's appearance changes from a small full phase around superior conjunction on the far side of the Sun, to a large, slim, crescent close to inferior conjunction (Figure 6.1).

Of alls the planets, Venus is by far the brightest and easiest to observe with the naked eye. It is best placed for observation when at its greatest elongation (angular distance) of 47° from the Sun in the sky. Venus then appears above the horizon for about 5 ½ hours after sunset or before sunrise. However, it is actually at its brightest when seen as a broad crescent at an elongation of 39°, about 36 days before and after inferior conjunction. At its closest, the

---

[1] The oblateness of Venus is zero. This means that, unlike other planets, there is no appreciable difference between the equatorial and polar diameters of Venus. This is due to its extremely slow rotation.

Exploring the Solar System, Second Edition. Peter Bond.
© 2020 John Wiley & Sons Ltd. Published 2020 by John Wiley & Sons Ltd.
Companion Website: www.wiley.com/go/Bond-Solar-System2e

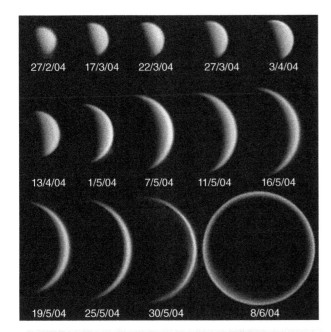

Figure 6.1 The phases of Venus. In this sequence, it appears fully illuminated, but very small, when it is at superior conjunction, on the opposite side of the Sun from Earth. As the distance between the two planets shrinks, Venus seems to grow in size and moves through its phases. Around inferior conjunction, when the crescent is narrowest and Venus is nearest Earth, its apparent diameter is about seven times larger. (Statis Kalyvas – VT-2004 programme/ESO)

Table 6.1
Transits of Venus 1631–2125

Date	Start Time (UT)	Duration
1631, December 7	03.49	3 h 00 m
1639, December 4	14.56	6 h 58 m
1761, June 6	02.01	6 h 35 m
1769, June 3-4	19.15	6 h 20 m
1874, December 9	01.50	4 h 36 m
1882, December 6	13.57	6 h 13 m
2004, June 8	05.15	6 h 13 m
2012, June 5–6	22.22	6 h 31 m
2117, December 11	00.03	5 h 41 m
2125, December 8	13.19	5 h 35 m

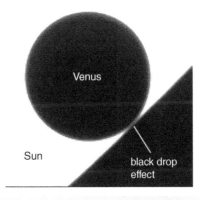

Figure 6.2 Venus at the western limb of the Sun, imaged by NASA's TRACE spacecraft during the transit of June 8, 2004. Unlike airless Mercury, Venus is outlined by a faint, glowing ring during transits, caused by scattering of sunlight in its thick atmosphere. The "black drop" effect was observed for about four minutes. It is thought to be caused by the "point-spread function" of the telescope and the light-scattering properties of Venus' atmosphere. (NASA/Stanford-Lockheed Institute for Space Research)

planet's apparent diameter is more than 65 arc seconds – providing easily the largest telescopic disk of any planet.

Transits

From time to time, both Mercury and Venus appear to cross the face of the Sun, as seen from Earth, at the time of inferior conjunction (see Chapter 5). If the conjunction occurs when the orbit of Venus crosses the ecliptic (the ascending or descending node), then it appears as a small dark spot moving from east to west across the Sun's disk. However, the slight inclination of the planet's orbit to the ecliptic means that such transits are extremely rare. The most recent transit took place on June 5–6, 2012, and the next will occur on December 11, 2117.

As Table 6.1 shows, transits of Venus are only visible from Earth in June or December. However, the duration of the transit – typically over 6 hours – varies according to the distance to be traversed across the Sun's disk.

The Venus transits have played an important role in the history of astronomy. In the 18th and 19th centuries many international scientific expeditions were dispatched to remote corners of the world in order to note the time the transit began and the apparent position of Venus on the solar disk from widely separated locations. By careful analysis of observations and the use of trigonometry, astronomers attempted to determine the distances of Venus and the Sun. Perhaps the most famous observation was made from Tahiti in 1769, during Captain James Cook's expedition to the South Pacific and circumnavigation of the globe.

Unfortunately, the accuracy of the measurements was adversely affected by a phenomenon known as the "black drop," which occurred at the beginning and end of each transit (see Figure 6.2). Observers found that the silhouette created by Venus appeared to lengthen and link to the solar limb. By the time the effect disappeared, the transit was already well under way. The black drop is now generally regarded as an optical effect, linked to the quality of a telescope's optics and its resolving power.

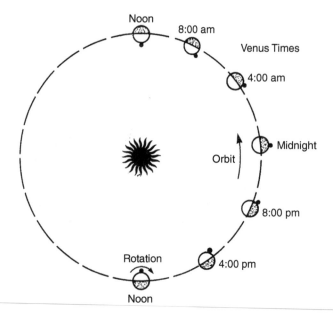

Axial Rotation: 243.01 Earth days
Orbital period: 225 Earth days (= Venus year)
Noon to Noon: 116.8 Earth days (= Venus year)
Venus year: 1.93 Venus year

Figure 6.3 Venus spins in a retrograde (east to west) direction, completing one full turn every 243.01 days. Since each orbit of the Sun takes 224.7 Earth days, a solar day on Venus – from one sunrise to the next – lasts for 116.75 Earth days. (NASA)

Table 6.2
Venus and Earth Comparison

	Venus	Earth
Mass	4.87×10^{24} kg	5.98×10^{24} kg
Equatorial radius (km)	6,052	6,378
Mean density	5,250 kg/m^3	5,520 kg/m^3
Average distance from Sun (million km)	108.21 (0.7 AU)	149.60 (1 AU)
Rotation period	243 Earth days (retrograde)	23 hours 56 minutes
Length of year (orbital period)	224.7 Earth days	365.2 days
Axial inclination (degrees)	177.3	23.5
Mean surface temperature	460°C	15°C
Highest point on surface	Maxwell Montes (11 km)	Mount Everest (8.8 km)
Reflectivity (albedo)	0.76	0.37
Surface gravity (at equator)	8.9 m/s^2	9.8 m/s^2
Surface atmospheric pressure	90 bars	1 bar (sea level)
Major atmospheric components	96% carbon dioxide, 3% nitrogen	78% nitrogen, 21% oxygen, 1% argon

Rotation Period

Since Venus comes closer to Earth than any other planet, it should be extremely easy to observe. The first telescopic observers recorded pronounced markings that suggested a cloud-free view of Venus' surface, but it was soon realized that it is covered by an impenetrable blanket of cloud. Even with the most modern optical telescopes, it is only possible to record gray, nebulous markings that change from day to day.

In the absence of any visible surface features, it was impossible to determine its rotation period. The first clue came in 1956, when Doppler measurements – small shifts in the wavelengths of light reflected from Venus – indicated a very slow retrograde rotation. If true, this meant that it must be spinning from east to west – the opposite direction to Earth and most other planets.

Further Doppler studies in the early 1960s supported the "backward" rotation, but suggested that Venus was spinning remarkably quickly – once every four days. This was later shown to be due to hurricane force winds that swept high-level clouds around the planet every four days (see Acid Clouds).

The first information about the hidden surface came in 1961, when ground-based radar was used to pierce the clouds. By measuring the Doppler shift of the radio waves reflected from Venus, the period of spin was finally revealed. In contrast to the high-level

winds, the solid planet rotates very slowly, completing one retrograde revolution in 243 Earth days. Venus has by far the longest rotation period of any planet in our Solar System.

When combined with a 224.7-day orbital period, this sluggish spin has a remarkable effect. The Venusian "day" from sunrise to sunset lasts for 116.75 Earth days. If the Sun could be seen from the surface, it would rise in the west and set in the east.

A number of theories have been put forward to explain Venus' strange behavior. Perhaps Venus was struck a glancing blow by a large impactor early in its history, causing it to topple over, so that the direction of its rotation appeared to reverse.

Alternatively, the planet may have simply slowed to a halt and then begun to rotate in the opposite direction. This could result from the chaotic effects created by several factors: gravitational tidal forces generated by the Sun combined with atmospheric thermal tides; friction between the core, mantle, and crust; and the angle of obliquity between the planet's equator and the plane of its orbit around the Sun. In this scenario, the Sun's gravitational pull on the solid body of the planet (possibly working in tandem with internal core–mantle friction), caused the spin to slow down. As the planet slowed and the influence of the atmospheric tides grew, its direction of rotation reversed.

In 2001, Alexandre Correia and Jacques Laskar (CNRS Institute of Celestial Mechanics) simulated the evolution of Venus' rotation and concluded that this type of reversal mechanism was most likely if Venus had a slow initial spin and a modest axial inclination.

Another possibility is that precession of the axis of rotation and gravitational perturbations from the other planets might cause chaotic shifts in its axial inclination, reaching a variety of positions throughout the planet's evolution (see also Chapters 3 and 5). If Venus reached a high obliquity, i.e. an equatorial inclination of more than 70° in relation to the plane of its orbit, and it was also spinning rapidly, the axis could have eventually flipped completely over. (Without the stabilizing influence of the Moon, Earth might also have experienced a similar trauma.)

It is interesting to note the relationship between the orbits of Venus and Earth and the rotation of Venus. At every inferior conjunction, the planet presents the same hemisphere to Earth. (This was obviously an obstacle to those who wanted to study its entire surface with ground-based radar.)

Atmospheric Composition

Prior to the Space Age, there was some uncertainty over the composition of Venus' atmosphere. Spectroscopic data indicated that carbon dioxide was a major constituent, but there were conflicting views over the surface temperature and the amount of water that was present.

In 1918, the Swedish Nobel prize winner Svante Arrhenius declared that Venus was a wet planet covered in swamps and tropical vegetation. Others argued that the presence of plentiful carbon dioxide and the uncertainty in the amount of water vapor favored a desert planet swept by dust storms.

By the 1960s, observations of radio emissions from Venus clearly showed that its surface temperature is extremely high, removing any possibility of a watery environment. We now know that Venus is the hottest planet in the Solar System, with an even higher surface temperature than Mercury, despite its greater distance from the Sun.

Spacecraft measurements confirmed that carbon dioxide makes up about 96% of Venus's atmosphere, compared with 0.03% on Earth (see Table 6.3). Apart from carbon dioxide and water vapor, the atmosphere consists primarily of inert gases, particularly nitrogen and argon. Other minor constituents include sulfur dioxide, carbon monoxide, hydrogen chloride, and hydrogen fluoride.

As is the case with the other rocky planets, the present-day atmosphere of Venus is almost certainly very different from the primordial atmosphere which existed more than 4 billion years ago (see Chapter 3). The original gaseous blanket was probably lost long ago when the young Sun increased in surface activity and luminosity.

Today, the atmosphere is remarkably massive and dense. It contains 100 times as much gas as Earth's atmosphere, with a surface pressure of 92 bars. (One bar is roughly equivalent to sea-level air pressure on Earth.) This crushing pressure is equivalent to what a diver would experience at a depth of more than 900 m in Earth's oceans.

The present atmosphere is the product of outgassing from the interior, a process which almost certainly continues today through active volcanism, and the influx of cometary and meteoritic material – rich in carbon, nitrogen, and water. With similar

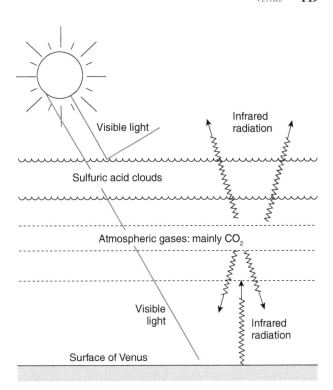

Figure 6.4 About 75% of the sunlight arriving at Venus is reflected by the high-level clouds. However, some light filters through to the surface, where it is absorbed. The warmed ground emits infrared radiation, most of which is absorbed by carbon dioxide and other greenhouse gases in the atmosphere. (Peter Bond)

Table 6.3

Composition of the Terrestrial Planet Atmospheres (% volume)

	Venus	Earth	Mars
Carbon dioxide	96	0.03	95
Nitrogen	3.5	77	2.7
Argon	0.007	0.93	1.6
Water vapor	~0.01 (variable)	~1	~0.03
Oxygen	0.13	21	~ 0

Some minor constituents are not shown.

sources of volatiles, Earth and Venus would be expected to have similar atmospheric compositions. However, at first sight the two "sister" worlds appear to possess very different amounts of carbon and water.

Runaway Greenhouse

One of the most intriguing mysteries about Venus is the reason for its completely hostile environment, with a sulfurous atmosphere, sweltering surface temperature (460°C), and crushing atmospheric pressure.

Why is Venus so hot and arid? The amount of water vapor and liquid water bound up with sulfuric acid and other compounds in the clouds is 100,000 times less than exists in Earth's oceans and atmosphere. On the other hand, the amount of deuterium (chemical formula ^{2}H or D, a "heavy" form of hydrogen) is about 100 times greater on Venus than on Earth.

Although both deuterium and normal hydrogen escape from the atmosphere while there is free water on the surface, the heavier isotope escapes less efficiently – hence its relative abundance today. This suggests that Venus initially had much more water than is observed today, but most of it has been lost.

What about the present abundance of carbon dioxide on Venus, compared with our planet? Both worlds probably started out with similar amounts of CO_2, but on Earth most of it was dissolved in the oceans and eventually ended up in carbonate-rich rocks, such as limestone (see Chapter 3). In contrast, any hot ocean that existed on the young Venus would soon have evaporated.

Estimates of how long the proposed ocean may have survived vary considerably. One factor is the output of solar energy, which was perhaps 30% lower in the early history of the Solar System than it is today. Eventually, as the Sun became more active and surface temperatures inexorably rose, the ocean began to disappear. In the absence of liquid water, the carbon dioxide remained in the atmosphere, trapping heat like the windows of a greenhouse. This resulted in a "runaway greenhouse" effect, which would have caused the ocean to totally evaporate, possibly within 600 million years of Venus' birth.

An alternative scenario suggests that that the surface water could have persisted for up to two billion years. According to this theory, the global cloud blanket would have reflected a lot of incoming sunlight back into space, thereby limiting the rate of surface heating. At the same time, as the oceans slowly evaporated, perhaps 50% of the atmosphere was composed of water vapor.

Under those conditions, the vapor pressure was so high (i.e. the atmosphere was saturated with water vapor) that further evaporation was inhibited – rather like water in a pressure cooker. The continuing loss of hydrogen at the top of the atmosphere was steadily compensated by further evaporation from the ocean. Only after a very long period, perhaps several billion years, would the ocean have disappeared altogether. This is known as the "wet greenhouse theory."

Both scenarios are based on the fact that, although carbon dioxide is transparent to incoming solar radiation, enabling sunlight to warm the surface of the planet, it is also an efficient absorber of infrared radiation which is emitted from the hot ground (see Chapters 3 and 7). Together with other greenhouse gases, such as water vapor and sulfur dioxide, the carbon dioxide built up in the atmosphere and formed a heat barrier which transformed the troposphere into a sweltering greenhouse.

Water vapor (H_2O) was carried to high levels, where it combined with sulfur compounds or was dissociated (split) into hydrogen and oxygen atoms by solar ultraviolet radiation. The light hydrogen gas then escaped into space, or was carried away by interaction with the solar wind, while the oxygen combined chemically with sulfur dioxide or oxidized surface rocks, a process similar to rusting on Earth.

Over billions of years the atmosphere was affected by widespread, massive volcanic eruptions and changes in solar output, but today it has reached a state of equilibrium, so that Venus is losing as much heat as it gains from the Sun and the surface temperature is no longer increasing.

Acid Clouds

Venus is blanketed by yellowish clouds, with no glimpse of the surface. It is the ability of these clouds to reflect sunlight which accounts for its brilliance in the night sky, along with its relative proximity to Earth.

For centuries, astronomical sketches of Venus merely displayed the planet's phases, upon which were superimposed vague, transient markings. The most noticeable variations were the horns or cusps of the crescent Venus, which appeared brighter than the rest of the planet.

Photography made little difference – pictures taken in visible light resembled views of a fog bank on Earth. However, images taken at near-ultraviolet wavelengths did reveal clearly defined cloud features, presumably at high level, that were only hinted at in ground-based views. These features survived for some time, enabling astronomers to calculate that they traveled around Venus at high speed, completing one circuit in a mere four days.

Spacecraft UV images show a complex pattern of cloud features. The most noticeable of these is a symmetrical, V-shaped

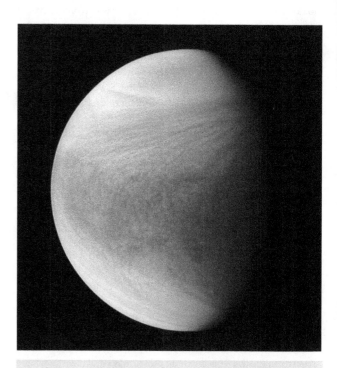

Figure 6.5 This false color ultraviolet image taken by Japan's Akatsuki orbiter shows turbulent, convectional clouds at low latitudes and less complex cloud patterns near the poles. Note also the dark, V-shaped cloud feature centered on the equator. (JAXA/ISAS/DARTS/Damia Bouic)

cloud pattern, centered on the equator. This occurs because the zonal (east–west) wind speed is fastest near the equator, slowing at higher latitudes and reducing to zero at the poles.

Broad bands of cloud diverge from the sub-solar point, where the Sun is overhead and heating is greatest, towards thick "collars" of cloud that are located at about 60° latitude and surround the poles.

Until the early 1970s, the composition of the clouds remained the subject of intense speculation. One of the favorite explanations was dust, stirred up by global, hurricane-force winds. However, measurements by NASA's Mariner 10 spacecraft, together with infrared spectroscopic studies, showed that the clouds resemble the output from a hazardous chemical plant.

The dominant constituent is highly concentrated sulfuric acid (H_2SO_4), together with traces of hydrochloric acid (HCl) and hydrofluoric acid (HF). The tiny droplets (about 1 micron across) of sulfuric acid are in a concentrated (75%) acid solution. Such corrosive clouds can exist over a much wider range of temperatures than the water clouds of Earth.

Where did the sulfur come from? One of the major gaseous products of volcanic eruptions on Earth is sulfur dioxide (SO_2). Since Venus is known to possess many hundreds of volcanic structures (see Volcanic Activity), it seems likely that these are the original sources of the sulfur. Volcanic eruptions would also replace any sulfur lost during chemical reactions with surface rocks.

When combined with water vapor (H_2O) in the upper atmosphere, the sulfur dioxide eventually produces sulfuric acid. However, any sulfuric acid rain falling beneath the main cloud deck soon evaporates in the high temperatures, producing acid vapors and water vapor.

When these constituents return to high levels, they are affected by photochemistry – chemical reactions involving ultraviolet sunlight. The UV photons break up SO_2, starting a series of chemical changes which results in the creation of H_2SO_4 once more. Some sulfur is also produced, and this may account for the yellowish color of the clouds. The substance that causes the ultraviolet cloud markings remains unknown.

Changes in Mariner 10's radio signals as they passed through the atmosphere showed the presence of at least two separate zones of cloud at altitudes between 35 and 60 km. Subsequent data from the Soviet Venera landers and the American Pioneer Venus probes have refined this view of the vertical cloud structure.

Above the main cloud deck, extending from about 90 to 70 km, is a layer of sulfuric acid haze. There are three main cloud layers between 70 and 48 km. The upper cloud deck comprises one-micron-sized sulfuric acid droplets, but the particle size is slightly larger in the middle and lower levels. At the level of the lower clouds, the temperature has already reached 95°C and atmospheric pressure is comparable to sea level on Earth.

Beneath the clouds the atmosphere is once again fairly hazy, but there is good visibility below about 30 km, except for the bottom 10 km, where conditions become increasingly murky due to scattering of light by the dense atmosphere, bathing the landscape with a reddish hue (see Figure 6.24). An observer on the surface would not be able to see the Sun through the dense, dreary blanket overhead.

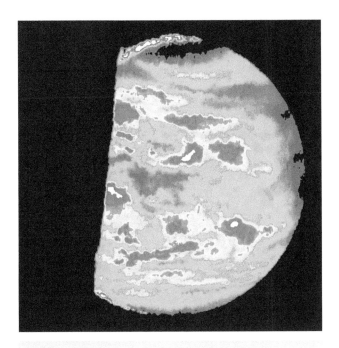

Figure 6.6   This near-infrared image, taken by the Galileo spacecraft en route to Jupiter, shows atmospheric "windows" in red and white – regions where heat escaping from lower levels is able to penetrate the clouds. This type of observation reveals the cloud morphology all over the planet, sometimes all the way to the surface. Patterns of thick and relatively thin clouds suggest that large clouds condense and evaporate in rising and falling air associated with weather systems. (NASA-JPL)

Atmospheric Circulation and Super-rotation

Ultraviolet images taken seven hours apart by Mariner 10 showed that the upper clouds speed around Venus at 360 km/h, completing one circuit of the planet every four days. This motion from east to west – the same direction as the planet's spin – is 60 times faster than the rotation of Venus. Scientists refer to it as an example of super-rotation.

This remarkably rapid retrograde motion of the upper winds has yet to be explained satisfactorily. The only comparable winds on Earth are the ribbon-like jet streams, but they are limited to narrow zones and much closer to the surface. In contrast, the Venusian zonal winds cover almost the entire planet, although their velocity quickly decreases towards higher latitudes.

Up to three east–west jets have also been identified: one above the clouds, between 50 and 55° latitude, with a maximum speed of 140 m/s; a second with a speed of 95 m/s, centered at 70° latitude and 60 km altitude; and a possible third jet at 15° latitude and 65 km altitude, with a speed of 100 m/s. A near-equatorial jet was also reported in 2017, after analysis of cloud movements imaged by Japan's Akatsuki orbiter in the low and middle cloud region, at an altitude between 45 and 60 km.

The only viable source of energy for these high-velocity jets appears to be the absorption of solar energy by the clouds, hazes,

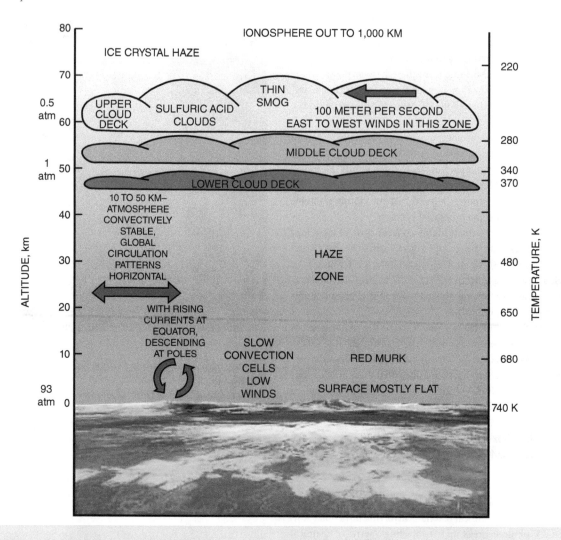

Figure 6.7 The horizontal structure of Venus' atmosphere, based on data from the Venera and Pioneer Venus missions. 360 km/h winds sweep the higher clouds around the planet in four days. Three layers of sulfuric acid cloud occur 48–70 km above the surface. In the gloom below, the atmosphere is stable, becoming increasingly dense and hot towards the surface, where the temperature is about 460°C. Surface pressure is equivalent to that found almost 1 km down in Earth's oceans, so low-level winds are slow. (NASA)

and relatively thin upper atmosphere, perhaps combined with turbulence.

A much more leisurely, meridional (north–south) circulation – only 5–10 m/s – also occurs within the troposphere. This Hadley cell circulation (see Chapter 3) takes the form of heated gas rising at the equator and spreading out at an altitude of about 50–65 km. At latitudes of about 60° N and S, the gas then sinks before returning toward the equator.

On Earth, the Hadley cells only extend to mid-latitudes (approx. 30°N and S) before being disrupted by Earth's rapid rotation (see Chapter 3). On Venus, the Hadley circulation is much simpler because of its small temperature variations, the absence of oceans and the slow planetary rotation. However, several additional north–south circulation cells may also occur at lower levels.

The polar edges of the Hadley cells are both marked by a "collar," where a wide (~1,000 km), shallow (~10 km) jet stream of fast-moving cold air circles the poles at about 70° latitude. The

clouds that mark each collar are colder and higher than their surroundings.

The troposphere, where most convective motion takes place, extends to an altitude of about 60 km, about four times higher than on Earth. Numerous cellular cloud structures occur in the region of maximum solar heating, where the Sun is overhead. These indicate that large convective plumes are thrusting upwards from the main cloud deck on columns of hot, rising gas.

Closer to the surface, where the atmosphere becomes extremely dense and there are no major variations in temperature, the atmospheric motion is fairly sluggish, generating only light breezes of about 1–2 m/s, in contrast to the super-hurricanes raging up above.

Although these modest surface winds should have enough momentum to move dust and fine sand, the thick air will certainly inhibit saltation – the bouncing motion of sand grains across the surface – and the erosion of rocks by abrasion or sandblasting.

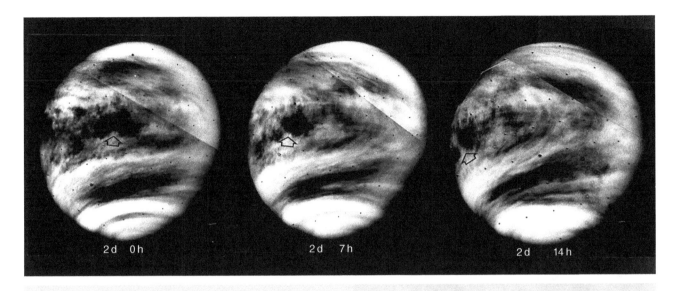

Figure 6.8 This series of ultraviolet photomosaics was taken at 7-hour intervals on February 7, 1974, two days after Mariner 10 flew past Venus. They show the rapid retrograde (east–west) rotation of light and dark markings at the top of the cloud deck. The bright feature marked with an arrow is about 1,000 km across. (NASA-JPL)

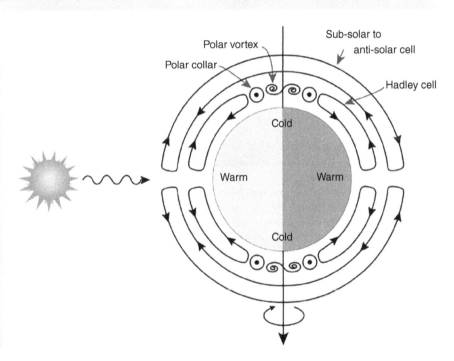

Figure 6.9 The general meridional (north–south) circulation of the atmosphere. The main feature is a convectively driven Hadley cell, in which the atmosphere moves away from the equator at an altitude of about 60 km, then sinks at about 60° latitude before returning towards the equator at a lower level. Above about 100 km, the circulation changes completely, with a sub-solar (midday) to anti-solar (midnight) pattern. Smaller vertical motions are associated with convection close to the subsolar point, where atmospheric heating is greatest. The overriding motion at the level of the cloud deck is the super-rotation from east to west (not shown). (ESA)

By tracking the movements of distinct cloud features in the cloud tops some 70 km above the planet's surface over many Venusian years, scientists have been able to monitor patterns in the long-term global wind speeds.

One discovery has been that, although the super-rotation is present on both the day and night sides of Venus, it seems more uniform in the day and becomes more irregular and unpredictable on the night side.

Another surprise was that the average cloud-top wind speeds between latitudes 50°N and 50°S increased from roughly 300 km/h to 400 km/h between 2006 and 2014. The reason for this dramatic increase is unknown.

On top of this long-term increase in the average wind speed, studies have also revealed regular variations linked to the local time of day and the altitude of the Sun above the horizon.

There were also dramatic variations in the average wind speeds observed between consecutive orbits of Venus Express around the planet. In some cases, winds at low latitudes circled the planet in 3.9 days, while on other occasions they took 5.3 days.

Polar Vortices and Gravity Waves

One of the most unusual features of Venus' atmospheric circulation is the presence of polar collars of cloud that surround swirling

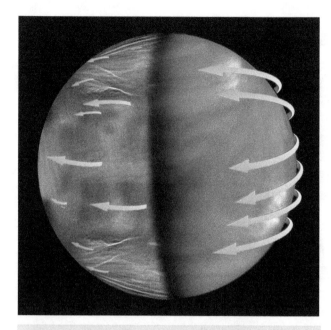

Figure 6.10 Atmospheric super-rotation occurs in the upper clouds, on both the day and night sides of Venus. However, it seems more uniform in the day (right side), whereas at night it becomes more irregular and unpredictable. The day image of the clouds was taken by Japan's Akatsuki in UV light, while the night image is a composite of thermal images obtained by ESA's Venus Express orbiter. (ESA, JAXA, J. Peralta and R. Hueso)

vortices above each geographical pole.[2] These permanent vortices, more than 2,800 km across, follow the direction of the atmosphere's super-rotation.

Sometimes, each vortex has two eye-like centers of rotation, which are like separate "tubes" that are connected by a distinct S-shaped cloud structure (Figures 6.11 and 6.12). The complex double "eye" rotates every 2.5–2.8 Earth days. However, this double, rather than single, vortex formation, seems to be just one of various configurations.

These vortices probably resemble the winter polar anticyclones that occur on Earth, especially over Antarctica. Air sinks in the center, possibly to an altitude of 50 km or lower, and rises in the polar collars. The temperature at the cloud tops in each polar vortex is −23 °C, some 13 °C warmer than in the nearby polar collars.

The centre of the southern vortex has been shown to have a highly variable shape and internal structure, and its morphology is constantly changing on timescales of less than 24 hours, as a result of differential rotation. Wind speeds are higher (35–50 m/s) near their outer edges, reducing to zero at the poles, so the vortex is being pulled and stretched continually.

Its center of rotation is offset from the geographical south pole by about 3°, or several hundred kilometers, and the center of rotation drifts right around the pole over a period of 5–10 Earth days. It is likely that the northern polar vortex has a similar structure and behaves in a similar way.

It might be expected that there is little connection between the baking atmosphere close to the ground and the upper atmosphere, some 60–70 km above. However, spacecraft observations over several decades indicate that the relationship more resembles an "ocean-like" lower atmosphere, topped by an opaque cloud layer which acts like the surface of the ocean. Ripples and air currents visible at the cloud tops provide hints about processes and influences far below.

Spacecraft observations have found evidence for the upward propagation of atmospheric waves from the surface to the main cloud deck and above (Figure 6.13). This process generates so-called gravity waves.

A gravity wave is an undulation caused by the interaction of gravity with a fluid or gas, rather like water in a stream flowing over a submerged boulder. They can only exist in a stably stratified atmosphere. Such waves are common on Earth, for example on the lee side of mountains, where they frequently reveal their presence through cloud formations. They often take the form of wave trains – a series of wave-like clouds travelling in the same direction and spaced at regular intervals.

On Venus, they appear to be linked to atmospheric turbulence created when horizontal air flow is disturbed by mountains, which cause the air to rise on one side and then subside on the lee side.

Early evidence of atmospheric waves being generated by air flowing over major topographic features came in 1985, when two Soviet Vega balloons flying at an altitude of 54 km experienced a bumpy ride above the southern uplands of Aphrodite Terra, the largest of Venus's three continent-sized highland regions (see Box 6.1).

Numerous, small, wave-like cloud features, some straight and some irregular, were later observed at the cloud tops, 62–70 km above the surface, by ESA's Venus Express orbiter. Most of them occurred at high latitudes (60–80°N) in a region of high cloud known as "the cold collar," and they were concentrated above the continent-sized highland of Ishtar Terra.

In December 2015, Japan's Akatsuki orbiter obtained infrared and UV images of a huge, bow-shaped wave for the first time in the planet's upper atmosphere (Figure 6.14). Although such a feature would be expected to be carried along by the fierce winds, the curved wave remained firmly in place, keeping pace with the planet's rotation. This suggested a link between the surface of Venus and the upper air, where the troposphere meets the lower stratosphere.

The wave extended for more than 10,000 km and stretched nearly from pole to pole. It was marked by the presence of slightly warmer air in the upper portion of the planet's thick atmosphere, some 65 km above the surface. The wave lasted for at least four days but when the orbiter returned to observe the area later in 2016, the feature had mostly disappeared.

The longitude of the huge cloud feature corresponded with the western slope of Aphrodite Terra. Mountainous features on the

[2] Several planets in the Solar System, including Earth, possess hurricane-like vortices, where clouds and winds rotate rapidly around the poles. Some of these have strange shapes, such as the hexagonal structure on Saturn, but none of them are as variable or unstable as the southern polar vortex on Venus.

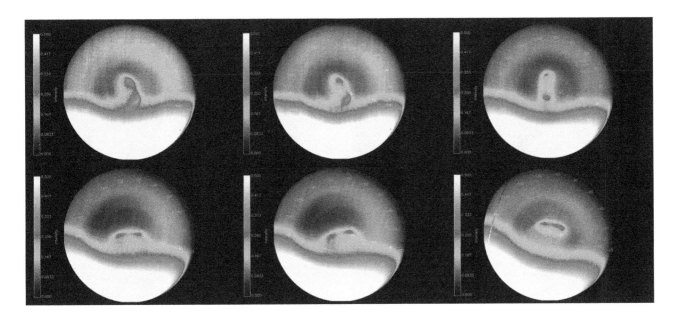

Figure 6.11 A series of false-color infrared images of the south polar region taken by Venus Express between April 12 and 19, 2006. The south pole is at the centre of each image. A peculiar double-eye vortex structure (red dots) is clearly visible. The sequence of images shows the rotation and the shape variation of the double vortex over time. It is also possible to see the rotation of the terminator: the day side is shown in yellow and the night side is blue. A collar of cold air (dark blue) is visible around the vortex structure. The vortices complete a full rotation in only three days. (ESA/VIRTIS/INAF-IASF/Obs. de Paris-LESIA)

surface may be forcing winds into the upper atmosphere, where they slow down enough to create a lasting bow wave.

Physical Characteristics

Venus is a rocky planet with a bulk density of 5.24 g/cm³, and a mass 81% that of Earth. This indicates that Venus formed in the inner region of the solar nebula, where it was too hot for volatiles such as ice to condense. As a result, the region was dominated by refractory elements, including iron, nickel, silicon, and aluminum, which condense and solidify at high temperatures.

These physical characteristics suggest that the internal structures of Venus and Earth are very similar, though the lower bulk density indicates that Venus has a slightly smaller iron-rich core (Figure 6.15). This core may be divided into a solid, iron-rich inner region, surrounded by a molten outer core of iron mixed with nickel or sulfur.

Modeling suggests that the core of Venus also differs from that of Earth by being entirely molten, with a temperature at the center of about 5,000°C – perhaps 1,000°C cooler than Earth's inner core. The smaller temperature difference between the core–mantle boundary and the surface, together with smaller gravitational acceleration, indicate that mantle convection is less powerful, which may help to explain the present lack of plate tectonics on Venus.

Another significant difference seems to be that Venus has a thicker crust than Earth. Comparisons of gravity data with surface topography show that the crust seems to be typically 25–40 km deep, increasing to perhaps 50 or 60 km under the older "tesserae" regions and the larger volcanoes. This thick, dry, rigid crust prevents Earth-like plate tectonics from taking place.

The Hidden Surface

The most promising technique for observing the surface is to transmit radar signals which penetrate the dense clouds and are then reflected by the rocky terrain. The echoes give important information about topography, surface texture, and composition, e.g. the more the return radar signal is scattered, the rougher the surface (see Box 6.2).

By the late 1970s, pioneering work with ground-based radar made it possible to detect features 10–20 km across on the hemisphere centered on 320°E. Particularly notable were the isolated, radar-bright regions named Alpha, Beta, and Maxwell. These were considered to be rough terrain, probably volcanic highlands or mountains. Also visible was a large canyon and a few craters, but most of the surface seemed fairly smooth and featureless.

The major breakthrough came with the arrival of NASA's Pioneer Venus orbiter in 1978. More detailed radar maps have since been provided by the U.S. Magellan spacecraft.

The satellite data make it clear that Venus is very smooth, with few uplands or low-lying basins compared with Earth (or Mars). Eighty percent of the surface mapped by Pioneer Venus lay within 2 km of the mean radius – the equivalent to sea level – and 60%

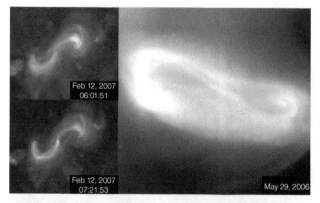

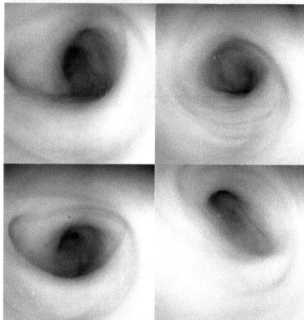

Figure 6.12 (a) A Venus Express infrared view of a double-eyed vortex at Venus' south pole, linked by an S-shaped cloud structure. Its overall diameter is about 2,800 km. (b) Four views of the single south polar vortex seen in thermal infrared light. The images show a drain-like structure where air is sinking in the center. White regions show cooler, higher cloud, with a darker, warmer center. (ESA/VIRTIS/INAF-IASF/Obs. de Paris-LESIA)

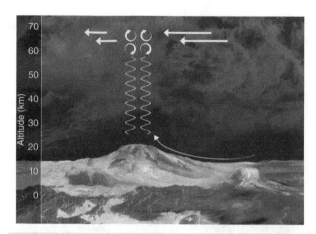

Figure 6.13 Possible behavior of gravity waves near mountainous terrain on Venus. Horizontal winds moving slowly over mountainous terrain generate gravity waves – atmospheric phenomena that form when air is forced over steep, undulating surfaces. The waves propagate vertically upwards, growing larger in amplitude until they break just below the cloud top, like sea waves on a shoreline. As the waves break, they push back against the fast-moving, high-altitude winds and slow them down. (ESA)

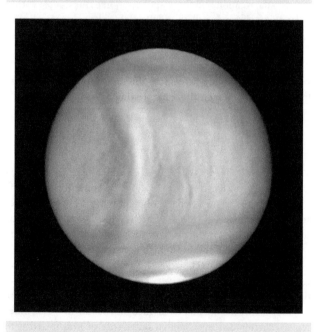

Figure 6.14 A 10,000 km-long wave feature was imaged in the upper atmosphere of Venus by Japan's Akatsuki orbiter in December 2015. This infrared image shows that the bow-shaped feature – probably created by a huge gravity wave – spanned most of the planet. It is marked by the presence of slightly warmer air in the upper portion of the planet's thick atmosphere, some 65 km above the surface. The stationary wave feature was probably associated with turbulence generated by air flowing over the mountains of western Aphrodite Terra. (Planet-C team/JAXA-ISAS)

within 500 m of this level. Although there are several upland regions which rise 2 km or more above the general level, these only cover 5% of the total area (Figure 6.16). Most of Venus is made up of flat, rolling plains, such as Guinevere Planitia and Kawelu Planitia.

This is not to say that there are no extremes of relief. The highest topography, in a northern region known as Maxwell Montes, has peaks that rise 11 km above the average level, dwarfing Mt. Everest on Earth. At the other extreme is a rift valley in the southern hemisphere known as Diana Chasma, which is 2.9 km below the average level – about one-quarter the depth of Earth's Marianas ocean trench.

Box 6.1   The Vega Balloons

Both of the Soviet Vega spacecraft dispatched to explore Halley's comet also carried payloads that were designed to study Venus. During flybys in June 1985, each Vega released a spherical capsule into the planet's atmosphere, eventually deploying a surface lander and a balloon – the only time meteorological balloons have so far been used in Solar System exploration.

Constructed of Teflon fabric, they weighed 12 kg and, when inflated, measured 3.4 m in diameter. Each balloon carried a payload of 6.9 kg on a gondola suspended beneath a 13-m tether. The gondola carried pressure, light, and temperature sensors, a nephelometer cloud sensor, and an anemometer (wind speed gauge).

The balloons alternated between 25 minutes of data gathering (often taking measurements every 90 seconds in order to conserve power) and 5 minutes of data relay to Earth. They were released in the night hemisphere, and, after filling with helium, floated to their equilibrium altitude of 53.6 km. This location was chosen in part for its benign temperature – it was feared that they would explode in higher daytime temperatures.

Tracking by an international array of 20 radio telescopes showed that their flights began at 180° E and 7° N and S of the equator, respectively. Both balloons traveled more than 11,000 km and operated for 46.5 hours, passing into the daylight hemisphere, 33 hours after deployment, before their lithium batteries failed.

The Vega 1 balloon was blown around the planet at speeds up to 250 km/h. Strong vertical winds bobbed the craft up and down 200–300 m through most of the journey. The atmospheric temperature averaged 40°C and pressure was 0.5 Earth atmospheres. The nephelometer found no clear regions in the surrounding clouds.

The second balloon began in a rather placid environment, but turbulence increased later. When it passed over a 5 km-high mountain on the "continent" of Aphrodite Terra, it was pulled 2.5 km towards the surface by a powerful downdraft. The temperature measured by the Vega 2 balloon was consistently 6.5°C cooler than found by its counterpart. Unfortunately, its nephelometer failed to function, so no cloud data were returned. No evidence for lightning was sent back by either balloon.

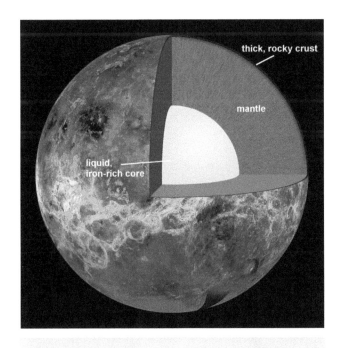

Figure 6.15 Venus probably has a very similar internal structure to Earth, with a large, iron-rich core, a mantle, and a rocky crust. The core may be rather smaller than that of Earth, but similarly divided into a solid inner region and a liquid outer core. (After Wikimedia)

Ishtar Terra

The most prominent features on the radar maps are two large upland regions: Ishtar Terra in the north and Aphrodite Terra along the equator.[3*] Ishtar, the size of Australia, is made up of three major units. At its core is Lakshmi Planum, a plateau which lies 4–5 km above the mean level – about the same height as the Tibetan plateau on Earth, but twice as large. On its northern and western flanks are the mountain ranges of Akna Montes and Freyja Montes, while to the south is Danu Montes.

However, the most spectacular mountains of Ishtar are Maxwell Montes, which lie to the east. Within these towering, radar-bright mountains is a 100 km-diameter, double-ring impact basin known as Cleopatra. Other notable features on Ishtar are a large shield volcano known as Sacajawea, and Colette Patera, another volcanic feature in the west of the plateau. The southern edge of Ishtar drops very steeply, suggesting possible faulting and tectonic activity on a large scale. The rough surface seems to be covered in loose debris.

Aphrodite Terra

Although Aphrodite is generally lower than Ishtar, it covers twice the area and is comparable in size to Africa. It contains several highland regions to the east and west separated by a broad, lower area. The most notable of these mountains are Ovda Regio and Thetis Regio.

However, the most intriguing features lie further east in the "scorpion's tail," where the surface is marked by a series of enormous, northeast-trending curved ridges and rift valleys, known as

---

[3] Almost all of the surface features on Venus are named after famous women and mythological goddesses, e.g. Ishtar was the Babylonian goddess of love.

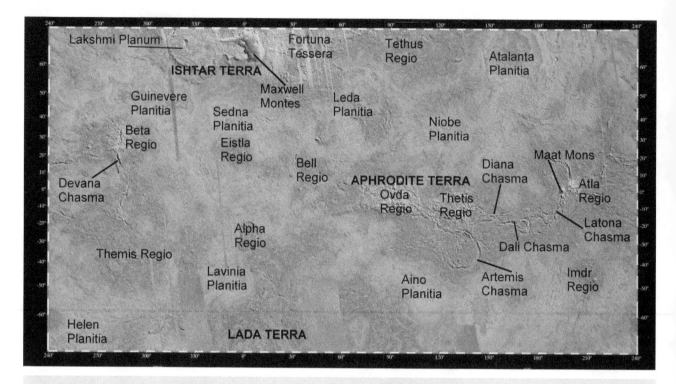

Figure 6.16   A topographic map of Venus obtained with the Magellan radar instrument. Highest land is red and yellow, lowlands are blue. Apart from the two upland regions of Ishtar (top) and Aphrodite, Venus is remarkably flat and low-lying. Most of the surface is within 500 m of the planet's mean radius. Zero degrees longitude on Venus is measured from a small feature known as Eve Corona, on the southwest slopes of Alpha Regio. Blank areas show where no data were obtained. (NASA-Ames)

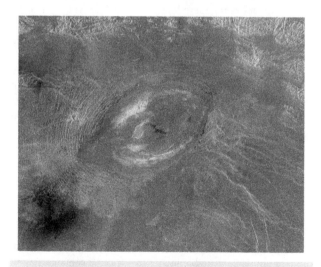

Figure 6.17   Sacajawea Patera is a 233 km-wide caldera located in western Ishtar Terra. The caldera is 1–2 km deep, 120 × 215 km in diameter, and bounded by a zone of graben and fault scarps. Extending up to 140 km toward the southeast is a system of linear structures thought to be a rift zone along which the injection and eruption of magma may have occurred. (NASA-JPL)

Diana Chasma, Dali Chasma, and Latona Chasma. Diana Chasma, for example, measures 2,500 km long and 280 km wide.

These valleys extend to a 3 km high upland region called Atla Regio. Also associated with these "chasmata" are impressive volcanic structures, such as 8 km-high Maat Mons, and circular structures, called "coronae," which are marked by radial or concentric fractures. The origin of these features is a matter of debate (see Coronae and Arachnoids).

One of the most intriguing, and mysterious, features of this region is Artemis, which lies between the rugged highlands of Aphrodite Terra to the north and relatively smooth lowlands to the south. Its interior topographic high is surrounded by a 2,100 km-diameter, nearly circular trough, called Artemis Chasma. The trough measures 25–200 km wide and is 1–2 km deep, with a raised outer rim. It fades away at its north western end and is deepest in the east. Despite its circular shape, Artemis in generally thought to be far too large to be classified as a corona, neither does it resemble a large impact basin.

Artemis is also associated with an outer trough (> 5,000 km in diameter), a swarm of fractures and volcanic dykes that radiate outward (12,000 km in diameter), and a suite of concentric wrinkle ridges (13,000 km in diameter) which represent modest crustal contraction.

Its origin is unknown, but it is thought to have formed in a region of relatively thin lithosphere. The trough with its raised rim would seem to suggest crustal extension, where the crust and lithosphere were pulled apart to create a graben or rift valley feature.

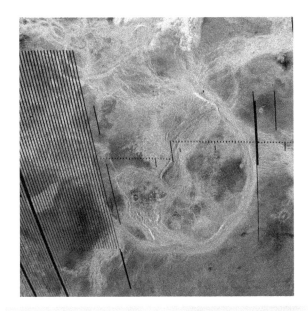

Figure 6.18   A Magellan radar image of the near-circular trough of Artemis Chasma. Artemis contains complex systems of fractures, numerous lava flows and small volcanoes, and two impact craters, the larger of which is located at lower left. The ring of fractures that defines Artemis forms a steep trough up to 200 km wide, with raised rims. The height from the rim crest to the bottom of the trough is up to 2.5 km. Radial fractures and concentric wrinkle ridges occur outside the trough. Artemis may be an extensional trough caused by the crust pulling apart due to the upwelling of a plume of hot material from the interior of Venus. Artemis has many features similar to coronae, but with a diameter of about 2,100 km, it is far larger than any of those. Black stripes are caused by lack of data. (NASA-JPL)

The unusual size and circularity of Artemis led to the suggestion that it may be a zone of intense compression and crustal underthrusting, similar to an oceanic subduction zone on Earth. If this was the case, Artemis would represent a failed attempt to initiate plate tectonics along a subduction zone on Venus. However, there is no evidence that surface crust was recycled into the mantle and this hypothesis is now generally discounted.

It may have been formed by a deep, rising mantle plume that resulted in the formation of an interior spreading center and newly formed crust. Rising and flattening of the plume head would lead to uplift and doming, which resulted in radial fracturing and a curved, exterior trough. As the plume head collapsed vertically and spread laterally, it likely caused outward migration of the trough, as well as fracturing and wrinkle-ridge formation outside the trough.

Weathering and Erosion

The absence of water and seasonal temperature variations limits the amount of weathering that takes place on the surface of Venus. Since wind speeds are low near the surface (1–2 m/s), the rapid movement of loose material and subsequent erosion through

sand-blasting is also minimal. Erosion and resurfacing through meteorite impacts is also reduced by the protective blanket of the atmosphere.

On the other hand, the dense atmosphere does possess sufficient momentum to transport loose material, as evidenced by radar images of dune fields and wind-blown streaks. Images from the Venera landers also showed pebbles and fine-grained debris, as well as large individual slabs, suggesting that solid rock can be broken down by mechanical or chemical processes. (An alternative theory is that this debris originated from a large volcanic eruption.)

The most prominent wind-related features in the Magellan images are streaks that form in the lee of large impact craters and other topographic obstacles as a result of deposition or removal of sand and dust. They can be used to calculate the direction of the most powerful or prevailing winds. However, some may be caused by the violent blasts generated during an impact event or a slower redistribution of fine impact debris by the wind.

Landslides are also quite common, particularly in the "scorpion's tail" region between Atla Regio and Aphrodite. Significant rock falls have occurred on the steep slopes of large volcanoes. Other landslides occur on the steep slopes that flank the major rift valleys or "chasmata." They are typically larger than terrestrial landslides but smaller than their Martian counterparts. Possible causes of these events may be oversteepening of slopes, perhaps by intrusion of magma, or Venusian quakes.

Impact Craters

About 1,000 impact craters have been found on Venus. Although this far exceeds the known impact features on Earth, it is well below the number found on most of the other solid bodies in the Solar System. This is not too unexpected, since most incoming meteorites would be expected to burn up in the dense atmosphere, while any ancient craters should be masked by billions of years of weathering, erosion, and resurfacing. However, their modest number and fairly regular distribution also suggests that the entire surface may be geologically young.

Although they are widely scattered over the surface, fewer impacts are found on the uplands. As expected, the least common, but largest, impact structures are multi-ringed craters. They have smooth floors and are typically more than 100 km across. Largest of all is 275 km-diameter Mead crater, located on the plains north of Aphrodite (Figure 6.22).

More than one-third of the craters are intermediate in size, with radar-bright central peaks or mounds and smooth floors. No impact craters smaller than 3 km across have been observed, presumably because any small incoming meteorites burn up or break apart in the dense atmosphere.

Volcanic Activity

About 1,100 volcanic centers have been detected on Venus. They are much more widely distributed than on Earth, where the majority of volcanoes are restricted to linear zones near plate boundaries (Figure 6.23). However, very few of the large volcanoes are found in the lowland plains or the highland regions, with higher

Box 6.2   Revealing Radar

Ground-based and spacecraft radar instruments have been powerful tools for studying the hidden Venusian surface. These observations took advantage of the fact that radio waves can penetrate the dense clouds and bounce back from the solid surface.

The echoes received on Earth give important information about the roughness and composition of the terrain, e.g. the more the radar signal is scattered in all directions, the rougher the surface. Slopes that reflect the signal back in the direction from which it came appear bright, while slopes that face a different direction appear dark. Topographic information is obtained by accurate measurement of each signal's return time – an early echo shows an area of highland.

An early success for this radar technique was the discovery in 1962 that Venus spins very slowly, so that one complete rotation takes 243 Earth days, in a retrograde (east to west) direction. It was also possible to build up a picture of one hemisphere from the radar echoes picked up by the Arecibo radio telescope in Puerto Rico.

Figure 6.19   This topographic map of Venus, centered at 180°E, was composed from Magellan radar images taken 1990–1994. Magellan imaged 98% of the surface at a resolution of about 100 m. Gaps were filled with images from the Earth-based Arecibo radar. Brown areas are high, generally rough terrain; dark blue areas are low and usually smooth. To the west (left) is the upland of Aphrodite, with huge shield volcanoes such as Maat Mons in the "scorpion's tail" just right of center. At the top (north) is Ishtar Terra. (NASA-JPL)

It eventually became possible to detect features 10–20 km across on the hemisphere centered on 320°E. Particularly notable were the isolated, radar-bright regions named Alpha, Beta, and Maxwell (in honor of British physicist James Maxwell). However, most of the surface seemed fairly smooth and featureless.

One limitation was the inability to accurately measure altitude. The breakthrough came through the radar altimeter on board the Pioneer Venus orbiter, which was capable of height measurements accurate to within 0.2 km. Over a period of two years, the instrument surveyed most of the planet between 73°N and 63°S, with a spatial resolution of 75 km (see Box 6.4).

Even before Pioneer Venus concluded its survey, both NASA and the Soviet Union launched spacecraft to map the surface in even greater detail. First to arrive were the Soviet Veneras 15 and 16, each equipped with a side-looking radar, which operated at a wavelength of 8 cm and was able to reveal structures less than 2 km across. Unfortunately, their elliptical orbits meant that only the northern hemisphere above 30°N could be mapped.

The images clearly showed linear ridges and folds skirting the edge of the Lakshmi Planum plateau, with Maxwell Montes dominating eastern Ishtar. Most controversy focused on strange circular features. Were they craters of volcanic or impact origin? Some, dubbed "arachnoids" because of their spider-like shape, were a total mystery.

Clearly, images with higher spatial resolution were required to clear up the mysteries. These were provided by NASA's Magellan spacecraft, which entered orbit around Venus in August 1990 (see Box 6.5).

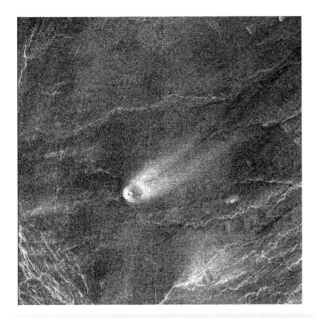

Figure 6.20 A Magellan radar image showing a bright wind streak next to a 5 km-diameter volcano located at the western end of Parga Chasma (9.4°S, 247.5°E). The northeast-trending streak was probably formed by deposition of material on the lee of the volcano. The streak is 35 km long and 10 km wide. (NASA-JPL)

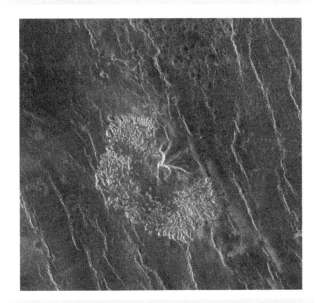

Figure 6.21 A Magellan radar image, centered at 45.2°S, 201.4°E, showing landslide deposits 70 km across which apparently result from the collapse of a volcanic structure. The bright deposit probably consists of huge blocks of fractured volcanic rock up to several hundred meters across. The remnant of the volcano, about 20 km across, is seen in the center. Numerous small volcanic domes can be seen in the northern half of the image. Bright linear features trending to the north-northwest are ridges caused by tectonic deformation of the upper crust. (NASA-JPL)

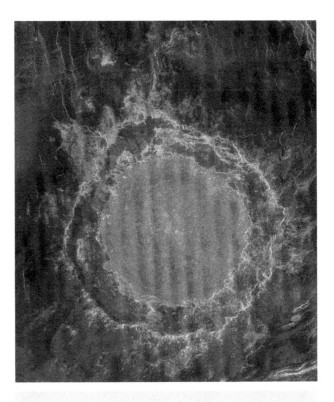

Figure 6.22 Multi-ringed Mead crater, the largest impact feature on Venus, is located north of Aphrodite Terra (12.50°N, 57.20°E). Its flat, radar-bright floor has several fractures visible as brighter lines. The patchy outer ejecta deposits are rougher than the crater floor, so they appear brighter. The surrounding plain is covered by fine debris that appears darker in the image. (NASA-JPL)

concentrations to the east and north east of Aphrodite, where there are major crustal fractures. The reason for this grouping is not fully understood, but it may reflect a cluster of closely spaced hotspots in a region of thinner, weaker crust.

The most notable volcanic region, and the first to be identified, is the radar-bright Beta Regio, which is centered at 27°N, 282°E. This is made up of two, 5 km-high shield volcanoes known as Theia Mons and Rhea Mons.

Theia Mons is the largest volcano on Venus, reaching over 4 km high, with lava flows covering an area more than 800 km wide. At its center is an oval caldera roughly 75 km long and 50 km wide. The volcano also lies at the junction of three major rifts. The most notable of these, Devana Chasma, is over 200 km wide in places, and up to 3 km deep on the flanks of the volcano. Another rift cuts through Rhea Mons.

Like most volcanic centers on Venus (and the Hawaiian Islands and Iceland on Earth), Beta Regio is thought to be associated with a hot spot in the mantle. According to this scenario, the region was uplifted and pulled apart by a plume of rising magma that could have formed at a depth of around 3,000 km, which approximately corresponds to the core–mantle boundary.

There are more than 150 large shield volcanoes on Venus. With diameters of between 100 and 600 km and rising up to 5.5 km

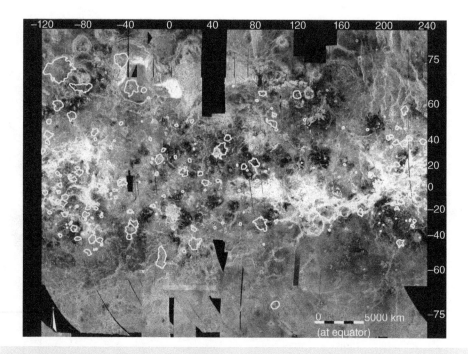

Figure 6.23 Most of the large volcanoes and calderas on Venus are located near the equator, particularly to the east of Aphrodite, where there are major fault structures and rift valleys. The major uplands have very few large shield volcanoes. (LPI)

Figure 6.24 Maat Mons is the highest shield volcano on Venus, rising 5 km above the surrounding terrain. It may still be active. Lava flows spread out from the volcano for hundreds of kilometers. Vertical scale has been exaggerated 10 times. Simulated hues are based on color images obtained by Venera 13 and 14. (NASA-JPL)

above the surrounding terrain, they are generally broader and flatter than their counterparts on Earth, such as Mauna Loa (see Chapter 3). On the other hand, although some of the Venusian shields cover nearly the same area as Olympus Mons, which has a basal diameter of approximately 800 km, the Martian volcano

towers to a much greater altitude and contains far more lava and ash.

Like their terrestrial counterparts, the Venusian shields have broad, gentle slopes and summit calderas, indicating that they are composed of non-viscous, basaltic lavas which flowed long distances from the central vent before they solidified.

The highest shield volcano on Venus is Maat Mons, in eastern Aphrodite. At its summit is a caldera about 30 km across. Within the giant caldera there are at least five smaller collapse craters, up to 10 km in diameter. A chain of small craters 3–5 km in diameter extends some 40 km along the southwest flank of the volcano. They may also have been formed by surface collapse, due to withdrawal of subsurface magma.

The vast majority of Venus' volcanoes are small shields, less than 20 km across. Numbering more than 100,000, these dome-shaped hills are often found in swarms on the lowland plains and lower uplands. Although they may have contributed significantly to the resurfacing of the planet in the relatively recent past, a large proportion has been partly buried by later lava flows. Many of the shield fields cluster around larger shield volcanoes, reflecting a concentration of volcanic activity in these regions.

The variety of volcanic landforms is evidence that the viscosity of lava on Venus is quite variable, although the high temperature of the atmosphere may extend the cooling time of surface flows, enabling them to travel further than on Earth or Mars.

At one extreme are numerous channel-like features which were eroded by fluid lava, rather than liquid water. One to three kilometers wide and hundreds or thousands of kilometers in length, they resemble mature rivers with meanders, cutoffs, and abandoned channel segments. However, Venus' channels are not as tightly

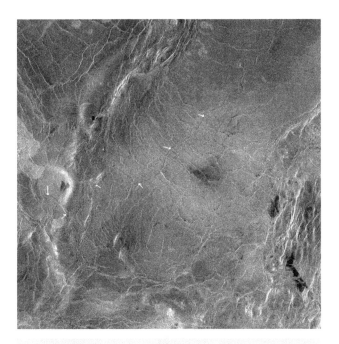

Figure 6.25 This Magellan radar image shows a 600 km-long segment of Baltis Vallis, the longest sinuous lava channel on Venus (marked with arrows). Both ends of the channel are obscured, so its original length is unknown. (NASA-JPL)

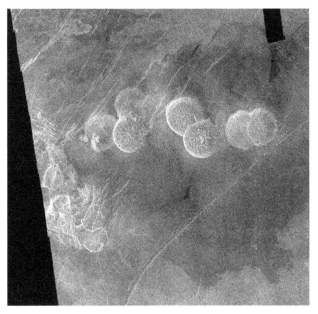

Figure 6.26 A chain of pancake domes located near the eastern edge of Alpha Regio (30°S, 11.8°E). The seven circular hills average 25 km in diameter, with maximum heights of 750 m. They are thought to have been formed by very viscous lava flows. The bright margins possibly indicate rock debris on their slopes. Some linear fractures on the plains cut through the domes, while others appear to be covered by them. (NASA-JPL)

sinuous as terrestrial rivers, and most are partly buried by younger lava plains, making their sources difficult to identify.

A few are associated with vast, radar-dark plains, suggesting large flow volumes. These appear to be older than other channel types, as they are crossed by fractures and wrinkle ridges, and are often buried by other volcanic materials. In addition, they appear to run both up and down slope, suggesting that the plains were warped by regional tectonism after channel formation.

The largest of all is Baltis Vallis, which is approximately 1.8 km wide and extends for about 6,800 km across the northern lowlands. Several hundred kilometers longer than the Nile, Earth's longest river, Baltis is the longest known channel in the Solar System.

Its remarkable length and almost constant width imply that the channel was carved by huge volumes of fast-flowing, fluid lava. Even in the hot Venus environment, the lava must have had an unusual composition, perhaps a high temperature form of basalt, or lava containing a lot of sulfur or calcium carbonate.

On the other hand, a few volcanic features seem to have formed from silica-rich, viscous lavas, which solidified close to their magma source. They fall into three types.

Pancake domes are circular in shape and typically 750 m high. Complex fractures on top of them suggest that a solid outer layer formed and then further lava flowing in the interior stretched the surface. Each dome also shows a small central pit. These pits look like the central vents found on shield volcanoes, but they seem to have formed after the domes themselves. After the domes grew during one large, slow eruption of viscous lava, the central pits seem to have formed as gases escaped from the magma or as the

dome cooled and shrank. Many domes lie near corona structures in the lowland plains.

Peculiar "tick" shaped structures are often much larger, but have fairly flat, concave summits from which ridges and valleys radiate. The origin of these lateral ridges and troughs is unknown. They may mark the scars of landslides, or possibly volcanic dykes running away from the central summit. Like the pancake domes, most ticks occur in the lowland plains far from the major shields. Many are found near the lowland coronae, and some are quite close to one or more pancake domes.

A few volcanoes have thick, fan-shaped or banded flows. Since most basalt lava flows are very fluid and fairly thin, these volcanoes seem to be made from other material, perhaps granitic or rhyolitic lavas rich in quartz. Most examples are located in one area of the southern plains.

Whether any of the volcanoes on Venus are active remains uncertain, although there is a great deal of circumstantial evidence to support this premise.

In 2010, studies of the amount of infrared radiation emitted by different surfaces (their emissivity) showed that Idunn Mons, and two other suspected volcanic areas in Themis and Dione Regio, are markedly different from the surrounding terrain. This may indicate that fresh lava flows spread across the surface only a few thousand or tens of thousands of years ago. The analysis is supported by high resolution radar maps which reveal five fresh-looking lava flows on the summit and eastern flank of Idunn Mons.

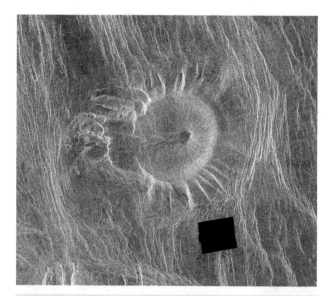

Figure 6.27 Volcanic features such as this are dubbed "ticks," since they resemble tiny insects. This example lies near the edge of Eistla Regio (18°S, 5.5°E) and is 66 km across. It has a broad, flat, slightly concave summit with a central vent. The sides are characterized by radiating ridges and valleys of uncertain origin. To the west, the rim has been breached by dark lava flows. (NASA-JPL)

Thermal measurements by Venus Express have also shown evidence of possible volcanic hot spots (see Figure 6.28). The data include transient spikes in temperature at several spots on the planet's surface. The hotspots, which were found to flare and fade over the course of only a few days, appear to be generated by active flows of lava on the surface.

The data showed spikes in temperature of more than 100°C in spots ranging in size from 1 sq. km to over 200 km wide. The spots were clustered in a large rift zone called Ganiki Chasma. Such rift zones are formed by stretching of the crust due to internal forces and hot magma that rises toward the surface.

Indirect evidence is provided by marked variations in the amount of sulfur dioxide in the atmosphere that have been detected from time to time by orbiting spacecraft (Figure 6.29). These changes have generally been attributed to large volcanic eruptions, but they could also be caused by changes in atmospheric circulation.

Coronae and Arachnoids

Some enigmatic landforms, known as coronae and arachnoids, are apparently unique to the lowlands of Venus, with no counterparts on any other world.

Coronae are circular or oval features that are surrounded by rings of ridges and troughs. (The name is derived from the Latin term for crown.) Radial fractures extend outwards, sometimes from the central regions. Coronae range in diameter from 75 km to more than 1,000 km, but they differ greatly in size and detailed

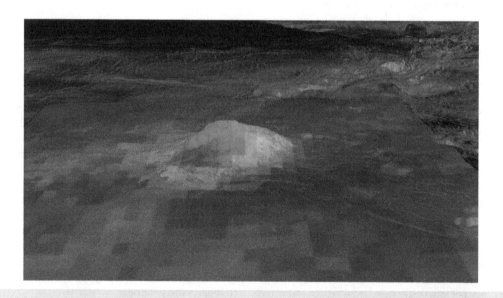

Figure 6.28 Idunn Mons volcano in Imdr Regio (46°S, 214.5°E). The brown area is 3D terrain data generated by the Magellan radar instrument, vertically exaggerated 30 times. Bright areas are rough or have steep slopes, dark areas are smooth. The colored overlay shows heat patterns derived from surface brightness data collected by Venus Express. Temperature variations due to topography were removed. The brightness indicates the composition of the minerals. Red-orange is the warmest area and purple is the coolest. The warmest area – probably recent lava flows – is centered on the summit, about 2.5 km above the plains. Idunn Mons has a diameter of about 200 km. (ESA/NASA-JPL)

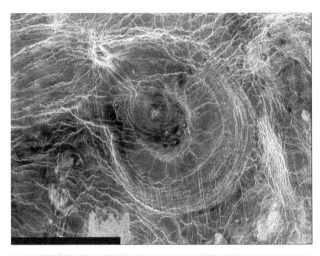

Figure 6.29 A graph showing changes in sulfur dioxide (SO_2) abundance in the upper atmosphere of Venus, measured in units of parts per billion by volume (ppbv). The data on the left are mostly from NASA's Pioneer Venus orbiter. The data on the right are from ESA's Venus Express. A large, sudden increase in sulfur dioxide can be interpreted as evidence for volcanic activity or for decadal-scale variations in the circulation of Venus' vast atmosphere. In the background is an artist's impression of volcanic terrain beneath the thick, cloud covered atmosphere. (E. Marcq et al./Venus Express, L. Esposito et al, ESA/AOES Medialab)

Figure 6.30 As the name suggests, "arachnoids" are spider-like, circular or oval structures, with a central dome or depression surrounded by a complex network of radial and concentric fractures. They range from 50 to 230 km across and may be linked with volcanic domes that collapsed, forming cracks in the surface. This example is part of an arachnoid cluster in Fortuna Regio. (NASA-JPL)

morphology – they may be domes, plateaus, plateaus with interior depressions, or rimmed depressions.

There may be a central depression which lies above or below the surrounding plains, and a possible raised rim or a moat outside the rim. Many coronae have been affected by volcanism, ranging from small vents to large, regional lava flow fields.

More than 500 coronae have been observed on Venus. Artemis Corona, to the south of Aphrodite, was once considered to be the largest of these, but it has a far larger diameter than any other coronal feature and it is almost surrounded by a huge trough, Artemis Chasma. It is now thought to be a distinct, though geologically related, type of landform (see Aphrodite Terra).

The largest example is now the 1,060 km-wide Heng-O Corona, which lies near the equator in Guinevere Planitia.

The coronae tend to occur in linear clusters along the major tectonic belts. The majority are found in three distinct geological environments: on topographic rises (11%), as relatively isolated features in the plains (25%), and along chasmata (62%). The few exceptions include three coronae located on Lakshmi Planum (a volcanic highland plateau) and five in tessera terrain.

There is no generally accepted explanation for their formation. However, most of them are thought to form over localized upwellings, or plumes, of magma, which develop as the result of partial melting at relatively shallow depths. When the subsurface magma retreats, the crust subsides, forming a depression surrounded by radial cracks.

Arachnoids are similar, but smaller, features which resemble spiders' webs in radar images. 265 arachnoids are known, with diameters between 50 and 175 km. They take the form of central depressions surrounded by a radial system of ridges and concentric ridges or fractures.

Most of them occur in the lowlands of the northern hemisphere, often in large clusters, e.g. in Bereghinya Planitia, where they are linked by a pattern of ridge-like linear features.

They may be precursors to corona formation, the result of small plumes of magma rising to the surface. Cooling and contraction of the magma may have caused the centers of volcanic mounds to collapse, cracking the crust and producing a distinctive pattern of radial and concentric fractures around a central depression. In some cases, lava may be extruded onto the surface as radial dykes.

In some cases, radial fracture patterns can be seen without a central dome or depression. Called "novae" these may represent an even earlier phase of corona formation, created when magma rises toward the surface, bulging it upward. Fractures radiate out from the center of the bulge, but subsidence due to cooling and withdrawal of magma has not yet created a central depression.

However, half of the novae are located in the inner part of coronae and seem to postdate the corona formation. Compared with coronae and arachnoids, novae are more concentrated near the equator. They are relatively scarce on the plains regions, but common near chasmata, where crustal deformation has occurred.

Tectonics

The crust of Venus is mainly basaltic, although the major uplands may bear some resemblance to Earth's continents, which contain granitic rocks high in silicon and oxygen. The surface is dominated by volcanic processes, but radar images also reveal extensive ridged plains, faulting, large rift valleys, and towering fold mountains.

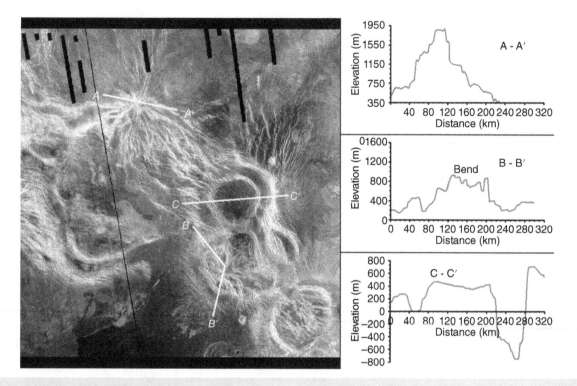

Figure 6.31 This area, which covers 945 sq. km is centered at 29.8°S, 274.3°E. It contains three features along a rift zone that illustrate a possible evolutionary sequence of corona structures. With the rise of a hot magma body a few hundred kilometers across in the mantle, the surface domes and cracks in a radial fashion (profile A-A'). As the domed structure begins to collapse, concentric faults and a surrounding trough begin to form (profile B-B'). When collapse is complete, it creates a central plateau or depression with a raised rim, surrounded by a circular trough (profile C-C', Gertjon Corona). (LPI)

The most widespread geological terrain consists of ridged plains, volcanic areas which have been deformed by wrinkle ridges. These are concentrated in the lowlands and cover about 60% of the surface. Since they are overlain by younger material in some areas, the ridged plains must have once occupied an even larger area.

Few impact craters have been deformed by the ridges, implying that the formation of the plains and most of the wrinkle ridges took place soon after the apparent resurfacing of the planet about 600 million years ago (see below). They were created over a fairly short period – no more than 100 million years – after which they were pockmarked by new impact craters.

As is the case with similar structures on the other terrestrial planets, wrinkle ridges are thought to be the products of horizontal shortening of the lithosphere, the outer layer of the planet. This would account for the consistent alignment of most wrinkle ridges over areas covering thousands of kilometers and the tendency for the ridges to encircle the major highlands.

The plains also contain regions of more marked tectonic deformation, known as "ridge belts." These often take the form of broad, linear hills rising a few hundred meters above their surroundings.

About 8% of Venus's surface is made up of tesserae. These are areas of very rough, elevated terrain which show evidence of a complex deformational history – ridges, fractures, and rift valley structures (graben). Several large elevated areas, such as Alpha, Ovda, and Thetis, are composed largely of tesserae. In some cases, they display parallel ridges and troughs that cut across one another at many different angles.[4]

The deformation of tesserae is sometimes so complex that it is difficult to determine whether horizontal compression of the crust followed or preceded extension. However, it is widely accepted that they precede the postulated major volcanic resurfacing event and represent the oldest terrain on Venus.

Tesserae are sometimes bordered by mountainous fold or thrust belts, further evidence of crustal convergence. The most impressive of these is Maxwell Montes, which rises to an altitude of 12 km above the mean level (Figure 6.34). Mountain ranges also occur elsewhere around the edge of the Lakshmi plateau. The slopes of these mountains are very steep, with gradients of up to 30° over distances of 10 km, despite the searing heat which would tend to weaken the rocks and reduce the slopes.

It seems that these towering mountains are only able to persist through continuing compression of the crust. If this was removed, the mountains would quickly subside. One possibility is that convection currents in the mantle are dragging two regions of crust towards each other. Where the currents converge and sink, the

[4]The term "tessera" comes from the Latin term for "mosaic tiles."

Box 6.3   The Venera Landers

Most of the in situ observations of the Venusian atmosphere and surface have been provided by the Soviet Venera probes. After a number of failures, Venera 4 succeeded in transmitting data during its October 1967 descent, recording a maximum temperature of 277°C and a pressure of 22 Earth atmospheres. However, Soviet claims that it had reached the surface had to be withdrawn after subsequent analysis of Mariner 5 data. The first direct sampling of the atmosphere found 90–95% carbon dioxide, plus small amounts of oxygen and water vapor.

The first successful landing took place on December 15, 1970, when Venera 7, equipped with a smaller parachute, was able to make a more rapid descent on the planet's night side, eventually touching down at 5°S, 351°E. Unfortunately, contact was lost almost immediately after arrival.

In 1972, Venera 8 landed on an upland plain not far from its predecessor. 50 minutes of data from the surface confirmed the dense carbon dioxide atmosphere and 470°C temperature. Measurements of natural gamma radiation indicated a surface made of granite-like rock. This implied the presence of water long ago, since granite is formed on Earth when basaltic rocks are driven down into the mantle by plate movements, mix with water, and then slowly re-emerge.

A simple light meter indicated that lighting was equivalent to pre-dawn illumination on Earth. It was later determined that the gloom was caused by the Sun lingering only 5° above the horizon.

ВЕНЕРА-10   25.10.1975   ОБРАБОТКА ИППИ АН СССР   28.2.1976

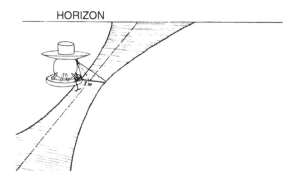

Figure 6.32   TV cameras on Venera 10 produced a 180°, black-and-white panoramic scan of a rock-covered, rolling plain to the southeast of the Beta Regio volcanic rise. The sketch shows the area scanned. (NASA)

The next generation of landers was better equipped to survive the hostile environment. Fitted with an internal cooling system and a partially collapsible "donut" base attached to shock absorbers, Veneras 9 and 10 completed their descent by aerobraking in the dense atmosphere, rather than using parachutes all the way down. Their instruments studied the cloud decks and atmosphere all the way down to the surface.

Both spacecraft touched down on the slopes of Beta Regio. Powerful floodlights on each craft were not needed, since illumination was comparable to a cloudy day on Earth. In contrast to the hurricanes raging high above, surface wind speeds were only 1.4 to 2.5 km/h. Suggestions that Beta is a giant volcanic complex were supported by gamma ray spectrometer readings indicating a basalt lava composition. However, the most notable breakthrough was the return of the first images from the surface.

The 180° black-and-white panoramic view from Venera 9 showed a possible rockslide, with 30–40 cm flat, angular boulders as far as the horizon, about 100 m away. Venera 10, which landed some 2,200 km away, revealed an older, flat plain covered with numerous rocky slabs, possibly evidence of weathered lava flows. Between the slabs were pebbles, sand, and dust.

The next pair of landers did not return any images, and few details of the surface conditions were returned. The most intriguing observations involved low-frequency radio bursts that were thought to be associated with extensive lightning activity. (Venera 9 had previously recorded a brief glow in the night sky, suggestive of lightning.)

In March 1982, Veneras 13 and 14 landed about 1,000 km apart, to the east of an upland area known as Phoebe. Venera 13 set down on rolling plains, recording a temperature of 457°C and a pressure of 89 atmospheres, while Venera 14 landed in a low-lying basin at 465°C and 94 atmospheres. They sent back the first (and so far, only) color pictures from the surface, revealing rust-colored landscapes draped beneath a lurid orange sky.

Venera 13 landed with a tilt of about 8°. Its 360° panorama showed a flat area with scattered angular rocks and pebbles of all sizes. A new instrument, known as an X-ray fluorescence spectrometer, was used to analyze a small sample of rock drilled from a depth of 3 cm. The results indicated the presence of an unusual type of basalt that is usually associated with continental volcanism on Earth. Venera 14 came down on a tabular terrain marked by continuous layers of bedrock divided into slabs. Analysis indicated a composition similar to the fluid basalt lavas found on Earth's ocean floors.

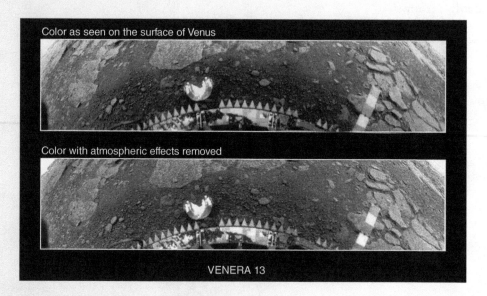

Figure 6.33   Venera 13 carried a twin camera system that imaged a 360° panorama, half of which is shown here. The color images showed a surface with an orange hue as a result of the atmosphere absorbing light at blue and green wavelengths. When this effect is eliminated, the rocks appear gray. In the foreground is a discarded lens cap. The saw-toothed object at the bottom of each image is the base of the lander. The teeth are about 5 cm apart. Also visible is a color bar for image calibration. Note the flat slabs separated by small stones and other loose debris. (James Head, Brown University/USSR Academy of Sciences)

The final Soviet landings took place in June 1985, when Vegas 1 and 2 dropped off capsules and balloons en route to comet Halley (see Box 6.1). No surface images were returned.

crust piles up into a thick, compressed mass, forming the mountain ranges.

Data based on perturbations of orbiting spacecraft show that the high gravity of Maxwell Montes correlates with thick crust and high topography. This suggests that the mountains of Ishtar Terra are made of low-density rock that rests on the denser mantle and is not uplifted by magma plumes thrusting upwards from beneath.

Uncertainty also exists over the origin of the deep, narrow canyons called "**chasmata**." These troughs often have a raised rim and may be many thousands of kilometers long, up to 100 km wide and up to 10 km deep. Some chasmata occur in major volcanic highlands, such as Atla Regio and Beta Regio. Many are associated with volcanism and are located around the margins of major coronae, such as Artemis and Latona.

The shapes and sizes of these features suggest that they are rift valleys caused by the crust pulling apart – rather like the East African Rift Valley on Earth. However, some chasmata appear to be associated with crustal compression – more like oceanic trenches on Earth.

A One Plate Planet?

On Earth, fold mountains, volcanoes, and rifts are generally located in linear zones along plate boundaries. No such pattern exists on Venus. It seems evident, therefore, that Earth-like plate tectonics, involving destruction of crust along subduction zones and creation of crust along ocean ridges, do not occur on Venus.

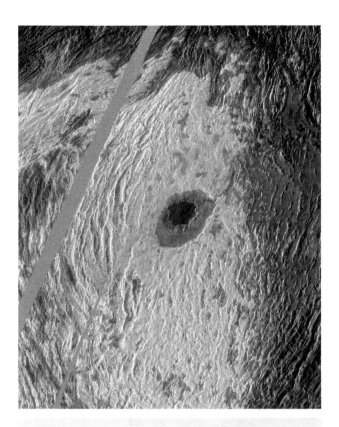

Figure 6.34 A Magellan radar image of western Maxwell Montes, the highest mountains on Venus, which rise up to 12 km above the mean planetary surface. The mountains were formed by compression of the crust. Although they reflect radar strongly, this may be due to the presence of an unusual radar-reflective mineral, such as lead sulfide, rather than a rough surface. The 105 km-diameter Cleopatra impact crater can be seen in the center of the image. The distance from Cleopatra to the steep south western slopes of Maxwell Montes is about 350 km. (NASA-JPL)

Indeed, Venus has been described as a one-plate planet, in contrast to the dozen or so rigid lithospheric plates that exist on Earth.

Assuming that Venus has an Earth-like, mobile mantle, the absence of plate tectonics could be accounted for by a thicker crust which has difficulty in sliding over the mantle. It may also be linked to the fact that Venus is much drier, deprived of the water that lubricates Earth's plate tectonics. (Water lowers the melting point of lithospheric rocks and makes them more malleable.)

Scientists hoped that gravity and magnetic field measurements by orbiting spacecraft would provide definitive evidence about the thickness of the crust. Unfortunately, the data are open to different interpretations, so that estimates of the thickness of the Venusian lithosphere range from perhaps 60 km to about 300 km.

However, radar images have provided strong evidence that the average age of the surface is 300–750 million years, indicating that the entire planet may have been resurfaced by volcanic eruptions in the geologically recent past. Not only are the volcanic landforms

and impact craters uniformly distributed, but all signs of any older, heavily cratered, terrain have been destroyed, with the possible exception of the major upland regions.

This has led scientists to propose a number of different scenarios:

1. Venus may undergo episodic resurfacing. For a long period of time, creation of new crust and destruction of existing crust ceases, so that heat builds up in the interior. Eventually, the surface becomes disrupted and catastrophically sinks into the mantle. The result is rapid overturning of the interior and virtually planet-wide volcanism, creating a vast magma ocean. This extracts so much heat from the interior that the mantle cools and the activity dies down, enabling a solid surface to be re-established. After a period of at least 500 million years, the process repeats itself. In this scenario, the visible surface is simply a record of the last resurfacing episode, some 600 million years ago.
2. Venus was continuously active, experiencing plate tectonics similar to those found on Earth. However, the interior then cooled sufficiently to shut down the crustal recycling process about 600 million years ago. In this case, convection in the mantle will continue to exert stresses on the lithosphere, but these are relieved in hot spots or relatively small compression zones instead of being concentrated at plate boundaries.

Magnetic Field

Spacecraft data show that the planet has a magnetic field no more than 0.09% the strength of Earth's. Since, like Earth, Venus almost certainly has a liquid, iron-rich core, the explanation must lie in its very slow rotation, which prevents the generation of an internal dynamo.

This means that there is no magnetic shield opposing the incoming solar wind. Instead, the solar protons and electrons strike the upper atmosphere and spacecraft data have shown that oxygen ions – a component of water – are stripped away by the solar wind. This may help to explain the planet's lack of water at the present time.

At the same time, solar UV and X-ray radiation splits neutral atoms to produce a region of positively charged particles known as the ionosphere. During the continuous battle with the solar wind, this region of the upper atmosphere is able to slow and divert the flow of particles around the planet, creating a small magnetosphere shaped rather like a comet's tail.

The height of the ionosphere's upper boundary, the ionopause, varies with the strength of the solar wind, typically dropping to 250 km or rising to a height of 1,500 km. However, during periods of very low solar wind pressure, it can extend outward at least 15,000 km from the planet, forming a long tail.

Although there is a well-developed bow shock in Venus' outer atmosphere, there is no evidence of charged particles being trapped. As a result, there are no radiation belts above the planet. The negligible magnetic field also means that only feeble auroras occur on the planet's night side.

Box 6.4   Pioneer Venus

In contrast to the sustained efforts of the Soviet Union to send landers to Venus, NASA concentrated its early efforts on the Pioneer Venus mission, which involved an orbiter and five atmospheric probes.

The drum-shaped orbiter, sometimes known as Pioneer 12, carried a dozen instruments to explore the planet's interaction with the solar wind, the composition of the clouds and the upper atmosphere, the shape of its gravitational field and the topography of its hidden surface. Its most innovative experiment was a radar mapper/altimeter to reveal the nature of the surface, after centuries of speculation.

In contrast, the Multiprobe mission of Pioneer 13 was designed to deliver four atmospheric probes which were attached to the top of a 2.5 m-diameter "Bus." After separating from the Bus, the probes, protected by heat shields, were designed to send back data during their descent to the surface.

Figure 6.35   Pioneer Venus comprised two separate missions. The Multiprobe mission (left) included the main Bus, which was not designed to survive atmospheric entry, and four atmospheric probes, each protected by a heat shield. The large Sounder Probe carried seven instruments, compared with three on the North, Day and Night Probes. The Pioneer Venus Orbiter (right) entered a near-polar, elliptical orbit, observing the atmosphere and mapping most of the surface with radar. (NASA-Ames)

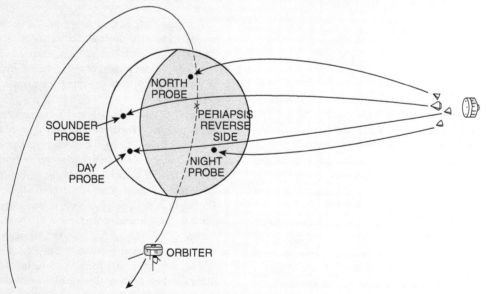

Figure 6.36   The Pioneer Venus Multiprobe mission began on November 15, 1978, with the release of the large Sounder Probe. Its three small companions followed four days later, when the Bus was 9.3 million km from Venus. They were targeted to investigate different regions of the Earth-facing hemisphere – two on the night side and two on the day side. The Sounder Probe slammed into the atmosphere on December 9, followed a few minutes later by the other probes. Each hit the surface at 32–35 km/h, but only the Day Probe sent back any meaningful data after impact. (NASA-Ames)

The orbiter set off from Cape Canaveral on May 20, 1978, following a long, looping trajectory that minimized the amount of onboard propellant required. Pioneer 13 took a faster route, so although the probes left Earth 2½ months after the orbiter, they slammed into Venus' atmosphere only 5 days after Pioneer 12 arrived on December 4.

The Sounder Probe entered the atmosphere at 42,000 km/h. Once its main parachute opened, the probe continued to descend for 16 minutes before the chute was cut loose. The final free fall lasted 39 minutes, ending with a loss of contact on impact.

The smaller probes followed diverging paths: one of them plunged into the atmosphere on Venus' day side, while the other two headed into the night. They survived their ballistic descent, and by the time they hit the ground, the dense atmosphere had slowed them to a speed of 35 km/h. The only useful data from the surface came from the Day Probe, which continued operating for 67 minutes.

Meanwhile, Pioneer 12 went into a near-polar orbit, which enabled it to survey almost the entire planet. The spacecraft survived for 14 years before it ran out of fuel and was destroyed in the dense atmosphere. During its long life, it mapped most of the surface, studied the clouds and upper atmosphere, and confirmed that Venus has a negligible magnetic field.

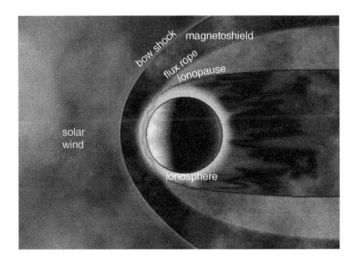

Figure 6.37 Venus has no internal magnetic field, so the solar wind interacts directly with the upper atmosphere. The solar particles are slowed when they reach the ionosphere, with a bow shock forming upstream. The solar wind then flows around the planet, forming a long magnetotail. This process slowly removes electrified (ionized) gases from the upper atmosphere. The ionosphere is usually confined close to the planet, but it may move outward to form a long tail when solar wind pressure is very low. (After UCAR)

Box 6.5   Magellan

NASA's Magellan orbiter, which entered a near-polar orbit around Venus in August 1990, provided the most detailed radar images and surface maps of the planet. During each orbit, its synthetic aperture radar (SAR) pierced the cloud blanket, imaging the landscape below in narrow strips about 24 km wide and 16,000 km long. By studying radar echoes from the surface, scientists were able to determine the nature of the terrain and map the topography.

Over a period equal to three Venus rotations (roughly two Earth years), Magellan mapped 98% of the surface with a spatial resolution of 120–250 m. During repeat passes, images taken from different viewing angles made it possible to obtain stereo views and search for surface changes. In September 1992, a fourth observation cycle was devoted to mapping the planet's gravity and internal structure.

By the end of its mission in October 1994, Magellan had returned more data than all previous U.S. planetary missions. Perhaps the most notable discovery was the importance of volcanic activity on Venus. There were volcanoes of all shapes and sizes, as well as unique features termed "coronae" and "arachnoids." The relative absence of impact craters suggested that Venus had been almost entirely resurfaced by lava flows about 600 million years ago.

Figure 6.38   Magellan followed an elliptical, near-polar orbit around Venus between 1990 and 1994. When the orbit was close to Venus, the synthetic aperture radar imaged a swath between 16 and 27 km wide, beginning near the north pole and continuing to the southern hemisphere. Subsequent swaths slightly overlapped, enabling the instrument to map 98% of the planet. Near apoapsis (its furthest distance from Venus), the high-gain antenna was turned toward Earth and data were transmitted. Magellan acquired radar data for about 37 minutes during each three-hour orbit. (NASA-JPL)

Questions

- Venus and Earth are often described as "twin" or "sister" planets. Summarize the reasons for this description.
- Briefly describe the ways in which the physical characteristics of Earth and Venus are different, giving reasons for these differences.
- Why is Venus said to have experienced a "runaway greenhouse" effect? Give reasons why such an effect could or could not take place on Earth.
- Describe the main characteristics of Venus' atmosphere and clouds.

- What evidence supports the theory that Venus was almost completely resurfaced about 600 million years ago? How might this resurfacing have taken place?
- Is Venus still volcanically active? Give reasons for your answer.
- Briefly describe the appearance and possible formation process of these Venusian landforms: coronae; tesserae; arachnoids; wrinkle ridges; chasmata.
- Explain why Venus has no magnetic field. How does this influence its near-space environment?
- Summarize the major difficulties to be overcome when planning a future space mission to Venus.

SEVEN
Mars

Mars is the fourth planet from the Sun and one of Earth's closest neighbors. Seen with the naked eye as a bright, reddish-orange "star," Mars is named after the ancient Roman god of war. However, its color is not caused by human blood, but the presence of weathered, iron-rich minerals in its surface rocks and dust. In the late 19th and early 20th centuries, science fiction writers

(and some serious observers) believed that the planet might harbor vegetation, possibly cultivated by intelligent beings, but the advent of space probes has made it clear that, if any life forms do exist on the red planet, they can only take the form of simple microrganisms.

As one of the "wandering stars" identified in prehistoric times, Mars appears in the night sky every 26 months or so, moving fairly rapidly through the constellations and increasing in brightness until it outshines everything apart from the Moon, Venus, and Jupiter. Observations of its looping path in the sky around these times of opposition, caused by Earth overtaking Mars on the inside track, helped to confirm the Copernican system of a heliocentric Universe.

Since it lies beyond Earth's orbit, Mars travels more slowly around the Sun, with an orbital period (its year) of 687 Earth days. The average distance from Mars to the Sun is about 227,920,000 km, roughly 1½ times the Earth–Sun distance. The orbit is tilted to the ecliptic by less than 2°.

The times of opposition and the brightness of the planet at such times are quite variable, because Mars follows a more eccentric orbit than most of the planets. The Sun–Mars distance can range from 206.6 million km to 249.2 million km. Consequently, the minimum distance between Earth and Mars at opposition also

varies considerably, ranging from 55.7 million km to 101 million km. During the most favorable oppositions, Mars reaches visual magnitude–2.9, outshining Jupiter in the night sky.[1]

The apparent size of the planet's disk changes correspondingly. It is largest (more than 25 arcsec) during close oppositions, when considerable surface detail is visible with modest telescopes, and smallest (less than 14 arcsec) during distant oppositions, when ground-based observations are seriously limited (Figure 7.1).

Since it is further from the Sun and receives less than half as much solar radiation as Earth, Mars is much colder. Surface temperatures vary from –130°C near the poles during the winter to over 30°C at midday near the equator. The average temperature is –63°C, comparable to severe conditions in Antarctica.

Sols and Seasons

Like most of the planets, Mars rotates on its axis from west to east. Visitors would see a much smaller, dimmer Sun in the pinkish sky,

---

[1] The closest opposition of historic times took place on August 28, 2003, when Earth and Mars were only 55.758 million km apart. We will have to wait until August 28, 2287, for a closer encounter.

Exploring the Solar System, Second Edition. Peter Bond.
© 2020 John Wiley & Sons Ltd. Published 2020 by John Wiley & Sons Ltd.
Companion Website: www.wiley.com/go/Bond-Solar-System2e

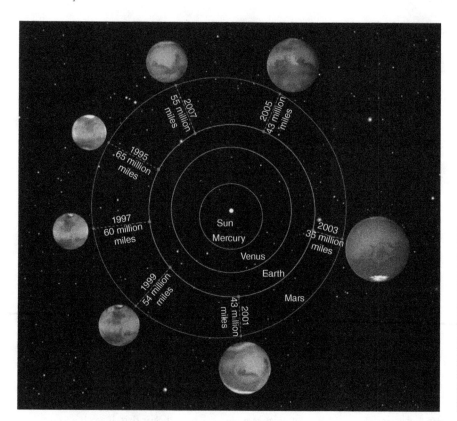

Figure 7.1 Mars follows a much more elliptical orbit than Earth. As a result, the Earth–Mars distance at each opposition varies considerably. In the period 1995–2007, the gulf between them ranged from 55.76 million km in 2003, when Mars was near perihelion, to 101.08 million km in 1995, near aphelion. The images compare the planet's apparent size at each opposition. (Z. Levay / STScI)

but otherwise conditions of illumination would be quite familiar. The Martian solar day (sol) – the time in which the planet rotates once with respect to the Sun – lasts 24 hours 39 minutes 35 seconds, slightly longer than a day on Earth.

Since the axes of both planets are tilted by similar amounts (see Table 7.1) relative to the plane of their orbits, they also have seasons. The axial inclination causes the amount of sunlight falling on certain parts of the planet to vary widely: regions near the poles experience continuous sunlight in summer, but nonstop darkness in winter.

The situation is complicated, however, by the eccentric orbit of Mars. On Earth, spring, summer, autumn, and winter are all similar in length, because the orbit is nearly circular and the planet moves at an almost constant speed around the Sun. However, the elongated orbit of Mars causes its seasons to vary significantly in length (see Table 7.2).

Since it travels faster along its orbit near perihelion, the southern spring and summer are noticeably shorter than the southern autumn and winter (and *vice versa* in the northern hemisphere). Martian months are defined as spanning 30° in solar longitude, but, due to the eccentricity of Mars' orbit, their length varies from 46 to 67 sols (Box 7.1).

It also means that Mars receives 44% more insolation at perihelion than at aphelion. Hence, the pole which is tilted toward the Sun at perihelion has warmer summers than the other pole. At present, the south has warmer summers, but this situation is not permanent. As a result of precession – a slow change in the

Table 7.1
Comparison of Mars and Earth

	EARTH	MARS
Diameter (km)	12,756	6,792
Rotation period	23 h 56 m	24 h 37 m
Density (water = 1)	5.52	3.91
Mass (Earth = 1)	1	0.11
Surface gravity (Earth = 1)	1	0.38
Inclination of axis to orbit	23.45°	25.19°
Orbital period (days)	365.24	686.93
Average distance from Sun (million km)	149.6	227.9
Average temperature (°C)	7	–63
Satellites	1	2

Table 7.2
Length of Seasons on Mars and Earth (northern hemisphere)

	EARTH (days)	MARS (approx. Earth days/sols)
Spring	93	199/194
Summer	94	184/178
Autumn	89	146/142
Winter	89	158/154

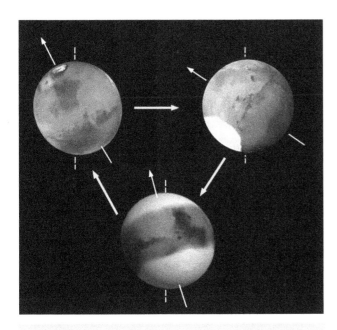

Figure 7.2   The evolution of Martian surface ice during an obliquity cycle. The angle between the white arrows and the dotted line denotes the obliquity (axial tilt). At high obliquity, the northern cap sublimates, losing a few centimeters of ice each year. The water vapor is then deposited as snow or frost in the cooler equatorial zones. When the obliquity decreases, ice sublimates from the lower latitudes and freezes out at high latitudes, resulting in growth of the polar caps. (ASD/IMCCE-CNRS, adapted from Jim Head/Brown University and NASA/JPL)

direction of tilt of the rotational axis – the hot and cold poles switch places every 51,000 years.

The obliquity of Mars is subject to large, chaotic changes, whereas Earth experiences only minor changes. The axial tilt of Mars generally ranges between 15° and 35°, but it may occasionally reach 60°, when the planet is almost lying on its side, relative to its orbital plane. Summer temperatures at the poles change dramatically during this obliquity cycle, becoming much warmer when the axial tilt is high than when it is fairly small. This means that Mars undergoes major climatic changes, shifting from ice ages to relatively warm interglacials.

At present, the fairly low obliquity means that it is coming out of an ice age which occurred between 500,000 and two million years ago. During this interglacial period, the ice cover has been degrading in the mid-latitudes as water ice sublimates and freezes out at the poles.

Physical Characteristics

With an equatorial diameter of 6,792 km, Mars is considerably smaller than both Earth and Venus, but larger than Mercury

(Figure 7.4). Its fairly rapid rotation causes it to bulge outward at the equator, so that its diameter is about 40 km less at the poles than at the equator.

It is also quite lightweight compared with Earth, which is about 10 times more massive. With an average density about 3.9 times that of water, Mars has roughly 70% of Earth's density and is clearly made of rocky material.

Because Mars is so much smaller and less dense, its surface gravity is only about 38% of that on Earth. Objects would fall to the ground more slowly than on Earth and lifting of heavy objects would be much easier.

Interior

Theories about the nature of the interior are based on modeling, Martian meteorites, comparisons with Earth, and the shape, mass, and moment of inertia of the planet.[2] It is almost certainly layered as the result of differentiation of lighter and denser materials very early in its history (Figure 7.5).

The average thickness of the crust is probably about 50 km – more than on Earth – since there is no evidence of craters that have punched through to the mantle, or of surface deformation due to the weight of overlying ice caps. On Earth, the weight of a similar stack of ice would cause the surface to sag, so the Martian lithosphere – its crust and upper mantle – must be markedly thicker and colder.

This is not surprising, since the outer layers of the smaller planet should cool faster and solidify to a greater depth. However, the crust is far from uniform in thickness, being thicker and older in the south than in the north (see North–South Divide).

The mantle is probably similar in composition to Earth's mantle, and composed mainly of peridotite, a rock made up chiefly of silicon, oxygen, iron, and magnesium. The most abundant mineral in peridotite is olivine. The average temperature of the mantle may be about 1,500°C, with the radioactive decay of elements such as uranium, potassium, and thorium as the main source of internal heat.

The dense core is probably composed of iron, nickel, and sulfur. Models of how the planets condensed from the solar nebula suggest that Mars, being farther from the Sun, should have incorporated a higher proportion of moderately volatile elements, such as sulfur, sodium, and potassium, than Earth.

The relatively low density of the planet, compared with Earth, indicates that the core makes up a smaller proportion of the interior. If it is mainly composed of iron, then the minimum core radius would be about 1,300 km. If it comprises less dense material, such as a mixture of sulfur and iron, the maximum radius may be up to 2,000 km.

Until recently, it was widely assumed that – since Mars has a relatively small volume – its core has largely cooled and solidified, but recent modeling suggests that the outer core may still be liquid, with plumes of hot material rising from the core–mantle boundary.

---

[2] The moment of inertia is a measure of the way in which the planet rotates on its axis, which is influenced by the distribution of material within it. The measurements reveal that the central core is dense while the outer regions are less dense.

Box 7.1 The Martian Calendar

A Martian year lasts for 668.6 sols (Martian days). Unlike on Earth, there is no large satellite that can be used to create a calendar based on "months." Scientists mark the progress of the year on Mars by referring to solar longitude (Ls). Martian months are defined as spanning 30° in solar longitude. Ls is 0° at the vernal equinox (beginning of northern spring), 90° at summer solstice, 180° at autumnal equinox, and 270° at winter solstice.

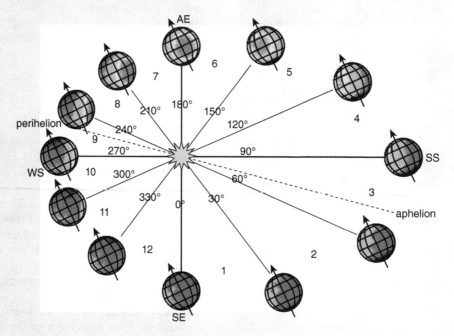

Figure 7.3   Martian seasons are based on the solar longitude, abbreviated Ls. Ls is 0° at the spring equinox (SE – beginning of northern spring), 90° at summer solstice (SS), 180° at autumnal equinox (AE), and 270° at winter solstice (WS). The orbit is quite eccentric, so the planet's distance from the Sun is 206.6 million km at perihelion and 249.2 million km at aphelion. *(Laboratoire de Météorologie Dynamique)*

Mars currently reaches aphelion at Ls = 71°, near the northern summer solstice, and perihelion at Ls = 251°, near the southern summer solstice. The Martian dust storm season usually begins just after perihelion, at around Ls = 260°.

Of course, this means that the dates when the Martian seasons begin and end vary from year to year on our calendars.

Figure 7.4   Mars and Earth shown to scale. With a diameter of 6,792 km, Mars has a surface area comparable to all the land masses on our planet. Whereas Earth is largely covered by oceans, Mars is covered by ochre deserts. (NASA)

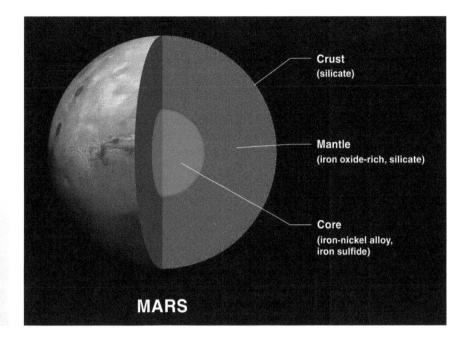

Figure 7.5 Mars has a layered interior, with a solid crust, a mantle, and a dense, iron-rich core. The outer core may still be at least partially liquid and conductive (though incapable of generating a global magnetic field), while the inner core may be solid. (LPI)

Impact Features

Mars has many surface features that are common on Earth: plains, canyons, volcanoes, valleys, gullies, and polar ice. The main exception is impact craters. When the Mariner spacecraft sent back the first images of its surface, they showed a heavily cratered landscape rather like the Moon. The density of cratering is much higher in the southern hemisphere, where most of the large impact features are found.

The largest of these structures is the Hellas basin, in the southern uplands (Figure 7.8). Its diameter from rim to rim is about 2,300 km, while its depth is more than 8 km – making it the deepest place on the planet. Overall, Hellas is only slightly smaller than the South Pole–Aitken basin on the Moon – the largest recognized impact structure in the Solar System. It is surrounded by a huge volume of excavated material, which, distributed evenly, would cover the continental US to a depth of 3 km.

Many smaller craters are found on its floor and in the surrounding uplands, indicating that it dates from the heavy bombardment of the Noachian era, just over 4 billion years ago. The mountains of ejecta along its rim are surprisingly low and dissected by valleys, indicating that they have been destroyed by erosion. Any surface drainage would be expected to travel down slope, away from the rim, but many valleys flow into the basin.

The basin floor is covered with layers of sediments deposited by wind and water during the Hesperian and Amazonian periods. These weakly consolidated deposits of sand and rock have been carved into remarkable curvilinear shapes. Some of the strange landforms may be created by glaciers of water ice buried beneath layers of dirt and rock (see Glaciation).

Hellas appears bright even in Earth-based telescopes, not just because of its great size and sand-colored sediments, but also because of its proximity to the south pole. Its low temperature in winter results in the formation of surface frost and low-level, morning haze.

The second-largest impact basin, also located in the southern hemisphere, is Argyre. Once again, the rim of this ancient structure has been eroded and dissected (Figure 7.9). One of the most striking features is an apparent dry river channel that runs from the northern rim toward the equatorial lowlands. Water may once have filled the basin until it overflowed, allowing floodwater to travel northwards until it eventually disgorged onto the northern plains (see Outflow Channels).

There are hundreds of thousands of craters on Mars, many of them without names. The vast majority of these have similar morphologies to those on the Moon and other planets, the main exception being pedestal craters (see Pedestal Craters). Small craters, no more than 8–10 km across, are simple, bowl-shaped depressions with constant depth-to-diameter ratios. Larger craters, up to about 140 km across, are more complex, with central peaks and a lower depth-to-diameter ratio. Largest of all are the multi-ringed basins.

Erosion rates at low latitudes for most of Mars' existence seem to have been very low. However, rates of erosion (and deposition) were much higher early in its history, when craters were modified by running water and glaciers, as well as wind. As a consequence, in the southern uplands, they range from sharply delineated circular structures to barely discernible, rimless, circular depressions. In contrast, almost all of the craters on the volcanic plains of the equatorial regions appear fresh, even though they may be billions of years old.

Frequent cratering is still taking place (see Figure 7.11). Many of these small craters are excavated in hidden surface ice. The bright hollows soon darken as freshly exposed ice sublimates into the atmosphere.

North–South Divide

One of the most surprising surface features is the dichotomy between the less cratered, lowland plains that cover much of the northern hemisphere and the ancient, heavily cratered, highlands

Box 7.2   Dating Mars

Mars appears to have experienced a similar cratering history to the Moon, although we do not know if the timing coincides precisely. As with all rocky planets, the age of the surface is estimated by counting craters: a higher number and density of craters indicates older terrain (see Chapter 4). Complications arise because Mars has undergone periods of large-scale volcanism, as well as erosion by glaciers and running water, and widespread deposition of sediments that may bury older craters.

Based on the presence of the largest impact structures, the highest crater densities, and the impact history of the inner Solar System, the southern highlands of Mars coincide with the oldest crust. They probably formed prior to 3.8 billion years ago. The more sparsely cratered northern plains must be younger, since they have fewer and smaller craters, having formed after the end of the great bombardment.

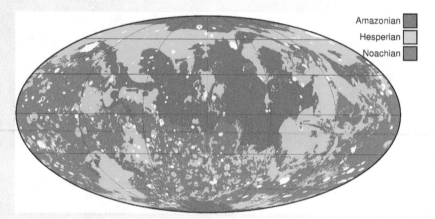

Figure 7.6   A map showing the ages of the major surface units. Most of the younger (Amazonian) surface is in the north, while the oldest terrain, the Noachian, is concentrated in the southern highlands. Major impact craters and their ejecta are marked in white. *(Sean C. Solomon et al.)*

The geological history of Mars has been divided into four main periods: pre-Noachian, Noachian, Hesperian, and Amazonian.
The geological record of the earliest events in Martian history, the so-called pre-Noachian era, has been largely erased by subsequent impacts and other resurfacing events.

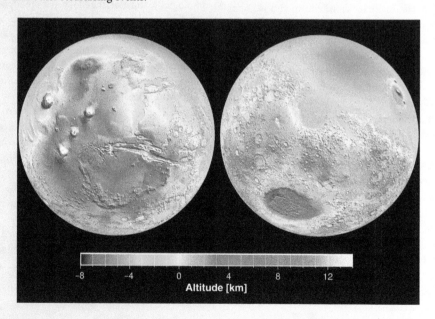

Figure 7.7   Global topography, based on data from the Mars Orbiter Laser Altimeter. Note the dichotomy between the heavily cratered southern highlands (yellow/orange) and the smooth, sparsely cratered northern lowlands (blue/green). Vertical accuracy is less than 5 m. Note the giant Hellas impact basin at right (purple, surrounded by mountains). Also, the Tharsis rise (red and white) and Valles Marineris (left image). *(David Smith/NASA/JPL)*

The Noachian period probably lasted from about 4.1 billion years ago until about 3.8 billion years ago. This was a period of heavy bombardment, intense volcanic activity, and warmer climate, with lakes, or possibly a shallow ocean. The Hellas and Argyre basins in the southern highlands date from this era. The Tharsis bulge and the Valles Marineris are also thought to have formed in this period. The atmosphere may have been thick enough for liquid water to survive for lengthy periods on the surface, and even for rain to fall from the sky. It is named after the ancient uplands of Noachis Terra in the southern hemisphere.

The Hesperian period (3.8 to perhaps 3 billion years ago) is considered to be a time of increasing cold and dryness as the greenhouse gases dispersed and thick permafrost formed. However, the planet may have remained warm enough for lakes – perhaps covered by ice – to have persisted until at least 3 billion years ago. There were also periodic and localized, but catastrophic, floods, caused by melting of ice. Most of the large outflow channels were probably carved and then abandoned at this time. The period was also marked by increased volcanic activity (including the birth of Olympus Mons). It is named after the plains of Hesperia Planum, not far from Noachis in the southern hemisphere.

The Amazonian period includes most of Martian history, extending from the Hesperian era to the present day. Landscapes of this age have few meteorite impact craters, all of them small. The climate was generally dry and cold. Some landscape changes occurred, e.g. eruptions on Olympus Mons and widespread lava flows elsewhere, landslides in Valles Marineris, and formation of broad plains and sand dunes near the poles. The period is named after Amazonis Planitia, a broad area of plains in the northern hemisphere.

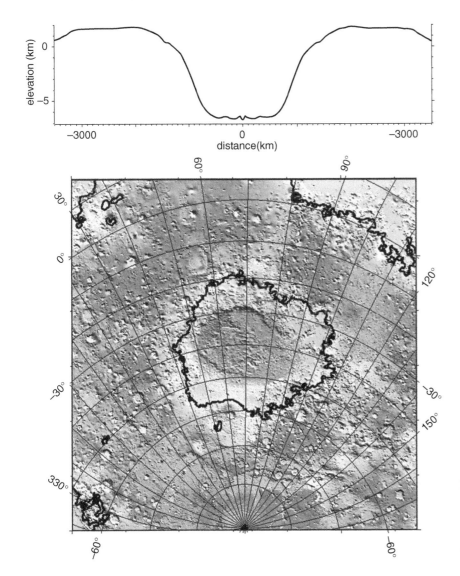

Figure 7.8 A topographic map of the Hellas impact basin, the largest on Mars, based on laser altimetry data. The upper panel shows a cross section through the basin. Hellas is about 2,300 km wide and over 8 km deep. (NASA-JPL)

Figure 7.9   A view of the southern part of the Argyre basin, taken by Viking Orbiter 1. The fragmented, eroded rim (right) may have been breached when water filled the basin and overflowed. Note the orange, dust-laden atmosphere above the horizon. (NASA-JPL)

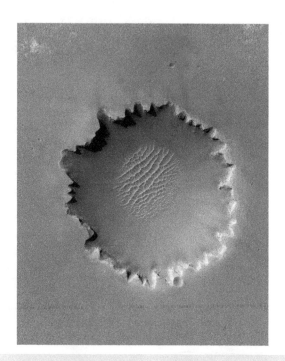

Figure 7.10   Victoria crater, in Meridiani Planum, is about 750 m in diameter and 75 m deep, with a 5 m-high raised rim. This HiRISE image shows a degraded, scalloped rim and sand dunes on the crater floor. Layered sedimentary rocks are exposed along the inner wall, and boulders have fallen into the crater. It has been infilled by sediments and widened by erosion over billions of years. (NASA/JPL/University of Arizona)

which dominate the southern hemisphere (Figures 7.7 and 7.12). The much lower number of craters in the north indicates that the plains are probably hundreds of millions of years younger than the highlands.

Altimetry maps show a difference in altitude of a few kilometers between the two regions, with a clearly delineated, fairly steep, boundary between them. Gravity data obtained by studying minute changes in the paths of orbiting spacecraft indicate that the mean crustal thickness is about 60 km in the south and 35 km in the north.

The reason for the dichotomy remains uncertain. One suggestion is that a huge impact early in the planet's history removed much of the original crust in the north. The absence of an obvious impact basin can be explained by the subsequent superimposition of the huge Tharsis volcanic bulge, and the fact that the 2,000 km-wide impactor may have arrived at a low angle, thus excavating an elliptical depression.

This theory is supported by gravity data from the Mars Reconnaissance Orbiter, which show signs of a huge, but indistinct, impact basin. 8,500 km across and 10,600 km long – the size of Asia, Europe, and Australia combined – it seems to cover about 40% of the planet's surface, making it far larger than any other known impact basin in the Solar System.

Other studies have suggested that a number of separate impacts could have combined to reproduce the irregular dichotomy boundary, with evidence for numerous hidden craters (known as

"quasi-circular depressions") provided by gravity and altimetry data. Likely subdued impact basins in the northern plains include Chryse Planitia, Acidalia Planitia, and Utopia Planitia.

Tectonics

The Martian crust also bears scars of large-scale tectonic activity early in the planet's history, such as the Tharsis bulge and the Valles Marineris canyon system. However, there is little evidence for global plate tectonics, in which the motions of separate slabs of lithosphere drive the formation of fold mountains, volcanoes, etc. (see Chapter 3).

Crustal spreading may have begun when Mars was very young, but then shut down quite rapidly when the interior cooled. Some of the oldest rocks are weakly magnetized, evidence that they cooled and solidified in the presence of a strong magnetic field that was generated by an internal dynamo (Figure 7.82). The banded pattern of strongly magnetized crust in the southern highlands bears some resemblance to the magnetic stripes imprinted on Earth's ocean floors by spreading on either side of a mid-ocean ridge.

The existence of a strong, reversing magnetic field is seen as good evidence that Mars once had a sufficiently high internal temperature to drive a dynamo in the iron-rich core. This would also be consistent with a vigorously convecting mantle, a thin

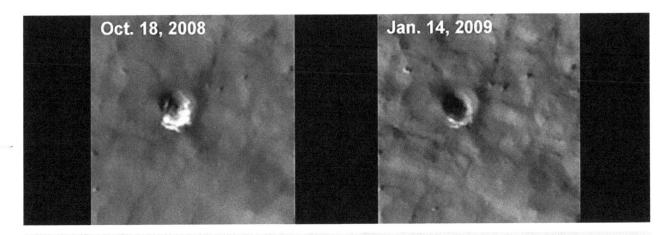

Figure 7.11   HiRISE images of a recent, 6 m-wide impact crater, taken on October 18, 2008, (left) and January 14, 2009. Each picture is 35 m across. The bright, freshly exposed ice darkened rapidly as it sublimated into the thin atmosphere. (NASA/JPL-Caltech/University of Arizona)

Figure 7.12   A cross section showing the thickness of the crust – about 60 km in the south and 35 km in the north – based on altimetry data from Mars Global Surveyor. It also shows the difference in altitude above the mean datum level, with the southern highlands (red-orange) separated from the lowland plains of the north (green-blue) by a sharply delineated boundary near the equator. (NASA-JPL/MOLA)

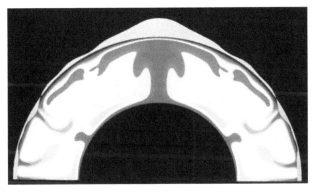

Figure 7.13   A computer simulation of mantle convection, modeling the internal processes that may have produced the Tharsis bulge. Hot regions are red, cold regions are blue and green, with a difference of around 1,000°C. Thermal expansion causes hot rock to have a lower density than cold rock, so hot mantle material rises toward the surface whilst colder material sinks, creating large-scale convectional circulation. The rising plume pushes up the crust, while the sinking mantle tends to pull the surface down. Deformation of the surface, shown in gray, is highly exaggerated: actual uplift in central Tharsis was about 8 km. This uplift also stretched the crust, forming fractures, grabens, and Valles Marineris. The hot, rising mantle material melted as it approached the surface, producing volcanic activity. (Walter S. Kiefer, Amanda Kubala – LPI)

lithosphere, and – possibly – plate tectonics. However, Mars lost its internal heat very quickly, so that the lithosphere grew too thick and rigid to easily break up into tectonic plates.

The absence of widespread plate tectonics means that almost all of the familiar deformational features on Earth are rare or absent. Mountains on Mars are associated with large impact craters, rather than plate movements. One possible exception is the Thaumasia mountain range, located to the south east of the Tharsis volcanic province and south of the Valles Marineris. These curved uplands may have been created by powerful crustal forces associated with the formation of the major tectonic features.

The lack of lateral plate motion means that any hot spots in the mantle stay in a fixed position relative to the surface. This, along with the lower surface gravity, may account for the huge Tharsis bulge and its enormous volcanoes.

The Tharsis bulge dominates the tectonics of Mars. The huge dome is thought to have been created by a long-lived, stationary hot spot in the mantle, where a rising superplume of magma

Figure 7.14 Acheron Fossae was once a region of intense tectonic activity and is part of a network of extensional fractures that radiates outward from the Tharsis bulge. This image from Mars Express shows curved depressions up to 1.7 km deep, created by faulting. When the crust dropped between parallel faults, a graben was formed. Fossa is the Latin word for trough. (ESA/DLR/FU-G. Neukum)

Figure 7.15 A Viking Orbiter mosaic of the north eastern Tharsis bulge showing an area of intense faulting to the west of the volcanoes Uranius Tholus (top) and Ceraunius Tholus. The presence of impact craters, particularly on Uranius Tholus, indicates that they are quite ancient and not active today. The summit crater of Ceraunius Tholus is about 25 km across. The true bases of the volcanoes are submerged beneath a flood of lava which produced the surrounding plain. The main field of radial fractures is known as Tractus Fossae, and the major double fault (left) is Tractus Catena. (NASA-JPL)

pushed the crust upward. The massive, widespread uplift of the crust has caused Tharsis to be the center of a vast array of radial fractures that spread across almost a third of the planet. To the east, several enormous interconnected canyons are aligned along these radial faults. The closely spaced, parallel grabens (rift valleys) are evidence of tensional deformation where the crust has pulled apart.

Many of these fractures are buried by younger lava flows and are visible only on older, higher terrain. To the north of Tharsis, the fractures are diverted around the large volcano Alba Patera, forming a ring of faults. At least some of the faulting is associated with volcanic dikes, where magma rose to the surface through cracks in the crust. Chains of pits are also seen in places where the surface has collapsed along the faults, e.g. Tractus Catena.

Similar, curved faults occur at Acheron Fossae, north of Olympus Mons, where cracks in the crust formed when magma rising from deep in the mantle pushed up the overlying lithosphere. Eventually, the brittle crust broke along zones of weakness. When the crust dropped between parallel faults, it created deep grabens, whilst the remnants of the previous surface were left upstanding as block mountains or "horsts."

Circular fractures around large volcanoes such as Elysium Mons and Ascreus Mons probably formed in a similar way. On the other hand, curved faults around the Isidis and Hellas impact basins were created as the crust adjusted to the formation of these giant depressions and the sudden removal of huge amounts of crustal material.

There is also evidence of crustal compression, where linear or curved wrinkle ridges occur, e.g. in the ridged plains of Lunae Planum, to the east of Tharsis.

Volcanism

Superimposed on the boundary between the cratered southern uplands and northern, low-lying plains are two large crustal rises (Figure 7.18). The Tharsis bulge, centered on the equator at 100° W, is 5,000 km across and 10 km high. The smaller Elysium bulge, centered at 30° N, 210° W, is 2,000 km across and 4 km high.

Both regions display raised gravitational signatures, indicative of large masses of dense material close to the surface. They were probably formed through a combination of large-scale, static convective plumes in the mantle, and isostatic adjustment due to shifts in the mass of the crust, as well as intrusive (subsurface) and extrusive (surface) igneous activity. The likelihood of periodic volcanism in recent times suggests that the domes may still be supported by rising plumes of mantle material.

Both of these regions have been sites of volcanic activity throughout most of the planet's history. Although they seem to have originated in the Noachian era, there is evidence of episodic volcanic activity over successive eons. Low numbers of craters on the lava flows and evidence that Martian meteorites which have fallen to Earth have crystallized within the last 150 million years, suggest that the planet may still undergo periodic volcanic activity.

On Mars, volcanic landforms created by the eruption of fluid, basaltic lava are far more common than those formed through explosive activity and production of ash. The most notable

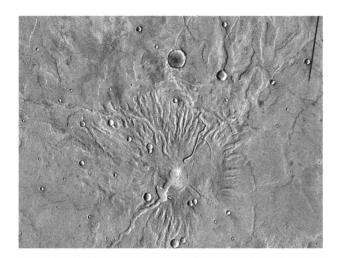

Figure 7.16 Tyrrhena Patera is about 200 km across but has very shallow slopes, deeply dissected by broad channels, probably carved by running water, that radiate from the summit. The low relief and easily eroded flank materials indicate that most of the volcano is composed of pyroclastic (ash) deposits. This is a mosaic of infrared images taken by the Thermal Emission Imaging System (THEMIS) on NASA's Mars Odyssey orbiter. (NASA/JPL-Caltech/Arizona State University)

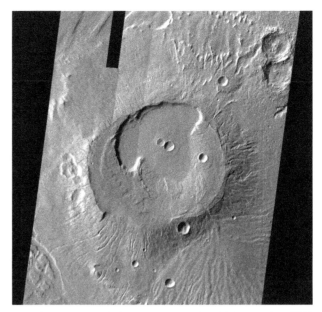

Figure 7.17 A THEMIS mosaic of Apollinaris Patera, an isolated, ancient volcano about 400 km across, located in the southern highlands (9°S, 186°W). To the west of the 80 km-wide central caldera is a cliff, marked by landslides, and chaotic terrain, formed by the collapse of ice-rich materials. Extending south from a breach in the caldera wall is a fan of material that flowed out of the volcano after the main cone was built. To the north and east are the younger (ash?) deposits of the Medusae Fossae Formation, incised by numerous gullies. (NASA/JPL/Arizona State University)

examples are the large shield volcanoes, located primarily in Tharsis and Elysium, which have many similarities with those of Hawaii. They display broad domes with gently sloping flanks of 6° or less. Each has a summit caldera, formed by collapse, following eruptions on the volcano's flanks.

Alba Patera (40.5°N, 109.9°W) is unlike any of the other shield volcanoes. It is very large (over 1,500 km across) and has a central caldera, but rises no more than 3 km above the surrounding plains. It appears to have been formed by numerous flows of non-viscous lava that were either much higher in output or much longer in duration than flows occurring on other volcanoes. Alba Patera is surrounded by a huge system of curved fractures, called "catenas."

Volcanoes located within the southern highlands have a very different morphology from the Tharsis or Elysium volcanoes. Two ancient, but very similar, examples are Tyrrhena Patera (21.9°S, 253.2°W) and Hadriaca Patera (30.6°S, 267.2°W).[3] Probably born around 2.5 billion to 3 billion years ago, each mountain is about 200 km across but has a gently sloping cone only a few hundred meters high. They represent the earliest central vent volcanism identified on Mars and may reflect a transition from the flood eruptions of very fluid lava which dominated early Martian history.

Their slopes have been dissected by radial gullies, hundreds of kilometers long, probably carved by streams of water flowing downhill and eroding weak layers of volcanic ash and lava. Massive clouds of steam billowing above the newborn volcano may have condensed into thick clouds from which heavy rain descended.

Not all Martian volcanoes are the result of quiet effusion of fluid lava from a single hot spot in the mantle. There are many other types of volcanic landforms, which range from small, steep-sided cones to plains covered in flows of solidified lava.

Volcanoes represent only a small fraction of the planet's volcanic activity. For example, surrounding the shield volcanoes in Tharsis and Elysium are extensive plains that were clearly built up from numerous lava flows superimposed on one another. These were probably fed by large fissures that are now buried beneath the flows.

Tharsis Giants

The most striking volcanic features on Mars are the three huge shield volcanoes on the Tharsis bulge – Arsia Mons just south of the Martian equator, Pavonis Mons on the equator, and Ascraeus Mons, 10 degrees north of the equator – and their larger, but isolated neighbor, Olympus Mons.

Each member of the Tharsis Montes chain is huge compared to terrestrial volcanoes – about 350 to 450 km across, towering to about 15 km above the surrounding plains.[4] Their location on the crest of a broad crustal dome means that they reach about the

[3] Patera is the Latin term for shallow dish or saucer. It is used to describe a shield volcano of broad areal extent but little vertical relief.

[4] The largest terrestrial shield volcano, Mauna Loa in Hawaii, is 120 km across and stands 9 km above the ocean floor.

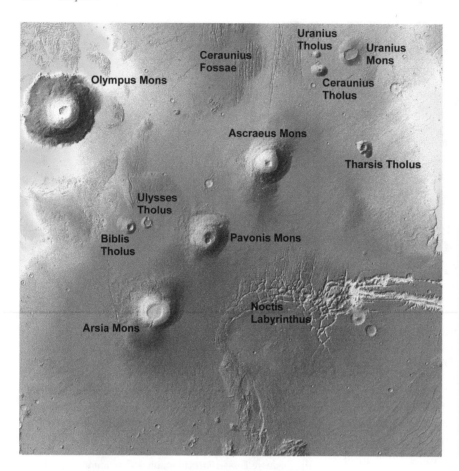

Figure 7.18 A MOLA relief map showing the three huge shield volcanoes that form a chain of mountains on the Tharsis bulge. To the west is the even larger shield structure, Olympus Mons – the largest volcano in the Solar System. A number of smaller volcanic and fault structures are also shown. Noctis Labyrinthus lies at the western end of the Valles Marineris canyon system. Highland is shown as red, pink brown, and white. (NASA/JPL-Caltech/Arizona State University)

same elevation as the summit of the much taller Olympus Mons. The fractured region southeast of Pavonis Mons, named Noctis Labyrinthus, merges with the Valles Marineris canyon system to the east.

The flanks of the Tharsis volcanoes rise gradually, evidence that they formed from eruptions of fluid lavas that could flow for long distances before solidifying. Each has a large summit caldera, caused by a monumental collapse of the surface when the subsurface magma supply was withdrawn. The flanks have a striated pattern caused by long linear flows, some with central leveed channels (i.e. with natural embankments of solidified lava on either side). Each has vents on its northeast and southwest flanks which fed vast lava flows that extend far across the surrounding plains.

All three volcanoes were built from countless thin sheets of runny lava flowing from natural pipes or tubes that develop within the flows. These carried lava long distances from its source to the spreading flow fronts. The roofs of some tubes have collapsed in places, forming an open pit. Where a longer section of roof has collapsed, the pit becomes an oval. If the flow in the tube eroded the roof sufficiently, many pits and ovals merged to form an open channel.

The Tharsis trio is aligned in a southwest–northeast direction, suggesting that, like the Hawaiian Islands, each one was formed by frequent, prolonged eruptions above a plume of rising magma. Studies of their surface features indicate that the oldest volcano, Arsia Mons, is in the south, and Ascraeus Mons is the youngest.

Since there is no evidence of plate movement, it seems likely that the plume moved beneath the stationary crust, instead of the other way round (Figure 7.19). Another possibility is a stationary plume that originated near Arsia Mons, then spread out toward Ascraeus as it neared the base of the crust.

Observations of surface roughness and crater counts indicate that they may currently be dormant, having erupted within the last few million years. It may be that future eruptions could contribute significant amounts of water and carbon dioxide to the Martian atmosphere.

Olympus Mons (18°N, 133°W) lies 1,500 km northwest of the Tharsis Montes, away from the main Tharsis rise. It is the largest volcano in our Solar System, with about 100 times the volume of Mauna Loa, the largest volcano on Earth (Figures 7.20 and 7.21). It is 25 km high (three times the elevation of Mount Everest above sea level) and about 620 km in diameter, or the size of the state of Arizona. The summit caldera alone is 80 km in diameter and 3 km deep.

Olympus Mons is so big that it would be impossible for an observer on the surface to see the entire edifice. The volcano is surrounded by a steep scarp, 6 km high in places, over which lava flows tumbled in the distant past. Extending several hundred kilometers from the base of the cliff is an aureole, consisting of several huge lobes of ridged terrain. The origin of the cliff and the aureole on the nearby plains is unknown – similar features are absent on the other Tharsis mountains.[5]

[5] Aureole is Latin for "circle of light."

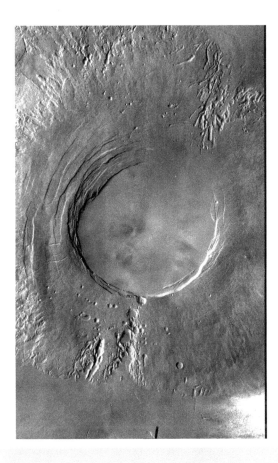

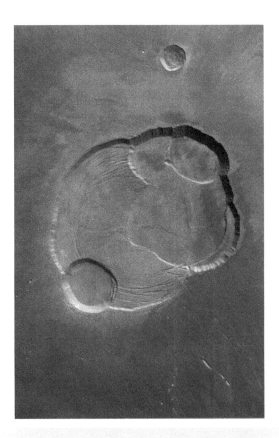

Figure 7.19   A THEMIS infrared mosaic of Arsia Mons – the southernmost, and possibly the oldest, of the Tharsis volcanoes. Indentations and collapsed lava tubes on the southwest and northeast sides align with Pavonis Mons and Ascreaus Mons to the northeast – evidence that a large fracture/vent system fed the eruptions that formed all three volcanoes. Note the curved fractures around the caldera. (NASA/JPL/ ASU)

Figure 7.21   An overhead view of the complex caldera at the summit of Olympus Mons. The striations are tectonic faults. After lava production ceased, the caldera collapsed over the emptied magma chamber, forming curved, extensional fractures. The level on which these fractures are observed represents the oldest caldera collapse. Later lava production and withdrawal produced new collapses, forming the other circular depressions. The image is 102 km across with a resolution of 12 m per pixel. South is at the top. (ESA/DLR/FU Berlin – G. Neukum)

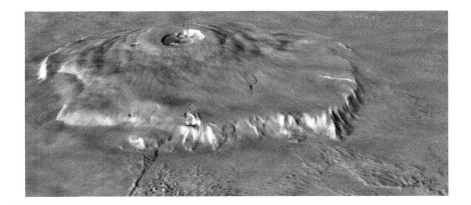

Figure 7.20   Olympus Mons is the largest volcano in the Solar System. The shield volcano is 624 km in diameter, 25 km high, and rimmed by 6 km-high cliffs. Note the 80 km-wide central caldera. A massive aureole deposit was possibly produced by flank collapse. The vertical exaggeration is 10:1. (NASA/MOLA Science Team)

Valles Marineris

Around the turn of the 20th century, some astronomers drew maps of linear channels – which they called "canali"– on the Martian surface. We now know that they were almost entirely optical illusions. One exception is a huge system of interconnected canyons, on the eastern flanks of the Tharsis bulge, which runs just south of the equator for about 4,000 km.

This great rift in the crust was discovered in 1972 by the Mariner 9 spacecraft – hence its name, Valles Marineris (Mariner Valleys). It dwarfs all similar features in the Solar System, and would swallow up thousands of valleys like the Grand Canyon in Arizona.

Valles Marineris is formed out of several parallel, connecting troughs. Individual canyons are generally 50–100 km wide, reaching a depth of 8–10 km in places, six to seven times deeper than the Grand Canyon.

Their characteristics change from west to east. At the western end is a complex graben system known as Noctis Labyrinthus, with numerous, intersecting linear depressions. The depressions are generally aligned parallel to faults in the surrounding plateau.

To the east, the valleys become deeper, wider, and more continuous, forming canyons aligned roughly east–west. Three large canyons merge in the central section to form a huge depression up to 600 km across. Still further east, the individual canyons become shallower, and fluvial features become more common, both on their floors and on the surrounding plateau. At the eastern end, they lead into large areas of chaotic terrain.

The canyons probably formed during the Hesperian era, about 3.5 billion years ago, as the result of tectonic activity that accompanied the growth of the nearby Tharsis volcanoes (see Tharsis Giants) (Box 7.2). The tectonic stress caused the

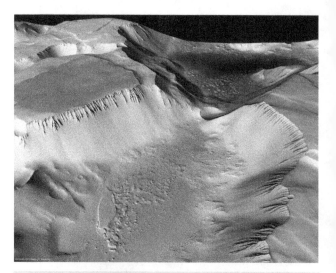

Figure 7.23 Noctis Labyrinthus, at the western end of the Valles Marineris, forms part of the graben system from which the canyons formed. The elongated, trench-like features, bounded by parallel faults, are some 5 km deep. They are greatly eroded, as seen from debris on the valley floors. Younger rock formations can be seen on the upper slopes. This view was based on High-Resolution Stereo Camera images from Mars Express. Vertical height is exaggerated. (ESA/ DLR/ FU Berlin – G. Neukum)

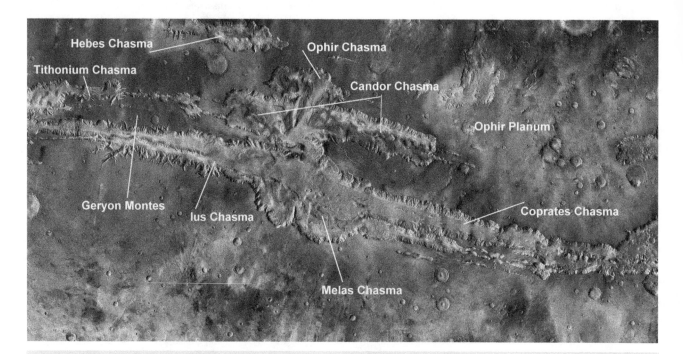

Figure 7.22 A mosaic of Valles Marineris images obtained by the THEMIS camera on Mars Odyssey. The broad valley floor, parallel canyons, and lateral landslides are clearly shown. Spatial resolution is about 100 m. (NASA/JPL/ASU)

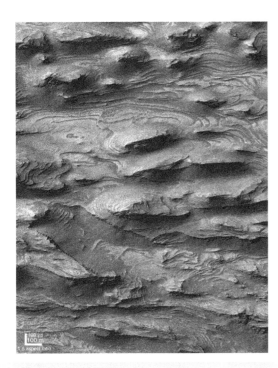

Figure 7.24 Layered sedimentary rock in western Candor Chasma. The area covered is 1.5 × 2.9 km. The MOC image reveals over 100 uniform beds on the canyon floor, each about 10 m thick. Each layer has a relatively smooth upper surface and is hard enough to form steep cliffs at its margins. This suggests that the deposition was not continuous, but interrupted at regular intervals. The layers may have been deposited on a lakebed. (NASA-JPL/MSSS)

surrounding crust to stretch. As the crust thinned and fractured, blocks of rock sank between parallel cracks. The result was a series of grabens – elongated, trench-like features bounded by parallel normal faults.

The faulting also opened paths for subsurface water to escape. As the surface broke apart, the water escaped, washing away material and causing the ground to collapse. The steep, newly exposed valley walls became unstable, causing numerous landslides, up to 100 km long, that widened the canyons further. The main extensional activity probably came to an end roughly 2 billion years ago, although landslides undoubtedly still occur.

Today, we can see where faults have opened into collapse pits, some of which then join together to form larger depressions. The process steadily ate away at the surface until the land between two neighboring faults was largely destroyed.

Massive floods seem to have escaped through several outflow channels at the valley's eastern end. The ground here was so saturated that when the water escaped, the surface collapsed almost completely, leaving only isolated mesas and hills which are termed "chaotic terrain."

In the northwest of Valles Marineris, a similar torrent emerged from a depression called Echus Chasma, forming the Kasei Valles outflow channel (Figure 7.38). Again, the floodwaters eventually poured into the northern lowlands. It seems most likely that such

floods occurred in several stages, rather than as single, catastrophic events.

Not all of the erosion within the canyons resulted from major floods. In places such as Louros Valles, numerous deep, tributary valleys have been cut into the canyon's southern rim, contributing to the overall widening of Ius Chasma. These minor tributaries were probably created by the release of local groundwater from springs.

As would be expected where flood water (and possibly glaciers or ice sheets) once existed, the canyons have also been shaped by large-scale deposition of sediments. These have since been eroded and are now exposed as multiple layers, e.g. in Candor Chasma (Figure 7.24).

Plains

Much of Mars, particularly in the northern hemisphere, consists of low-lying plains – some of the smoothest surfaces in the Solar System. Like the plains of Mercury and the Moon, they have been constructed from innumerable thin lava flows. Laser altimeter measurements show that, apart from the craters, the plains typically vary in height by 50 m or less. Shallow grooves and broad hollows are filled with windblown dust and dunes, separated by low rises. In places, the more resistant surfaces appear as higher land.

Wind is not the only agent of erosion and deposition to have modified the plains. There is ample evidence that large volumes of

Figure 7.25 NASA's Phoenix spacecraft landed on the northern plains in a region covered by polygons, 1.5 to 2.5 m across, that stretched to the horizon. Similar in appearance to surface features in Earth's Arctic regions, the polygons are caused by continual expansion and contraction of ice wedges within the permanently frozen surface (permafrost). (NASA/JPL-Caltech/University of Arizona/Texas A&M University)

water carved deep channels and terminated in lowland basins or spread out onto the plains, where sedimentary layers of rock and sand were deposited (see Valley Networks, Lakes, and Deltas).

Glaciation almost certainly modified the plains in the distant past, as ice sheets spread onto the plains from the polar ice caps and upland regions, even at low latitudes. The flowing ice sheets would have eroded and smoothed the terrain, eventually burying much of it under deep blankets of glacial moraine and outwash deposits (see Glaciation).

Even today, the plains contain a lot of water ice at shallow depths, as confirmed by the Phoenix spacecraft which landed in the Martian Arctic in 2008. Its panoramic images revealed vast plains corrugated by polygonal ice wedges.

Polar Regions

At each pole, extending out to 80° latitude, is a permanent ice cap composed of finely layered deposits of ice and dust a few kilometers thick. In the north the cap rests on lowland plains, in the south on cratered uplands. The small number of superimposed impact craters suggests that they are relatively young, possibly only a few hundred million years old.

The residual northern cap covers an area bigger than Texas. Its volume of 1.2 million cubic km is less than half that of the Greenland ice sheet, and about 4% of the Antarctic ice sheet. In comparison, the residual southern ice cap is estimated to contain 1.6 million cubic km, making it the largest source of water on the planet's surface. Radar data indicate that it contains approximately 15% dust – significantly less than the northern cap.

Both polar caps possess a perennial core composed of water ice. This is overlain by a seasonal blanket of frozen carbon dioxide (CO_2) mixed with dust. Extending for tens of kilometers around this is a vast area of dirty water ice, containing varying amounts of dust.

Some 30% of the carbon dioxide in the planet's atmosphere is cycled between the polar caps as the seasons change. As the warming summer cap shrinks due to sublimation of carbon dioxide ice, CO_2 condenses out on the other polar region, which is permanently in winter darkness, to form a seasonal ice cap. The depth of the seasonal ice layer has been estimated to be no more than a meter or two, less at lower latitudes.

As summer approaches, the temperatures rise from the freezing point of CO_2 (−120°C) to around the freezing point of water (0°C). Not only does the CO_2 cap vaporize, exposing the underlying water ice, but the latter also begins to sublime, thus increasing the content of water vapor in the atmosphere (Figure 7.28).

There are some significant differences between the two ice caps. In the northern hemisphere, the seasonal cap can extend toward the equator as far as 60°N, while its southern counterpart extends to 50–60°S.

The northern CO_2 cap burns off completely in summer, leaving a residual water ice cap. At the south pole, the CO_2 cap does not completely disappear, since temperatures remain close to −120°C. Only a modest increase in water vapor is detected, although a permanent cap of water ice has been detected beneath the dry ice.

The southern cap is several kilometers higher – and therefore, cooler – than its northern counterpart, but the main reason for the colder summer conditions in the south may be the effect of dust in the atmosphere. Dust storms and dust devils are much more common during the southern summer, and the fine particles that are carried into the upper atmosphere block some of the incoming solar radiation, keeping the surface cooler than expected.

Whereas the winter cap is symmetrical about the south pole, the summer residual cap is offset by some three to four degrees. This is caused by much more snow falling on one side than the other. On the western hemisphere side of the south pole the winds are changed by the giant Hellas basin. A low-pressure system forms, resulting in greater precipitation of snow. On the other side, the drier air results in less snow and more frost. Snow tends to reflect

Figure 7.26 The northern polar cap is approximately 1,000 km across. The dark, spiral-shaped bands are deep troughs. In the center of the image is a large canyon, Chasma Boreale, which almost bisects the ice cap. Chasma Boreale is about the length of the Grand Canyon and up to 2 km deep. (ESA/DLR/FU Berlin; NASA MGS MOLA science team)

Figure 7.27 The residual south polar ice cap lies on top of smooth layered deposits that overlie the cratered southern highlands. Nearby is a large area of dunes. (NASA/MOLA Science Team)

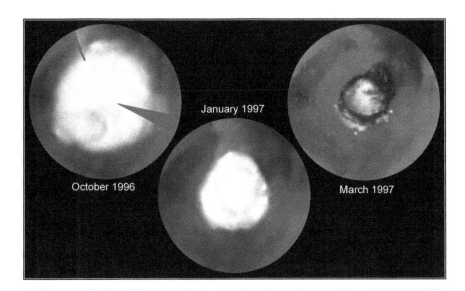

Figure 7.28 Seasonal changes in the north polar cap imaged by the Hubble Space Telescope. In October 1996 (early spring), the almost circular cap extends to 60°N, nearly its maximum extent. Bluish areas on the edge of the cap are clouds. Some of the area was still in shadow. By January 1997, increased warming has sublimated the carbon dioxide ice and frost below 70°N. The faint, darker circle within the cap's boundary marks circumpolar sand dunes. These dark dunes are warmed more quickly than their brighter surroundings, so their frost coating sublimates earlier. By March 30, 1997 (early summer), the cap has retreated, leaving a small core of water ice, almost cut in half by the Chasma Borealis canyon. Outliers of ice persist south of the sand sea. Bright circular features are ice-filled craters. The edge of the images is at 50°N latitude. 0° longitude is at the top. (Phil James/Univ. Toledo, Todd Clancy/Space Science Inst., Boulder, CO, Steve Lee/Univ. Colorado, and NASA)

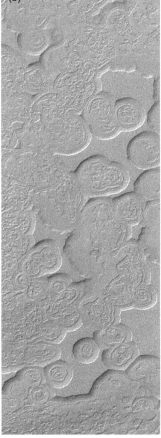

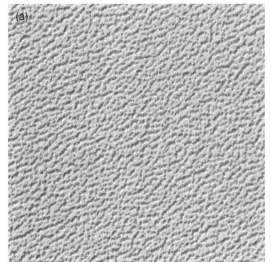

Figure 7.29 (a) "Cottage cheese" terrain on the north polar residual ice cap. The flat surface has many closely spaced pits up to 2 m deep and 10–20 m across. These form as water ice sublimes, growing slowly over thousands of years. This 1 × 1 km picture is illuminated from upper left. (b) The south polar cap is colder and coated with carbon dioxide ice. Large circular depressions resembling Swiss cheese are separated by flat-topped mesas and buttes up to 4 m high. The area covered measures 3 × 9 km. (NASA-JPL/MSSS)

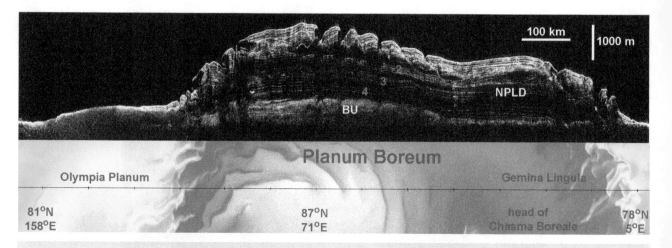

Figure 7.30   A cross section through the north polar ice cap (top), taken by the radar instrument on Mars Reconnaissance Orbiter, reveals layers of ice, sand, and dust. The colored map below shows surface topography (red and white = higher ground, green and yellow = lower). The radar reveals four thick layers of ice and dust separated by layers of nearly pure ice. The boundary between the ice layers and the underlying surface is relatively flat, suggesting that the lithosphere is thick and rigid. NPLD = north polar layered deposits. BU = basal unit, an ice-sand deposit that lies beneath parts of the north polar layered deposits. (Courtesy of NASA/JPL-Caltech/University of Rome/SwRI)

more sunlight in the summer, reducing the surface temperature and the amount of sublimation. Frost, on the other hand, has a rougher surface and tends to trap more sunlight, resulting in greater sublimation.

The north polar cap has a flat, pitted surface that has been likened to cottage cheese or the surface of a sponge (Figure 7.29). In contrast, the south polar cap has roughly circular depressions that are more like Swiss cheese, with flat floors and steep-sided walls. As the layered CO_2 ice is removed, it leaves a landscape of pits, troughs, flat-topped mesas, and buttes.

Spacecraft images and laser altimeter data show that the polar caps are composed throughout of multiple horizontal, or near-horizontal, layers. These polar layered deposits are stacked on top of each other, reaching a maximum thickness of about 2.7 km in the north, and 3.1 km in the south.

Numerous individual layers of ice are visible, particularly in the walls of valleys cut into the sediments, such as Chasma Boreale. They are generally believed to be the result of variable amounts of dust and water ice being laid down over many climate cycles in the last four million years or so. The primary cause of the banding may be periodic changes in the temperature and winds at the poles, induced by variations in the planet's orbit and rotation. These cycles affect the stability of water ice and CO_2 ice at the poles and the dynamics of the atmosphere.

The poles act as a cold trap for water. Any water entering the atmosphere as a result of geologic processes, such as volcanic eruptions or floods, is ultimately frozen out at the poles. They may also be a trap for dust, since dust nuclei can be taken out of the atmosphere as CO_2 freezes onto the pole in the fall and winter. The layered deposits sit on top of a much older ice-sand deposit, known as the "basal unit."

One of the most unusual features of both caps is the deep, curved valleys that dissect the residual summer ice. They are unlike any terrain on Earth, and various hypotheses for their formation have been put forward, including subsurface volcanic activity, glacial erosion, aeolian processes (such as local katabatic winds[6]) and sublimation, outflow of meltwater, or a combination of these processes.

The largest canyon in the northern cap is Chasma Boreale, which cuts into the layered polar terrain and crosses several smaller, curved troughs (Figure 7.26). It extends 500 km into the residual cap, has a width of up to 100 km, and a depth of about 2 km. The head of Chasma Boreale ends in a steep, icy cliff more than 1 km high. The origin of the canyon is uncertain, but it seems to be associated with an old, linear depression that underlies the layered deposits. The depression was later modified by meltwater, katabatic winds, or local variations in the rate of sublimation.

Both short and longer-term changes occur in the polar caps. Frequent avalanches have been observed on the scarps at the edges of the ice caps during spring (Figure 7.31). The falls appear to be composed mostly of carbon dioxide frost, mixed with loose, dusty material. Frost builds up on the scarp during the long winter, then begins to sublimate when sunlight returns. This results in avalanches that create dust clouds tens of meters high.

The carbon dioxide ice at the south pole is rapidly disappearing. Comparisons of images taken a few years apart show circular pits that have enlarged, shrunken mesas, and small buttes that have vanished. The average rate of retreat is about 3 m per Martian year, though it varies from place to place. Studies suggest that an 8 m-thick layer of CO_2 ice is being removed to reveal a permanent layer of water ice beneath. It seems that the planet was markedly

---

[6]**Katabatic winds occur when cold, dense air, slides downslope, becoming warmer and drier. Famous examples on Earth are the Chinook ("snow eater") and the Föhn.**

Figure 7.31 The HiRISE camera imaged an avalanche in progress on a steep scarp in the northern polar region. The 700 m-high cliff is made of layers of water ice with varying dust content. On top is white carbon dioxide ice and frost. The dust cloud rises up to 50 m high. (NASA/JPL/University of Arizona)

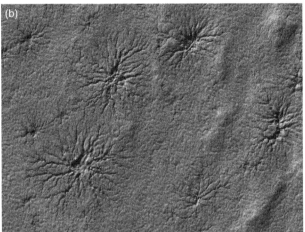

Figure 7.32 (a) Sand-laden jets shoot into the southern polar sky in this artist's view. These leave a delicate, spider-like pattern on top of the residual polar cap, after the seasonal carbon dioxide ice has disappeared. (Arizona State University/Ron Miller) (b) Spider-shaped channels in the south polar region are carved by vaporizing dry ice. This HiRISE image was taken in the southern summer, after the winter frost had disappeared. The channels carved into the ground are typically 1 to 2 m deep. The area shown is 1.2 km wide. Carbon dioxide gas coming from the bottom of the ice builds up pressure and carves channels into the ground as it flows toward a point where it escapes into the atmosphere. Often the channels are radial in nature, with the escape point for the gas at the center. (NASA/JPL-Caltech/University of Arizona)

colder only a few centuries to a few tens of thousands of years ago.

The residual southern cap also displays dark spots, typically 15 to 45 m wide and spaced up to 100 m apart. They appear every spring and last for three or four months before vanishing, only to reappear the next year, after a fresh layer of CO_2 ice has been deposited during the winter. Most of the spots recur at the same locations year after year, and some of them display spider-like grooves that extend outward in all directions.

These features are created when powerful jets of trapped carbon dioxide erupt through weaknesses in the overlying ice as the temperature rises in spring. As the gas rushes toward the vents, it picks up loose particles of sand and erodes the ice, carving networks of spidery grooves that converge on the vents. The "geysers" climb high into the thin air, carrying fine, dark sand and spraying it for many meters around each vent.

Permafrost

On Earth, much of the ground surface adjacent to the polar ice caps is permanently frozen to a depth of many meters, a condition known as permafrost. On Mars, the area of permafrost is much more extensive. The subsurface ice is so cold that it resembles solid rock.

The precise depth of the Martian permafrost is uncertain, but it probably extends down more than a kilometer at the equator and several kilometers at the poles. At latitudes between 40° and 80°, water ice on the surface sublimes during summer, but it is very stable only a few meters below the surface. Only near the

equator is ice unstable at all depths. This may enable liquid water to accumulate in porous aquifers deep underground. It may even exist close to the surface if the water contains a high concentration of salts, which alters its freezing point.[7]

The extent of the ice, which is covered by a thin regolith, has been determined by measuring high energy neutrons and gamma ray emissions from the surface, which provide information on the hydrogen content. High concentrations of hydrogen detected in Mars' circumpolar regions are consistent with plentiful water ice in the shallow subsurface.

Visual support for the widespread presence of ground ice comes from spacecraft images which show fresh impact craters excavated

---

[7]The presence of salt causes sea water to freeze at a lower temperature than pure water. This is also why common salt (sodium chloride) is used to de-ice roads and pavements in winter.

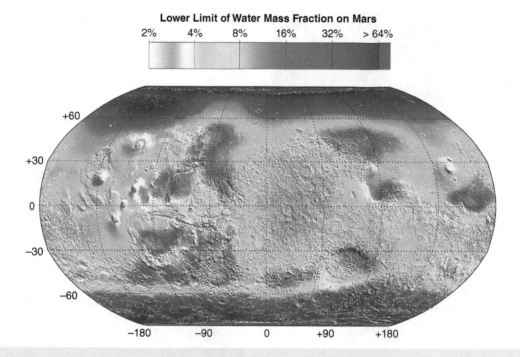

Lower Limit of Water Mass Fraction on Mars

Figure 7.33   A map showing the estimated lower limit of the water content of the upper meter of Martian soil. The estimates are based on the hydrogen abundance measured by the neutron spectrometer on Mars Odyssey. The highest water mass fractions, exceeding 30% to well over 60%, are in the polar regions. Significant concentrations also occur in the uplands east of Valles Marineris (center), and to the southwest of Olympus Mons (extreme left and right). (Courtesy of NASA/JPL/Los Alamos National Laboratory)

in ground ice (see Figure 7.11) and thick deposits of subsurface ice that are exposed on scarp faces that are being eroded. The ice was likely deposited as snow long ago. The deposits are exposed in

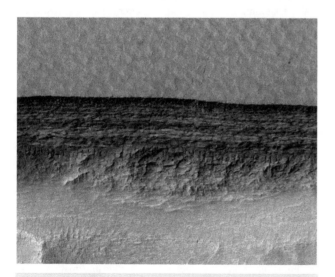

Figure 7.34   A cross-section of underground ice is exposed at the steep slope that appears bright blue in this enhanced-color view from the HiRISE camera on Mars Reconnaissance Orbiter. The scene is about 550 m wide. The scarp exposes a layer of about 80 m of water ice. The scarp is located at 56.6°S, 114.1°E (Courtesy of NASA/JPL-Caltech/UA/USGS)

cross section as relatively pure water ice, capped by a layer one to two meters thick of ice-cemented rock and dust.

Other visual evidence for surface water ice is the existence of numerous pedestal craters, each surrounded by what look like mud flows. These craters are created when a meteorite hits a region containing substantial subsurface ice (see Pedestal Craters).

Another key piece of evidence for permafrost is the presence of a network of small polygonal features on flat, debris-covered plains (Figure 7.25). Such polygons have been mapped from the edges of the polar caps to mid-latitudes in both hemispheres. In many cases, the pattern comprises large rectangular features, about 160 m across, which are subdivided into less prominent sections, each separated into several dozen smaller polygons.

The polygons form when the ground contracts in the winter, creating small spaces. If they are infilled with dust, the ground ice has no room to expand again when the temperature rises in summer. As a result, the ground buckles, producing low mounds at the centers of polygons, and shallow troughs at the edges.

Not all polygons on Mars are associated with permafrost. Some large craters contain large polygons subdivided into smaller polygons. These may be desiccation cracks caused by the drying out of the floors of evaporating lakes (see Valley Networks, Lakes, and Deltas).

Pedestal Craters

The most distinctive impact structures on Mars are pedestal craters (also called rampart craters), which are particularly common at higher latitudes. Instead of linear rays of ejecta radiating

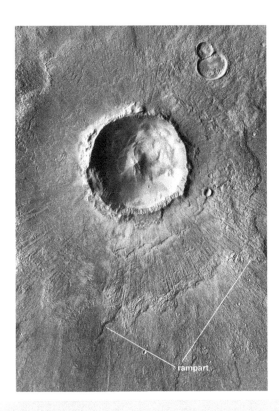

Figure 7.35 20 km-wide Bacolor crater lies in Utopia Planitia (33.0°N, 118.6°E). The pedestal crater formed where the subsurface held a lot of water or ice. This false color view from Mars Odyssey shows a thick, inner layer of ejecta, surrounded by a thinner layer that extends off the image. The inner debris layer came to rest about one crater diameter from the rim, piling up a rampart 50–80 m high. Note the two small craters, 14 km northeast of Bacolor, which are surrounded by debris. (Courtesy of NASA/JPL/ASU)

from the impact site, they are notable for a raised platform (pedestal) that is piled up against the crater rim and lobes of debris. The unusual ejecta field is produced when water-rich material is blasted out of the ground and dropped onto the surrounding terrain.

A thick, hot, slurry of mud, water vapor, and rock fragments flies away from the impact site at a steep angle and falls to the ground close to the newly formed crater. This inner ejecta apron forms a muddy blanket that flows sluggishly away from the crater rim.

Where the viscous mud comes to rest, about one crater diameter from the rim, it piles up to form a debris rampart composed of broken rocks and smaller fragments, mixed with groundwater and pieces of ice. A thinner, outer region of ejecta forms when the tall column of debris that rose over the impact site begins to collapse. At the surface, the debris column races outward until it loses its momentum and halts, roughly two crater diameters from the rim.

Only large pedestal craters are found near the equator, whereas much smaller examples can be found at higher latitudes. This makes sense if the permafrost is buried at great depth in equatorial regions, but close to the surface nearer the poles.

Glaciation

Since the temperature rarely rises above 0°C, water ice is more or less permanent on Mars. Any water ice that is warmed above its vaporization point sublimates, changing almost instantly from ice to gas. On the other hand, carbon dioxide ice is much less stable. As already mentioned, CO_2 gas freezes out over much of the planet in winter, but the ice disappears through sublimation in summer (except in most of the residual south polar cap).

On Earth, regular glacial cycles have resulted in advances of ice sheets from the polar regions to lower latitudes. These cycles are usually linked to changes in the Earth's axial tilt and orbit (see Chapter 3). Mars has also undergone numerous ice ages, but these are more extreme than on Earth, since the variations in insolation caused by periodic changes in its obliquity and orbit are more pronounced.

Under present conditions, there is little precipitation on Mars, so the formation of glaciers and ice sheets is unlikely. However, its climate has not always been so equable. During warmer periods, vaporization of water ice increases the atmospheric pressure and allows substantial amounts of water to circulate before eventually falling as snow. When the climate switches back, the

Figure 7.36 This 5 km-long, lobed feature on the inner wall of a large crater was imaged by Mars Global Surveyor. It appears to be a rock glacier, a mass of glacial ice overlain by rocky debris. The feature is located at 38°S, 113°E, a region famous for "softened terrain" which may involve flowing ice. (Courtesy of NASA/JPL/Malin Space Science Systems)

Figure 7.37 Three craters in the eastern Hellas region which appear to contain concealed glaciers. On the left is how the surface looks today; on the right is an artist's concept, based on radar data, showing subsurface ice. The thickness of the buried ice ranges from about 250 m in the upper crater to 300 m and 450 m in the middle and lower craters respectively. Each image is 20 km across and 50 km deep. (Courtesy of NASA/ Caltech/JPL/UTA/UA/MSSS/ESA/DLR)

glaciers become inactive or retreat. There is evidence that some glaciers still exist, protected from sublimation by a blanket of rocky debris.[8]

The walls of many mid-sized impact craters, for example in the Arabia Terra region of the northern hemisphere and the area east of the Hellas basin, display lobed features that resemble rock glaciers, with pitted surfaces and raised ridges, similar to moraines, on their lateral margins. Some major outflow channels, such as Kasei Valles, may also have been modified by subsequent glacial activity, which deepened and widened them to create U-shaped valleys.

Other terrain features between latitudes 30–45° may also be related to glaciation. Small arcuate (arc-shaped) ridges enclose depressions at the base of crater walls, often below gullies. These may be glacial moraines, remnants of unsorted glacial deposits from which the ice has since evaporated. Debris aprons – pitted and lineated deposits on crater floors – are similar to debris-covered glaciers or ice-rich landslides on Earth. Various types of narrow, sinuous channel are seen crossing crater floors, often breaching the crater walls and extending onto the terrain beyond. These channels may have been formed by flowing glacial meltwater.

An hourglass-shaped crater in Promethei Terra, near the Hellas basin, seems to contain debris-covered ice that flowed from one side of a nearby massif. The "block glacier" almost filled a bowl-shaped impact crater, approximately 9 km wide, then flowed 500 m downhill, through a breach in the crater rim, into a 17 km-wide crater below. The lack of impact craters shows that the surface is only a few million years old.

Similar examples in the Hellas region and in the northern hemisphere have been documented by the SHARAD radar on Mars Reconnaissance Orbiter. The concealed glaciers extend for tens of kilometers from the edges of mountains or cliffs and are up to one kilometer thick. Such evidence supports the theory that snow and ice accumulated on higher topography within the last few million years and then flowed downhill.

Spacecraft images also indicate that glaciers covered the flanks of volcanoes and mountains only a few million years ago. Each of the Tharsis volcanoes has an unusual, fan-shaped deposit on its western or north western flank which may be formed by glacial processes.

Outflow Channels

One of the most surprising discoveries of the 1970s was the presence of numerous dry channels, some of which closely resemble river networks on Earth. Crater counts show that the sinuous channels, hundreds or thousands of kilometers in length, date from the Hesperian era, around 3.5 billion years ago, a time of transition from a wet Mars to a dry Mars.

Careful study of the dry channels reveals features typical of erosion and deposition by water: terracing of channel walls, teardrop-shaped "islands," hanging channels that dead-end, and braided channels or oxbows that separate and then reconnect to the main branch. Lava is generally thought to be unable to create such a wide variety of features, although studies of a channel system on the flanks of Ascraeus Mons have shown that this is not always the case.

These features are divided into two types: outflow channels and valley networks. The outflow channels occur mainly on the young surfaces of the northern lowlands (from the Amazonian period of Martian history), whereas the valley networks occur throughout the older, heavily cratered terrains of the Noachian and Hesperian periods. This suggests that the outflow channels were formed after the valley networks.

Most of the outflow channels occur in lowlands north of Valles Marineris and to the west of the Chryse Planitia region. They

---

[8]In theory, it is cold enough for glaciers to exist at the Martian equator today, but the prevailing low atmospheric pressure would cause ice to sublimate unless it was protected by a coating of surface dust.

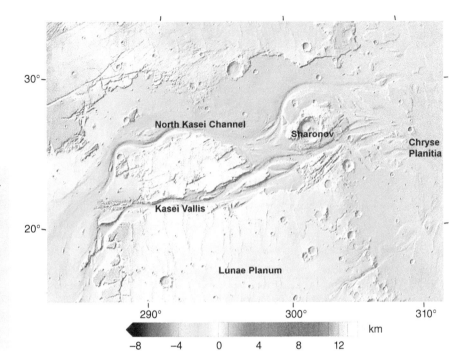

Figure 7.38 A topographic map of Kasei Valles, the largest outflow channel system on Mars. Kasei Vallis is around 3,500 km long, more than 400 km wide, and exceeds 2.5 km in depth. Huge floods carved the channels as the water flowed into the Chryse Planitia lowland. The channel system extends another 1,200 km south to Echus Chasma. (After Wikimedia)

often begin abruptly, and are sometimes more than 100 km across and several thousand kilometers in length. Some of the outflow channels dwarf anything on Earth. The most dramatic examples flow north from the Argyre basin and eastward from the Valles Marineris, cutting around and through impact craters, before disgorging onto the Chryse Planitia lowland.

The outflow channels appear to be fed from a fairly localized source, often beginning in a box canyon or an area of collapsed or chaotic terrain that may be hundreds of kilometers across. Some in Elysium and around Hellas start adjacent to large volcanoes; others appear to originate at faults. Tributaries are few in number, and any that do exist are usually short and stubby. They typically have scoured floors, often with teardrop-shaped "islands," and convex walls.

The sudden emergence of full-sized channels and the lack of tributaries indicate that these features were likely carved by massive flash floods that eroded the surface over a short period of time. The floodwaters may have been fed by an abrupt release of meltwater, caused either by sudden heating of surface/near-surface ice – possibly caused by a large meteorite impact or volcanic activity – or the collapse of a barrier that was damming up a huge lake. Another possibility is that groundwater became trapped beneath a thick layer of permafrost and, as the pressure built up, it episodically burst onto the surface.

In the case of Mars, the water source may have been an underground "reservoir" of meltwater, rather than a surface lake. When the softened ground collapsed, the torrent of water was released.

The dimensions of these channels show that the volume of water must have been enormous, with discharges in the range of 100 million to one billion cubic meters per second, 1,000 to 10,000 times greater than the Mississippi. These discharges are so large that flow could have continued even under the present freezing

climatic conditions. Thousands of cubic kilometers of material would have been removed by such floods.

However, such torrents could not have lasted for long, perhaps a matter of days or weeks. It is likely that the outflow channels were created by at least one catastrophic flash flood, possibly augmented by spasmodic, smaller events, rather than a prolonged, continuous discharge.

Although the temperature was frigid by Earthly standards, the water could have survived for weeks or months if it was briny (hence with a lower freezing point) and the air pressure was higher than it is today. The rate of evaporation would also have been reduced by a surface layer of ice, beneath which the river flowed.

One of the most dramatic examples of an outflow channel is Ares Vallis, which begins in a region of chaotic terrain known as Arram Chaos and flows 1,600 km north to Chryse Planitia (Figure 7.40). The main channel is typically 2.5 km wide and varies in depth from 100 m to one kilometer.

Ares Vallis is one of the few areas of Mars where a spacecraft has conducted surface measurements (Figure 7.39). Mars Pathfinder landed downstream from the channel mouth and sent back panoramic pictures of a dry, desert landscape littered with many dark, basaltic rocks separated by patches of sand and dust.

Ares Vallis probably dates back to Hesperian times. A more youthful example is the Athabasca Vallis/Marte Vallis system, which seems to cut through more recent, Amazonian lavas. It begins on the south eastern side of the Elysium volcanic region, in the Cerberus Fossae region, where there are numerous parallel fractures aligned from north west to south east.

Athabasca Valles resembles a flood-carved landscape, with branching channels, streamlined "islands" and dunes. However, its floor appears remarkably uneroded in high resolution images. What carved these relatively recent valleys? One theory is that they were eroded by a catastrophic flood originating from the

Figure 7.39   The Sojourner rover, which landed near the mouth of the Ares Vallis outflow channel, using its X-ray spectrometer to analyze the composition of a basaltic rock nicknamed Yogi. The rock's two-toned surface may be the result of accumulated windblown dust, or evidence of a break from a larger boulder as it was deposited in an ancient flood. Just beyond the ramp, on the left, is a silica-rich rock known as Barnacle Bill – a possible andesite. Patches of red or dark gray dust and sand lie between the rocks, with white material exposed by the rover's wheels. In the distance are the Twin Peaks, smooth-sloped hills that may have been shaped by fast-flowing water. (NASA-JPL)

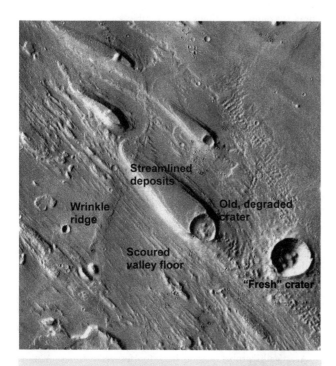

Figure 7.40   A THEMIS image of part of Ares Vallis, near its channel mouth. The teardrop-shaped "islands," several hundred meters high, may have been deposited where obstacles such as craters produced a zone of turbulence that allowed debris to settle and collect behind them. The water was moving from lower right to upper left. Smaller, parallel ridges were created by erosion of the valley floor. After the channel bed became dry, tectonic activity raised a wrinkle ridge that runs across the valley. (Courtesy of NASA-JPL/ASU)

Cerberus Fossae fractures or from a large lake in the Elysium Basin. The floodwaters which eventually spread out on the northern plains now take the form of broad, frozen "sea" known as Cerberus Palus.

Valley Networks, Lakes, and Deltas

Valley networks are present over almost half the planet, mostly in the ancient, heavily cratered southern highlands.

There are two main types: long, winding valleys with few tributaries, and smaller valley networks, often with complex, multiply branched patterns of tributaries. A good example of the first is Nirgal Vallis, south of the eastern part of Valles Marineris.

The dendritic (branching) valley networks that dissect much of the southern uplands are much more like terrestrial river valleys than the huge outflow channels. Not only are they narrower – up to a few hundred meters across – and much shallower, but many of them have tributaries (although these may be stunted) and increase in size downstream because they are fed by numerous, small channels.

The general consensus is that they formed as a result of slow erosion by water. Some of this water may have been derived directly from precipitation or from the melting of surface snow and ice, although searches have failed to find the tiny rivulets that would be expected if precipitation was the cause. However, any rivulets may have been blanketed and hidden by later deposits, and the dendritic networks remain the best evidence of surface runoff from rain or snowfall on young Mars. On the other hand, many of the networks have characteristics that suggest they were formed by seepage of groundwater, possibly from melting ice.

It is unlikely that these valleys could form under present climatic conditions, since even brine, which has a lower freezing point, is not stable for long in the low-pressure environment. The

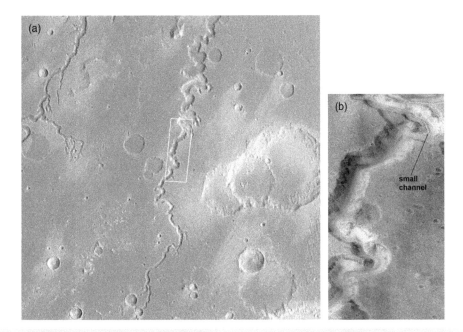

Figure 7.41 (a) 2.5 km-wide Nanedi Vallis is one of the valleys cutting through cratered plains in the Xanthe Terra region. The valley lacks lengthy tributary channels and has short, box-headed tributaries that suggest formation by collapse. A close-up view of the region in the white box is shown in (b). (NASA-JPL) (b) Rocky outcrops occur along the upper walls and weathered debris lies on the lower slopes and the valley floor. A 200 m-wide channel in the valley bottom (upper right) is partially covered by dunes and debris. It suggests that a small river flowed in the valley after the main channel was created. The valley may also have been widened by slumps and landslides, wind and groundwater flow. This picture spans 9.8 x 18.5 km. Spatial resolution is 12 m. (NASA-JPL/MSSS)

networks probably formed when Mars was warmer and wetter, and possessed a thicker atmosphere. This is supported by the concentration of the channels on old, cratered uplands, e.g. in Terra Tyrrhena, although the presence of a few relatively young examples indicates that cyclical climatic changes may have allowed surface drainage in more recent times.

One of the most persuasive pieces of evidence for prolonged surface drainage is meandering channels, which are created where streams flow across a flood plain that is resistant to erosion. Erosion takes place on the outside of the bend, where the flow is faster, with deposition on the inside, where it is slower. Over time, the meander loops increase in size and migrate downstream, widening the flood plain. On Mars, meandering river sediments have been found in many places, suggesting that much of the planet experienced lengthy periods of surface run off. Low features within the bends, known as scroll bars, show where the meanders migrated over time.

Many of the meanders and dry channels are now seen as ridges rather than valleys. These examples of inverted relief are the result of wind erosion removing finer, less cemented sediment from the surrounding flood plain (Figure 7.42).

Rivers on Earth often enter lakes or the sea through deltas. The river splits into numerous distributaries as it crosses the flat, low-lying area which is created by deposition of a large sediment load when the rate of flow decreases. Deltas are also found on Mars, often at the margins of craters. In these cases, the channels probably disgorged into a lake which filled the impact basin.

Figure 7.42 A dendritic valley network at 42°S, 92°W. The small, branching channels merge to form larger channels, although the network is less dense than typical drainage systems on Earth. They may have been eroded primarily by groundwater flow – perhaps beneath a layer of ice – rather than by runoff from rain. The area shown is about 200 km across. (NASA-JPL)

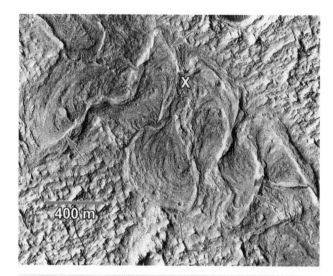

Figure 7.43 A fossil flood plain containing a meandering channel, indicated by red arrows. This is an example of inverted relief: the channel bed is now a ridge, rather than a valley, because wind erosion has removed finer sediment from the flood plain. The meanders were created by erosion of the bank on the outside of bends. A cut off may have occurred shortly before flow ceased above location "X," resulting in abandonment of the loop downstream. (NASA-JPL/MSSS)

The most famous example is a fossil delta in the Eberswalde crater at 24.0°S, 33.7°W. After its formation, the delta material was further buried by other materials – probably sediments – that are no longer present. The entire area became cemented and hardened to form rock. Subsequent erosion stripped away the overlying rock, re-exposing the delta. Today, the hardened delta sediments are preserved as ridges, an example of inverted relief.

Occasionally, fans of sediment are found where an ancient channel emptied into a large crater that was presumably filled by a lake. When the river slowed down upon entering the lake, its load of material was dropped, eventually building up into a wide fan crossed by a number of distributary channels. Many of these are fossil landforms, in which the channels through which sediment was transported are no longer present.

Another notable delta is found in Jezero crater, which has been selected as the landing site for NASA's 2020 rover mission. The rover, which is scheduled for launch in July 2020, will seek signs of ancient habitable conditions – and past microbial life – and collect rock and soil samples that will be retrieved and returned to Earth at some future date.

Jezero crater is located on the western edge of Isidis Planitia, a giant impact basin just north of the Martian equator. More than 3.5 billion years ago, the 45 km-wide crater contained a lake that could have collected and preserved ancient organic molecules and other potential signs of microbial life from the water and sediments that flowed into the crater billions of years ago.

Another piece of evidence for former crater lakes is networks of polygons up to 250 m in diameter. Although polygons on Mars are associated with permafrost (see Permafrost), some large craters

Figure 7.44 The Eberswalde delta lies in a crater in the southern highlands. It was formed when a river entered a crater lake. Over billions of years, the delta sediments hardened to form rock, while the softer surrounding material was eroded by the wind. Today, the fossil delta is raised above the rest of the crater floor. This detailed MOC image shows meandering channels, a cut-off meander, and crisscrossing channels at different elevations. (NASA-JPL/MSSS)

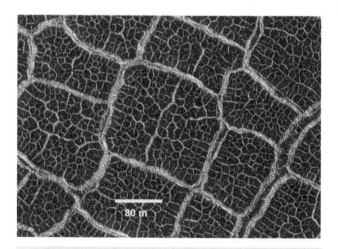

Figure 7.45 Large-scale polygons on crater floors may be caused by a desiccation (drying out) process, with smaller polygons caused by thermal contraction inside. The central polygon in this image is 160 m in diameter, smaller ones are 10 to 15 m wide, and the cracks are 5–10 m across. (NASA/JPL)

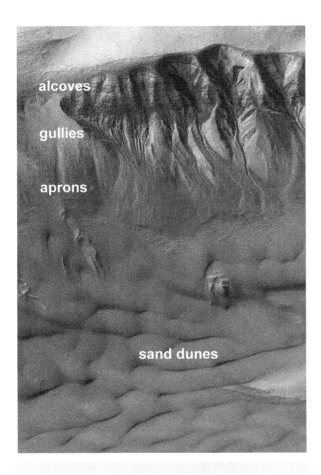

Figure 7.46 Gullies in an impact crater in Newton Basin, Sirenum Terra (42.4°S, 158.2°W). Spanning a few kilometers, this image from Mars Global Surveyor shows large alcoves near the top of the crater wall, leading into narrow gullies with aprons of loose material at their lower ends. Broad sand dunes cover the crater's floor. Frost is visible near the top and on the dunes below. (NASA-JPL/MSSS)

contain large-scale polygons subdivided into smaller polygons. These may be desiccation cracks caused by the drying out of the floors of evaporating lakes.

Unexplained Gullies

No evidence of hot springs or geothermal activity has yet been detected on Mars, not even in volcanic fissures. However, many impact craters display fresh gullies on their steep walls, often with deposits at their lower ends. Similar gullies have also been found on sand dunes. Clearly, some erosional process involving fluids or gases is active at these sites.

A typical gullied slope is smooth and fairly steep, but topped by a rocky outcrop (Figure 7.46). Many gullies originate from an indentation or "alcove" which is typically a few hundred meters across and located just beneath the upper rocky layer. On Earth, such alcoves form as water comes out of the ground and undermines the material from which it is seeping. The erosion of material causes

rock and debris on the slope above to collapse and slide downhill, creating the alcove.

In the center, the gullies are fairly straight and follow the down slope. They are typically about 10 m wide and hundreds of meters in length. At the bottom, some gullies fade away and disappear, whilst others end in a delta-shaped fan or apron of material – possibly a mixture of ice and loose debris. The aprons do not extend very far onto the crater floors, indicating that a limited amount of loose material (and possibly fluid) reaches the bottom of the slope.

The most obvious explanation is that occasional flows of groundwater have eroded the crater wall and carried away loose material. However, it is very difficult for liquid water to penetrate the frozen crust and survive on the surface in the present Martian environment.

If the source is melting ice, then the gullies should be most frequent near the equator – but this is not the case. They are more common at mid-latitudes – particularly in the southern highlands – and almost extend to the fringes of the permanent ice caps. This may be explained by the fact that ground ice is deep below the surface in equatorial areas, whereas it is within a few meters of the surface in higher latitudes.

Another curiosity is the predominance of gullies on shaded, polar-facing slopes at mid-latitudes, though there is little preference nearer the poles. One possibility is that groundwater moving toward a cold, shaded slope freezes near the surface. Its passage is blocked until the barrier is warmed sufficiently for the dam to burst, enabling a sudden rush of water down the slope. The trapped water could more easily remain in a fluid state if it contained dissolved salts – like sea water on Earth.

Other agents have been proposed, e.g. sublimating carbon dioxide ice which escapes, causing sand to become fluidized and flow downhill, eroding a gully. A similar scenario may apply to dunes that display gullies, e.g. in the Hellespontus and Noachis regions.

Several newly formed deposits have been observed in existing gullies. The lower end of the flow may split into several branches, and the material appears to be diverted around low obstacles. Such flows must have moved slowly, since they do not over-top some of the low obstacles in their path.

The gullies are distinct from another type of linear feature on Martian slopes, dark streaks called recurring slope lineae (RSL). These features grow in length, fade when inactive and recur annually during each summer, when temperatures may be warm enough for salty ice to melt. They are mostly found on steep, rocky slopes in regions such as the southern mid-latitudes, Valles Marineris, and in Acidalia Planitia on the northern plains.

Their appearance and growth resemble those of landforms created by seeping liquid water, but how they form remains unclear. Research has shown that they are probably flows of granular material like sand and dust. However, small amounts of water could still be involved in their initiation, as hydrated minerals have been detected at some RSL locations.

A Mars Ocean?

Did oceans ever exist on Mars? Scientists continue to debate over whether such a large body of surface water ever existed billions of years ago.

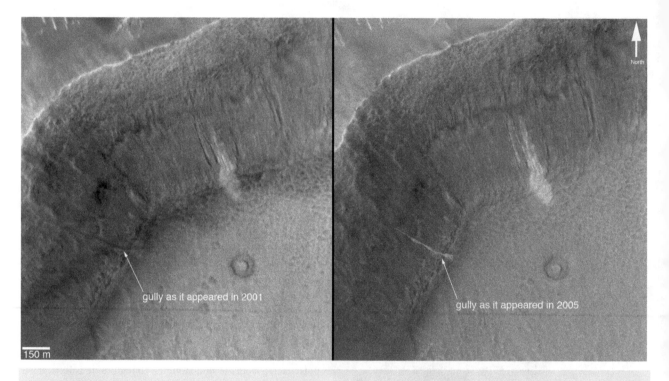

Figure 7.47   Images taken by Mars Global Surveyor in December 2001 (left) and September 2005 show new, light-toned deposits that appeared in an existing gully on the southeast wall of an unnamed crater in Terra Sirenum. The shallow, debris-laden flow was diverted around obstacles in its path and spread out into fingers at its lower end. (NASA-JPL/MSSS)

Figure 7.48   Dark, seasonal linear features emanate from bedrock exposures at Palikir crater in this HiRISE image. These recurring slope lineae form and grow during summer, when surface temperature is warm enough for salty ice to melt, and then fade or completely disappear in winter. The crater is located at 41.6°S, 202.3°. Three arrows point to bright, smooth fans left behind by flows. The ground slopes down toward top left. North is to the left. (Courtesy of NASA/JPL-Caltech/Univ. of Arizona)

There is certainly a large area of flat lowland in the northern hemisphere onto which numerous channel networks converge from the southern highlands (see Valley Networks, Lakes, Deltas). They must have delivered large volumes of water to the lowlands, although the discharge was probably more episodic than continuous.

What happened to the floodwaters? Did they supply enough water to fill a Martian ocean? Studies of the polar caps, ground ice, and numerous channel networks between the equator and mid-southern latitudes suggest that a plentiful supply of liquid water could have been available in the Noachian era, and favorable conditions may have recurred periodically in the Hesperian and Amazonian.

The fate of the water that flowed down the channels is uncertain. Any terminal lakes that pooled at the ends of the outflow channels may have quickly frozen, leaving thick bodies of ice that may still be present. After freezing, the ice would have slowly sublimated into the atmosphere, unless the frozen lakes were quickly covered by insulating dust.

Such a frozen lake, covered with surface pack ice, may have been identified in southern Elysium (5°N, 150°E). The feature, presumably fed by water from the Cerberus Fossae fissures, measures about 800–900 km across and may be up to 45 m deep – similar in size and depth to the North Sea. (The "lake" has also been interpreted as a hardened lava flow.)

Is there any evidence of a much larger body of water – a northern ocean? One clear indicator would be the presence of ancient shorelines: many former beaches can be seen on Earth. However,

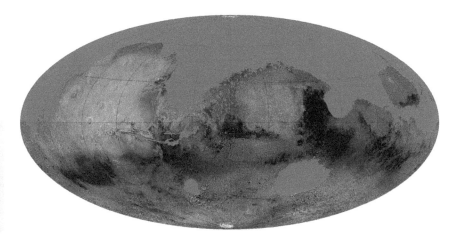

Figure 7.49 How Mars might have appeared some 3.5–4 billion years ago if an ocean covered the northern plains, fed by run off from the southern highlands. (Brian Hynek, LASP, University of Colorado)

there is no reason to believe that such clearly demarcated shorelines would be visible on Mars. First, there are no large moons to create powerful ocean tides. Second, any Martian oceans may have been ice covered, which would prevent wave action.

A controversial analysis of Viking orbiter images supported the presence of two ancient shorelines. However, when Mars Global Surveyor mapped the topography to a resolution of 300 m, it was found that the so-called Arabia "shoreline" varied in elevation by about 2.5 km, while the Deuteronilus shoreline varied by about 0.7 km, rising and falling like a wave with peaks several thousand kilometers apart. Since shoreline elevations, measured relative to sea level, are typically constant, the idea that Mars once had an ocean was widely rejected.

One attempt to overcome this objection suggests that the undulating shorelines can be explained by a surface displacement of Mars' poles by nearly 3,000 km within the last 2–3 billion years. Because spinning objects bulge at their equator, this so-called "true polar wander" could have caused shoreline elevation shifts similar to those observed on Mars.[9]

It is argued that, since the solid lithosphere behaves elastically, its surface deforms differently than the sea surface, creating a non-uniform change in the topography. If a flood of water had filled the Arabia ocean to a depth of several kilometers, about 3 billion years ago, the additional mass at the pole might have been enough to shift the pole 50° to the south. Once the water disappeared, the pole could have moved back, then shifted again by 20° during the deluge that created the Deuteronilus shoreline.

An alternative scenario proposes that the two shorelines represent the receding edge of a large sea, rather than separate inundations, caused when the Arabia ocean retreated to the Deuteronilus shoreline, shifting the pole from 50 to 20°. Once the ocean disappeared entirely, the pole returned to its current position.

A study released in 2018 argues that the irregular elevations of these shorelines can be explained by the growth of the Tharsis bulge with its huge shield volcanoes some 3.7 billion years ago. This would have deformed the topography and misaligned the shorelines.

The presence of a northern ocean that covered one third of the planet's surface is also supported by detailed, global maps of channels and deltas. The regions that are most densely dissected by valley networks lie between the equator and mid-southern latitudes, consistent with past precipitation and the presence of an ocean. Rain would be mostly restricted to the area over the ocean and nearby land surfaces, whilst the southernmost regions, located far from the water, would be arid and without any valleys. This would also explain why the valleys become shallower from north to south.

The distribution of ancient deltas around the margins of the northern plains has also been claimed as a marker for a planet-wide equipotential surface (i.e. a Martian sea level) along the margins of the northern lowlands. In other words, the levels of the deltaic landforms act as tracers of an ancient shoreline, where the Noachian and Hesperian rivers entered the ocean. Such data also support the existence of large seas in the Argyre and Hellas basins.

Chemical data have also been used to support the argument in favor of a northern ocean. The gamma-ray spectrometer on Mars Odyssey was used to detect elements buried 0.3 m underground by tracing their gamma-ray emissions. Concentrations of potassium, thorium, and iron were found along what appeared to be two ancient shorelines – one that covered a third of Mars' surface, and another that bordered a younger, smaller ocean (Figure 7.50). The elements were apparently leached out of the rocks on the neighboring uplands and carried by rivers into the oceans, where they were deposited.

Studies of the chemical signatures of two slightly different forms of water in Mars' atmosphere have been used to assess how much water the red planet lost over billions of years (see Solar Wind Interaction.) The conclusion was that, perhaps 4.3 billion years ago, Mars had enough water to cover its entire surface in a liquid layer about 137 m deep. In practice, the water would have formed an ocean occupying almost half of the northern hemisphere, in some regions reaching depths greater than 1.6 km.

To conclude, there is certainly plenty of circumstantial evidence that the northern, circumpolar plains may once have supported a

---

[9] Other possible causes of polar wander include a major shift of mass involving thermal convection within the mantle or a major asteroid impact.

Figure 7.50 This 3D image superimposes gamma-ray data from Mars Odyssey onto topographic data from Mars Global Surveyor. The red arrow indicates the shield volcanoes of Elysium, seen obliquely to the south east. Blue-to-violet colors in Elysium and the highlands mark areas poor in potassium. Red-to-yellow mark potassium-rich sedimentary deposits in lowlands below the Mars Pathfinder landing site (PF) and Viking 1 landing site (V1). (NASA/JPL-Caltech/University of Arizona)

large, shallow body of water (Figure 7.49). The proposed ocean would probably have resembled Earth's Arctic Ocean, with a surface layer of ice that waxed and waned with the seasons and helped to protect the underlying body of water from evaporation.

Hydrated Compounds

Apart from the physical evidence for running water and lakes, there is compelling chemical evidence that groundwater was widespread on ancient Mars (Figure 7.51). Thousands of outcrops of hydrated minerals, which have been chemically altered by exposure to water, have been found in the southern hemisphere, and others have been discovered in craters and plains in the northern hemisphere. However, there is still considerable debate over the amount in different localities.

Some of the debate centers around the presence (or absence) of carbonate compounds and a common mineral called olivine. Earth has enormous amounts of carbon locked up in its rocks in the form of carbonates. This occurred as the result of atmospheric carbon dioxide being dissolved in water and then reacting with other elements in the rocks, such as calcium.[10] These carbonates were washed into the oceans and eventually formed thick layers of chalk and limestone.

If Mars once possessed large lakes or oceans, similar carbonate beds might be expected to be widespread. However, these compounds have proved to be very elusive, even though many Martian

meteorites contain small amounts of carbonate, as well as altered minerals, such as clays, salts, and gypsum (see Chapter 13).

Carbonates originate in wet, near-neutral conditions, but are easily dissolved by acids, and many of the surface water and groundwater sources on the young Mars may have been sufficiently acidic to destroy any carbonates that formed.

However, recent observations have revealed some surface exposures of carbonates which have apparently been buried for billions of years. Mars Reconnaissance Orbiter has detected calcium or iron carbonate which was excavated from depths of up to 5 km by the giant impacts that created Huygens and Leighton craters. The largest known exposure is in Nili Fossae, a region of large rift valleys (grabens) in the northern hemisphere.

Exposed magnesium carbonate ($MgCO_3$) has also been found in the Nili Fossae region, where it is closely associated with rocks that contain phyllosilicates (clays) and olivine. It probably formed during the Noachian or early Hesperian era from the alteration of olivine by hydrothermal fluids or near-surface water. The presence of carbonate with accompanying clays suggests that the waters were neutral to slightly alkaline at the time of its formation, so that acidic weathering, thought to be characteristic of Hesperian Mars, did not destroy the carbonates in this particular environment.

Small exposures of carbonate have also been found by the Spirit rover in the Columbia Hills of Gusev crater (Figure 7.52). Magnesium iron carbonate makes up about one-quarter of the measured volume in the Comanche outcrop, a carbonate concentration 10 times higher than previously identified for any Martian rock. However, the presence of olivine indicates that neutral conditions may not have lasted very long.

Olivine, a common mineral in volcanic rocks, breaks down quickly when exposed to water. If olivine is still present today, then the implication is that it has not been exposed to wet conditions for any length of time. The mineral is widespread in areas such as Syrtis Major, but it has also been found in deep canyons, such as Ganges Chasma (Figure 7.53).

In some places, such as Nili Fossae, clays appear beneath – and therefore were deposited earlier than – fresh, unweathered rock rich in olivine. It seems that such sites were initially wet, but since the volcanic rocks were laid down, they have remained dry for hundreds of millions, or even billions, of years.

Another key mineral is hematite. On Earth, this iron-rich mineral is often associated with hot springs, where geothermally heated water dissolves iron-bearing minerals. After it evaporates, the iron compounds are left behind as colorful deposits. However, it can also be made in other ways, e.g. from crystals that grow in volcanic lava.

Exposures of hematite have been identified at various sites. The most abundant exposures include Aram Chaos, near Valles Marineris, and Meridiani Planum, the landing site of NASA's Opportunity rover (Figure 7.55). The rover came to rest in a small crater whose walls contained plenty of hematite in the form of small, spherical concretions which scientists dubbed "blueberries." Many more hematite/blueberry deposits were later found, far beyond the landing site.

---

[10] Carbonates are also formed by living organisms, but there is no proof of their existence on Mars, so they will not be discussed here.

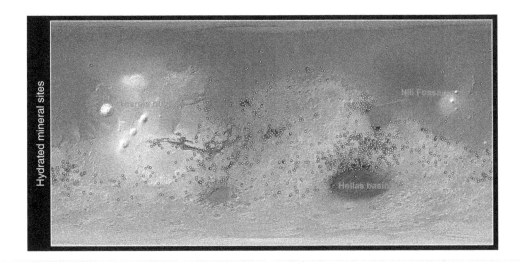

Figure 7.51 A map showing individual sites where hydrated minerals, formed in the presence of water, have been detected by Mars orbiters. More than 10 of these minerals have been found, with several thousand individual exposures detected. Over 50% occur in crater walls, floors, or central peaks, where older material has been excavated from below. Sedimentary deposits and crustal outcrops account for most of the remainder, with ancient river deltas accounting for 5% of the sites. They are primarily found in the most ancient terrain, dating back to over four billion years ago, when Mars may have sustained surface and subsurface liquid water. (ESA/CNES/CNRS/IAS/Université Paris-Sud, Orsay)

Figure 7.52 A false-color image of an outcrop, named Comanche, in the Columbia Hills of Gusev crater. When found, it had 10 times more carbonate than any previously identified Martian rock. Spectrometers on the Spirit rover found that about one-quarter of the rock is composed of magnesium iron carbonate, a mineral that formed when the environment was wet and non-acidic. (Courtesy of NASA/JPL-Caltech/Cornell University)

The blueberries – which are actually more gray than blue – were found in layers, which suggested that they had formed due to leaching of iron minerals in water, or the slow evaporation of iron-rich water in the rock's pore spaces – rather like stalactites and stalagmites in caves. Many blueberries had been released by weathering and wind erosion of softer rock, so that they gathered in small piles. Further support for an aqueous history was provided by cross bedding and possible concave patterns formed by underwater ridges within the rocks.

The Curiosity rover also found resistant bands of hematite in a region known as Vera Rubin Ridge, on the lower slopes of Aeolis Mons – a 5 km-high mountain at the center of Gale crater.

The Spirit rover made its own water-related discoveries, halfway around the planet. Apart from the detection of carbonate in the Columbia Hills, one of the most surprising of these was a measurement of about 90% pure silica (SiO_2) in a patch of fine-grained soil that was churned up by a stuck wheel on the rover (Figure 7.56). Indeed, wherever, the wheel dug into the surface, a white layer rich in sulfur or silica was revealed. The silica may have been laid down as the result of hot spring activity or when acidic, volcanic vapors interacted with water at the surface. High sulfate levels were also found in the Columbia Hills.

The picture that is now emerging is of a planet that was fairly warm and wet more than 4 billion years ago. Atmospheric carbon dioxide interacted with ancient bodies of water, producing calcium- and iron-rich carbonates. Prolonged contact with water of low acidity – either on the surface or underground – meant that volcanic rock minerals such as olivine were rapidly hydrated, creating widespread phyllosilicate (clay) beds and magnesium carbonate.

As Mars began to cool and dry out, salt flats with intermittent pools between dunes became the norm. Planet-wide volcanic eruptions belched sulfur into the atmosphere, producing acid rain or highly acidic groundwater. Chemical reactions with this water again modified the surface rocks, creating different types of hydrated minerals, such as sulfates – especially where there was evaporation of salty, and sometimes acidic, water.

Figure 7.53 Almost 5 km deep, this section of Ganges Chasma reveals ancient rock layers in the canyon walls, as well as several enormous landslides. One landslide on the north wall flowed more than 17 km across the canyon floor, eventually coming to rest on top of an older landslide from the south wall. Outcrops of olivine-rich, basaltic bedrock appear as long, dark red ribbons at the base of the walls. They indicate that the region has remained relatively dry, because olivine is rapidly broken down in wet environments. The colors show night temperatures. Warmer, rocky areas are red and yellow. Cooler, dusty areas are green and blue. (Courtesy of NASA/JPL/ASU)

Figure 7.55 A near-true-color image of a rock in Eagle crater called "Berry Bowl." It shows numerous ball-shaped "blueberries." Rich in the iron compound hematite, the spherules are thought to have been deposited in water. The circular area (center) was brushed clean by the rover's rock abrasion tool. (Courtesy of NASA/JPL/Cornell)

Hydrated silica, found in and around the margins of Valles Marineris and dry river channels, is evidence that some low temperature, acidic water remained on the surface until about 2 billion years ago. Finally, Mars became largely dry. During the second half of the planet's history, the rocks have been altered by slow weathering, punctuated only by occasional relapses to warmer, wetter conditions.

A Windblown Desert

Orange-red dust and sand cover almost the entire surface of Mars. The color is the result of oxidation of iron minerals in the crust: essentially, the planet is "red" because its rocks have rusted. The

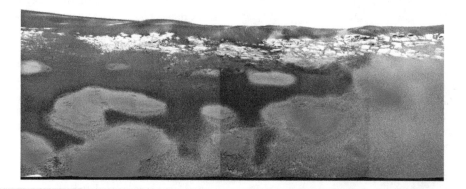

Figure 7.54 The Opportunity rover landed in a small hollow, named Eagle crater. Spectrometer data show the location of crystalline hematite, a product of past water activity. Red and orange indicate high levels of the iron-bearing mineral. Blue and green denote low concentrations. The crater was about 22 m across, and the bedrock outcrop (mainly white) was 30–45 cm deep. The northeastern outcrop on the rim of the crater does not appear to contain much hematite. Also lacking hematite are the rover's airbag bounce marks (foreground). (Courtesy of NASA/JPL/Arizona State University/ Cornell)

Figure 7.56 A shallow trench made by Spirit's wheel uncovered bright patches of almost pure, fine-grained silica – evidence for ancient, water-rich environments in Gusev crater. On Earth, such deposits can be created in warm, evaporating coastal waters or in hot, mineral-laden waters around geysers and hot springs. This false-color view was taken on May 17, 2007. (Courtesy of NASA/JPL-Caltech/Cornell)

iron component in the dust has been confirmed by spacecraft spectrometers and by magnets carried on surface landers and rovers, which have attracted particles of dust. The dust is dominated by silicate minerals.

Although the atmosphere is only about 1% as dense as Earth's at sea level, it is capable of picking up and transporting small, weathered debris on a massive scale. These shifting sands are largely responsible for the changing seasonal pattern of dark and light markings that have been recorded by astronomers for hundreds of years. (For several centuries they were thought to be associated with the seasonal expansion and dying off of Martian vegetation.)

In general, the classical bright features seen on pre-space-age maps are regions covered by a layer of dust up to 2 m thick with particles in the size range 2–40 microns. The dark features are relatively dust-free, though still coated with particulate materials ranging from 0.1 mm to 1 cm across, rather than bedrock.

Tiny particles of dust – comparable to cigarette smoke – are also suspended high in the sparse atmosphere, producing an orange sky, as well as spectacular sunrises and sunsets. These can remain aloft almost indefinitely, although they are periodically replenished by dust storms and mini-tornados known as dust devils (see Dust Devils).

The gradual build-up of fine dust on exposed surfaces was documented by the Mars Exploration Rovers, which experienced large drops in power when a coating of dust reduced the sunlight reaching their solar panels. However, Nature also intervened from time to time when gusts of wind removed much of the dust and boosted their power output.

As befits a desert world, Mars has many landforms which have been created by wind action, and dunes of all sizes are found. A huge dune field covers much of the north polar region, and there are many isolated sand seas, especially within old, degraded craters west of Hellas Planitia. In many ways these dunes resemble their counterparts on Earth. Some are dark, probably of basaltic composition, and others are bright.

Swarms of crescent-shaped barchans occur where the prevailing wind is mainly in one direction (Figure 7.60). Sand is moved up a gentle slope on the windward side. As the sand piles up at the crest, it collapses and forms a steep slope on the lee side. The entire dune is driven forward by this process. The horns of the crescent show the direction of movement and are created by sand moving around the dune's outer edge. Sometimes smaller, faster-moving dunes merge with larger, slower-moving dunes, resulting in single, large features over 500 m wide.

Seif dunes are linear in shape, with their long axis parallel to the wind. These are often seen extending from sand sheets next to eroding mesas and rocky knobs, *as well as from the horns of some barchans.* They form when wind that encounters the dune obliquely is deflected on its lee side, causing it to flow along that side parallel to the crest line.

Most dunes on Mars are transverse, lying at right angles to the prevailing wind direction. Like barchans, they are asymmetrical, with a gently sloping windward side and a steeply sloping leeward side. However, in many places there is a juxtaposition of fresh and degraded dunes, including star dunes and transverse dunes with differing alignments, indicating multiple wind directions.

Images taken over many years show few changes: the larger dunes appear to be stationary, implying that many of them are "fossil" features left over from a different climatic regime. Only a few small, dome-shaped dunes have been seen to diminish in size or disappear completely, although streaks have been seen to appear on the downwind sides of some dunes, indicating that sand is moving. Calculations show that, on average, it takes about 4,000 years for a Martian dune to shift by 1 m. However, it is likely that dune change is limited to occasional dust and sandstorms driven by strong winds.

Sand ripples that resemble small tranverse dunes are found almost everywhere (Figure 7.61). Over time, stationary ripples and small dunes develop a surface crust that is hard to identify. In March 2009, the Spirit rover became permanently stuck when its wheels broke through such a crusty surface and churned into soft sand hidden underneath. The rotating wheels revealed loose layers of basaltic sand, sulfate-rich sand and areas with silica-rich materials, possibly sorted by wind and cemented by chemical action.

The key to dune and ripple movement is saltation, a cascade effect which occurs when a sand particle starts to hop along the surface and dislodge other grains. The thin atmosphere means the wind has much less momentum and is much less likely to disturb a sand grain than on Earth. However, the lower gravity and air pressure enable a particle that has begun saltating to travel 10 times faster than it would on Earth. As a result, each Martian grain should eject about 10 times more sand upon impact than on Earth.

Other evidence that wind can move loose material around the surface is provided by "tails" of material on the leeside of craters

Box 7.3   Curiosity in Gale Crater

The largest and most advanced roving vehicle to explore the surface of Mars is NASA's Curiosity, the heart of the Mars Science Laboratory mission. The nuclear-powered rover carries 10 instruments designed to continue NASA's "follow the water" strategy, in order to discover potential habitable environments capable of supporting microbial life.

Launched on November 26, 2011, the spacecraft began a ballistic descent through the Martian atmosphere on August 6, 2012 (GMT). During the final stages of the descent, retrockets were fired to slow the rate of fall, then the 899-kg rover was lowered toward the Martian surface on nylon cords, using a new "skycrane" technique. The touchdown occurred in 154 km-wide Gale crater. Curiosity's primary mission was intended to last one Mars year (687 Earth days), but it is still operational.

One reason that Gale crater was selected as the target was its low altitude and near-equatorial location (4.5° S), which made it a relatively easy place to land. Another crucial factor was evidence from orbiting spacecraft that it once contained substantial amounts of water. In the center of the crater is Aeolis Mons (unofficially dubbed Mount Sharp by NASA), a 5 km-high "mound" with sides which are gentle enough to be climbed by Curiosity and easy to sample.

The payload includes mast-mounted instruments to survey its surroundings and assess potential sampling targets from a distance; instruments on a robotic arm for close-up inspections; laboratory equipment inside the rover to analyze samples of rocks, soils and atmosphere; and instruments to monitor the environment around the rover.

Some of the tools, such as the laser-firing Chemcam for checking the elemental composition of rocks, are the first of their kind on Mars. In addition to the science payload, engineering sensors on the heat shield gathered information about Mars' atmosphere during its descent.

Observations from orbit indicate the presence of water-related clay and sulfate minerals in the lower layers of Aeolis Mons, and textures higher on the mountain indicate that mineral-saturated groundwater filled fractures and deposited minerals. Stratification in the mountain suggests that it is the surviving remnant of extensive deposits that were laid down after the impact that excavated the crater, billions of years ago.

Scientists decided first to investigate closer outcrops, including the first streamed pebble deposits ever examined up close on Mars. The images showed rocks containing rounded stones eroded by flowing water and subsequently cemented into conglomerate rock.

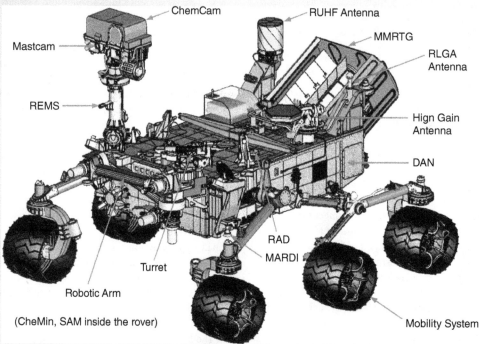

Figure 7.57   The 900-kg Curiosity rover is the largest, most sophisticated vehicle yet to land on Mars. It carries 10 instruments to explore Gale crater. These include 17 cameras to survey its surroundings and assess potential sampling targets from a distance; a rock-zapping laser; chemical analysis instruments inside the rover; and a package to monitor local weather, radiation, etc. A turret mounted on the 2.1 m-long robotic arm carries a drill to obtain powdered samples from interiors of rocks; an Alpha Particle X-ray Spectrometer (APXS) to identify chemical elements in rocks and soils; a sample processing subsystem named Collection and Handling for In-Situ Martian Rock Analysis (CHIMRA), which includes a scoop for acquiring soil samples; a Dust Removal Tool (DRT) for brushing rock surfaces clean; and the Mars Hand Lens Imager (MAHLI) for magnified, close-up views of rocks. Power is provided by a radioisotope thermoelectric generator which converts heat from plutonium-238 dioxide into electricity. *(NASA)*

Evidence of a past environment suited to support microbial life came within the first eight months. Analysis of the first sample material ever collected by drilling into a rock on Mars identified sulfur, nitrogen, hydrogen, oxygen, phosphorus, and carbon.

The rover's Sample Analysis at Mars (SAM) and Chemistry and Mineralogy (CheMin) instruments indicated that the Yellowknife Bay area was at the end of an ancient river system or an intermittently wet lake bed. The rock was made up of a fine-grained mudstone containing clay minerals, sulfate minerals, and other chemicals. This ancient wet environment, unlike some others on Mars, was not harshly oxidizing, acidic, or extremely salty.

Meanwhile, the ChemCam team detected veins of gypsum running through the rocks of Yellowknife Bay, some 700 m away from Curiosity's landing site. The veins were composed mainly of hydrated calcium sulfate, evidence that water once circulated in rock fractures. The accumulated evidence indicates that the area once was home to ponds created by runoff or subsurface water that had percolated to the surface,

Mission scientists have concluded that habitable conditions in the Yellowknife Bay area may have persisted for millions to tens of millions of years. During that time, rivers and lakes probably appeared and disappeared. Even when the surface was dry, the subsurface likely was wet.

Curiosity also found evidence of finely layered bedrock that is thought to result from sediments laid down on a lake bed, as well as mineral veins deposited by salty fluids moving through the rock. By late 2017, the rover had progressed to Vera Rubin Ridge, a region where iron seems to have precipitated out of fluids and become cemented to form a weather-resistant rock layer rich in hematite.

Other interesting results include the first radiation measurements from the surface and the detection of variable amounts of methane in the atmosphere. The radiation dose equivalent rate was 0.67 millisieverts per day from August 2012 to June 2013. Long-term population studies have shown that exposure to radiation increases a person's lifetime cancer risk; exposure to a dose of 1 Sv is associated with a five percent increase in fatal cancer risk.

By mid-February 2020, Curiosity had driven a total of 21.79 Km from its landing site, and it was continuing to climb the layered foothills of Aeolis Mons.

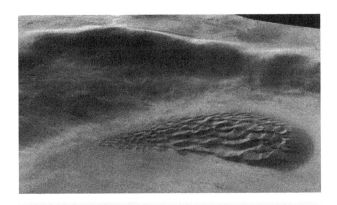

Figure 7.58 Many craters and canyons contain sand dunes. This Mars Express image shows a crater that contains a small field of dark dunes which measures 7 × 12 km. The crater, located at 43°S, 303°E, is about 45 km wide and 2 km deep. The dunes' asymmetric profile, with a gentle slope on the wind-facing side, suggests an easterly wind direction. The dark sands could be basaltic. (ESA/DLR/FU Berlin – G. Neukum)

which are probably accumulations of bright material shaped by wind currents.

Wind is an agent of erosion, as well as transportation and deposition (Figure 7.62). Perhaps the most obvious example of aeolian erosion are yardangs, parallel ridges and grooves which are aligned in the direction of the prevailing wind. They are common in southern Amazonis and south of Elysium Planitia, sometimes reaching several kilometers in length, though they are often only a few tens of meters high.

Sand blasting of the lower parts of rocks and wind erosion of harder bedrock, such as lava flows, seems to be rare. This may be because wind speeds are generally low, and fine dust, rather than larger, more abrasive particles, is picked up and moved by the thin air.

Dust Storms

Astronomers have long been aware of regional dust storms that obscure the planet's surface. These are seasonal events, confined to spring and summer in either hemisphere, when the seasonal carbon dioxide ice cap sublimates as the temperature increases. This causes the atmospheric pressure to rise, thus enabling more dust to be suspended and remain aloft for a longer time.

Most are local or regional events, driven by strong surface winds, turbulence, and rapid convection. They are very common near the poles in spring, powered by large temperature differences between the polar ice and the nearby regions of dark sand and rock which are heated and defrosted by the springtime Sun. The dust may obscure the ice caps as it is picked up and rapidly spread by the polar jet stream.

An even more important source region is the Hellas basin in the southern hemisphere, which is effectively a huge dust bowl. Because the basin is up to 8 km deep, air at the bottom is about

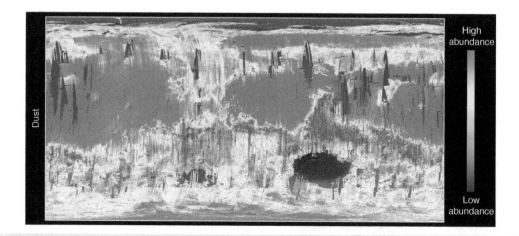

Figure 7.59   A global map of dust cover, based on data from the OMEGA instrument on Mars Express. The dust is in the form of ferric oxide nanoparticles, at most a few tens of microns in size. The particles are thought to be the result of chemical reactions with the atmosphere, causing rocks to "rust" slowly over billions of years, giving Mars its distinctive red hue. The dust is closely linked to iron-rich terrains, including the volcanic Tharsis region. Weathering and erosion from past glacial activity and impact events, as well as dust storms, winds, and freeze-thaw cycles today, contribute to continued production of fine-grained dust. Grey areas are caused by a lack of data, notably in the Hellas basin at lower right. (ESA/CNES/ CNRS/IAS/ Université Paris-Sud, Orsay)

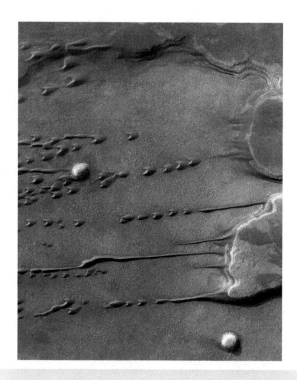

Figure 7.60   Barchan and linear dunes that formed on the floor of a crater and extended from an eroding mesa to the west of Hellas Planitia, near 41.8°S, 44.5°W. The prevailing wind blows from right to left. The location of the linear dunes coincides with areas where supply of loose material is greatest. The linear dune breaks up into barchans with distance from the flow obstruction. (Courtesy of NASA-JPL/University of Arizona)

10°C warmer than at the top. This temperature gradient creates major updrafts and strong, turbulent winds, which can carry dust above its rim and out of the crater. On rare occasions, the dust from Hellas spreads far and wide, triggering other long-lived storms.

A dozen planet-encircling dust storms have been reported since 1877. Perhaps the most famous example erupted in September 1971, whilst NASA's Mariner 9 and several Soviet spacecraft were en route to the planet. When they arrived in orbit, Mars was completely shrouded in dust. Only the south polar cap and four dark spots, eventually revealed as the tops of the Tharsis volcanoes, were visible. The dust sometimes rose to a height of 70 km, and the storm did not disappear until the end of the year.

The evolution of a similar storm was closely monitored in 2001, at the onset of the Martian southern spring. Studies showed a complex sequence of events which culminated in the dust eventually covering the entire planet.

It began as a small dust cloud inside the Hellas basin. At first there was little change. The storm only "exploded" after it crossed the equator on June 26. In less than 24 hours, additional dust clouds were being raised in Arabia, Nilosyrtis, and Hesperia, far from Hellas.

During the following week, dust lifted into the stratosphere was carried eastward by the south circumpolar jet stream. At the same time, a turbulent wind front moved across Mars, favoring the creation of many other dust storms, sometimes thousands of kilometers from Hellas. By early July, dust had enveloped the entire planet. The storms largely subsided by late September, but the atmosphere remained hazy into November.

Dust storms play an important role in modifying the weather and climate of Mars. Smaller storms, especially near the poles, affect the rate at which seasonal frost evolves, and control local and regional weather patterns. The global storms temporarily alter the entire planet's heat balance, promote variations in seasonal frost

Figure 7.61 Two sizes of wind-sculpted ripples are visible on the top surface of a Martian sand dune. The area shown is part of "Namib Dune" in the Bagnold Dune Field, along the northwestern flank of Aeolis Mons in Gale crater. Sand dunes and the smaller type of ripples also exist on Earth, but the larger ripples – roughly 3 m apart – do not occur on Earth. The mosaic was taken by the Curiosity rover on December 13, 2015. The sand is very dark, both from morning shadows and from the darkness of the minerals that dominate its composition. (Courtesy of NASA/JPL–Caltech/MSSS)

Figure 7.62 Parallel ridges and grooves known as yardangs are sculpted by windblown sand. They occur where the winds blow in the same direction for a long period. Weaker material is eroded faster than more resistant material. This Mars Express image was taken south of Olympus Mons at 6°N, 220°E. More resistant rock may occur in the three flat regions, which measure about 17 × 9 km. (ESA/DLR/FU Berlin – G. Neukum)

This was hardly surprising, since images from the Viking landers showed how the sky darkened during even local dust events, and this was confirmed by Spirit and Opportunity, which occasionally recorded a 99% drop in direct sunlight, enough to stop all surface operations (Figure 7.65).[11]

In contrast, the loss of heat from the surface at night was reduced by the dust blanket, so the nighttime average actually increased by 2–3°C. At the same time, the dust absorbed solar radiation and heated the upper atmosphere by some 30°C. The warmer atmosphere ballooned outward, causing increased aerodynamic drag on orbiting spacecraft.

The most recent major storm began at the end of May 2018 and became a global event on June 19. This storm was unusual because it began in Mare Acidalium in the northern hemisphere, and started much earlier than usual, only nine days into southern hemisphere's spring. (All of the historical dust storms of this size began in the southern hemisphere, in the areas of Hellas, Noachis, or Argyre.) It rapidly grew in size as it spread south and west toward the Chryse and Sinus Meridiani regions – and kept going.

The storm caused a drop in insolation which obliged NASA to suspend science activities with the solar-powered Opportunity rover in an effort to preserve the rover's batteries. The prolonged lack of sunlight eventually caused the demise of the long-lived vehicle.

Dust Devils

formation and dissipation, and greatly affect the distribution of water vapor.

During the 2001 storm, the fine airborne dust blocked a significant amount of sunlight, causing a drop in the daytime average temperature on the surface of almost 3°C (Figures 7.63 and 7.64).

Loose surface material is also transported by much smaller atmospheric disturbances, known as dust devils, which form when the ground heats up during the day. Much larger than their counterparts on Earth, these column-shaped vortices resemble miniature tornadoes. As the warmer, lighter air nearest the ground begins to

---

[11] Before dust storms began blocking sunlight in June 2007, Opportunity's solar panels produced about 700 watt hours of electricity per day, enough to light a 100-watt bulb for seven hours. By July 18, the solar panel output had dropped to 128 watt hours.

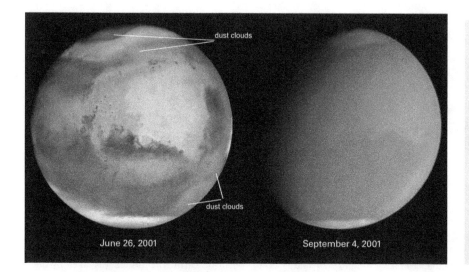

Figure 7.63 The evolution of a global dust storm. (Left) On June 26, 2001, one large dust storm was visible above the northern polar cap (top), with a smaller dust cloud nearby. Another large dust storm was spilling out of the Hellas basin in the southern hemisphere (lower right). White water ice clouds are seen near the polar caps. (Right) Over two months later, the entire planet was still blanketed by dust. (Courtesy of James Bell/Cornell Univ., Michael Wolff/Space Science Institute, Hubble Heritage Team)

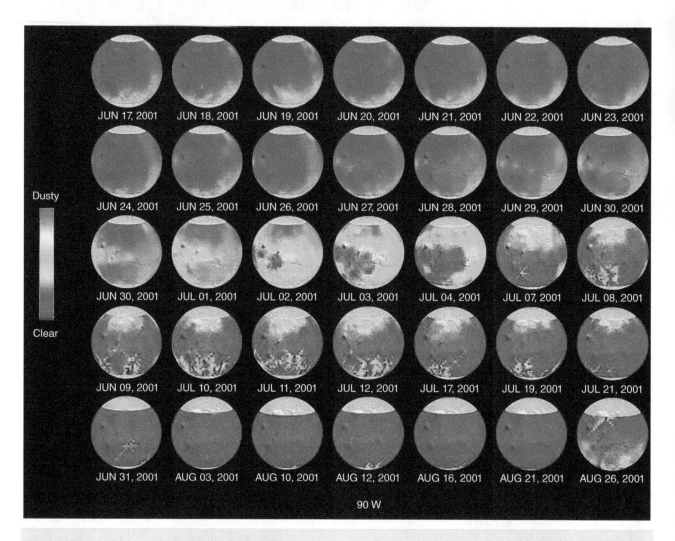

Figure 7.64 The evolution of the June 2001 dust storm that began in the Hellas basin (near bottom), as measured by Mars Global Surveyor. The cloud grew until, by late July, it enveloped the whole planet. It then slowly subsided over the next few months. Blue represents a fairly clear atmosphere and red shows dense concentrations of dust. (Courtesy of NASA-JPL/Arizona State University)

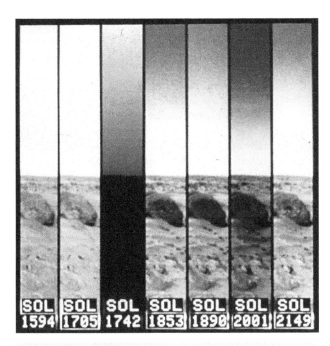

Figure 7.65 The impact of a dust storm at the Viking 1 landing site (22°N, 48°W) on sol 1,742 of the mission, compared with illumination on other days. The relative levels of illumination show how atmospheric dust affects the brightness of the sky. The images have been computer enhanced. (Courtesy of NASA-JPL/Mary A. Dale-Bannister, Washington University, St. Louis)

rise, it spins. When the spinning column moves across the surface it picks up dust, making the vortex visible.

The peak time for dust devil formation is around 2 p.m. local time, the warmest time of day in midsummer. On Earth, dust devils typically last for only a few minutes, but the Martian versions can survive much longer. Sometimes they grow to the size of terrestrial tornadoes, spiraling up to 9 km above the surface (Figure 7.67). (The height of a dust devil can be calculated from the length of its shadow and the height of the Sun above the horizon.)

The Mars Exploration Rovers observed hundreds of dust devils, often appearing in small swarms. Gusev crater alone probably experiences about 90,000 dust devils per sol, and these whirlwinds

collectively hoover up and redeposit an estimated 4.5 million kg of sediment per sol. Clearly, these modest storms reshape the landscape over much of the planet and contribute to the build-up of larger dust storms.

Atmosphere

From the surface, the sky usually appears light brown or orange, due to light scattered from suspended dust particles, roughly 1.5 microns across. These can be raised to high altitude by dust devils and seasonal dust storms, remaining aloft for long periods before they slowly drift back to the surface.[12]

The atmosphere is about 100 times less dense than on Earth. It is dominated by carbon dioxide, which accounts for more than 95% of the atmosphere by volume. During the winter, when the poles are in continual darkness, the surface gets so cold that up to 25% of the atmospheric CO_2 freezes out at the polar caps. When the poles are again exposed to sunlight in summer, the CO_2 ice sublimates back into the atmosphere. There are also seasonal variations in the amount of water vapor in the atmosphere as the polar caps wax and wane.

The surface pressure ranges from about 14 mb in the deepest parts of the canyons to 0.3 mb on top of the highest volcanoes. It also changes with the seasons, as a result of sublimation of ices in the summer hemisphere and freezing out of volatiles at the winter pole. This cycling of the polar ices causes the average atmospheric pressure to change by 20% as the seasons change. Variations of this magnitude are unknown on Earth, where even a hurricane causes a drop in pressure of only a few percent.

Since Mars is between 30% and 67% further from the Sun than Earth, it is correspondingly colder. The maximum solar irradiance on Mars is about 590 W/m², compared to about 1,000 W/m² at Earth's surface – equivalent to the intensity on an autumnal day when the Sun is a mere 36° above the horizon.

Without a thick atmosphere to trap warmth by night or minimize the Sun's heat by day, Mars experiences extreme day–night temperatures. Local temperature variations are related to factors such as the reflectivity and thermal properties of surface materials, and the slope and orientation of the ground.

Latitude also plays a part (see Chapter 3). The Spirit rover recorded greater temperature swings than Opportunity because it was farther from the equator, where there were greater variations

Figure 7.66 An image by the Spirit rover showing two dust devils crossing the dusty interior of Gusev crater. Small dust devils such as these appeared each afternoon during the summer months. (NASA-JPL)

[12] The first color picture returned by Viking Lander 1 was hurriedly processed to produce a blue sky. Mission scientists soon realised their error, and all subsequent images showed the dust-laden, orange sky.

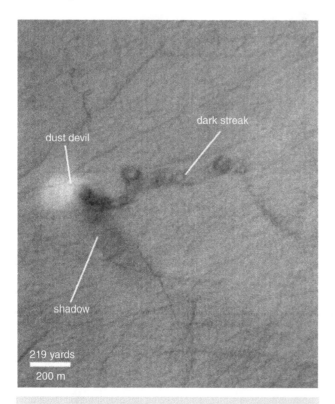

dust devil

dark streak

shadow

219 yards

200 m

Figure 7.67 A dust devil seen from above as it creates a curly, dark streak in Promethei Terra. Traveling from right to left (east to west), the rotating, dust-laden column throws a shadow. The dust devil is less than 100 m wide. The picture covers an area of 1.5 × 1.7 km. (NASA-JPL/MSSS)

in the Sun's height above the horizon and the length of the day.

Near-equatorial temperatures in summer typically range from –90°C at night to above 20°C at midday, but this variation increases if the surface consists of low density, fine-grained material (Figure 7.73). However, at depths of several centimeters below the surface, temperatures even out with a diurnal mean of about –50 to –60°C. The air temperature is also quite different only a few meters above ground level.

Conditions are even more extreme at the poles in winter, where temperatures drop below −120°C. At this point, carbon dioxide condenses out of the atmosphere to form a seasonal ice layer which covers the permanent polar cap and extends to lower latitudes.

Clouds of water ice often gather above the slopes of the giant volcanoes and in the bottoms of the canyons and deep basins, such as Hellas, where they resemble blankets of fog or low-lying stratus (Figures 7.75 and 7.76). Wispy cirrus clouds also appear at high altitudes, both during the day and night (Figure 7.77). The daytime clouds have been found up to 100 km above the surface – the highest clouds ever detected on any terrestrial planet. Circular clouds are common in summer near the north pole, and may result from merging air currents.

Nocturnal clouds are five times thicker than their daytime counterparts and hover close to the ground. Despite their insubstantial

nature, they do trap some heat and help to keep the surface warmer when temperatures plummet after sunset. When the air becomes dustier, the upper atmosphere is warmed and the area of water ice clouds dwindles.

The Viking Landers sent back images of a frost-covered surface in winter, but snow was not thought to be likely (Figure 7.74). However, the Mars Phoenix lander detected ice crystals falling from clouds at an altitude of about 4 km, so this form of precipitation may also be an important part of the water cycle. However, the total amount of water in the atmosphere is very small. If all the water in the Martian atmosphere simultaneously fell as rain, it would make a global ocean less than 0.01 cm deep.

On Earth, warm air rises in the tropics, travels poleward at high altitude, then cools and descends in the subtropics before returning toward the equator (see Chapter 3). Computer simulations suggest that a similar, though less complex, circulation also occurs on Mars, with a single, trans-equatorial Hadley cell which reaches an altitude of 50 km, compared with about 20 km on Earth. This cell is most prominent around the solstices.

The atmospheric flow associated with the Hadley cell transports surface air from the cold winter hemisphere to the summer hemisphere, e.g. at the northern winter solstice, the rising air is in the warmer southern hemisphere and the sinking air is in the northern hemisphere.

Because the equator rotates faster than the rest of the planet, it creates a Coriolis force that deflects the zonal winds. The result is a tradewind-like pattern of easterlies in the winter hemisphere and westerlies in the summer hemisphere. A strong westerly jet also occurs near the winter pole, while an easterly jet occurs in the southern subtropics in summer.

Winds associated with the Hadley circulation can reach very high speeds: models indicate that mean winds in the subtropics during the southern midsummer exceed 160 km/h about 200 m above the surface. Surface winds can also be very variable. At the Viking landing sites, typical wind velocities were from zero to 12–16 km/h in summer, with rare gusts in winter of 48–65 km/h. Wind patterns were also fairly predictable, with a repetitive pattern of wind direction and velocity for specific times of the day. However, during southern spring and summer, when dust storms are common, gusts of up to 144 km/h have been recorded.

Large scale variations in atmospheric pressure occur in the winter hemisphere, as high and low pressure systems travel eastwards – just as they do on Earth (Figure 7.71). Associated frontal systems – boundaries between air masses of contrasting temperature – also occur on both planets.

The sparse atmosphere responds almost instantly to solar heating. This results in a daily pressure cycle, with minimum pressures near 4 a.m. and 6 p.m. local time, and maximum values near midnight and 10 a.m. It also causes strong winds across the day–night terminator. Winds are also modified by local topography, especially near the large volcanoes and in large basins such as Hellas, creating stationary eddies. Pathfinder, for example, found that winds at night came from the south, consistent with air flowing down Ares Vallis.

Despite its low density, the atmosphere does offer some protection from incoming interplanetary debris. The glow from meteors

Box 7.4   The Vikings

NASA's Viking 1 and 2, each consisting of an Orbiter and a Lander, were the first missions to obtain high resolution images of the Martian surface, characterize the structure and composition of the atmosphere and planetary surface, and conduct in situ biological tests.

Figure 7.68   Scientist Carl Sagan with a model of the Viking lander. The upright cylinder to the right of the dish-shaped high gain antenna is a TV camera. The meteorology boom is visible at right. *(NASA-JPL)*

Each Orbiter was equipped with two vidicon cameras, an infrared spectrometer, and an infrared radiometer. Their task was to image the entire planet at high resolution.

Each 663-kg Lander was sterilized before launch and housed inside a protective, pressurized shell on the bottom of the Orbiter. It was 2.1 m tall and 3 m wide, with three legs and a 3 m-long robot arm to dig trenches and collect samples. Power came from two radioisotope thermoelectric generators, which used plutonium-238 as fuel, and four batteries.

The Landers carried two TV cameras, a biology lab with three experiments, a gas chromatograph mass spectrometer, and X-ray fluorescence spectrometer, pressure, temperature, and wind velocity sensors, a three-axis seismometer, a sampler arm, and magnets. Other sensors measured air pressure, temperature, and acceleration during the descent.

Viking 1 was launched on August 20, 1975, and arrived at Mars on June 19, 1976. The Viking 1 Lander touched down at Chryse Planitia (22.48°N, 49.97°W) on July 20, 1976. Viking 2 was launched on September 9, 1975, and entered Mars orbit on August 7, 1976. The Viking 2 Lander arrived on Utopia Planitia (47.97°N, 225.74°W) on September 3.

Originally designed to function for 90 days, the Vikings continued collecting data for more than six years. The Landers accumulated 4,500 high-resolution images of the Martian surface, whilst the Orbiters provided more than 50,000 images, mapping 97% of the planet.

Viking 1 landed on rocky plain marked by occasional ridges and impact craters. The gently rolling landscape was yellowish-brown and strewn with rocks in the centimeter to meter size range. The rocks were volcanic and probably deposited as ejecta from distant large impacts. Between the rocks were drifts of fine-grained material, with bigger dunes of windblown sand in the distance. The surface was loosely cemented, probably by water-soluble salts such as sulfates, to form a crust.

Viking 2 also landed on a level, rock-strewn, but otherwise featureless plain. The source of the rocks was believed to be a 90 km-diameter impact crater situated 170 km to the west. Although no chemical analyses were obtained from rocks at either Viking site, they were believed to be basaltic. The Landers scooped up surface material in order to measure the chemical composition of the soils. They also performed a variety of chemical experiments on the soils, as part of their search for life (see Life on Mars?).

burning up in the atmosphere was captured by the Spirit rover's panoramic camera, and meteor showers, similar to those experienced on Earth, are almost certainly visible on Mars.

Several spacecraft in orbit around Mars on October 19, 2014, observed the aftermath of the close passage of Comet Siding Spring. Their observations of the upper atmosphere revealed that debris from the comet added a temporary, but enhanced, layer of ions to the Martian ionosphere. This suggests that an observer on the planet's surface would have seen a spectacular meteor shower, with several thousand meteors per hour.

Figure 7.69   The Viking 1 landing site on Chryse Planitia. Orange-red sand dunes and rocks covered most of the surface, interspersed by patches of darker material. The 1 m-high rock (extreme left) was named "Big Joe." Shallow trenches dug by the sampler arm (center) are visible in the foreground. Some of these were dug deeper to study soil which was not affected by solar radiation and weathering. The orange sky was caused by dust suspended in the lower atmosphere. The boom holding the meteorology sensors is at left. *(NASA-JPL)*

Figure 7.70   Viking Lander 2 touched down on Utopia Planitia, further north than Lander 1. Although they were about 6,400 km apart, the sites looked broadly similar, with a boulder-strewn surface reaching to the horizon nearly 3 km away. Fine particles of dust settled on the spacecraft. Three color calibration charts for the cameras are visible to the right of the Stars and Stripes. The circular high-gain antenna (top) was the main communication link with Earth and the Orbiters. *(NASA-JPL)*

Figure 7.71   The Opportunity rover took this image of vast plains of sand beneath an orange tinted, dusty sky, and wispy clouds of water ice. (Courtesy of NASA-JPL/Cornell University)

Magnetic Field

Mars does not possess a global magnetic field, although there is a patchy, localized surface field where the crust retains the magnetism implanted when Mars still had a global field (Figure 7.82). An induced magnetic field is also present on the dayside.

It seems that Mars was a magnetic world soon after its interior differentiated and its core formed in the pre-Noachian era. Today, the strongest magnetic anomalies are associated with late Noachian and early Hesperian terrain, but surface magnetism is absent from much of the northern plains, and younger volcanic or impact structures.

Over billions of years, most of the atmosphere has escaped into space or been frozen into the surface as permafrost. Much of the carbon dioxide may be tied up in the regolith or in the formation of carbonate rocks. Some of the water has been broken up by solar ultraviolet light and the resultant hydrogen has escaped from the planet.

Table 7.3

Atmospheric Composition of Mars (by volume)

Major gases	carbon dioxide (CO_2) – 95.32%;
	nitrogen (N_2) – 2.7%;
	argon (Ar) – 1.6%;
	oxygen (O_2) – 0.13%;
	carbon monoxide (CO) – 0.08%
Minor gases (ppm)	water (H_2O) – 210
	nitrogen oxide (NO) – 100
	neon (Ne) – 2.5
	hydrogen-deuterium-oxygen (HDO) – 0.85
	krypton (Kr) – 0.3
	xenon (Xe) – 0.08

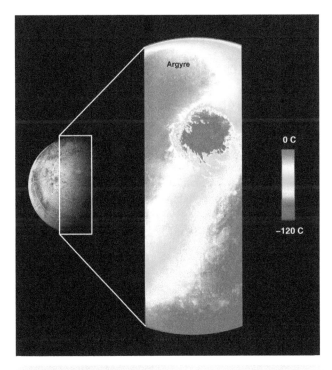

Figure 7.72 A thermal infrared image acquired by Mars Odyssey on October 30, 2001 – late spring in the southern hemisphere. It shows a temperature range of more than 120°C from the pole to the equator – a distance of over 6,500 km. The near-circular blue feature is the southern polar cap of carbon dioxide ice, where the temperature was about –120°C. The cap was more than 900 km in diameter and shrinking as summer approached. Cool air blowing off the ice is shown in orange, to the left of the cap. The cold region at lower right is the night hemisphere. The warmest regions (red) occur near local noon. The 900 km-diameter Argyre impact basin can be seen in the early afternoon. The thin blue crescent along the planet's upper limb is the atmosphere. (Courtesy of NASA/JPL/ Arizona State University)

The patchwork magnetic field on the surface resembles scattered umbrellas that act as barriers to erosion of the atmosphere by the solar wind (Figure 7.81). They are produced by regions of highly magnetized rock up to 1,000 km wide and 10 km deep.

The surface fields are comparable to Earth's magnetic field – a few tenths of a Gauss – and they can extend to an altitude of 1,300 km. Where the anomalies are stronger, the ionosphere reaches to a higher altitude, an indication that the solar wind is being kept at bay.

The anomalies with the largest amplitude are in the southern highlands. Weaker anomalies occur on parts of the northern lowlands, particularly near the boundary with the southern highlands. The major impact basins (Utopia, Hellas, Isidis, and Argyre) have no apparent magnetic signatures, probably because the shock and heating that accompanied the massive impacts demagnetized the crust. This implies that Mars no longer possessed an active dynamo in its core by the time they were created: otherwise, the rocks would have been remagnetized once they cooled and solidified.

Why did the magnetic dynamo in the core shut down within the first billion years of Mars' existence? One theory blames a series of major impacts that took place during the period of heavy bombardment, about 3.9 billion years ago. One or more massive collisions in the northern hemisphere may have warmed the planet's mantle sufficiently to disrupt core convection. This could cause the dynamo to – at least partially – grind to a halt, thus shutting down the global magnetic field and enabling the solar wind to strip the atmosphere away.

The induced magnetic field is very different in origin. It is formed when the magnetic field that is carried by the solar wind is compressed at Mars and draped around the planet (Figure 7.83). In the absence of a strong global magnetic field, the solar wind interacts directly with the ionosphere, inducing electrical currents. This interaction results in the formation of two boundaries: the bow shock and the magnetic pile-up boundary (MPB).

At the bow shock, the solar wind is decelerated from supersonic to subsonic speed as its flow is obstructed by Mars. Closer to the planet's surface is the MPB, which is the top of the induced magnetic field.

The location of the bow shock – the boundary where the solar wind slows suddenly as it begins to plough into the planet's outer atmosphere – varies according to the position of Mars in its orbit. This is in addition to other factors such as the dynamic pressure of the solar wind, and the amount of extreme ultraviolet solar radiation.

On average, the bow shock is closer to Mars near aphelion and further away near perihelion. Its average distance from the planet, when measured from above the terminator (the day–night boundary) has a minimum of 8,102 km around aphelion, while its maximum distance of 8,984 km occurs around perihelion – a variation of approximately 11% during each Martian orbit.

This rather surprising discovery indicates that the bow shock's location is more sensitive to variations in the solar EUV output than to changes in the solar wind's dynamic pressure.

This may be largely due to the impact of EUV on the density and thermal pressure of the ionosphere, and the expansion of the exosphere. These processes create buffers against the solar wind.

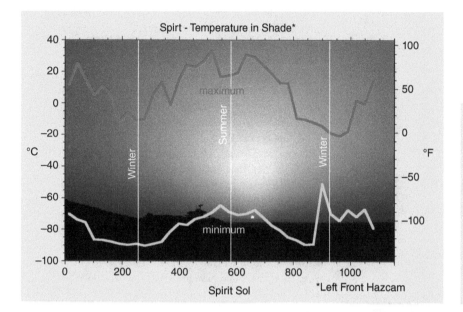

Spirt - Temperature in Shade*

Spirit Sol

*Left Front Hazcam

Figure 7.73 Shade temperatures for the Spirit rover ranged from about 35°C in summer to −90°C in winter. The horizontal axis shows the number of Martian days (sols). The data were obtained by a sensor on the hazard avoidance camera, which was in shade beneath the solar panels. The background image shows a sunset. (Courtesy of NASA/JPL-Caltech/Cornell /NMMNH)

Figure 7.74 A frosty morning on Utopia Planitia, imaged by Viking Lander 2 on May 18, 1979. A similar coating of water ice on the rocks and soil appeared almost exactly one Martian year earlier, evidence of the planet's regular annual climate cycles. The frost remained at this mid-latitude site (48°N) for about 100 days. Dust particles in the atmosphere may act as nuclei around which water and carbon dioxide freeze. The particles become heavy enough to fall to the ground. Warmed by the Sun, the carbon dioxide sublimates and returns to the atmosphere, leaving behind water ice and dust. The ice may be less than 1 mm thick. (NASA/JPL)

There may also be a link with annual changes in the amount of dust in the Martian atmosphere: the Martian dust storm season occurs around perihelion, when the planet is warmer and receives more solar radiation.

Solar Wind Interaction

Like Earth, Mars possesses an ionosphere – a layer of ionized (electrically charged) particles in its upper atmosphere. In this region, high-energy solar radiation and particles in the solar wind split the atoms and molecules in the upper atmosphere, releasing free electrons.

There are two main layers, at altitudes of about 110 and 130 km. A third, non-continuous layer has been detected between 65 and 110 km. This lower layer may be created by the interaction of the atmosphere with fast, incoming meteors. This may involve an exchange of electrical charge between the atmospheric gases and magnesium and iron atoms present in meteors.

The ionosphere appears to vary in shape and composition over time scales of only a few minutes, and it is not always able to prevent the solar wind from penetrating. The solar particles then energize and ionize the upper atmosphere, causing the escape of hydrogen and oxygen ions and ionized molecular species, such as carbon dioxide.

One of the most surprising results from Mars Express was the discovery of an ionosphere above the night hemisphere. This can be explained if the solar wind flows around the planet and produces a population of free electrons in the upper atmosphere. The regions of high electron density above the night side are associated with strongly magnetized areas, especially south of the equator, where the local magnetic field lines are perpendicular to the surface.

The interaction between the solar wind and the ionosphere is responsible for erosion of Mars' upper atmosphere. This erosion is particularly marked during periods of high solar activity, when a strong solar wind and coronal mass ejections compress the ionosphere, pushing it to a lower altitude. This exposes more hydrogen

Figure 7.75 Thin cloud cover is widespread on a northern summer day in April 1999. This global view shows bluish-white water ice clouds, e.g. above the Tharsis volcanoes (left). The bright white ice clouds (lower right) lie above the Hellas basin. The equator runs across the center. The high southern latitudes (above 60°S) were in winter darkness. (Courtesy of NASA/JPL/Malin Space Science Systems)

Figure 7.76 Four images of an annular (circular) cloud that formed over the same terrain in the north polar region during successive summers. The first, taken in April 1999, was obtained by the Hubble Space Telescope. The others were taken by Mars Global Surveyor. Despite its apparent resemblance to a cyclone, the cloud does not rotate. It forms when different air currents merge in the morning; by afternoon, it usually dissipates or breaks up into smaller clouds. (Courtesy of NASA/JPL/Malin Space Science Systems)

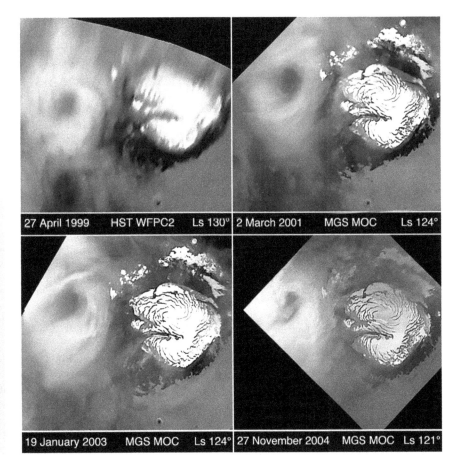

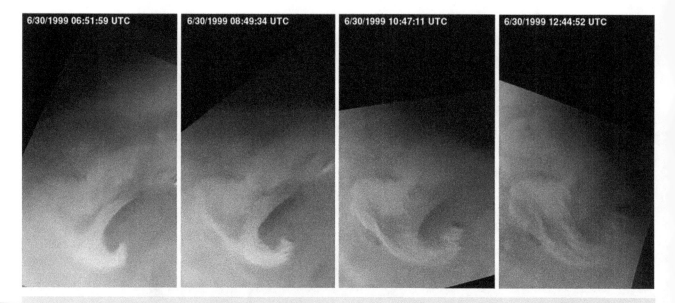

Figure 7.77   Four images, taken approximately two hours apart, show the evolution of a storm near the north pole on June 30, 1999 – late summer on Mars. The polar ice cap is at the center. White clouds are mainly water ice, orange/brown clouds contain dust. The "hook" in the clouds indicates a vortex associated with a storm front moving toward upper right. High surface winds raised dust and mixed it with water vapor over the polar cap to create this effect. Such storms are common in late summer – early autumn. (Courtesy of NASA-JPL/Malin Space Science Systems)

and oxygen ions to the solar wind, which picks them up and carries them away. Such erosion is thought to have increased around 3.5–4 billion years ago, when the planet's internal dynamo shut down. Erosion by the solar wind since that time could account for a loss of water equivalent to a global Martian ocean about 10 m deep. Today, the only regions to receive any protection from the solar wind lie above the stronger magnetic anomalies that still exist on the surface, where the altitude of the ionopause may exceed an altitude of 400 km.

Auroras

In the absence of a global magnetic field, auroras were not expected to occur on Mars. However, spacecraft have shown that they are very common. Hundreds have been recorded on the night side, particularly close to local midnight. Their locations coincide precisely with patches of strong magnetic field. Just as Earth's auroras occur where the magnetic field lines dive toward the surface at the north and south magnetic poles, Mars' auroras occur at the borders of magnetized areas where the field lines arc vertically toward the planet.

Earth's auroras light up when charged particles from the Sun are diverted down magnetic field lines and collide with atoms in the atmosphere to create an oval of light around each magnetic pole. The process on Mars is probably similar. Solar wind particles are funneled around to the night side of Mars where they interact with crustal field lines. The aurora is produced when the ions and electrons hit carbon dioxide molecules in the atmosphere.

NASA's MAVEN orbiter has discovered that some solar wind protons can slip past the bow shock by first bonding with electrons from the planet's upper atmosphere to form hydrogen atoms. Because these hydrogen atoms are electrically neutral, they can pass through the bow shock and collide with gas molecules in the Martian atmosphere, causing them to emit ultraviolet light. The result is an ultraviolet proton aurora on the dayside of Mars.

Increased solar wind activity and solar storms tend to make the auroras brighter and stronger. When a solar storm hits the Martian atmosphere, it can trigger auroras that illuminate the entire planet in ultraviolet light. An aurora on Mars can envelop the entire planet because, unlike Earth, the planet has no strong magnetic field to concentrate the aurora near polar regions. The energetic particles from the Sun also can be absorbed by the upper atmosphere, increasing its temperature and causing it to swell outward.

Such major storms can also significantly increase radiation levels on the surface. The increased radiation also interacts with the atmosphere to produce additional, secondary particles, which would be a threat to future human explorers.

Meteorites from Mars

Missions to bring samples of Martian material back to Earth for analysis have been under consideration for many years, but they have always been delayed by the huge cost and technological complexity. However, there are samples of Martian rock already available for study in terrestrial laboratories. These are meteorites that originated on the Red Planet, were ejected into space by impacts, and then fell to Earth after millions of years in interplanetary space.

Box 7.5 The Methane Mystery

One of the most intriguing minor constituents of the atmosphere is methane (CH_4), the first organic molecule found on Mars. The global average abundance is tiny, around 10 ppb, although some measurements have claimed concentrations of up to 250 ppb.

The methane levels change with the seasons: they are highest in autumn in the northern hemisphere and lowest in winter.

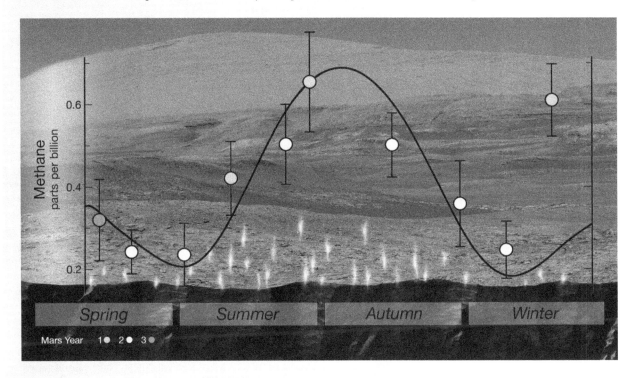

Figure 7.78 NASA's Curiosity rover detected seasonal changes in atmospheric methane in Gale Crater. The graph shows the methane concentration over nearly three Martian years (nearly six Earth years), with a peak each summer. *(Courtesy of NASA/JPL-Caltech)*

Measurements by the Curiosity rover confirm that methane levels on Mars vary with the seasons. Background levels of methane in Gale Crater have a mean value of 0.41 ± 0.16 parts per billion by volume (ppbv) and show a strong, repeatable seasonal variation of 0.24–0.65 ppbv, peaking near the end of summer in the northern hemisphere.

Mission scientists noted that the marked seasonal variation and occurrences of higher temporary spikes (about 7 ppbv) are consistent with small localized sources of methane released from surface or subsurface reservoirs.

The spatial distribution is also variable. There are three regions in the northern hemisphere where methane concentrations are high: Tharsis and Elysium, the two main volcanic provinces, and Arabia Terrae, which has plentiful underground water ice.

Figure 7.79 A map of methane distribution during the northern autumn, based on Mars Global Surveyor data. Highest concentrations (orange) occur in the Tharsis and Elysium volcanic regions, and Arabia Terra. Lowest concentrations are purple. *(Courtesy of NASA-JPL, University of Salento)*

Methane is an unstable gas which is soon broken down by natural processes, so, unless there is a source of continuous replenishment, it should have disappeared long ago from Mars.[13] How is the methane being generated?

Methane can be produced by microbes and by processes that do not require life, such as reactions between water and olivine (or pyroxene) rock. Ultraviolet radiation (UV) can induce reactions that generate methane from other organic chemicals produced by either biological or non-biological processes, such as comet dust falling on Mars.

Methane generated underground in the distant or recent past might be stored within lattice-structured methane hydrates called clathrates, and released at a later time, so that methane being released to the atmosphere today might have formed in the past.

Winds on Mars can quickly distribute methane, reducing localized concentrations of methane. Methane can be removed from the atmosphere by sunlight-induced reactions (photochemistry). These reactions can oxidize the methane, through intermediary chemicals such as formaldehyde and methanol, to produce carbon dioxide, the predominant gas in Mars' atmosphere.

At the time of writing, only 124 of the 60,000 or so meteorites discovered on Earth have been identified as originating from Mars.[14] All of them are made of igneous rock, and the vast majority are between 1.3 billion years and 165 million years old, much younger than typical meteorites, which are 4.5 billion years old (see Chapter 13). They also have higher contents of volatiles than igneous meteorites and oxygen isotope ratios that are distinctively different from terrestrial ratios.

The only plausible source that could have been volcanically active so recently is Mars. Conclusive evidence of their origin came from the measurement of gases trapped inside the EETA 79001 meteorite, which match those measured in the Martian atmosphere by the Viking Landers. Since the oxygen isotope content of the other SNC meteorites is the same, they must all come from the same parent object.

There are half a dozen sub-groups of these meteorites, representing rocks that formed in different locations on or beneath the planet's surface. Their varying mineral content and chemistry show that they cannot all have originated from a single impact event. Some contain shock-produced glass formed by partial melting as the result of a large impact event. Some have igneous minerals with little water, but many contain evidence of interaction with liquid water, especially carbonates, sulfates, and clays.

The most unusual, and oldest, example is ALH 84001, which seems to have crystallized 4.5 billion years ago.[15] It displays evidence of a long, complex history of shock and thermal alteration, and also contains carbonates, evidence that it has been in contact with Martian water (Figure 7.85). However, the lack of hydrated minerals in ALH 84001 favors limited exposure to moisture in a region of restricted water flow, such as an evaporating saline pool. The meteorite made headlines around the world in 1996 when a team of U.S. scientists announced that it may contain evidence of primitive life (see Life on Mars?).

Life on Mars?

Since life as we know it requires liquid water, the history of water on Mars and its current distribution are regarded as the key factors in determining whether the planet ever supported life and where it may have flourished.

It now seems certain that a denser atmosphere very early in Mars' history supported a warmer, wetter environment, creating conditions that could have been favorable for life. Hydrated mineral deposits suggests that the liquid water on or near the surface was widespread and of low acidity in the Noachian. However, even if life began within the first few hundred million years, its evolution may have been severely restricted.

The fairly benign conditions changed in the Hesperian, to be replaced by a more acidic environment. Huge volcanic eruptions coincided with the shutdown of the global magnetic field and the depletion of the atmosphere as gases escaped into space. Apart from occasional warmer spells and flash floods, the surface conditions have generally been far more extreme than anywhere on Earth for billions of years.

However, this does not necessarily exclude the survival of any primitive organisms that may have evolved long ago. Liquid water can exist deep below the surface, where temperatures are higher and there is no exposure to solar ultraviolet light.

Today, although permafrost predominates at high latitudes, only a few centimeters below the surface, numerous environments with liquid water probably exist a few kilometers below the surface, over most of the planet. This might enable organisms to survive in or beneath ice sheets, in subterranean aquifers, or in protected habitats such as lava tubes, caves, and fissures. The potential for life could be enhanced in warmer, volcanic regions, such as Tharsis and Elysium.

In 2018, the first discovery of a subsurface body of liquid water was announced by scientists who had analysed radar data returned

---

[13] The lifetime of CH_4 in the atmosphere of Mars is estimated to be 300–600 years, based on the rate of its destruction by photochemistry, or shorter if strong oxidants such as peroxides are present in the surface or on airborne dust grains.

[14] They are often called SNC meteorites after the names of three typical examples: Shergotty, Nakhla and Chassigny.

[15] The designation ALH indicates that it was found in the Allan Hills in Antarctica. The number 84001 indicates that it was the first meteorite found at that location in 1984.

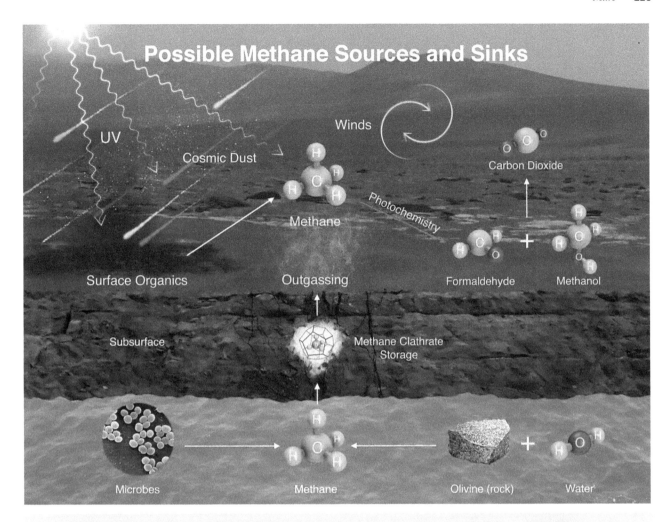

Figure 7.80 Various possibilities have been proposed as sources and sinks of Martian methane. Potential non-biological sources for methane include comets, degradation of interplanetary dust particles by ultraviolet light, and interaction between water and rock. A potential biological source would be microbes, if they ever lived on Mars. Potential sinks for removing methane from the atmosphere are photochemistry in the atmosphere and loss of methane to the surface. *(Courtesy of NASA/JPL-Caltech, SAM/GSFC)*

by Mars Express. The data indicated the presence of a 20 km-wide water layer at least several tens of centimeters thick. The water was detected about 1.5 km beneath the layers of ice and dust that comprise the south polar cap of Mars. It may be comparable to subglacial lakes on Antarctica, which are home to various forms of microbial life.

Furthermore, recent discoveries on Earth of extremophiles – organisms that can survive in conditions of extreme heat or cold, acidity or alkalinity, prolonged dryness, lack of sunlight, and even high radiation – prove that life forms are remarkably resilient.

It is, therefore, possible that life may have evolved to cope with alternating climatic cycles or different environments, so that dormant microbes could be present close to the surface whilst active variants survive in protected environments.

Since most potentially habitable environments are out of reach for the foreseeable future, the hunt for Martian organisms is currently focused on seeking evidence of past life, preserved as fossil biosignatures in surface rocks. This will require microscopic

imagers capable of viewing rocks in many wavelengths as well as seeing details only a hundredth of a millimeter across. Also needed are organic chemistry laboratories to analyze promising rocks.

The first attempt to identify life on Mars was made in the 1970s by the Viking Landers, which carried four experiments to search for indirect evidence of living organisms.

The biology lab on the Landers comprised three experiments to analyze gases released by soil samples under different conditions (Figure 7.86). The Pyrolytic Release Experiment was looking for organisms that could assimilate and reduce carbon dioxide or carbon monoxide (rather like the photosynthesis carried out by plants on Earth). Samples were tested in darkness and light, as well as in dry or humid conditions. After they were exposed to radioactively tagged carbon dioxide, the "labeled" gas was removed and they were heated so that any gases that were released could be analyzed.

The fixation of gaseous carbon was detected, but similar results occurred even in darkness, and the reaction remained stable at a

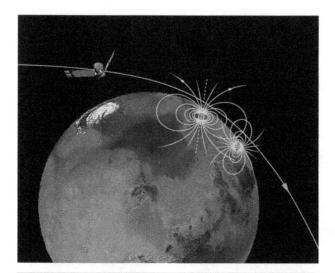

Figure 7.81 A schematic representation of localized magnetic sources in the Martian crust, buried beneath the surface, and revealed by observations of the magnetic field (blue) at the satellite's orbital altitude of 120 km. (Courtesy of NASA/JPL/GSFC)

temperature of 90°C, suggesting a chemical, rather than biological cause.

The Labeled Release Experiment also used radioactive carbon-14 as a tracer, but in this case it was in the food supply – an inorganic nutrient that was added to the soil. It was hoped that any organisms would consume the nutrient and then exhale radioactive carbon dioxide or carbon monoxide.

Once again, there was an unexpected result. Part of the organic mixture was rapidly converted to CO_2, as would be expected in a biological reaction, but then the conversion largely stopped. A second injection of nutrient had no effect. This contrasted with the behavior of terrestrial organisms, which continue to take up nutrients until they are completely consumed. However, strong heating seemed to destroy whatever was converting the nutrient, a possible sign of organisms being killed during the sterilization process.

In the Gas Exchange Experiment, a very rich organic nutrient "broth" was added to the soil after it was initially dampened with water vapor. The atmosphere in the chamber was then monitored for changes that would indicate biological activity. Although a burst of oxygen was released after the sample was humidified, few changes occurred when the broth was added. This suggested that the soil contained a strong oxidizing substance, such as a peroxide or superoxide, rather than living organisms.

The landers also carried a Gas Chromatograph – Mass Spectrometer experiment. Several samples of material were heated in steps to 500°C, and the released gases were then analyzed for organic compounds. The experiment was capable of detecting simple carbon compounds in concentrations of about 1 part per million, as well as larger molecules in the parts per billion range.

Carbon dioxide and water absorbed in the soil were detected, as were some terrestrial contaminants, but no organic molecules were recorded, a remarkable result in light of the continuous delivery of carbon-based material by falling meteorites. After this negative result, it was generally concluded that any positive indications from the biology experiments were suspect. The argument against biological interpretations was reinforced by the apparent presence

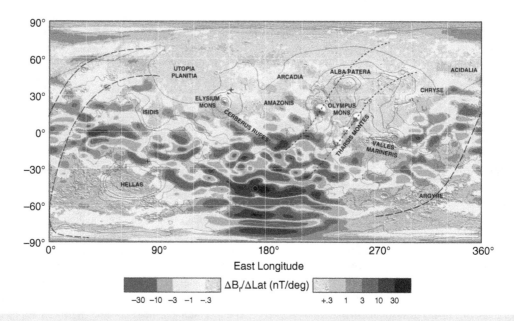

Figure 7.82 The locations of magnetic anomalies, based on data from Mars Global Surveyor. The strongest crustal magnetization is largely confined to east–west "stripes" of alternating polarity in the ancient southern highlands. Regions of extensive volcanism (e.g. Olympus Mons, Tharsis Montes) are non-magnetic, as are much of the northern plains and regions surrounding the largest impact basins. (Courtesy of NASA-JPL/J.E.P. Connerney et al.)

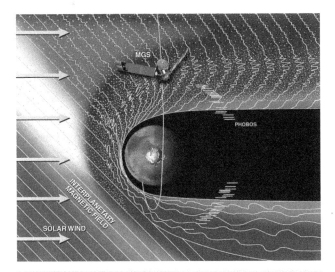

Figure 7.83 The solar wind of electrically charged particles (ions and electrons) is slowed to subsonic speeds at a bow shock, upstream of the planet. Here, the magnetic field fluctuates wildly and the flow of the solar wind becomes chaotic. The absence of a global magnetic field allows the solar wind to interact directly with the ionosphere and upper atmosphere. Solar ultraviolet radiation ionizes gases such as oxygen and hydrogen, which can then be removed by the magnetized solar wind. Also shown is the near-polar orbit of Mars Global Surveyor. (Courtesy of NASA/JPL/GSFC)

of a powerful inorganic oxidizing agent that decomposes organic compounds.

One of the principles behind the Viking experiments – the detection of life through the release of gases produced by metabolic processes – is also related to a more recent detection of localized methane sources on Mars (see Box 7.5). On Earth, more than 90% of atmospheric methane arises from current or past biological activity, but the question of whether the Martian methane is produced by living organisms or some inorganic process remains unresolved.

Other evidence is available from the small collection of Martian meteorites, most notably **ALH 84001** (see Meteorites from Mars) (Figures 7.87 and 7.88). In August 1996, NASA and Stanford University scientists announced that it contained probable evidence of past microscopic life.

The group cited four main arguments, all of which were associated with unusual orange globules of carbonate minerals, up to a few millimeters across:

1. Layered calcium, magnesium, and iron carbonates were formed due to contact with non-acidic, liquid water on Mars.
2. Complex organic chemicals in the carbonates could be the decay products of microbes.[16]

3. The carbonate globules contain tiny grains of magnetite, an iron oxide, and iron sulfides. These particles were similar to those associated with certain bacteria on Earth.
4. The globules contain segmented, tubular objects, less than one micron in length, similar to fossilized nanobacteria found on Earth.

Although each piece of evidence can be accounted for by inorganic processes, it was argued that biogenic activity is the simplest explanation to account for all of the observed features.

As a result of the furore, ALH 84001 became the most closely examined piece of rock in history. Although the meteorite was clearly very ancient, forming when liquid water was almost certainly present on its surface, and the carbonates almost certainly formed at that time, the other lines of evidence were not generally accepted.

Instead, it was argued that some of the features originally identified as possible fossil organisms may have been caused during sample preparation, while some others may be the result of weathering of its surface. The magnetite grains could be non-biogenic, while the organic compounds may be the result of contamination after the meteorite landed in the Antarctic.

Phobos and Deimos

After years of speculation that Mars may have some small moons, they were eventually found by American astronomer Asaph Hall during the opposition of 1877. He named the inner satellite Phobos (Fear) and its outer companion Deimos (Panic), after the mythological companions of the Greek war god, Ares (the Roman god Mars).

The average distance of Deimos from the center of Mars is 23,459 km – only 6% of the Earth–Moon distance. It follows a near-circular path, tilted at an angle of 1.8° to the planet's equator. With an orbital period of 30 hours 21 minutes, Deimos takes a little over one Martian day to complete a single orbit. Consequently, an observer on Mars would see it track slowly westwards across the sky.

Phobos orbits only 9,378 km from the planet's center – so close that it would be permanently below the horizon for anyone above latitude 69°. In order to avoid crashing into Mars, it sweeps around the planet in 7 hours 39 minutes (about one-third of a Martian day). This means that an observer near the equator would see Phobos rise above the western horizon roughly every 11 hours. It would then cross the sky "backwards" (west to east) in 4½ hours, eventually setting below the eastern horizon. During this brief appearance, it would go through more than half its cycle of illuminated phases.

Both Phobos and Deimos are irregular in shape, with average diameters of about 22 and 12 km respectively. From Mars' surface, Phobos appears about one-third the diameter of the Moon, as seen from Earth, whereas Deimos appears as a bright star. Phobos is the denser of the two, though it still has less than one-thousandth the surface gravity of Earth.

---

[16]An "organic" compound is not necessarily associated with biology. It simply comprises one or more carbon–hydrogen (C–H) bonds, or a carbon–carbon (C–C) bond.

Box 7.6   Mars Orbiters

The modern era of Mars exploration began on July 14, 1965, when the Mariner 4 spacecraft flew by the planet and sent back 22 pictures, with resolutions of several kilometers. Scientists had speculated that Mars might be more like Earth than the Moon, but the images revealed an ancient, cratered surface. Confirmation was provided by the next two Mariner flybys.

Perceptions then changed dramatically in 1972, when Mariner 9 became the first Mars orbiter. Despite an initial delay caused by a global dust storm, the spacecraft mapped 70% of the planet and made the first detailed study of the atmosphere and the surface.

As the mission progressed, huge volcanoes, deep canyons, enormous dry riverbeds, and extensive dune fields came into view, and a complex geological history became apparent. It also sent back the first close-up images of Phobos and Deimos, and provided information on the shape of Mars and its gravity field.

Encouraged by the evidence that Mars was once much warmer and wetter, NASA sent two Viking spacecraft to survey the planet in much greater detail. By the end of the mission, almost all of the surface had been imaged from orbit at a resolution of about 250 m/pixel and smaller areas with resolutions as high as 10 m/pixel. The Viking Landers photographed their landing sites on the northern plains and successfully carried out a variety of experiments directed mostly toward detecting life and understanding the chemistry of the soil.

After the failure of the Vikings to detect any organic compounds, no more U.S. spacecraft were dispatched to the planet for 16 years. Unfortunately, contact with Mars Observer was lost shortly before arrival at Mars, and further studies had to wait until Mars Global Surveyor (MGS) entered orbit on September 12, 1997.

Propellant consumption was reduced by aerobraking, using friction with the planet's upper atmosphere to move into the operational polar orbit. MGS returned data for more than 9 years, and its Mars Orbiter Camera returned more than 240,000 images which revolutionized understanding of the planet's surface and weather. Discoveries included recent gully erosion, multi-layered sedimentary rocks, and retreat of ice on the south polar cap.

Some of the most important breakthroughs came from the Mars Orbiter Laser Altimeter (MOLA), which provided the first accurate topographic maps. Other important results came from the magnetometer, which discovered local magnetic anomalies in the crust, and the Thermal Emission Spectrometer, which mapped surface minerals and discovered deposits of hematite – a clue to a watery past.

The next spacecraft to arrive was Mars Odyssey, which entered orbit on October 24, 2001. After aerobraking was completed, it entered a 400-km polar, nearly Sun-synchronous mapping orbit. Its gamma-ray suite recorded significant quantities of hydrogen close to the surface – evidence of large deposits of underground water ice. Analysis of temperature data enabled scientists to distinguish between solid rock and a variety of loose surface materials. Odyssey was still operational in March 2019.

Figure 7.84   Mars Express was Europe's first planetary orbiter. Although its piggybacking Beagle 2 lander was lost during atmospheric entry, the orbiter continued to operate well beyond its design life. Among its instruments were a radar sounder, three spectrometers, a stereo camera, and a particle instrument. *(ESA)*

Europe's first Mars orbiter was Mars Express, which comprised an orbiter and the Beagle 2 lander. Although Beagle 2 failed to survive its landing on December 25, 2003, the orbiter was successfully placed in a near-polar orbit. Its radar experiment sent back the first subsurface soundings of Mars and the High Resolution Stereoscopic Camera provided a series of stunning 3D images of the surface. Another breakthrough came with the discovery of phyllosilicates and other hydrated minerals by the OMEGA spectrometer, evidence of a warmer, wetter planet billions of years ago. Mars Express was still operational in March 2019.

Mars Reconnaissance Orbiter, the largest U.S. spacecraft ever sent to the Red Planet, arrived on March 10, 2006. The spacecraft sends data to Earth at about 10 times the rate of any previous Mars mission. It carries five main instruments to characterize the surface, subsurface and atmosphere of Mars, particularly the history of water, past and present. The High Resolution Imaging Science Experiment (HiRISE) is the most powerful camera ever flown on a planetary mission, capable of detecting features one meter across. MRO's Context Camera (CTX), which has a resolution of about 6 m per pixel, has imaged more than 99% of Mars. The Italian-built Shallow Subsurface Radar probes to a depth of up to one kilometer to study layers of ice and rock, including the polar caps. By early 2017, it had returned 300 terabits of data, many times the amount sent back from all other deep space missions combined.

In recent years, other countries have become interested in exploration of the Red Planet. India's first venture was the Mars Orbiter Mission (MOM), also known as Mangalyaan, which entered orbit around the planet on September 24, 2014. It was still operational in January 2019, having returned about 1,000 color images of Mars and its satellites.

The European Space Agency has teamed with Russia to develop its ExoMars program. The first phase was only partially successful, after the Schiaparelli lander was lost during atmospheric entry and descent on October 19, 2016. However, the Trace Gas Orbiter (TGO) arrived safely on the same day, and its science mission began in April 2018. TGO's main goal is to make a detailed inventory of rare gases that make up less than 1% of the atmosphere's volume, including methane, water vapor, nitrogen dioxide and acetylene.

NASA's most recent orbiter is the Mars Atmosphere and Volatile Evolution Mission (MAVEN), which is designed to investigate how the planet lost its atmosphere and abundant liquid water. MAVEN has been operational since its arrival on September 21, 2014.

Figure 7.85 ALH 84001 is the oldest known Martian meteorite, which is thought to have formed about 4.5 billion years ago. Since then, it has undergone shocks and metamorphism, probably associated with violent impacts on Mars. Carbonate minerals indicate that it was once in contact with ground water, possibly an evaporating pool of brine. (Courtesy of NASA-JSC)

Each satellite has been affected by tidal forces over billions of years, so that its long axis is aligned with the planet and the same hemisphere always faces Mars.

Studies of their orbits show that Phobos is slowly accelerating and spiraling in towards Mars. About 100 million years from now, the moon will be destroyed when it collides with the planet. Deimos, on the other hand, is slowly moving away from Mars.

Phobos has a heavily cratered, ancient surface which is dominated by a 9 km-wide crater called Stickney, the maiden name of Asaph Hall's wife (Figure 7.93). Also present are two 5 km-wide craters – one of which is named after Hall himself.

Radiating away from Stickney are linear chains of small craters and parallel striations that are only 5–10 m deep but up to 15 km in length. These have often been regarded as features related to internal fractures caused by the large impact that excavated Stickney (and almost destroyed Phobos).

Also visible are linear grooves about 500 m wide that may be evidence of lines of weakness produced by tidal forces or rolling boulders ejected from Stickney and other sizeable craters. Recent, multi-wavelength images show a coating of ejecta on the flank of Stickney and landslides on its inner walls.

Measurements made during close spacecraft encounters indicate that it has a lower density than expected. With 25–35% of its interior as empty space, it is now believed to be little more than a jumbled "rubble pile."

Deimos also has an irregular shape, with several small craters, the two most prominent being Swift and Voltaire (Figure 7.94). However, its surface is much smoother than that of Phobos. There are no visible grooves or striations, suggesting that tidal forces and large impacts have played a lesser role in its evolution, and most of the craters appear to be partially filled or covered with loose regolith. Close-range images also reveal a surface strewn with boulders that were presumably ejected during low-energy collisions with small meteorites.

Both objects are very dark, with albedos of only 5 to 7%. Early spectral measurements in visible and near-infrared light indicated that they resemble carbonaceous chondrite meteorites (see Chapter 13). This led to the belief that the moons are captured asteroids, although it was difficult to explain their near-circular, near-equatorial orbits.

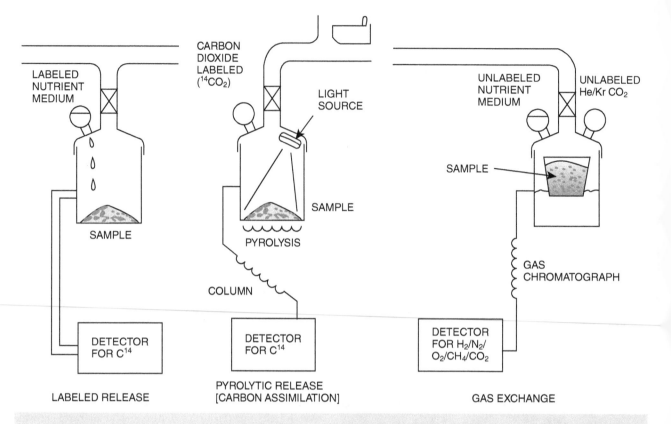

Figure 7.86 The biology lab on the Viking landers comprised three separate experiments. In the Labeled Release Experiment, a radioactively tagged liquid nutrient was added to soil and emitted gases were analyzed to detect any uptake by microbes. The Pyrolytic Release Experiment heated soil samples that had been exposed to radioactively tagged carbon dioxide to see if it had been used by organisms to make organic compounds. The Gas Exchange Experiment analyzed gases released by soil to which a nutrient was added. (NASA-JPL)

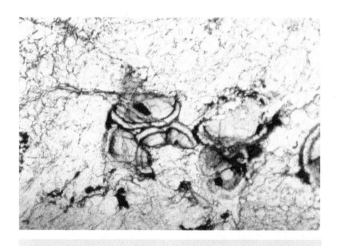

Figure 7.87 Rounded globules in ALH 84001 are made of siderite (brownish iron carbonate) and magnesite (clear magnesium carbonate), while the dark, outer rinds are rich in iron oxide and sulfides. The globules formed after ALH 84001 solidified from lava billions of years ago, but their age is uncertain. Possible bacterial nanofossils were found in these globules. This thin-section view is about 0.5 mm across. (Allan Treiman/Lunar and Planetary Institute)

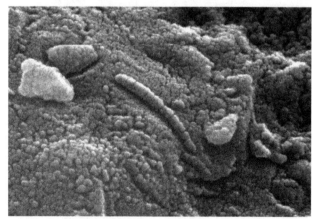

Figure 7.88 This segmented, worm-like object, seen in a scanning electron microscope image of ALH84001, has been interpreted as a possible microfossil bacterium. The image, 1 micron across, shows the surface of orange carbonates in the rock. (Courtesy of NASA-JSC)

Box 7.7   The Mars Exploration Rovers

Two U.S. Mars Exploration Rovers, Spirit and Opportunity, were launched in 2003 to search for evidence of liquid water, a possible precursor for the evolution of Martian life.

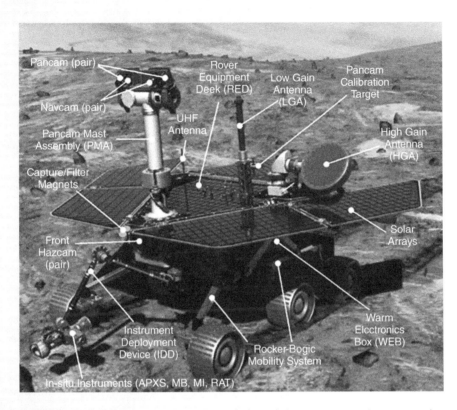

Figure 7.89   The rovers were equipped with navigation cameras, two high-resolution stereo panoramic cameras, and a spectrometer to study rocks at infrared wavelengths. The robotic arm carried a microscopic imager, for detailed views of rocks, and two spectrometers to determine surface composition. Also on the arm was an abrasion tool to remove the weathered outer crust from rocks. The six-wheeled rovers received power from a deck of solar panels. Eight radioisotope heaters enabled them to survive the cold nights and long winters. *(NASA-JPL)*

After considerable debate about where to send them, it was eventually decided to drop Spirit in Gusev crater, a 166 km-wide impact structure on the edge of the southern highlands. Orbiter images indicated that it once contained a lake that was fed by an enormous outflow channel, Ma'Adim Vallis. However, little evidence of the lake was found, perhaps because any sediments were subsequently buried by lava.

Opportunity's target was Meridiani Planum, a broad plain on the other side of the planet. This region was known to have deposits of gray hematite, an iron mineral that is usually produced in the presence of liquid water.

Spirit arrived on the surface, encased in airbags, on January 3, 2004. Its first images showed a flat, relatively featureless plain, strewn with small rocks. Some low hills were just visible 3 km away. Analysis of the soil showed that it was rich in silicon and iron, with significant levels of chlorine and sulfur. One unexpected finding was the presence of olivine, a mineral which is easily weathered. The surface material was found to be fine-grained basalt.

Opportunity touched down 10,000 km away on January 24, coming to rest inside a 20 m-diameter crater which featured a small outcrop of layered bedrock. There were thousands of gray, iron-rich spherules – dubbed "blueberries" – scattered across the surface and embedded in the exposed rock.

Over the coming months, both rovers found evidence that confirmed the presence of liquid water at their landing sites in the distant past. This was particularly true at Meridiani Planum, where the data indicated that finely layered rocks in Eagle crater and elsewhere were probably deposited on a salt flat, or playa, which was sometimes covered by shallow water and sometimes dry.

Eventually, the rovers began to explore further afield. Spirit passed by the 200 m-diameter Bonneville crater in mid-March 2004 and set off on a 2.3-km trek to the 120 m-high Columbia Hills, where it found layered rocks containing minerals that were either formed in, or altered by, water. Some of the rocks contained high concentrations of sulfates and carbonates.

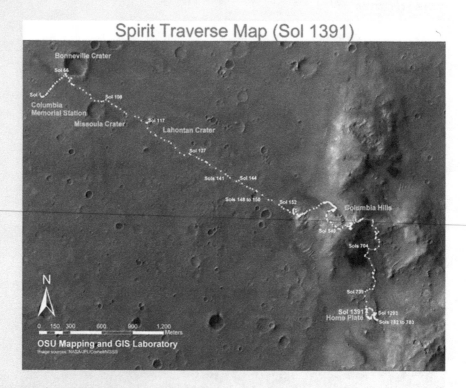

Figure 7.90   The route followed by Spirit during the first 1,391 sols, showing the main exploration sites. The traverse ended on the edge of a low plateau called Home Plate. The landing site is 3.8 km away to the north west. Sol numbers refer to when Spirit was at that location. Spirit reached the summit of Husband Hill on sol 619 (Sept. 29, 2005). *(Courtesy of OSU Mapping and GSU Laboratory/NASA/JPL/Cornell University/MSSS)*

During its descent from the summit, Spirit came across an unusual low plateau called "Home Plate," a possible site of ancient volcanic explosive activity. It continued to explore this area until April 2009, when its wheels broke through a surface crust and dug more than hub deep into sand. It was unable to escape from the sand trap and failed to survive the following winter. During its 6-year mission, it traversed more than 7 km.

Figure 7.91   A wide-angle view of Burns Cliff inside Endurance crater, with an image of Opportunity superimposed. The approximately true-color rendering spans more than 180° side to side. The cliff walls appear to bulge outward. In reality, they form a gently curving surface. Note the various horizontal layers, which were probably modified by surface streams and groundwater. *(Courtesy of NASA/JPL-Caltech/Cornell)*

Opportunity initially examined sedimentary rock outcrops and sandy, windblown regolith within Eagle crater. It then spent six months inside Endurance crater, where it made a close inspection of rock layers exposed in Burns Cliff, part of the crater wall. En route to its next target, Erebus crater, Opportunity found the first meteorite ever identified on another planet.

Despite becoming temporarily stuck in a sand dune, Opportunity arrived at the 300 m-wide Erebus crater in late September, where dunes and rifts lay between outcrops of rock. It continued south through etched terrain toward the 800 m-wide Victoria crater, where a substantially thicker sequence of layered rock was found.

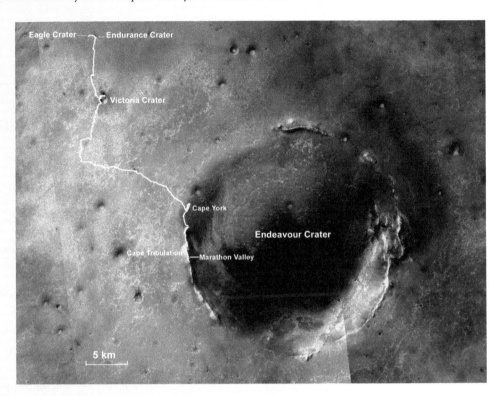

Figure 7.92 The route followed by Opportunity up to March 2015 from its landing site in Eagle crater. After venturing inside Victoria crater, it set off south toward the ancient, 22 km-wide Endeavour crater. It then traveled south along the western rim of Endeavour crater. *(Courtesy of NASA/JPL-Caltech/MSSS)*

After entering the crater on September 13, it experienced slopes of up to 25°, but was able to study the crater wall. After 11 months, Opportunity backed out the way it had come and headed for 22 km wide Endeavour crater, about 12 km away.

Opportunity reached the rim of Endeavour crater on August 9, 2011. For the rest of its mission, it explored the more ancient rocks exposed along the crater's western rim. Not long after its arrival at a site named Cape York, found a bright mineral vein identified as hydrated calcium sulfate. This was described as "the clearest evidence for liquid water on Mars that we have found in our eight years on the planet."

Subsequent results suggested that water of low acidity flowed in this region before Endeavour crater formed, whereas the water was much more salty and acidic afterward.

The solar-powered rover lost contact during the 2018 dust storm, when it was about one-third of the way down Perseverance Valley, a shallow, fluid-carved channel incised into the rim's crest. During its 14-year life it has travelled more than 45 km and returned about 225,000 images.

More recent measurements have shown that they contain minerals that are common on Mars, including phyllosilicates that are created by interaction of silicate materials with liquid water and particles of volcanic basalt. This has led to the theory that they formed relatively near their current orbits through re-accretion of material blasted into space by a catastrophic Martian impact about 4 billion years ago. Such an origin would account for their rubble pile composition.

Recent computer simulations suggest that Phobos and Deimos formed after an oblique impact with Mars by an object similar in size to the largest asteroids Vesta and Ceres – a much less massive object than previously considered. Such an event would

Figure 7.93   A false color image of Phobos, taken on October 23, 2007, by Mars Reconnaissance Orbiter, shows features 20 m across. Ejecta around the rim of Stickney crater (right) appear pale blue. Linear grooves and crater chains radiate from Stickney. Landslides are visible on the internal walls of large craters, despite Phobos' weak gravity. (NASA/JPL-Caltech/University of Arizona)

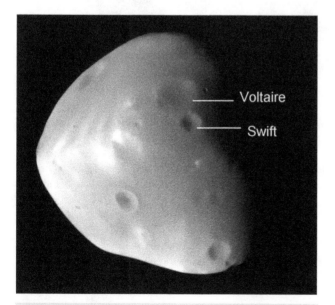

Figure 7.94   An enhanced-color view of the Mars-facing hemisphere of Deimos, taken with the HiRISE camera on February 21, 2009. At that time, only two craters (labeled) had been given names. North is up and the equator is roughly horizontal. (Courtesy of NASA-JPL/UA)

have created a large impact feature on the Martian surface, such as the Borealis, Utopia, and Hellas basins.

Material ejected from the impact site would form an equatorial disk of debris. The outer regions of the disk would accumulate into Phobos and Deimos, while the inner regions would accumulate into larger moons that eventually spiralled inward and crashed into Mars.

Questions

- Mars is often described as the most Earthlike of all the planets in the Solar System. Discuss the validity of this statement by discussing the major similarities and differences between the two planets.
- How is the Martian environment modified by variations in the planet's orbit and axis of rotation?
- Explain the importance of (a) dust, (b) water, and (c) carbon dioxide in the Martian environment today.
- Summarize the main phases in the geological evolution of Mars.
- Summarize the evidence that suggest Mars may have once been partially covered by an ocean.
- (a) Why do scientists believe that Mars was once warmer and wetter than it is today? (b) What processes caused Mars to be an arid world today?
- Summarize the similarities and differences between the two Martian polar regions.
- Compare the magnetic environments of Mars and Earth.
- Is Mars a likely location for the evolution of life? Explain your answer.
- Summarize how (a) orbiters and (b) landers have revolutionized our understanding of Mars in the past half century.
- Briefly summarize how and why the natural satellites of Mars differ from Earth's Moon.
- Summarize the difficulties that will face the first human colonists on Mars.

EIGHT

Jupiter

Beyond the main asteroid belt is Jupiter, the largest planet in the Solar System, with more than twice the mass of all the other planets combined. Its gravitational field is so powerful that it can hold on to 79 known satellites and thousands of Trojan asteroids, while removing asteroids from parts of the main belt. Its magnetic influence extends all the way to Saturn and the major moons are bathed in lethal doses of radiation. Yet, despite its huge size, Jupiter spins faster than any other planet, completing each rotation in less than 10 hours.

Jupiter's existence has been known since prehistoric times. This is hardly surprising, since it can reach magnitude –2.9 around the time of opposition, every 399 days or so. For about six months of each year, Jupiter is an unmistakable beacon, often dominating the night sky after midnight, when only the Moon (and, very occasionally, Venus) rivals its brilliance.

Early observers were aware that Jupiter took almost 12 years to progress through the constellations of the zodiac – longer than any planet apart from Saturn. From its stately movement against the stellar background and prolonged illumination of the night sky, they deduced that the planet was further away than Mercury, Venus, or Mars. Nevertheless, it was remarkably bright, so the Romans named it after Jupiter, the king of the Olympian gods.

Orbit and Physical Characteristics

Jupiter lies beyond the outer edge of the main asteroid belt, the fifth planet from the Sun. Like most of the others, it follows an almost circular orbit which is very close to the ecliptic.

After the introduction of the telescope, it soon became obvious that Jupiter is the dominant body in the Solar System, after the Sun. With a volume equivalent to 1,300 Earths, it is large enough to swallow all of the other planets. Although it comes no closer to the Earth than 599 million km at times of opposition, its visible disk dwarfs every planet apart from nearby Venus, offering a magnificent sight in telescopes of even modest aperture.

Jupiter has more than 2½ times the mass of all the other planets combined – or 318 times Earth's mass (Figure 8.2). As a consequence, its gravitational pull is enormous. To escape Jupiter's gravity, a spacecraft must travel away from the planet at a speed of 59.6 km/s, compared with 11.2 km/s for Earth. However, since its mean density is not much greater than that of water, it is clear that the bulk of the planet is made of gases and ices, rather than rock.

Despite an equatorial diameter 11 times that of Earth, Jupiter rotates once every 9 hours 50 minutes – faster than any other planet.[1] This means that the equatorial cloud deck is traveling from west to east at more than 45,000 km/h.

Such rapid rotation of a largely gaseous planet causes a very noticeable equatorial bulge. The outward centrifugal force of rotation opposes the effect of gravity. Since the speed of rotation is greatest at the equator and least at the poles, the planet expands into an oblate shape that is elongated along the equator. Jupiter's equatorial diameter is actually 9,270 km larger than its polar diameter.

[1] The equatorial rotation period of about 9 hours 50 minutes is about 5 minutes shorter than the rest of the planet. Radio emissions from Jupiter's interior give a period of 9 hours 55 minutes 30 seconds.

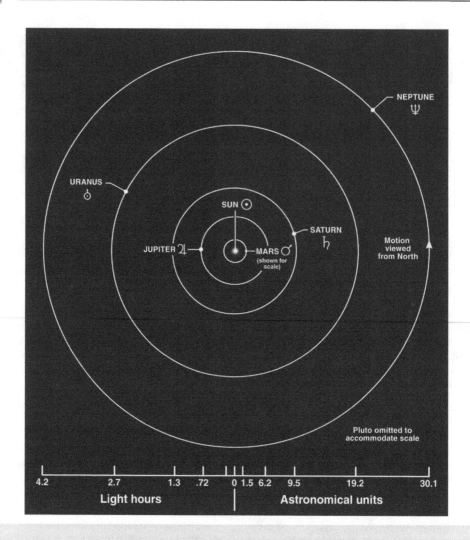

Figure 8.1 Jupiter, Saturn, Uranus, and Neptune are sometimes known as the jovian (Jupiter-like) planets. This diagram shows their approximate distances from the Sun. Note the enormous gaps between their orbits, shown in astronomical units and light hours – the time light from the Sun takes to reach them. (LPI)

Table 8.1

Jupiter Summary

Mass	1.9×10^{27} kg (317.8 Earth masses)
Equatorial Diameter	142,984 km
Polar Diameter	133,708 km
Axial Inclination	3.13 degrees
Semimajor Axis	778,330,000 km (5.2 AU)
Rotation Period (internal)	9.92 hours
Sidereal Orbit Period	11.86 years
Mean Density	1.33 g/cm^3
Number of known satellites	79

Zones and Belts

Jupiter is a dynamic, cloud-covered world with no visible solid surface. Most of the planet is dominated by reddish-brown "belts" and whitish "zones," linear cloud features that are parallel to the equator. These are associated with alternating east–west wind currents, with the fastest jets blowing from west to east around the equatorial region. Spacecraft data show that the zonal winds are remarkably constant over time (Figures 8.3 and 8.4).

In general, the strongest winds on Jupiter blow toward the east. Two jets, one at about 24°N and another at about 10°S, move clouds to the east at 150 m/s (540 km/h). Most of the faster jets move along the boundaries between belts and zones.

Although they are ever-present, the belts and zones are not unchanging (Figure 8.5). Amateur astronomers eagerly record the latest streaks, plumes, and spots that form and evolve in the highly turbulent atmosphere. Some of the spots and plumes survive for several days, months, or even decades. However, even Jupiter's turbulent atmosphere occasionally goes through a quiet spell, when the belts become fainter, or even disappear, and fewer cloud variations are visible, e.g. the 2010–2011 disappearance of the South Equatorial Belt.

The cloud patterns also reflect the vertical motion of the atmosphere in which the clouds form. The light-colored zones

Figure 8.2   Jupiter has more than 2½ times the mass of all the other planets combined. Although 1,300 Earths would fit inside it, gaseous Jupiter has only 318 times Earth's mass. Despite its enormous bulk, Jupiter rotates faster than any other planet, causing its equatorial region to bulge outward. The clouds are arranged in light and dark bands aligned parallel to the equator. (Courtesy of Bob Kanefsky, NASA-Ames)

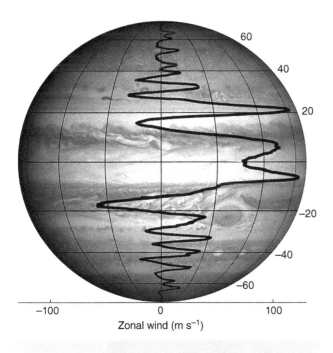

Figure 8.3   Zonal wind speeds superimposed on an image of Jupiter taken by the Cassini spacecraft. The strongest winds on Jupiter blow toward the east. Two jets, one at about 20°N and another at about 10°S, move clouds to the east at 150 m/s (540 km/h). (Y. Kaspi et al./Nature)

are regions of warmer, rising gas. When this reaches the cloud tops, it spreads out and sinks back in the neighboring dark belts. The direction in which the gas moves is also influenced by the extremely strong Coriolis force (see Chapter 3).

The Juno orbiter's measurements have revealed that Jupiter's gravitational field is asymmetric north–south, a signature of the planet's atmospheric and interior flows. The observed jet streams, revealed by the colourful bands at cloud level, are now known to take the form of nested cylinders that extend down to a depth of about 3,000 km. At greater depth the flows decay, possibly slowed by Jupiter's strong magnetic field.

Above 50° latitude it is a different story. The familiar bands are replaced by a mottled pattern of reddish-brown clouds in the form of numerous rotating cells. Images taken by the Juno orbiter have revealed giant cyclones, organized in persistent polygonal patterns, which occur at both of Jupiter's poles.

Infrared images show that the cloud colors coincide with altitude (Figure 8.6). Occasional bluish regions are warmest of all, indicating that they are the deepest of the visible clouds and only visible through holes in the overlying blanket. Above these are brown clouds, followed by white. The highest clouds of all are reddish.

All of the major cloud layers predicted by models are white in color, so the precise composition of most of these colorful compounds remains uncertain. Elemental sulfur, which can assume many different colors, depending on its temperature and molecular structure, is one possibility. Phosphine is another.

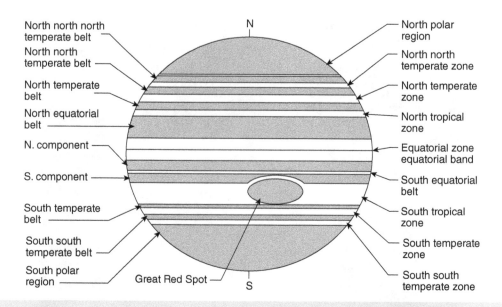

North north north temperate belt
North north temperate belt
North temperate belt
North equatorial belt
N. component
S. component
South temperate belt
South south temperate belt
South polar region
Great Red Spot

North polar region
North north temperate zone
North temperate zone
North tropical zone
Equatorial zone equatorial band
South equatorial belt
South tropical zone
South temperate zone
South south temperate zone

Figure 8.4 Jupiter's dark cloud bands are called belts, and the light bands are known as zones. The rest of each name is based on the corresponding latitudes, e.g. equatorial or tropical. North is at the top. (NASA)

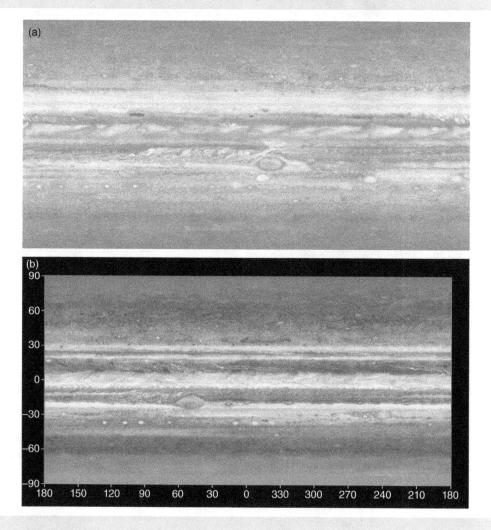

Figure 8.5 Two cylindrical mosaics of Jupiter obtained during the flybys of (a) Voyager 2 in July 1979 and (b) Cassini in December 2000. Note how the parallel belts and zones are replaced by numerous rotating cells above 50° latitude. The Great Red Spot and numerous white ovals are visible in the southern hemisphere. (NASA-JPL)

Box 8.1   Mass and Weight

"Mass" and "weight" are often interchangeable in our daily speech, but they have very different scientific meanings. Scientists measure MASS in kilograms, whereas they measure WEIGHT in Newtons (N). So, instead of saying that someone weighs 100 kg, we should strictly say that their weight is 980 N.

The mass of a body is a measure of how much matter it contains. Mass is also a measure of an object's inertia, i.e. how difficult it is to move from a resting position, or to slow down and stop if it is moving. Objects at rest typically want to remain at rest. Once they are in motion, they want to keep moving. This quality of matter is its inertia.

An object's weight is not only related to its mass, but also to how strongly the force of gravity is pulling on that object. Every object in the universe attracts every other object, but the amount of attraction depends on their masses and the distance between them. An apple has a very small mass, so it falls toward the center of the Earth, rather than the other way around.

The gravitational attraction between two apples is far too small to be measured. However, the pull between a very large object, such as Earth, and another object can be measured easily by using a weighing scale. Scales measure the amount of physical attraction between an object and Earth. This attraction is called weight.

Astronauts in orbit have to adapt to the fact that objects with mass can have no measurable weight. Everything on a spacecraft floats because it is (almost) weightless – a scale placed beneath an anvil in space would read zero. However, if you try to shake the floating anvil, you will soon note that it needs a hefty push to get it moving and another, equal force to stop its motion. This is because the anvil still has mass and inertia.

The pull of gravity is directly related to an object's mass. If its mass is doubled, its gravitational pull also doubles. Similarly, if the planet you are standing on is twice as massive as another planet of the same size, its surface gravity is twice as strong.

On the other hand, the pull of gravity decreases quite rapidly with distance from the center of a planet. The force drops off with the square of the distance. Hence, if a spacecraft doubles its distance from a planet, the pull of the planet's gravity is reduced to one-fourth of its previous value. At 10 times the distance, the planet's gravitational attraction is 100 times weaker.

Sir Isaac Newton (1642–1727) was the first person to describe this relationship in his law of gravitation. This is shown by the equation: $F=GMm/r^2$. F stands for the effective force of gravity between any two masses. G stands for the gravitational constant – a value which applies everywhere in the Solar System (6.67259×10^{-11} $m^3/kg/s^2$). "M" and "m" are the masses of the planet and another nearby object. "r" is the radius, i.e. the distance of the object from the planet's center. Note that the force of gravity never becomes zero, no matter how far you travel from a planet.

When applied to giant planets such as Jupiter, the equation throws up some surprises. Since Jupiter has 318 times the mass of Earth, an astronaut standing on its visible surface might expect to weigh 318 times more than at home. This would be true if Jupiter and Earth were the same size, but Jupiter is much larger, so the astronaut would actually be over 11 times further from the center. This reduces the gravitational pull by a factor of 125.4, so Jupiter's surface gravity is "only" about 2.53 times stronger than on Earth. In other words, an astronaut "weighing" 10 kg on Earth would weigh 25.3 kg on the surface of Jupiter.

On smaller, less massive planets, the opposite is true. For example, surface gravity on the Moon is about one-sixth that on Earth, so heavy lifting is not a problem for lunar explorers. A person with a "weight" of 100 kg would weigh about 16 kg on the Moon. Note that it is also possible for someone to weigh less on the surface of a very large planet whose mass is relatively low for its size. One example is Uranus, where the surface gravity is about 90% that on Earth (see planets table in appendices).

However, the clouds are certainly the products of active physical and chemical processes taking place in the atmosphere. Unstable compounds may form in a variety of ways: bombardment by charged particles; lightning discharges; rapid ascent and condensation; or solar ultraviolet light causing photodissociation of compounds such as methane.

Atmospheric Composition

Like all of the giant planets, Jupiter has an atmosphere dominated by the two lightest gases, hydrogen and helium. However, many more constituents have been found by spectroscopic observation.

The first major advances came in the 1930s, when dark lines in its spectrum were identified as ammonia (NH_3) and methane (CH_4). Around 1960, the presence of molecular hydrogen was confirmed, leading to the conclusion that Jupiter's overall composition must be very similar to that of the Sun – which is not too surprising if the planet formed from gas in the solar nebula.

Table 8.2

Atmospheric Compositions of Jupiter and Sun (%)

Molecule	Sun	Jupiter
Hydrogen (H)	84	89.8
Helium (He)	16	10.2
Water vapor (H_2O)	0.15	0.0004
Methane (CH_4)	0.07	0.3
Ammonia (NH_3)	0.02	0.026

This opinion is still held today, although there are some small differences that are hard to explain (see Table 8.2).

The first detailed radio and infrared measurements of Jupiter's atmosphere were obtained during the Voyager flybys of 1979. This

Figure 8.6   False color, near-infrared images from the Galileo orbiter show clouds and hazes at various altitudes. The top left and right images, taken at 1.61 microns and 2.73 microns, respectively, show the deep atmosphere, with clouds down to a level of about three Earth atmospheres. In the 2.17 micron image (top center), light is strongly absorbed by hydrogen. Only the Great Red Spot, the highest equatorial clouds, a small feature at mid-northern latitudes, and thin, high photochemical polar hazes can be seen. At 3.01 microns (lower left), deeper clouds are just visible, despite gaseous ammonia and methane absorption. The 4.99 micron image (lower middle), shows heat rising from the deep interior through thin cloud cover. In the image at lower right, red areas denote warmth from the deep atmosphere; green shows cool tropospheric clouds; blue shows the cold upper troposphere and lower stratosphere. Polar regions appear purplish, because small-particle hazes allow leakage and reflectivity. Yellowish regions at temperate latitudes may indicate tropospheric clouds with small particles which also allow heat to escape. The blue color of the Great Red Spot and equatorial region is caused by high- and low-altitude aerosols. (NASA-JPL)

made it possible to produce pressure-temperature profiles to a depth of several hundred kilometers. Theoretical studies, along with the known gas abundances, indicate that there should be three cloud layers at varying depths.

At the top, where the atmosphere is coldest, the clouds are composed of ammonia ice crystals. A little lower should be a layer of ammonium hydrosulfide (NH_4SH) – a mixture of ammonia and hydrogen sulfide. Lower still, where the atmospheric pressure approached 5 bars (five times sea level pressure on Earth), water ice clouds should form. Scientists reported in 2018 that they had found evidence for all three cloud layers by studying thermal emissions leaking through the cloud-covered Great Red Spot.

The only attempt to confirm conditions in these outer layers by *in situ* measurement took place in December 1995, when the Galileo Probe made a suicidal entry into the atmosphere (Figure 8.7). Surprisingly, hazes predominated and no thick layers of cloud were detected. One cloud structure – presumably ammonia ice – was found near the 0.6 bar pressure level. Another

well-defined cloud feature was found above the 1.6 bar level, presumably composed of ammonium hydrosulfide ice crystals. In both cases, the clouds seemed to be broken up and patchy.

Why were the models so wrong? Images of the Probe's atmospheric entry point showed that – against all the odds – it had descended within a rare "hot spot." Such holes in the main cloud decks allow heat to escape from deep inside the planet, resulting in major changes to the normal atmospheric conditions.

The Probe also found that the atmosphere is easily mixed in the vertical direction – a so-called "neutrally stratified atmosphere" – at depths below the 0.5 bar level. This confirmed that Jupiter's internal heat escapes principally by convective motion of the atmosphere at these levels.

The strong eastward winds were studied by the Doppler Wind Experiment, which measured changes in precisely tuned radio transmissions to the Galileo orbiter, 215,000 km above the probe (Figure 8.57). Winds speeds of 540 km/h were recorded just below the cloud tops, persisting and increasing with depth until they reached 720 km/h toward the end of the Probe's transmissions.

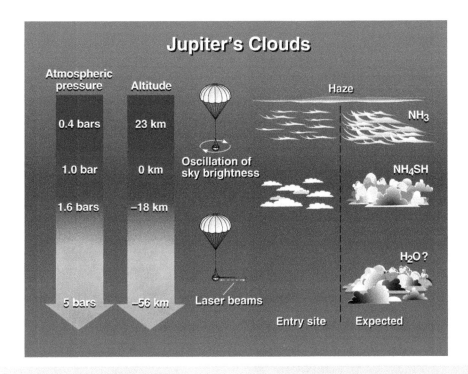

Figure 8.7 The Galileo Probe detected clouds during its descent into Jupiter's atmosphere. One instrument observed variations in brightness and detected a cloud layer near the 0.6 bar pressure level. Another instrument observed laser light scattered by passing cloud particles. It found one well-defined, tenuous cloud structure with a base at the 1.6 bar level, though the higher cloud layer was not seen. This may have been because the upper cloud layer is patchy and the Probe went through a clear area. Theoretical models (right) predict layers of ammonia (NH_3), ammonium hydrosulfide (NH_4SH), and water (H_2O) clouds. The explanation for their absence may be that the Probe descended in an atmospheric "hot spot." (Courtesy of NASA-Ames)

The presence of such strong winds at depth supported previous data that solar heating is supplemented by a powerful internal heat source.

The orbiter made its own contribution by detecting the first discrete cloud of pure ammonia ice found on the planet. The small, bright feature, perhaps 15 km thick, was named the Turbulent Wake Anomaly because it lay in the disturbed clouds downstream from the Great Red Spot, where powerful wind currents moving in opposite directions drew ammonia gas from below. The ammonia then cooled rapidly to form ice in the −120°C atmosphere.

A different view of the atmosphere was obtained by the Juno orbiter, which carried a microwave radiometer that could study atmospheric variations down to a depth of about 350 km. Juno found that Jupiter appears to have a high abundance of ammonia around its equator – contrary to expectations that ammonia would be uniformly mixed at depths greater than 100 km.

The convective nature of the atmosphere, driven by solar radiation and heat from inside the planet, also accounts for the numerous thunderstorms on Jupiter. The Galileo orbiter identified many thunderheads towering above the main cloud deck to the north of the Great Red Spot.

One image sequence showed a dense, white convective storm, which measured 1,000 km across and rose 25 km above most of the surrounding clouds. The cloud base nearby was shown to be at least 50 km below the general cloud level, so deep that it could only be made of condensed water.

Many lightning flashes, hundreds of times brighter than electric storms on Earth, have been recorded on the planet. The Juno orbiter detected peak rates of four lightning strikes per second (similar to the rates observed in thunderstorms on Earth). Interestingly, the Galileo Probe did not observe any optical lightning flashes during its descent, although many discharges were observed at radio frequencies. The radio signals indicated that the discharges were many thousands of kilometers away.

Juno also found that lightning distribution on Jupiter is the opposite to Earth's, with a lot of activity near Jupiter's poles but none near the equator. This is thought to be because solar warming at Jupiter's equator is just enough to create stability in the upper atmosphere, inhibiting the rise of warm air from within. The poles, which do not have this upper-level warmth, and therefore no atmospheric stability, allow warm gases from Jupiter's interior to rise to the cloud tops, driving convection and creating favorable conditions for lightning.

The Great Red Spot

All of the giant planets display discrete storm systems. In the case of Jupiter, these take the form of about 80 huge ovals – some more than half the diameter of Earth – that drift around the planet within the main jet streams.

Most of these are distinguished by white clouds, although some take on a reddish hue – most notably the famous Great Red Spot

Figure 8.8   A false-color mosaic of the northern hemisphere derived from near-infrared images taken by the Galileo orbiter. Atmospheric circulation between 10°N and 50°N is dominated by alternating eastward and westward jets. Light blue clouds are high, while thin, reddish clouds are deep, and white clouds are high and thick. The high clouds and haze over the whitish ovals extend into Jupiter's stratosphere. Dark purple probably represents haze overlying a clear atmosphere. Also visible are large white ovals, bright spots, dark spots, interacting vortices, and turbulent cloud systems. The north–south diameter of the two interacting white vortices is about 3,500 km. (NASA-JPL)

(GRS). This enduring atmospheric feature may have existed for well over 300 years.[2] Its discovery is often credited to Englishman Robert Hooke, who sketched a spot on Jupiter in May 1664, although the first confirmed sighting was in 1831.

The huge anticyclonic storm is located at about 22.4°S, within the South Tropical Zone. Although its distance from the equator is stable, the Spot is continually moving westward – the opposite direction to Jupiter's rotation – and drifting in longitude. During the 20th century, it wandered a total of about 1,200°, equivalent to more than three complete laps of Jupiter.

Observations made in the late 1800s determined that the GRS spanned about 41,000 km – wide enough to fit three Earths comfortably side by side. By the time of the NASA Voyager flybys

in 1979 and 1980, the Spot had shrunk to 23,335 km. In 2017, the feature measured 16,350 km across. Although its dimensions have fluctuated over the past 150 years – sometimes increasing, but mostly decreasing – the Spot is still large enough to swallow the entire Earth (Figure 8.12). However, no change in internal wind speed has occurred, despite it becoming more compact.

The Spot's reddish coloration first became noticeable in the 1870s. Since then, its red coloration has frequently changed in intensity. Since 2014 its color has been deepening, becoming intensely orange. One possibility is that the chemicals which color the storm are being carried higher into the atmosphere as the spot grows taller. At higher altitudes, the chemicals would be exposed to more UV radiation and take on a deeper color.

It has also reduced its rate of drift around the planet (though with considerable variation). From 1940 onwards, the average time it took to travel once around the planet was 9h 55m 42s. However, its westward drift has increased from ~0.26° per day in the 1980s to ~0.36° per day currently.

No one knows precisely why it is red. The leading theory is that deep under Jupiter's clouds, a colorless ammonium hydrosulfide layer could be reacting with cosmic rays or UV radiation from the Sun. However, observations show that it represents an enormous anticyclone which towers 8 km above its surroundings.

Infrared images confirm that it has a cold center and warmer surroundings, suggesting that a powerful updraft is carrying material from deep beneath Jupiter's cloud tops (Figure 8.11). When it is exposed to solar ultraviolet radiation, the gas is altered by some unknown photochemical reaction, possibly involving phosphorus or sulfur, which results in the familiar brick color.

The movement of nearby cloud features shows that winds circulate around the GRS at up to 430 km/h. In Voyager images the clouds on the outskirts of the spot took about 7 days to complete one circuit, whereas there was little horizontal motion in the center of the storm. Amateur observations made in 2006 revealed that the rotation period had reduced to 4.5 days, indicating a substantial increase in wind speed.

The anticyclonic circulation around the GRS seems to be driven by powerful wind currents moving from east to west on its northern flank, and in the opposite direction on its southern edge. Like an island in a stream, the Spot acts as a barrier, causing considerable turbulence in the surrounding clouds, particularly on its western side. However, the clouds that enter its domain suffer various fates.

One bright feature imaged by Voyager circled the Spot for 60 days without any appreciable change. Some clouds swept around the northern edge and kept on going. Some were captured and consumed. Yet others split in two when they reached either end of the giant oval, and then traveled in opposite directions.

Cloud patterns around the Spot's edge also change fairly rapidly. In early 1979, Voyager 1 saw the GRS partially covered to the north by a veil of high ice cloud – probably ammonia. Four months later, a more marked ribbon of white cloud had developed around its northern edge, blocking the motion of small spots coming from the east and forcing them to reverse direction.

---

[2] There is an observational gap of 118 years after 1713, leading some to suggest that the current spot first appeared as a long, pale "hollow" in 1831, with the prominent red color first developing in the 1870s.

Figure 8.9   Juno's microwave radiometer examined Jupiter's atmosphere to a depth of about 350 km. In the cross-section (right), orange signifies high ammonia abundance and blue shows low ammonia abundance. Jupiter appears to have a band around its equator with high ammonia levels – contrary to expectations that the gas would be uniformly mixed. (Courtesy of NASA/JPL-Caltech/SwRI)

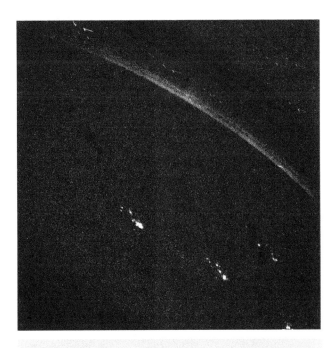

Figure 8.10   A Voyager 1 image showing lightning on Jupiter's night side, together with a faint aurora above the horizon. (NASA-JPL)

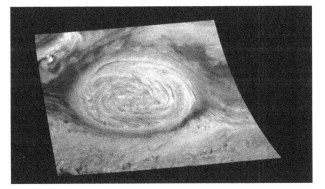

Figure 8.11   A false-color infrared mosaic of the Great Red Spot taken by the Galileo orbiter. High, thick clouds are white; high, thin clouds are pink; low-altitude clouds are blue and brown. Clouds altitudes vary by more than 30 km, and the Great Red Spot rises about 8 km above the main cloud deck. The deepest clouds are in the collar surrounding the anticyclonic storm, and in the northwest corner of the image. White cumulus clouds to the north appear to be thunderheads. The smallest clouds are tens of kilometers across. (Caltech)

Juno data indicate that the Spot's roots penetrate about 300 km into the planet's atmosphere, where warmer gases rise toward the surface and power the giant storm.

The upper atmosphere, high above the GRS, is also much warmer than anywhere else on the planet. Infrared data show that the atmosphere 800 km above and near the storm is about 370°C warmer than normal.

Researchers have concluded that hot region above the GRS may be caused by turbulent, energetic waves that rise from below, then

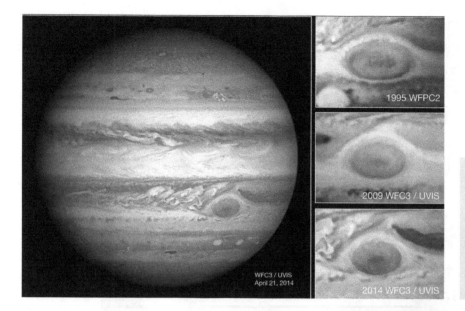

WFC3 / UVIS
April 21, 2014

1995 WFPC2

2009 WFC3 / UVIS

2014 WFC3 / UVIS

Figure 8.12 Hubble Space Telescope images show the shrinking size of Jupiter's Great Red Spot (right). In 1995 its diameter was just under 21,000 km; by 2009 it was just under 18,000 km, and in 2014 its diameter was about 16,000 km. (Courtesy of NASA/A. Simon, ESA)

collide and heat the upper atmosphere. The two possibilities are gravity waves and acoustic (sound) waves created by compression of the air.

Merging Storms

The white ovals tend to survive for very long periods, but mergers between storms drifting at similar latitudes do occur. One remarkable sequence of mergers took place between 1997 and 2000, when three of them coalesced into one. Each of the ovals was a swirling, high-pressure vortex, marked by colder temperatures than its surroundings – evidence of upwelling at the center – and winds that rotated counterclockwise at about 470 km/h. The largest of them was about 9,000 km across, the others slightly smaller.

One of the storms had been in existence for at least 90 years, while the youngest had first appeared in 1939, during a period of enhanced atmospheric activity. Over the following decades, they ploughed inexorably around the planet, on the edge of the South Temperate Belt at 31–35°S (south of the Great Red Spot). Although they sometimes approached each other, no collisions took place. However, by early 1998, two of the ovals were approaching each other as Jupiter went into conjunction, behind the Sun. Between the ovals was a darker, warmer region of sinking gas – a low-pressure system.

When the planet became visible once more, the pair had combined to form a single feature, while the pear-shaped, reddish storm had been weakened and disappeared. Curiously, the area of the newcomer did not match the combined size of its two components, suggesting that the new oval might have lost some material during the merger or grown vertically.

The second merger was not long in coming. By October 1999, the newly merged oval and the last of the original trio were approaching each other. Between them was a dark red storm, a low-pressure system where clouds swirled clockwise. The darker oval was pushed out of the way and torn apart as the three pressure systems skirted past the Great Red Spot.

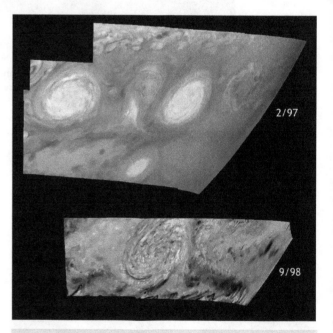

2/97

9/98

Figure 8.13 Near-infrared images from the Galileo orbiter show the "before" and "after" of a merger between two anticyclonic storms in February 1998. The top panel shows two white ovals separated by a darker, pear-shaped cyclonic region. Below, the low-pressure system has disappeared and the ovals have coalesced to form a single, larger storm. Light blue clouds are high and thin, reddish clouds are deep, and white clouds are high and thick. The clouds and haze over the white ovals extend into Jupiter's stratosphere. High haze is largely absent above the cyclonic feature. Dark purple probably represents haze overlying a clear atmosphere. (NASA-JPL)

The disappearance of the cyclone cleared the way for the two white ovals to meet and merge. The process began in March 2000 and lasted about three weeks. At the cloud tops, the storms circled counterclockwise around each other, then consolidated into an enormous oval about 12,000 km across – becoming the second largest storm on the planet. (A similar merger centuries ago may have created the Great Red Spot itself.)

Such events may be significant on a planetary scale, perhaps as part of a recurring climate cycle that causes Jupiter to lose most of its spots before the atmosphere is sufficiently destabilized to create new vortices.

Red Junior

The huge merged oval which was born in the year 2000 (see Merging Storms) continued to drift around the planet over the next few years. At first, the storm remained white in color, but by December 2005 it was slowly turning brown and by early 2006 its color was similar to the Great Red Spot.

Officially named Oval BA, but popularly nicknamed Red Junior, the changeling was about the same size as Earth – or about half the diameter of its famous rival. When viewed at near-infrared wavelengths which absorb methane gas, Oval BA was about as prominent as the GRS, suggesting that it, too, protruded above the main cloud deck. However, its darker appearance indicated that its top deck might be a little lower or less reflective. On the other hand, cloud studies showed that the highest wind speeds around Red Junior had increased from 430 km/h to 640 km/h – comparable to the GRS.

When the two storms passed close to each other in July 2006, there was little interaction, apart from a fairly normal stretching of Red Junior in an east–west direction. Since then, Oval BA has not only persisted, but it has retained its color. As long as the storm can maintain its high wind speeds and continue to tower above

its surroundings, continually pulling up material from deep in the atmosphere, it seems likely to continue to thrive.

In contrast, small atmospheric features known as brown barges are often seen on Jupiter. Barges are small cyclonic features where sinking air clears a break in the clouds, allowing us to see down to a strongly colored cloud that could be ammonium sulfide or ammonium hydrosulfide in the deeper atmosphere.

Barges are most frequently found in the North Equatorial Belt, although they may occur in other belts, such as the North Temperate Belt.

Turbulent Poles

Until NASA's Juno spacecraft entered a near-polar orbit around Jupiter in 2016, astronomers had never had a good look at the planet's polar regions (Figure 8.16). The spacecraft discovered families of cyclones around both of the poles, tightly packed into remarkably stable geometric patterns around their central storms. At the south pole, five cyclones form a pentagon; at the north pole, eight cyclones form a modified octagon or "double square."

The eight north polar cyclones have diameters of 4,000–4,600 km, and they are separated from the central cyclone by a narrow chaotic zone with a westward flow and some vortices. The octagonal arrangement did not change substantially in over

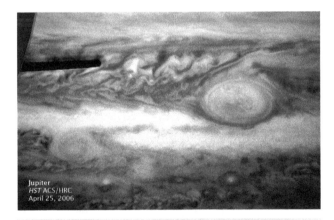

Figure 8.14 This Hubble Space Telescope image, acquired on April 8, 2006, shows Jupiter's two red spots. Oval BA (left) was drifting eastward while the Great Red Spot was moving westward. "Red Junior" passed to the south of the Great Red Spot in July 2006. Note a halo of high-altitude cloud around Oval BA and a small area of pale cloud in its center. A large thunderstorm appears as a bright white cloud in the turbulent region west (left) of the Great Red Spot. (Courtesy of NASA-ESA, A. Simon-Miller/NASA-GSFC, Glenn Orton/NASA-JPL, and Nancy Chanover/New Mexico State University)

Figure 8.15 A false color, near-infrared view of the two red spots – shown here in white – brushing past each other in Jupiter's southern hemisphere in July 2006. White cloud features are at relatively high altitudes; blue indicates lower clouds and red represents still deeper clouds. Also prominent is an orange/red polar stratospheric haze, which is bright near the pole. Other white spots are regions of high cloud, such as thunderheads. (Travis Rector/University of Alaska Anchorage, Chad Trujillo and Gemini ALTAIR adaptive optics team)

Figure 8.16 Juno's camera has revealed persistent, polygonal patterns of cyclones around both of Jupiter's poles. The north pole (a) has eight storms forming an octagon or "double square" around the central storm. The south pole (b) has five even larger cyclones arranged in a pentagon around a central cyclone. Almost all of the cyclones are so densely packed that their spiral arms come in contact with adjacent cyclones. The colors in these infrared composites indicate radiant temperature – the amount of heat escaping into space. The white clouds are thicker, higher, and colder. (Courtesy of NASA/JPL-Caltech/SwRI/ASI/INAF/JIRAM)

a year, with only limited lateral motions around the pole. The central cyclone is located only 0.5° of latitude off the true pole.

The diameters of the south polar cyclones range from 5,600 to 7,000 km, and the five forming the pentagon are mostly in contact with each other and with the central storm. The pentagon is not entirely symmetrical. It is approximately centered on the middle cyclone, whose center is offset from the pole by 1.0° to 2.5° of latitude. Juno images show that this offset has varied cyclically. On the side furthest from the pole, there is a gap between two cyclones in the pentagon, and this gap was seen to grow and shrink during Juno's first eight close flybys.

The cyclonic wind speeds are mostly about 200–350 km/h at a radius of 1,500 km. Closer to their centers, some of them have even faster speeds, exceeding 450 km/h near the center of the south polar cyclone. These speeds are similar to terrestrial hurricanes, or even faster, though the storms are several times larger.

The polygonal clusters are remarkably stable, having persisted with little change during the Juno mission, despite the furious winds which blow within them, and the buffeting from the chaotic, rapidly changing storm systems around them.

It is not yet clear how these polygons remain stable, but they may be examples of patterns called "vortex crystals," which were previously known only from laboratory modeling. These are created when small vortices form and persist as the material in which they are embedded continues to flow. For reasons which are not clear, the patterns are different in nature and appearance from the north polar hexagon on Saturn, which is a kinked jet stream that flows around a single giant cyclone (see Chapter 9).

Interior

Jupiter's visible clouds are only the thin outer layer of an "ocean" of compressed gases thousands of kilometers deep. Data from the Juno orbiter suggest that Jupiter's zonal winds decay slowly with depth until about 3,000 km below Jupiter's surface (5% of the planet's radius), a point at which the pressure is about 100,000 times that of the atmosphere at Earth's surface. At this depth, hydrogen becomes conductive enough to be dragged into near-uniform rotation by the planet's powerful magnetic field.

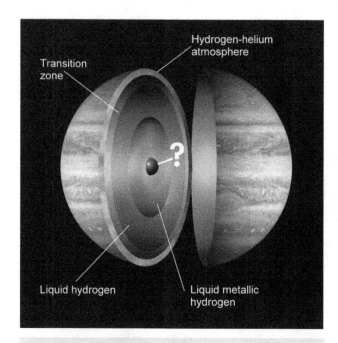

Figure 8.17 Jupiter's cloud layer extends to a depth of no more than 50 km. Beneath the cloud decks is a hydrogen–helium atmosphere about 3,000 km deep. The gas then changes into liquid hydrogen, which extends to a depth of perhaps 15,000 km. When the pressure exceeds one million bars and the temperature exceeds 6,000 K, the hydrogen starts to change phase, forming liquid metallic hydrogen. At the center there is probably a rock-ice core. (NASA)

Until the 1930s, astronomers had very little idea about the nature of the planet's deep interior. Then Rupert Wildt (1905–1976) discovered methane and ammonia in the atmospheres of Jupiter and Saturn. He went on to propose a model which included a solid, rocky core overlain by a layer of ices and topped by an atmosphere that was largely made up of hydrogen and helium, with small amounts of ammonia, methane, and other gases.

Later researchers realized that the planet's low density probably indicated a hydrogen-rich composition all the way down to the core. However, under the pressure cooker conditions deep inside Jupiter, the highly compressed hydrogen ceases to behave like a gas and turns into a liquid.

Various models have been developed in recent years, based on theoretical calculations and laboratory experiments. It is now believed that hydrogen gas begins to change into a hot liquid at a pressure exceeding 100,000 bars. A further transition begins at a depth of perhaps 15,000 km, when the pressure reaches 1 million bars (1 megabar) and a temperature of about 6,000 K – the same as the surface of the Sun.

The molecular and atomic bonds begin to break down under such extreme conditions, creating a new form of hydrogen which behaves like a metal. This liquid "metallic" hydrogen, which is made up of ionized protons and electrons, makes up most of Jupiter. Since metallic hydrogen is an electrical conductor, slow-moving convection currents flowing through this layer would be expected to generate a magnetic field, and this is exactly what is found (see Magnetic Field).

The vast layer of metallic hydrogen is believed to have a depth approaching 50,000 km. At its base, the temperature reaches 20,000 K, with a pressure of around 20 megabars. Models indicate that there may be a gradual transition to a layer dominated by "ice," a liquid mixture of water, methane, and ammonia. Finally, at the center there may be a rock or rock-ice core of up to 10 Earth masses. If the core exists, it is probably molten, although it is possible that the tremendous pressure compresses it into a solid.

Some refinements to this overall picture have been proposed, based on the results sent back by the Galileo Probe. Elements such as argon, krypton, xenon, carbon, and nitrogen were shown to be 2–3 times more abundant on Jupiter than on the Sun. This enrichment can be largely explained by the addition of heavy elements from other planetesimals in the later stages of Jupiter's formation.

On the other hand, the abundance of neon was only 10% of the solar figure. This may be due to the removal of neon from the upper atmosphere by helium "rain" at a depth of about 10,000 to 13,000 km. According to this theory, helium does not mix with metallic hydrogen, but it does mix easily with neon. When helium droplets separate out and fall toward the interior, they carry neon with them.[3]

One other characteristic of Jupiter is an internal heat source which causes it to radiate more than twice as much heat as it receives from the Sun. The heat is thought to be generated by the slow gravitational compression of the planet. It flows outward

Box 8.2   Metallic Hydrogen

Normally, hydrogen is a non-metallic element. Each hydrogen atom is composed of a single, negatively charged electron in orbit around a single, positively charged proton. However, as a result of the tremendous pressures deep inside Jupiter and Saturn, hydrogen is compressed so much that many atoms occupy the space normally filled by only a single atom.

Although there is still plenty of space in each hydrogen atom for the electron to circle the proton, high-pressure laboratory experiments and theoretical calculations suggest that the molecular and atomic bonds begin to break at a pressure of about 1 million bars and a temperature of about 5,700°C. As a result, electrons begin to migrate from atom to atom – the protons and electrons are said to be ionized. Some atoms contain so many electrons that the outermost one can easily be detached.

Under these conditions, hydrogen takes on the properties of a metal. As such, it becomes an electrical conductor and is opaque to visible radiation. Currents circulating within the metallic hydrogen are thought to be responsible for the powerful magnetic fields of Jupiter and Saturn.

In the case of Jupiter, the transformation from liquid hydrogen to metallic liquid hydrogen may occur at a depth of about 15,000 km. Saturn is considerably smaller, so the layer of metallic hydrogen is deeper and thinner. Internal pressures and temperatures are much lower for Uranus and Neptune, so metallic hydrogen is unlikely to exist.

from the core, driving the convectional circulation in the liquid hydrogen mantle and the outer atmosphere. Hot gas rises from the interior, then cools and flows back down.

Magnetic Field

The first planetary magnetic field to be discovered (other than Earth's), and by far the largest, was that of Jupiter. The breakthrough came in early 1955, when radio astronomers discovered a conspicuous radio source which shifted position from day to day. The signals coincided with the presence of Jupiter in the sky. It was the first radio emission detected from another planet.

At first it was thought that these decametric (tens of meters) wavelength radio emissions were caused by electrical disturbances in Jupiter's atmosphere. However, the regular variations in the intensity of radio waves eventually led to the conclusion that they were produced by electrons trapped in a strong magnetic field, i.e. radiation belts. This theory was confirmed in 1973, when the Pioneer 10 spacecraft flew past Jupiter and detected very intense radiation belts within an enormously powerful magnetic field.

[3] Helium rain has also been proposed for the interior of Saturn (see Chapter 9). However, Saturn's helium rain was predicted because the planet is warmer than expected. The falling rain releases heat, accounting for the temperature difference. Jupiter's temperature is in accord with models of its cooling rate and age, so the hypothesis of helium rain was only introduced to account for neon depletion in the atmosphere.

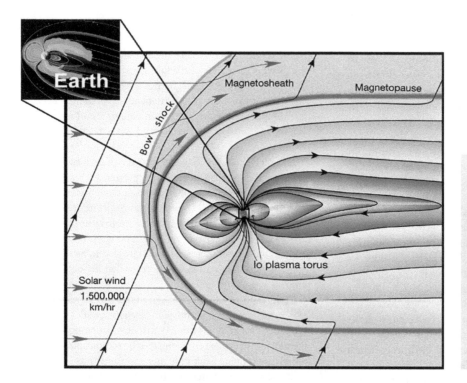

Figure 8.18 Jupiter is surrounded by an enormous magnetosphere which deflects the charged particles of the solar wind. The flow of solar particles shapes the magnetic bubble, creating a magnetotail that stretches all the way to Saturn's orbit. Trapped plasma is concentrated in a disk-like plasma sheet aligned with the planet's equator. The magnetic axis is inclined 9.6° to Jupiter's rotational axis. Earth's magnetic field (inset) is shown for size comparison. (SWRI)

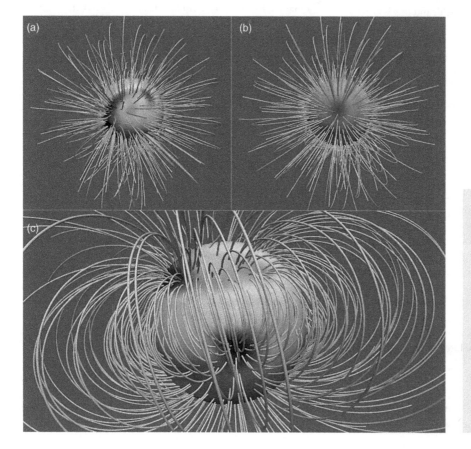

Figure 8.19 Three views of Jupiter's magnetic field lines. (a) north polar view; (b) south polar view; (c) equatorial view. The non-dipolar nature of the magnetic field in the northern hemisphere and the dipolar nature in the southern hemisphere is apparent. The equatorial view shows the linkage of magnetic field lines that enter through that location. The contoured surface on which the field lines start and end is at 0.85 Jupiter radii. Red marks outward flux, blue inward flux. (Courtesy of Kimberley M. Moore et al./Nature)

As with most planets, the Jovian magnetosphere is shaped by the solar wind so that it resembles a giant tadpole, with a blunt head and a long tail. However, Jupiter's magnetic "bubble" is the largest single feature in the Solar System. If our eyes could see it, the magnetosphere would appear twice the width of a full Moon in the night sky.

Despite the pressure of the solar wind, the outer edge of the magnetosphere extends 3–5 million km toward the Sun (Figure 8.18).[4] On the planet's night side, the magnetotail stretches at least 740 million km (more than 100 Jupiter radii) – beyond the orbit of Saturn. All of Jupiter's major satellites are immersed in the magnetosphere – unlike the Moon, which dips occasionally into Earth's magnetotail.

Jupiter's magnetic field is also the strongest in the Solar System (except for fields associated with sunspots and other active regions on the Sun), with a dipole moment about 19,000 times that of Earth. However, the planet is so large that the magnetic field at its cloud tops is a mere 14 times stronger than at Earth's surface (about 4.2 gauss on the equator and 14 gauss at the poles).

The magnetic field is thought to be generated in the zone of liquid metallic hydrogen deep inside Jupiter, where the great pressure changes the properties of hydrogen, so that it conducts much like a metal.

Jupiter's field was thought to be a simple dipole, like Earth's, which generally resembles that generated by a bar magnet. The dipole's axis (which joins the magnetic poles) is offset from the planet's center and inclined by 9.6° to the rotation axis. The polarity is opposite to Earth's, so a north-seeking compass would point south on Jupiter.

However, data from the Juno orbiter have shown that the magnetic field is highly unusual (Figure 8.19). It is non-dipolar in the northern hemisphere, but dipolar in the southern hemisphere. While flux lines emanate from its north pole, there are two return points, rather than just one: one is located near the south pole, the other close to the equator.

The discovery of a third magnetic pole near the equator is hard to explain, but researchers suggest that Jupiter's dynamo, unlike Earth's, does not operate in a thick, homogeneous shell. Instead, it may be influenced by variations in internal layering, density, and/or electrical conductivity.

There are other differences. Jupiter's trapped plasma seems to co-rotate with the planet all the way out to the dayside boundary with the solar wind. At Earth, this co-rotation ceases well inside the magnetosphere. The Jovian plasma is composed not only of protons and electrons, but also of sulfur and sodium ions that originate in volcanic eruptions on the moon Io.

In the middle magnetosphere, from 20–60 Jupiter radii, ionized particles form a sheet of electric current. This "ring" current produces a magnetic field which is actually stronger than the planet's magnetic field. The current is concentrated near the equator, creating a "magnetodisk" where the magnetic field is much more flattened than Earth's.

The major satellites sweep high-energy particles from the radiation belts, so that total radiation near Jupiter is reduced by as much as 100 times. By far the largest number of particles is removed by Io, the innermost of the large moons. When Io passes into Jupiter's shadow, a faint auroral glow appears in the satellite's sparse atmosphere due to high speed collisions with ions from the magnetosphere. The "sponge effect" makes it possible to fly spacecraft close to the planet without major radiation damage.

Jupiter's Auroras and Io

Jupiter displays brilliant auroras which are caused when electrically charged particles trapped in the magnetic field spiral inward at high energies toward the north and south magnetic poles. When these particles collide with the rarefied upper atmosphere, they excite atoms and molecules, causing them to glow (similar to fluorescent streetlights).

Some of these particles come from solar eruptions, but many originate from Jupiter's large satellites (Figures 8.21 and 8.22). As the plasma trapped in the planet's rotating magnetic field passes by, it sweeps charged particles off the surface of Io and the other moons. Io also contributes about one tonne of material per second through its continuous volcanic eruptions.

Figure 8.20   A composite, ultraviolet image of Jupiter's polar auroras, taken with the Hubble Space Telescope in September 1997. The bright ovals are created by highly energetic particles which have been accelerated by the planet's powerful magnetic field. These electrons and ions flow along magnetic field lines toward the magnetic poles, generating auroras when they collide with the upper atmosphere. Narrow, curved trails outside the main ovals show where the powerful electric current from Io's flux tube enters the atmosphere. (Courtesy of NASA-ESA)

---

[4] As usual, the magnetosphere expands and contracts under the fluctuating pressure of the solar wind.

Figure 8.21 Io's yellow cloud of neutral sodium gas, imaged by the 0.6 m telescope at Table Mountain Observatory, California. The images of Jupiter and Io's orbit were superimposed. The sodium is ejected from Io's volcanoes. Io's size and location are shown by the orange dot within the crossbars. (Courtesy of NASA-JPL/Bruce Goldberg and Glenn Garneau)

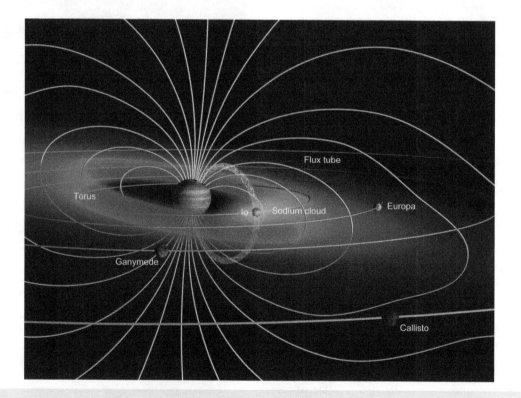

Figure 8.22 The main features of the inner Jovian magnetosphere. The plasma torus (red) of ionized sulfur and oxygen is trapped in the rotating, tilted magnetic field, so it co-rotates with the planet. The neutral sodium cloud (yellow), also derived from Io, is mainly concentrated close to the satellite. The Io flux tube (green) is a flow of electrical current connecting Io with Jupiter's ionosphere (upper atmosphere). (John Spencer, SwRI)

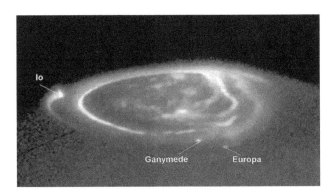

Figure 8.23 A Hubble Space Telescope ultraviolet image of the Jovian aurora, which is centered on the magnetic north pole, as well as more diffuse emissions across the polar region. Also visible are the magnetic "footprints" of three of Jupiter's largest moons, Io, Europa, and Ganymede. These emissions, produced by electric currents generated by the satellites, flow along Jupiter's magnetic field lines. (Courtesy of NASA-ESA/John Clarke, University of Michigan)

Figure 8.24 This false-color, infrared image by the Hubble Space Telescope shows Io and Ganymede. Also visible are their shadows and the shadow of Callisto, which is out of view. (Courtesy of NASA, ESA, Erich Karkoschka/Univ. of Arizona)

This enormous amount of material is stripped of its electrons (ionized), so that the particles become trapped in Jupiter's magnetic field. They are then carried around Jupiter once every 10 hours, creating a donut-shaped ring or "torus" – predominantly ionized oxygen and sulfur – that occupies Io's orbit.

A cloud of neutral sodium gas also surrounds Io and trails around its orbit, as well as chlorine in the plasma torus. The source of these elements, gaseous sodium chloride, was confirmed by radio astronomers in 2002. Although the salt constitutes only about 0.3% of the moon's tenuous atmosphere, it may be present in much larger concentrations near the active volcanoes.

As Io sweeps through Jupiter's magnetic field during each orbit of the planet, the moon acts as an electrical generator, producing about 10 million volts of electricity and electrical currents of 10 million amps (Figure 8.23). This enormous electrical current flows along the magnetic field lines to Jupiter's ionosphere. When the energetic ions hit the ionosphere, at the end of the flux tube, they generate a narrow, linear auroral trail about 250 km above the main cloud decks. Similar auroral footprints are visible for the weaker flux tubes of Ganymede and Europa.

The Galilean Moons

In January 1610, Galileo Galilei found four bright "stars" near Jupiter (see Chapter 1). Observations over the following weeks led him to conclude that they were actually satellites, with orbital periods that varied from about 42 hours to 17 days. The so-called Galilean moons – visible to anyone with a modest pair of binoculars – are named Io, Europa, Ganymede, and Callisto.

These satellites provide a rich variety of visual phenomena – eclipses, when they disappear and reappear from Jupiter's shadow; transits, when they cross the planet's disk; occultations, when they disappear and reappear from behind the planet; and transits of their shadows across the cloudy atmosphere. On rare occasions, the moons pass in front of each other or are eclipsed by the shadow of a neighbor.

Until the Space Age, little was known about the satellites, although it was noticed that a resonant configuration exists among the inner three Galileans, so that innermost Io completes four orbits to every two of Europa, and every one of Ganymede.

Their relative sizes and densities were determined by direct observation and calculations based on their motions and mutual gravitational perturbations (Figure 8.25). Ganymede, Callisto, and Io are all larger than the Moon, while Europa is only a little smaller. Indeed, Ganymede is the largest satellite in the Solar System, and noticeably larger than the planet Mercury. Callisto is only fractionally smaller than Mercury.

Despite their sizes, the moons have lower densities than the rocky terrestrial planets, and only Io has a higher bulk density than the Moon. This implies that they are a mixture of rock and ices. Their densities decrease with distance from Jupiter, probably a reflection of conditions within the nebula of dust and gas that surrounded the planet, their birthplace more than 4 billion years ago. Like the solar nebula, temperatures would have been much higher near the center, so that fewer volatiles condensed out. Further away from Jupiter, the colder conditions would have allowed more ices to condense.

Like our Moon, all of them have been influenced by tidal interactions with their planet, so they keep the same face towards Jupiter. Furthermore, since the satellites are embedded within its radiation belts, they are bombarded by energetic particles that cause chemical reactions and physical erosion of their surfaces (see Jupiter's Auroras and Io).

Figure 8.25   The four largest moons of Jupiter, shown to scale. From left to right, they are Ganymede, Callisto, Io, and Europa. Ganymede, the largest satellite in the Solar System, is bigger than Mercury. Io is larger than Earth's Moon. (Courtesy of NASA-JPL/DLR)

Io

Until the Voyager flybys of 1979, Io remained an enigma. Radio astronomers knew that it was linked to the bursts of radio noise from Jupiter and that it was surrounded by a cloud of sodium (Figure 8.21). Visual observations showed a yellowish object with reddish polar caps. When the moon emerged from Jupiter's shadow, it was frequently (but not always) a few percent brighter than usual for about 15 minutes. The mystery deepened in 1973, when Pioneer 10 detected a layer of ionized particles about 100 km above Io's surface. Clearly, Io had a tenuous atmosphere.

Voyager 1 sent back pictures of a "pizza" world, marked by irregular patches of white, yellow, orange, and black. The absence of any impact craters indicated that Io's surface was extremely young and global resurfacing was still ongoing.

Then an optical navigation engineer named Linda Morabito noticed a strange, crescent-shaped cloud extending beyond the satellite's limb. After eliminating all obvious possibilities, she concluded that it must be a cloud of material rising hundreds of kilometers above the surface.

Morabito had stumbled upon Io's secret. The satellite is the most volcanically active world in the Solar System. Further observations showed at least 8 active volcanoes, with mushroom-shaped plumes rising up to 250 km into space (Figure 8.28). Six of these eruptions were still taking place when Voyager 2 arrived four months later, although the largest plume – from a volcano named Pele (after the Hawaiian goddess of the volcano) – had ceased.[5]

Independent confirmation of the nature of Io's volcanoes and surface material has come from measurements of thermal emission and spectral data. The volcanic centers are hundreds of degrees hotter than their surroundings. The plumes of sulfur, sulfur dioxide (SO_2) gas, and solid pyroclastics coat the surface in vivid hues of yellow, orange, red, and white. They are also the source of the sulfur and oxygen ions in Jupiter's magnetosphere.

Figure 8.26   Io, the most volcanic body in the Solar System, as imaged by the Galileo spacecraft. The moon's colors are caused by sulfur deposits at different temperatures: the "normal" color of sulfur is yellow, but it changes to orange and red when exposed to heat. Several of the dark, flow-like features correspond to hot spots, and are probably active silicate lava flows. The whitish areas are coated with volcanically deposited sulfur dioxide frost. There are no impact craters because volcanism rapidly covers the surface with new deposits. The image is centered on the side of Io that always faces away from Jupiter. The smallest features that can be seen are 2.5 km across. (NASA/JPL/University of Arizona)

[5] Place names for Io are taken from various mythological associations with fire and volcanoes.

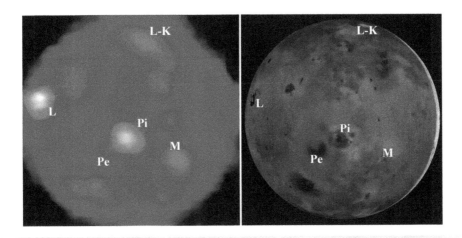

Figure 8.27 Volcanoes and the previous day's sunshine warm the nighttime surface of Io in a thermal image from Galileo (left). Several volcanoes are identified: L-K is Lei-Kung Fluctus, L is Loki, Pi is Pillan, M is Marduk, and Pe is Pele. Much of the heat comes from a few discrete volcanoes. The brightest is Loki, which radiates roughly 15% of Io's total volcanic heat. Second brightest is Pillan, where the heat is radiated by extensive, cooling lava flows. Although Pele produces Io's largest plume, its small crater emits very little heat. Blue indicates the coldest temperatures, near 90 K (–183°C). Oranges and yellows show the highest areas, above 170 K (–103°C). Small volcanic areas exceed 1,500 K (1,227°C), but they are not visible in this low-resolution view. Io's overall temperature varies little with latitude. (Right) A visible light image of the same night hemisphere. A thin crescent marks the area in sunlight. (NASA-JPL)

The presence of sulfur had been suspected even before the Voyager flybys, since its color changes dramatically according to temperature. Although the "normal" color of sulfur is yellow, the element changes to orange, red, and black when exposed to heat.

The colorful surface revealed by the Voyager spacecraft led scientists to believe that lava flows on Io's surface were made of sulfur, but this idea is now largely rejected. Only in a few cases does sulfur seem to be the primary form of molten lava.

A more likely scenario is a thin veneer of sulfur or sulfur compounds on top of a crust of silicate rocks. The presence of cliffs and mountains, together with the large size of the calderas on Io, requires the crust to be strong and thick. Silicate volcanism probably results in high volume, low viscosity (basalt) lava flows erupted from low shields and possible fissures.

The presence of hot spots – active or recently active volcanic regions – also supports the presence of silicate volcanism. High resolution images of hot spots are interpreted to be caldera floors filled with lava lakes and/or lava flows. These relatively small areas have temperatures of 725°C or higher – exceeding the boiling point of sulfur in a vacuum – and thus supporting the existence of silicate lavas.

Extensive lava flows spread out from these basins, either as meandering rivers or broader fans and lobate features. Extensive white patches are attributed to deposits of sulfur dioxide frost or snow.

Io's surface is continually changing. In the 17-year interval between the Voyager and Galileo missions, a dozen areas the size of Connecticut were resurfaced. At least 120 volcanic hot spots were recorded, compared with about 600 on the much larger Earth. Although Io's larger centers of activity could remain active for many months, or even years, some of the smaller volcanoes were found to turn off and on over a period of a few weeks.

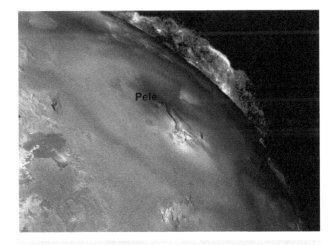

Figure 8.28 A dramatic Voyager 1 view of an umbrella-shaped plume rising 300 km above Pele, one of Io's largest volcanoes. Pele's vent is marked by a dark spot just north of the triangular-shaped plateau. The plume fallout covered an area the size of Alaska, but the cloud was so sparse that it could hardly be seen against the surface. Note the slab-like mountains near the limb (right). (Courtesy of NASA-USGS)

Galileo sent back the first direct images of ongoing surface activity, capturing a 60 km-long curtain of lava erupting along the edge of a 200 km-wide caldera named Tvashtar Catena. In another center, known as Loki, a gigantic block of solidified sulfur dioxide was floating on the dark crust that covered a lake filled with molten rock. Along one shoreline, a ribbon of glowing lava had broken through the overlying crust (Figure 8.30).

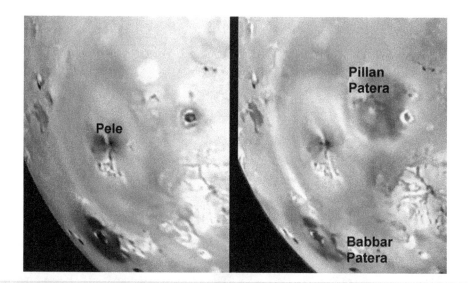

Figure 8.29   Galileo images showing major changes caused by volcanic activity between April 4 (left) and September 19, 1997 (right). The appearance of a new, 400 km-wide dark spot around Pillan Patera was caused by its eruption of a 120 km-high plume. The pinkish deposits of sulfur around Pele were partially covered and other subtle changes are visible. Some apparent differences are due to variations in illumination, emission and phase angles, notably at the dark volcano named Babbar Patera. (Courtesy of NASA-JPL/Univ. of Arizona)

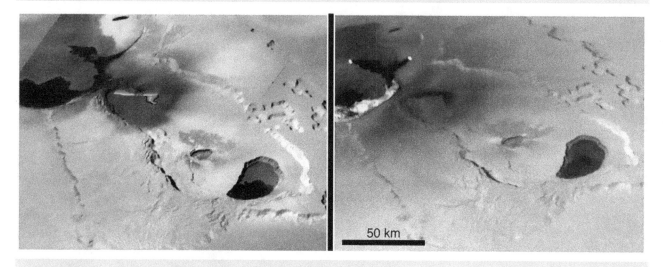

Figure 8.30   Two images of volcanic eruptions at Tvashtar Catena, a chain of calderas centered at 60°N 120°W. The view on November 25, 1999 (left), is a drawing produced after the camera was saturated by extremely hot lava. It shows an orange-red lava curtain (a chain of lava fountains) and surface flows. The lava fountains are up to 1.5 km high. Several other lava flows can be seen on the caldera floors. The darkest flows are probably the most recent. In the February 22, 2000, image (right), the November outburst appears as a dark, L-shaped feature. White and orange areas are fresh, hot lava. Two small, bright spots show where molten rock is exposed at the ends of lava flows. The orange and yellow ribbon is a cooling lava flow over 60 km long. Dark, diffuse deposits around the active lava flows were also recent. The main caldera is almost surrounded by a mesa about 1 km high. In places the mesa's margins are scalloped, due to an erosional process called sapping. (NASA-JPL)

In another volcanic center, Prometheus, the active vent had migrated 70 km since 1979. Images indicated the presence of a subsurface lava flow which moved west before eventually breaking through the overlying crust and interacting with sulfur dioxide frost to produce a new umbrella-shaped plume.

In many respects, Io's plumes resemble high-temperature geysers on Earth, such as Old Faithful in Yellowstone National Park. However, the low gravity of Io (about 1/6 that of Earth) and the extremely low atmospheric pressure allow the sulfur dioxide to soar to much greater heights.

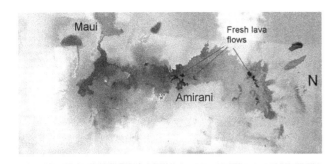

Figure 8.31 The Amirani lava field, 500 km long and 180 km wide, is a patchwork of separate, overlapping flows. The most recent lavas are the darkest because they are too hot to be covered by deposits from sulfur dioxide plumes. Fresh lava is leaking from at least five areas at the northern end and three places in the middle. The lava probably originated in the south, then traveled north beneath a solid crust of older lava before breaking out onto the surface hundreds of kilometers from the vent. Diffuse, white halos around the darkest flows are probably sulfur dioxide snows and frosts. (Moses Milazzo, Lunar and Planetary Lab., University of Arizona)

When subsurface pools of liquid sulfur dioxide come into contact with hot magma, some of the SO_2 becomes a superheated vapor.[6] This high-pressure mixture of liquid and gas opens up a fracture in the crust, enabling it to move toward the surface. As the SO_2 rises, the pressure decreases and it vaporizes completely, eventually erupting at the surface as a high-velocity plume of gas. In the extreme cold of Io, where the typical temperature is about −146°C, the gas immediately starts condensing into snowflakes which fall onto the frigid surface.

The largest known active lava field in the Solar System lies in the Amirani region on Io, where 23 distinct new flows were visible (Figure 8.31). The area spans more than 300 km, and individual flows are several kilometers long. Galileo images showed that new lava flows at Amirani covered about 620 km² in less than five months.[7] Despite the huge amount of lava disgorged – over 100 tons every second – the eruptions were relatively tranquil, with few explosive outbursts.

The extreme vulcanicity of Io was reinforced once more in February 2001, when the Keck II telescope in Hawaii detected the largest eruption ever observed in the Solar System. The high-temperature event involved incandescent lava fountains which were kilometers high, propelled by expanding gases and accompanied by extensive lava flows. Its thermal output almost matched the entire energy output from all of the other volcanoes on Io.

For many years, the volcanoes on Io were thought to be driven by a sulfur cycle, but this theory was disproved when infrared data indicated temperatures of 1,700°C, making it the hottest lava ever recorded. Since liquid sulfur would vaporize at this temperature, it seems certain that silicate magma is associated

with such eruptions – although this does not eliminate the possibility that some lava flows on Io are composed primarily of sulfur.

The arguments favoring silicate volcanism are further supported by the presence of plateaus and mountains, some more than twice as high as Everest. The mountains and deep, steep-sided calderas must be underlain by a crust of considerable rigidity and strength, something which cannot be explained by a sulfurous surface.

About two percent of the surface is occupied by block- or slab-shaped mountains. Io does not have tectonic plates such as those on Earth, but vertical movements seem quite common, as well as occasional rifting. Uplift and thrust faulting have caused some mountains, such as Euboea Montes, to reach heights of up to 13 km.

How these mountains form is not clear. One theory suggests that crustal recycling may be responsible. Since Io is being resurfaced continuously at a rate of about 1 cm per year, the surface may be buried by up to one kilometer of material over a period of 100,000 years. Under the weight of these deposits, the crust may sink and merge with the molten mantle. The resulting compression may force large crustal blocks upward along deep faults, at the same time enabling magma to seep to the surface. Another possibility is that some of the mountains have been pushed up by intrusions of rising magma.

The remarkable volcanoes and tectonic activity are sustained by heat generated by tidal flexing, similar to the tides on Earth, but on a huge scale. Io's shape is periodically distorted by the gravitational

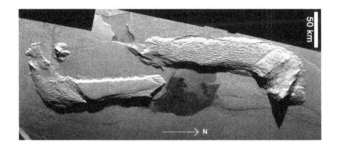

Figure 8.32 Like many of the mountains on Io, these raised blocks alongside Hi'iaka Patera (the irregular, dark depression at the center) seem to be in the process of collapsing. Huge landslides have left piles of debris at the bases of the mountains. The ridges parallel to their margins also indicate that material is sliding downhill due to gravity. By measuring the length of the shadows, it has been estimated that the sharp peak in the north is about 11 km high, while the elongated plateaus are about 3.5 km high. The two blocks have similar shapes which appear to fit together – possible evidence of horizontal rifting which left a depression that has been filled by lava flows. (Moses Milazzo, Lunar and Planetary Lab, University of Arizona)

[6] Superheating means that the liquid is heated beyond its normal boiling point, but remains largely in a liquid (metastable) state.
[7] Amirani is huge even when compared to other Ionian lava flows. The Prometheus lava flow field covered only about 60 km² during the same period.

Box 8.3   Ocean Worlds

Until the last few decades, the only world in our Solar System known to possess an ocean was Earth, although there were indications that large areas of surface water may once have existed on Mars, and even on red-hot Venus. The objects that orbit beyond the asteroid belt were assumed to be too cold and inert for liquid water to exist.

Since then, observations by spacecraft have revealed that many outer Solar System bodies probably harbor liquid water oceans beneath their ice shells. The prime candidates include Europa, Ganymede, and Callisto, which are described in this chapter, and two satellites of Saturn: Titan and Enceladus. (Large lakes of liquid ethane and methane have been found on Titan, the only object other than Earth to have large bodies of liquid on its surface.) Theorists have also suggested that subsurface oceans may exist on some extremely frigid worlds, notably Pluto and Neptune's largest satellite, Triton.

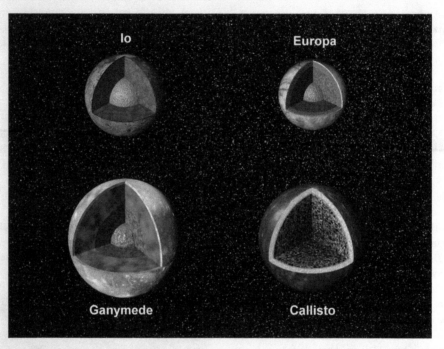

Figure 8.33   The possible internal structures of the Galilean satellites, shown to scale. Three of the satellites have iron-nickel cores (gray) and rocky mantles (brown). Io's rock (silicate) shell extends to the surface, while the mantles of Ganymede and Europa are overlain by water ice or liquid oceans (white/blue). Callisto has a relatively uniform mixture of ice and rock, perhaps with a shallow ocean beneath the icy crust. (NASA-JPL)

The clearest evidence for the existence of liquid water oceans is surface activity and an absence of impact craters which is indicative of recent resurfacing. In the case of Europa, the smooth icy surface is crisscrossed by enormous linear fractures and apparent ice floes. Enceladus is much smaller, but it, too, displays a large area of smooth, uncratered, icy terrain.

Evidence for hydrated salts has been found on the surfaces of moons such as Ganymede and Europa, implying the presence of liquid water that has seeped to the surface.

Related features are cryovolcanoes – ice volcanoes that erupt slurries of volatile compounds such as water, methane, or ammonia from an ocean hidden in the moon's interior, instead of rocky lava. They can erupt violently or flow gently, just like volcanoes on Earth.

Cryovolcanoes are thought to occur on some of the Solar System's coldest objects, such as Titan, where internal warmth is capable of melting an icy mantle. On Titan, spacecraft images have shown landforms that resemble volcanic features on Earth, such as lava flows and mountains with central craters.

The most remarkable feature of Enceladus is its geyser-like plumes of water ice particles which erupt though fractures near the moon's south pole. Recent observations have suggested that Europa also ejects plumes of water ice into space. Triton has been observed releasing dark columns of nitrogen-rich material into its sparse atmosphere, though these may result from solar heating of the surface ices.

Indirect evidence of the internal structure of icy satellites is generally deduced from measurements of their bulk density, gravity field, magnetic field, and shape. Bulk density can be used to infer the approximate rock-ice ratio: a higher density implies more rock. Shape or gravity measurements can be used to infer whether a satellite has undergone differentiation (separation into rock-dominated and ice-dominated layers).

Secondary magnetic fields detected at satellites such as Europa and Callisto are thought to be induced in salty, subsurface oceans – which act as conductors – as the moons travel through Jupiter's powerful magnetic field.

It is generally assumed that ocean-bearing bodies are differentiated. This is because melting of water ice will lead to efficient downward segregation of denser silicates. A typical icy satellite, such as Enceladus, will therefore consist of a water layer sitting on top of a silicate (rocky) core and below a solid, icy shell.

For a larger, ice-rich body like Ganymede, different types of ice will form according to the internal pressure. If high-pressure ices are present, there will be a "water sandwich" with the ocean sitting between the crystalline ice shell and the denser, higher-pressure variants.

How can oceans exist on worlds that orbit so far from the Sun? It has long been recognized that heat released during the decay of radioactive elements in a satellite's core can provide a source of warmth over millions or billions of years. However, radioactive decay alone is probably insufficient to produce subsurface oceans in large moons.

Perhaps the most significant factor is tidal forces, generated when satellites' orbits and rotations are modified by gravitational interactions between two or more objects. The tidal force varies with distance, so, if a satellite's orbit is eccentric (elliptical), the satellite will be stretched and squeezed during each orbit. This mechanism applies to a moon such as Enceladus. Similar interactions can occur between satellites that have orbital resonances with each other, such as Europa.

The repeated changes in shape – known as tidal flexing – generate heat inside the satellite that may result in partial melting of an icy mantle and a subsurface ocean, perhaps with cryovolcanism.

Survival of a subsurface ocean over billions of years may partly depend on its composition. The presence of an antifreeze such as ammonia, which can lower the freezing point of water by almost 100°C, will reduce the likelihood of an ocean freezing out. Ammonia ice has been identified in some of the Saturnian satellites but not in the Jovian moons.

pull of Jupiter, Europa, and Ganymede.[8] This tug of war creates tidal bulges up to 100 m high in Io's solid crust. This squeezing and stretching strongly heats the interior, with the heat being released at the surface by global volcanism.

Unlike the other Galilean satellites, Io seems to be lacking in water. Its extremely dry surface (and, presumably, interior) can be explained by the high rate of volcanic activity and resurfacing throughout much of its history. The silicate-sulfur volcanism, possibly involving magnesium and iron silicates, indicates a rocky lithosphere at least 30 km thick.

The uniform, global distribution of volcanic centers suggests the presence of a convective, partially molten asthenosphere beneath the solid crust, with magma rising and sinking in cells several hundred kilometers apart. Models of the interior suggest a silicate mantle overlying a core of molten iron, or a mixture of iron and iron sulfide, which fills 35 to 60% of Io's radius.

Europa

Beyond Io is Europa, an icy world which is lacking in volcanoes, mountains, or craters, but instead is covered by enormous, linear fractures – rather like a cracked eggshell. Instead of closely packed impact craters, Europa displays a young, smooth surface – evidence that resurfacing is still taking place.

Its most striking characteristic is a global network of narrow, crisscrossing, linear features. Many of these are straight, but some are curved or irregular. Ranging from a few hundred kilometers to 3,000 km in length, and varying from several kilometers to 70 km in width, they appear to be evidence of tectonic processes.[9]

Although the dark "streaks" are the most obvious features, closer examination shows a comparable web of smaller, bright

Figure 8.34 Long, linear cracks and ridges crisscross the surface of Europa, interrupted by regions of disrupted terrain where the ice crust has broken up and re-frozen into new patterns. The image approximates how Europa would appear to the human eye. Areas that appear blue or white contain relatively pure water ice. Reddish and brownish areas include non-ice components in higher concentrations. North is at right. (Courtesy of NASA/JPL-Caltech/SETI Institute)

lines. No more than about 10 km across, these curved features can be seen when the Sun is low above Europa's horizon. Largely invisible when the Sun is overhead, their vertical height is estimated to be only a few hundred meters. However, the pattern of curves tends to repeat regularly every 100–300 km.

---

[8] The other Galilean moons are also affected by tidal heating, but to a much lesser extent, since they are further from Jupiter.

[9] The markings only appear dark in comparison with the brightness of their surroundings: the difference in albedo is only about 10%.

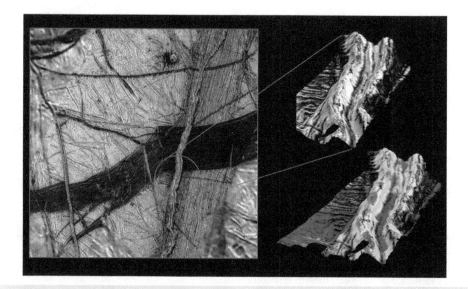

Figure 8.35   Galileo images of a double ridge on Europa that cuts across older plains and a darker, wedge-shaped band. (North is to the right.) The parallel ridges are separated by a valley about 1.5 km wide. A perspective view (upper right) shows that bright material, probably pure water ice, lies on the ridge crests and slopes, while most of the dark material (perhaps ice mixed with silicates or hydrated salts) is confined to lower areas. The northernmost slope is darker than the south-facing slopes. A color-coded relief model (lower right) shows that the ridges (red) rise more than 300 m above the surrounding furrowed plains (blue and purple). The crust seems to have pulled apart, allowing dark material to well up from below. (DLR, NASA-JPL)

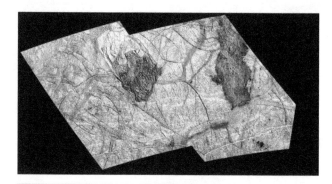

Figure 8.36   An enhanced color mosaic of Thera and Thrace Macula, two dark, reddish "chaos" regions found on the ridged plains of the southern hemisphere. Thera (left) measures about 70 x 85 km and appears to lie slightly below the level of the icy plains. Curved fractures along its boundaries suggest that collapse may have been involved in its formation. Thrace (right) is hummocky and appears to lie slightly above the plains. On its southern edge, material from Thrace overlies a gray band known as Libya Linea. The image covers an area of 525 x 300 km. (Courtesy of NASA-JPL, PIRL/University of Arizona)

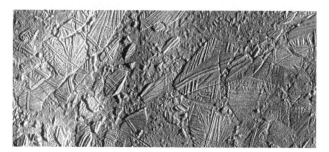

Figure 8.37   An enhanced color view of thin, disrupted, ice crust in the Conamara region (9°N 274°W) that resembles Arctic ice floes on Earth. White and blue areas are blanketed by fine ice particles from the Pwyll impact crater about 1,000 km to the south. A few secondary craters, less than 500 m across, can be seen. The reddish-brown color is due to mineral contaminants released from below the crust when it was disrupted. The image covers an area of 70 x 30 km. (Courtesy of NASA-JPL, PIRL/University of Arizona)

Most of the satellite is classified as smooth or ridged plains. Smooth plains appear to be among the youngest surfaces on Europa. One possibility is that they have been created by a form of cryovolcanism (ice volcanism), in which fluid erupted from below floods a low-lying area and freezes.

Europa's ridged plains are notably older and lack the bland, ice rink appearance of the smooth plains. They generally contain multiple, overlying ridges that have formed at different times. The linear ridges range in complexity from simple parallel pairs to intricate, braided triple bands where two ridges, about 10 km wide, lie either side of a steep-sided, narrow valley. Color images show that dark material – perhaps in the form of mineral-rich liquid water or warm ice – appears to have welled up, filling the gaps created by fracturing along these ridges.

Figure 8.38 An enhanced color image showing the region around Pwyll, one of the few impact craters on Europa. The bright rays – made of fine water ice ejecta – indicate that the 26 km-diameter crater is geologically young. Its central peak, not visible here, stands about 600 m above the floor and is much higher than its rim. The rays extend for over 1,000 km in all directions. Also visible are dark linear features known as "triple bands." Pwyll is on the trailing hemisphere, at 11°S 276°W. The area shown is about 1,240 km across. (Courtesy of NASA-JPL, PIRL/University of Arizona)

More difficult to explain are the numerous dark patches, surrounded by older ridged and fractured plains (Figure 8.36). Typically 50–100 km across, they are named "lenticulae," Latin for "freckles." Although many of the features are domes, others are small depressions. Some are smooth, others are rough.

Comparisons of their irregular shapes and the alignment of nearby ridges and fractures indicate that these regions are not controlled by pre-existing cracks. One model for their formation is complete melt-through from an ocean below. Another possibility is that warm ice welled up from below, resulting in partial melting and disruption of the surface.

Close-up images of chaos regions such as Conamara show that they are made up of a disrupted matrix which resembles a mixture of frozen slush interspersed with rafts of pre-existing crust (Figure 8.37). Some blocks of ice appear to be shifted, rotated, tipped or partially submerged. This chaotic pattern is thought to be the result of blobs of warm ice moving up through colder, near-surface ice. Rough terrain was created if the surface was broken up and partially melted. Smooth lenticulae were formed if meltwater spread onto the surface and then quickly froze.

There has been considerable debate about the composition of the dark materials. Both sulfuric acid and compounds such as sulfates and carbonates – presumably originating in a subsurface ocean – have been suggested.

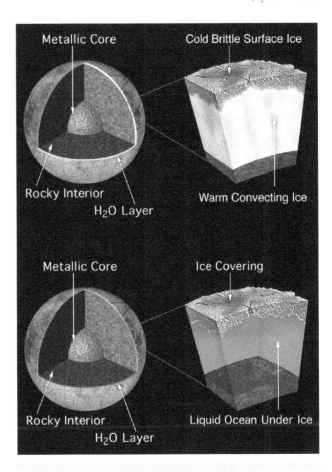

Figure 8.39 Two different models of Europa's internal structure. (Top) A warm, "slushy" convecting ice layer located several kilometers below a cold, brittle ice crust. (Bottom) A layer of liquid water, possibly more than 100 km deep, beneath the icy crust. Most of Europa is made up of a rocky mantle with an iron or iron-iron sulfide core. (Courtesy of NASA-JPL, SETI Institute)

This theory was supported by Galileo's discovery of marked fluctuations in the magnetic field around Europa. This strongly implies that a magnetic field is being created – or induced – within the icy moon by a deep layer of electrically conductive fluid beneath the surface. The most likely explanation for the induced magnetic field is a global ocean of salty water.

The thickness of Europa's crust remains uncertain, as does the presence of liquid water. Estimates for the thickness of the crust range from 3 to 40 km. However, images of the few sizeable impact craters, such as Pwyll, show central peaks and sharp topography that can only be explained by a fairly thick, strong crust. The impacts that formed them clearly did not break through to the moon's postulated ocean.

One scenario favors an ocean up to 100 km deep – 10 times deeper than any ocean on Earth. If so, it would contain twice as much water as Earth's oceans. However, whereas Earth's oceans contain sodium chloride brought down by rivers and concentrated by evaporation, Europa's ocean would contain salts such as magnesium sulfate which have been derived from warm water circulating

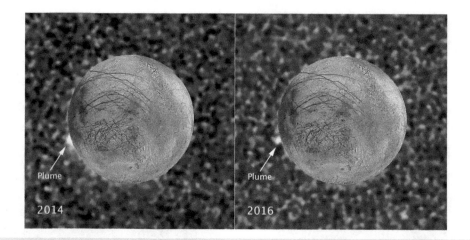

Figure 8.40 These composite images show a suspected plume of material from the same location on Europa in 2014 and 2016. These images, taken in ultraviolet light by the Hubble Space Telescope, support the existence of intermittent plume eruptions. The plumes were seen in silhouette as the moon passed in front of Jupiter. The 2016 plume seems to rise about 83 km above Europa's surface, while the other plume is estimated to be about 48 km high. The plumes correspond to the location of an unusually warm spot on the moon's icy crust. They provide circumstantial evidence for water venting from an ocean beneath the frozen crust. (Courtesy of NASA, ESA, W. Sparks [STScI], and the USGS Astrogeology Science Center)

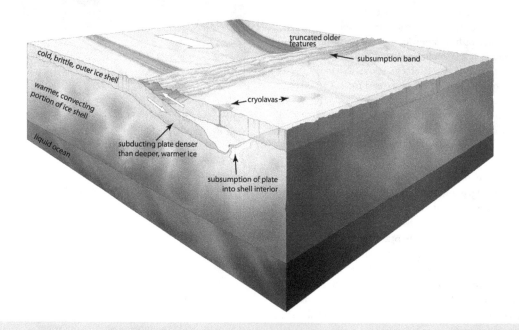

Figure 8.41 Evidence of icy plate tectonics has been found on Europa. This would involve subduction (where one plate is forced under another) when a cold, brittle, outer slab of Europa's 20–30 km-thick ice shell sinks into a warmer ice shell and is ultimately subsumed. A low-relief subsumption band is created at the surface in the overriding plate, alongside which cryolavas may erupt. (Courtesy of NASA/Noah Kroese, I.NK)

through its rocky mantle. Alternatively, Europa's outer layer may simply consist of warm, convecting ice.

Observations of Europa by the Hubble Space Telescope have revealed further evidence for a subsurface ocean. Images taken two years apart show probable plumes of material erupting from the moon's surface at precisely the same location.

The images, taken on March 17, 2014, and February 22, 2016, were obtained when the moon was silhouetted against Jupiter. They show a dark feature protruding slightly above Europa's equatorial region. The 2016 plume seems to rise about 83 km above Europa's surface, while the 2014 plume is estimated to be about 48 km high.

Box 8.4   Voyagers at Jupiter

The Voyager missions to the outer planets were made possible by a planetary alignment that occurs once every 176 years. On such rare occasions, the four gas giants are suitably placed in their orbits for spacecraft to use gravity assists from Jupiter and Saturn in order to reach Uranus and Neptune.

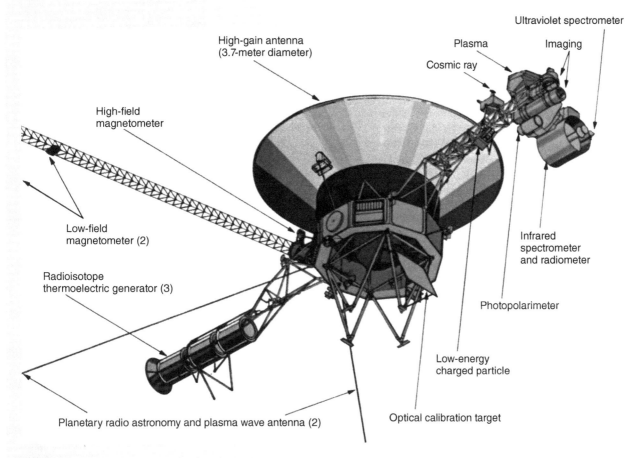

Figure 8.42   The Voyagers carried instruments to investigate atmospheres, satellites, and magnetospheres. The TV cameras were located on a movable scan platform alongside other experiments that required tracking. Color images were obtained by combining pictures taken through blue, orange, and green filters. The Voyagers were powered by three radioisotope thermoelectric generators. (NASA-JPL)

Voyager 2 was the first to leave Earth, blasting off from Cape Canaveral on August 20, 1977, followed 16 days later by the identical Voyager 1. Leaving Earth faster than any previous spacecraft, it crossed the Moon's orbit in less than 10 hours and overtook Voyager 2 soon after entering the main asteroid belt.

During the cruise to Jupiter, the scan platform on Voyager 1, which carried the cameras and other tracking instruments, temporarily stalled. More serious was the failure of Voyager 2's primary radio receiver, followed by a malfunction in the back-up receiver. Concerned about a total breakdown in communications, the mission team uploaded a command sequence that would ensure a minimum science return at Jupiter.

Voyager 1 flew past Jupiter on March 5, 1979, followed by its sister craft on July 9. The dual flybys made it possible to arrange complementary observations of the Jovian system. Voyager 1 was able to study 5 of Jupiter's 13 known satellites, obtaining remarkably detailed images of Io, Ganymede, Callisto, and Amalthea after it passed by the planet. Voyager 2 encountered the satellites on the inward leg, enabling it to image the anti-Jupiter hemispheres of Callisto, Ganymede, and Europa, with a long-range reconnaissance of Io.

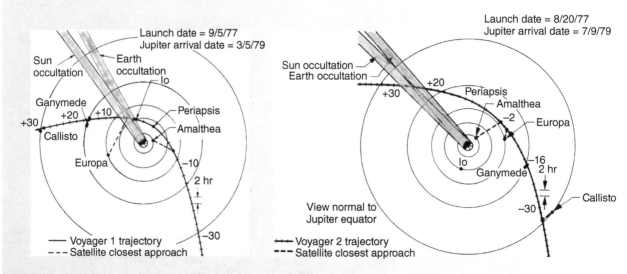

Figure 8.43 The Voyager flybys of Jupiter took place four months apart. In both cases, they flew very close to the planet's equatorial plane. Voyager 1 imaged the major satellites on its outward leg, while Voyager 2 observed the opposite hemispheres on the inward leg. Voyager 1 was able to study Io at close quarters, while Voyager 2 obtained the most detailed images of Europa. (NASA-JPL)

Each spacecraft took more than 15,000 pictures of Jupiter and its moons, which were used to create movies of the planet's cloud motions and to map the surfaces of the major satellites. The glows of auroras and lightning were detected for the first time, as was a dark ring system. Three small satellites were also discovered close to Jupiter. Perhaps the most exciting finds were the sulfur-laden volcanic eruptions and flux tube of Io, together with the "eggshell" surface of Europa.

The plumes correspond to the location of a thermal anomaly – an unusually warm spot – on the moon's icy crust, detected in the late 1990s by NASA's Galileo spacecraft. This region, roughly 320 km across, was found to be a few degrees warmer than the surrounding terrain. This is considered to be circumstantial evidence for a water geyser erupting from the moon's subsurface.

Two possible scenarios have been proposed. The warmer area could be caused by the heat from liquid water, located more than 1.5 km beneath Europa's crust. The water is pushed upward and cracks the surface, venting as a plume.

Another idea is that water ejected by the plume falls onto the surface as a fine mist. This process could change the structure of the surface grains, allowing them to retain heat longer than the surrounding landscape.

If Europa has a mineral-rich ocean, there is the tantalizing possibility that it may be able to support some form of life. Protected from harmful radiation by the overlying ice, it is possible to envisage ecosystems flourishing alongside hydrothermal vents, where the water is warm, there is an abundance of chemical compounds, and chemical reactions provide the energy required for each organism to function.

A number of studies have suggested that a form of plate tectonics may exist on Europa which is comparable to the processes that take place on Earth (Figure 8.41). Many parts of Europa's surface show evidence of extension, where the surface has pulled apart and fresh icy material from the underlying shell moved into the newly

created gap – a process similar to sea floor spreading on Earth. This would explain why Europa's surface is so young.

However, the moon is not expanding, because this crustal extension seems to be offset by subduction, where two cold, brittle "plates" converge, with icy crust moving down beneath a neighboring slab. Ice volcanoes occur on the overriding plate, possibly formed through melting and absorption of the slab as it dived below the surface, and a lack of mountains at the subduction zone implies that material was pushed into the interior rather than crumpled as the two plates collided.

This subduction could be eased if the outermost ice shell incorporates salts, which are denser than ice. Even if a slab warmed up as it sank, reducing its density to some degree, the added density of the salt would allow it to continue sinking. The slab would be incorporated into the ice shell, but it seems unlikely that the downward moving slab would reach as far as the postulated ocean.

Europa has an extremely thin atmosphere which is created by "sputtering," when fast-moving particles in Jupiter's magnetosphere hit the moon's surface and dislodge water molecules.[10] There may also be some sublimation of ice due to solar heating. However, its low gravity means that Europa cannot hold onto an atmosphere for very long. Water molecules lost from the surface are quickly separated into neutral oxygen and hydrogen. The molecules can also be quickly ionized by ultraviolet radiation and charged particles in the vicinity. As

---

[10] **Europa's surface pressure is barely one hundred billionth that of the Earth.**

a result, Europa has a neutral atmosphere as well as an ionosphere. The relatively lightweight hydrogen gas escapes into space, while the heavier oxygen molecules accumulate to form an atmosphere which may extend 200 km above the surface. The oxygen slowly leaks into space and must be replenished continuously.

This "leakage" means that, like Io, Europa is the source of a huge torus, a donut-shaped cloud of hydrogen and oxygen atoms that stretches for millions of kilometers around Jupiter.

Ganymede

With a diameter of 5,262 km, the third Galilean satellite, Ganymede, is the largest moon in our Solar System. The Voyager spacecraft revealed a satellite which had undergone considerable modification and resurfacing during its 4.5 billion-year history. Although there was a lot of tectonic activity during the first billion years of Ganymede's history, crater counts show that relatively little has changed since then.

About one-third of its surface is covered by dark, heavily cratered terrain. The most notable example is a roughly circular region named Galileo Regio, which measures about 3,200 km across. Located in the northern hemisphere on the hemisphere facing away from Jupiter, the region is pockmarked by lighter

impact features and partially blanketed by a thin layer of frost. Its circular outline suggests it was created by an enormous impact early in the satellite's history. Superimposed on it is a series of slightly curved, concentric markings – again presumably the result of an ancient impact, although the center of the former basin has since been erased by resurfacing.

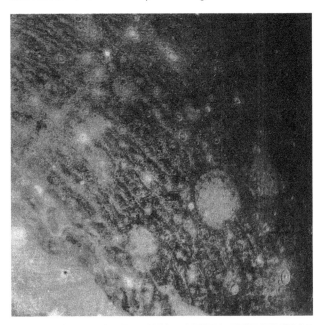

Figure 8.45 A Voyager 2 view of the western edge of Galileo Regio. The dark, fairly ancient terrain contains numerous small impact craters, many with central peaks. The larger, bright circular features – known as "palimpsests" – show little relief and are the remnants of old craters which have been filled by the inflow of icy material. Slightly curved lineaments in the dark region (center) suggest the presence of a large impact basin to the south west which was later hidden by the formation of the grooved terrain of Uruk Sulcus (bottom left). (NASA-JPL)

Figure 8.44 A near-global view of Ganymede taken by the Galileo orbiter. The large, circular, dark region is Galileo Regio, centered at 36°N, 138°W. On its western boundary is a curved region of parallel ridges and troughs known as Uruk Sulcus. The dark areas are older and more heavily cratered, whereas the light areas are younger, tectonically deformed regions. The brownish-gray color is due to mixtures of rocky materials and ice. Bright spots are geologically recent impact craters and their ejecta. (NASA-JPL)

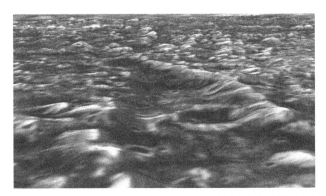

Figure 8.46 Topographic detail of Galileo Regio is seen in this 3D view. Note the deep furrows and impact craters. Bright walls show where cleaner ice may be exposed, and deposits of dark material fill the lower terrain. (NASA-JPL)

Like other regions of ancient, icy crust, Galileo Regio displays a variety of impact craters and furrows. More than 4,000 craters larger than 10 km in diameter have been identified. The oldest impact features take the form of lighter colored palimpsests or "ghosts" which show no present-day topography.[11] This may be because they formed when the crust was warmer and malleable.

In contrast, the younger, usually smaller, craters display more typical impact features, such as deep basins and uplifted rims. The largest preserved impact basin, named Gilgamesh, is about 800 km across. It was created after the formation of nearby grooved terrain, suggesting that the crust eventually became rigid enough to preserve the impact structures. Other evidence for relatively recent impacts comes in the form of a number of bright ray craters.

Much of Ganymede shows evidence of a complex geological history, with a mottled pattern of light and dark regions. Some of the bright regions have a lower density of impact craters, indicating a more recent origin; others appear (based on the crater count) to be only a little older than the dark terrain.

Bright, grooved regions, such as Uruk Sulcus, which marks the western boundary of Galileo Regio, cover more than half of Ganymede. These regions of linear ridges and troughs have a vertical relief of a few hundred meters and extend for thousands of kilometers, but most of them seem to have a different origin from the grooved terrain on Europa. Instead of new material welling up from below, they seem to have been created by crustal tension in the dark terrain.

Once the icy surface was fractured and pulled apart, it was broken into many parallel ridges and valleys. This tectonic process was associated with parallel faulting, which resulted either in the formation of graben separated by raised blocks of crust, or by tilted blocks that resemble toppled dominoes.

In some cases, the surface is seriously disrupted by multiple fractures and grooves running in different directions – presumably created at different times. This creates a broken landscape of separate, small hills known as "reticulate terrain."

The brightness of the grooved terrain may be accounted for by cryovolcanism which blanketed the surface with fresh ice. This theory is supported by a number of impacts which have punched through the bright surface layer to dredge up dark material from beneath. Also present are peculiar, caldera-like depressions which seem to have been formed by volcanic flows of liquid water and slush (Figure 8.47).

Scientists have suspected that Ganymede has a massive ocean under a thick, icy crust since the 1970s. Supporting evidence comes in the form of hydrated salt minerals similar to those seen on Europa, possibly the result of brine making its way to the surface by eruptions of through cracks.

Bright, broken ridges and troughs, disrupted dark plains, and features such as Arbela Sulcus, which appear to have formed by complete separation of Ganymede's icy crust, also suggest that Ganymede may once have experienced some form of tectonics and cryovolcanism.

A third piece of evidence is the presence of a magnetic field that would require the presence of something more electrically conductive than ice. The scenario that best fits the data is a melted

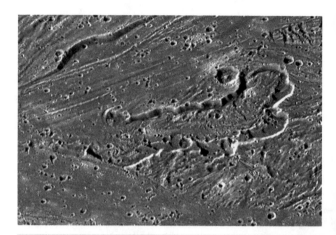

Figure 8.47 An unusual, caldera-like depression in the Sippar Sulcus region, possibly the result of water ice volcanism. This depression, marked by scalloped walls and internal terraces, is about 55 km long and 17–20 km wide. On the floor of the inner depression is a lobate, flow-like deposit 7–10 km wide, with ridges that are curved outward (and apparently downslope) toward a junction of smooth terrain and a grooved band. The morphology suggests that volcanic eruptions created a channel and eroded the surface. The mosaic, centered at 31°S 189°E, covers an area of 91 x 62 km. (Courtesy of Brown University, NASA-JPL)

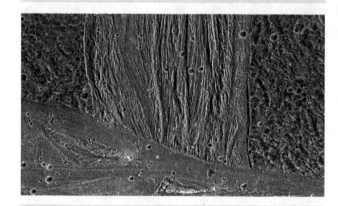

Figure 8.48 Not all bright terrain on Ganymede shows parallel ridges and troughs. In this Galileo view, an 85 km-wide grooved feature, known as Erech Sulcus, runs north–south between two blocks of the older, dark terrain of Marius Regio. The numerous grooves probably formed when tectonic forces pulled apart the icy surface. In the south it meets a much smoother region known as Sippar Sulcus, which may have been flooded by ice volcanism. (Courtesy of/NASA-JPL)

layer of water as salty as Earth's oceans and several kilometers thick, starting about 200 km down.

Some models show an ocean sandwiched between a top and bottom layer of ice. Another model published in 2014 and based on laboratory experiments that simulate salty seas, shows that the

[11]The term "palimpsest" comes from reused parchment on which previous writing was poorly erased.

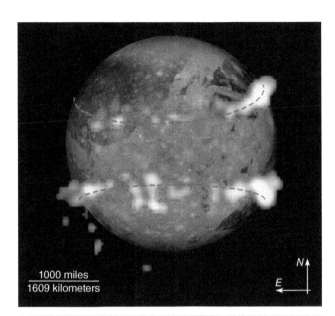

body
Figure 8.49 Hubble Space Telescope images of Ganymede's auroral belts (shown in blue) are overlaid on a Galileo orbiter image of the moon. The amount of rocking of the moon's magnetic field suggests that it has a subsurface saltwater ocean. The images were made in ultraviolet light. (Courtesy of NASA/ESA and J. Saur, University of Cologne)

1000 miles
1609 kilometers

N
E

The magnetic field causes auroras – ribbons of glowing, hot electrified gas, in regions circling the north and south poles of the moon (see Chapter 3). Because Ganymede is quite close to Jupiter, it is also embedded in the planet's magnetosphere. When Jupiter's magnetic field changes, the auroras on Ganymede also change, "rocking" back and forth.

If a saltwater ocean were present, Jupiter's magnetic field would create a secondary magnetic field in the ocean that would counter Jupiter's field. This "magnetic friction" would suppress the rocking of the auroras. This ocean counters Jupiter's magnetic field so effectively that it reduces the rocking of the auroras to 2°, instead of 6° that would be expected if the ocean was not present.

These observations, combined with Ganymede's fairly low mean density, suggest that the moon has a crust of water ice lying on top of a salty ocean, and a thick rocky (silicate) mantle, with a core of iron and iron sulfide. However, this generalized model of uniform layers hides a more complex reality, as exemplified by Galileo tracking data which revealed the presence of irregular "lumps" beneath the surface. These masses may be rock formations which have been supported by Ganymede's icy shell for billions of years.

Like Europa, Ganymede has an extremely sparse exosphere of atomic hydrogen and molecular oxygen which is formed when charged particles trapped in Jupiter's magnetic field rain down onto the icy surface. As the particles penetrate the surface, molecules of water are disrupted, leading to production of ozone (O_3). This gas is then broken down to form oxygen.

Callisto

Of the four Galilean moons, only the outermost, Callisto, lived up to expectations by revealing a heavily cratered, ancient surface. Callisto is the only body with a diameter greater than 1,000 km that shows no signs of undergoing any extensive resurfacing since the end of the period of heavy bombardment, about 4 billion years ago.

Its surface seems saturated with 100 km-diameter craters, making it the most heavily cratered object in the Solar System (Figure 8.50). However, only a few huge impact structures are visible, suggesting that the remainder were erased by the subsequent bombardment of smaller impactors.

The largest of these structures is Valhalla, a bright, circular feature about 600 km across (Figure 8.51). The flat central region is surrounded by numerous concentric rings which extend about 1,300 km from the center. The absence of high ridges, ring mountains, or a large central depression suggests that the impact resulted in melting of the surface and the outward spread of shock waves. Refreezing took place in time to preserve the concentric shock rings.

Some minor resurfacing of Callisto's surface is taking place, since there are many fewer small impact craters than expected. Instead of deep, saucer-shaped basins, the craters are very shallow, suggesting that they have been gradually filled and partially obliterated over billions of years (Figure 8.52).

The entire satellite seems to be coated with a blanket of fine, dark material of uncertain origin – so Callisto is the least reflective

ocean and ice may be stacked up in multiple layers, more like a club sandwich.

These layers may arise because ice exists in different forms depending on pressures. "Ice I," the least dense form of ice, is the type that floats in chilled beverages. As pressures increase, ice molecules become more tightly packed and increase in density. Since Ganymede's oceans may occur up to 800 km deep, they would experience much more pressure than Earth's oceans. The deepest and most dense form of ice thought to exist on Ganymede is called "Ice VI." It has been suggested that, in total, Ganymede's oceans may have 25 times the volume of those on Earth.

Ganymede is the only moon to generate its own magnetic field. It was something of a surprise when Galileo's instruments showed that the moon creates a magnetic field about 1% the strength of Earth's. (At Ganymede's orbital distance, the moon's intrinsic field is actually stronger than the field of Jupiter.)

Although the satellite probably has a molten, iron-rich core, this might be too small or solid to generate the required dynamo effect. Natural radioactivity, possibly combined with modest tidal heating, might provide enough warmth to maintain this molten core.[12]

In addition to the intrinsic magnetic field, Ganymede has an induced dipole magnetic field that is similar to those of Europa and Callisto. This induced field is created in a subsurface, briny ocean with a high electrical conductivity as the satellite moves through the variable plasma environment and Jovian magnetic field.

[12] At the present time, Jupiter and the other Galilean moons cause only a small amount of tidal heating on Ganymede. However, tidal effects were probably much more pronounced when Ganymede's orbit was more elliptical, perhaps 3 billion years ago.

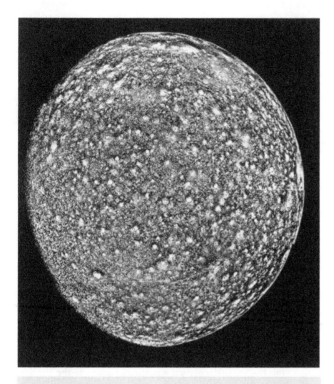

Figure 8.50   This Voyager 2 image of Callisto shows a uniform distribution of medium-sized impact craters on the hemisphere facing away from Jupiter. Typically about 100 km across, many of these craters have bright rims, evidence of exposed water ice. Some relatively fresh ray craters are visible, as well as several larger impact structures. One of these, Asgard, is visible near the limb (top) and has about 15 concentric rings around a bright central spot. The smooth limb confirms the absence of high topography. (NASA-JPL)

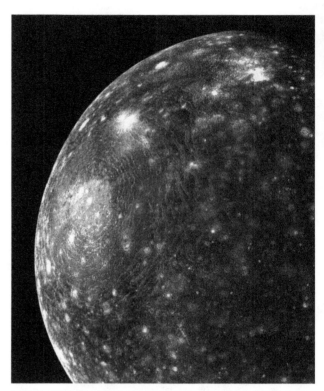

Figure 8.51   The largest impact feature on Callisto is the multi-ringed Valhalla (left). Unlike craters on rocky planets, Valhalla is quite flat, with no encircling mountains or central peak. The bright central plains were possibly created by the excavation and ejection of "cleaner" ice from beneath the surface, with fluid impact melt filling the crater. The bright central region is around 600 km across, while the surrounding rings extend 1,500 km from the center. Valhalla is on the hemisphere facing Jupiter. (NASA-JPL)

of the Galilean moons.[13] On steep slopes, this dark coating slides downhill, leaving bright crater rims where underlying water ice is exposed. Intercrater plains typically appear quite dark and smooth.

Callisto is the third-largest satellite in the Solar System. Although it is only a little smaller than Ganymede, the two moons display major differences. Callisto's bulk density ($1.86\,g/cm^3$) is the lowest of the Galileans, indicative of a world which is composed of a mixture of about 40% ice and 60% rock and iron.

There is very little tidal heating of Callisto at the present time and it seems that the satellite has never undergone the strong heating experienced by its neighbors. Tracking data of the Galileo orbiter suggest that little, if any, internal differentiation has taken place (Box 8.3). Models indicate the absence of a rock-metal core, with most of the interior filled with a mixture of ice and rock. The surface crust may be composed of ice and rock less than 350 km thick.

In addition to the expected water ice and hydrated minerals, four unusual spectral features have been detected. One of these appears to show carbon dioxide trapped in the surface. Two others

may represent sulfur derived from Io's volcanic eruptions. The fourth corresponds to carbon-nitrogen compounds – possibly complex organic molecules known as tholins.

Callisto may also possess a subsurface ocean. As was the case with Europa and Ganymede, the first evidence came from measurement of magnetic fluctuations around the satellite. However, subsequent studies showed that, unlike Ganymede, but like Europa, Callisto is not generating its own magnetic field. Instead, an electrically conducting layer beneath the surface makes the moon act like an enormous electromagnet. A magnetic field is induced by its interaction with Jupiter's magnetosphere.

How can Callisto be hot enough to support an ocean without melting its interior and producing separate layers of ice and rock? Some feeble heating of the interior may still occur through radioactive decay of elements. The very cold, rigid crust also acts as a lid, keeping in the heat. This may be sufficient to melt the ice at a depth of at least 150 km below the surface, creating a subterranean ocean. An ocean at least 10 km thick would have the required conductance if the water was comparable in salinity to Earth's oceans.

---

[13] **Callisto's surface is the darkest of the Galileans, but it is still twice as reflective as the Moon.**

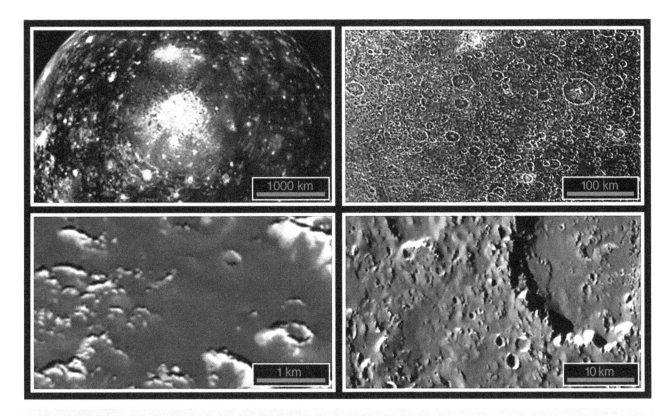

Figure 8.52 Four Galileo views show how increasing resolution modifies interpretation of Callisto's surface. In the broad view (top left) there are many small, bright spots around the Valhalla impact structure. The regional view (top right) shows the spots are large craters with bright rims. The local view (bottom right) shows some smaller craters and bright material on crater rims amid a smooth, dark layer that covers much of the surface. In the close-up frame (bottom left), where spatial resolution is 30 m, the surface is very smooth between the craters. (DLR/NASA-JPL)

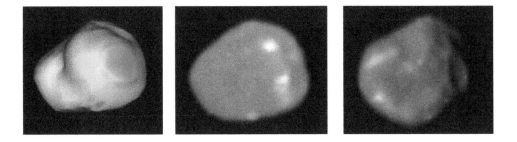

Figure 8.53 The left image shows escape velocities color-coded on a shape model of Amalthea. Blue areas have the lowest escape velocity, barely 1 m/s near the anti-Jupiter end; red shows the tiny regions of much higher escape velocity, nearly 90 m/s. The low escape velocities are due to its low density and rapid rotation. The middle image also shows Amalthea from the anti-Jupiter side. The visible area is about 150 km across. Since the Sun is overhead, there are no shadows, but brighter markings on the ends of a ridge are prominent. The image at right shows bright ridges on the satellite's leading side and a 40 km-wide crater with two nearby ridges. (Courtesy of Cornell University, NASA-JPL)

Callisto's sparse exosphere seems to be mostly carbon dioxide, presumably derived from dry ice on the surface. A few other constituents, notably sulfur dioxide, have been tentatively identified. Hydrogen atoms escaping from Callisto are derived from water ice on its surface. Solar ultraviolet radiation striking the rock-hard ice is probably the primary mechanism for separating the hydrogen and oxygen atoms. Sputtering is less likely, since Callisto orbits outside the radiation belts and does not interact as strongly as Ganymede with the charged particles in Jupiter's magnetosphere.

Box 8.5   Birth of the Jovian Satellites

It is generally accepted that the Galilean satellites were created in orbit around Jupiter. Most models assume that they formed within a gaseous disk, in which temperatures were highest near the young, rapidly growing planet, and lowest on the outskirts. This interpretation is supported by the bulk densities of the moons, which decrease with distance from the planet.

This trend indicates an increasing proportion of low-density ice and a decreasing proportion of rock and iron, so that the outer moons, Ganymede and Callisto, contain up to 50% ice by mass. The combined mass of the Galilean satellites indicates that the nebula was equal to about 2% of Jupiter's mass.

This view has been challenged by researchers who argue that temperatures in such a massive, gas-rich disk would be too high to retain ices, even in its outer reaches. In addition, the gravitational interaction of the satellites with the disk material would have caused their orbits to decay rapidly inward, probably leading to their destruction.

The alternative model does not require that all of the mass needed to form the satellites was present in the disk from an early stage. Instead, it suggests that the disk grew slowly over a long period of time. As the inflow of gas and dust from solar orbit fed the growth of Jupiter, it also supplied material for the disk. In this "gas-starved" disk, temperatures remained low enough for ice to exist in the region of Ganymede and Callisto. According to this model, the satellites formed fairly slowly, over 100,000 to 1 million years.

The model accounts for the moons' bulk compositions, the unusual internal structure of Callisto, and the ability of the satellites to avoid rapid inward migration of their orbits due to interaction with the disk.

It also predicts that the satellites would migrate inward during their formation, each at a rate proportional to its mass. This inward migration could lead to the establishment of an orbital resonance, so that Io completes four orbits to every two of Europa, and every one of Ganymede. The existence of the resonance leads indirectly to internal heating of the satellites, causing Io's extensive volcanism, for example.

However, the discovery that Jupiter's fifth-largest moon, Amalthea, which lies inside the orbit of Io, has a low density and probably contains a lot of water ice, is not explained by any of the disk models. Temperatures would have been too high at Amalthea's current position for an icy moon to form. This suggests that Amalthea was born in a colder environment. One possibility is that it formed later than the major moons. Another is that it formed beyond the orbit of Europa or in a cold region of the solar nebula. It would then have migrated or been captured in its current orbit around Jupiter.

The multitude of tiny satellites that lie close to Jupiter or follow eccentric orbits far from the planet are likely to be asteroids that were captured by the giant planet's gravity.

Inner Satellites

For almost three centuries, the only known Jovian satellites were the four Galileans. Then, in 1892, American astronomer Edward Barnard (1857–1923) discovered Amalthea. It was the last planetary moon to be found without the aid of photography.

Amalthea orbits only 181,300 km (2.5 Jupiter radii) from the center of the planet (Figure 8.53). At this distance, within Jupiter's rings, the satellite is bombarded by intense radiation and micrometeorites. It has an irregular shape, pockmarked with numerous impact craters. Two of these are quite large. Pan, located in the northern hemisphere, is about 90 km in diameter and at least 8 km deep. Gaea, which straddles the south pole, is about 70 km across but probably twice as deep. Overall, the potato-shaped moon measures 270 x 165 x 150 km, and, like most satellites of the giant planets, it is in a synchronous rotation, with its long axis always pointing towards Jupiter.

One of Amalthea's most surprising characteristics is its dark red color – in fact, the satellite is the reddest object in the Solar System.

This is thought to be the result of its surface being sprayed by sulfur and sodium derived from Io's volcanoes.

Precise tracking of the Galileo orbiter indicated that Amalthea's overall density is well below that of water ice. Instead of being a solid chunk of rock and ice, the satellite seems to be a loosely packed mixture of rubble in which the voids take up more of its volume than the solid material. It seems that Amalthea was once shattered by a collision and the pieces are now held together through their mutual gravitational attraction.

The density is so low that even its solid rubble is apparently less dense than Io. This contradicts models suggesting that moons forming close to Jupiter would be made of denser material than those farther out (see Box 8.5).

Three more moons lie close to the planet. The orbits of Adrastea and Metis are less than 1,000 km apart, and within 130,000 km of the center of Jupiter. Both lie deep within the Jovian rings. **Thebe,** the largest of the trio with a maximum diameter of 110 km, orbits at the outer edge of the rings (see Jupiter's Rings). All three are irregular in shape and images show Thebe has a number of sizeable impact craters (Figure 8.54).

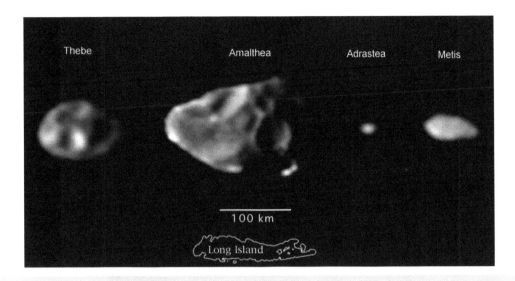

Figure 8.54 Jupiter's inner satellites compared in size to Long Island, which measures 190 km E–W. These small satellites have very low surface gravities: a person weighing 68 kg on Earth would weigh about 0.2 kg on Amalthea, and approximately 0.01 kg on Adrastea. Impact craters 35–90 km across are visible on the larger satellites. (NASA-JPL)

Box 8.6   The Galileo Mission

NASA's Galileo mission, named in honor of Galileo Galilei, comprised two spacecraft: an orbiter and an atmospheric probe. The orbiter was designed to study Jupiter's atmosphere, satellites, and surrounding magnetosphere, while the probe would make the first in situ measurements of a giant planet's atmosphere.

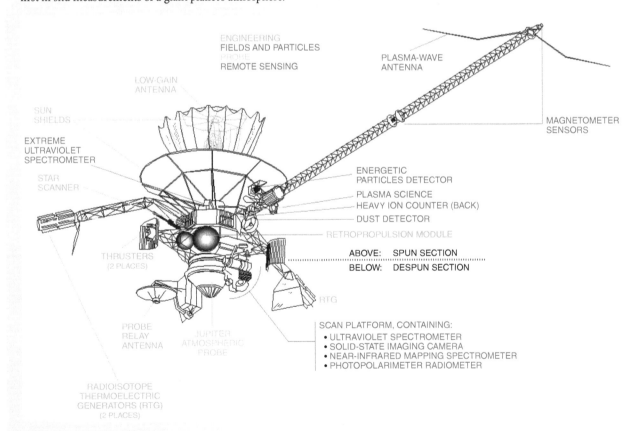

Figure 8.55 The Galileo spacecraft weighed 2,223 kg at launch and measured 5.3 m from the top of the high gain antenna to the bottom of the probe. It was the first dual-spin planetary spacecraft. The upper, spinning section rotated at about 3 rpm and contained the power supply, propulsion module and most of the computers and control electronics, as well as instruments to study magnetic fields and particles. The despun section was counter-rotated to provide a fixed, stable orientation for cameras and other remote sensors. Power was provided by two radioisotope thermoelectric generators. (NASA-JPL)

The spacecraft was carried into Earth orbit by Shuttle Atlantis on October 18, 1989. After release, a two-stage booster sent it toward Jupiter along a complex trajectory which involved one flyby of Venus (February 10, 1990) and two flybys of Earth (December 8, 1990, and December 8, 1992). It also made the first flybys of main belt asteroids, discovered a satellite around Ida, and observed the impacts of fragments of comet Shoemaker-Levy 9 with Jupiter.

Figure 8.56   The Galileo probe, with its heat shield (below) and parachute (above), entering Jupiter's upper atmosphere at a speed of about 48 km/s. Data from its six instruments were relayed to Earth via the orbiter before it was destroyed by extreme temperature and pressure. (NASA-JPL)

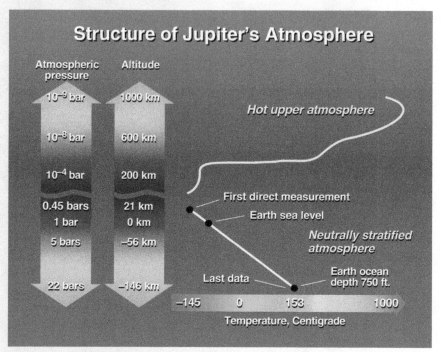

Figure 8.57   The Galileo Probe measured the variation of temperature and atmospheric density with pressure and altitude to a depth of about 20 km. The probe stopped transmitting at a pressure of 22 bars – equivalent to the pressure at an ocean depth of 230 m on Earth – and a temperature of 153°C. (Courtesy of NASA-Ames)

The piggybacking Probe was released from the mother craft on July 12, 1995. The 339-kg Probe slammed into Jupiter's atmosphere on December 7, 1995, at a speed of 170,000 km/h. For the next 58 minutes it sent back data on lighting levels, heat flux, pressure, temperature, winds, lightning, and atmospheric composition. By the time high temperatures silenced its transmitters, it had reached a pressure 22 times the sea level pressure on Earth.

The main craft also arrived on December 7, 1995, becoming the first spacecraft to orbit an outer planet. Over the next 8 years it completed 34 orbits of Jupiter and 35 flybys of the major satellites – 7 of Io, 8 of Callisto, 8 of Ganymede, 11 of Europa, and 1 of Amalthea.

The orbiter carried 11 scientific instruments, 7 of which were located on a section that provided pointing stability by spinning. The four remote sensing instruments, which sent back images and data on composition and temperature, were on a despun section. Its radio link to Earth and the probe-to-orbiter radio link were also used to conduct scientific investigations, such as mapping of gravity fields.

Unfortunately, the amount of data sent back from the mission was severely curtailed when the high-gain antenna failed to deploy. Although far fewer images could be returned than originally anticipated, the mission team estimated that 70% of the original science objectives were accomplished. In particular, Galileo provided major advances in our understanding of Jupiter's magnetosphere, and the nature of its large moons, including evidence for saltwater oceans on Europa, Ganymede, and Callisto.

In December 2000, Galileo participated in a unique dual examination of the Jovian system which coincided with the distant flyby of the Cassini spacecraft en route to Saturn.

Outer Satellites

Jupiter has 79 known satellites – second only to Saturn. 71 of these are small objects that orbit far beyond Callisto.[14] All of these follow eccentric, highly inclined orbits. These orbits are stable over the age of the Solar System, although they are strongly influenced by solar and planetary perturbations.

There are five distinct groups, each dominated by one relatively large body. In general terms, the innermost irregular satellites follow prograde orbits. They are named the Himalia and Themisto groups after the largest object in each. Most of the more remote moons travel in the opposite, retrograde direction. There appear to be at least three groups (named Pasiphae, Ananke, and Carme).

The groups were probably formed through collisional shattering of larger precursor objects after they were captured by Jupiter. At present, it is very difficult for Jupiter to capture satellites permanently because no efficient means of changing an object's orbital momentum and energy exists. However, satellite capture could have occurred more easily in the past, for example due to gas drag from the planet's bloated atmosphere and/or a higher probability of collision with nearby small bodies. This would suggest that most of the outer satellites were captured towards the end of Jupiter's formation, around 4.5 billion years ago.

The sizes of the outer satellites can be estimated by assuming geometric albedos (reflectivity) of 4%. This makes Himalia the largest irregular satellite of Jupiter with a diameter of about 160 km. Each retrograde group contains one or two large objects a few tens of kilometers in diameter, but the remainder are mere fragments, typically 2–4 km across.

One "oddball" satellite that was announced in 2018 has an orbit like no other known Jovian moon. Named Valetudo, it follows a path that is more distant and more inclined than the prograde group of moons and takes about one and a half years to orbit Jupiter. Unlike the closer-in prograde moons, this new prograde moon has an orbit that crosses the outer retrograde moons. As a result, head-on collisions are much more likely to occur between the newcomer and the retrograde moons, which are moving in opposite directions. It is also likely Jupiter's smallest known moon, being less than one kilometer in diameter.

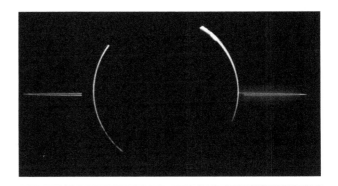

Figure 8.58   A mosaic of Jupiter's ring system taken when the Galileo spacecraft was in Jupiter's shadow and looking toward the Sun. This reveals tiny, dust-sized particles in the ring and upper atmosphere of Jupiter. Only the main ring and a hint of the surrounding halo can be seen in this image. (Courtesy of Cornell University, NASA-JPL)

Jupiter's Rings

Like the other giant planets of the Solar System, Jupiter is circled by a ring system. It was discovered when Voyager 1 crossed the planet's equatorial plane and its narrow angle camera captured light reflected from a ring of tiny particles as it looked back toward the Sun.

Unlike the spectacular disk that surrounds Saturn, the Jovian rings are extremely dark, with an albedo of about 0.05 (5% of the sunlight falling on the rings is reflected). They would probably be invisible from the Galilean moons, except when backlit by the Sun. The rings are made of billions of microscopic specks of dust, similar in size to particles of cigarette smoke. Unlike the rings of Saturn, they contain little or no ice.

Four distinct regions have been identified. The innermost of these is a faint torus of material, about 10,000 km thick, that extends down to the planet's cloud tops. Its particles are very small – perhaps 100 times smaller than the width of a human hair. Particles this size should survive for only a short time, so they must

---

[14]The rapid rate of discovery in recent years, made possible by the introduction of large format charge-coupled devices (CCDs), indicates that many more have yet be found. One estimate suggests that there may be about 100 irregular satellites larger than 1 km in diameter.

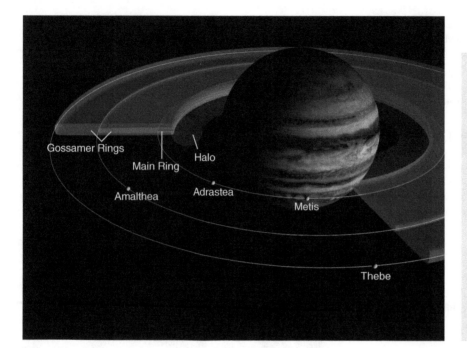

Figure 8.59 The geometry of Jupiter's rings in relation to the four inner satellites, the sources of the ring material. The innermost and thickest ring (gray) is the halo that ends at the main ring. The thin, narrow main ring (red) is bounded by Adrastea and shows a marked decrease in brightness near the orbit of Metis. It is composed of fine particles ejected by meteoroid impacts on Adrastea and Metis. (The orbits of Adrastea and Metis are about 1,000 km apart.) Thebe and Amalthea, which follow inclined orbits, supply dust to the thicker gossamer rings (yellow and green). (Courtesy of Cornell University, NASA-JPL)

be replenished by material derived from other parts of the ring system.

Outside this is the brighter, denser, main ring, which begins at a distance of about 122,500 km and ends at about 129,000 km, very close to the orbit of Adrastea. The brightness of this ring drops markedly at about 127,850 km, near the orbit of another moon, Metis. The outer ring segment, which is very faint, is known as the gossamer ring, and extends all the way out to the orbit of Thebe. It is divided into two distinct sections.

Jupiter's four inner satellites, Metis, Adrastea, Amalthea, and Thebe, affect the structure of these tenuous rings. The main ring is supplied by fine particles knocked off Adrastea and Metis by small meteoroid impacts. Similarly, Amalthea is the source of material for the inner gossamer ring, while Thebe feeds the outer gossamer ring.

Instead of tailing off to nothing, the gossamer ring is unusually thick, with a sharp outer edge. The depth of this faint, outer ring is accounted for by the inclined orbits of Thebe and Amalthea, which spread out the ring debris as they travel above and below Jupiter's equator.

Questions

- Why is Jupiter sometimes described as a failed star?
- Compare and contrast the atmospheric features imaged during the Voyager 2 and Cassini flybys, shown in Figure 8.5.

- Make a list of the similarities and differences between the atmospheric conditions on Earth and Jupiter.
- Why does Jupiter display belts and zones?
- Describe and explain the main features of the red spots and white ovals on Jupiter.
- What are the main characteristics of Jupiter's magnetic field? (b) How does this field differ from those of other planets?
- The Galilean satellites are comparable in size to Mercury. Describe the major differences and similarities between each of these satellites and the innermost planet.
- Briefly explain the scientific rationale for believing that at least three of the Galilean satellites may have subsurface oceans.
- Why is Europa often cited as a high-priority target for future space missions?
- The surfaces of Earth and Europa are both thought to be shaped by plate tectonics. Explain the similarities and differences between both systems.
- Describe the relationship between the Jovian satellites and the ring system.

NINE

Saturn

Saturn is the outermost of the naked eye planets. As such, it was long regarded as the furthest outpost of the Solar System. In many ways, Saturn may be regarded as a smaller version of Jupiter, with a hydrogen-helium composition and parallel bands of cloud driven by high-speed jets. What make the Saturnian system unique are the planet's enormous rings and its many-faceted satellites. Most notable of all is Titan, the only satellite with a dense atmosphere and a world that has been likened to a primitive Earth in deep freeze.

The oldest written records to mention Saturn date back to the Assyrians, around 700 B.C. A few centuries later, the Greeks named it in honor of Kronos, their god of agriculture and the father of Zeus. The Roman equivalent was Saturnus. This name turned out to be very appropriate, since Saturn was later found to be a giant, second only to Jupiter among the Sun's family.

By studying its movement against the "fixed" stars, early astronomers discovered that Saturn took 29½ years to complete one circuit of the Sun. However, every 13 months or so, it would brighten as it neared opposition. From its slow motion through the constellations and modest brightness, it was clear that Saturn was the furthest of the known planets. We now know that Saturn never comes closer to Earth than about 1.2 billion km, or 8 AU. Its average distance from the Sun is about 1.4 billion km (9.5 AU).

With a diameter more than 9 times that of our world, 764 Earths would fit inside Saturn (Figure 9.1). Despite its impressive size, it is only 95 times more massive than Earth. Clearly, Saturn is a largely gaseous object, with an average density 0.7 times that of water. If a sufficiently large ocean could be found, it would easily float. This combination of great size and low density means that the pull of gravity at its visible surface is lower than on Earth.

The planet has the most oblate ("squashed") shape of all the planets, with a marked outward bulge at the equator: the equatorial diameter is almost 12,000 km greater than the polar diameter – a difference of 10%. This is explained by its surprisingly rapid rotation and gaseous composition. By observing occasional long-lived cloud features, it is possible to calculate that Saturn rotates once every 10 hours 14 minutes (Box 9.1). This is a little longer than Jupiter's rotation period, but shorter than the all the other planets.[1]

Saturn's orbit is inclined about 2.5° to the ecliptic and its axis of rotation is inclined almost 27° to the orbital plane. As a result, different hemispheres are tilted towards us during Saturn's journey around the Sun. (Our views of the rings also vary dramatically – see Discovering Saturn's Rings.) As on Earth, the marked axial

[1] The period of rotation was shown to about 20 minutes longer at higher latitudes – the result of different wind speeds at cloud level. Similar variations occur on Jupiter.

Exploring the Solar System, Second Edition. Peter Bond.
© 2020 John Wiley & Sons Ltd. Published 2020 by John Wiley & Sons Ltd.
Companion Website: www.wiley.com/go/Bond-Solar-System2e

Figure 9.1   Saturn is the second-largest planet in the Solar System, with an equatorial diameter of 120,000 km. If the planet and its main ring system were placed between Earth and Moon, there would not be much room to spare. Earth would fit inside Saturn 764 times, but the gas giant's bulk density is well below that of water, so its mass is only 95 Earth masses. (Courtesy of NASA)

Figure 9.2   When Cassini arrived at Saturn in 2004, its northern hemisphere appeared pale blue, much like the deep, clear atmospheres of Uranus and Neptune, rather than the usual tan color. Since it was winter in the northern hemisphere, there was less sunlight reaching the atmosphere – particularly since it was largely covered by the ring shadow. The cooler conditions seem to have caused the tan clouds to sink to depths where they are no longer visible. The upper atmosphere became relatively haze- and cloud-free, so the blue color was probably caused by Rayleigh scattering of sunlight (like the blue skies on Earth). The thin rings are barely visible, but they cast broad, dark shadows. (Courtesy of NASA/JPL/Space Science Institute)

Table 9.1
Saturn Summary

Mass: 5.68 x 10²⁶ kg (95.16 Earth masses)
Equatorial Diameter: 120,536 km
Polar Diameter: 108,728 km
Axial Inclination: 26.73°
Semimajor Axis: 1,433,530,000 km (9.58 AU)
Rotation Period (internal): 10.55 h
Sidereal Orbit Period: 29.46 years
Mean Density: 0.69 g/cm³
Number of known satellites: 82

tilt produces noticeable seasonal variations, exacerbated by the growth and decline of the shadow cast by the broad rings.

Winds and Cloud Bands

At first glance, Saturn's atmosphere seems to be a pale imitation of Jupiter (Figure 9.3). It appears as a fairly bland, tangerine disk,

marked only by muted bands and occasional spots, rather than turbulent, colorful ribbons of cloud and huge, long-lived storm systems.

Saturn's more quiescent version of Jupiter is probably explained by the fact that it receives only 25% as much heat from the more distant Sun, resulting in cloud condensation at lower levels. Its weaker gravitational pull also leads to a smaller pressure gradient, with greater potential for the scattering of light by a deep layer of high-level haze above the main tropospheric cloud deck.

Curiously, the pattern of belts and zones seems to have little connection with the zonal wind pattern. Bands of cloud running parallel to the equator are surprisingly common, with 24 counted in the southern hemisphere alone. A number of undulating, ribbon-like streamers are also seen within unstable, high-speed jet streams. The major seasonal variations caused by Saturn's pronounced axial tilt and the umbrella effect of the huge ring system may account for this apparent anomaly.

When Cassini probed below the haze and into the troposphere, it revealed that the width of Saturn's bands alternates with latitude: narrower ones are darker and coincident with rapid jet streams, and the wider bands tend to be brighter, aligned with jets that are slower and maybe even stationary, relative to the general rotation of the planet.

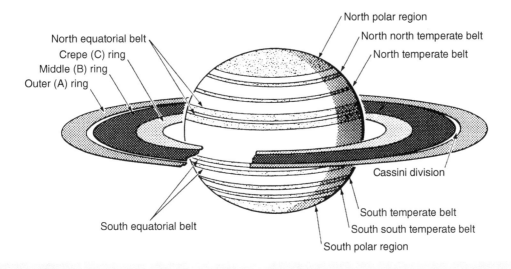

Figure 9.3 Saturn has a regular pattern of belts and zones aligned parallel to its equator. These are less prominent and more difficult to observe than the clouds on Jupiter because they are obscured by a thick, high-level haze. The bands disappear at high latitudes. (Courtesy of NASA)

Observations of occasional long-lived cloud features enable astronomers to recognize differences in rotation rates and wind speeds at various latitudes. As on all other planets with dense atmospheres, the winds are modified by the Coriolis force and arranged symmetrically on either side of the equator (see Chapter 3). However, the overall pattern is more Earthlike and less complex than on Jupiter. Only above 40° latitude does Saturn's atmospheric circulation mimic that of its larger neighbor.

On Saturn, almost all of the zonal winds blow from west to east, in the same direction as the planet's rotation.[2] Most notable is a broad equatorial jet located between 35° N and S, where winds speed around the planet at up to 1,720 km/h. This is more than four times the velocity of Jupiter's equatorial jet stream and much faster than any wind measured elsewhere in the Solar System. However, Cassini data show that the winds at the equator vary by ~200 m/s, which is 10 times larger than the variation of Earth's jet streams. Whether this is a variation with time or a variation with altitude is still an open question.

Analyses of Hubble Space Telescope and Cassini images show a marked slowdown in Saturn's equatorial jet at the visible cloud level since the Voyager flybys of the early 1980s. Since Cassini data suggest that winds decrease with altitude in the equatorial region, the apparent reduction of the equatorial wind speed may be due to the fact that the "tracer" cloud features are now at higher altitudes, rather than an actual decrease in velocity at a given level.

Only four easterly wind currents were identified from Voyager images, and all of them were very weak. Despite the planet's large axial tilt, there are no major seasonal differences in local weather between the northern and southern hemispheres. Wind speeds tail off dramatically at high latitudes, causing the parallel cloud bands to disappear.

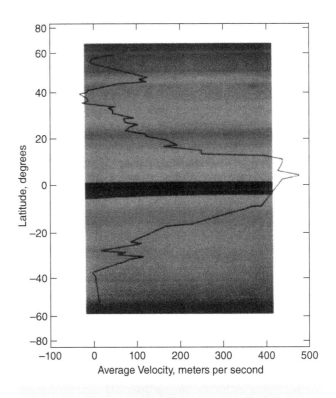

Figure 9.4 Saturn's wind speeds were calculated by studying Voyager images showing the movement of long-lived spots and eddies. The planet's huge equatorial jet propels the clouds toward the east at up to 480 m/s (1,720 km/h) – the fastest winds on any planet in the Solar System. (NASA-JPL)

[2] Meteorologists refer to these as "westerlies" because they come from the west.

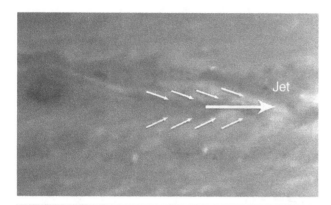

Figure 9.5 Small-scale cloud features are associated with turbulent eddies and wind shear close to one of Saturn's westerly jet streams. The jet, located at 27.5°S, is indicated by the horizontal arrow. Winds in this jet blow continuously at speeds close to 320 km/h. By tracking the movements of the cloud features for about 10 hours (roughly one Saturn day), it was shown that eddies on either side of the jet transfer energy and momentum to help sustain the jet. The small arrows indicate the direction in which this transfer takes place. The strong winds that accomplish this combine to stretch out the eddies into bright, tilted streaks parallel to the arrows. (Courtesy of NASA/JPL/Space Science Institute)

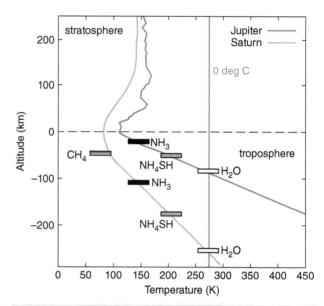

Figure 9.6 As on Jupiter, Saturn's upper atmosphere is thought to have three main cloud decks: water ice (H_2O), ammonium hydrosulfide (NH_4HS), and ammonia (NH_3). They are found at lower altitudes on Saturn because of the planet's lower temperature. Saturn's cloud decks are also more widely separated because of Saturn's lower gravity and pressure gradient. In the troposphere, temperature increases steadily with depth at the so-called "adiabatic lapse rate." In the stratosphere, above the tropopause, the temperature rises with altitude, probably due to absorption of solar radiation by gases and haze particles. Zero altitude marks the level where atmospheric pressure is 1 bar. (Peter Bond)

How are such remarkably rapid jet streams created? The primary sources of heat energy are the Sun and the planet's interior. Observations indicate that Saturn emits a similar amount of heat to that received from the distant Sun. The resultant convection produces an alternating pattern of belts and zones, aligned parallel to the equator, where gas is rising or sinking. The parallel pattern is the result of the Coriolis effect created by the planet's rapid rotation.

As with Jupiter, it has generally been assumed that the bright cloud bands (zones) are regions where warm gas rises and the dark bands (belts) are where cooler gas sinks. However, detailed observations show that towering plumes of white cloud associated with thunderstorms, generated by condensation of water and powerful updrafts, occur only in the dark belts. Further evidence that the belts are regions of convection is provided by the interaction between small eddies (rotating storms) and the atmospheric jets.

Clearly, the atmosphere cannot be rising everywhere. One possibility is that the link between zones/belts and rising/sinking gas might not apply on Saturn. Alternatively, it may be that any storm clouds in the belts are local phenomena, surrounded by other regions in the belts that are sinking. Upwelling may still be occurring in the zones, though obscured by clouds.

Boundaries between eastward- and westward-flowing jet streams create turbulent, eddy-filled regions that pump energy into the neverending gales. The eddies apparently power the horizontal jet streams, rather than the other way round. For example, Cassini pictures have confirmed that eddies on either side of a jet at 27.5°S supply its energy and momentum (Figure 9.5).

Fed by heat that rises from deep within Saturn, the spinning storms and occasional thunderstorms help to maintain the powerful eastward and westward currents in much the same way as rotating gears power a conveyor belt. The eddies themselves are created by temperature differences in the atmosphere caused by localized condensation of water. Water vapor condenses in some places as air rises and releases heat as it makes clouds and rain – just as it does on Earth.

Cloud Decks

Like Jupiter, Saturn is completely covered by dense cloud, although its turbulent activity is better hidden by high-level hazes (Figure 9.6). On Saturn, the cloud decks occur at lower levels and higher pressures than on Jupiter, since the low temperatures required for cloud formation occur at greater depths.

Calculations based on the known composition of the atmosphere indicate that there are at least three main cloud layers. The predicted altitude of these cloud decks is based upon the temperature at which a particular gas will condense into droplets. The outer atmosphere consists of about 97% hydrogen and 3% helium, with minor amounts of methane, ammonia, water, acetylene, ethane, and phosphine. However, since hydrogen and helium will not condense at Saturn's temperatures, it is the minor constituents that are the key to cloud formation in the troposphere.

Box 9.1   How Fast Does Saturn Spin?

For many centuries, astronomers studied the motions of major cloud features in order to determine the length of Saturn's day. The results indicated that there were different zones, each with its own periodicity. At the equator, clouds complete one circuit of the planet every 10 hours, 14 minutes – the result of a high-speed westerly jet stream. This was known as System I. At higher latitudes, the typical rotation period (System II) is around 10 hours, 38 minutes. But what about the rotation of the planet's interior (System III)?

Saturn is a source of radio emissions, although these are considerably weaker than Jupiter's. It has a broad band of emission in the frequency range 20 KHz to 1 MHz, with maximum intensity between 100 and 500 KHz. Saturn's kilometric radio emissions show a regular variation in intensity, measured by the Voyagers to occur every 10 hours, 39 minutes, 24 ($\pm$7) seconds. This was taken to be the planet's internal rotation period, although the cause of the emissions was hard to understand, since no asymmetry of the internal magnetic field has been detected. (In the cases of Jupiter, Uranus, and Neptune, the internally generated magnetic field is asymmetric about the spin axis, so the direction of the field seems to nod up and down as the planet spins. This means that variations in the intensity of the observed radio emissions are directly linked to the rotation period.)

The question of Saturn's rotation rate has been thrown wide open by recent data, which show that the periodicity of the radio signals is surprisingly variable over long and short time scales. The Cassini orbiter found that Saturn apparently rotates in 10 hours, 45 minutes, 45 seconds, $\pm$36 seconds. It is inconceivable that a planet could slow its rotation by 6 minutes in a few decades. Furthermore, Cassini's observations of radio emissions have shown that the rotation rate seems to vary by as much as one percent in a week.

Since the planet's spin cannot account for the irregular radio signals, some other process must be the cause. One possibility is that gusts in the solar wind cause the emission to vary when they impact the magnetic field. Studies show that there is a characteristic variation in the behavior of the short-period radio emission every 25 days. This period happens to coincide with the Sun's rotation rate, as seen from Saturn.

Another influence on the radio emissions is Saturn's icy moon Enceladus. The geysers on Enceladus may have become more active than a few decades ago. Large numbers of gas molecules ejected into space by the satellite become electrically charged, and are captured by Saturn's magnetic field, forming a torus – a disk of plasma, which "weighs down" the magnetic field. As a result, the rate of rotation of the plasma disk slows down slightly.

Meanwhile, scientists still struggled to deduce Saturn's true internal rotation rate. In 2015, a team of Israeli scientists reported a period of 10 hours, 32 minutes, 45 ($\pm$46) seconds, using Saturn's gravitational field and limits on the planet's observed shape and possible internal density profiles.

Any tilt to the magnetic field would make the daily wobble of the planet's deep interior observable, thus revealing the true length of Saturn's day, but – unique in the Solar System – the field's inclination coincides with the rotation axis, so its precise duration remained elusive.

In January 2019, scientists used data obtained by the Cassini spacecraft during its final, close Saturn flybys to study the planet's structure and its rings with unprecedented detail. They determined that the interior of Saturn vibrates at frequencies that cause variations in its gravitational field. The rings respond to these vibrations within the planet, acting rather like the seismometers that are used to measure movement caused by earthquakes. The variations in its gravitational field, in turn, are displayed in the ring structure.

The outcome of this analysis was that Saturn's interior rotates once every 10 hours, 33 minutes, and 38 seconds.

As with Jupiter, the lowest cloud layer is probably made up of water ice. Some 10 km thick, it occurs about 250 km below the tropopause (the boundary between the troposphere and stratosphere) where the temperature is about –23°C and pressure is greater than 10 bars. (Deeper layers of cloud presumably exist, but they cannot be seen.)

Some 70 km above this is a cloud layer composed of ammonium hydrosulfide ice (NH_4SH), where the temperature is –93°C and pressure is about 5 bars. 70–80 km higher still are the upper clouds, composed of white crystals of ammonia (NH_3). At this level, about 100 km below the tropopause, the temperature is –153°C and the pressure is between 0.5 and 1 bar.

Above these are found occasional clouds that are propelled upward by convection, with a stratospheric haze layer, probably composed of ammonia crystals and photochemical products – hydrocarbons produced by the effect of sunlight on methane.

Saturn Storms

As on Jupiter, cyclonic and anticyclonic storms sometimes appear through the haze, particularly away from the equator, where wind speeds are relatively modest. Warmer, rising gas from the interior creates sizeable storm systems, with numerous white or brown spots and peculiarly shaped convective clouds that are associated with gigantic thunderstorms. Their color is clearly related to cloud composition and altitude.

The smaller eddies last only a few days, although some of the larger features found by Voyager 1 were recovered in Voyager 2 images, demonstrating that they could survive for nine months or more.

The most impressive storms resemble smaller versions of Jupiter's white ovals. Like their Jovian counterparts, they are anticyclonic in nature, rotating in a clockwise direction in the

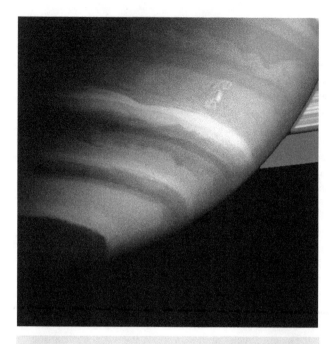

Figure 9.7 A false-color, near-infrared, image of the "Dragon Storm," a large, complex convective feature that appeared in the southern hemisphere in 2004. The image was taken through filters that reveal different amounts of methane gas. Regions with an abundance of methane above the clouds are red, indicating clouds that are deep in the atmosphere. High clouds are gray and brown clouds are at intermediate altitudes. Powerful radio emissions, similar to bursts of static generated by lightning, indicated that it was a giant thunderstorm. (Courtesy of NASA/JPL/Space Science Institute)

northern hemisphere. Nothing comparable to Jupiter's Great Red Spot has been seen, although Voyager 1 did discover a much smaller red feature, about 16,000 km across, at latitude 55°S.

Cassini monitored Saturn's cloud patterns more or less continuously after it entered orbit around the planet in June 2004. Some of the most interesting features were occasional white clouds, often irregular in shape, which appeared in a "storm alley" at about 30–35°S. This area lies between two jets moving at different speeds. The resultant wind shear seems to generate tremendous turbulence, allied to strong updrafts of warmer gas from the interior.

One of the best documented of these active systems was the so-called "Dragon Storm," a complex feature which arose in the same region that had earlier produced large, bright convective storms (Figure 9.7). This suggested that it was a long-lived feature that originated deep in the atmosphere and periodically flared up to produce dramatic white plumes.

During July and September 2004, the storm was shown to be associated with powerful radio emissions that resembled short bursts of static generated by lightning on Earth. The radio bursts started while the storm was below the horizon on the night side and ended when the storm was on the day side. This led to the conclusion that it was a giant thunderstorm whose precipitation generates electricity as it does on Earth.

Images taken over many months showed that the Dragon Storm spawned three small, dark ovals that broke away. Two of these subsequently merged while the wind current to the north carried the third oval off to the west. Such small storms are generally stretched out until they merge with opposing currents to the north and south.

Every few decades, the equatorial region hosts even more dramatic atmospheric eruptions, usually in the form of a huge white spot – a planet-encircling thunderstorm complex that produces intense lightning and enormous cloud disturbances. The head of one of these storms can be as large as Earth. Unlike Jupiter's Great Red Spot, which is calm at the center and has no lightning, the Saturn spots are active in the center and have long tails that eventually spread right around the planet.

Notable disturbances which lasted for several months were recorded in 1876, 1903, 1933, 1960, 1990, and 2010. They occur approximately every 20–30 years – roughly once every Saturnian year.

The most recent of these storms, known as the Great Northern Storm, erupted at 35°N in December 2010 (soon after the spring

Figure 9.8 This Cassini image shows the Great Northern Storm that erupted in 2010 and lasted for months. The storm affected the clouds, temperatures, and composition of the atmosphere for more than three years. This true-color view was taken on February 25, 2011, about 12 weeks after the storm began. It shows the turbulent patterns within the storm. By this time, the westward-moving storm head had merged with its fainter tail, having caught up with itself. Some of the cloud south and west of the storm head is tinged blue as it interacts with other currents in the atmosphere. (Courtesy of NASA/JPL-Caltech/Space Science Institute)

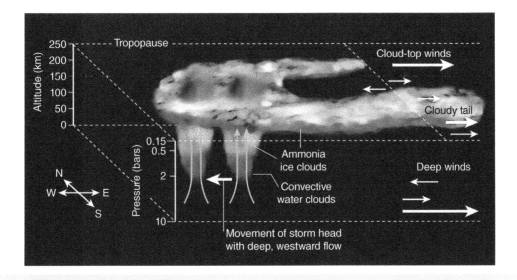

Figure 9.9 The white clouds of the 2010–2011 storm are thought to be produced by convective upwelling of warm, moist gas. Deep water clouds (originating at pressure levels of about 10 bars) rise into the upper troposphere. As the upwelling material approaches the tropopause (where the temperature stops decreasing with height and begins increasing into the stratosphere) at pressures of 0.1–0.5 bars, the bright, white, ammonia ice clouds spread horizontally (like "anvil" thunderstorm clouds on Earth). These are carried along by strong cloud-top eastward jets, producing an elongated "tail." The cluster of convective clouds moves westward relative to the winds at the cloud tops, suggesting that westward zonal winds are stronger at deeper levels, where the convection is initiated. (Nature)

equinox) and continued until August 2011. During that time, the storm exploded through the clouds. As it drifted westward, traveling faster than the weak jet stream at cloud level, it caught up with its tail, having circled the planet (Figure 9.8).

The effects of this storm continued until 2013. The wave activity associated with the storm headed towards the equator, affecting the clouds, temperatures, and composition of the atmosphere tens of thousands of kilometers away. As a result, the entire equatorial region cooled dramatically.

The fundamental trigger for these enormous eruptions remains uncertain, but most of them seem to be generated around mid-summer in the northern hemisphere after a sudden upwelling of warmer gas.

Observations of the 2010 storm showed that water vapor was carried upward from great depth, condensing, then freezing as it rose (Figure 9.9). The release of latent heat during condensation increased the temperature and buoyancy of the rising gas column. The water ice crystals then appeared to become coated with more volatile materials like ammonium hydrosulfide and ammonia as the temperature decreased with their ascent. This may explain how the storm was generated.

Research published in 2015 focused on the water molecules in the planet's atmosphere. These are heavy compared to the hydrogen and helium that make up most of the planet's atmosphere, so they form a discrete wet layer which suppresses convection.

Over time, the absence of warm, rising gas leads to a cooling of the upper atmosphere. However, that cool air eventually becomes dense enough to sink, overriding the suppressed convection and allowing warm moist air to rise rapidly and trigger a huge

thunderstorm complex. The upper atmosphere is so cold and so massive that it takes 20 to 30 years for this cycle to be completed, triggering another storm.

The absence of similar planet-encircling storms on Jupiter could be explained if its atmosphere contains less water vapor than Saturn's atmosphere.

Polar Hot Spots and a Hexagon

Saturn's considerable distance from the Sun means that the visible cloud layers are cold, well below the freezing point of water.[3] The average temperature of the cloud tops is about −140°C. The absorption of sunlight by high-altitude hazes means that temperatures increase in the stratosphere, rising to −130°C at a height of 200 km above the tropopause. Unlike Jupiter, no large hot spots pierce the all-embracing cloud blanket.

An axial tilt of almost 27° and huge shadows cast by the rings either side of the summer or winter solstices inevitably cause some differences in atmospheric temperature, although seasonal effects are quite small. The immense atmosphere is so massive that it only responds sluggishly to the modest changes resulting from summer heating and winter cooling.

Nevertheless, ground-based measurements taken in 2005 during the southern midsummer showed that there was a general warming towards the south pole, which had been exposed to continuous, direct insolation for many years (Figure 9.10). In general, the southern atmosphere was up to 10°C warmer than the north.

Less expected was a temperature increase of about 2°C in the upper troposphere between 69°S and 74°S, together with another

[3] On average, Saturn receives about 1% as much solar radiation as Earth.

Figure 9.10 An infrared mosaic obtained by the Keck I telescope on February 4, 2004, showing temperatures at Saturn's tropopause, well above the main cloud decks. Lighter shades show warmer temperatures. The bright hot spot (bottom) is located above the south pole. Temperature increases toward the pole, rising from −185°C near 70°S to −182°C at the pole. Note the cooler region in the ring shadow. The rings (orbiting clockwise in this image) also show temperature variations. Ring particles are coldest just after leaving Saturn's shadow (lower left) and become warmer until they pass behind Saturn again (lower right). (Courtesy of NASA/JPL)

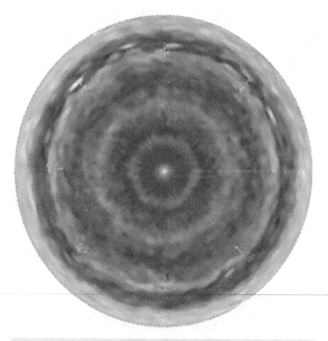

Figure 9.11 A false-color thermal image from Cassini, showing night temperatures from 30°N to the north pole in the center. Data relate to temperatures near the 100 millibar tropopause, the top of the convective layer. At the north pole is a relatively hot, cyclonic vortex very similar to that found at the much sunnier south pole. Colors show temperatures from −201 to −189°C. The polar hexagon is evident in the warm belt between 75 and 80°N. (Courtesy of NASA/JPL/GSFC/Oxford University)

marked rise of some 2.5°C between 87°S and 90°S. These elevated temperatures correlated with regions of lower visible reflectivity.

Most notable was a dark spot surrounded by high walls of clouds at the south pole that coincided with the highest temperatures measured in the atmosphere.[4] The cyclonic vortex at Saturn's south pole has a 4,200 km-diameter, cloud-free eye.

The eye has a warm core extending from the troposphere into the stratosphere, a concentric eye wall of clouds extending 70 km above the internal clouds, and numerous external clouds whose anticyclonic rotation suggests a convective origin. The eye rotation speeds reach 150–190 m/s, and probably strengthen with depth. The Saturn polar vortex has features in common with terrestrial hurricanes and with the Venus polar vortex.

Surprisingly, Cassini infrared measurements obtained in 2007 showed that the north pole possessed a hot spot similar to the southern vortex, even though it had been in winter darkness for more than a decade (Figure 9.11). Temperatures at the polar tropopause, the top of the atmosphere's convective layer, peaked at −189°C.

Both hot spots coincide with large, hurricane-like cyclonic circulations in the atmosphere. The centers of both vortices are depleted in phosphine gas, an imbalance probably caused by a downward motion of the atmosphere. It seems that the hot spots are the result of gas from equatorial regions moving poleward, then being compressed and heated as it descends over the poles.

The main difference between the polar regions is the presence of a unique hexagonal feature encircling the north pole at 78°N, which is unlike anything seen on the other planets (Figures 9.11 and 9.12). First imaged by the Voyagers, it has persisted to the present day, and is visible in thermal images taken at night by

Cassini. The reason for the asymmetry between the north and south polar structures remains unknown.

Infrared images show that the hexagon is largely a clearing in the thick clouds that extends some 100 km below the cloud tops. The structure, which surrounds the cyclonic polar vortex at its center, is about 32,000 km wide. Inside the six-sided cloud feature are numerous small storms.

Although Saturn's northern aurora lies nearly overhead, its considerable depth indicates that the hexagon is not linked to Saturn's radio emissions or to auroral activity (see Figure 9.18).

The hexagon appears to have remained fixed with Saturn's rotation rate and axis since first glimpsed by Voyager several decades ago. The points of the hexagon rotate around its center at almost exactly the same rate that Saturn rotates on its axis. Moreover, a jet stream, which flows eastward at up to 360 km/h, follows a path that appears to match the hexagon's outline.

Until recently, the jet stream was regarded as a lower-atmosphere phenomenon, restricted to the clouds of Saturn's troposphere. However, research published in 2018 indicated that the high-speed belt of wind extends about 300 km above the cloud tops, up into the stratosphere, at least during the warmer period of northern spring and summer.

---

[4]Neptune has a similar south polar hot spot.

Figure 9.12 A false-color Cassini image of Saturn's north pole. The eye of a hurricane-like storm appears dark red while the fast-moving hexagonal jet stream is yellowish green. Clouds circling inside the hexagon at lower altitudes have a muted orange color. A second, smaller vortex at lower right is pale blue. The rings of Saturn are bright blue at top right. (Courtesy of NASA/JPL-Caltech/SSI/Hampton University)

The origin of the unique hexagon is not clear. However, computer simulations of an eastward jet flowing on a curved path near Saturn's north pole indicate that small perturbations in the jet – possibly caused by interactions with other air currents on either flank – could make it meander into a hexagonal shape. It seems that the hexagon may be an example of an unusually strong planetary wave.[5]

The changing seasons may also influence the color of the polar region, which was seen to change from a pale blue to a golden-brown hue in Cassini images taken between 2012 and 2016. This is thought to be the result of an increased exposure to sunlight as the planet moved from northern winter toward the summer solstice in May 2017. The increase in insolation led to greater production of photochemical hazes in the polar atmosphere.

Interior

The exact nature of Saturn remained uncertain for a long time, though it was thought to be largely gaseous with a fluid interior. Then came the introduction of the spectroscope, which led to the 1932 discovery of methane and ammonia – simple compounds that combine hydrogen with carbon and nitrogen.

This led Yale University's Rupert Wildt to suggest that Saturn had formed by pulling in a huge amount of hydrogen, helium, and other gases from the original solar nebula. He envisaged an Earth-like, rocky core surrounded by a thick layer of water, ammonia, and methane ices, topped by a hydrogen-rich atmosphere. More recently, it has been suggested that the hydrogen atmosphere is compressed so much that it changes into metallic hydrogen at a depth of about 20,000 km (see Chapter 8).

Today, our knowledge of Saturn's interior is still largely based upon its size and bulk density, together with measurements of its atmospheric composition compared with the elemental composition of the Sun. Scientists also use information on the distortion of the planet's gravitational field, as measured from its effects on the motion of nearby spacecraft, to gain insights into the amount of mass concentrated near Saturn's center.

The most obvious characteristic is Saturn's remarkably low bulk density, approximately $0.7\,g/cm^3$, or 70% the density of water. When combined with the fact that 97% of the atmosphere (by molecular abundance) is hydrogen, with almost all of the remainder accounted for by helium, it is clear that Saturn is made almost entirely of the lightest gases in the universe.

Additional constituents include ammonia (NH_4) and methane (CH_4), a small amount of water vapor, and trace quantities of compounds such as carbon monoxide (CO), phosphine (PH_3), ethane (C_2H_6), acetylene (C_2H_2), methyl acetylene ($CH_3C\equiv CH$), and propane (C_3H_8).

In the upper atmosphere, most of these are probably produced by photochemical reactions in which solar ultraviolet radiation breaks down methane. In the cloudy troposphere, lightning discharges also contribute to these chemical processes. Lower down, a water/methane reaction may result in the formation of carbon monoxide.

Jupiter and Saturn are believed to have similar internal structures, with a thick layer of liquid hydrogen (plus some helium) beneath their thick, gaseous atmospheres (Figure 9.13). This changes into liquid metallic hydrogen at a depth of about 30,000 km, when the increasing pressure exceeds about two megabars. This highly compressed, conductive form of hydrogen behaves like a metal and convectional currents in this fluid layer probably generate Saturn's magnetic field.

Pressures and temperatures are inevitably lower in the smaller of the two gas giants, so Saturn's metallic hydrogen zone is relatively thin and located at much greater depth. On the other hand, its liquid hydrogen layer is likely much thicker.

At the center, there may be a fairly small core – perhaps with a radius of about 16,000 km – made of water, methane, and ammonia ices overlying rocky materials such as silicates and iron. (Another model has only a rocky core with the ices dispersed through the overlying hydrogen-helium layers.) The core temperature is in the region of 13,000°C, with pressure about 18 megabars.

Infrared data show that Saturn has a strong internal heat source. It actually radiates 1.8 times more heat than it receives from the Sun, compared with 1.7 times for Jupiter (Figure 9.14). Since this internal powerhouse cannot be explained by emission of heat

---

[5] A planetary wave (or Rossby wave) is a large, long-lived wave in a planet's atmosphere (like a meander in a jet stream) caused by temperature differences and modified by the Coriolis effect produced by the planet's rotation.

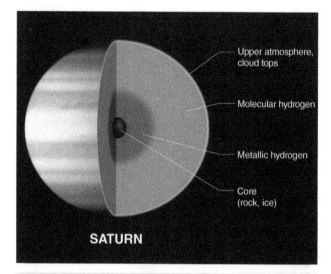

Figure 9.13 Saturn is composed mainly of hydrogen, with a substantial amount of helium. Beneath the 1,000 km-thick atmosphere is a layer of liquid hydrogen that changes to liquid metallic hydrogen as temperature and pressure increase. The liquid hydrogen layer is probably thicker than that of Jupiter, while the metallic hydrogen layer is thinner. There may be a core composed of rock and ice. (LPI)

left over from its formation and subsequent contraction (Saturn should have completed this long ago), some other process must be at work.

Supporting evidence is provided by the depletion of helium in Saturn's atmosphere, compared with Jupiter and the Sun. The favored theory is that metallic (ionized) hydrogen and neutral helium atoms – once thoroughly mixed – are separating deep inside the planet. Some 2 billion years ago, the helium began to form tiny liquid droplets which are raining toward the center of the planet. This process of differentiation would provide a source of internal heat.

Magnetic Field

Although Saturn's radio emissions indicated the presence of a magnetic field, the existence of a magnetosphere was not confirmed until the flyby of Pioneer 11 in September 1979. The data indicated that Saturn's magnetosphere is, in many ways, a smaller version of Jupiter's, with a blunt sunward side and a long downwind tail that is shaped by the solar wind. Since then, further details have been revealed by the Voyagers and the Cassini orbiter.

The first interaction between the solar wind and Saturn's magnetosphere takes place at the bow shock, where the incoming particles are dramatically slowed. The average distance from the planet of this outer boundary is about 1.8 million km (30 Saturn radii). However, changes in solar wind pressure mean that the bow shock oscillates between 20 and 35 Saturn radii. During its insertion into orbit around Saturn in 2004, Cassini crossed the boundary 17 times.

Figure 9.14 A false-color mosaic taken above the northern, unlit side of the rings over a period of 13 hours. On the nightside (right), Saturn's internal thermal radiation is shown (wavelength 5.1 microns). The speckled and banded patterns are created by variable, thick clouds, which block some of the heat deep in the atmosphere. These clouds are likely made of ammonium hydrosulfide and cannot be seen in reflected light on the dayside (left), since they are hidden by overlying hazes and ammonia clouds. The icy ring particles are highly reflective at 2.3 microns (blue). (Courtesy of NASA-JPL/University of Arizona)

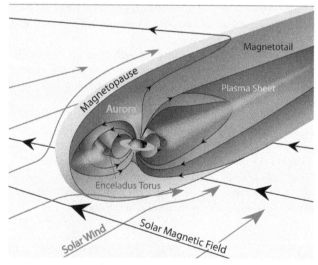

Figure 9.15 Saturn is surrounded by a "normal" tadpole-shaped magnetosphere, with many similarities to those of Earth. The solar wind is slowed at the bow shock and diverted around the magnetopause, the outer edge of the region dominated by Saturn's magnetic field. Saturn is the only planet whose magnetic field is almost exactly aligned with its rotation axis. (Fran Bagenal/University of Colorado, Boulder)

After crossing the bow shock, the solar wind changes direction and is forced around the magnetosphere. The magnetopause typically lies about 1.2 million km above Saturn's sunlit hemisphere, although the distance varies with the level of solar wind activity. The magnetotail likely stretches hundreds of Saturn radii downwind.

Of the planets in our Solar System, Saturn's magnetic field strength is second only to that of Jupiter. Its magnetic moment is 600 times Earth's, but only about 3% that of Jupiter. Although its magnetosphere dwarfs that of Earth, Saturn is so large that the strength of the magnetic field at its cloud tops is actually lower than on Earth's surface: 0.2 gauss, compared to 0.5 gauss. At present, the polarity of the field is opposite to that of Earth, but like our planet, the field is dipolar.

Like Jupiter, the key factors in the generation of Saturn's field are thought to be its rapid rotation, an internal heat source and the presence of fluid, metallic hydrogen, a conductive material. These combine to drive a dynamo involving moving currents deep in the interior.

Saturn's field resembles a dipole with secondary features. In addition to the main dynamo action in the metallic hydrogen layer, deep zonal flow (differential rotation) and small-scale convective motion in the semiconducting region of Saturn could lead to a secondary dynamo action.

Curiously, Saturn is the only planet whose magnetic field axis is almost exactly aligned with its rotation axis, even though the center of the field is displaced northwards along the axis by about 2,800 km. The existence of the magnetic field is hard to explain because theories for the generation of planetary magnetic fields require the field axis to be inclined to the planet's rotation axis. Without any tilt, the dynamo's currents should eventually subside and the field would disappear.

Saturn's magnetosphere is populated by numerous heavy ions (charged particles) which are largely derived from geysers on Enceladus that eject water ice particles and dust, sputtering by charged particles on the surfaces of Saturn's icy satellites and meteoroid bombardment of ring material. Smaller contributions come from the planet itself and Titan's nitrogen-rich atmosphere. Disks of neutral and ionized hydrogen and nitrogen surround Saturn near the orbit of Titan, and plumes of heavy ions may also extend beyond the satellite's orbit.

The geysers on Enceladus are thought to deliver as much as 300 kg/s of water to Saturn's inner magnetosphere. The neutral water molecules are split by sunlight and turned into ions, which Saturn's magnetic field captures and shapes into a torus of energized plasma. The rotating plasma disk slightly slows down the rate of rotation of the planet's magnetic field.

Saturn's plasma torus, similar to the one around Jupiter, is the biggest plasma structure associated with any planet. It extends out to 15–20 Saturn radii. The torus contains approximately 3,000 particles per cm³.

The rotating plasma torus is also associated with an electrical current (the "equatorial ring current") of about 10 million amps. Ring currents are caused when plasma becomes trapped between magnetic field lines and gradually drifts around the planet. The aggregate motion of the hot ions distributed around the equator generates an electrical current. The ring current occupies a region

of the equatorial plane between 540,000 and 1,080,000 km from the center of Saturn.

Whereas Earth's ring current is associated with hydrogen and appears during solar flares, Saturn's is largely associated with ionized oxygen and is always present. Unlike Earth's fairly stationary ring current, Saturn's rotates with the planet and is asymmetric, with the side facing the solar wind showing evidence of squeezing.

Auroras and Radiation Belts

Saturn's magnetosphere traps ionized particles in radiation belts where the number of particles is similar to the terrestrial magnetosphere (Figure 9.16). The main belts extend outward from the edge of the major rings to the orbit of Tethys, with the most heavily populated belts occurring between 139,000 km and 362,000 km from Saturn's center. They contain particles with energies up to tens of MeV.

The belts are largely composed of oxygen and water-derived ions produced by the bombardment of the rings and icy moons by solar ultraviolet light and particles trapped in the magnetic field.

The belts are more intense on the night side of the planet, and there are marked gaps created as the trapped ions collide with small, embedded moons, the tenuous E ring, and gas.

Until recently, the inner edge of the radiation belts was thought to be marked by the main rings, which absorb any particles that encounter them. However, Cassini found another radiation belt that stretches between Saturn's cloud tops and the inner region of the D ring (the closest ring to the planet). The outer sector overlaps with the extended D ring, and its intensity is reduced compared with that of the inner sector, owing to proton losses on ring dust.

This inner radiation belt, which coincides with the equatorial plane, is dominated by protons with energies from 25 MeV up to the gigaelectron volt range – much higher than observed outside the main rings. These protons are among the ß-decay products of neutrons, which are released through galactic cosmic ray collisions with Saturn's rings.

Another phenomenon associated with energetic particles trapped in the magnetosphere is the aurora (Figure 9.17). On Earth, auroral storms may develop in about 10 minutes and last for a few hours, whereas Saturn's auroras always appear bright and may last for several days. The auroral zones typically measure about 9,000 km across and rise more than 1,600 km above the cloud tops, but their extent and brightness vary over time. These changes indicate that the aurora is primarily shaped and powered by a continual interaction between Saturn's magnetic field and the solar wind.

As on Earth, the auroras occur when charged particles from the Sun spiral down the magnetic field lines, making the molecules in the upper atmosphere glow. This results in circular or oval auroras, shaped by the magnetic field lines that converge like a narrowing funnel towards the magnetic poles. Unlike Earth, Saturn's auroras are only "visible" in ultraviolet light.

Studies of interplanetary shock waves that strike Saturn's magnetosphere show that they result in a brightening of auroral emissions, with accompanying increases in radio emissions. This suggests that there is a direct correlation between auroral intensity and the state of the solar wind – similar to the situation on Earth. (Jupiter's auroras are largely controlled by processes taking place

Figure 9.16 Saturn's main radiation belts extend outward to the orbit of the moon Tethys. They are segmented because of particle absorption by moons and rings. The innermost radiation belt (inset) threads through Saturn's D ring and contains protons with energies up to several gigaelectron volts, much higher than observed outside the main rings. These protons are among the ß-decay products of neutrons, which are released through galactic cosmic ray collisions with Saturn's rings. The most heavily populated regions are shown in red. (Science/E. Roussos et al)

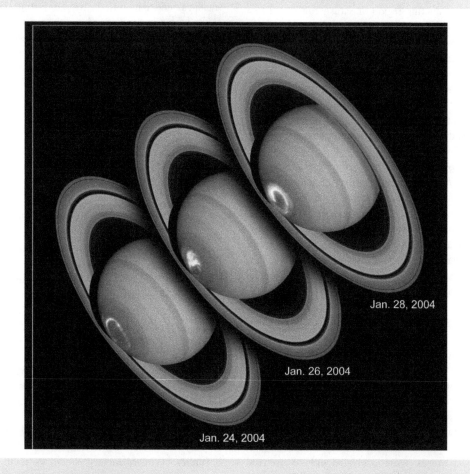

Figure 9.17 A sequence of images of Saturn's southern hemisphere taken by the Hubble Space Telescope, January 24, 26, and 28, 2004. The ultraviolet aurora is superimposed on visible images. The brightening on January 28 (right) corresponds with the arrival of a large disturbance in the solar wind. When Saturn's auroras become brighter (and more powerful), the ring of light encircling the pole shrinks in diameter. (Courtesy of NASA, ESA, J. Clarke/Boston University, and Z. Levay/STScI)

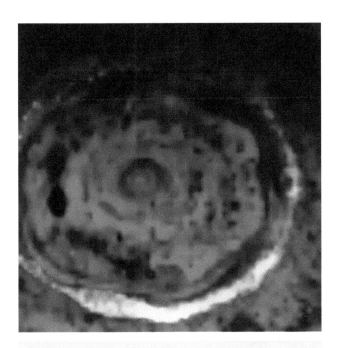

Figure 9.18   A Cassini infrared view of Saturn's north pole at night reveals heat rising from a depth of at least 75 km below the upper clouds. Surrounding the north pole is a nearly stationary hexagonal feature, possibly an unusually strong planetary wave that extends deep into the atmosphere. The blue color shows auroral emissions caused by charged particles entering the atmosphere. The aurora seems to coincide with the hexagon, though it lies at a much higher altitude. (Courtesy of University of Arizona–LPL, NASA-JPL)

inside its huge magnetosphere, rather than the conditions in the solar wind – see Chapter 8.)

However, further study indicates that the main factor controlling Saturn's auroras appears to be compression of the magnetopause as a result of increased solar wind dynamic pressure. Unlike Earth, the orientation of the interplanetary magnetic field (i.e. its north–south polarity) and the resultant magnetic reconnection that links the solar and planetary magnetic fields is less important. This may be because, during Saturn's southern summer, the solar magnetic field is generally aligned at right angles to the planet's field, making it less easy for magnetic reconnection to take place.

Discovering Saturn's Rings

The rings of Saturn were discovered in 1610 by Galileo Galilei. Using a homemade "spyglass" that could magnify a mere 20 times, he noticed that Saturn seemed to be sandwiched between two identical, but smaller planets. When he returned to the scene two years later, the companions had vanished and Saturn was now in splendid isolation (Figure 9.20). Subsequent observers produced drawings of detached *ansae* ("handles") attached to either side of the planet.

Confusion reigned for more than 40 years, until a Dutch astronomer named Christiaan Huygens announced that he had discovered a satellite – now known as Titan – and found an explanation for Saturn's *ansae*. Curiously, he decided to disguise the answer in the form of an anagram and waited three years to unveil the secret in his book, *Systema Saturnium* (The Saturn System).[6]

Huygens declared that the mystery appendages were actually a thin, solid ring that circled above its equator. Since the planet was inclined to the plane of its orbit, the ring, too, was tilted by the same amount. This meant that the ring's appearance changed during each 29½ year-long orbit of Saturn. When the ring was edge-on to Earth, during Saturn's spring and autumn, it seemed to virtually disappear (Figure 9.21). During summer or winter, when the poles were tilted toward the Sun and Earth, the ring opened out and appeared in all its glory. Then the entire cycle operated in reverse until the disk was viewed edge-on once more.

It soon became clear that the ring was not solid. In 1675, Giovanni Domenico Cassini noticed a continuous dark region, about two-thirds of the way out from Saturn, which he correctly surmised to be a gap – now known as the Cassini Division – that separates an outer A ring from an inner B ring. Cassini also stated that the ring was composed of a large number of small satellites orbiting Saturn.

As telescopes improved, further structure was observed in the ring system. The faint, diaphanous, C or "Crepe" ring was discovered in 1850, extending inward from the B ring. A sizeable gap, now known as the Encke Division, was also noticed near the outer edge of the A ring.

Meanwhile, the true nature of the rings was revealed when James Clerk Maxwell proved mathematically that a very thin, solid ring circling the planet would be destroyed by gravitational forces. His prize-winning paper noted that the rings must be composed of small particles in separate orbits, with the inner ring particles obeying Kepler's laws and traveling faster than those further out.

For many years, it was thought that the rings were formed at the same time as the planet, coalescing out of swirling clouds of interstellar gas, 4.5 billion years ago. An alternative explanation was provided by French mathematician Edouard Roche, who calculated that an object would be torn apart by tidal forces if it approached too close to Saturn. The distance at which this would occur – the so-called Roche limit – more or less corresponded to the position of the rings.

The Roche limit is an imaginary boundary around a planet within which a "rubble pile" satellite will be torn apart by gravitational forces. At the same time, particles in this region cannot gravitationally coalesce to form a satellite because of the disruptive effect of planetary tidal forces.

Today, the origin and age of the main rings is still unknown, but opinion seems to be shifting toward the idea that they were formed at least 4 billion years ago. This may have occurred when comets or small, icy satellites approached too close to Saturn and were broken apart by tidal forces or by collisions with each other after they had been captured by Saturn's gravity. Alternatively, they may have formed after an icy moon about the size of Mimas was broken up by an impact with a comet or asteroid.

[6]**The solution to the anagram (translated into English) was: "It is surrounded by a thin flat ring, nowhere touching, and inclined to the ecliptic."**

Box 9.2   Voyagers at Saturn

NASA's Voyager spacecraft were not the first visitors to Saturn. A gravity assist from Jupiter enabled the much simpler Pioneer 11 to fly past Saturn on September 1, 1979. Pioneer's TV camera showed very little cloud detail on the planet but there was clear evidence of structure in the main rings. Its major discoveries included the thin F ring, about 3,600 km beyond the A ring, and the presence of a small satellite between the new ring and the orbit of Mimas. Pioneer also made the first survey of Saturn's magnetosphere and particle environment.

The Voyager spacecraft were designed to carry out a comprehensive exploration of Saturn's rings and satellites. Furthermore, the second spacecraft could be reprogrammed to take advantage of any discoveries by its predecessor.

Titan was the key to the Voyager 1 encounter. In order to observe this mysterious moon, while obtaining a good view of the rings and images of some smaller satellites, the spacecraft was targeted to arrive from above the ring plane. It then swung over Saturn's southern hemisphere at a height of 124,200 km and returned northward, using the planet's gravity to head toward interstellar space.

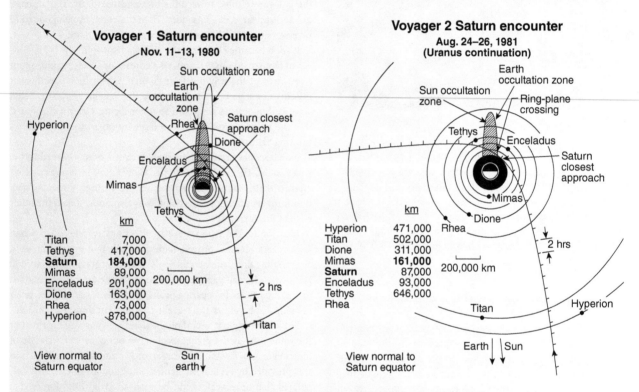

Figure 9.19   The flight paths of Voyager 1 and 2 through the Saturn system, as seen from above the planet's north pole. Voyager 1 was targeted to fly close to Titan. Saturn's gravity then diverted it above the ecliptic to send it out of the Solar System. This successful flyby meant that Voyager 2 could be redirected toward Uranus. (NASA-JPL)

During the Saturn approach, Voyager 1 skimmed past Titan on November 11, 1980, just 4,520 km above the orange clouds. It then sent back close-ups of Tethys before plunging beneath the rings. The camera turned toward Mimas, Enceladus, and Dione before Voyager recrossed the ring plane and obtained detailed images of Rhea. On the outward leg, as seen from the Earth, the spacecraft disappeared behind the planet and the main ring system.

Although the flight from Titan to Rhea lasted little more than 24 hours, Voyager's instruments monitored the Saturn system continuously for almost four months. The rings were found to contain hundreds of ringlets, some elliptical, while the F ring comprised several braided strands.

Another puzzle involved shadowy "spokes" in the B ring that survived despite different rotation rates within the ring material. The existence of the D and E rings was confirmed, and a seventh main ring (G) was discovered. The theory that some narrow rings were confined by nearby satellites was confirmed by the discovery of two tiny "shepherds" on either side of the F ring and another just outside the A ring.

Although Voyager's cameras were unable to pierce the smog that shrouded Titan, new data were returned about the temperature, pressure, and composition of its atmosphere, and the high-altitude haze layers. Particularly intriguing was the dense, nitrogen-rich atmosphere in which complex organic compounds were created.

Voyager 2 arrived more than nine months later. Equipped with a more sensitive camera and an approach trajectory well above the ring plane, it was ideally placed to study the rings. The northern side of the rings was now receiving more direct sunlight and so appeared much brighter.

From early June to late September 1981, Voyager 2 sent back more than 18,500 images of Saturn, its rings and satellites. Closest approach occurred on August 25, 101,000 km above the cloud tops.

Thousands of ringlets were detected, though a search for small satellites within them was unsuccessful. Saturn's bland clouds came alive with images of brown and white spots, and undulating, ribbon-like jet streams. Titan was only viewed from afar, with Hyperion, Enceladus, Tethys, and Iapetus the satellites that gained most attention. Voyager 2 crossed the equator only once, providing an opportunity to look back at the shaded side of the rings.

Two-thirds of the Enceladus pictures were lost through jamming of the scan platform, which carried the cameras, 45 minutes after the ring passage. However, the platform was operational again nine days later, when Voyager made a distant pass of Phoebe, the outermost satellite then known.

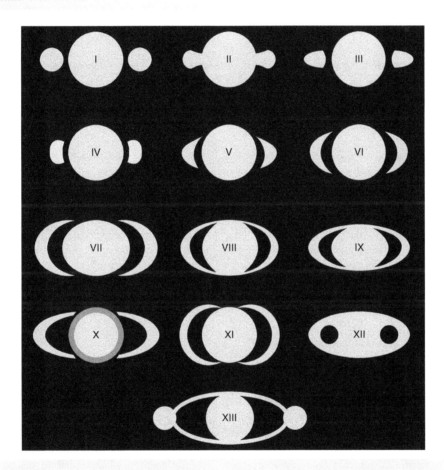

Figure 9.20 Sketches of Saturn made by various observers during the first half of the 17th century. Galileo's observation of 1610 is labeled I. Not until 1659 did Christiaan Huygens reveal that the "handles" were actually a ring system above the planet's equator. (NASA)

In one model of the ring formation process, Kuiper Belt objects cross the Roche limit, the distance at which a giant planet's gravity is strong enough to cause tidal disruption of a passing object. The objects are destroyed during these close encounters and some fragments are captured into orbits around the planet. Repeated collisions between the fragments cause the captured debris to break down, their orbits become gradually more circular, and the current rings are formed.

The thinner, dustier rings are produced by collisions between small objects within the rings, or by meteorite erosion of these bodies. Geysers on Enceladus provide tiny grains of water ice that generate the huge E ring, while an even larger outer ring is fed by material removed from the moon Phoebe.

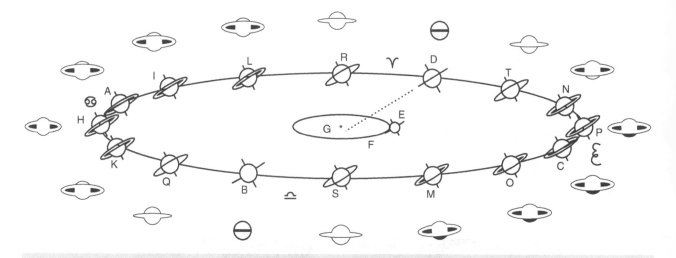

Figure 9.21   In this diagram from his book, *Systema Saturnium,* Christiaan Huygens showed how its large axial tilt resulted in seasons on Saturn and caused different amounts of its ring to be visible from Earth. (NASA)

Multiple Rings

Today, there are seven recognized major rings, which are labeled in the order in which they were discovered. From the planet outward, they are D, C, B, A, F, G, and E. Other faint rings or ring arcs are associated with Janus, Epimetheus, Anthe, Methone, and Phoebe.

From outer edge to outer edge, the main system measures 374,000 km, so it would almost span the gap between Earth and the Moon. Despite their vast width, they are generally only a few tens of meters thick – the reason why they "disappear" when they lie edge-on to the Earth.

The rings are best envisaged as thousands of ringlets, each comprising innumerable particles and small chunks of icy debris in separate orbits. The particles are composed primarily of water ice mixed with silicates, and range in size from microns to tens of meters. Above this size range, their number drops sharply. A typical ring particle is more like a fluffy snowball than a hard chunk of ice.

The main rings are thought to be made up of boulder-size snowballs, but the size of the ice crystals on their surfaces ranges from very small, like powdery snow, to larger grains, similar to more granular snow.

Subtle color variations reveal differences in the composition and particle sizes within the main rings, with the largest objects furthest from Saturn. The cleanest ice occurs in the A ring, whereas the C ring is noticeably "dirtier," containing larger amounts of silicate-containing dust that may have been delivered by a rocky intruder within the last 100 million years. The Cassini Division and other gaps also contain thin ringlets composed of an unidentified "dirty" substance which appears remarkably similar to the surface material seen on Saturn's remote moon, Phoebe.

Until recently, it was assumed that the rings were ephemeral, and would disperse over time. However, it is now thought that they may be a permanent feature.

Cassini observations show that the A ring is primarily empty space, with ever-changing clumps of material that range in size from 2–13 m. The gaps between the clumps are larger than the clumps themselves (Figure 9.25).

The individual clusters are largest near the middle of the ring and became smaller toward the edges of the ring. These clusters are torn apart and reassembled by Saturn's gravity. Even when they fragment, they tend to come back together, thereby maintaining the overall ring structure.

Similarly, the B ring contains numerous clumps that are 30 to 50 m wide and very flat, like big sheets. Once again, there is a constant destruction process as Saturn's gravity pulls them apart, followed by reformation. The particles move from clump to clump as they are destroyed and new ones reformed. Such recycling may enable the rings to persist almost indefinitely.

An atmosphere of molecular oxygen and neutral hydrogen extends 60,000 km above and below the main rings and some distance beyond the A ring. Water ice is the likely source for this cloud. Water molecules are driven off the ring particles by solar ultraviolet light. They are then split into hydrogen, as well

Figure 9.22   A cross-section of the main rings showing how they differ in structure, color and density. The natural-color mosaic was acquired as the Cassini spacecraft soared above the shaded side of the rings. Major gaps are named at the top. Scale in the radial (horizontal) direction is about 6 km per pixel. (Courtesy of NASA/JPL/Space Science Institute)

as molecular and atomic oxygen – a process known as photodissociation. Much of the hydrogen gas escapes to space, while the atomic oxygen and any remaining water are frozen back into the ring material due to the low temperatures. This leaves behind a concentration of oxygen molecules.

A and B Rings

Both the A and B rings are very bright and easily observed from Earth. However, the B ring is much denser and more opaque than the other rings, hence its dark appearance when illuminated from behind.

The A ring, the outermost of the main rings, is subdivided by two "empty" regions. Each of these gaps is cleared of particles by a small satellite embedded in the ring (Figure 9.26). The 42 km-wide Keeler Gap is swept clear by 8 km Daphnis, while the 330-km Encke Gap is occupied by 26 km-diameter Pan.

The A and B rings are separated by the Cassini Division, the largest gap in the rings. The Division is cleared by a 2:1 gravitational resonance with the moon Mimas. Particles orbiting at this distance from Saturn travel twice around the planet for every single orbit of Mimas. This regular gravitational interaction pushes them into different orbits, opening up a gap.

Once thought to be empty, spacecraft images have shown that the Division, which measures about 5,000 km across, actually contains five broad rings that display additional structure. The effective particle diameter in these rings, derived from radio occultation measurements, is about 8 m.

At least four faint, dusty ringlets have also been observed in the Encke Gap. One of them occupies almost the same orbit as Pan, suggesting that the satellite is the source of the material. The presence of the other narrow ringlets indicates that there may be other unseen moonlets in the Encke Gap. The ringlets vary in brightness and appear to move in and out along their length, resulting in notable "kinks" which are similar in appearance to those in the F ring (see F Ring).

Both Daphnis and Pan perturb the particles along the borders of the gaps they inhabit. When the satellite passes, its gravity generates a series of visible stripes or "wakes" in the ring material on either side of it. These move diagonally away from the gap's edge. In the faster-moving ring material closer to Saturn, they drift ahead of the moon as leading wakes. Similar trailing wakes are created as Pan overtakes the slower-moving ring material beyond its orbit. Since Pan is larger than Daphnis, it creates correspondingly larger wakes.

Cassini images of the middle A ring also show numerous "propeller-shaped" disturbances that are thought to be caused by embedded objects approximately 40–120 m in diameter. These moonlets are not large enough to create a continuous gap in the ring, but they can make a partial clearing which is shaped like an airplane propeller (Figure 9.27). Their gravity perturbs the surrounding ring material, creating wavelike features in their wakes.

Cassini imaged swarms of these features and it is likely that millions of them exist within the A ring alone. During the final years of its mission, six of the most noticeable propellers, each popularly named after famous aviators such as Blériot, were regularly imaged and tracked by Cassini scientists.

Cassini also detected numerous wavy features called "density waves" in ring A. Many of these were in its outer region, and were generated by repeated gravitational interactions (resonances)

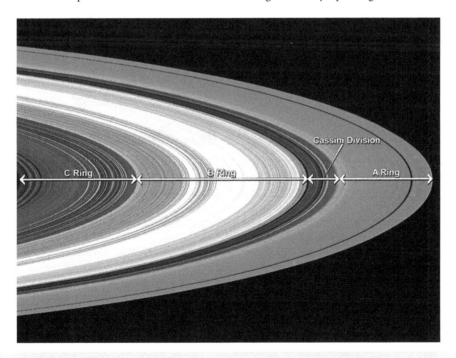

Figure 9.23 A Cassini false-color image showing the size and density of ring material, based on observations of the changing brightness of stars as they pass behind the rings. The brightest areas have the greatest density of ring particles. The B ring is densely packed with clumps, called self-gravity wakes, separated by nearly empty gaps. These clumps, 30 to 50 m across, are constantly colliding. The formation of wakes is strongest in the bluer regions. Particles in the central, yellow regions are too densely packed for starlight to pass through. (Courtesy of NASA/JPL/University of Colorado)

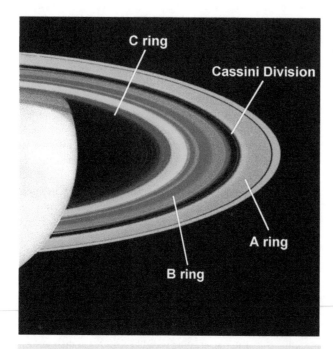

Figure 9.24 The temperatures of the unlit side of Saturn's rings are shown in this false-color image from Cassini. Red is about –173°C, green –183°C, and blue –203°C. Opaque regions are cooler, e.g. the outer A ring and the middle B ring. More transparent sections are relatively warm, e.g. the Cassini Division or the inner C ring. (Courtesy of NASA/JPL/GSFC/Ames)

Figure 9.25 An artist's concept of Saturn's ring material. The mini-satellites are composed mostly of ice, but are not uniform. They clump together to form elongated, curved aggregates, continually forming and dispersing. The space between the clumps is mostly empty. The largest individual particles shown are a few meters across. (Courtesy of NASA-JPL/ University of Colorado)

with the small moons orbiting just outside the ring. Spiral bending waves generated by Mimas produce corrugations in the ring with vertical amplitudes as large as 1 km.

One unexpected discovery was an object nicknamed "Peggy," a moonlet that appeared to be forming within the outer edge of Saturn's A ring.

The B ring has a very complex structure. It contains rings that are hundreds of kilometers wide and vary greatly in the amount of material they contain. A thick, 5,000 km-wide core contains several bands of material that are nearly four times as dense as ring A and nearly 20 times denser than ring C. Organized into uncountable ringlets, many of which are eccentric (not circular), the B ring is a dynamic, constantly changing region. Once again, the ringlets may be formed by the presence of numerous unseen moonlets.

Since there are so few gaps, most of the radial structure in the B ring must be due to variations in its optical thickness due to density waves, gravitational instabilities, or dynamical instabilities. Density waves are generated by the gravitational effects of Saturn's satellites, propagating out from the resonant orbits, e.g. at the 2:1 resonance with Janus, where particles travel twice around the planet for every single orbit of the satellite.

The outer edge of the B ring is elliptical, with a difference of some 140 km in its maximum and minimum diameters. This eccentricity is accounted for by the 2:1 resonant interaction with Mimas.

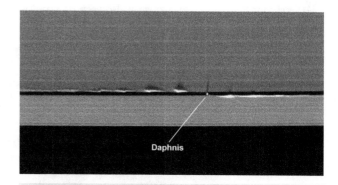

Figure 9.26 The 8 km-wide Daphis occupies an inclined orbit within the Keeler Gap in Saturn's A ring. The moon's gravity perturbs the orbits of the particles forming the gap's edge, creating horizontal and vertical waves. Here, the vertical waves cast dark shadows on the rings. Measurements show that they are 0.5–1.5 km high, i.e. up to 150 times taller than the thickness of the ring. Daphnis also casts a shadow onto the nearby ring. (Courtesy of NASA-JPL)

Cassini images also revealed shadows cast by vertical structures up to 3.5 km high in the variable outer edge of the B ring. Scientists concluded that this region contains independently orbiting moonlets embedded near the ring's edge. These moonlets are likely big enough to cause ring material streaming past them to be excessively compressed and piled up vertically as a result. The existence

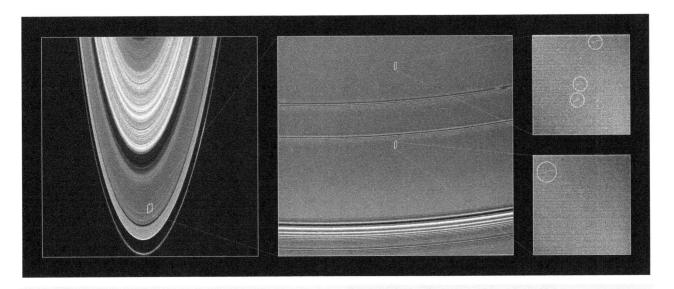

Figure 9.27 Propeller-shaped disturbances in the middle of the A ring. These indicate the presence of at least four small, embedded moons. This was the first evidence of the objects intermediate in size between the ring moons Pan and Daphnis and much smaller particles that comprise the bulk of the rings. The moonlets may be remnants of the body whose break-up produced the rings. (Courtesy of NASA/JPL/Space Science Institute)

Figure 9.28 Analysis of radio signals sent through Saturn's rings to Earth provided a profile of ring material distribution. This simulated image depicts the ring structure at a resolution of about 10 km. Color is used to represent ring particle sizes, based on the measured effects of the radio signals. Purple indicates regions where there are few particles smaller than 5 cm across. Green and blue shades indicate regions where there are numerous particles smaller than 5 cm and 1 cm, respectively. The broad white band near the middle of ring B is the densest region, where most of the radio signals were blocked. Such observations also show that all main ring regions are populated by a broad range of particle sizes, including boulders that are sometimes many meters across. (Courtesy of NASA/JPL)

of one 300 meter-wide moonlet embedded in this region of the B ring was confirmed because it cast a shadow over the ring.

Theoretical modeling has shown that, without forces to confine them, the rings would spread out over hundreds of millions of years. This spreading would happen because faster-moving particles that orbit closer to Saturn occasionally collide with slower particles on slightly more distant orbits. When this takes place, some of the faster particles' momentum is transferred to the slower particle, speeding the latter up in its orbit and causing it to move farther outward. The inverse happens to the faster, inner particle.

Gravitational tugs from Mimas are solely responsible for halting the outward spread of Saturn's B ring. Ring scientists had thought that Janus was responsible for confining the outer edge of the A ring, but a new modeling study, led by Radwan Tajeddine of Cornell University, shows that the A ring's outward transfer of momentum is kept in check by the cumulative effect of tugs from numerous moons, including Pan, Atlas, Prometheus, Pandora, Janus, Epimetheus, and Mimas. They combine to create the sharp outer edge of the A ring.

Spokes

The B ring is also characterized by sporadic radial markings or "spokes." In Voyager images these faint, wedge-shaped features typically extended outward 10,000 to 20,000 km across the B ring (Figure 9.29). The spokes were clearly visible to the Voyagers when Saturn was near its equinox, but in 1998 they faded from view and were not seen by Cassini when it arrived during Saturn's southern summer in 2004. It was not until 2005 that the first spokes reappeared.

When seen at low phase angles, they appeared dark; at high phase angles they appeared bright – indicative of tiny dust

Figure 9.29   A Voyager 2 image of numerous, dark "spokes" in the B ring. Often more than 10,000 km long, they could appear within a matter of minutes. The spokes survived for several hours, even though the particles in the B ring orbit at different rates. They are probably created by electromagnetic forces acting upon micron-sized ice particles. (NASA-JPL)

particles.[7] They formed at a distance from Saturn where the rotational speed of the ring particles roughly coincides with that of the planet's magnetic field lines.

The dark spokes appeared very quickly after the ring particles entered sunlight and then dissipated gradually as they moved toward the planet's shadow. Broader, less radial spokes that showed normal orbital behavior were also visible, probably older remnants that had survived. In some cases, new spokes appeared to be superimposed on older ones.

The importance of electrostatic charging effects in spoke formation and dynamics was supported by the detection at radio wavelengths of electrostatic discharges from the rings. These discharges loosely correlated with Saturn's rotation.

The spokes are made up of particles less than a micron across that collect electrostatic charges in the plasma environment of the rings and become subject to electrical and magnetic forces. Under the right conditions they gain an extra electron, allowing them to briefly levitate en masse from the ring's surface to a height of perhaps 80 km above the B ring.

Their visibility is now believed to depend on the elevation of the Sun above the rings and the changing light-scattering geometry: the less sunlight, the more likely they are to form. It seems that the spokes are visible for a period of about eight years, when the rings are nearly edge on to the Sun, and then disappear for six or seven years.

C and D Rings

The C ring is fainter and more transparent than the A and B rings – hence its other title, the "Crepe ring." The fine dust particles that make up the ring scatter light forward. It contains a number of dense ringlets whose locations are unrelated to orbital resonances with the larger satellites.

Within the C ring is the 270 km-wide Maxwell Gap, which contains an eccentric ringlet whose width varies from about 30 to 100 km. A more diffuse ring may indicate the presence of unseen moonlets in the gap.

The still fainter D ring extends from the C ring to within 7,000 km of Saturn's cloud tops. Details are difficult to observe, except with long exposures taken at favorable viewing angles. Cassini discovered that, although some ringlets within the D ring had stayed in the same place since the Voyager flybys, one of its brighter ringlets has dimmed considerably and moved 200 km closer to Saturn during that 25-year period.

Another example of dynamism in the D ring is a series of vertical or corrugated ripples, possibly generated by a recent collision with a comet or asteroid.

Towards the end of its mission, Cassini made numerous dives between the planet and the inner edge of the D ring, obtaining the first-ever samples of the planet's atmosphere and main rings. The spacecraft found no large particles, but many nanometer-size ring particles in the gap.

Cassini found a lot of water, which wasn't surprising because water makes up about 90% of the rings. However, it also sampled a variety of organic molecules in and just above Saturn's atmosphere, including hydrocarbons similar to propane, plus some methane and sulfur-bearing molecules.

The types of molecules became less well-mixed as the spacecraft looked deeper into Saturn's atmosphere, which is what would happen if the particles came from the rings and sank at different speeds. The material is thought to be especially raining down from the D ring, and, as a result, the innermost ring is slowly being eroded as it loses mass.

F Ring

Beyond the A ring is a gap known as the Roche Division, which contains two diffuse rings: R/2004 S1 is a 300 km-wide feature that lies along the orbit of Atlas, while R/2004 S2 is very similar and located immediately interior to the orbit of Prometheus. The latter's orbit is elliptical, swinging inward to the outer edge of R/2004 S2 and outward to the inner boundary of the F ring. The region occupied by its orbit has been cleared of material through the gravitational sweeping action of the 102 km-wide satellite.

The association of R/2004 S1 with Atlas and the main Encke Gap ringlet with Pan (plus two other diffuse rings either side of the G ring) suggests that all of these ringlets are fed by material derived from their orbital companions. The weak gravity of these small moons means they cannot retain any loose material on their surfaces. When they are struck by meteoroids, this material is blasted into space, creating diffuse rings. Collisions among

---

[7] The D, E, F, and G rings all show similar forward scattering properties to the B ring spokes, evidence that they are made of very small particles.

Box 9.3   Cassini

NASA's Cassini was one of the largest and heaviest scientific spacecraft ever launched. Together with the European Space Agency's piggybacking Huygens probe, the spacecraft weighed about 5.7 tonnes at lift off on October 15, 1997. Despite the use of a powerful Titan IV-B/Centaur launch vehicle, Cassini had to rely upon gravity assists from Earth, Venus, and Jupiter before it reached Saturn on June 30, 2004.

During the Saturn orbit insertion, the spacecraft passed through the gap between the F and G rings, looped above the ring plane, then descended between the F and G rings. The closest approach to Saturn prior to its final atmospheric entry occurred shortly before burn completion, some 20,000 km above the cloud tops. To protect the spacecraft from small particles, Cassini was turned so that its high-gain antenna acted as a shield.

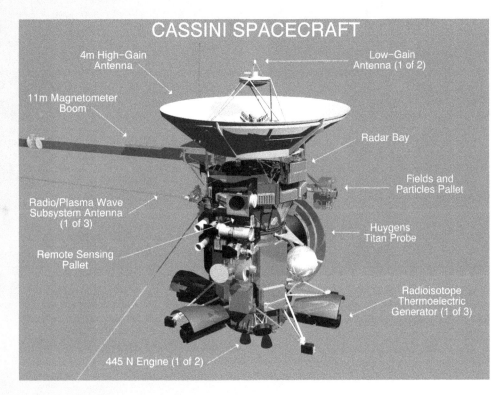

Figure 9.30   The nuclear-powered Cassini orbiter carried 12 remote sensing instruments to study the Saturn system. The European Huygens probe piggybacked on the ride to Saturn. (NASA-JPL)

During its four-year prime mission, Cassini completed 76 orbits of Saturn, including 45 close flybys of the Titan and numerous flybys of other moons. Titan's gravity was used to change Cassini's course and navigate around the Saturn system. The mission was subsequently extended until 2017.

The 12 remote sensing experiments included cameras, various spectrometers, an imaging radar, a magnetometer, a magnetospheric imaging instrument, a dust counter, and a radio science experiment.

Many of the most surprising results came from investigations of the satellites. The first of these to be seen at close quarters was Phoebe, which proved to have a very dark, ancient surface. However, the main focus of attention was Titan, with its orange clouds, dense atmosphere, and hidden surface. The radar instrument, in particular, revealed smooth plains, icy uplands, impact craters, dry riverbeds, and methane-filled lakes.

Perhaps the greatest surprise was Enceladus. Cassini discovered a tortured surface of "tiger stripes" where tidal forces have fractured and buckled the icy crust. An underground heat source at the south pole feeds geysers that propel a plume of water vapor, ice, and dust particles thousands of kilometers into space. These vents are the source of the minute particles that make up Saturn's enormous E ring.

The most bizarre, cratered landscape was found on Hyperion, an irregularly shaped rubble pile whose sponge-like surface displays closely packed and deeply etched pits. The cratered landscapes of Rhea and Dione also show signs of past geological activity in the form of fractures and ice cliffs.

Cassini's instruments also studied the smaller moons and the rings, e.g. the ring shepherd, Prometheus, which creates knots and kinks in the F ring. Studies of clumps, propellers, and waves in the rings showed an ever-changing population of ice-rich particles.

Saturn's cloudy atmosphere was also scrutinized, showing high-speed jet streams, long-lived oval storms, and thunderclouds, as well as a hexagonal cloud feature at the north pole.

After more than 13 years at Saturn, Cassini's mission ended on September 15, 2017, with an intentional plunge into the planet's atmosphere. During its final months, the spacecraft's cameras captured views from within the gap between the planet and the rings and made the first-ever direct measurements of the particles and gaseous molecules in this previously unexplored region. It completed 294 orbits of Saturn and made 162 targeted flybys of its satellites.

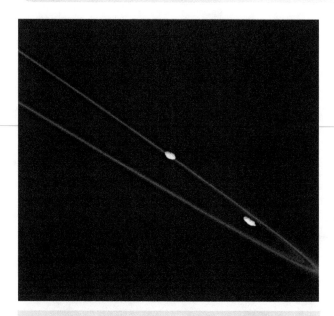

Figure 9.31 Two shepherd moons, Prometheus (right) and Pandora (left), are seen on either side of the F ring. Pandora is closer to the spacecraft in this view. Prometheus is slightly closer to Saturn, so it travels faster around its orbit, periodically overtaking Pandora. The moons' gravitational interactions disturb the ring particles, creating complex braids and other structures. (Courtesy of NASA/JPL/ Space Science Institute)

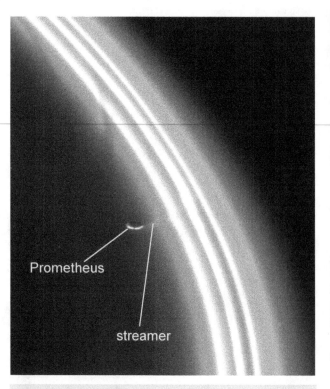

Figure 9.32 Prometheus creates complex structures in the F ring, shown here as five distinct strands. The inner strands are disrupted during the moon's regular passes on the inside of the ring. Prometheus is pulling material from the ring, as shown by the faint streamer. Discontinuities or "kinks" in the ringlets are seen, as well as gaps in the diffuse inner strands. (Courtesy of NASA/JPL/Space Science Institute)

several moonlets, or clumps of boulder-sized rubble, may also lead to debris trails. Many of the particles in these ringlets may be micron-sized dust, implying that non-gravitational forces also affect the ringlets' dynamics.

On the outer edge of the Roche Division is the F ring, the most chaotic member of the entire ring system. The narrow ring is shepherded on either side by two tiny moons, Prometheus (closer to Saturn) and Pandora. Prometheus sometimes enters the F ring, leaving a "gore" or channel where its gravitational influence clears out some of the smaller ring particles. The moon then draws ring material with it as it exits the ring, leaving streamers in its wake.

Discovered by Pioneer 11 in 1979, the F ring displays kinks, clumps, and strands, as well as braids. Voyager images showed one bright "core" strand and four fainter ones, each about 70 to 100 km across. During a stellar occultation, the core was seen to be subdivided into many narrower strands about 3 km wide, possibly as a result of larger particles in the ring. Clumps of material in the

ring were shown to be long-lived, and fairly uniformly distributed at roughly 9,000-km intervals around the ring.

The clumps and some of the other ring patterns are undoubtedly formed as a result of gravitational tugs from the nearby shepherding satellites. Since encounters between a moon and a ring particle recur during each orbit of the planet, the overall effect is a pattern that roughly repeats, with a spacing proportional to the difference in orbital speeds between the ring particle and the moon. The length scale of most F-ring structures indicates that Prometheus, the larger and closer of the two moons, is responsible for most of the braiding.

However, perturbations by the shepherds cannot explain many of the ring's other features. Voyager images showed that the

Figure 9.33 The F ring shows a variety of phenomena in this Cassini image. Near the lower right are two "fans" of material radiating out of the main strand (or "core") of the ring. Kinks are apparent all along the core, and dark "channels" in the core can be seen in places, the result of a recent interaction with the shepherd moon Prometheus (not visible in this image). Image scale is 6 km per pixel. (Courtesy of NASA/JPL-Caltech/Space Science Institute)

brightest clumps are not regularly spaced. Some of them, called "bursts," appear suddenly and then spread out over periods of days to weeks. Others evolve more slowly but fade away after a few months. One hypothesis is that the most rapid "bursts" are dust clouds arising when a meteoroid hits the ring; the other is that all the clumps arise from mutual impacts among ring bodies.

The ghostly ringlets flanking the ring's core are arranged into a spiral structure that wraps around the planet three times. Comparable structures in the main rings are associated with density and bending waves initiated by the gravitational influence of a large moon. However, the F ring's spiral structure contains very little mass and appears to originate from material that is somehow ejected periodically from the core strand and then sheared out due to the different orbital speeds followed by the constituent particles.

The spiral may be caused by one or more tiny moons crossing the F ring and spreading particles around, or by a rare, large impact with the ring. Cassini found at least two moonlets or sizeable clumps of material that appear to cross the F ring periodically. One of them, S/2004 S6, can intersect the F ring at very high speeds and may be responsible for disrupting the ringlets and forming the spiral. S/2004 S6 may be a particularly large shard left over from the impact.

How was the F ring formed, and why it is so strange? First, it orbits at the edge of the Roche limit, where some re-accretion is possible, so ring bodies can continuously break up and join back together. Second, it is faint and narrow, so that small injections of new dust are quite noticeable. The ring and its shepherds all follow eccentric orbits, which means that the ring is perturbed quite markedly as Prometheus and Pandora approach and recede during each orbit.

On the other hand, it seems to have sufficient mass to maintain a fixed eccentricity, overcoming the tendency of ring particles to

precess at different speeds. The overall picture favors a ring that arose from the disruption of a small moon – perhaps the size of Prometheus – but subsequently suffered too many external disturbances to ever settle down into a uniform, circular feature.

Outer Rings

Beyond the F ring are two much wider, but less noticeable, rings, labeled G and E, plus a number of faint rings that are co-orbital with small satellites. All of these are many times more tenuous and transparent than the inner rings.

The Janus-Epimetheus ring is visible between the G ring and the bright main rings and is about 5,000 km wide. Another recent discovery, the Pallene ring, is about 2,500 km across, and located between the G ring and the E ring (and between the moons Mimas and Enceladus). The satellites are the source of material for the rings. Impacts on Janus, Epimetheus, and Pallene knock particles off their surfaces and inject them into Saturn orbit.

The G ring, first imaged by Voyager, lies more than 15,000 km from the nearest sizeable satellite and about 170,000 km from the center of Saturn. Like the D ring, it contains very little material, and is largely made of tiny, icy particles.

Its existence at this location was a long-standing mystery. However, Cassini observations showed that it contains a bright, long-lived, arc of material. The arc is very tightly confined to the

Figure 9.34 This view, acquired on the shaded side of Saturn, reveals two previously unknown faint rings. One of these is a broad, diffuse ring coincident with the orbits of Janus and Epimetheus. It lies between the F and G rings and is about 5,000 km wide. The other, coincident with the orbit of Pallene, is about 2,500 km across and located between the E and G rings. Pallene orbits Saturn between Mimas and Enceladus. The bright dot in the Pallene ring is a background star. (Courtesy of NASA/JPL/Space Science Institute)

Figure 9.35 The entire inner ring system seen when Cassini was in Saturn's shadow so that microscopic ring particles appear brighter. The mosaic shows one faint ring along the shared orbit of Janus and Epimetheus, and another along Pallene's orbit. Color contrast is greatly exaggerated. Reddish, radial "flares" are artifacts caused by scattered light within the camera's optics. (Courtesy of NASA/JPL/Space Science Institute)

inner edge of the ring. It extends one-sixth of the way around Saturn and is about 250 km wide, much narrower than the full 7,000 km width of the ring.

Cassini also discovered a moon, called Aegaeon, which is estimated to be 0.5 km in diameter, orbiting within the G ring arc. This seems to be the major source of the ring material. Similar to the other dusty rings, debris escaping from the moon forms the arc, then spreads to form the rest of the ring.

Encircling the entire Saturn system is the diffuse E ring, which extends about 1 million kilometers from the orbit of Mimas as far out as Rhea. Discovered by ground-based telescopes in 1967, the thick disk is made of ice and dust particles only one micron across. Five of Saturn's seven largest moons are embedded within it: Mimas, Enceladus, Tethys, Dione, and Rhea.

The E ring has a very different origin from its neighbors. Its material is largely derived from geysers on the surface of the geologically active moon Enceladus (see Tiger Stripes on Enceladus). The vents eject clouds of water ice particles which are spread into a broad torus around the planet by the motions of the moons and Saturn's magnetic field.

An even larger ring was discovered by the Spitzer Space Telescope in February 2009. Made of ice and dust particles, the sparse ring lies in the plane of Saturn's orbit. Its inner edge is known to start about 7.7 million km from the planet and extends out to at least 12.5 million km, but it may be even larger (Figure 9.36). Its remarkable depth of about 2.4 million km is accounted for by the presence of the heavily cratered moon Phoebe within the ring.

Continuous impacts with Phoebe produce the particles that feed the ring. However, Phoebe has an elliptical orbit whose plane is inclined to Saturn's orbit. Bodies that are ejected into Phoebe's orbit share this inclination, so they travel above and below the main ring plane. Over time, their orbits precess at different rates, depending upon their average distance from Saturn, so the ring particles spread out over a huge area.

Moreover, the particles are affected by re-radiation of absorbed sunlight, which exerts an asymmetric force that causes them to

spiral in towards Saturn. Most material from 10 microns to centimeters in size ultimately collides with Iapetus, with smaller amounts striking Hyperion and Titan (see Spongy Hyperion).

Satellites

Saturn has 82 known satellites (not counting the numerous moonlets in the rings), more than any other planet. One of these, Titan, is a planet-sized object, comparable to Jupiter's four Galilean moons. The family also includes seven intermediate-sized worlds (Figure 9.37). Mostly between 400 and 1,500 km across, they have very low bulk densities, evidence that they are mostly composed of water and other ices. All of these "regular" satellites move along nearly circular, prograde orbits in the planet's orbital plane (Box 9.5).

However, by far the majority of the satellites are much smaller. They can be subdivided into two main groups: the inner moons associated with the ring system (which also follow regular orbits) and a retinue of captured objects that occupy eccentric, often retrograde, orbits far from Saturn.

The first to be discovered was the largest and brightest – Titan. Found by Christiaan Huygens in 1656, it was only the sixth known planetary satellite. Iapetus, Rhea, Dione, and Tethys were subsequently found by Giovanni Domenico Cassini between 1671 and 1684.

More than a century passed before William Herschel discovered Enceladus and Mimas in 1789. Saturn's eighth satellite (Hyperion) was discovered by William Lassell and William Bond in 1848. Phoebe, found by William Pickering in 1898, was unusual, since it followed a retrograde, eccentric path with an average distance of 13 million km from the planet – far beyond the other known satellites.

More recent additions to the satellite list have come from spacecraft observations and dedicated searches carried out using ground-based telescopes. The most favored times to search were once considered to be when the rings were edge-on, but the latest discoveries have been made far from the rings and the ring

Figure 9.36 In February 2009, the Spitzer Space Telescope discovered a huge ring between 7.7 million and 12.5 million km from Saturn. It is tilted about 27° to the main ring plane and encompasses the orbit of Phoebe. Both the ring and Phoebe orbit in the opposite direction to the other rings and most of Saturn's moons. Material blasted off Phoebe by small impacts remains in its neighborhood for millions to billions of years. Some of it spirals in toward Saturn and is swept up by Iapetus, Titan, and Hyperion. (NASA-JPL)

Figure 9.37 The relative sizes of Saturn's nine major satellites in order of distance from the planet (left to right: Mimas, Enceladus, Tethys, Dione, Rhea, Titan, Hyperion, Iapetus, and Phoebe). Most of them are between 400 and 1,500 km across and have very low bulk densities, evidence that they are mostly composed of water and other ices. Titan is similar in size and density to Jupiter's largest satellites, Ganymede and Callisto. (Paul Schenk, Lunar and Planetary Institute)

plane. Discoveries of these small, outer satellites require some of the world's largest telescopes in combination with a sensitive, wide-field digital camera.

Orange Titan

Long before Voyager 1 flew past Titan in November 1980, ground-based observations had shown that the satellite was unique, with an intriguing orange hue and spectroscopic evidence of methane and hydrogen gas. Clearly, Titan was large enough to have a significant atmosphere, and speculation was rife that, beneath its clouds, it might have a nitrogen-rich atmosphere and an icy surface coated with tarry organics or oceans of liquid methane.[8]

Stellar occultations and direct measurements had indicated that Titan is the largest satellite in the Solar System, but more accurate radio occultation data obtained using Voyager revealed that it

[8] Titan's average distance from Saturn is 20.3 Saturn radii, so it is sometimes inside, and sometimes outside, the planet's magnetosphere.

Figure 9.38   A Voyager 2 image showing Titan completely shrouded in a high-level, orange smog made up of complex hydrocarbons. The main features visible are a relatively bright southern hemisphere and a dark northern polar hood. The image was taken on August 22, 1981, from a range of 4.5 million km. Spring was just beginning in its northern hemisphere. (NASA-JPL)

is actually a little smaller than Ganymede, with a diameter of 5,150 km. The reason for the error was that Titan's atmosphere was unexpectedly thick – some 10 times deeper than Earth's.

With a bulk density almost twice that of water, the moon is made up of equal amounts of ice and rock. In that sense, it is a near twin of Ganymede and Callisto, with an undifferentiated core of silicate rock and ice that extends about two-thirds of the way to the surface.

Recent models suggest that the overlying layers comprise various high-pressure forms of ice. These are topped by an ocean of liquid water mixed with ammonia (which acts as an antifreeze) and a surface crust of rock-hard "ordinary" ice a few tens of kilometers thick. Despite the possible presence of a subsurface ocean, no evidence of a magnetic field has been found.

The methane that plays such a key role in Titan's atmospheric and surface evolution is locked in the crust of water ice. This may be in the form of methane clathrate – methane trapped in the crystal structure of the ice or in pores within the crust. The gas would then be released by seasonal warming or localized cryovolcanism.

Like many satellites, Titan is tidally locked to Saturn, and hence always presents the same face toward the planet. To view all hemispheres of Titan from Earth therefore requires observations during almost one entire orbital period, 16 days.

Today, Titan's atmosphere has a surface pressure of about 1.5 bars, 50% greater than on Earth. It is likely to have been much more massive in the past. Why does Titan have such a thick atmosphere when Ganymede and Callisto have essentially none?

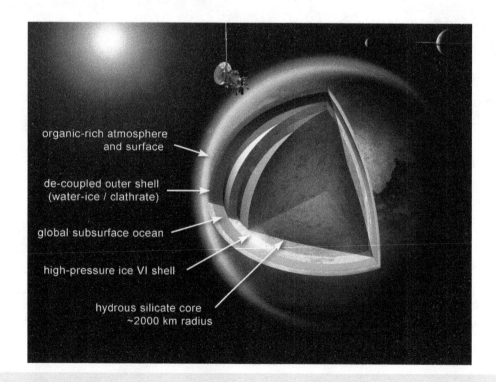

Figure 9.39   A cutaway view of Titan showing a possible subsurface ocean of liquid water and ammonia, sandwiched between layers of water ice. At the center is an undifferentiated core of silicate rock and ice. (Courtesy of A.D. Fortes/UCL/STFC)

The low temperature of the nebula that surrounded the young Saturn allowed Titan to acquire and hold onto the moderately volatile compounds methane and ammonia (NH$_3$), in addition to water. As sunlight split the ammonia, it was converted to nitrogen and hydrogen. While most of the hydrogen escaped into space, the heavier nitrogen remained. The higher temperatures of the Jovian moons, which were closer to the Sun, prevented them from retaining such an atmosphere.

Today, about 95% of the satellite's atmosphere is nitrogen, with most of the remainder in the form of methane. In fact, Earth and Titan are the only places in our Solar System with a dense atmosphere of molecular nitrogen.

The dominance of nitrogen on both worlds has led to descriptions of Titan as an early Earth in deep freeze. Titan's present environment could be similar to that on Earth billions of years ago, before life forms began pumping oxygen into the atmosphere. However, a surface temperature of –180°C suggests that the emergence of life – at least on the surface – is most unlikely.

Another intriguing aspect of the atmosphere is the substantial amount of methane (CH$_4$) –second only in abundance to nitrogen. At the temperatures present on Titan, methane can take the form

of a gas, liquid, or solid – just like water on Earth. This means that Titan has a methane cycle, with ethane-methane lakes that slowly evaporate, eventually resulting in cloud condensation that leads to methane rain or snow.

However, the entire methane content of the atmosphere would be destroyed in only 10 million years by the action of solar ultraviolet light. In an irreversible process, the liberated hydrogen escapes to space while organic photolysis products, such as ethane, drizzle down to the surface (see Chemical Factory).

So how can the gas have survived in large quantities to the present day? The only explanation is that some source – perhaps an underground ocean, a large aquifer, or cryovolcanism – is continually replenishing the methane supply. (An unlikely alternative is that biology may also play a role in creating Titan's methane.) Outgassing from the interior may have pumped methane into Titan's atmosphere over billions of years.

Chemical Factory

Methane and nitrogen are the raw materials for many chemical reactions that lead to the creation of complex organic molecules in Titan's atmosphere and on the surface.

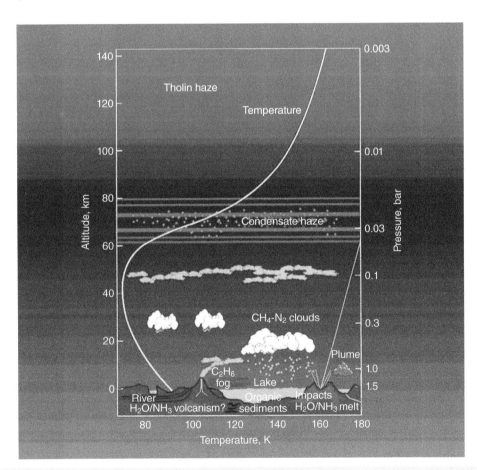

Figure 9.40 A cross section through Titan's atmosphere. High altitude haze is made of complex organic molecules known as "tholins." Lower down is a denser layer of tiny frozen condensates, including hydrocarbons such as ethane. In the troposphere are localized clouds of liquid methane which occasionally produce heavy rainstorms. The methane cycle is completed by evaporation from surface lakes and cryovolcanism. (NASA-JPL)

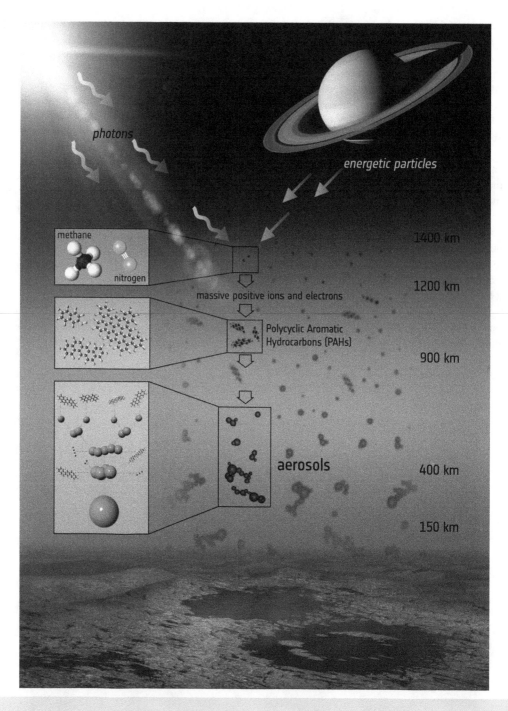

Figure 9.41   The stages in the formation of the aerosols that make up Titan's high level orange haze. When sunlight (solar photons) or highly energetic particles from Saturn's magnetosphere hit the atmosphere above 1,000 km, they split the nitrogen and methane molecules. This creates massive positive ions and electrons, which trigger a chain of chemical reactions. These produce various hydrocarbons, including polycyclic aromatic hydrocarbons (PAHs) – large carbon-based molecules that form from the aggregation of smaller hydrocarbons. As they grow, PAHs become heavier and tend to sink into the lower atmosphere. This eventually leads to the production of large, carbon-based aerosols in the lower haze layers, below about 500 km. (Courtesy of ESA/ATG medialab)

Much of the chemistry takes place at high altitude (Figure 9.41). Several tenuous haze layers occur up to 700 km above the hidden surface. These and the main brownish haze layer at an altitude of 200 km were created by interactions between the nitrogen-methane atmosphere, ultraviolet light from the Sun, cosmic rays, and charged particles trapped in Saturn's magnetic field. The hydrocarbons produced by this methane destruction form a photochemical smog similar to that found over large cities, only much thicker.

Many minor constituents have been detected in the smog layer, including ethane (C_2H_6), acetylene (C_2H_2), and propane (C_3H_8). Methane interacting with nitrogen atoms also forms "nitriles" such as hydrogen cyanide (HCN). Lower down is a denser layer of tiny, frozen condensate particles, possibly nitriles and hydrocarbons such as ethane.

The photochemical reactions result in large amounts of liquid and solid hydrocarbons raining onto the frigid surface. The heaviest rainstorms come from low-level methane-ethane clouds created by localized convection or gas rising over mountains. However, the solid smog particles, including acetylene and other compounds, also fall very slowly through the atmosphere, taking about a year to reach the surface. There they accumulate over millions of years, possibly forming a layer of hydrocarbon sludge hundreds of meters thick.

Seasonal Changes

Since its rotation axis is tilted at an angle of 26.7° to the Sun, Titan has seasons. With a year on Titan lasting almost 30 Earth years, each hemisphere experiences winters and summers lasting more than 7 years. The southern summer ended with the autumnal equinox in late 2009, after which the northern hemisphere started to tilt towards the Sun.

Voyager's cameras revealed a satellite surrounded by an impenetrable cloud blanket, relieved only by a dark hood over the north pole and high-level hazes (see Figure 9.38). Information on atmospheric density and temperature was provided by studying radio signals transmitted to Earth, revealing the profile all the way to the surface.

Subsequent observations have shown that Titan's climate changes with the seasons. The winter hemisphere typically displays more high-altitude haze, making it darker at shorter wavelengths (ultraviolet through blue) and brighter at infrared wavelengths (Figure 9.42). The switch between dark and bright occurred over the course of a year or two around the 2009 equinox. The boundary between the two areas appears to run directly east–west near the equator, but its position is actually offset from the equator by about 10° of latitude.

Titan experiences other regular climatic changes during the different seasons. When Cassini arrived at Saturn in 2004, Titan's north pole was in midwinter, while the south pole was enjoying a long summer.

Apart from the hood of high-altitude haze over the winter pole, Cassini observed a swirling northern polar vortex which was associated with a localized hot spot and a downwelling of gas. The meridional (north–south) atmospheric circulation was dominated by a single pole-to-pole cell that ascended at warmer southern and equatorial latitudes and subsided over the northern winter pole.

Strong vortex winds formed a mixing barrier and effectively isolated the polar air mass, permitting a distinct temperature and composition within the vortex. This vortex was

Figure 9.42   An artist's impression of seasonal atmospheric changes on Titan. In each image a detached haze layer (blue) extends around the moon (blue). During the southern summer, a hood of dense gaseous haze (white) existed in a vortex above its north pole, along with a high-altitude hot spot (red). At equinox, both hemispheres received equal heating from the Sun. After equinox, there was still a significant build-up of trace gases over the north pole, even though the vortex and hot spot had almost disappeared. Meanwhile, similar features began developing at the south pole. This probably signified a large-scale reversal in the single pole-to-pole atmospheric circulation cell of Titan immediately after equinox, with upwelling of gases in the summer hemisphere and a corresponding downwelling in the winter hemisphere. (Courtesy of ESA)

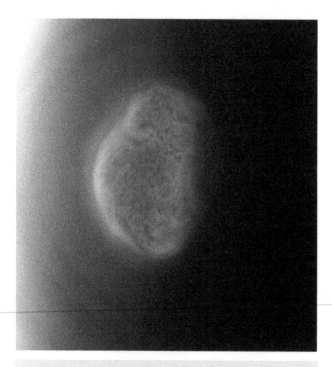

Figure 9.43   A high-altitude haze associated with a swirling vortex appeared over the south pole of Titan in 2012, during the southern winter. This natural-color image shows a region where Titan's air is sinking in the center and rising around the edges. One rotation of the haze feature took about 9 hours. (Courtesy of NASA/JPL-Caltech/Space Science Institute)

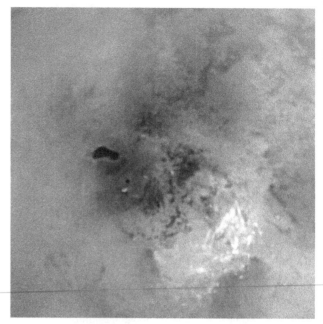

Figure 9.44   This dark feature in Titan's south polar region marks a former or present lake of liquid hydrocarbons. The brightest features are convectional methane clouds. A red cross below center marks the pole. This composite of three Cassini infrared images was taken on June 6, 2005, during Titan's southern summer. Resolution is approximately 3 km per pixel. (Courtesy of NASA/JPL/Space Science Institute)

characterized by trace gas enrichment due to subsidence from upper-atmosphere photochemical source regions; a cold lower stratosphere due to radiative cooling and a lack of insolation; and a hot stratopause/mesosphere due to subsidence.

After the 2009 equinox, Titan's south pole began to enter winter and the pattern of circulation began to reverse almost immediately. The resulting south polar subsidence created a hot spot (–93°C at 300–400 km altitude) and high-altitude enrichment of trace gases.

The reversal was accompanied by a two-cell transitional global circulation, with upwelling around the equator and subsidence at both poles. This persisted for two years before a fully reversed single circulation cell was established.

Methane Clouds

As is the case with water on Earth, Titan has a methane cycle: the gas evaporates from the surface, then cools and condenses as it rises, releasing latent heat and causing cloud formation. Occasional rainstorms then return liquid methane to the surface, feeding rivers and lakes.

Ground-based infrared observations and images from Cassini revealed clouds in three distinct latitude regions of Titan. Widespread ethane clouds at altitudes of 30–50 km occur over the winter pole, due to a constant influx of ethane and aerosols from the stratosphere.

There is also patchy, convectional methane cloud at the summer pole, which is in permanent sunlight. At the height of the summer, large, convectional clouds occur within 30° of the pole.[9] These clouds have lifetimes of several weeks and result from heating of the surface, possibly combined with rising gas where the summer Hadley cell meets cooler polar air. When the fall arrives, these clouds disappear.

Short-lived (typically one terrestrial day), often elongated clouds also appear regularly in mid-latitudes during the summer. Solar heating alone may be the cause, although other explanations have been put forward, including a seasonal shift in global winds or enhanced surface activity which disturbs the atmosphere.

Severe convective storms, accompanied by intense precipitation, sometimes occur during summer. The strongest storms grow when the methane relative humidity in the middle troposphere is above 80 percent, producing updrafts with maximum velocities of 20 m/s that reach as high as 30 km before dissipating in 5 to 8 hours. Raindrops 1–5 mm across produce torrential downpours.

Clouds are rare in Titan's equatorial regions, unlike on Earth. At Titan's equator, the atmosphere is so thick that the temperature at night is the same as during the day. This is because the thermal response of the atmosphere at such a great distance from the Sun is

[9]Titan's southern summer solstice took place in October 2002.

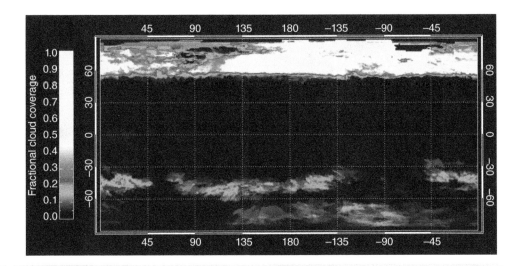

Figure 9.45 The percentage of cloud cover on Titan, July 2004 to April 2010, based on data from Cassini. Black is cloud free, yellow shows complete cloud cover. During the northern winter, clouds of ethane formed in the northern troposphere. In the warmer southern hemisphere, rising convection currents resulted in the condensation of methane, forming mid- and high-latitude clouds. (Courtesy of NASA/JPL/University of Arizona/University of Nantes/University of Paris Diderot)

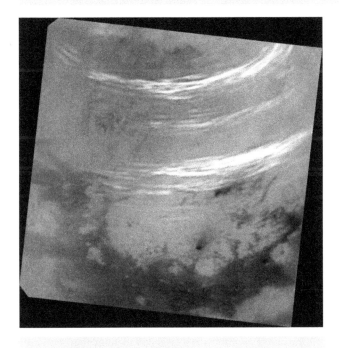

Figure 9.46 With the northern summer solstice only a few weeks away, Cassini observed an outburst of bright bands of methane cloud at Titan's mid-northern latitudes. This was the most extensive cloud outburst since clouds reappeared there in early 2016. A few isolated cloud streaks are also visible outside these bands. The image was taken on May 7, 2017. (Courtesy of NASA/JPL-Caltech/Space Science Institute)

much longer than the Titan day (16 Earth days). In the equatorial regions of Titan, the temperature has been largely unchanged in 20 years.

Significant changes took place between July 2004 (early southern summer) and April 2010, the start of northern spring

(Figure 9.45). Whereas both polar regions had been heavily overcast during the late southern summer, the clouds at the south pole completely disappeared just before the equinox in August 2009 and the clouds in the north began to thin out.

Observations of the northern polar regions were difficult during their long, dark winter. However, in September 2006, Cassini imaged a huge cloud formation at a height of 40 km that extended almost to latitude 60°N. This cloud, over 2,400 km in diameter, was probably composed of frozen ethane, rather than methane droplets, since the particles were only 1–3 microns across. This may produce rain (or snow) as strong downdrafts drive organic particles towards the surface. This precipitation would feed the numerous methane/ethane lakes found on the surface.

As hoped, Cassini observed the reappearance of clouds at northern mid-latitudes and near the polar lakes in early 2016, although they were fewer than climatic models had predicted. However, by May 2017, with the summer solstice only a few weeks away, three broad bands of methane clouds appeared at mid-northern latitudes (Figure 9.46). These were some of the brightest clouds Cassini had ever observed on Titan.

The southern band was located between 30°N and 38°N – a region where few clouds had previously been seen. A fainter, middle band between 44°N and 50°N occurred where cloud bands had been observed fairly regularly over the past year. A northern band was also seen between 52°N and 59°N.

A few isolated cloud streaks were imaged outside these bands, including some near 63°N, while other streaks were observed as far south as 23°N.

Less spectacular is the bank of high, layered cloud that causes a widespread, persistent morning drizzle or mist of methane over the western foothills of Xanadu, the main continent on Titan. This drizzle may be formed as moisture-laden winds travel upslope, resulting in condensation. The drizzle or mist seems to dissipate after about three Earth days, when the Sun is fairly high in the sky.

Figure 9.47   A mosaic of Titan's surface taken from an altitude of 16 km during the descent of ESA's Huygens probe on January 14, 2005. The bright ridge is marked by numerous black, branching channels that appear to have been formed by flowing liquid – probably methane loaded with organic sediment. The darkness of the channels and the nearby lowland suggests that hydrocarbon dust has drifted from above onto the surface, then been washed into the dry riverbeds during a rainstorm and transported to the low-lying terrain – possibly an evaporated lake. (Courtesy of ESA/NASA/JPL/University of Arizona)

Figure 9.48   The Huygens landing site on Titan was imaged after touchdown by the probe on January 14, 2005. The surface was darker than expected, and covered with numerous pebble-sized stones made from a mixture of water ice and hydrocarbons. The largest boulder, 85 cm from the camera, was about 15 cm across. The ice pebbles rest in small hollows like those caused by erosion on riverbeds. The hazy image suggested the presence of methane or ethane mist. (Courtesy of ESA/NASA/University of Arizona)

Such observations support Huygens data indicating widespread clouds of frozen methane at a height of 25–35 km, plus liquid methane clouds below 20 km, with rain or drizzle at lower levels. On Titan, methane droplets are predicted to be at least one millimeter across, 1,000 times larger than water droplets in terrestrial clouds.

Rivers and Lakes

The Voyager flybys led to speculation that large expanses of liquid ethane and other hydrocarbons might exist on Titan. Infrared and radar observations made by the Hubble Space Telescope and ground-based instruments seemed to confirm this by revealing dark and bright patches – possibly oceans separating large, icy continents.

The Cassini orbiter and Europe's Huygens probe discovered a surface that bears some remarkable similarities to Earth (Figure 9.47). Bright, icy uplands are drained by branching river systems which flow towards much darker plains that likely flood from time to time, although they are largely dry at present. The plains seem to be covered with carbon-rich sediment that has

probably been removed from nearby uplands by rainstorms or seasonal springs, collected by streams and rivers, then transported downstream to lowland basins.

The evidence for fluvial erosion, transportation, and deposition was supported by images of the Huygens landing site (Figure 9.48). The surface had the consistency of soft sand and was covered with numerous rounded stones and pebbles made from a mixture of water ice and hydrocarbons. The largest boulder measured about 15 cm across. Hollows at the base of these ice pebbles resemble those caused by erosion on riverbeds.

Cassini's radar instrument also found numerous extremely smooth, dark patches around Titan's north pole (Figure 9.49).[10] These were assumed to be lakes – at least some of which contained liquid – but confirmation had to wait until Cassini obtained an image of sunlight reflected from a liquid surface.

There are dozens of these features. Some are less than a kilometer wide, but others are many times larger. Ontario Lacus, the largest lake in the southern hemisphere, is about 235 km long, although it may be only partly wet. However, it is dwarfed by Kraken Mare, which lies near the north pole. The largest hydrocarbon lake on Titan covers nearly 400,000 sq. km – more than the Caspian Sea on Earth.

---

[10] Dark regions in radar images generally mean smoother terrain, while bright regions mean a rougher surface.

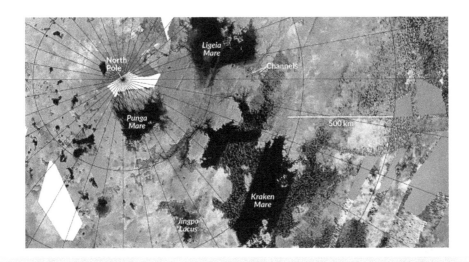

Figure 9.49 Numerous dark patches resembling lakes are visible in radar images of Titan's north polar region. With a surface temperature of about −180°C, the lakes probably contain a mixture of liquid methane and ethane. In this false color view, lakes and seas are blue. Radar-bright regions of dry land are tan. Black regions are caused by lack of data. The north pole is at the center. The view extends down to 50°N latitude. (Courtesy of NASA/JPL-Caltech/ASI/USGS)

The lakes, probably filled with liquid methane, ethane, and nitrogen, are the source of the hydrocarbon smog. In some cases, rims can be seen, suggesting deposits that form as liquid evaporates. Nearby channels seem to have been carved by liquid. The surface is also covered with alluvial features, suggesting that methane rain sometimes feeds surface flow, resulting in widespread erosion, transportation, and deposition.

An unusual phenomenon detected in Cassini's radar images was bright regions in Ligeia Mare and Kraken Mare that appeared and then disappeared. One suggestion is that these "magic islands" may be the result of bubbles which form as a result of mixing between liquids at the surface and at depth.

Scientists have theorized that part of the nitrogen-rich surface mixture flows down to deeper regions, which contain more ethane. The liquids then separate under the effect of pressure, and the gaseous nitrogen bubbles back up to the surface. These bubbles are formed at a depth of 100 to 200 m, and can reach a diameter of four cm.

This phenomenon can spread over several hundred square kilometres of sea, although the "magic island" is a transient, sporadic phenomenon.

In 2018, a simpler explanation was put forward. The brightening could simply be the glint of sunlight reflecting directly off giant waves on the lakes, like ocean ripples tinted with gold at sunset. Simulations of Titan's dense atmosphere suggest these waves could be raised by winds as slow as 0.5 m per second, which would barely move a wind vane on Earth.

Dunes, Mountains, Volcanoes, and Craters

Radar also reveals dune fields, mainly concentrated within 30° of the equator. They are thought to be composed of small hydrocarbon or water ice particles – probably about 250 microns

Figure 9.50 A Cassini synthetic aperture radar image of long, dark ridges – longitudinal dunes created by prevailing winds. They may be composed of solid organic particles or ice coated with organic material. Up to 3 km apart, the dunes curve around bright features that may represent high topography. The image is centered at 8°N, 44°W, and covers approximately 160 x 325 km. Spatial resolution is about 500 m. (Courtesy of NASA/JPL)

in diameter, similar to sand grains on Earth. The "sand" may have formed when methane rainstorms produced flash floods that eroded particles from the icy bedrock. Alternatively, it may come from organic solids that are produced by photochemical reactions in the atmosphere and then fall onto the surface.

The huge longitudinal dunes are 100 m high and run parallel to each other for hundreds of kilometers. Up to 10 km apart, they

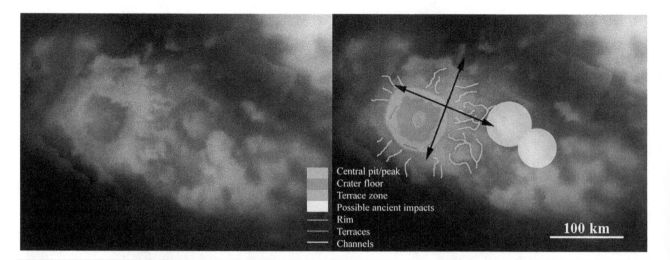

Figure 9.51 A Cassini image of 90 km-diameter Selk impact crater (left) and a simple geology map of the area (right). The crater's rim is in the middle of the optically bright area and the interior walls are terraced. The crater rim is polygonal, indicating some possible planes of weakness in the crust (black arrows). The exterior is cut by drainage channels (blue lines), and the crater may be superposed on two ancient, degraded craters (yellow circles). (Courtesy of NASA-JPL/UA/Soderblom et al.)

form and grow through the action of fairly strong (1 m/s or more) west-to-east surface winds.

Data from Cassini also revealed what appear to be giant dust storms in equatorial regions of Titan. The discovery makes Titan the third Solar System body, in addition to Earth and Mars, where dust storms have been observed.

The storms appeared as unusual equatorial brightenings in infrared images taken by Cassini around the 2009 equinox. The features were visible for periods between 11 hours and five weeks. Analysis indicated that they were about 10 km high, well below where methane clouds would be expected at that time of year. Modeling showed that they must be tiny solid organic particles lifted aloft by strong winds blowing over the dune fields around Titan's equator.

Unlike most satellites, impact features are rare on Titan. Although radar has mapped little more than half of the surface, only about 60 impact craters have been found. This is partly due to comets and meteoroids breaking up and incinerating during descent through the thick atmosphere, but it likely also indicates relaxation of the icy surface over time, as well as weathering by methane rain, fluvial erosion, and burial by photochemical sediments.

The craters were all within about 30 degrees of the equator, a relatively dry region on Titan. Craters on Titan are, on average, hundreds of meters shallower than similarly sized craters on Jupiter's giant moon Ganymede, suggesting that some process on Titan is filling its craters. Images show craters at all stages of filling; some just beginning to be filled in, some half filled, and some that are almost completely full. This suggests the presence of windblown sand, which would fill the craters and other features at a steady rate.

Cassini's radar has imaged one large, possibly multi-ringed crater – known as Menrva – with an outer diameter of 440 km. The feature lacks a central peak, suggesting that it has been eroded or otherwise modified since formation.

Another, known as Selk, measures roughly 180 km in diameter, and is surrounded by bright material that is probably an ejecta blanket covering the surrounding plains (Figure 9.51). The inner part of the crater is dark and may represent smooth deposits. There is also evidence of a central peak.

Cassini also discovered several mountain ranges near Titan's equator. The highest peaks rise to a height of about 2 km, comparable to the Appalachian Mountains of the USA. The main range, which runs southeast to northwest, is about 150 km long and 30 km wide. To the east, in a region known as Xanadu, are three parallel ridges known as Mithrim Montes, which include Titan's tallest peak, with an elevation of 3,337 m (Figure 9.52). However, Titan's crust is icy rather than rocky, so its bedrock is softer and its mountains cannot rise as high as on Earth.

The parallel ridges are spaced about 50 km apart and they appear to be tilted or separated blocks of broken or faulted crust. Their regular spacing is typical of regions that have been compressed or extended over large areas. In the case of Titan, crustal compression resulting in folds and thrust faults is considered to be more likely.

Along the south sides of the ridges are prominent cliffs, or scarps, up to a few hundred meters high, that are likely associated with linear faults. This indicates that tectonic forces have acted in a north to south direction across Titan's equatorial region. However, there is no suggestion that Titan is shaped by plate tectonics, which is a process unique to Earth.

The origin of Titan's tectonic forces is unknown. One suggestion is that the mountains may result from cooling of the interior and resultant shrinkage, which has caused the outer ice crust to compress, fold, and fracture. Other possibilities are changes in Titan's rate of rotation and tidal forces from Saturn.

Cryovolcanism (ice volcanism) may be fairly widespread on Titan, involving emissions of a slurry of ice and liquid hydrocarbons, instead of rocky lava. However, proof is hard to find. Radar

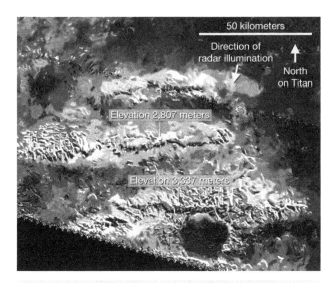

Figure 9.52 Cassini's radar instrument imaged three parallel ridges, known as Mithrim Montes, in a mountainous region called Xanadu. They include Titan's highest known peak (3,337 m). These may be folds or tilted blocks of faulted crust. Along the south sides of the ridges are prominent cliffs, or scarps, visible as thin, radar-dark lines trending west-to-east. The ridges are spaced about 50 km apart. They were probably formed by compressional tectonic forces acting in a north–south direction in Titan's equatorial region. Bright regions indicate materials that are rough or scatter the radar beam; dark regions are relatively smooth or absorb radar waves. (Courtesy of NASA/JPL-Caltech/ASI)

Figure 9.53 A false-color, 3D radar image of Doom Mons, the best known candidate for a cryovolcano on Titan. The landform includes two peaks, both more than 1,000 m high, and a nearby depression (possibly a caldera) known as Sotra Patera, which is about 1,700 m deep and 30 km wide. Topography is vertically exaggerated by a factor of 10. Dunes appear brown-blue. Possible ice-lava flows appear bright yellowish-white, like the mountain and caldera. (Courtesy of NASA/JPL-Caltech/USGS/University of Arizona)

images of apparent domes, cryovolcanic flows, and changes in surface brightness can be interpreted in various ways.

The most likely candidate found to date is 70 km-wide Doom Mons[11], which is estimated to stand about 1,450 m high (Figure 9.53). Nearby is a depression (possibly a caldera caused by surface collapse) known as Sotra Patera. The deepest depression yet found on Titan, the non-circular Sotra Patera is about 1,700 m deep and 30 km wide. On the side of the mountain are finger-like flows, named Mohini Fluctus, which may be less than about 100 m thick. This is the most likely example yet found of an ice volcano (cryovolcano) on Titan.

At least two more mountains, one with another big crater were observed nearby, forming a chain hundreds of kilometers long,

Another possible cryovolcano is Erebor Mons which is located near the equator, about 470 km to the north-northeast of Doom Mons. It is 40 km across, more than 1 km high, and has lobate flow features to its north and east.

The cryovolcanoes may eject "lava" composed of water and ammonia. Since liquid water is denser than ice, the process that enables it to rise to the surface is unclear, although the presence of ammonia could substantially lower its density and prevent it from freezing. Methane gas is also a likely ingredient, helping to explain the persistence of that gas in Titan's atmosphere.

Tiger Stripes on Enceladus

Even more surprising than Titan is Enceladus, a small icy moon that orbits much closer to Saturn, inside the planet's enormous E ring. The satellite has long been known to have an extremely reflective surface, evidence that it is coated with almost pure water ice. Ground-based observers also noted that Enceladus appears brighter when its south pole is visible. If the moon had frosted over, it must have happened recently or radiation would have changed the chemistry of the ice and darkened it.

The Voyager spacecraft confirmed that Enceladus has one of the youngest surfaces in the Saturn system, with sizeable expanses of relatively crater-free terrain. Some areas were moderately cratered with hollows up to 35 km across, some had numerous small craters, while others were smooth, grooved plains.

Many of the craters were softened in outline, suggesting some relaxation of the surface, possibly due to warming of the ice. There was also evidence of fairly recent geological activity in the form of sinuous mountain ridges 1 to 2 km high and linear fractures where the crust had apparently shifted. No giant craters were seen – evidence that the entire satellite had been resurfaced at least once.

Dynamic processes were expected on planet-sized Titan, but there was little reason to anticipate anything other than ancient, heavily cratered surfaces on the smaller satellites. Then, in 2005, the Cassini magnetometer discovered evidence of an extended, asymmetric atmosphere around the 504 km-diameter moon, Enceladus. The sparse gas was acting as an obstacle to the planet's magnetic field and plasma.

---

[11] By convention, mountains on Titan are named for mountains from Middle-earth, a fictional setting in fantasy novels by J. R. R. Tolkien.

Box 9.4 Huygens

In June 1982, less than a year after Voyager 2 flew past Saturn, a joint USA-Europe working group recommended development of a Saturn orbiter and a Titan probe. Named in honor of Christiaan Huygens, the probe was eventually selected as part of ESA's Horizon 2000 long-term science plan.

The probe was carried to Saturn by the U.S. Cassini spacecraft, which would relay data from Huygens during its descent through Titan's atmosphere. After launch, a design flaw was discovered in the Huygens communications system, so the mission plan had to be revised. It was decided to shorten Cassini's first two orbits around Saturn and insert an extra orbit. This meant that Cassini would fly over Titan's cloud tops at an altitude of about 65,000 km, more than 50 times higher than originally anticipated.

Huygens was released toward Titan on December 25, 2004, for an entry into the moon's atmosphere on January 14, 2005. The 319-kg, wok-shaped probe was equipped with six instruments and 39 sensors that would enable it to investigate Titan's haze layers and atmospheric composition, search for evidence of lightning, and measure wind speed, air pressure, and temperature. If it survived the landing, Huygens carried a Surface Science Package that its creators hoped would send back the first in situ measurements of an alien sea.

Figure 9.54 After a journey lasting almost seven years, Cassini-Huygens entered orbit around Saturn on June 30, 2004. The piggybacking Huygens probe was released on December 25, and entered Titan's atmosphere on January 14, 2005. This artist's view shows the descent sequence of the 2.7 m diameter probe after separation from Cassini (top left). Saturn was not visible through the dense clouds. (Courtesy of ESA/David Ducros)

After a 4 billion-km piggyback ride that lasted almost 7 years, Huygens separated from Cassini and headed for a near-equatorial landing site on Titan's sunlit hemisphere. On January 14 (European time) it slammed into the upper atmosphere at a speed of 6 km/s.

Protected by an ablative heat shield, Huygens' descent rapidly slowed, enabling it to deploy a 2.6 m pilot chute at an altitude of about 160 km. After 2.5 seconds, the main 8.3 m parachute was released to stabilize its swinging motion. Once the front shield was released, its instruments began to sample the atmosphere, take pictures, and study the strange environment. Microphones recorded the sound of wind rushing by the probe.

The descent lasted for 2 hours 28 minutes, but transmissions from the surface continued long after Cassini dropped below the horizon and was no longer able to receive the data. More than 474 megabits of data were received, including some 350 pictures taken during the descent and on the ground. The atmosphere was probed and sampled from an altitude of 160 km all the way to the surface, revealing a uniform mix of methane with nitrogen in the stratosphere.

Methane concentration increased steadily in the troposphere, with clouds of methane at about 20 km altitude and methane or ethane mist near the surface. There were some indications of rain falling from clouds. The measurement of atmospheric argon-40 indicated that Titan may have experienced cryovolcanic activity associated with water ice and ammonia.

Although some Doppler data and images were lost, the probe's signal was monitored by a global network of radio telescopes on Earth, making it possible to analyze variations in Titan's winds by reconstructing the descent trajectory with a remarkable accuracy of 1 km.

Results showed that the winds generally blow west to east above 10 km altitude, but in the opposite direction nearer the ground. They were weak near the surface, less than 5 m/s, increasing slowly up to an altitude of about 60 km. Higher in the atmosphere, Huygens was given a bumpy ride, probably due to significant vertical wind shear. The maximum wind speed of roughly 430 km/h was measured at a height of about 120 km. This powerful, high-altitude wind is an example of super-rotation, which is also seen in Venus' atmosphere.

Panoramic images returned during the descent showed a complex network of narrow, black, steep-sided drainage channels running from brighter highlands to lower, flatter, dark regions.

The branching channels merged into rivers that seemed to run into lake beds featuring offshore "islands" and "shoals." Although any surface liquid seemed to have evaporated or seeped underground, leaving the rivers and lakes dry, it was thought rain may have occurred in the recent past.

Bright hills rose about 200 m above the dark plains. The uplands were made of water ice (which is rock-hard at Titan's temperature of about –180°C), covered by organic deposits that drift down from above. Methane rains apparently wash the hydrocarbon dust off the hillsides into rivers, eventually depositing it on the lowlands.

Huygens landed at 10° S, 192° W. A penetrometer on the probe recorded the moment of landing at a speed of 4.5 m/s. The surface resembled wet sand or clay, composed of a mixture of water ice and hydrocarbons, and overlain by a thin crust. Bursts of methane gas boiled out of the damp surface as the result of heat generated by Huygens.

Bathed in subdued orange light, the landing site resembled a dry riverbed, overlain by small, rounded pebbles between 3 mm and 15 cm across. The absence of sharp edges and hollows at the bases of some rocks suggested that the area was sometimes inundated by flash floods of liquid methane (see Figure 9.49).

Subsequent flybys detected dust particles around Enceladus, and a huge cloud of gas over its south pole. Water vapor comprised about 65% of the cloud, with molecular hydrogen at about 20%. The rest was mostly carbon dioxide and other simple, carbon-based molecules.

Close-up images revealed a landscape near the south pole almost entirely free of impact craters and littered with house-sized ice boulders. The most noticeable features were continuous, ridged, slightly curved, and roughly parallel faults, nicknamed "tiger stripes" due to their distinct appearance when viewed in false color (Figure 9.55). These unique tectonic fractures are approximately 150 km long and about 40 km apart. Compounds found in the tiger stripes include carbon dioxide and hydrocarbons such as methane, ethane, and ethylene.

Enceladus is one of the coldest places in the Saturn system because its surface reflects more than 80% of the incoming sunlight (Figure 9.56). The temperature near the equator of Enceladus is –193°C, similar to predictions. However, Cassini's infrared spectrometer showed the south pole to be much warmer than expected. Although the poles should be even colder because the Sun shines so obliquely there, the average figure near the southern fractures is a remarkable –93°C. Clearly, the satellite has a strong internal heat source – unprecedented for an object of Enceladus' size.

Then, on November 27, 2005, Cassini observed the nightside of Enceladus (Figure 9.57). Clearly illuminated above the limb was a series of at least eight tenuous plumes erupting from geysers close to the south pole. The jets of fine particles reached hundreds of kilometers into space. All of the plumes were found to be associated with the "tiger stripes," or sulci.

The geysers are venting water vapor and fine water ice particles that have crystallized on Enceladus' surface within the last 10–1,000 years, as well as carbon dioxide, carbon monoxide, and organic compounds. One theory suggests that they are driven by liquid water under pressure below the surface, possibly containing

methane that has been locked up inside the satellite and is now escaping through the vents.

Studies suggest that the large number of ice particles and the steady rate at which these are produced require temperatures close to the melting point of water ice. Vapor evaporating from the water expands and cools as it rises to the surface through cracks in the ice crust. The resultant ice grains are carried along by the vapor plumes as they blast upwards at speeds of 300–500 m/s.

On reaching the narrow vents at the surface, the faster-moving water vapor shoots high above Enceladus, becoming trapped in Saturn's magnetosphere. However, most of the condensed ice particles fail to reach the escape velocity of 240 m/s and fall back to the surface. Only about 10% escape to form Saturn's E-ring.

The inescapable conclusion is that the geysers are supplied by a subsurface reservoir of liquid water between the moon's porous, rocky core, and icy crust (Figure 9.58). At first, researchers thought it existed only as a large sea beneath the south polar region, but it is now believed that the ocean is global, with a likely depth of about 50 km.

It is thought that strong tidal forces generated by Saturn affect the "loose" rock in the moon's core. Frictional heat is transferred very efficiently to the water circulating through the core, heating it to more than 90°C.

It seems possible that Enceladus' ocean floor could include hydrothermal vents similar to those on Earth, which are known to support life on the sea floor (see Chapter 3). This has led to speculation that habitable conditions could exist beneath the moon's icy crust.

Scientists have discovered hydrogen gas and very complex organic molecules in the plumes of material erupting from Enceladus. Analysis indicates that the hydrogen and carbon-based molecules are likely produced through chemical reactions between the moon's rocky core and warm water from its subsurface ocean.

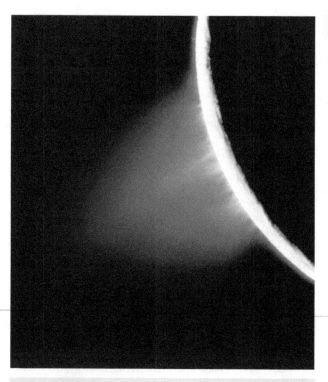

Figure 9.55 The anti-Saturn hemisphere of Enceladus imaged on July 14, 2005. The false-color view includes images taken at wavelengths from ultraviolet to infrared. Softened craters and complex, fractured terrains are visible. Many long fractures in the south (shown in blue) are markedly different in color from the surrounding terrain. These walls of these "tiger stripes" may expose coarse-grained ice, whereas powdery material covers the flatter surfaces. These linear fractures are the source of powerful geysers that blast icy plumes into space. (Courtesy of NASA/JPL/Space Science Institute)

Figure 9.57 A false color Cassini image show numerous gaseous plumes erupting into space from the south polar region of Enceladus. This view was processed to enhance the individual jets. Some artifacts due to the processing occur in the image. (Courtesy of NASA/JPL/Space Science Institute)

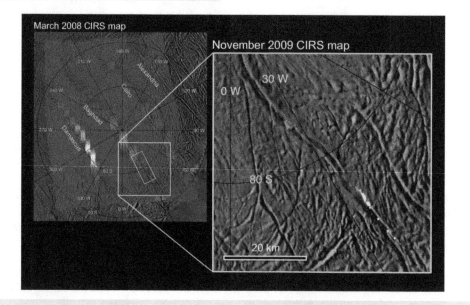

Figure 9.56 A false-color, infrared view of Enceladus' south polar region (left) shows heat radiating from the 150 km-long "tiger stripe" fractures in March 2008. Temperatures of at least 180 K (−93°C) were recorded along Damascus Sulcus. The higher-resolution infrared image (right), obtained on November 21, 2009, shows the warm regions are less than 1 km across within the Baghdad Sulcus fracture. The intensity of heat radiation increases from violet to red, orange, and yellow. (Courtesy of NASA/JPL/GSFC/SwRI/SSI)

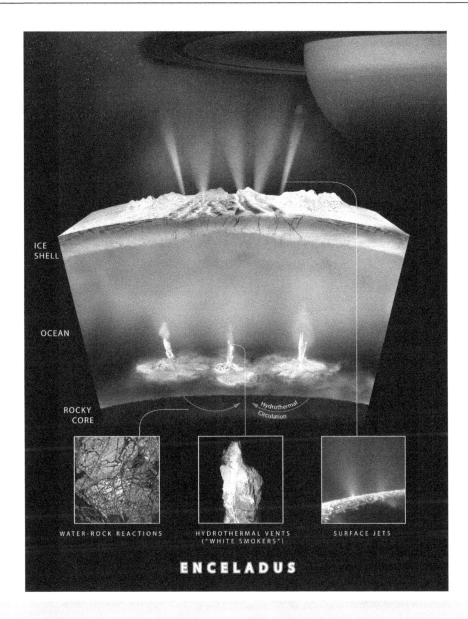

Figure 9.58   Various models have been proposed for the geysers of water vapor and ice particles on Enceladus. This model shows a subsurface ocean of water that circulates through the sea floor, where it is heated and interacts chemically with the rock. This warm water, laden with minerals and dissolved gases, then re-enters the ocean though chimney-like hydrothermal vents. These vents may thin the overlying ice, enabling water to rise toward the surface through cracks in the crust. Vapor evaporating from the water cools, creating ice grains that are carried aloft by the vapor plumes as they blast upwards at speeds of 300–500 m/s. (Courtesy of NASA/JPL-Caltech/Southwest Research Institute)

The ascending hydrothermal fluids probably trigger local melting in the ice layer of the polar region. This would explain why the ice layer at the poles is only 3–10 km thick, compared with 35 km near the equator. At the south pole, the water is able to rise through fissures almost to the moon's surface.

How can a satellite which should have frozen solid long ago retain enough internal heat to energize such powerful geysers? And why are they only found near the south pole? Enceladus is unlikely to contain enough internal rocky material for substantial radioactive heating or enough ammonia to lower its melting temperature significantly.

Tidal heating may be the main source of energy, even though the satellite's orbit is almost circular. Perhaps the moon is stretched between Saturn's powerful gravity and the opposing pulls of Tethys and Dione. In particular, Enceladus is near the 2:1 mean motion resonance with Dione, which is the main cause of its small orbital eccentricity. Dione may also tug on Enceladus sufficiently to cause a regular widening and closing of the tiger stripe fractures, explaining why the plumes periodically brighten and dim.

Such heating would then be due to localized tidal dissipation within either the ice shell or the underlying silicate core. This may

Box 9.5   Formation of Saturn's Moons

Saturn's satellite system is unusual in that it contains one huge moon, Titan, and half a dozen medium-sized objects (see Figure 9.38). How did this family come about?

It seems likely that Saturn was surrounded by a disk of gas and dust during its early period of formation. Like a miniature Solar System, this would have resulted in snowball-like growth and accretion through collisions, in which rock and volatile elements, particularly water ice, gradually grew into large satellites in orbit around the gas giant.

According to this scenario, the Saturn system started out resembling that of Jupiter, with several large moons. However, after a few collisions and mergers involving these objects, only one large satellite (Titan) remained. Meanwhile, icy material from the outer layers of the colliding satellites would create enough debris to form satellites such as Mimas, Enceladus, Tethys, Dione, Rhea, and Iapetus.

Models suggest that material in ice-rich spiral arms would clump together through gravitational attraction and eventually create Saturn's diverse family of medium-sized, ice-rich moons. This scenario may also explain why Titan, Iapetus, and Mimas have fairly high orbital eccentricities.

The moon mergers probably took place billions of years ago, perhaps as the result of a gravitational nudge from Uranus or Neptune as they migrated through the outer Solar System.

Most of Saturn's satellite retinue of 82 moons is made up of small objects measuring just a few kilometers across. Most of these "moonlets" are captured asteroids, or their remnants, many of which follow highly inclined, retrograde orbits (see Small Satellites).

be sufficient to cause geological activity, resulting in shear heating caused by movement along polar faults.

A different theory suggests that the current polar location of the hot spot can be explained by reorientation of the satellite's rotation axis. A plume of relatively warm ice at any point within its icy shell could have caused Enceladus to reorient itself so that the warm region becomes the new south pole. As the warmer ice expands and becomes less dense, the resulting uneven distribution of mass shifts the rotation axis.

Other Major Moons

All of the other medium-sized moons have ancient surfaces that are heavily cratered, although they also display evidence of varying degrees of tectonic activity in the past.

Mimas, which orbits between the G ring and the E ring, has a flattened or oblate shape with an equatorial diameter of 397 km, nearly 10% larger than its polar diameter (Figure 9.59). Its low density suggests that it is primarily composed of ice. It has one of the most heavily cratered surfaces in the Saturn system.

Its most distinctive feature is the 130 km-diameter Herschel impact crater, which lies near the equator on the leading hemisphere. Herschel has a large central peak caused by crustal rebound during the impact (see Chapter 4). The collision which created the ancient crater may have almost destroyed Mimas, spreading ejecta over much of the surface.

Most of the surface displays numerous, ancient, impact craters of various shapes. These often overlap – evidence of the intensity of the bombardment. Many of the larger craters appear to be filled in with landslides that may have been triggered by subsequent impacts elsewhere.

Mimas also has a series of grooves and canyons, some more than 1 km deep and over 100 km in length, that are generally aligned away from Herschel. These features may be related to the giant impact that created Herschel. Alternatively, they may indicate that the moon once had a warm, active interior.

Beyond Enceladus is Tethys, something of a Mimas look-alike, though much larger, with a diameter of more than 1,000 km

(Figure 9.60). Once again, the leading hemisphere is dominated by a huge impact feature. Known as Odysseus, the 450 km-wide, multi-ringed basin is remarkably well-preserved, even though a superimposed peppering of craters indicates its ancient origin.

Another striking feature is Ithaca Chasma, a huge, ancient rift, marked by parallel canyons running north–south, on the Saturn-facing hemisphere. It is over 1,000 km long, averages 100 km in width, and is up to 4 km deep.

Dione is slightly larger than Tethys, and, after Titan, the densest of Saturn's moons – implying that it has a fairly large rock core under the layers of ice (Figure 9.61). However, its terrain is surprisingly varied. Although much of the surface is heavily cratered, other regions display only smaller craters, while the trailing hemisphere displays a broad area of lightly cratered, smooth terrain which may have been resurfaced by cryovolcanism.

Voyager images showed broad, bright wisps spreading out across this smoother region. These are now known to be numerous, roughly parallel, cliffs and grooves. Cutting across older craters, many of the cliffs are sharp and fresh looking. Other fractures in the southern polar region and some of the craters have a softened appearance, suggesting that they are older. Dione has clearly been tectonically active for much of its existence, presumably driven by internal heating.

Cassini sent back other indirect evidence that both Dione and Tethys may still be geologically active. Streams of plasma are apparently coming from the moons and becoming trapped in Saturn's magnetosphere. A cloud of methane and water ice has also been detected around Dione, suggesting possible cryovolcanism, similar to that on Enceladus, but on a much smaller scale. It has also been suggested that Dione may have a global ocean beneath a 100 km-thick crust, based on its topography and how strongly it pulled on the Cassini spacecraft.

Just as tidal heating involving Dione may be the cause of Enceladus' activity, so the reverse may also be true. The satellites may have been hot early in their lives and remained warm for billions of years through continued heating caused by Saturn's strong gravitational pull. The 1:2 orbital resonance between Dione and Enceladus may also have played a role, so that they are gravitationally

Figure 9.59 The clear-filter image of Mimas (left) has been processed to enhance the brightness and sharpness of visible features. The false-color, multi-spectral image (right) shows subtle variations in the composition or granular texture of its surface. Shades of blue and violet reveal materials that are bluer in color and have a weaker infrared brightness than more typical materials, shown in green. Herschel crater is visible at upper right. The bluer color may represent material that was excavated from depth when the Herschel impact occurred. (Courtesy of NASA/JPL/Space Science Institute)

Figure 9.60 Multi-ringed Odysseus crater, 450 km in diameter, covers much of northern Tethys in this view of the leading hemisphere. (Courtesy of NASA/JPL/Space Science Institute)

Figure 9.61 Dione's bright cliffs diverge across the smooth plains on its trailing hemisphere. At lower right is a feature called Cassandra, which is surrounded by thin, linear rays. (Courtesy of NASA/JPL/Space Science Institute)

squeezed. Since Dione is further from Saturn, it experiences less severe flexing than Enceladus. Radioactive decay of rocky elements may also have played a part.

Beyond Dione is Rhea, second in size and mass only to Titan among Saturn's satellites (Figure 9.62). Rhea is heavily cratered, with two large impact scars on its leading hemisphere. One of them, Tirawa, is about 360 km across and slightly elliptical in outline. Its shape and elongated central peak complex infer an

oblique impact. Tirawa overlaps a larger and more degraded basin to its southwest. Both are covered in smaller craters, indicating that they are quite ancient.

Another striking feature is a bright ray crater on the leading hemisphere. About 50 km across, the relatively recent impact feature has sprayed fresh ice for hundreds of kilometers across the older terrain.

Figure 9.62   A natural color view (left) of Rhea's trailing hemisphere show bright, wispy terrain that resembles the heavily fractured landscape on Dione. The false-color image (right) shows that the wisps cut across an older, cratered surface. A series of thin, north–south lineaments (also likely fractures) is visible at left. The false colors reveal subtle differences in surface composition or grain size. Rhea's diameter is 1,528 km. (Courtesy of NASA/JPL/Space Science Institute)

The Voyagers revealed some intriguing bright streaks that run for hundreds of kilometers across Rhea's trailing hemisphere – in a region that is generally darker and less cratered than its surroundings. Cassini showed that these resemble the bright, icy canyons seen on Dione, and they are probably due to a combination of crustal extension and shearing. However, its greater distance from Saturn and weaker tidal effects mean that internal heating is much lower than on Dione or Tethys.

Rhea's low density indicates that it is mainly water ice, with perhaps 25% rock. Models of Rhea's interior suggest that it is almost undifferentiated, and composed of a rock-ice mixture, surrounded by a relatively thin ice shell.

Cassini's Magnetospheric Imaging Instrument found indirect evidence for a broad debris disk, possibly extending up to 4,800 km from the moon's center, and several rings around Rhea. The debris appeared to be shielding Cassini from the usual rain of electrons. Brief, sudden drops in electrons on both sides of the moon also suggested the presence of three rings or extended arcs of material within the disk. However, no visual evidence for the disk or rings has been found.

Another surprise was the discovery of a sparse atmosphere of oxygen and carbon dioxide around Rhea, the first direct evidence of oxygen in the atmosphere of any body other than Earth. Oxygen could be released as the surface is irradiated by ions (charged particles) from Saturn's magnetosphere. The source of the carbon dioxide is less clear, but could be the result of similar irradiation, or from impacts by comets.

Spongy Hyperion

Some of the satellites orbiting beyond Titan also possess unique characteristics (Figure 9.63). Hyperion measures about 270 km × 201 km × 336 km and is one of the largest irregularly shaped moons in the Solar System. Its axis of rotation tumbles unpre-

dictably. Its slightly eccentric orbit makes it subject to varying gravitational forces from Saturn, so that its rotational period differs from one orbit to the next. The 4:3 orbital resonance between Titan and Hyperion may also contribute to the random motion.

It has one major, degraded crater and several others that are 120 km in diameter and 10 km deep. However, most of the satellite is peppered with medium-sized, well-preserved craters, typically 2–10 km across and 400 m deep. They are unlike any craters seen on other bodies, creating a strange, sponge-like appearance. Hyperion has a reddish surface, coated with water and carbon dioxide ices, as well as dark material that fits the spectral profile of hydrocarbons – particularly on the crater floors.[12]

The crucial factor in explaining the weird landscape seems to be Hyperion's low density – only about 55% as dense as water. This can only be explained if the moon is a rubble pile with large empty spaces. This means that impactors striking the porous outer layers form craters more by compressing the surface than by blasting out material. Furthermore, its weak gravity allows much of the ejected material to escape completely, rather than re-impact the surface. As a result, the craters are sharper and less blanketed by debris than on other bodies.

Two-Toned Iapetus

Iapetus, with a diameter of 1,436 km, is the outermost of Saturn's large, icy satellites (Figure 9.64). Like most of Saturn's moons, it always keeps the same hemisphere toward the planet. Its two-toned nature has been recognized since 1671, when Giovanni Domenico Cassini noted that it appeared bright to one side of Saturn and much dimmer on the opposite side. The great observer deduced that the moon must have a lighter, trailing hemisphere abutting a much darker leading surface.

His deductions have been confirmed by spacecraft, revealing one coal-black hemisphere, known as Cassini Regio, alongside

---

[12]The dark hydrocarbon material on the crater floors seems to be similar to material on the dark side of Iapetus and the surface of Phoebe.

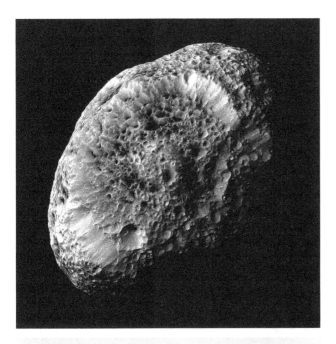

Figure 9.63 A high-resolution, false-color mosaic of Hyperion. At the center is a possible circular impact feature upon which numerous smaller craters have been superimposed. Its walls display fresh ice that has been exposed quite recently. Dark organic material covers many crater floors. Numerous bright-rimmed craters indicate that the dark material may be only tens of meters thick with brighter material beneath. Downslope movement causes craters to fill with debris. (Courtesy of NASA/JPL/Space Science Institute)

Figure 9.64 Iapetus has a dark leading hemisphere and a bright, icy, trailing hemisphere. The dark, heavily cratered area, called Cassini Regio, is centered on the equator at roughly 90°W. Within Cassini Regio, deposits with an albedo of only 4% coat nearly everything. Above about 40°N the surface gradually becomes much brighter. An ancient, 400 km-wide impact basin is visible near the center. Also shown near the equator is a mountain ridge, at least 1,300 km long and 13 km high. (Courtesy of NASA/JPL/Space Science Institute)

a region of clean water ice more than five times as reflective. The dark material is thought to be composed of nitrogen-bearing organic compounds called cyanides, hydrated minerals, and other carbonaceous minerals.

Iapetus displays at least nine large impact basins, including several on the bright hemisphere. Cassini Regio also displays a 600 km-wide feature alongside a 400 km-diameter basin, both overlain by smaller, more recent, craters. Around their rims are steep scarps up to 15 km high and altered in places by major landslides. Many of the scarps appear bright, probably due to exposed outcrops of relatively clean ice. The scarps and walls of nearby craters are brightest on north-facing slopes, while the opposite scarps are stained with darker material.

Even though the Sun is over 1.4 billion km away, its warmth is able to vaporize ice from sunward-facing surfaces, leaving behind dark material that is mixed with the ice. Subsequent downslope motion slowly transports this to the crater floors and other low lying terrain.

Although the surface is heavily cratered and ancient, the carbon-rich, dark coating is obviously much more recent. However, the presence of very small craters that reveal bright ice beneath indicates that the dark layer is generally thin. The material is almost certainly dust which spirals in from Phoebe and is deposited on the forward-facing hemisphere.

Dark surfaces absorb more sunlight and warm up faster than bright, reflective regions. On Iapetus, infrared observations show that the dark material is warm enough (–146°C) to enable a very slow release of water vapor from water ice. The gas then condenses on the nearest cold spot, such as the moon's poles and icy areas at lower latitudes on the trailing side of the moon. According to this model, the dark hemisphere loses its surface ice and becomes even darker, while the brightness of the trailing face is enhanced by fresh deposits of ice.

The most surprising feature is an elongated mountain ridge that stretches at least 1,300 km along the satellite's equator, bisecting the entire dark hemisphere. No more than 20 km wide, the ridge rises at least 13 km above the surrounding terrain, higher than Mt. Everest. The ridge is pitted with craters, indicating that it is old, but its origin remains a mystery. Is it a mountain belt that has somehow been folded upward, or does it mark a fracture through which viscous material erupted onto the surface and then piled up, rather than spreading outward?

Iapetus' rotation period and shape, and its 15° inclination to Saturn's equator, indicate strong early heating. As Saturn's gravity slowed the moon's spin rate to match its 79-day orbit, gravitational tides within the satellite would dissipate the rotational energy (see Chapter 4). In addition, Iapetus' early rotation may have been sufficiently rapid – once every 17 hours or less – and its surface warm enough to be easily deformed at the equator. As the moon cooled, the bulge "froze" in place.

Phoebe

For many years, Phoebe was thought to be outermost moon of Saturn. Unlike the other sizeable satellites, it follows a highly eccentric and retrograde path at an average distance of 13 million km from the planet. The orbit is inclined approximately 27° to Saturn's equator.

Phoebe is almost four times farther from Saturn than its nearest major neighbor, Iapetus, and substantially larger than Saturn's other outer moons. It rotates every nine hours and 16 minutes and completes one orbit of Saturn in about 18 months. Phoebe is also unusually dark, reflecting only 6% of the sunlight it receives.

Images reveal an irregular chunk of ice, about 220 km across and heavily potholed with craters, large and small. Also visible are bright streaks in the ramparts of the largest craters, bright rays emanating from smaller craters, and long, uninterrupted grooves. The plethora of 50-km craters suggests that debris from past impacts created many of the tiny outer moons that follow eccentric orbits much like Phoebe's.

The moon's low mean density indicates that it is mainly composed of water ice and carbon dioxide ice with a little rock, overlain by a layer of darker material – probably carbon-rich compounds – perhaps 300 to 500 m thick. This primordial mixture suggests that Phoebe may have formed as a Kuiper Belt object in the outer Solar System before it was nudged inward and captured by Saturn (see Chapter 12). Material ejected from its surface feeds a huge ring and coats some of the inner satellites, most notably Iapetus (see Two-Toned Iapetus).

Small Satellites

As with Jupiter, there is a mixed collection of small, irregularly shaped moons orbiting quite close to the planet. As mentioned above, they are closely linked with the ring system.

Pan and tiny Daphnis orbit within the main rings and are responsible for clearing gaps in the ring particles. Pan keeps the 330-km Encke Gap open while Daphnis is responsible for the

Figure 9.65 Cassini's view of Phoebe, taken from a distance of 32,500 km. 220 km across, Phoebe is an ice-rich body coated with a thin layer of dark material. Small, bright impact craters reveal fresh ice and are probably fairly young. On some crater walls, dark material has slid downwards, exposing brighter ice. The largest crater, about 100 km across, is named Jason (top). Some areas that appear particularly bright – especially lower right – are over-exposed. (Courtesy of NASA-JPL/Space Science Institute)

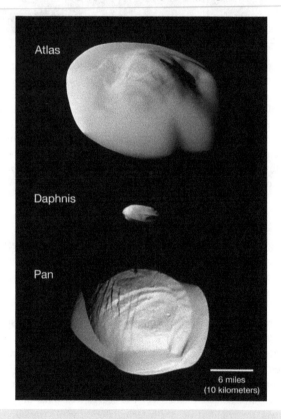

Figure 9.66 A montage of Cassini images shows three of Saturn's small ring moons, Atlas, Daphnis and Pan, at the same scale. Pan and Atlas have distinctive "flying saucer" shapes, caused by equatorial ridges. Pan's equatorial band is much thinner and more sharply defined, and the central mass of Atlas (beneath the smooth equatorial band) appears to be smaller than that of Pan. The images of Atlas and Pan were taken using infrared, green, and ultraviolet spectral filters to highlight subtle color differences across the moons' surfaces. Pan measures 33 x 21 km. The dimensions of Atlas are 39 x 18 km. (Courtesy of NASA/JPL-Caltech/Space Science Institute)

narrow Keeler Gap within the A ring. Atlas lies just beyond the A ring and is probably responsible for confining the ring particles and creating a well-defined edge to the ring.

It is believed that these moons began as leftover shards from larger bodies that broke apart. These acted as cores that have grown through the accretion of ring material. The result is a moon with a small, dense ice core that is covered by a thick shell of porous, icy ring material.

The equatorial ridges of Pan and Atlas, which make them resemble flying saucers, may have been created during a secondary stage of accretion that occurred after their main growth was completed and after the rings became extremely thin and flat. This would have allowed the rings to form small accretion disks above the satellites' equators.

The multi-stranded F ring is shepherded on either side by Prometheus and Pandora (see F Ring). Pandora keeps the narrow ring from spreading outwards, while Prometheus shapes its inner edge. Observations show that they behave unpredictably – Pandora was found to have moved far ahead of where it should be in its orbit, while Prometheus was lagging far behind. Their motions can be explained by chaotic gravitational interactions between them each time Prometheus overtakes Pandora, about once every 28 days. Since their orbits are elliptical, the distance of closest approach – and the strength of the gravitational push – varies on each occasion.

Other pairs of small satellites closely interact in the Saturn system. Epimetheus and Janus follow almost identical paths between the F and G rings. Every four years the inner moon overtakes the outer one and swaps positions – one of them moves closer to Saturn and the other moves farther away. Janus was the innermost of the pair 2006–2010, then it switched positions.

Saturn is the only planet known to have "Trojan" moons, so called because they resemble Jupiter's Trojan asteroids by orbiting near stable "Lagrange points." In Saturn's case, both Tethys and Dione are accompanied by pairs of co-orbital satellites with positions centered roughly 60° ahead and behind them. All four Trojans are very small and irregular in shape.

Telesto is located 60° ahead of Tethys, at the L4 Lagrange point, whilst Calypso orbits behind Tethys at L5. Helene and Polydeuces are co-orbital with Dione. Polydeuces is much more of a wanderer than Helene, coming as close as 39° to Dione and then drifting up to 92° away. It takes over two years to complete its journey around the L5 point.

The vast majority of Saturn's small moons orbit millions of kilometers beyond Phoebe. Most of them have been discovered in recent years during dedicated searches by Scott Sheppard, David Jewitt, and Jan Kleyna, among others, using some of the world's biggest ground-based telescopes and large, sensitive CCD detectors.

The satellites generally follow eccentric, inclined, and retrograde orbits, although a minority occupy prograde orbits with inclinations of less than 50°. These remote moons are probably icy remnants of much larger objects that were destroyed during cataclysmic impacts and then gravitationally captured by Saturn.

Questions

- Both Jupiter and Saturn are described as "gas giants." Summarize the main similarities and differences between them.
- Why has it been difficult to determine Saturn's sidereal rotation period?
- Why do major storms occur periodically on Saturn? What are the main characteristics of these storms?
- How were the various rings of Saturn probably formed?
- Describe and explain these features of the main rings: propellers, spokes, gaps, and density waves.
- What are the main characteristics of these rings: (i) A and B; (ii) C and D; (iii) F; (iv) E; and (v) Phoebe?
- Titan is comparable in size to Jupiter's satellite Ganymede, but otherwise very different in nature. Describe and briefly explain the main differences between these planet-sized satellites.
- Describe Titan's methane cycle. What are the main similarities between Titan and Earth?
- Summarize the evidence for geyser/plume activity on Enceladus. What factors contribute to making this activity possible on such a small satellite?
- Briefly describe the following features of Saturn's system: (a) magnetosphere; (b) auroras; (c) Trojan moons.
- Summarize the main features of Saturn's major moons (other than Titan and Enceladus) and explain their similarities and differences.

TEN
Uranus

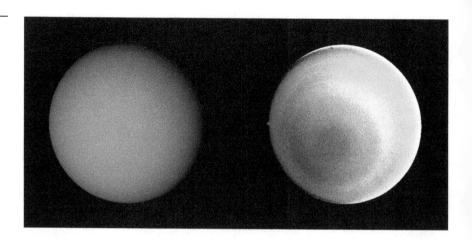

The third of the Solar System's giant planets is Uranus, a blue-green world which seems to be mainly composed of liquid water, methane, and ammonia. Orbiting twice as far from the Sun as Saturn, Uranus is a frigid world where the feeble solar radiation is the major driver of occasional storms in the deep atmosphere. Many mysteries remain, not least the reason for the planet's 98° inclination, which causes the most unusual seasons in the Solar System.

Discovery

Until the late 18th century, studies of the Solar System were restricted to the Sun, Moon, six planets, and a number of planetary satellites. No one was prepared for the major upheaval that occurred on March 13, 1781: the first recorded discovery of a planet.

The revolution began with a routine sky survey by a German-English musician who also happened to be an enthusiastic amateur astronomer and telescope maker. Working from his house in Bath, with the willing assistance of his sister, Caroline, William Herschel (1738–1822), was using a homemade 6.2 inch (15.7 cm) reflecting telescope when he noted in his journal: "In the quartile near Zeta Tauri the lower of the two is a curious either nebulous star or perhaps a comet."

It soon became clear that the object had no tail and was not following a comet-like orbit. The mystery was eventually solved when calculations showed that it was following an almost circular path around the Sun.

The seventh planet turned out to be a giant, though smaller than either Jupiter or Saturn. Located more than twice the distance of Saturn from the Sun, it was obviously a faint, cold world. At a stroke, Herschel had doubled the size of the Solar System.

In gratitude for an appointment as King George III's personal astronomer (and a royal pension of 200 pounds a year) Herschel suggested that the new planet be named Georgium Sidus (George's Star). Not surprisingly, this title was completely unacceptable to other nations, and many years passed before the name Uranus was officially recognized.[1]

The Seventh Planet

Herschel was not the first person to see Uranus. John Flamsteed had noted it on six separate occasions between 1690 and 1715 without recognizing its true nature. In fact, the planet is just visible with the unaided eye if you know exactly where to look. Nevertheless, with a disk measuring only four arc seconds in diameter, it was difficult to find out more than the basic information about the planet's size and orbit.

With a semi-major axis of 2,871 million km and an average orbital velocity of 6.8 km/s, Uranus takes 84 years to complete one revolution (Table 10.1). Since its discovery, the planet has only completed two full circuits of the Sun. The orbit is only slightly elliptical, ranging between 3,005 million km and 2,734 million km, so its apparent size and brightness vary little. Uranus last reached aphelion on February 27, 2009.

For many years, there was uncertainty about its size, density, and rotation period. Today Uranus is known to be the third-largest planet in the Solar System, with an equatorial diameter of 51,118 km – four times greater than Earth (Figure 10.1). 64 Earths would fit inside Uranus. Like the other giant planets, its rapid rotation causes the disk to appear noticeably flattened at the poles – evidence of a deep atmosphere and the absence of a solid surface. But where are the poles?

[1] In Roman mythology, Uranus was the father of Saturn and grandfather of Jupiter.

Table 10.1
Uranus Summary

Mass: 86.8 x 10²⁴ kg (14.5 Earth masses)

Equatorial Diameter: 51,118 km

Polar Diameter: 49,946 km

Axial Inclination: 97.77 degrees

Semi-major Axis: 2,872.46 million km (19.2 AU)

Rotation Period (internal): 17.24 hours

Sidereal Orbit Period: 84.011 years

Mean Density: 1.27 g/cm³

Number of known satellites: 27

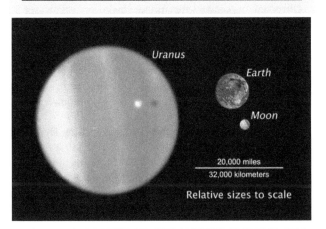

Figure 10.1 The relative sizes of Uranus, Earth, and Earth's Moon. Uranus is approximately 50,000 km in diameter, or about four times the diameter of Earth. (Courtesy of NASA, ESA, and L. Sromovsky/University of Wisconsin)

A key clue to the axial inclination of Uranus came in 1787, when Herschel discovered two satellites (Titania and Oberon) whose orbits were tilted almost 90° to the plane of the ecliptic. If the moons were orbiting above the planet's equator and in the same direction as the planet's rotation – as most regular moons do – then Uranus must have an extreme axial inclination. Further confirmation came in the 19ᵗʰ century when spectral observations of the planet's upper atmosphere showed that it appeared to be rotating in a retrograde direction – the opposite of most of the other planets.

We now know that Uranus spins like a top that has been knocked over onto its side, with the rotation axis lying in almost the same plane as the orbit. The actual axial inclination is 97.77°, so what would normally be regarded as the north pole actually lies below the ecliptic plane.[2]

This unusual arrangement means that the poles endure extreme seasonal variations (Figure 10.2). For about 21 years, the south pole is in permanent sunlight, while the north pole endures continuous darkness – a situation that reverses when Uranus has moved halfway around its orbit. In between, the equatorial regions have more sunlight and conditions are less extreme.

The south pole was last aligned toward the Sun in 1985, since when the planetary aspect has been changing. In 2007, the rings and equator were aligned with the Sun and Earth (as they were in 1964 and 1922), and by 2028 the north pole will be facing the Sun.

Even with this knowledge of its obliquity, the bland appearance of the bluish disk made it extremely difficult to determine the rotation period. Before the Voyager flyby of 1986, the best estimates, based on observations of occasional, isolated clouds, indicated a period of about 16 hours. Today, the accepted rotation period – based on radio signals generated by particles trapped in the magnetic field – is 17 hours 14 minutes, the slowest of the giant planets.

Why is Uranus spinning on its side? The usual explanation is that a large planetary embryo, several times the size of Earth, was involved in an off-centre collision with the young Uranus (Figure 10.3). The positioning of the planet's rings and inner moons above the equator suggests that this took place more than 4 billion years ago, perhaps when Uranus was still migrating out to its present orbit (see Chapter 1).

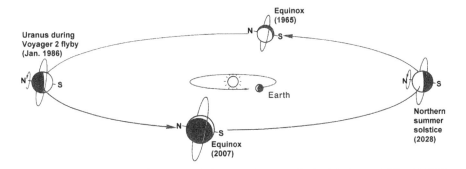

Figure 10.2 The axis of Uranus is tilted almost parallel to the plane of the planet's orbit. As a result, one pole is in permanent sunlight for 21 years, while the opposite pole is in permanent darkness. In between, when the equator receives more sunlight, conditions are less extreme. At such times, the rings and the orbits of the satellites are seen almost edge-on. (Courtesy of NASA/Peter Bond)

[2] There has been some contradiction over the naming of Uranus' poles. In 1982 the International Astronomical Union declared that all poles above the ecliptic should be regarded as north poles. Under this system, the south pole of Uranus was in sunlight during the Voyager 2 flyby in 1986. However, the Voyager scientists preferred to refer to it as the north pole.

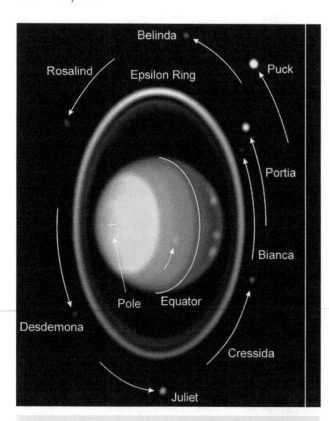

Figure 10.3   From the late 1960s, Uranus' south pole was aligned with the Sun and Earth. However, when this Hubble Space Telescope image was taken in November 1997, the planet had moved far enough around its orbit to make part of its northern hemisphere visible for the first time since the early 20th century. (Courtesy of Erich Karkoschka/NASA)

Computer simulations suggest a modified scenario, involving two or more smaller collisions rather than a single, major collision early in the Solar System's history, when Uranus was still surrounded by the disk of dust and gas that would eventually form its moons.

The simulations show that, after a single, massive collision, the disk would have reformed above the equator of the toppled planet, and the satellites that eventually condensed from the disk material would also be in the new equatorial plane. However, they would orbit in a retrograde direction, the opposite of what they do today. In order to recreate the present, prograde, configuration, the tilt of Uranus and its disk would have had to be rearranged by a series of two or more smaller collisions.

Atmosphere

Like the other giant planets, Uranus has an atmosphere composed largely of hydrogen and helium. Spectroscopic data show that molecular hydrogen (H_2) is the most abundant gas on Uranus,

implying that helium must also be a major atmospheric constituent. Confirmation came from the Voyager 2 spacecraft, which measured 83% hydrogen, 15% helium, and 2% methane.

The pale blue color of Uranus indicates that its atmosphere differs from those of Jupiter and Saturn. For many years, the reason for this coloration was unclear. Then, in 1932, Rupert Wildt announced that dark lines in the spectra of all four giant planets could be attributed to absorption of sunlight by methane gas (CH_4). Subsequent measurements show that Uranus has 10 times more methane in its upper atmosphere than Jupiter or Saturn. The gas acts as a filter, scattering blue wavelengths but absorbing red light. As a result of this filtering effect and the presence of a widespread photochemical haze at high level, the underlying cloud features appear washed out and fairly featureless.

Other minor constituents include hydrogen sulfide (H_2S), deuterated hydrogen (HD), ethane (C_2H_6), and acetylene (C_2H_2). The last two are created by photochemistry – the action of sunlight on methane gas. Despite their scarcity, these minor ingredients do have a subtle influence on their surroundings: they condense to form high level hazes which absorb and reflect sunlight, raising the temperature of the upper atmosphere while lowering the temperature of the deeper atmosphere.

The atmosphere of Uranus can be divided into three layers. The lowest is the troposphere, which is about 350 km thick. Pressures range from 100 to 0.1 bar (10 MPa to 10 kPa). Here, the temperature ranges from −153°C at the base to −218°C at the upper boundary – the tropopause. This means it is the coldest atmosphere in the Solar System.

Models suggest that the troposphere has several cloud layers similar to those found at depth on the other giant planets (Figure 10.4). Water clouds condense at the highest pressures, with ammonium hydrosulfide clouds above them. Ammonia and hydrogen sulfide clouds come next, topped by thin methane clouds where the temperature is at its lowest. (See Chapters 8 and 9.)

Above the tropopause is the stratosphere, where pressures are between 0.1 and 10^{-10} bar (10 kPa to 10 µPa). The stratosphere is about 3,500 km thick.

In the stratosphere, temperatures range from −218°C to −153°C, thanks largely to heating caused by solar radiation. The stratosphere contains ethane smog, which may contribute to the planet's dull appearance. Acetylene and methane are also present, and these hazes help warm the stratosphere.

The outermost layer, the thermosphere/exosphere, extends as far as 50,000 km into space. This region has a uniform temperature of 520–570°C, although the reason is not clear, because the amount of heat and UV radiation coming from the distant Sun is insufficient to generate such high temperatures and auroral activity cannot provide the necessary energy.

Since Uranus orbits more than 19 AU from the Sun, it receives 360 times less light and heat from the Sun than Earth. As a result, its atmosphere is extremely cold, with a temperature of about −214°C at the 1 bar pressure level (equivalent to the air pressure at sea level on Earth).

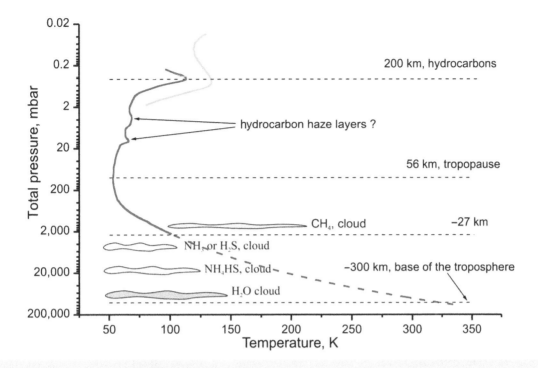

Figure 10.4 Temperature profile of the Uranian troposphere and lower stratosphere, showing presumed cloud and haze layers. (Based on Wikipedia/Ruslik0)

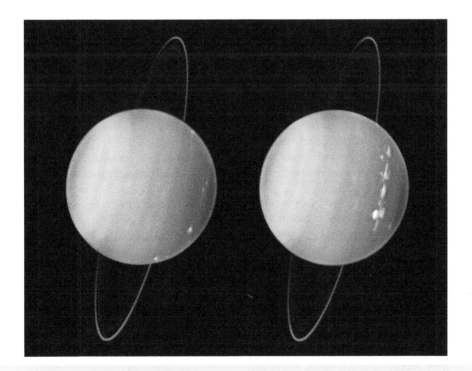

Figure 10.5 These two near-infrared images, taken with the Keck II telescope on July 11–12, 2004, show both hemispheres of Uranus, with its north pole at 4 o'clock. The rings appear red because of processing to reveal methane absorption in the atmosphere. Thirty-one cloud features are visible (13 in the left image and 18 at right), more than the total observed up to the year 2000. The bright white feature (right) reached a high altitude and was very variable. The isolated bright green cloud (lower left limb in the right hand image) was a long-lived feature that oscillated in latitude. Note the line of bright methane clouds in the northern hemisphere, which was coming out of a long winter. The 28,800 km-long band of clouds dissipated completely within one month. (Lawrence Sromovsky, UW-Madison Space Science and Engineering Center/Keck Observatory)

Voyager 2 found that temperatures were almost identical everywhere on the planet, evidence that the small amount of heat received from the Sun is rapidly spread by winds around the planet. However, Voyager instruments did detect a slightly colder band between 15 and 40°S, where temperatures were 2 to 3°C lower.

Clouds and Storms

Before modern technological advances were introduced, ground-based telescopes often showed no features at all on Uranus, although some observers recorded two faint belts on either side of a bright equatorial zone and dark hoods over the poles (Figure 10.14). This absence of cloud features persisted throughout the Voyager 2 flyby of 1986, when even the most extreme computer processing of images revealed only 8 discrete clouds and a high layer of photochemical haze around the sunlit pole.

Only since 1994 have individual cloud structures become visible from the Hubble Space Telescope and ground-based telescopes equipped with adaptive optics (Figures 10.3 and 10.5). As the planet's seasons changed and the northern hemisphere emerged from its prolonged winter darkness, the banded structure near the south pole slowly evolved, while bands of bright, high-altitude clouds were seen for the first time in the northern hemisphere.

Images taken at infrared wavelengths (not possible with Voyager) show a global blanket of methane ice crystals with bands and spots. Many of these spots are thought to be convectional storms – comparable to continent-sized thunderheads on Earth – that develop at the top of huge columns of warmer, rising gas. (Since there is very little temperature contrast, only small inputs of energy are required to drive the weather on Uranus.)

However, others appear to be companion clouds of possibly orographic nature, associated with uplift around vortex circulations which results in local condensation as the flow is displaced above the methane condensation level. These may be similar to companion clouds associated with the Great Dark Spot on Neptune, but at a much smaller size scale, spanning only a few degrees of longitude at their greatest extents.

Given the pressures and temperatures of the brighter clouds (about 1 bar and 77K), it seems certain that they mainly comprise frozen methane particles. The composition of the deeper clouds – at pressures near 1.6 bar – are less certain. Hydrogen sulfide and ammonia (for the even deeper layers) might be components of these clouds.

In recent years, the fairly bland face of Uranus has become increasingly changeable. Although large, bright cloud features have become common in mid-latitudes, none of the dark spots common on Neptune had ever been confirmed on Uranus before 2006 – as the equinox approached (Figure 10.6). In that year, observations from both the Hubble Space Telescope and Keck Telescope revealed a dark spot in the northern hemisphere, which was just emerging from winter. It was located at a latitude of about 28°N and measured approximately 1,300 km in latitude and 2,700 km in longitude.

The feature, called the Uranus Dark Spot (UDS) moved in the prograde direction relative to the planet's rotation with an average speed of 43.1 m/s, which is almost 20 m/s faster than the speed of clouds at the same latitude. The latitude of UDS was approximately constant, but the feature was variable in size and appearance and it was often accompanied by bright white clouds.

Since then, other dark spots have been confirmed on Uranus. Their behavior and appearance are similar to dark spots and their bright cloud companions on Neptune, although they are significantly smaller. This similarity suggests that they have the same origin. Neptune's dark spots are thought to be anticyclonic vortices which cause air to rise around and above them, like orographic clouds on Earth, causing condensation of methane to create the bright companions.

There were some differences in the infrared brightness of the UDS and the Neptunian dark spots, evidence that they occur at somewhat different pressure levels – the Uranian feature probably lay near the 4 bar level (four times Eath's sea level pressure). However, the dark color of the spots on both ice giants may be caused by thinning of the underlying hydrogen sulfide or ammonium hydrosulfide clouds.

The emergence of dark spots and numerous bright cloud features on the hemisphere of Uranus that was in darkness for many years are evidence that, near equinox, Uranus experienced a period of elevated weather activity. Instead of a near-instantaneous increase in activity caused by greater insolation, followed by a dying down of storm activity, the planet exhibited a delayed reaction.

This activity continued to increase in the ensuing years, resulting in enormous cloud systems so bright that, for the first time ever, amateur astronomers were able to see details in the planet's hazy blue-green atmosphere. Astronomers detected eight large storms in Uranus's northern hemisphere during August 2014.

One of these, located at 30–40° N, was the brightest storm ever seen on Uranus at 2.2 microns, a near-infrared wavelength that senses clouds just below the tropopause, where the pressure ranges from about 300 to 500 mbar, or half the pressure at Earth's surface (Figure 10.7). The storm accounted for 30 percent of all light reflected by the rest of the planet at this wavelength.

Later in the year, amateur observers detected a different storm at 1.6 microns, infrared light that was emitted from deeper in the atmosphere, which meant that this feature was below the uppermost cloud layer of methane ice in Uranus's atmosphere. The colors and morphology of this cloud complex suggested that the storm might be associated with a vortex in the deeper atmosphere, similar to two large cloud complexes seen during the equinox. Other observations at a variety of wavelengths by the HST revealed multiple storm components extending over a distance of more than 9,000 km, and clouds at a variety of altitudes.

Some cloud features evolve rapidly, others persist over a number of years. They vary greatly in size. Some are dim and diffuse, others bright and sharp. One large, bright feature near 34°S was first recorded in 1986 and was still visible in 2004. During that period its latitude oscillated between 32° S and 36.5° S over a period of about 1,000 days. It also varied its longitudinal drift rate between −20° per day and −31° per day as it drifted in the strong zonal wind, behavior similar to that of the DS2 feature observed on Neptune by Voyager 2.

In 2004 it became much brighter; and in 2005 it started to migrate towards the equator, turning into a very powerful storm system. In 2009, when it came to within a few degrees of the equator, it dissipated. This sizeable storm was thought to be the result of vigorous convection created as the southern hemisphere shifted to day–night illumination instead of permanent sunlight. As already mentioned, a comparable storm has since erupted in the northern hemisphere.

Like the other giant planets, the winds on Uranus blow parallel to its equator, resulting in faint cloud belts that are pale imitations of those on Jupiter and Saturn. The zonal pattern is apparently controlled by the planet's rapid spin rather than by direct heating from the distant Sun.

Voyager 2 enabled scientists to track the rare clouds and determine the wind speeds in the southern hemisphere. The images showed that mid-latitude winds blow in the direction of the planet's rotation at 40–160 m/s (145–575 km/h). Radio science experiments found winds of about 100 m/s blowing in the opposite direction at the equator. (This is the opposite of Jupiter and Saturn, where equatorial winds are prograde.) The atmospheric circulation in the dark northern hemisphere remained a mystery.

With the shifting of the seasons, it is now possible to observe the cloud motions to the north of the equator (Figure 10.8). Observations made in October 2003 showed wind speeds of 107–111 m/s, reaching a peak of 218 m/s (785 km/h), the highest wind velocity seen yet on Uranus. At northern mid-latitudes (20° to 40°), the winds appeared to have accelerated when compared to earlier HST and Keck observations. However, the comparable wind speeds at southern mid-latitudes had not changed since the Voyager measurements in 1986.

As the northern hemisphere has become visible, it has been possible to confirm the symmetrical pattern of zonal winds, with a band of fast, westerly winds peaking at around 60° latitude in both hemispheres (Figure 10.8).

The equatorial region exhibits a subtle wave structure, indicated by diffuse patches roughly every 30° in longitude (Figure 10.9). The largest discrete cloud features show complex structure and nearby streaks extending over tens of degrees, reminiscent of activity seen around Neptune's Great Dark Spot in 1989. There has been no sign of a northern "polar collar" resembling that seen in the south, though it has been suggested that a number of discrete features at high latitudes might mark the early development of such a collar. However, by 2014, the first detection of a widespread polar haze had been made over the northern polar region.

Detailed images have revealed what looks like a field of cumulus convective features around the north pole, extending down to 55°N. These do not seem to be very dense clouds, although they are quite large (typically 450 km across). This finding was unexpected because such features had never been seen in the south polar region of the planet. They may be fairly transient, arising as the northern hemisphere begins to warm up, then fading as midsummer arrives.

The outer planets are monitored yearly for changes, and an HST image obtained in November 2018 showed a vast, bright cloud cap across the north pole of Uranus (Figure 10.10). With the planet approaching the middle of its summer, when the Sun shines almost directly onto the north pole and never sets, the polar cap was

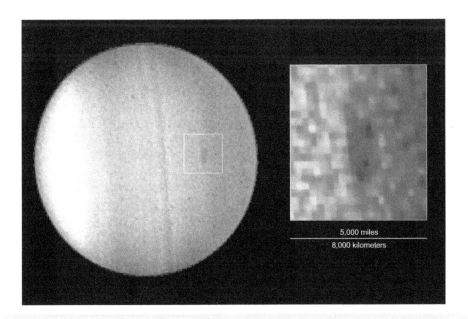

Figure 10.6   The first confirmed image of a dark spot on Uranus was obtained by the Hubble Space Telescope in 2006, as the planet approached its equinox – when the Sun is overhead at the equator. The spot was detected at latitude 27 degrees in Uranus' northern hemisphere, which was just becoming fully exposed to sunlight after many years of winter darkness. This three-wavelength composite image was taken with Hubble's Advanced Camera for Surveys on August 23, 2006. The inset image shows a magnified view of the spot with enhanced contrast. Uranus' north pole is near the 3 o'clock position in this image. The bright band in the southern hemisphere is at 45 degrees south. (Courtesy of NASA, ESA, L. Sromovsky and P. Fry/University of Wisconsin, H. Hammel/Space Science Institute, and K. Rages/SETI Institute)

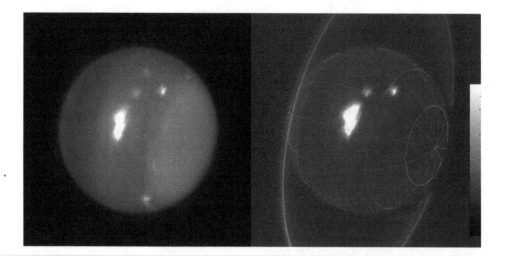

Figure 10.7   Infrared images of Uranus (1.6 and 2.2 microns) obtained on August 6, 2014, with adaptive optics on the 10-meter Keck telescope. The white spot was an extremely large storm that was brighter than any feature ever recorded on the planet in the 2.2-micron band. The cloud rotating into view at the lower right limb grew into a large storm that was seen by amateur astronomers at visible wavelengths. (Imke de Pater/UC Berkeley and Keck Observatory)

becoming more prominent. This polar hood may have formed by seasonal changes in atmospheric flow.

Near the edge of the cloud cap was a large, bright, methane-ice cloud formation. A narrow cloud band could be seen encircling the planet north of the equator. It is a mystery how bands like these are confined to such narrow widths, because Uranus has very broad westward- blowing jets.

Interior

Even the most basic physical data indicate that Uranus must be very different in make-up from the larger gas giants. With a volume less than 5% that of Jupiter and 8% that of Saturn, Uranus is clearly dwarfed by those two planets. At the same time, its overall density is not much higher than water – twice that of Saturn, but very similar to that of Jupiter. Uranus' relatively modest size and bulk density indicate that pressure levels in the planet's interior are low compared with its larger cousins.

Low-density elements must account for a substantial fraction of the planet's mass (Figure 10.11). However, to be consistent with the bulk density, Uranus cannot be composed largely of hydrogen and helium. Instead, roughly 80–90% of the total mass of Uranus must be ice and rock (compared with 10% for Jupiter). The high temperatures and pressures in the planet's interior mean that the "ices" are actually in the form of a compressed liquid. In other words, Uranus is a fluid planet, whose gaseous atmosphere gradually merges with the liquid interior.

The formation of Uranus (and Neptune) is not well understood. One reason for its different composition is probably the planet's slow growth in the outer regions of the solar nebula. The relatively slow orbital motion of the proto-Uranus would have limited its rate of accretion from the dusty disk. By the time a significant rocky core had grown, much of the gas in its neighborhood may have been cleared by a powerful outflow of radiation from the young Sun. The smaller core and limited supply of gas meant that the planet captured a less massive atmosphere.

Various models for the internal structure of Uranus have been proposed to account for these properties, and even today a number of questions remain. In 1934, Rupert Wildt suggested that it has a rocky core overlain by a thick layer of ice and a deep atmosphere of hydrogen, methane, and ammonia. An alternative model put forward in 1951 by W. Ramsey preferred a world largely composed of ammonia, methane, and water.

By the early 1980s, this was refined to a three-layered planet with a fairly small, rocky core made of a mixture of silicate, magnesium, and iron-nickel, overlain by a liquid mantle of water, methane, and ammonia, and topped by a thick hydrogen-helium atmosphere.

Current models suggest that the hydrogen-helium atmosphere extends about 7,500 km below the visible cloud tops. There is then a layered mantle of rock and "ices" – water, ammonia, and methane in solar proportions – with traces of other substances. In this ice layer, pressure and temperature increase from 20 GPa and 1,700°C at the upper boundary to 600 GPa and 7,700°C at the mantle–core boundary. It is unclear how these "ices" are distributed within the mantle – as homogeneous mixtures, with continuous concentration gradients, or as well-separated layers of specific composition.

Shock wave experiments indicate that, under such extreme conditions, methane dissociates – possibly into molecular hydrogen and carbon. Some scientists have suggested that the high pressures and temperatures result in the carbon taking the form of tiny, solid diamonds.

Recent studies suggest that ammonia-rich hydrates can precipitate out of any ammonia–water mixture at sufficiently high pressures.

At the very center of the planet may be an Earth-sized rocky core with a radius of around 6,000 km and a temperature of about 8,000°C.

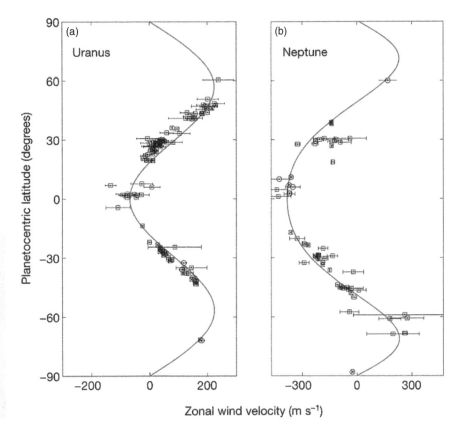

Figure 10.8 Average zonal wind speeds observed at cloud level on Uranus and Neptune. Both planets have similar zonal wind patterns with two eastward (prograde) blowing jets at about 60° N and S, and a westward (retrograde) jet at the equator. Most storm activity occurs around 30°N and S. The horizontal lines show the possible range of error in the various data points. (Y. Kaspi et al/Nature)

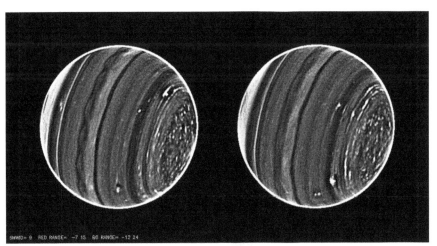

Figure 10.9 These specially processed images, taken with the Keck telescope, reveal an astounding level of detail in Uranus' clouds. The images are composites of near-infrared images from July 25, 2012 (left), and from July 26 (right). White features are high-altitude, optically thick clouds (like Earth's cumulus clouds), and bright blue-green features are at high altitude but optically thin (like cirrus clouds on Earth). Reddish tints indicate deeper cloud layers. Very few of the discrete features are deep. Most are probably between 1 bar and 2 bars, the typical cloud-top range for Uranus. The narrow bright blue arc on the left of each image is due to the main (Epsilon) ring of Uranus. The planet's bright perimeter is an artifact of the filtering process. The north pole is at the right and slightly below center. Small "convective" spots appear mainly from about 55°N to the pole, and were not seen in the south polar region when it was imaged by Keck at similar wavelengths in 2003. They are highly reminiscent of features seen on Saturn's pole by the Cassini spacecraft. The broad bright band just to the left of the disk's center covers the latitude range from the equator to about 10°N. Just south of the equator, and never before seen on Uranus, is a scalloped wave pattern, similar to instabilities that develop in regions of horizontal wind shear. Near the bottom of the left image is a small dark spot with bright companion clouds. (Courtesy of NASA/ESA/L.A. Sromovsky/P.M. Fry/H.B. Hammel/I. de Pater/K.A. Rages)

Missing Heat

All of the planets should have started as hot objects, when gravitational energy and kinetic energy from impacts were transformed into heat during planetary accretion. Over the age of the Solar System, the smaller planets have lost most of this heat, but the massive giant planets store heat well and radiate poorly from their cold surfaces. Consequently, they should have retained much of the heat produced during their formation and this should still be escaping today.

The total heat output of a planet is determined from its infrared emission, while the heat input is determined from the fraction of incident sunlight that is absorbed, i.e. not scattered back into space. The fraction that is scattered is called the albedo, which for Uranus is about 0.3 (30%).

By studying the far-infrared emissions and albedo of Uranus, its internal heat output can be estimated. The surprising conclusion is that Uranus is the only giant planet that does not radiate a substantial amount of heat from its interior. Voyager data confirmed that the internal heat source – if it exists – produces no more than 6% of the energy it receives from the Sun, though the degree of uncertainty meant that there could be no excess at all. (The equivalent figures for the other giant planets exceed 70%.)

Since Uranus and Neptune appear to be virtual twins, the reasons for this energy balance are not readily apparent, though the planet's relatively bland appearance seems to be one obvious consequence. The lack of vertical motion and storm activity is the result of limited interaction between heat escaping from the interior and the atmosphere.

One possibility is that some form of layering within the mantle is preventing large-scale convection. If layers of differing composition, density, and temperature exist, they may confine the regions where convective motions occur. Instead of being allowed to rise rapidly to the surface by convection, Uranus' heat flow may occur thru a much slower diffusive process in regions in the upper mantle which have a high molecular weight gradient. Such an interpretation is supported by the presence of a highly unusual magnetic field (see Magnetic Surprise).

An alternative hypothesis is that the large tilt of the rotation axis may have been responsible for a higher rate of heat loss, resulting in the current temperature being much lower than expected. The possible massive impact with a planetary embryo soon after its formation may also have caused it to lose most of its primordial heat, leaving it with a low core temperature.

Magnetic Surprise

Although its fluid interior and rapid rotation indicated that Uranus possesses a magnetic field, it was not possible to directly measure it from the Earth. The best indirect evidence came in the early 1980s when the International Ultraviolet Explorer spacecraft detected "aurora-like" emissions from hydrogen in the planet's upper atmosphere.

Everything changed in 1986, when Voyager 2 flew past the planet. Radio emissions detected several days before closest approach provided the first conclusive evidence that Uranus possesses a magnetosphere. The magnetic field turned out to be 50 times stronger than Earth's (though at Uranus' visible surface the field is generally weaker than on our planet).[3]

The alignment of the magnetic field was a major surprise (Figure 10.12). The field is not only tilted at an angle of 58.6° to the rotational axis (compared with 11° for Earth), but it is also offset from the planet's center by about 8,000 km, or one-third of the distance to the cloud tops. This means that the magnetic poles are nowhere near the geographical poles. Instead they lie at 15.2°S and 44.2°N on the visible surface. This unusual offset is thought to be caused by the field being generated by an electrically conductive fluid layer far above the core.

The source of the magnetic field remains uncertain, although it has been suggested that it may arise from nearly complete ionization of water, the most abundant component of the icy mantle.

One important clue is the "roughness" of the magnetic field. Instead of being predominantly dipolar (like a bar magnet), the field also has a quadrupole component (as though produced by a combination of two bar magnets, with two north and two south poles). However, small-scale components of a field die out rapidly above the electrically conducting region. When combined with the

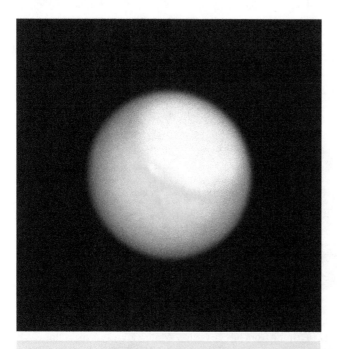

Figure 10.10   This Hubble Space Telescope image of Uranus, taken in November 2018, reveals a vast, bright, stormy cloud cap across the planet's north pole. Also visible are a large oval cloud of methane ice and a narrow band of bright cloud near the equator. (Courtesy of NASA, ESA, A. Simon/NASA-GSFC, M.H. Wong and A. Hsu/University of California, Berkeley)

---

[3]The intensity of the magnetic field at Uranus' visible surface varies considerably from place to place because of its large offset from the planet's center. Earth's field strength is about 0.3 gauss everywhere on the surface, whereas on Uranus it varies from 0.1 to 1.1 gauss.

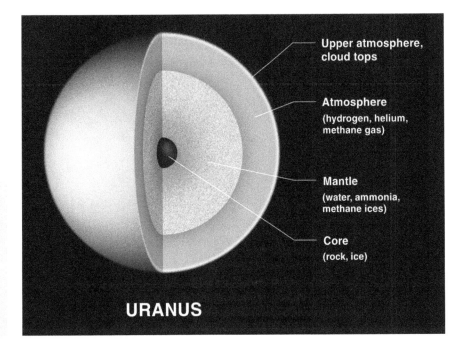

Figure 10.11 The overall density of Uranus is only a little higher than that of water, less than all the planets except Saturn. Current models of the interior favor a hot, rocky core, surrounded by a mantle of water, methane, and ammonia and a thick atmosphere of hydrogen, helium, and methane. (LPI)

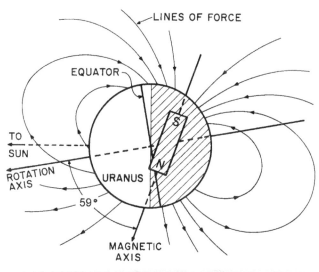

Figure 10.12 The magnetic field of Uranus, shown as a dipole tilted about 59° from the rotation axis and offset 8,000 km toward the dark north polar region. (NASA-JPL)

During the late 20th century, the magnetic dipole axis appeared to rotate in the anti-clockwise direction as seen from the Sun, with the solar wind approaching Uranus more or less from above the south pole. With the changing seasons, this alignment changes, so that the solar particles arrive at latitudes close to the equator. This latter configuration causes the magnetic poles to gyrate toward and away from the Sun every 17 hours (Uranus' rotation period).

The magnetosphere has many of the characteristics of those around other planets. The field repels the stream of charged particles in the solar wind, creating a huge magnetospheric cavity, filled with trapped charged particles, that surrounds the planet and extends downstream. The magnetotail extends at least 10 million km (400 radii) beyond Uranus's night side.

On the upstream (sunward) side, the boundary between the magnetosphere and the solar wind extends to approximately 25 planetary radii. A bow shock at about 33 planetary radii deflects the supersonic flow of the solar wind in front of the magnetosphere.

The magnetosphere of Uranus is fairly empty, since it encompasses only a few, relatively small, icy satellites and solar wind-driven convection moves particles through the region in a matter of days. There is no evidence of helium, which might originate from the solar wind, and no evidence of heavier ions (e.g. oxygen) that might be sputtered from the surfaces of the moons.

The main rings of Uranus extend to two planetary radii and sweep away much of the material in that region. This means there is only a small plasmasphere containing trapped particles, and a lot less plasma can be forced from the magnetosphere down into the atmosphere to create auroras.

The geometry of the interaction between the solar wind and Uranus' magnetosphere dramatically evolves over timescales ranging from a quarter of a rotation (hours) to seasons (decades). The planet's unique axial inclination means that, around the time of Uranus' equinoxes, one magnetic pole can be aimed almost straight into the solar wind and then almost directly away a half day later.

unusual tilt and offset location from the center, this evidence suggests that the field is generated at shallow depths within the planet.

Since water and ammonia dissociate into positive and negative ions at relatively low pressures and temperatures, the field could easily be generated close to the surface where these moderate conditions exist. According to current models, the dynamo that creates the magnetic field occurs in a thin shell of convecting, ionized fluid that lies above a non-convecting, ionized "ocean."

The planet's highly inclined rotation produces an unusual corkscrew motion in the field lines as the cylindrical magnetotail, shaped by the movement of the solar wind past the planet, rotates in space.

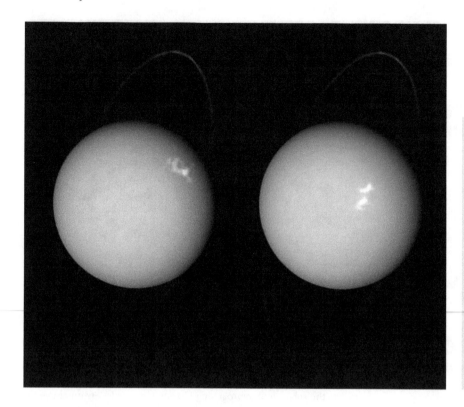

Figure 10.13 A composite image of Uranus by Voyager 2 and two different observations made by the HST – one for the rings and one for the far-ultraviolet auroras (bright white spots) in the planet's southern hemisphere. These images show the first direct evidence that Uranus' auroras rotate with the planet. The detected emissions occur close to the expected arrival of interplanetary shocks. They also enabled Uranus' magnetic poles, which were lost shortly after their discovery by Voyager 2 in 1986, to be rediscovered. (Courtesy of ESA/Hubble and NASA/Laurent Lamy)

Box 10.1   Voyager 2

Most of what we know about Uranus comes from the brief Voyager 2 flyby in January 1986. Although the Voyagers were not optimized to operate beyond Saturn, the mission team hoped that it would be funded to include visits to Uranus and Neptune – completing the "Grand Tour" of the outer Solar System. A rare alignment of the outer planets made it possible to redirect the spacecraft by a gravitational "slingshot" from each planet.

Since Uranus orbits about twice as far from the Sun as Saturn, the spacecraft's cameras had to cope with light levels that were four times lower. When combined with the high speed of the flyby, this increased the likelihood of image smear, particularly if the spacecraft vibrated while the shutters were open. Imaging was further complicated by the darkness of Uranus' rings and moons.

Voyager 2 was already suffering from a number of malfunctions, most notably the loss of its primary radio receiver and limited communication via the backup receiver. In addition, output from its nuclear power source had dropped to about 400 watts – not enough to operate all spacecraft subsystems simultaneously.

In order to obtain the maximum data return, the spacecraft's stability was improved by modifications to the attitude control system. To avoid image smear, the entire spacecraft was programmed to turn at the correct rate, like a camera panning on a fast-moving vehicle – even though this caused the main antenna to point away from Earth.

Greater distance also meant reduced rates of data return. In order to minimize the loss of information, the spacecraft was programmed to compress the image data, transmitting only the change in brightness between neighboring picture elements (pixels) rather than the absolute brightness values. The ground facilities were also upgraded by electronically combining the signals received by several antennas.

The encounter began on November 4, 1985, 81 days before closest approach, with a long range survey of the planet, its rings, and satellites. The bland, blue sphere of Uranus showed only a dark polar hood and the occasional high altitude cloud. By studying the rare methane clouds, it was found that mid-latitude winds were circling the planet in the same direction as its rotation at speeds of 40–160 m/s.

Uranus was found to be extremely cold, with an average temperature of –210°C. Temperatures were fairly uniform at similar levels in the atmosphere, with the exception of a slightly colder band between 15 and 40°S. Even the sunlit south pole was similar in temperature to the north pole, which had not seen the Sun for two decades.

As expected, the main constituents of the atmosphere were found to be hydrogen and helium, although the 15% helium abundance was much lower than some Earth-based studies had suggested. However, this figure was comparable to the amount of helium thought to be present in the original solar nebula.

Figure 10.14 Two views of Uranus returned by Voyager 2's narrow angle camera on January 17, 1986. The spacecraft was 9.1 million km from the planet, several days from closest approach. The picture at left – a composite of images taken through blue, green, and orange filters – shows the southern hemisphere in true color. False color and extreme contrast enhancement reveal a polar hood of high altitude methane haze (right). (NASA-JPL)

The mystery of the planet's internal rotation was solved by studying fluctuations in the magnetic field and associated radio signals. A day on Uranus was found to last for 17 hours 14 minutes, 1½ hours longer than anticipated. There was no significant internal heat source.

Spacecraft measurements showed that the magnetic tail extended at least 10 million km behind the planet. The strength of the field at Uranus' surface turned out to be roughly comparable to that on Earth. What was not anticipated was the unique magnetic field, which was tilted 59° to the spin axis and offset from the center of the planet by almost 7,700 km.

Since the magnetic poles were not expected to be so close to the equator, most scans by Voyager's instruments were aimed toward the wrong places. However, the ultraviolet experiment did find some evidence of an aurora on the night side, along with an "electroglow" caused by energized electrons above the dayside.

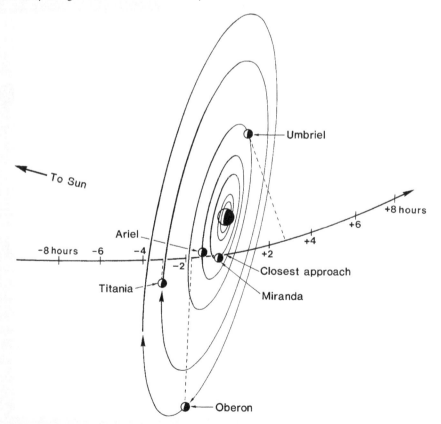

Figure 10.15 Voyager 2 flew to within 81,600 km of Uranus' cloud tops on January 24, 1986. Since the planet, its rings, and the satellite orbits were arranged like concentric rings on a target, the best photo opportunities were limited to a period of less than 6 hours around the time of closest approach. The trajectory was also influenced by the need for a gravitational assist toward Neptune. This meant that images of the larger moons were taken from fairly long range, while Miranda received the closest scrutiny. (NASA-JPL)

Voyager 2 also obtained brief glimpses of the individual rings and moons. One priority was a search for "shepherd" satellites that might be confining the narrow rings (only two were found). Altogether, 10 new satellites were discovered, nine of them orbiting between the epsilon ring and the orbit of Miranda. Of the previously known moons, Miranda was the star of the show, with a bizarre surface marked by three regions of parallel ridges, one of which resembled a chevron.

This means that each day Uranus' magnetosphere is "open" in one orientation, allowing solar wind to flow into it; as time goes by and the orientation changes, the magnetosphere closes, forming a shield against the solar wind and deflecting it away from the planet.

This behaviour inevitably influences the planet's auroral emissions at the magnetic poles. Ultraviolet (UV) auroras were first detected by Voyager 2 in 1986.

Some auroras seem to be produced by current systems which connect the moon Miranda to the magnetic poles of Uranus. However, many of the auroras are created from the precipitation of solar wind particles into the atmosphere and the arrival of interplanetary shock waves.

These create auroral ovals around the solstices, rather like the dynamic curtains of light found at the poles of Earth or Jupiter. However, short-lived elongated spots appear near the equinoxes. The near-equinoctial auroras may be triggered by large-scale compressions of the magnetosphere by the solar wind, as well the variable solar wind/magnetosphere geometry.

Hubble Space Telescope images have shown that the auroras rotate with the planet (Figure 10.13). The UV emissions were found to last for tens of minutes and varied on timescales down to a few seconds.

They also enabled researchers to get a better idea of the longitudes of both of Uranus' magnetic poles. Fitting the observations with model auroral ovals constrained the longitude of the southern and northern magnetic poles to $104 \pm 26°$ and $284 \pm 26°$ respectively in the Uranian Longitude System.

The longitudes of the magnetic poles were first measured in 1986 by Voyager 2, but the longitude was soon lost owing to the low-accuracy measurement of Uranus' rotation period. Now, with the newly determined longitudes and Voyager 2's original measurements, researchers hope to be able to reconstruct a longitude system for Uranus's magnetic poles for the last 30 years and determine the rotation period at high resolution.

Voyager 2 found radiation belts of similar intensity to those at Saturn, although they differ in composition. The radiation belts at Uranus appear to be dominated by hydrogen ions, probably derived from the planet's upper atmosphere and ionosphere. The escaping hydrogen atoms glow in ultraviolet light on the sunlit side, close to the magnetic poles.

The radiation belts are so intense that irradiation would darken any methane trapped in the icy surfaces of the inner moons and ring particles within 100,000 years. This process may have contributed to the darkened surfaces of the moons and ring particles.

Major Satellites

Prior to the Voyager 2 flyby, only five medium-sized moons were known to orbit Uranus (Figure 10.16). The two largest, Oberon and Titania, were discovered by William Herschel six years after he first observed Uranus. Ariel and Umbriel were found by William Lassell in 1851, while the smallest, Miranda, was discovered in photographs taken by Gerard Kuiper in 1948. All of these were located well outside the rings, following normal orbits of low eccentricity and inclination (except Miranda, whose orbit is inclined about 4° to the planet's equator).

The satellites were generally much darker than the moons of Saturn and all but Miranda appeared to be partially covered by spectrally neutral material similar to charcoal or carbonaceous chondrite meteorites. (The spectral reflectance of this dark material also closely resembled that of the rings).

Their bulk densities were higher than those of Saturn's icy satellites, but lower than the Galilean giants of Jupiter – suggesting a fairly even mixture of water ice and silicate. Since the moons were considered to be heavily cratered, ancient relics left over from the formation of the planet, no evidence of geologic activity was anticipated.

Like most satellites in the Solar System, the moons of Uranus are in synchronous rotation, always keeping the same face toward the planet. Normally, this would enable partial imaging of both the northern and southern hemispheres. However, in the case of Uranus' satellites the Voyager 2 spacecraft passed above their fully illuminated south poles, so imaging was limited to their southern hemispheres.

The other giant planets all possess at least one satellite comparable in size to Earth's Moon, but the main moons of Uranus are all quite modest in size. They range in diameter from 472 km for Miranda to 1,578 km for Titania.

Their mode of formation is unclear. Unlike Jupiter and Saturn, the young Uranus almost certainly was unable to draw in a large cloud of dust and gas. Hence, the satellites would not have formed during the infall of material from the solar nebula.

However, the accretion process would have caused the planet to be hotter and larger than it is today. As it cooled and shrank, it may have left behind a "spinout disk" of material that eventually condensed and stuck together. In the classical accretion process, the debris could then have grown into the satellites we see today.

Recent computer simulations suggest another possibility. If an Earth-sized protoplanet collided with the newborn Uranus, it may not only have given the planet its current tilt but also created a debris disk. The rubble from the two impactors could have come together to form moons with orbits and masses similar to those of Uranus' regular moons.

The larger satellites, Oberon and Titania, are the most distant from Uranus. They are remarkably similar in size, density, color, and albedo, although their surfaces are quite different.

Oberon has numerous large craters between 50 and 100 km in diameter (Figure 10.17). Spatial resolution in Voyager images was only about 12 km, but it was possible to identify a large mountain, probably the central peak of a large impact crater, rising 11 km

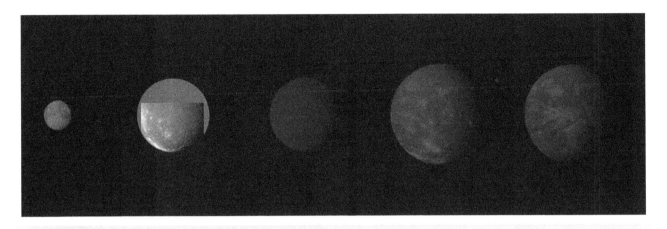

Figure 10.16   A "family portrait" of Uranus' five largest moons. From right to left, in order of decreasing distance from the planet are Oberon, Titania, Umbriel, Ariel, and Miranda. The image shows correct relative sizes and brightness. The two largest, Oberon and Titania, are about half the size of Earth's Moon, or roughly 1,600 km in diameter. All five satellites are generally gray, with only slight color variations on their surfaces. (NASA-JPL)

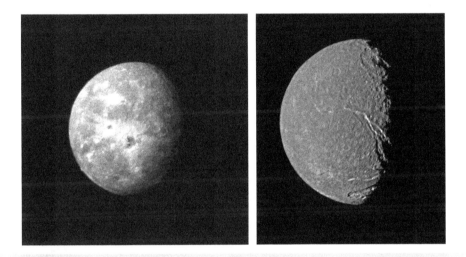

Figure 10.17   Oberon (left) and Titania (right) are the largest moons of Uranus. They are very similar in many respects, although Titania shows more small craters and seems to have undergone global extension of its crust after the main cratering phase ended. (NASA-JPL)

above the limb of Oberon. Also visible were several large craters with dark floors and bright rays of ejecta. The largest of these (Hamlet) had two dark spots on either side of a bright central peak. Perhaps isolated, icy volcanism had released a dark fluid in the distant past. Further evidence of past surface activity included linear and curved features that appear to be enormous faults.

Titania revealed a number of large impact basins (100–200 km in diameter) and far fewer craters than Oberon in the 50–100 km size range, with an abundance of 20 km-diameter craters. The size–frequency distributions of these smaller craters suggested that they were mainly secondary features excavated by debris from larger impacts.

Several patches of smoother terrain indicated a prolonged early period of resurfacing. There was also an extensive network of branching, intersecting faults, with a number of inward-facing

scarps on either side of grabens. Calculations based on length of shadows near the terminator showed that the scarps rise to heights of 2–5 km, while the grabens are 20–100 km across. The longest graben extends for at least 1,500 km and its stepped floor is several kilometers deep. The absence of superimposed craters suggests the faults are some of the youngest features on the satellite.

Closer to Uranus are Umbriel and Ariel. Again, they are almost identical in size and density, but with significant surface differences (Figure 10.18). Umbriel is the darkest of all the major Uranian satellites and has a weaker spectral signature for water ice. The black surface shows little variation in brightness, the most conspicuous feature being a ring, about 30 km wide, on the floor of a sizeable crater called Wunda. The population of large craters indicates that, along with Oberon, it has the oldest surface of Uranus' major satellites. The reasons why Umbriel is darker than its neighbors (particularly Ariel) and apparently unaffected by internal melting remain unknown.

On Ariel's relatively bright, complex surface, most of the larger, ancient craters have been erased, either by viscous relaxation of the icy surface over billions of years or by extrusion of fluid material (or both). The largest remaining crater appears flattened, with a gently domed floor partly surrounded by a shallow trough.

Voyager also revealed extensive faults and flat-floored rift valleys over 10 km deep and hundreds of kilometers long. Lacking craters, the spectacular canyons have been filled with relatively recent volcanic flows. The resurfacing appears to be too extensive to be accounted for solely by radiogenic heating (heat from the decay of radioactive elements in the moon's interior). Instead, tidal heating may have been involved.

Although it has no orbital resonances with other satellites today, couplings with Umbriel, and possibly Titania, may have occurred in the past, causing repeated gravitational tugs that heated Ariel's interior. As a result, a cryogenic mix of water ice, methane, and small amounts of ammonia could have risen to the surface and filled many of the valleys and impact craters.

In addition to the ubiquitous water ice, carbon dioxide ice has been discovered on the surfaces of Ariel, Umbriel, and Titania.[4] It may be primordial (from the solar nebula) or delivered by impacting comets. Alternatively, the dry ice could be the result of greater plasma irradiation on the trailing side as protons and electrons trapped in Uranus' magnetosphere overtake the moons from behind and impact their surfaces. The CO_2 ice is only found on the trailing side of the satellites, suggesting that the ice may have been eroded or buried by meteorite impacts on the more exposed forward-facing hemisphere.

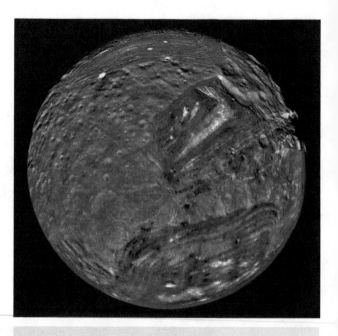

Figure 10.19 Although it is only 472 km across, Miranda displays a bizarre variety of landforms. About half of the surface is heavily cratered, but the most obvious features are the three coronae. The V-shaped "chevron" is known as Inverness Corona. The ridged polygon at the bottom is Arden Corona, while the similar feature on the opposite limb is Elsinore Corona. Also visible (right limb) is the 20-km cliff known as Verona Rupes. (Courtesy of NASA-JPL/USGS)

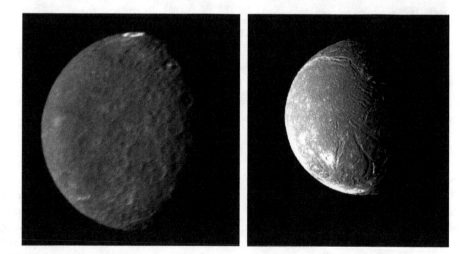

Figure 10.18 Umbriel (left) and Ariel (right) are similar in size and density. Umbriel has a dark, ancient surface covered in impact craters. Only a few brighter features are visible, most notably a ring on the floor of Wunda crater (top). Much of Ariel has been resurfaced, leaving few large craters. Also visible are numerous linear faults and smooth-floored grabens filled with volcanic flows. (NASA-JPL)

---

[4]Carbon dioxide ice has also been observed in the reflectance spectra of Jupiter's satellites Europa, Ganymede, and Callisto, Saturn's satellites Phoebe and Iapetus, and Neptune's moon Triton.

Miranda

Voyager 2 imaged almost the entire southern hemisphere of Miranda at fairly high resolution, and the smallest of the major satellites turned out to be the most intriguing (Figure 10.19). Its unique landforms, superimposed on more heavily cratered terrain, included a "chevron" and two huge polygonal features with rounded corners that were nicknamed "circi maximi" after Roman chariot-racing tracks.[5] All three features displayed parallel faults and ridges with a scattering of fresh impact craters. Clearly they had been formed in relatively recent geological times.

On the day–night terminator, close to the chevron (Inverness Corona), could be seen a 20 km-high cliff (twice the height of Mount Everest), now known as Verona Rupes. In the weak gravity of Miranda, an object would take almost 10 minutes to fall to the valley floor below.

Various theories have been put forward to explain this tortured terrain. One suggestion was that the satellite slowly reassembled itself after being shattered by a huge impact. Each corona was thought to represent a discrete fragment. However, this is difficult to reconcile with the much older terrain between the coronae.

Alternatively, the coronae may have formed during past episodes of tidal heating. Multiple eruptions of viscous fluid material, possibly associated with tidal heating of the mantle and stretching of the crust, could have created the broad lanes of parallel ridges.

Tidal heating could have occurred when Miranda was in an eccentric orbit – moving closer to and further from Uranus. This would cause the tidal forces from Uranus to vary, periodically stretching and squeezing Miranda and generating heat in its ice shell. As the warm, buoyant ice rose toward the surface, it caused concentric surface extension beneath the locations of the coronae, leading to the formation of extensional tectonic faults.

Bright patches on the surfaces of Arden and Inverness Coronae could be deposits of material erupted from explosive vents. The development of the coronae may have ceased prematurely when the satellite cooled before its internal differentiation into ice and rock was complete.

Rings

For many centuries, Saturn was assumed to be the only planet to possess a set of rings. Then, in March 1977, during a stellar occultation by Uranus, astronomers noticed brief dips in the starlight, both before and after the planet passed in front of the star (see Box 10.2). By combining observations from different locations, nine separate rings were recognized.

Since then, four more rings have been discovered, bringing the total to 13 (Figure 10.20). Two of them – lambda (λ) and zeta (ζ) (formerly 1986U2R) – were discovered in Voyager 2 images. Another pair, mu (μ) and nu (ν), (originally known as R/2003 U1 and R/2003 U2) was discovered in Hubble Space Telescope images in 2005.

Even before the Voyager 2 flyby, it was possible to gather some basic information about the rings from ground-based observations and analysis of data from various occultations. The nine known rings were shown to begin about 16,000 km above the planet, with their outer limit about 25,600 km from Uranus' cloud tops. Unlike most of the rings around Saturn, they had very sharp edges and were very dark, reflecting less light than coal dust.[6]

From the very brief blips in the light curves, it was obvious that the main rings are very closely spaced – often only a few hundred kilometers apart. They are also remarkably narrow, and, with the exception of the outer, epsilon ring, generally no more than 10 km across. Even more surprising was the fact that some of the rings are slightly elliptical, rather than circular. In particular, the distance of the epsilon ring from Uranus was found to vary by about 800 km.

The inner rings reflect only a small percentage of what little sunlight reaches them, so imaging the charcoal black rings against the dark background is extremely difficult. Occultation data at radio and optical wavelengths showed that particles smaller than tens of centimeters were absent from the epsilon ring, and the rings as a whole are remarkably free of dust. However, Voyager images taken looking back towards the Sun did reveal that most of the region inward of the lambda ring is filled with faint, broad bands and ringlets of micron-sized particles.

Since the effects of sunlight and drag from the tenuous hydrogen gas of Uranus' outer atmosphere would be expected to remove the smoke-sized particles within one million years, the particles must be slowly replenished – presumably by material derived from collisions involving ring boulders or larger moons.

Voyager also detected a sparse but broad band of dust 90,400 km above the cloud tops, between the epsilon ring and Miranda. Thirty hits per second were recorded from particles striking the spacecraft during this crossing of the ring plane – very modest compared with 600 hits per second during the ring plane crossing at Saturn.

Images obtained 20 years later indicated that large changes in the rings had taken place. At least two Uranian rings were found to have broad, optically thin components. The radial distribution of dust had changed quite dramatically – much more than Saturn's D and F rings, and the ring arcs of Neptune. Such changes are dynamically plausible because the dust populations that we see represent extremely tiny amounts of material, and the orbits of small dust grains evolve rapidly in response to non-gravitational forces and occasional impacts.

At least six of the main rings were inclined to the planet's equator, though only by a few hundredths of a degree. Several seemed to wobble and "pulsate" in and out, probably as the result of Uranus' slightly flattened shape, along with the gravitational influence of nearby moons.

The epsilon ring, the broadest and brightest of the family, was one of the few rings orbiting above the equator, but its orbit was also the most eccentric. Not only did its distance from Uranus vary,

---

[5] These polygonal features are known as "coronae," although they bear little resemblance to the ovoid coronae on Venus.

[6] Since the black rings are the same color as the nearby small moons, it seems likely that both are being darkened by the same process – the bombardment of organic molecules on their surfaces by high energy particles.

Box 10.2   Discovering Uranus' Rings

On March 10, 1977, several teams of astronomers observed a rare stellar occultation by Uranus. Their intention was to time very precisely the moments when a star known as SAO 158687 was extinguished by the intervening bulk of the planet and when it reappeared. Such measurements taken from many different locations would be used to refine the planet's size and orbit.

Since the precise time of the occultation was unknown, the teams' observations began well ahead of the predicted commencement. The resulting light curves caused a sensation. A series of blips in the light levels were visible, both before and after the star was eclipsed. Eventually, James Elliot, leader of the American team using the Kuiper Airborne Observatory over the Indian Ocean, realized that five sizable dips in the starlight appeared as a mirror image on either side of the planet. They had discovered a set of narrow rings around Uranus.

Similar dips in the starlight were recorded by groups in India, Japan, and South Africa. One team at Perth Observatory in Australia saw them, even though Uranus passed to the north of the star.

When the data from all of the sites were taken into account, it became clear that at least eight separate rings had been detected. The three rings closest to Uranus and first recognized by the Perth group, led by Robert Millis, were known as 6, 5, and 4. The five rings announced by Elliot's team were given Greek letters whose alphabetical order corresponded to their distance from the planet: alpha (α), beta (β), gamma (γ), delta (δ), and epsilon (ε). A less obvious ring that was identified at a later date was labeled eta (η). This complex, rather haphazard system of nomenclature has survived to the present day.

Figure 10.20   In this false color view, taken by Voyager 2, the nine original rings are shown. The fainter, pastel lines between them are the result of computer enhancement. Six 15-second narrow-angle images were used to image the extremely dark, faint rings. The brightest, or epsilon, ring (top) is neutral (gray) in color, while the eight fainter rings show slight color differences. The lambda ring is too faint to be visible. Spatial resolution is about 40 km. (Courtesy of NASA-JPL)

of comparable width, and a bright inner strip about 15 km across. Studies of the way light passed through the ring suggested that it was mainly made of boulders up to 1 m across.

The most important discovery about the epsilon ring was the presence of two shepherding satellites, Cordelia and Ophelia, on either flank of the epsilon ring – comparable to the moons that accompany Saturn's narrow F ring (Figure 10.22). However, no other shepherds were found.

Orbiting only 4,000 km apart, the moons are able to force wayward particles and boulders back into the epsilon ring. Ophelia, the outer shepherd, completes 13 orbits in the time a particle at the outer edge of the epsilon ring completes 14 orbits. As they slowly overtake Ophelia, the particles are attracted by the moon and lose energy. This causes them to "fall" closer to Uranus.

In contrast, particles at the inner edge of the ring complete 24 orbits in the same time that Cordelia completes 25 orbits. This is known as a 24:25 resonance. As Cordelia overtakes the ring particles, it gives them increased energy and forces them into higher orbits. The combined result is to give sharp edges at both the inner and outer boundaries of the ring.

The almost circular orbit of Cordelia and the slightly elliptical orbit of Ophelia complicate the gravitational interaction. This may explain the fact that the average eccentricity of the epsilon ring is between that of the orbits of the nearby shepherds.

Resonances with these companions of the outer ring may also modify the behavior of some of the other rings, though other unseen shepherds only a few kilometers across could also exist between the lanes of particles. If so, they would explain the narrow, confined nature of the main rings.

Two more rings located well outside the main system were found with the Hubble Space Telescope in 2003–2004 (Figure 10.24). Both of these faint rings are much wider than the previously known rings. They are very dusty, with particles predominantly in the sub-micron to micron size range. Such tiny grains cannot survive for long in planetary orbit because they are

but its width changed proportionately – at its closest to the planet, it was only 20 km across, whereas at its furthest the ring expanded to 100 km (Figure 10.21).

Voyager observed some fine structure when the epsilon ring passed in front of distant stars. Some camera images revealed a bright outer component about 40 km wide, a darker central region

[7] Although they were first noticed in 2004, the rings were subsequently found in HST images taken the previous year – hence their designation.

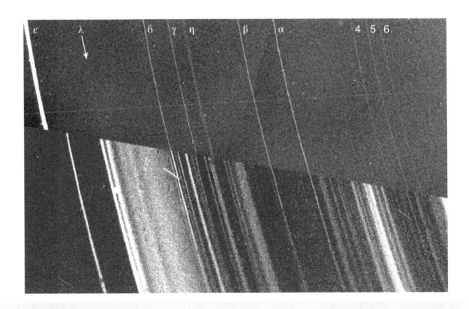

Figure 10.21 Two Voyager images of the main Uranian ring system compared. The top image, obtained one day before closest approach, shows 10 narrow rings from their sunlit side. The lower image, taken looking almost directly towards the Sun (phase angle 172.5°), reveals a sheet of micron-sized dust particles inward of the lambda ring. The differing distance of the elliptical epsilon ring is clearly shown. (Courtesy of NASA/Wikimedia)

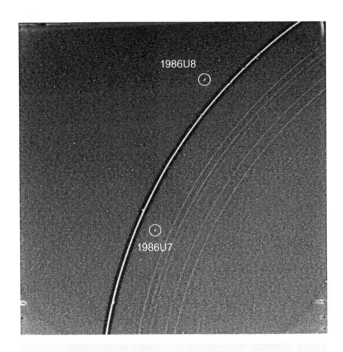

Figure 10.22 Two small moons, both about 40 km in diameter, were found by Voyager 2 on either side of the epsilon ring. Cordelia (1986 U7) and Ophelia (1986 U8) – the tiny specks inside the circles – are the only known "shepherd moons" in Uranus' ring system. Their gravitational influence confines the ring's boulder-sized debris. (NASA-JPL)

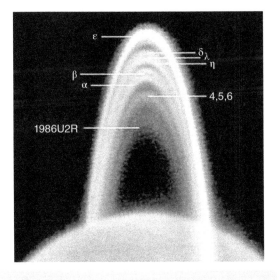

Figure 10.23 At 2.2 microns, methane and hydrogen in Uranus' atmosphere absorb most of the light falling on the planet, so little is reflected back into space. Imaging at this wavelength results in a dark Uranus, making the faint ring system clearly visible. The region interior to rings 4, 5, and 6 is a broad, dusty ring known as zeta (1986U2R). This near-infrared image was taken using adaptive optics with the Keck 2 telescope in July 2004. (Imke de Pater, Seran Gibbard, and Heidi Hammel/Science, April 7, 2006)

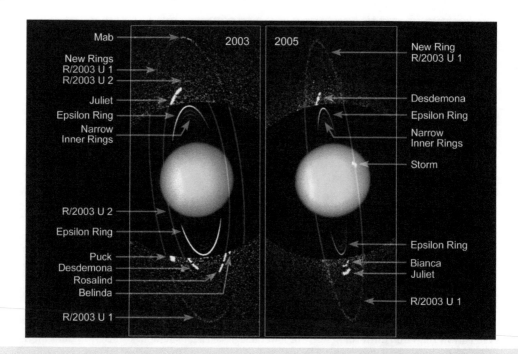

Figure 10.24 These Hubble Space Telescope images taken in 2003 and 2005 show the most recently discovered rings of Uranus. The outer mu (μ) ring (R2003 U1) lies almost twice as far from the planet as the outer edge of main ring system. Hubble also detected two small satellites, named Mab and Cupid. Mab shares an orbit with the outer ring and is probably the source of dust that replenishes it. The nu (ν) ring (R2003 U2) lies about 67,700 km from Uranus, in a region with no known satellites. (Courtesy of NASA, ESA and M. Showalter-SETI Institute)

subject to a variety of removal and destruction mechanisms. Continuous replenishment from larger source bodies is required.

The outer ring, now known as mu (μ), was discovered orbiting between 86,000 and 103,000 km from the planet's center.[7] The ring shares its orbit with a small satellite called Mab. Since it is brightest along Mab's orbit, it is probably replenished by material ejected during impacts with the moon. This is supported by its blue color, evidence that the ring is made of tiny water ice particles, mostly less than a tenth of a micrometre across.[8]

The other recently discovered ring, now designated nu (ν), is almost 4,000 km wide and squeezed between the orbits of Portia and Rosalind, which probably confine its limits. The ring is centered about 67,700 km from Uranus, in a region with no known satellites to provide source material. This red ring may be comprised of centimeter-sized or larger debris from the collisional disruption of a moon, as well as fine dust, since it has no known source bodies.

The appearance of the rings changed dramatically after the 2007 equinox, when the rings could be viewed from the dark (shaded) side for the first time. The brightest ring (epsilon) faded rapidly year by year, eventually disappearing from view, while the optically thin zeta ring became the brightest part of the ring system. Most of the rings interior to epsilon also brightened between 2004 and 2006.

Their considerable variability in width and opacity suggests that the rings are likely to be younger than the planet. Since the rings

all lie inside the Roche limit, one possibility is the collisional or gravitational break-up of a small moon (see Chapter 9). If so, it must have been quite small – the entire ring system could be compressed into an icy body only 30 km in diameter.

Chaotic Inner Moons

By the time the data from the Voyager 2 encounter were analyzed, the satellite count had trebled, with the discovery of 10 satellites between Uranus and Miranda (Figure 10.26). An eleventh moon, Perdita (S/1986 U10), was found 13 years later in a re-analysis of the Voyager images.

Since then, improvements in ground-based instruments have led to the discoveries of two more satellites inside the orbit of Miranda. One of them, Cupid, is probably about 8 km across. It comes to within 800 km of Belinda and lies in the midst of the so-called "Portia group," a family of nine moons (collectively named after their largest member), which occupy the relatively confined space between the orbits of Bianca (59,166 km) and Perdita (76,417 km).

The other recent discovery, Mab, is the only moon known to orbit between Puck and Miranda. It is also the only satellite known to be the primary source of material for one of Uranus' rings. Like the previously known moons, the newcomers follow nearly circular orbits above the planet's equator and are assumed

---

[8]The only other known blue ring is Saturn's E ring, which surrounds Enceladus. The E ring is fed by jets of water vapor spewing out of the moon.

Figure 10.25 A comparison of the outer rings of Saturn (top) and Uranus, with each system scaled to a common planetary radius. The recently discovered outer ring of Uranus, like that of Saturn, is blue because the material they contain is smaller than the material in the inner, red rings. Saturn's blue E ring is created by dust and ice particles ejected from the moon Enceladus. Uranus's blue mu ring is probably fed by material ejected during impacts with the icy surface of the small moon Mab. (*Imke de Pater, Heidi Hammel, Seran Gibbard, Mark Showalter/2006 Icarus issue 180, 186-200*)

to have a synchronous rotation. Mab is thought to be covered in ice, and around 6 km across, so it is probably Uranus' smallest regular satellite.

Puck was the first and largest of the moons discovered by Voyager (Figure 10.27). Orbiting at a distance of 86,000 km from the planet's center, it takes a little over 18 hours to complete one circuit. It is about 160 km across and roughly spherical, with a large impact crater visible on the limb. It reflects only 7% of the incoming sunlight – much less than the larger moons, but more than the rings.

Images of the other inner moons show only smudges, so few details of their physical nature, including their masses or densities, are known. However, estimates of their sizes can be derived from their brightness and distance. Assuming that they reflect as much sunlight as Puck, their diameters have been calculated to be between 20 and 135 km. Their dark surfaces suggest a mixture of water ice and organic (carbon-based) compounds that have been modified by interactions with charged particles trapped in the magnetosphere.

Most of the inner moons are more or less uniformly distributed between the rings and Puck, their orbits typically separated by a few thousand kilometers. However, Showalter and Lissauer have measured numerous changes to the orbits of these moons since 1994.

Their simulations reveal that there is a continual exchange of energy and angular momentum between the satellites. As a result, their orbits are chaotic and dynamically unstable over time scales of only one million to 100 million years.

One example is Cupid, which approaches the considerably larger Belinda. Calculations predict that that the pair will collide within a few million years. A similar process in the past could have led to the birth of Cupid and Perdita, which may be pieces that broke away when a comet smashed into nearby Belinda.

Probably the most unstable moon of all is Cressida, which calculations indicate may collide with its neighbor, Desdemona, about a million years from now. Desdemona orbits just 900 km outside Cressida's path.

Additional small moons, as yet unseen, may be shepherding and shaping the surprisingly narrow rings. Studies of Voyager 2 data made 30 years later have shown that the amount of material on the edge of the alpha ring varied along different parts of the ring. A similar, even more noticeable, pattern occurred in the same part of the beta ring. The pattern in Uranus' rings appears to be similar to structures in Saturn's rings, known as moonlet wakes.

The hypothesized moons would be 4 to 14 km in diameter, as small as some moons of Saturn, but smaller than any of Uranus' known moons. The moons, if they exist, may be acting as "shepherds," helping to keep the rings from spreading out. Ophelia

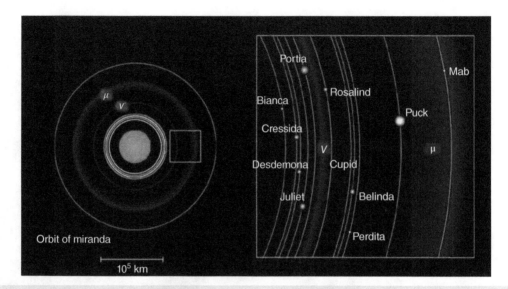

Figure 10.26 The crowded region within the orbit of Miranda contains 13 known rings and 13 small satellites. (Left) The mu (µ) and nu (ν) rings lie between the orbit of Miranda and the narrow rings detected in 1977. (Right) An illustration of the region near the outer rings, showing the satellites' orbits. The satellites are shown to their correct relative sizes, but exaggerated by a factor of ∼16 compared with the width of their orbits. (Carl D. Murray)

Figure 10.27 Puck is roughly spherical and about 160 km across, making it the sixth-largest of Uranus' moons. Its surface is almost as black as the planet's rings. A 45 km-wide impact crater is visible near the limb (right), with another large indentation above it. (NASA-JPL)

and Cordelia are currently known to act as shepherds to Uranus' epsilon ring (see Rings).

Remote Moons

Uranus has at least nine small satellites beyond Oberon, bringing the total number to 27 (Figure 10.28). All but one of these outer moons follow irregular, retrograde orbits. The orbits are also highly elliptical and inclined to the planet's equator – further evidence that they have been captured by Uranus.

Box 10.3   Naming the Moons of Uranus

The International Astronomical Union (IAU) has the task of assigning names to newly discovered objects in the Solar System. In the case of planetary satellites, the IAU usually follows the tradition of naming them after characters in Greek or Roman mythology. However, in the case of Uranus, the first four moons had been named in the mid-19th century after characters from the literary works of Shakespeare (Oberon, Ariel, and Titania) and Alexander Pope (Umbriel). Miranda and the more recent discoveries have also been named after Shakespearean characters, with the exception of Belinda (1986 U5), which was named after a character from Pope's "The Rape of the Lock."

The process of capture could have taken place in two different ways. The moons could have simply been trapped by Uranus' gravity as they came close to the planet, or the young Uranus could have been surrounded by a gaseous nebula that would have caused a drag on objects passing nearby.

Almost all of the outer satellites have been discovered since 1997 by ground-based teams of observers. Since they are very faint (below magnitude 20), large modern telescopes with adaptive optics are required to find them. The first to be recognized were Caliban and Sycorax. Both were noticeably red, resembling the Kuiper Belt objects that exist beyond Neptune.

Caliban, the smaller of the two moons, has a diameter of about 70 km and orbits Uranus at an average distance of about 7.2 million

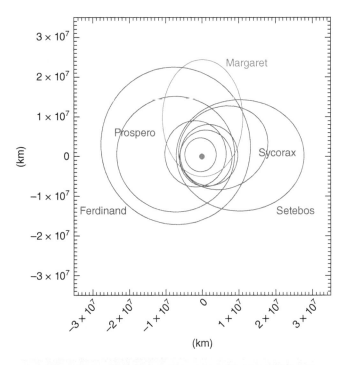

Figure 10.28   The orbits of the outer satellites. All are highly elliptical and steeply inclined to the planet's equator. They also travel in a retrograde direction (with the exception of Margaret, which is prograde). The satellites were probably captured by Uranus. (Adapted from Scott S. Sheppard, Carnegie Institution for Science)

km, taking 1.6 years to complete one revolution. Sycorax appears to be about twice the diameter and has an orbital period of 3.5 years, with a mean distance of about 12.2 million km. However, Sycorax has a much more elliptical orbit than Caliban, bringing it as close as 6 million km to the planet and as far away as 18 million km.

In 1999, three more irregular moons (Stephano, Prospero, and Setebos) were found with the Canada-France-Hawaii Telescope.

Considerably smaller than the moons discovered two years earlier, these are about 20 times fainter and orbit Uranus at average distances of 8 million, 16.2 million, and 17.4 million km. The orbit of Setebos is even more eccentric than that of Sycorax.

Francisco and Trinculo were found in 2001, followed by Margaret and Ferdinand in 2003. Francisco is the innermost irregular moon so far discovered, with a mean distance of about 4.3 million km from Uranus – more than seven times further out than Oberon. Ferdinand is the outermost member of the group, taking almost eight years to complete one orbit at an average distance of 20.9 million km. The odd one out is Margaret, which travels in a "normal" prograde direction, though it has the most eccentric orbit of all the Uranian satellites.

No missions to Uranus are planned for the foreseeable future.

Questions

- Uranus was the seventh planet to be discovered orbiting the Sun. Why did it take so long to find?
- Summarize the major differences between Uranus, Jupiter, and Saturn.
- Why does Uranus have such a high axial inclination and rotate on its side? Is this likely to have been an ancient or relatively recent event? Explain your answer.
- How does Uranus' high obliquity influence conditions on the planet?
- Describe and explain the differences in the appearance of Uranus during the Voyager 2 flyby and its appearance in recent years.
- Why were scientists surprised by the magnetic field of Uranus? Briefly explain how this field is thought to be generated.
- Compare and contrast the rings of Uranus with those of Saturn and Jupiter.
- What are the major similarities and differences between the satellites of Uranus and Saturn?

ELEVEN

Neptune

1,600 million kilometers beyond Uranus is its near-twin, Neptune, another blue-green ice giant. Despite their many similarities, the planets reveal a number of unexplained differences, notably their contrasting heat output and storm activity. Although most of what we know about this remote world was returned by the Voyager 2 spacecraft in 1989, its changing atmosphere and elusive ring system continue to be examined by the Hubble Space Telescope and state-of-the art ground-based instruments. These observations make it clear that Neptune is one of the most dynamic members of the Sun's family.

With the discovery of Uranus in 1781, the Solar System once again seemed complete (see Chapter 10). However, when astronomers began to refine their calculations of its orbit, Uranus refused to behave as expected. Until 1822 it moved faster than predicted, but after 1822 it seemed to slow down. By the mid-1830s, a number of scientists were wondering whether the unexpected errors in its position were caused by another, more remote, planet pulling on Uranus.

Would it be possible to calculate the position of such an object from its perturbations of Uranus, and then recognize it among the plethora of background stars? A number of mathematicians decided to attempt the incredibly difficult calculation. By 1845, John Couch Adams, a student at Cambridge University, England, informed James Challis, Professor of Astronomy at Cambridge, that he had come up with a solution to the problem, but little action was taken.[1]

By then, another brilliant mathematician was causing a stir on the other side of the English Channel. Unaware of Adams' work, Urbain Jean Joseph Le Verrier produced his own independent solution, which predicted that the new planet was located about 5° east of the star delta Capricorni – a very similar answer to that of his English rival.

Le Verrier wrote to Johann Galle at the Berlin Observatory, requesting that a search be carried out. With the backing of the observatory's director, Johann Encke, Galle and graduate student Heinrich d'Arrest immediately started to scan Le Verrier's predicted location with a 23 cm refractor. As Galle stared through the telescope and called out each star in the field of view, d'Arrest compared them with a new, detailed map of the heavens. Within one hour they found an object that was not on the chart.

The following night (September 24), they were able to confirm that the object had moved and that it displayed a tiny disk. The discovery had been made within one degree of Le Verrier's prediction, and was a great triumph for Newton's gravitational theory.

The solution to the Uranus problem was now clear (see Figure 11.1). Before 1822, Uranus was lagging behind Neptune

---

[1] Adams' solution was actually within two degrees of arc of Neptune's actual position. If a search had begun straight away, he would probably have gone down in history as the unchallenged discoverer of the eighth planet from the Sun. Instead, as a result of bad luck and incompetence by his seniors, the search by Challis did not begin until late July 1846.

Exploring the Solar System, Second Edition. Peter Bond.
© 2020 John Wiley & Sons Ltd. Published 2020 by John Wiley & Sons Ltd.
Companion Website: www.wiley.com/go/Bond-Solar-System2e

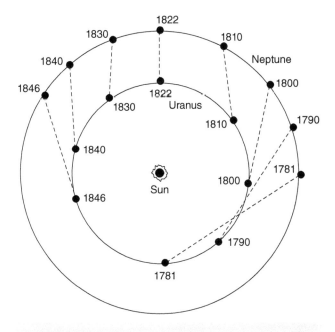

Figure 11.1 Unexpected perturbations in the motion of Uranus caused astronomers to recognize that the gravitational pull of an unknown giant planet could be influencing its orbit. Until 1822, Neptune's gravity caused Uranus to move faster than expected. After their opposition in 1822, Uranus overtook Neptune, but the pull of the unseen planet caused it to slow down. As the distance between the pair increased, Neptune's gravitational influence gradually diminished. (Peter Bond)

in its orbit, but slowly catching up with the outer planet. As a result, Neptune's gravity was accelerating Uranus' motion. In 1822 the two planets were at opposition. Thereafter, although Uranus began to pull ahead of Neptune, the outer planet's gravitational pull slowed it down, causing it to lag behind its predicted position.

The Eighth Planet

After the usual debate over what to call the newly discovered world, it was eventually decided to name it Neptune, after the Roman god of the seas and brother of Jupiter. Neptune is too faint to be seen with the naked eye, although it is often brighter than eighth magnitude. With an apparent diameter of little more than three arc seconds, it is 600 times smaller than a Full Moon. However, with a telescopic magnification of more than 100, the planet clearly shows a bluish disk.

The eighth planet turned out to be another giant, similar in size and outward appearance to Uranus. For more than a century, astronomers were unsure which of the pair was the larger, and it is only in recent years that Uranus has been given the nod. Neptune's equatorial diameter is about 49,528 km (almost four

Table 11.1

Neptune Summary

Mass: 102.43 x 10^{24} kg (17.1 Earth masses)
Equatorial Diameter: 49,528 km
Polar Diameter: 48,682 km
Axial Inclination: 28.32 degrees
Semi-major Axis: 4,495.06 million km (30 AU)
Rotation Period (internal): 16.11 hours
Sidereal Orbit Period: 164.79 years
Mean Density: 1.64 g/cm³
Number of known satellites: 14

times the diameter of Earth). Fifty-seven Earths would fit inside Neptune (Table 11.1).

On the other hand, it has a greater mass than Uranus and its bulk density is actually the highest of the Jovian planets. Neptune has a mass more than 17 times that of Earth, although the gravity at Neptune's cloud tops is only 1.19 times Earth's surface gravity. This is because it is such a large planet (the gravitational force a planet exerts upon an object at the planet's surface is proportional to its mass and to the inverse of its radius squared). Hence a 10-kg object would weigh 11.9 kg at Neptune's visible surface.

These physical properties suggest slightly different compositions and internal structures for the two blue giants. This seems to be confirmed by measurements indicating that – unlike Uranus, but similar to Jupiter and Saturn – Neptune radiates more than twice as much energy as it receives from the Sun.

The difficulty in observing long-lived cloud features meant that the planet's rotation period was uncertain for many years. (As with all gas giants, clouds swept along by strong zonal winds are unreliable markers for the rotation of the planet as a whole.) It was not until the Voyager 2 flyby of Neptune that the internal rotation period was established to be 16 hours, 6 minutes – considerably faster than that of Uranus – by measuring the rotation rate of the planet's magnetic field.[2] Neptune's rapid rotation causes an equatorial bulge, so the polar diameter is about 850 km smaller than the equatorial diameter.

Neptune has an axial inclination of 28.3°, only a little more than Earth and Mars. This means that it experiences seasons that last approximately 41 years during its 164-year circuit of the Sun. At the time of the Voyager 2 flyby the south pole was illuminated by early spring sunshine, with an absence of solar heating at the wintry north pole. Midsummer in the south (and midwinter in the north) will arrive in 2020.

Neptune's orbit is almost circular, with a semi-major axis of about 4.5 billion km (30 AU). At such a vast distance, the planet crawls around the Sun at 5.5 km/s, taking almost 165 years to complete one orbit. Neptune completed its first circuit of the Sun since its discovery on July 11, 2011.

---

[2] Studies of long-lived cloud features have led to a shorter estimate for one complete rotation: 15 hours 57 minutes 59 seconds.

Box 11.1   Voyager 2's Grand Finale

After the flyby of Uranus in January 1986, NASA's Voyager 2 spacecraft set course for Neptune on the final leg of its Grand Tour. In order to prepare for the high-speed flyby of Neptune, where light levels are 900 times lower than on Earth, the spacecraft's computers were reprogrammed to allow longer imaging exposures and greater spacecraft stability, and to compensate for the relative motion of the spacecraft and its targets.

Typical camera exposures at Neptune were to last for 15 seconds or longer, compared with 5 seconds at Uranus. However, exposures of up to 61 seconds were now possible, with additional extensions in multiples of 48 seconds.

In order to ensure that Voyager's weak signals were safely received, all of NASA's 64 m Deep Space Network antennas were enlarged to 70 m and made as efficient as possible. Other antennas and ground stations were enlisted, including the Parkes radio telescope in Australia and a dish in Usuda, Japan.

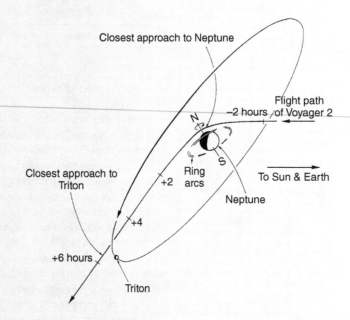

Figure 11.2   Voyager 2 was targeted for closest approach only 4,500 km above Neptune's northern hemisphere. This would bend its trajectory south of the ecliptic, enabling it to pass only 38,500 km from the centre of Triton less than six hours after the Neptune encounter. This flyby sent Voyager 2 heading south, out of the Solar System, with no possibility of redirecting it towards Pluto. It also meant that Voyager was unable to make a close approach to the other known moon, Nereid. *(NASA-JPL)*

The main objective was to find the closest approach to Neptune that would result in a grandstand view of Triton. On the date of the flyby (August 25, 1989) Triton would be south of Neptune's orbital plane, so Voyager had to sweep low over the north pole, allowing the planet's gravity to bend its path southward. In the event, Voyager passed a mere 4,500 km above Neptune's cloud tops.

By May 1988, images from Voyager were already showing more detail than any obtained by ground-based telescopes. Then, in January 1989, a huge dark spot was discovered in the planet's southern hemisphere. In early July a 400 km-diameter satellite was found inside the orbit of Triton. Five more small moons were found close to Neptune's rings in the weeks that followed.

The spacecraft crossed the bow shock – the boundary where the solar wind ploughed into Neptune's magnetosphere – about 9½ hours before closest approach. It eventually exited the magnetosphere 38 hours later, but multiple bow shock crossings were subsequently experienced almost 4 million km from the planet. The magnetic dipole was found to be highly inclined and offset from the planet's center.

One of the most intriguing puzzles solved by Voyager was the apparent presence of ring arcs – partial rings – around Neptune. Images and occultation observations showed that the arcs are actually elongated clumps within one of the five dark, narrow, ringlets. Voyager crossed the ring plane twice, about 60 minutes before and 85 minutes after its closest approach to Neptune.

Voyager's trajectory was bent about 48° south of the ecliptic, and, 110 minutes after leaving Neptune behind, its instruments began to observe Triton. The spacecraft eventually passed 38,500 km from the satellite's center, less than six hours after the Neptune encounter. Images of the sunlit southern hemisphere revealed an extremely cold, icy world, with a surface of pink ice, cantaloupe patterns, and dark streaks.

Voyager's farewell view of Neptune and Triton was taken on August 28, when the narrow-angle camera captured two pale crescents hanging in a black sky.

Atmosphere

Hydrogen and helium, the main constituents of Jupiter, Saturn, and Uranus, were also expected to be the major components of Neptune's atmosphere, but spectroscopic evidence was hard to obtain. However, by the late 1940s, techniques had improved sufficiently to recognize the spectral signature of molecular hydrogen (H_2) and methane (CH_4). The abundance of hydrogen inferred that helium, the second-most common gas in the Sun, was also present in large quantities.

This was confirmed by data from Voyager 2, which indicated that the upper atmosphere (above the methane ice clouds) is 84% molecular hydrogen, 14% helium, and 2% methane (Box 11.1). As is the case with Uranus, the scattering of blue light and absorption of red light by methane gives the planet its characteristic blue color (Figure 11.3).

There are also trace amounts of carbon monoxide (CO), hydrogen cyanide (HCN), acetylene (C_2H_2), and ethane (C_2H_6) – the last two being products of methane photochemistry. As is the case with Uranus, ultraviolet light from the Sun destroys the methane high in Neptune's atmosphere, converting it to acetylene and ethane. These gases diffuse down to a lower, colder part of the stratosphere, where they condense into tiny ice particles. On reaching the warmer region of the upper troposphere, the hydrocarbon ice particles evaporate and are converted back to methane gas.

Voyager's instruments were only able to probe the atmosphere down to a level where pressure reached a few bars. Beneath one

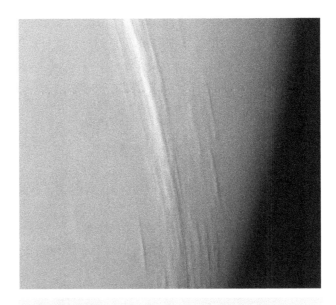

Figure 11.4 Voyager 2 imaged bright, linear clouds of methane ice casting shadows on a lower cloud deck of uncertain composition. These clouds were observed at latitude 27°N and extended parallel to the equator for thousands of kilometers. The width of the cloud streaks ranged from 50 to 200 km and the shadows were 30–50 km across. The clouds were above most of the methane, hence their whitish hue. Their height above the main cloud deck was estimated to be 50–100 km. The Sun is toward lower left. (NASA-JPL)

Figure 11.3 This false-color image from Voyager 2 shows a high-level haze (red), particularly over the south pole and near the limb, where the instruments were looking through a greater thickness of atmosphere. At the center of the disk, sunlight penetrates the haze and is absorbed by methane, producing a blue color. Highly reflective ice clouds appear as patches of white. (NASA-JPL)

or more layers of ethane–acetylene haze is a physically thick but optically transparent layer of methane ice haze. Discrete, denser, clouds of methane ice created by localized upwelling sometimes cast shadows on the main cloud deck 50–75 km below (Figure 11.4). This lower, blue-tinted, opaque layer has an unknown composition, but may be made of hydrogen sulfide (H_2S) ice or ammonia (NH_3) ice.

Below this, conditions may be suitable for the existence of ammonium hydrosulfide (NH_4SH) clouds – a mixture of water, ammonia, and hydrogen sulfide. The deepest hypothesized cloud layer is thought to exist between 40 and 400 bars, at a depth of several hundred kilometers.

Voyager also measured the temperature distribution in the atmosphere. The infrared data showed that the upper atmosphere is warmer near the equator, cooler in mid-latitudes, and warmer again above the south pole (which was enjoying its long summer). This pattern resembled that of Uranus, even though their axial inclinations are very different. In the troposphere, there was hardly any temperature difference between the south pole and the equator – a characteristic common to all four giant planets.

The temperature of the stratosphere fell from about –155°C at an altitude of 100 km above the 100 millibar level to a minimum of –220°C at the 100 millibar level. It then increased steadily through the troposphere, reaching a temperature of –146°C at the 5 bar level. The planet was notably warmer than expected, based on its distance from the Sun, with an average surface temperature of –220°C, compared with –214°C for Uranus.

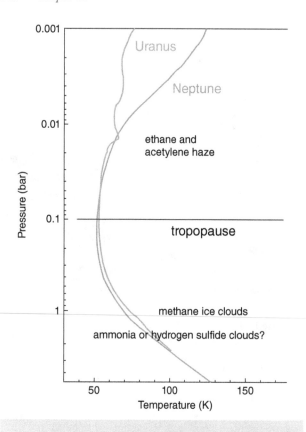

Figure 11.5 A comparison of the temperature and pressure of the upper atmospheres of Neptune and Uranus, showing probable haze and cloud layers. The region above the tropopause (the stratosphere) is dominated by methane photochemistry. Models suggest that other cloud layers exist beneath the methane ice. The uppermost of these may be composed of hydrogen sulfide. (Peter Bond)

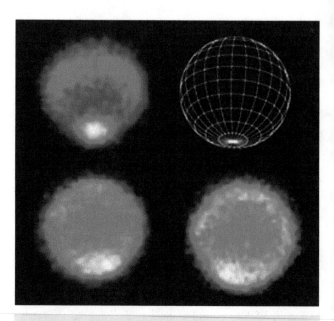

Figure 11.6 Thermal images show temperatures near the top of Neptune's troposphere (upper left image), with the hottest temperatures found at the south pole (graphic, upper right). The lower two images show temperatures at higher altitudes, in Neptune's stratosphere, or lower atmosphere. (Courtesy of VLT/ESO/NASA-JPL/Paris Observatory)

Neptune's warmer-than-expected temperature is accounted for by the fact that, unlike Uranus, it has a substantial internal heat source. Neptune emits about 2.5 times as much heat as it receives from the Sun. This internal source of warmth may explain why Neptune's winds are much stronger than those on Uranus.

Temperature maps of the planet's atmosphere obtained in 2007 confirmed that the upper troposphere above the south pole was about 10°C warmer than the rest of the planet, including the equator (Figure 11.6). This was explained by the fact that the southern hemisphere had been tilted toward the Sun for about 40 years and was nearing the end of its lengthy summer.

The temperatures at the south pole were high enough to enable methane gas, which should be frozen out in the upper part of Neptune's atmosphere (the stratosphere), to leak out – accounting for the planet's high stratospheric methane abundance.

By studying the motions of long-lived cloud features moving parallel to the equator, it was possible to determine that – as on Uranus – zonal winds in equatorial latitudes are retrograde – opposite to the eastward rotation of the planet (Figure 10.8). (On Jupiter and Saturn the equatorial jets are prograde, while retrograde winds are weak and scarce.)

At the equator, winds are blowing towards the west at speeds of up to 400 m/s (1,440 km/h), comparable to the near-equatorial winds of Saturn. Clouds caught up in this wind take more than 20 hours to travel around the planet, compared with Neptune's rotation period of a little over 16 hours.

In contrast, winds at latitudes of 70° peak at about 200 m/s in a prograde direction, sweeping clouds around the planet in a mere 11 hours (Figure 11.7). This means that the variation of wind speeds on Neptune far exceeds that on any other planet.

The most noticeable cloud features imaged by Voyager were in the southern hemisphere. The largest of these was the Great Dark Spot, located at about 22°S. Blown westward at about 30 m/s by the zonal wind, it completed one circuit in just under 18 hours.

In many ways, the oval storm, which measured some 15,000 km by 7,000 km, resembled Jupiter's Great Red Spot (Figures 11.8 and 11.9). The counterclockwise motion of clouds around its rim indicated an anticyclonic weather system, an area of rising gas that was creating a high-pressure system.

Bright, cirrus-like clouds of methane ice crystals were seen above and around the Great Dark Spot. They appeared to be similar to orographic clouds on Earth, which form when moist air is forced upwards over a mountain. In this case, the Spot was apparently forcing methane-rich gas to higher levels, resulting in cloud formation.

Two significant differences between the huge anticyclones of Jupiter and Neptune were their stability and longevity. Not only was the Great Dark Spot "floppy" – changing shape and orientation over a period of about 8 days – but it was drifting toward the equator at a rate of 15° per year. Scientists were not too surprised

Figure 11.7   Neptune is the windiest planet in the Solar System. The planet appears to have three broad jet streams: one that blows westward at the equator and two that blow eastward at higher latitudes, closer to the poles. Most of the cloud features are blown westward – the opposite direction to the planet's rotation – with peak wind speeds of up to 400 m/s near the equator. (Courtesy of Yohai Kaspi/Weizmann Institute of Science, NASA)

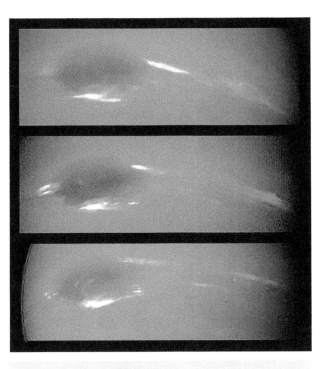

Figure 11.8   Voyager 2 images of the Great Dark Spot showed bright cirrus-like clouds above and around the anticyclone that formed and dissipated quite rapidly. The sequence spans about 36 hours, a little over two rotations of Neptune. (Courtesy of NASA-JPL)

when Hubble images taken in 1994 indicated that the Spot had disappeared.

Two more striking cloud features were seen in the Voyager images (Figure 11.9). One, a triangular white cloud at latitude 42°S, was nicknamed Scooter because every few revolutions it caught up with and overtook the Great Dark Spot, completing one circuit in about 16 hours.

At latitude 54°S was another dark spot, known as D2, which displayed a bright core of methane ice above a dark oval. Unlike its giant cousin, D2 appeared to rotate in a clockwise direction, indicating that it was a cyclone or region of low atmospheric pressure. Gases would, therefore, be expected to rise in the cloudy centre and descend in the dark oval around it. D2 almost matched the planet's rotation period, reappearing roughly every 16 hours.

Although the detailed observations made by Voyager are unique, an observation campaign has been undertaken since the early 1990s with modern ground-based instruments and the Hubble Space Telescope. These found evidence that the Great Dark Spot and Scooter had disappeared by 1993.

However, images taken in late 1994 did reveal a new dark spot in the northern hemisphere that appeared to resemble its southern counterpart in terms of size and appearance. Once again, high-altitude clouds around the edge of the spot were thought to be cooled clouds of methane ice crystals.

Several more dark ovals – most of them smaller than the Great Dark Spot – have been observed in mid-latitudes since then. The particles in the storm clouds – possibly hydrogen sulfide – are still highly reflective, but they are slightly darker than the particles in the surrounding atmosphere, so they appear as dark ovals. (H_2S is best remembered on Earth for its pungent smell of rotten eggs.)

None of these high-pressure atmospheric systems have survived more than a few years – unlike comparable anticyclonic storms on Jupiter. This may be related to the fact that Neptune seems to only have three broad jet streams: a westward one at the equator, and eastward jets around the north and south poles (see Figures 11.7 and 10.8). This is much less complex than Jupiter's atmospheric circulation, which has numerous alternating wind jets that are marked by cloud bands in its atmosphere (see Chapter 8). Neptune's rotating storms are, therefore, free to shift in latitude anywhere between the jets.

Storm activity also seems to have been increasing as the seasons slowly change. The planet has been growing brighter since the early 1980s, as it moves from the onset of southern spring toward southern midsummer. There is also some evidence that the planet's overall brightness varies over a period of about 10 years, possibly related to the cycle of solar activity.

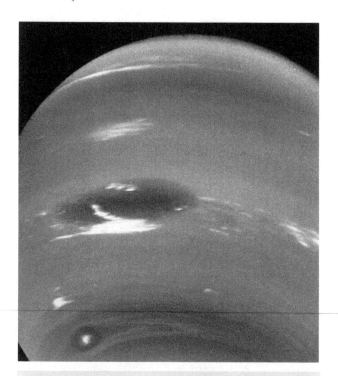

Figure 11.9   A Voyager 2 image showing features in Neptune's southern hemisphere. Most prominent is the Great Dark Spot and its associated cirrus ice clouds. Also visible are a bright cloud feature nicknamed Scooter (lower left) and a small dark spot (D2) with a bright core. Driven by zonal winds, all the features were moving around the planet at different speeds. The maximum diameter of the Great Dark Spot was about 15,000 km. (NASA-JPL)

In the early 1990s, cloud activity shifted from the southern to northern mid-latitudes. Since the mid-1990s, there has been a distinct increase in the area and brightness of banded cloud features, particularly in the planet's southern hemisphere.

Bright cloud features observed in 1996 were found consistently at latitudes of 45°S and somewhat less frequently at 30°S and 30°N. By 1998, the clouds at both of these southern latitudes were more numerous and a new cloud feature, similar to one seen in Voyager 2 images, was seen near 67°S (Figure 11.11). A much more dramatic change was apparent by 2002, when individual clouds appeared to be much brighter at 30°S, 45°S, and 67°S, and the spatial extent of the bright clouds was much greater.

The changes are especially prominent at near-infrared wavelengths, which are much more sensitive to high-altitude clouds. They are thought to be a response to seasonal variations in sunlight, particularly at higher latitudes, where the seasons tend to be more pronounced. The result is stronger upwelling of convectional currents, leading to more condensation and increased cloud cover. Since the response of the massive atmosphere to the small change in insolation is thought to be very slow, the planet is likely to continue brightening for another 15–20 years.

One notable recent change was the 2017 appearance of a larger bright cloud feature near Neptune's equator, a region where no similar feature had ever been seen before (Figure 11.12). The center of the storm complex was about 9,000 km across, roughly one-third of Neptune's radius. The storm, which spanned at least 30 degrees in both latitude and longitude, brightened considerably in only one week. The reason for its existence was not clear: it might have been caused by condensation as air flowed over a large, unseen, vortex, or it could have been a seasonal storm, rather like those that erupt from time to time on Saturn (see Chapter 9).

Meanwhile, a dark spot that appeared in the mid-southern latitudes in 2015 was seen to be shrinking and drifting toward the south pole in 2018 (Figure 11.10). Observers had expected it to drift toward the equator, but the presence of only three planetary jet streams allowed it to wander in an unpredictable fashion.

In later 2018, the HST imaged a new northern Great Dark Spot that was developing alongside a partially overlapping region of white cloud (Figure 11.13). This new dark storm was of a similar size and shape to the storm discovered in 1989 by Voyager 2. There was some image evidence that the storm had been forming from as early as 2015. This slow growth process indicated that the storm developed deep within Neptune's atmosphere, drawing up dark material from its depths, and only became visible once the top of the updraft reached higher altitudes.

Interior

The internal structure of Uranus and Neptune is difficult to determine (see Chapter 10) because their bulk composition seems to be more complex than for Jupiter and Saturn, with a smaller fraction of hydrogen and helium, and their gravitational moments are known with a lower accuracy (Figure 11.14). Both are thought to be "ice giants," but there are some differences.

Neptune's mass and density are higher than those of Uranus, despite it being slightly smaller. This is partly due to greater compression, but could also be the result of a slightly different composition, perhaps with a slightly higher proportion of rock to ice. (Assuming that the ice/rock ratio of both planets is similar to the protosolar nebula, their overall compositions, by mass, should be about 25% rock, 60–70% ice, and 5–15% hydrogen and helium.)

The stormy atmosphere has been estimated to be no more than 1,000 km deep. Beneath the atmosphere of hydrogen, helium, and methane, there is probably a layer of liquid hydrogen that also includes some helium and methane. Most of Neptune's interior is thought to be made up of an "ocean" of molten "ices," probably a mixture of water, methane, and ammonia. At the center may be a rocky core, perhaps slightly larger than the core of Uranus.

There has been considerable debate over the possible chemical reactions that take place at high temperatures and pressures inside Neptune and Uranus. One theory suggests that the outcome would be molecular hydrogen and pure carbon in the form of diamond crystals. The latter, if large enough, would irreversibly sink as sediment toward the center of the planet, whereas the hydrogen would rise into the atmosphere. Another theory favors the dissociation of methane into free hydrogen and ethane.

Why is Neptune releasing more than twice as much energy as it receives from the Sun, so that its temperature is comparable to that of Uranus? As with Jupiter and Saturn, much of Neptune's internal heat is thought to be produced by continuing, slow contraction

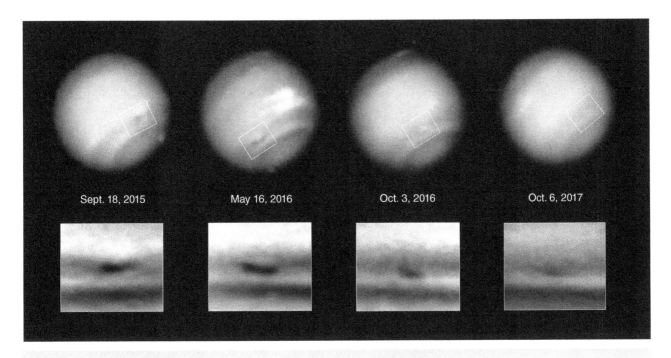

Figure 11.10 Occasionally, Neptune exhibits large, dark cloud features. These dark spots are high-pressure atmospheric systems. This feature, named SDS-2015 (Southern Dark Spot discovered in 2015), was first observed at 45°S. HST observations from 2015 to 2017 revealed that the spot's maximum extent reached 4,960 km across its long axis before it rapidly diminished in size. The spot drifted slowly toward the south pole as it evolved. Bright clouds alongside the spot formed when atmospheric flow was disturbed and diverted upwards over the spot, causing gases to freeze into methane ice crystals. (Courtesy of NASA, ESA, M.H. Wong, and A.I. Hsu/UC Berkeley)

because of its strong gravitational field. Differentiation in the icy mantle also allows heavier molecules to sink to lower levels, liberating heat energy.[3]

This heat seems to escape to the surface more easily in Neptune than in Uranus, possibly because convection is less inhibited by differential layering in the latter's interior. If this layering exists at deeper levels in Neptune, it would allow more heat to be transported outward.

Magnetic Field

Like the other Jovian planets, Neptune has a powerful magnetic field and a large magnetosphere. Voyager found that Neptune's magnetic field was 25 times stronger than Earth's, though the planet is so large that the field is actually weaker at its cloud tops than on the surface of our planet.

One of the biggest surprises of the Voyager encounter was the lopsided nature of the magnetic field, which bore many similarities to that of Uranus (Figure 11.15). The axis of the magnetic dipole is tilted 47° to the rotation axis and offset from the planet's center by 55% of Neptune's radius.

This means that the field originates in a region of the interior that is actually closer to the cloud tops than to the center, and the field is much stronger in one hemisphere than the other. In fact,

the field strength at the surface varies from more than 1 gauss in the southern hemisphere to less than 0.1 gauss in the northern hemisphere. (Earth's equatorial magnetic field at the surface is 0.32 gauss.)

The large tilt of the magnetic dipole axis and the planet's rotation result in a dynamic magnetosphere which changes from an Earthlike configuration (when the angle between the dipole axis and solar wind is about 90°) to a pole-on configuration and back every 16 hours or so. As a result, the cusp regions above the poles change location and size as the planet rotates while the magnetic equator and field lines appear to gyrate wildly in the solar wind as the planet rotates. When Voyager 2 passed the planet, one pole of the highly inclined magnetic field was pointing toward the Sun.

The field seems to be generated by slow-moving convectional currents of ionized material in the planet's icy mantle. Recent studies suggest that the magnetic fields of both ice giants are generated in a thin outer layer of the mantle, rather than in the lower regions where convection does not take place.

Compared with other planetary environments, the magnetosphere is actually quite empty, with fewer protons and electrons per unit volume than any of the other gas giant planets. The main source of nitrogen ions is Triton's sparse atmosphere or ionosphere, while the hydrogen ions come from a large, diffuse cloud that extends inward from the satellite's orbit for more than 150,000 km.

---

[3] Energy is released when mass sinks toward the center of an object, effectively converting gravitational potential energy into heat.

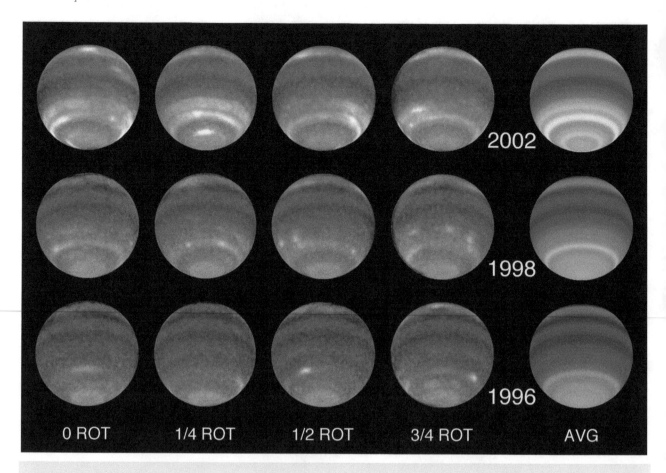

Figure 11.11 Changes in Neptune's appearance 1996–2002, as seen by the Hubble Space Telescope. The individual images were processed to reduce limb darkening and enhance contrast. The first four columns display views at 0, 1/4, 1/2, and 3/4 rotations to show the distribution of discrete features. The last column displays the longitudinal average, showing little change at the equator but dramatic increases in brightness near 30°S, 45°S, and 67°S. (Courtesy of L. Sromovsky, P. Fry/Univ. of Wisconsin, NASA, ESA)

The high inclination of Neptune's magnetic field means that charged particles trapped in the magnetosphere are repeatedly swept past the satellites and rings. Many of these charged particles are absorbed by the satellites and rings, effectively removing a large fraction of its charged particle population.

Weak auroral emissions were identified on the nightside of Neptune, and they are thought to occur close to both magnetic poles.

Triton

Prior to the Voyager 2 flyby, only two satellites of Neptune were known. The largest of these, Triton, was found by William Lassell in October 1846, only 17 days after the planet's discovery.

Triton was quite bright for its distance, suggesting that the satellite rivaled Titan and Ganymede in size. However, its precise diameter remained uncertain until the Voyager encounter, when it became apparent that ice-covered Triton acts like a mirror, reflecting 72% of sunlight at visible wavelengths. This unexpectedly high albedo explained why astronomers had seriously overestimated Triton's size. With a diameter of 2,710 km, Triton is smaller than the Galilean satellites, Titan, and Earth's Moon, but larger than Pluto.

Based on its size and density (derived from its effect on Voyager's trajectory), it is possible to model the moon's interior. With a density about twice that of water, Triton has approximately the same density as Ganymede and Callisto, suggesting that it is rocky rather than icy. Models indicate that there is a thick layer of water ice above a rocky central region, which occupies about two-thirds of its radius. There may also be a small iron-rich core.

Triton is unusual in other ways. It orbits about 355,000 km from Neptune, completing one circuit every 5.9 days in a retrograde direction – opposite to the planet's rotation.[4] A season on Triton lasts about 41 years, and Triton passed its southern summer solstice in 2000 (Figure 11.17). At the time of Voyager's flyby Triton's southern hemisphere was experiencing high summer, while the northern hemisphere was in the middle of a prolonged winter.

[4]Triton was the first satellite found to have a retrograde orbit. It is also the only large satellite in the Solar System to orbit a planet in a retrograde direction.

Neptune: 26 June 2017

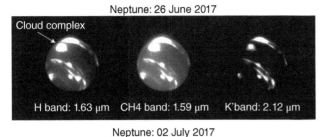

Neptune: 02 July 2017

Images displayed on log scale

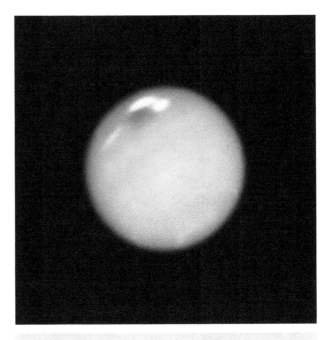

Figure 11.12   Images obtained with the Keck telescope in 2017 revealed an extremely large bright storm system (labeled) near Neptune's equator a region where no such bright cloud had ever been seen before. The center of the storm was about 9,000 km across, about one-third of Neptune's radius. The storm brightened considerably between June 26 and July 2, 2017. (N. Molter/I. De Pater, UC Berkeley/C. Alvarez, W.M. Keck Observatory)

Figure 11.13   A large dark spot to the left (west) of a partially overlapping region of white cloud was imaged by the Hubble Space Telescope in 2018. This new dark storm in the northern hemisphere, which had taken some three years to mature, was similar in size and shape to the Great Dark Spot discovered in 1989 by Voyager 2. (Courtesy of NASA, ESA, A. A. Simon/NASA-Goddard, and M.H. Wong and A.I. Hsu/University of California, Berkeley)

However, the satellite also experiences a much more complex, 688-year-long cycle of seasons, resulting from Neptune's 164-year passage around the Sun, an orbital inclination of 23° to the planet's equator, and the precession of Triton's orbit about Neptune's axis of rotation (Figure 11.16). This means that the seasons vary considerably in length and intensity in the long term.

Despite the seasonal variations in sunlight, solar heating appears to have very little impact on Triton, with only minor variations in temperature from one hemisphere to the next. The average temperature of –235°C means that Triton is so cold that most of the nitrogen in its tenuous atmosphere freezes out onto the surface, making it the only satellite in the Solar System known to have a surface made mainly of nitrogen ice.[5] (Infrared spectra have also revealed the presence of frozen methane, carbon monoxide, and carbon dioxide.)

The surface pressure determined by Voyager was about 16 microbars – about the same as that 70 km high in the Earth's atmosphere. Nevertheless, Voyager imaged discrete clouds up to 8 km above the surface, and haze was detected up to altitudes of 20–30 km.

With Triton's southern hemisphere beginning to receive more direct sunlight as summer approached, the thin layer of frozen gases on the surface sublimated into gas, thickening the atmosphere. However, the atmosphere of molecular nitrogen, with traces of methane and carbon monoxide, remained extremely thin.

Confirmation of seasonal variations came in November 1997 when Triton passed in front of a star. Astronomers took advantage of the rare occultation to study its atmospheric density by measuring the gradual decrease in the star's brightness as its light traveled through the satellite's sparse atmosphere.

The data indicated that Triton's atmosphere had doubled in density. Based on the unusually strong link between the temperature of its surface ice and atmospheric pressure, the moon had apparently warmed by almost 2°C since the Voyager encounter nine years earlier, resulting in a "balmy" temperature of –234°C.

By 2010, it was estimated that Triton's atmospheric pressure had risen by a factor of four since 1989, so that the atmospheric pressure had increased to between 40 and 65 microbars (still 20,000 times less than on Earth).

Triton is one of only four objects in the Solar System known to have a nitrogen-dominated atmosphere. The others are Earth, Pluto, and Saturn's giant moon Titan. Nitrogen, carbon dioxide, and carbon monoxide are known to exist in interstellar clouds, so it is likely that the very cold objects in the outer Solar System have probably retained a lot of the original molecular mix of the nebula from which the planets formed.

Voyager's cameras unveiled a strange landscape shaped by the seasonal variations in ice cover and by internal processes. The visible area was dominated by two dramatically different types of terrain: the so-called "cantaloupe" terrain close to the equator and a receding ice cap.

The bright southern polar cap appeared to be composed largely of nitrogen frost. There was some evidence that this ice was

[5] Dwarf planet Pluto also has a surface of nitrogen ice.

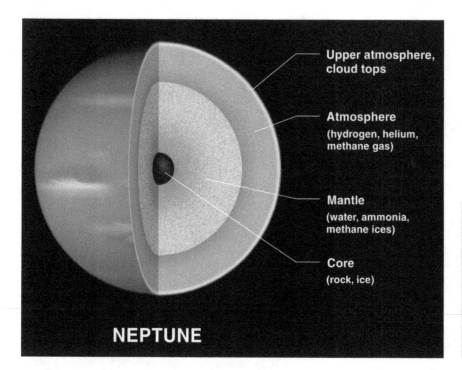

NEPTUNE

Upper atmosphere, cloud tops

Atmosphere
(hydrogen, helium, methane gas)

Mantle
(water, ammonia, methane ices)

Core
(rock, ice)

Figure 11.14 Like Uranus, Neptune is believed to consist mainly of a vast internal ocean of water, methane, and ammonia "ices" which become liquid at the high temperatures (several thousand degrees Celsius) and pressures deep in the interior. The presence of a rock-ice core about the size of Earth is inferred. (LPI)

sublimating – turning directly from ice to vapor – in the equatorial region, presumably to recondense in the dark northern polar region (Figure 11.18).

More than 100 dark streaks were visible on the ice cap, mainly l.l between 15° and 45°S. Ranging from 4 km to over 100 km

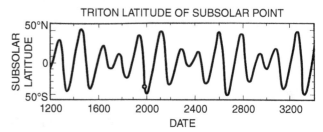

TRITON LATITUDE OF SUBSOLAR POINT

Figure 11.16 Triton experiences a remarkably complex succession of seasons as the position of the overhead Sun (the subsolar point) shifts from the southern to the northern hemisphere. At the time of the Voyager encounter in 1989, it was early summer in the southern hemisphere, whereas the northern hemisphere was in prolonged darkness. The subsolar point reached latitude 50°S in late 2000, the first time in over 350 years that it had been so far from the equator. (NASA-JPL)

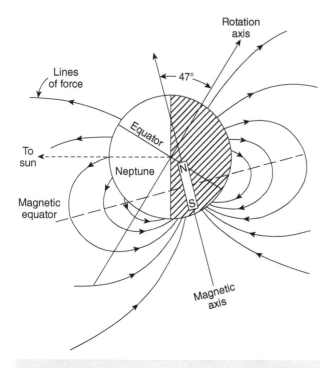

Figure 11.15 Like the other giant planets, Neptune's magnetic polarity is the opposite to that of Earth. The field is tilted 47° with respect to the rotation axis and offset 0.55 Neptune radii from the center. (NASA-JPL)

in length, they were often fan-shaped and aligned toward the northeast. Their appearance suggested that they were windblown deposits – despite the sparse atmosphere on Triton. This interpretation was confirmed when stereo images of some streaks revealed that they were associated with plumes that soared to an altitude of 8 km before being diverted horizontally by high-level wind.

The Mahilani Plume at 48°S 2°E appeared as a very narrow, linear cloud at least 90 km in length, whereas the Hili Plume at 57°S 28°E comprised several much broader streamers up to 100 km long (Figure 11.19). In both cases the clouds were moving toward the west. It seems that the wind direction shifts dramatically at different levels on Triton, with a northeastward flow close to the surface, an eastward flow above 1 km and a westward flow at the height of the plumes.

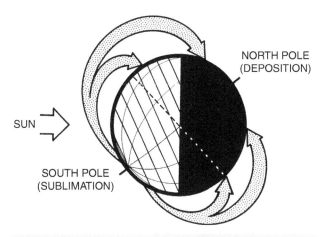

Figure 11.18 Sublimation of nitrogen frost in Triton's summer hemisphere is counterbalanced by condensation and deposition of ices in the winter hemisphere. (NASA-JPL)

Figure 11.17 A photomosaic of Triton centered near 20°N, 0°W. The equator lies approximately along the pale green region (centre). The surface is mainly nitrogen ice. Pinkish deposits in the southern hemisphere contain methane ice which has been modified by ultraviolet sunlight to form organic compounds. The dark streaks are probably carbonaceous dust deposited from geyser-like plumes. The pale green band that extends all the way around Triton is relatively fresh nitrogen frost. The greenish areas include the "cantaloupe" terrain and cryovolcanic plains produced by icy material erupted from Triton's interior. (USGS/NASA-JPL)

The origin of the plumes remains uncertain. Most models assume that they are somehow driven by the heat of the overhead Sun in summer. The weak solar radiation could penetrate the transparent nitrogen ice to a depth of several kilometers. Since the nitrogen ice is an excellent insulator, subsurface ices may melt or sublimate. Pressure building up over a period of time could suddenly be released when liquid nitrogen (and perhaps other materials) finds its way to the surface and erupts as a boiling geyser. Traveling rapidly upward, the plume could pick up dust from the surface, while nitrogen and other gases making up

the plume would freeze and condense. Such an eruption could last for at least a year before it subsided.

It is also possible that they are geysers powered by cryovolcanic activity involving localized internal heat sources, or that the thickness of the nitrogen ice cap plays a role. Another possibility is that the plumes are an atmospheric phenomenon, perhaps resembling some form of dust devil.

There is a certainly a strong case for arguing that many of the surface features seen beyond the edge of the polar cap are associated with cryovolcanism and internal activity. Voyager found that the western half of Triton's visible equatorial region comprises a unique surface of large, elliptical or kidney-shaped "dimples" crisscrossed by linear ridges. The depressions (known as cavi) that make up this "cantaloupe" terrain are typically 25–35 km across, with no overlapping.

The favored explanation is that these dimples are created by diapirs – blobs of warm, buoyant, less dense material rising to the surface through denser overlying material. This theory implies that Triton has distinct crustal layers which allow ammonia-water

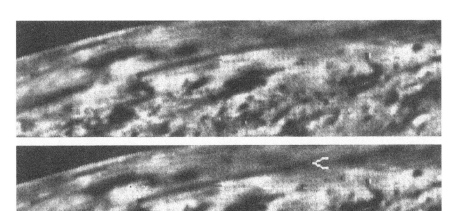

Figure 11.19 Two images of the dark Mahilani plume on Triton, imaged by Voyager 2. The plume (arrowed) rose rapidly to a height of about 8 km – presumably the top of the troposphere – and was then blown westward for about 100 km by winds of 10–20 m/s. (NASA-JPL)

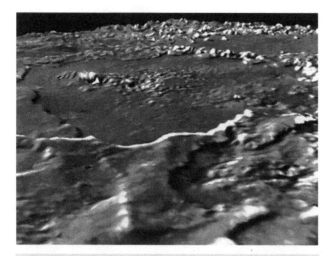

Figure 11.20 Ruach Planitia is a 175 km-wide walled plain bounded on all sides by rougher plains. The extremely flat floor may have been flooded by fluid, icy material. Terraces and benches around the edges may be evidence of previous flooding followed by collapse. In this computer-generated view, topography has been exaggerated roughly 30 times; the actual range of relief is about 1 km. (NASA-JPL)

ices to respond to heating from below and move toward the surface.

The network of ridges bears some resemblance to features on the surface of Jupiter's moon Europa. Some parallel ridges 6–8 km wide occur in pairs either side of a central trough; others are triple or in the form of single, narrow valleys 2–3 km in width. All of them appear to be tectonic features created as the result of extension of the crust. The medial ridges may be the result of subsequent intrusions of material, similar to volcanic dikes on Earth.

Most of Triton is fairly flat, with a difference in relief of no more than 1 km. This lack of topography is particularly noticeable to the east of the cantaloupe terrain. The smooth plains close to the eastern limb in Voyager's close-up images are thought to be the oldest, based on their higher crater count. They include Mozamba, the largest known impact crater on Triton, which has a diameter of 27 km and a prominent central peak.

Nearby are some smooth, relatively dark patches surrounded by brighter aureoles. Unlike any features seen on other objects, these "maculae" (spots) are of unknown origin. It may be that the central patches are made of carbonaceous material (presumably including some methane ice) bounded by brighter, nitrogen-rich ice.

Equally intriguing are four walled plains whose edges rise in steps (scarps) to the surrounding, terraced terrain (Figure 11.20). Their floors are the flattest places on Triton, implying that they were flooded by very fluid material. Clusters of small pits in their central regions may be associated with cryovolcanic vents.

Is this activity still ongoing? The absence of large impact craters excavated by comets from the Kuiper Belt, together with the relatively low crater density (only a few hundred craters more than 3 km across have been identified), indicate that the surface is relatively young. Estimates range from 100 million years to only 10 million years, but if the latter figure is accepted, surface activity is almost certainly continuing at the present day.[6]

Although there have been various theories concerning the origin of Triton, its unusual retrograde orbit suggests that it was captured by Neptune, rather than formed in the planet's vicinity. Remarkable similarities with Pluto – including size, density, surface composition, and atmosphere – suggest a common birthplace in the Kuiper Belt.

According to this view, Triton was a planetesimal in the inner Kuiper Belt that approached too close to Neptune. The proto-Triton could have been slowed by a collision with an existing satellite or interaction with a disk of material orbiting the young planet. As time went by, the moon's highly elliptical orbit became smaller and more circular due to tidal forces generated during each close passage. A periodic rise and fall of the bulge in Triton raised by Neptune's gravity at each approach would dissipate energy and modify the orbit.

Critics argue that the likelihood of a collision which is large enough to slow down Triton, but small enough not to destroy it, is extremely small. Similarly, aerodynamic drag from a disk of gas around Neptune would have to occur at a particular stage in the planet's early evolution. It would have to be surrounded by a disk that was dense enough to slow Triton down and bring it into a circular orbit, but the gas would have to disperse before Triton's orbital motion was slowed so much that it was sent crashing into Neptune.

An alternative scenario suggests that Triton may once have belonged to a binary pairing, similar to Pluto and its large moon Charon. In a close encounter with Neptune, the pair were pulled apart by the planet's gravitational forces and Triton was captured by the giant planet.

Tidal forces are still influencing Triton today, causing it to spiral slowly towards the planet. In 10 to 100 million years, it will be so close that Neptune's gravity will tear it apart, forming a new, spectacular ring.

Small Satellites

Before the Voyager flyby, the Neptune system was known to have one small moon, Nereid, which was found by Gerard Kuiper in 1949. Nereid follows one of the most elliptical orbits of any satellite, traveling between 1.4 million km and 9.7 million km from Neptune over a period of 360 days. The orbit is inclined about 7° to Neptune's equator and is prograde.

The fuzzy Voyager images suggest that Nereid is roughly spherical, with a diameter of about 340 km. It is brighter than the other small satellites and gray in color, suggesting an icy surface. This seems to be confirmed by the discovery of water ice in its spectrum. Unusually, the rotation period, based solely on changes in its light curve, is almost certainly not synchronous. One estimate suggests a period of 13.6 hours, but others have failed to find any

---

[6] All of the craters are on Triton's leading hemisphere, implying that it is sweeping up debris – possibly from a shattered satellite – as it orbits Neptune.

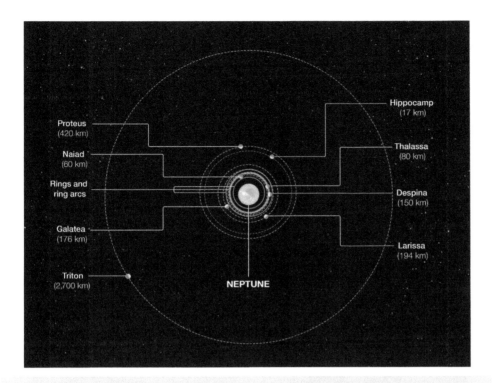

Figure 11.21   The orbits of Neptune's inner satellites and Neptune's rings. Hippocamp is thought to be a fragment which broke off Proteus after a collision with a comet. (M. Showalter/SETI Institute)

regular pattern, possibly pointing to a chaotic motion resembling that of Saturn's satellite, Hyperion.

Voyager revealed six more satellites inside the orbit of Triton. They were arranged in order of size, with the smallest (Naiad) closest to Neptune and the largest (Proteus) furthest away (Figure 11.21). The discovery of another moon, known as S/2004 N1 and now named Hippocamp, was announced in 2013. HST images show that it is only about 34 km wide, making it the smallest of Neptune's 14 known satellites. It is thought to be a fragment that broke away from Proteus during a collision with a comet. Hippocamp orbits between Larissa and Proteus, well outside the rings.

Proteus, orbiting at a distance of 117,650 km, is the second largest satellite of Neptune (Figure 11.22). With a diameter of more than 400 km, it has an odd, box-like shape, with a large impact crater, called Pharos. With a slightly higher mass, its gravity would have been sufficient to pull the body into a sphere.

Like Proteus, all of the other small, inner moons are dark and follow circular orbits with low inclinations. Apart from Proteus, they all scoot around the planet in less than one Neptunian day.

Perhaps the most important characteristic of the five inner satellites was their shepherding interaction with the planet's narrow rings (see Chapters 9 and 10). Naiad, Thalassa, and Despina orbit between the innermost ring (Galle) and the Le Verrier ring. Galatea patrols between the Arago and Adams rings, while Larissa shapes the outer edge of the Adams ring.

Naiad, in particular, proved difficult to observe, because it orbits so close to the much brighter planet. The 100-km moon was "lost" for many years until it was rediscovered in archived HST images obtained in 2004. The new observations showed that Naiad appeared to now be far ahead of its predicted orbital position, leading to speculation whether gravitational interactions with one of Neptune's other moons may have caused it to speed up.

With the advent of CCD cameras and ground-based telescopes fitted with adaptive optics, five more, irregular moons were discovered far beyond Triton and Nereid in 2002 and 2003, boosting the number of known satellites to 13. This was something of a surprise, since it was thought that cataclysmic events related to the capture of Triton would have dispersed any outer satellites.

Little is known about these moons, but they are all thought to be 60 km or less in diameter. Their average orbital distances are between 15 million km and 49 million km. Indeed, Neso is the furthest known satellite of any planet, taking about 25.6 years to complete one orbit.

The satellites follow highly elliptical orbits steeply inclined to Neptune's equator. One of them (Halimede) even crosses the orbit of Nereid. Two of the satellites follow prograde paths, while the other three have retrograde orbits. The irregular nature of their orbits indicates that they have been captured by Neptune, rather than forming in a disk around the young planet.

The outer satellites are regarded as remnants of a system that has been modified over billions of years by collisions and gravitational interactions with Triton and Nereid, together with solar tides (Figure 11.23). Dynamical studies show that the eccentricities and inclinations of their orbits undergo large variations over time. Collisions between Nereid and the innermost of the outer satellites, particularly Halimede, are quite probable over the lifetime of the Solar System.

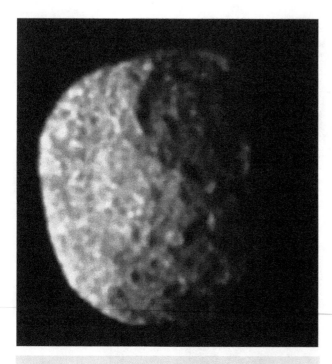

Figure 11.22   Proteus imaged on August 25, 1989, from a range of 146,000 km. Spatial resolution is about 2.7 km. The half-illuminated satellite has a diameter of more than 400 km. It is dark (albedo 6%) and spectrally gray. Apart from a large crater, hints of craters and groove-like features can be seen. (NASA-JPL)

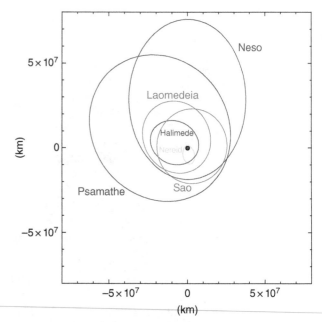

Figure 11.23   The orbits of Neptune's five outer moons are all highly elliptical and inclined to its equator. Two of them (Sao and Laomedeia) follow prograde orbits. The others are retrograde. (After Scott S. Sheppard/Carnegie Institution for Science)

Rings and Arcs

With the discovery of rings around Uranus in 1977, it seemed possible that similar, dark rings might exist around Neptune. The best chance of finding them was to examine light curve data obtained during a stellar occultation.

During the 1980s, many occultations were meticulously studied, but definitive proof that Neptune had a ring system was hard to find. Although more than 90% of the observations yielded no detection, at least five occultations seemed to provide evidence of material orbiting Neptune.

One observation indicated the presence of either a small satellite or an optically thick, incomplete ring with a radial width of about 80 km.[7] Other confirmed sightings also detected dimming on only one side of the planet. The puzzling data led to the theory that Neptune was orbited by several discontinuous ring arcs. These were thought to be narrow (15–25 km), at least 100 km in length, and located between 41,000 and 67,000 km from Neptune.

Voyager 2's cameras soon solved the mystery by discovering five charcoal black, but continuous, rings between 41,900 km and 62,933 km from the center of Neptune (Figure 11.25). Images confirmed the existence of elongated arcs of material as distinct sections of the outer, Adams ring. There was also some evidence

for a faint, partial ring sharing the orbit of Galatea – about 1,000 km inside the Adams ring.

The ring nomenclature is a little easier to memorize than for Uranus, since they were named after some of the leading characters involved in the planet's discovery. In order of distance from the planet, they are: Galle, Le Verrier, Lassell, Arago, and Adams (Figure 11.24).

Three of the rings are extremely narrow, with a radial width of less than 100 km. Only Galle and Le Verrier measure several thousand kilometers across. Radio and optical occultation data indicate that the Adams and Le Verrier rings are composed of material no more than a few centimeters across. There is a much higher proportion of micron-sized dust than in the rings of Uranus or Saturn. Their overall mass is about 10,000 times lower than the rings of Uranus. All the material gathered together would form a ball only a few kilometers across.

One similarity with the Uranian system is the low albedo (less than 5%), possibly due to the alteration of methane ice by ultraviolet light and high-energy particles, leading to a coating of elemental carbon similar to graphite.

The Adams ring is particularly interesting, not only because it is the densest and slimmest member of the ring system, but also because it contains five arcs that are noticeably brighter than the remainder of the 50 km-wide ribbon. Voyager 2 showed that all were contained within a 40° section of the ring.

At the time of the Voyager flyby, the arcs made up about one-tenth of the ring's circumference, sufficient to explain the

[7] Data obtained during the Voyager 2 flyby confirmed that this was a small satellite, Larissa, which orbits just outside the ring system.

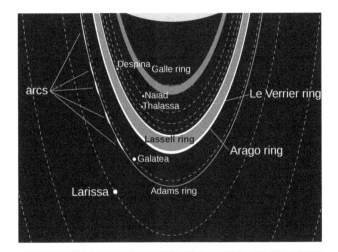

Figure 11.24 The five inner satellites orbit close to or within the rings of Neptune. Galatea may shape the arcs of the Adams ring. There may also be a partial ring which shares the orbit of Galatea. (Wikimedia)

puzzling occultation observations. Leading the orbital procession of "beads" was an arc called Courage, followed by Liberté, Egalité 1 and 2 (sometimes regarded as subdivisions of the same arc), and Fraternité.

The reasons for the formation and continuing existence of the rings and the arcs are not clear. The presence of a large amount of dust, together with the presumably ephemeral ring arcs, favors a

system that formed fairly recently. The most likely explanation is the destruction of a small satellite.

The evolution of the rings might be expected to be affected by the nearby presence of five small, icy moons. Naiad, Thalassa, and Despina travel between the innermost, Galle ring, and LeVerrier, while Galatea patrols between the Arago and Adams rings. A fifth satellite (Larissa) moves outside the Adams ring. (The four inner satellites and the rings all reside within the Roche limit for objects composed of water ice – see Chapter 9)

However, most of these moons do not behave like the shepherds that confine the rings of Uranus and some of Saturn's rings. Apart from Galatea, which shares its orbit with a partial ring, there is no obvious relationship between the inner moons and the narrow rings. The only orbital resonance that has been confirmed involves Galatea and the Adams ring, with the particles in the ring moving around the planet 42 times for every 43 orbits of Galatea. This repetitive resonance enables Galatea's gravitational influence to limit the longitudinal spread of the arcs and confines the radial width of the Adams ring.

However, this does not explain why the Adams ring does not migrate outward as energy is transferred from Galatea to the particles in the Adams ring, or the absence of arcs in the other rings. Neither does it explain why some of the other rings are so narrow and sharply defined. This has led to suggestions that one or more unseen moons, less than 10 km across, may be embedded either in the arcs or at a Lagrangian point in the ring, 60° ahead or behind the arcs.

Neptune's inner moons and ring arcs are difficult to observe because they are faint (below magnitude 20) and close to the bright planet. However, monitoring with the HST and ground-based

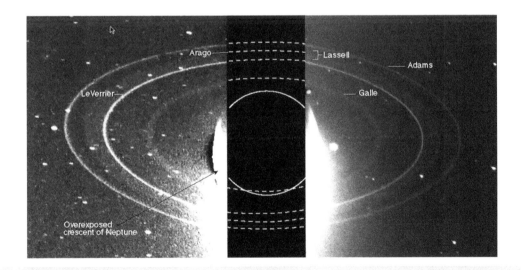

Figure 11.25 Two long exposures of Neptune's rings taken by Voyager 2 from a distance of 280,000 km. The period between exposures was one hour and 27 minutes. The arcs in the Adams ring were not visible, since they were on the opposite side of the planet during each exposure. The image shows the faint inner ring (Galle), about 42,000 km from the center of Neptune, and Lassell, a broad, faint band between the Le Verrier and Arago rings. The rings were backlit by the Sun, allowing fainter, dusty regions to be seen. The two gaps in the outer ring (left image) are due to computer processing. (NASA-JPL)

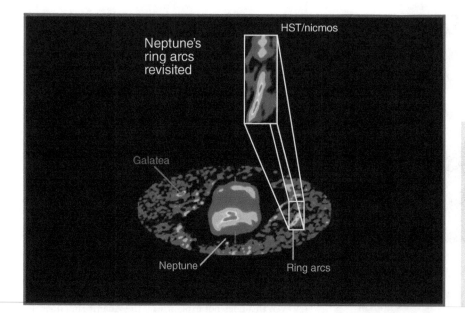

Figure 11.26 Near-infrared observations made with the Hubble Space Telescope have provided information about the orbital periods of the inner moons and subtle changes in the rings. The data indicated that the Liberté ring arc had edged forward, towards Courage. (Courtesy of Christophe Dumas, NASA-JPL, University of Arizona Steward Observatory/HST-NICMOS)

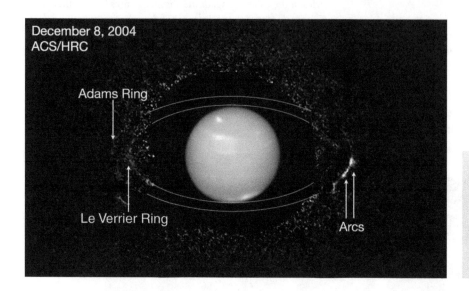

Figure 11.27 Reprocessed HST images taken in 2004 show the Adams and Le Verrier rings and two ring arcs. The Liberté and Courage arcs are not visible and may have dispersed altogether. (Mark Showalter/SETI Institute)

instruments has shown that the rings gradually evolve over time (Figures 11.26 and 11.27). The narrow Le Verrier ring brightened by a factor of four in the decade after 1989. By 2003 Galatea had moved 8° ahead of its predicted position, relative to the arcs. This led to the hypothesis that the arcs are massive enough to affect the satellite's orbit, possibly because of the presence of sizeable objects hidden in the ring, which act as the source of the visible dust.

Major changes in the arcs have also taken place since the Voyager flyby, both in their location and intensity relative to the trailing arc, Fraternité, which seems to be the most stable. In contrast, the leading arcs, Liberté and Courage, have shifted forward and decayed.

The most noticeable change has involved Liberté, which displayed a forward shift of about 2° and was considered to be on the verge of disappearing completely. At the same time, there was a slight lengthening of Egalité, possibly as the result of material being transferred from Liberté. The sub-arcs Egalité 1 and 2 appeared to have reversed in relative intensity, perhaps as the result of material migrating between them. A forward shift in location of 8° by Courage could also be associated with a movement of material over one full orbital resonance site.

Studies of archived HST images taken in 2004 showed that the two leading arcs were absent, while no changes were observed in the two trailing arcs.

No missions to Neptune are planned for the foreseeable future.

Questions

- The discovery of Neptune was said to be a triumph for gravitational theory. Do you consider this to be true? Explain your answer.
- Why is scientists' knowledge about the Uranus and Neptune systems much less than for the other planets?
- Why are Neptune and Uranus described as "ice giants"?
- How does the atmosphere of Neptune differ from the atmospheres of Jupiter and Saturn?
- Summarize the main similarities and differences between Uranus and Neptune. Suggest reasons for these similarities and differences.
- Compare and contrast the magnetospheres of Uranus and Neptune.
- Four planets possess ring systems. What are the main similarities and differences between the rings of Neptune and the other ring systems?
- How and why is Triton unique (as far as we know) among the major satellites of the Solar System?
- Why is a future mission to Neptune unlikely to take place for several decades?

TWELVE

Pluto and the Kuiper Belt

For more than 80 years, Neptune was considered to be the outer-most planet in the Solar System. Then, in 1930, a young research assistant at Lowell Observatory discovered Pluto, a small world in an eccentric orbit that crossed the orbit of Neptune. Forty-eight years passed before a large satellite of Pluto was found. Not until 1992 was the first of many objects that belong to the so-called Kuiper Belt discovered beyond Neptune. Today, the number of known Kuiper Belt objects larger than 100 km across is thought to exceed 200,000, though the total population could be as many as 100 billion. Some of these objects rival Pluto in size. Pluto and some of the other large Kuiper Belt objects are also classified as dwarf planets.

The Edgeworth-Kuiper Belt

For decades, the apparent emptiness of the region beyond Neptune was a puzzle. Apart from the occasional comet, there seemed to be few leftovers from the planetesimals that thronged the early Solar System.

The discovery of remote Pluto in 1930 (see Box 12.3) came as something of a surprise, since subsequent observations showed that this icy world was smaller than Earth's Moon and followed a much more elliptical orbit than the other planets.

For several decades, Pluto was regarded as a unique maverick, since it could not be classified either as a small, rocky planet or as a gas giant. In some ways it most closely resembled a periodic comet, since it follows an inclined, elliptical orbit, periodically warming and losing its atmosphere into space. However, Pluto is far larger and denser than any known comet nucleus. Until the discovery of Pluto's largest moon, Charon, the only object that seemed comparable in size and density was Neptune's satellite, Triton.

Was Pluto really a planet that accreted from the solar nebula about 4.5 billion years ago, or was it some kind of protoplanetary debris that happened to have been shifted into a strange orbit? Some astronomers found it hard to believe that Pluto was the outermost member of the planetary system, and there were various attempts to predict the locations of unseen planets beyond Pluto.

Clyde Tombaugh continued to search for such a planet for 14 years after he found Pluto, eventually concluding that no large worlds exist in the outer Solar System.

Although no other Pluto-like objects were revealed, studies of comets' orbits hinted at the existence of a swarm of icy objects in the outer reaches of the Solar System. The first suggestion that a reservoir of comets should exist beyond Neptune came from British astronomer Kenneth Edgeworth in 1943. He noted that collisions and coalescence between planetesimals would be so infrequent in the outer Solar System that only small bodies would be likely to form.

A Dutch-American astronomer, Gerard Kuiper, came to a similar conclusion in 1951, when he proposed that enormous numbers of icy bodies populate the region beyond Neptune, possibly extending as far out as 120 AU.

During the 1980s, various researchers suggested that a flattened, donut-shaped zone of icy planetesimals could be the source of the short-period comets that occasionally enter the inner Solar System. One key piece of evidence was that these comets travel around the Sun in the same direction as the planets, following orbits that are only slightly inclined to the ecliptic plane. Unfortunately, the instruments of the time were not sufficiently advanced to find any proof that these dormant comets actually exist.

The breakthrough came on August 30, 1992. After five years of searching the ecliptic with a new-generation CCD camera linked

Exploring the Solar System, Second Edition. Peter Bond.
© 2020 John Wiley & Sons Ltd. Published 2020 by John Wiley & Sons Ltd.
Companion Website: www.wiley.com/go/Bond-Solar-System2e

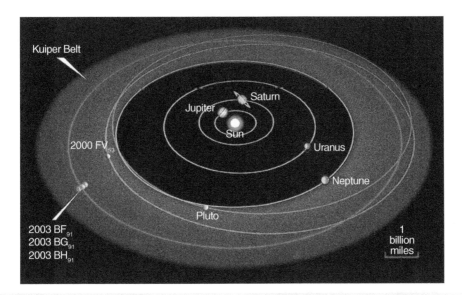

Figure 12.1 The Edgeworth-Kuiper Belt is named after Kenneth Edgeworth and Gerard Kuiper. They were the first to propose that enormous numbers of icy bodies exist beyond the orbit of Neptune. After many years of searching, the first object was found in 1992. Since then, more than 1,300 have been discovered. The orbits of four KBOs are shown on this diagram. (Courtesy of NASA and A. Field/STScI)

to the 2.2 m University of Hawaii telescope, David Jewitt and Jane Luu discovered 1992 QB1 drifting slowly among the background stars (Figure 12.2). The newcomer measured about 250 km across and took 296 years to complete one orbit. It was the first object in the Solar System known to spend its entire life beyond Neptune, and the first recognized member of the Kuiper Belt.[1]

Today, several thousand occupants of the belt have been found, although the total population is much larger. It is estimated that hundreds of thousands of icy bodies more than 100 km in diameter exist within the Kuiper Belt (compared with 200 objects that size in the main asteroid belt), along with a trillion or more comets. Nevertheless, the overall mass of the Kuiper Belt objects (KBOs) is probably less than 10 percent of Earth's mass.

Observations show that the KBOs are mostly confined within a thick, torus-shaped region that is centered within 10 degrees of the ecliptic, though some orbit far outside this plane (Figure 12.1). The vast majority of them exist between the orbit of Neptune at 30 AU and an apparent outer edge about 55 AU from the Sun. Beyond this outer limit there appears to be a sparse population of scattered KBOs with large orbital eccentricities.

The structure of the belt is linked to the gravitational influence of Neptune. The majority of the so-called classical KBOs (see The Classification of Kuiper Belt Objects) occur between the 2:3 and 1:2 resonances with Neptune, at approximately 42–48 AU. Any KBOs that are affected by these periodic gravitational interactions with Neptune over many millions or billions of years are destabilized. The ice giant ejects them into the inner Solar System or

out into the scattered disc or interstellar space. This causes the Kuiper Belt to have regions that are sparsely populated, similar to the Kirkwood gaps in the asteroid belt (see Chapter 13).

The Kuiper Belt population is thought to comprise the last survivors of the icy planetesimals that existed 4.5 billion years ago, when the planets were forming from the solar nebula. Whereas the inner, dense parts of the pre-planetary disk condensed into the major planets, probably within a few millions to tens of millions of years, the outer reaches were less dense, and collisional accretion progressed more slowly. Instead of coming together to form large planets, a multitude of small objects formed.

Some of this icy debris was probably ejected to their present remote location billions of years ago by gravitational interactions with the giant planets, particularly Uranus and Neptune. Models suggest that this could have happened during the planets' outward migration (see Chapter 1). Today, when they are occasionally disturbed by the outer planets, they enter Sun-approaching orbits and become short-period comets.

How the larger members of the Belt could have grown to their present size in the vast, near-empty region beyond Neptune has yet to be fully explained, although it has been suggested that turbulence and vortexes in the protoplanetary nebula might have enabled many tiny particles to coalesce and grow quite rapidly.

Collisions between KBOs may also have been involved, and an ancient collision involving Pluto is thought to account for the existence of Charon and the smaller Plutonian moons.

---

[1] Today, the name of the Edgeworth-Kuiper Belt is usually shortened to simply Kuiper Belt. Its members are called Kuiper Belt objects (KBOs), Edgeworth/Kuiper Belt objects, or trans-Neptunian objects (TNOs).

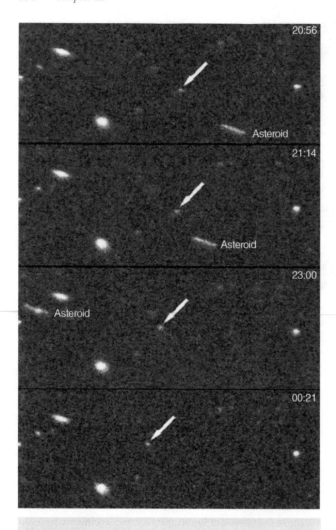

20:56

Asteroid

21:14

Asteroid

23:00

Asteroid

00:21

Figure 12.2 The first Kuiper Belt object, known as 1992 QB1, was found by David Jewitt and Janet Luu on August 30, 1992. These images taken with the University of Hawaii's 2.2 m telescope at Mauna Kea show its motion over a period of about 3½ hours. With a magnitude of about 23, 1992 QB1 was 6 million times fainter than the faintest star visible to the naked eye. Also visible is a streak caused by a faster-moving asteroid. (David Jewitt, UCLA)

The Classification of Kuiper Belt Objects

Studies of the orbits of KBOs have resulted in four main categories being recognized (Figure 12.3):

1. **Cold Classical KBOs.** The most numerous group so far detected, these occupy a narrow region which is about 7 AU thick and stretches between 42 and 48 AU from the Sun. They generally follow orbits of modest inclination, which suggests that they were born outside Neptune's orbit and have been largely unperturbed by the giant planet. They also tend to be smaller and browner than other KBOs. Examples include 1992 QB1, the first KBO ever discovered, which has a semi-major axis of 43.7 AU and an eccentricity of 0.07, and Quaoar, one of the largest KBOs yet discovered.

2. **Hot Classical KBOs.** These have similar average distances from the Sun to the Cold Classical KBOs, but their orbits are more eccentric and/or inclined. This suggests that their orbits have been disturbed and altered by Neptune's gravity. Their sizes and colors vary, but they include larger and grayer objects than Cold Classicals. Examples include 1996 RQ20 and 1997 RX9, which have inclinations greater than 30°.

3. **Resonant KBOs (Plutinos and Twotinos).** Like Pluto, Plutinos follow orbits that have been modified by Neptune's gravity, so they have 3:2 orbital resonances with the ice giant (i.e. they make two orbits for every three of Neptune). They include Orcus and Ixion. Another, more remote group ("Twotinos") has 2:1 resonances with Neptune. With semi-major axes at about 47.7 AU, they are close to the outer edge of the main Kuiper Belt. Less significant resonances also exist at 3:4, 3:5, 4:7, and 2:5.

4. **Scattered KBOs.** Although their perihelion distances of around 35 AU are not too unusual, they have inclined and eccentric orbits that may take them hundreds of AU from the Sun. Over billions of years, perturbations by Neptune around perihelion may gradually alter their orbits. They account for only 3–4% of the known Kuiper Belt objects, but this apparent depletion may be explained by the fact that they spend most of their time at large distances and are difficult to detect. Eris and Sedna are examples of this group.

Surfaces

Little is known about the surface properties of most KBOs, with the obvious exceptions of Pluto and its companions.

However, one small object, known as 2014 MU69 (and now named Arrokoth) has been observed from close range when the New Horizons spacecraft flew past at close range on January 1, 2019 (Figure 12.4). It is regarded as an example of a "cold classical" KBO, which has remained in the same region of the Solar System since its formation, 4.5 billion years ago.

The first images showed a dark, slightly reddish, contact binary that was assembled long ago. Both components are similar in color and probably consist of the same icy material. The larger lobe, nicknamed "Ultima," resembles a giant pancake, while the smaller lobe, nicknamed "Thule," is shaped like a dented walnut.

A narrow, bright "collar" separates the two lobes – evidence that it comprises two objects that gently collided and coalesced. The ancient surface shows a 6 km-wide circular depression, as well as numerous small pits up to about 0.7 km in diameter. These may have been created by impacts or past sublimation of subsurface volatiles.

Most KBOs are too small and distant for spectral data to be obtained, and broad spectra are available for a mere handful. Albedos and sizes have been estimated for a small sample, but thermal emissions are difficult to measure. Visible colors have been reported for a few hundred objects, with near-infrared colors for a smaller group.

Visible photometry shows that their colors cover a wide range, from neutral to very red, indicating that KBOs have very diverse surfaces. It seems unlikely that this diversity is related to widely differing chemical compositions, since few chemical reactions are likely in a region where the temperature is only 30–50°C above absolute zero.

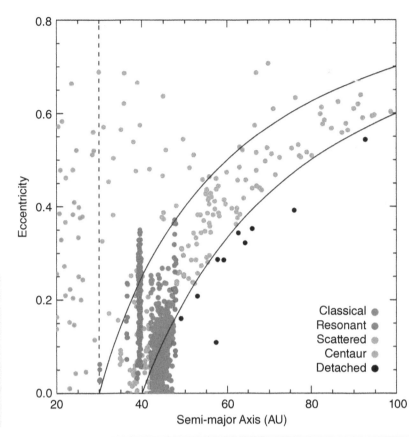

Figure 12.3 A graph showing the distribution of different classes of objects in the outer Solar System, plotting semi-major axis against orbital eccentricity. The orbit of Neptune is marked by a vertical dotted line. Objects in blue are resonant with Neptune. The most densely populated cluster of blue dots shows the 3:2 resonant objects, which include Pluto. (David Jewitt, UCLA)

Besides, there is no apparent correlation between color and current distance, as might be expected if location and temperature were important. However, the color range might be explained if some of the objects were born in different regions of the outer Solar System and subsequently relocated to their present positions.

One possibility is that the colors represent different evolutionary stages – as the surface ages, it becomes darker and more reddish. Many KBOs may have similar compositions, a mix of solid organic compounds, silicates, and volatile ices derived from the primordial solar nebula. Over time, irradiation by high-energy cosmic rays, the solar wind, and solar ultraviolet light, combined with sublimation of the volatiles and non-disruptive collisions with other objects, would change their surface compositions.

In this way, bright surfaces coated with fresh ices would be transformed by the loss of hydrogen atoms and carbonization. The end result would be a low albedo surface "crust" of red, organic-rich sludge. Grazing or non-disruptive collisions could remove some of this outer layer, revealing more reflective, neutrally colored, ices beneath.

There is also the possibility that resurfacing through deposition of fresh icy material, resulting from cryovolcanism, might also play a role.

The only "ground truth" data we have comes from the Pluto system (see various New Horizons sections). Pluto itself seems to be differentiated into a rocky core with a water ice mantle and crust (Figure 12.5). The frigid surface exhibits water, nitrogen, methane, and carbon monoxide ices which have created a remarkably varied landscape of icy plains, glaciers, mountains, and impact craters, with at least one likely cryovolcano.

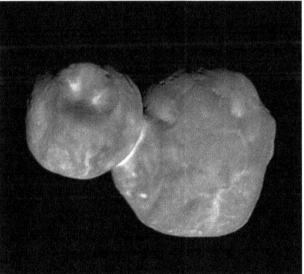

Figure 12.4 One of the first images of the small Kuiper Belt object 2014 MU69 (Arrokoth) sent back by the New Horizons spacecraft. The object, which measures 32 x 16 km, appears to be a contact binary which was assembled from a gentle collision more than 4 billion years ago. Note the bright collar where the two lobes are joined. Also visible are numerous small pits up to 0.7 km in diameter, and an apparent circular depression, about 7 km across, on the smaller of the two lobes. (Courtesy of NASA/Johns Hopkins University Applied Physics Laboratory/Southwest Research Institute)

Figure 12.5 This composite of enhanced color images of Pluto (lower right) and Charon (upper left) highlights striking differences in their color and brightness. Charon's surface is water ice, whereas Pluto also has nitrogen, methane, and carbon monoxide ices. They are shown with approximately correct relative sizes, but their true separation is not to scale. (Courtesy of NASA/JHU-APL/SwRI)

Even Charon, its largest moon, has undergone major tectonic activity in the past, generating a major canyon system which separates the more rugged, cratered, terrain of water ice in the northern hemisphere from smoother landscape in the south.

The Largest KBOs

Pluto, the largest known Kuiper Belt object, has a diameter about two-thirds that of our Moon. A handful of KBOs are almost as large (Figure 12.7). The largest of these Eris, which was first imaged on October 31, 2003, then rediscovered by Mike Brown (California Institute of Technology) and colleagues on January 8, 2005.

Eris is an example of a scattered KBO, and it follows a remarkably eccentric orbit. It was discovered near aphelion, 97 AU from the Sun, making it the most distant object ever observed in the Solar System. When it reaches perihelion in 2257, it will close to about 38 AU – so Eris is sometimes much closer to the Sun than Pluto, although never closer than Neptune.

Observations with the Keck Telescope in September 2005 led to the discovery of a satellite, named Dysnomia. Its diameter is uncertain, with estimates ranging from 100 km to 700 km. It orbits about 37,000 km from Eris over a period of about 15.8 days.

Like the Earth–Moon and Pluto–Charon systems, Eris and Dysnomia may have formed after a massive collision about 4.5 billion years ago. However, this does not tie in with recent observations which suggest that, in contrast to the extremely reflective Eris, Dysnomia may be darker than coal.

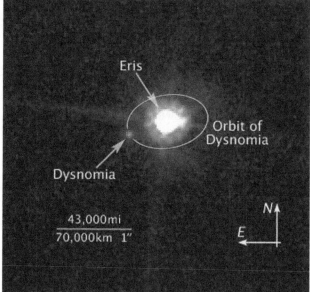

Figure 12.6 Eris is one of the largest known Kuiper Belt objects, similar in size to Pluto. It was discovered near aphelion – 97 AU from the Sun – and is one of the most reflective objects in the Solar System. A satellite called Dysnomia was discovered in 2005. The image was taken with the Hubble Space Telescope on August 30, 2006. (Courtesy of NASA, ESA, and M. Brown/California Institute of Technology)

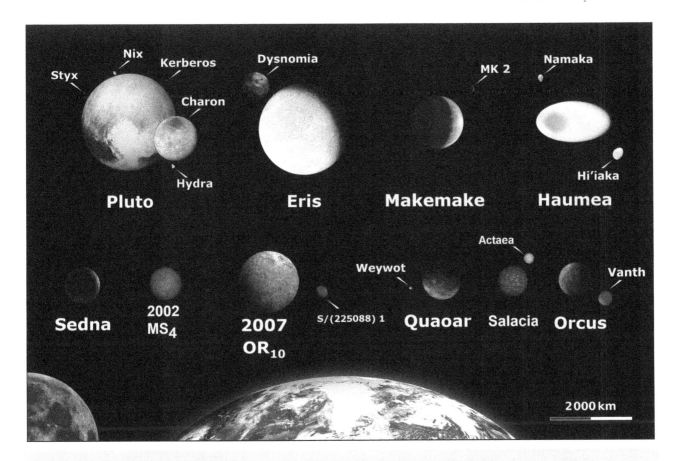

Figure 12.7 The largest known KBOs compared with Earth and the Moon. Eris is similar in size to Pluto. Haumea is thought to be elliptical in shape. Almost all of the largest KBOs have satellites. (Wikipedia)

Eris, which is thought to be almost identical in size to Pluto, is also classified as a dwarf planet. The orbital motion of Dysnomia was used to estimate that Eris is 27% more massive than Pluto. With a density of about 2.52 g/cm³, Eris is likely made up mainly of rock, with a relatively thin mantle of ice.

Despite its vast distance, Eris is the third brightest of the Kuiper Belt objects. Its surface is extremely reflective, with a visible albedo of 0.96. This is even brighter than fresh snow on Earth, making Eris one of the most reflective objects in the Solar System, along with Saturn's icy moon Enceladus.

As is the case with Pluto, this is almost certainly because its surface is composed of a nitrogen-rich ice mixed with frozen methane – as indicated by the object's spectrum – coating it with a thin and very reflective icy layer. These ices can exist because the surface temperature of the dwarf planet is estimated to be no higher than −238°C in the daytime, and even lower on the night side.

Since Eris takes about 560 years – twice as long as Pluto – to orbit the Sun and is now very close to aphelion, its atmosphere probably froze onto the surface hundreds of years ago. This could result in a layer of solid nitrogen several centimeters thick, or a layer of fresh methane ice tens of microns thick.

Beneath this seasonal ice cover may be a darker, methane-rich layer which is exposed around perihelion – at a distance of about 5.7 billion km from the Sun – and darkened when it is broken down by solar ultraviolet light.

Variations in surface ices and albedo are also apparent on Makemake, another dwarf planet discovered by Brown's team, whose diameter seems to be about 1,400 km.

Makemake takes 305 years to make one orbit of the Sun. With an average distance from the Sun of 45.8 AU, this classical KBO should have a surface temperature approximately midway between Pluto and Eris. Like Pluto, it seems to be a red-brown color, and covered by frozen methane.

In 2015, the Hubble Space Telescope discovered a moon around Makemake that is estimated to be 160 km wide. Provisionally designated S/2015 1 (136472), the moon orbits 20,800 km from the dwarf planet.

At present, the third-largest known member of the Kuiper Belt is 2007 OR10. Observations taken by NASA's Kepler Space Telescope revealed that 2007 OR10 has a slow rotation period of 45 hours, about double the period of most KBOs. Astronomers began to wonder if the slower rotation period could be caused by the gravitational tug of a moon. Sure enough, a satellite was announced in 2017.

Another large member of the Belt is Haumea, which is about one-third as massive as Pluto (Figure 12.8). Spectroscopic observations show a highly reflective, water ice surface similar to that of

Figure 12.8   Dwarf planet Haumea is the first Kuiper Belt object known to have a ring system. The dark, narrow ring was discovered when Haumea passed in front of a star. (IAA-CSIC/UHU)

Calculations based on its shape, mass, and spin rate indicate that it must be largely made of rock, with a thin veneer of ice.

The rapid rotation is thought to have been caused by a major impact that created its satellites, and possibly its ring. Haumaea is also followed in its orbit by a swarm of other small icy bodies, further evidence of a shattering collision long ago.

The stellar occultation also revealed that Haumea has a narrow ring system, similar to those of Uranus and Neptune – the first ring to be discovered around a Kuiper Belt object. The ring has a radius of about 2,287 km, a width of about 70 km, and an opacity of 0.5. It is well within Haumea's Roche limit. The ring plane approximately coincides with Haumea's equatorial plane and the orbital plane of its outer moon Hi'iaka.

Haumea has two known satellites, Hi'iaka and Namaka, which were discovered in 2005. Hi'iaka is roughly 310 km in diameter, and orbits Haumea in a nearly circular path every 49 days. Its bright surface seems to be coated in crystalline water ice.

Namaka, the smaller, inner satellite, is a tenth the mass of Hi'iaka. It orbits Haumea every 18 days in a highly elliptical orbit which is inclined 13° from that of the larger moon. This causes Hi'iaka to perturb the orbit of its little neighbor. These very different, inclined orbits argue in favor of a collisional origin, with the moons representing icy fragments of Haumea itself.

Another unusual object is Quaoar, which has an estimated diameter of about 890 km but an albedo of only 12%. It also has a small satellite, named Weywot, which follows an eccentric, 12.4 day orbit. Calculations based on the moon's orbital period indicate that Quaoar has a surprisingly high density of about 4.2 g/cm³, making it the most dense Kuiper Belt object currently known. This implies that Quaoar is mainly rocky, with only the thin veneer of water, methane, and ethane ices (which appear in its spectrum).

Pluto's moon Charon. Studies of its light curve show great variations in reflected sunlight that repeat every three to four hours – an indication of its rapid rotation rate.

Haumea is spinning so quickly that it has assumed an elongated, cigar shape. During a stellar occultation in 2017, Haumea was shown to be approximately the same diameter as Pluto along its longest axis and about half as much at its poles. Nothing else so large, so elongated, or so quickly rotating is known anywhere in the Solar System.

Box 12.1   Dwarf Planets

For 76 years, Pluto was regarded as the ninth major planet in the Solar System. However, the definition of a planet was not very precise, referring simply to a large, non-stellar object in orbit around a star. Then, in 2006, the International Astronomical Union (IAU) devised and introduced new classifications for Solar System objects, effectively demoting Pluto and reducing the number of planets to eight (see Chapter 1).

The problem arose partly because Pluto is not only smaller than the other planets, but also smaller than seven of their satellites, including the Moon. Furthermore, it is only one member of a vast population of Kuiper Belt objects. The 2005 discovery of Eris, a KBO originally thought to be even larger than Pluto, brought the issue to a head. If Pluto is a planet, then Eris must also be one, along with the numerous Pluto-sized objects yet to be discovered.

New IAU resolutions reclassified Pluto as a "dwarf planet," as well as one of the largest members of the Kuiper Belt. Ceres, the largest main belt asteroid, was also classified as a dwarf planet, along with the large KBOs Eris, Haumea, and Makemake.

The IAU subsequently decided that newly discovered trans-Neptunian objects with an absolute magnitude brighter than +1 (and hence with a minimum diameter of 838 km) will be classified (and named) as dwarf planets.[2]

No one knows how many dwarf planets exist. Astronomer Mike Brown, the discoverer of Eris, conjectures that there are 10 objects which are almost certainly dwarf planets, 26 objects which are highly likely to be dwarf planets, and hundreds of others that may also qualify.

The term "plutoid" was also introduced for trans-Neptunian dwarf planets. The IAU definition states: "Plutoids are celestial bodies in orbit around the Sun at a semi-major axis greater than that of Neptune that have sufficient mass for their self-gravity to overcome rigid body forces so that they assume a hydrostatic equilibrium (near-spherical) shape, and that have not cleared the neighborhood around their orbit." The currently recognized plutoids are Pluto, Eris, Haumea, and Makemake. Other candidates are 2007 OR10, Quaoar, Orcus, and Sedna.

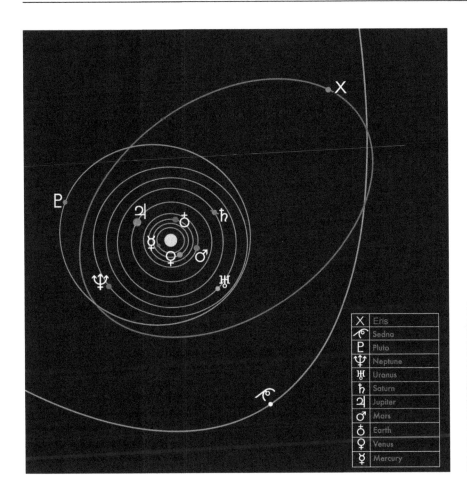

X	Eris
⚳	Sedna
P	Pluto
♆	Neptune
♅	Uranus
♄	Saturn
♃	Jupiter
♂	Mars
⊕	Earth
♀	Venus
☿	Mercury

Figure 12.9 The orbit of Sedna ranges between 76 AU and 975 AU, so that it is always far outside the main Kuiper Belt, but well inside the postulated Oort Cloud. It takes about 12,050 years to complete one orbit of the Sun and may belong to an inner Oort Cloud. Here the orbit of Sedna is compared with those of the planets and Eris. (Gemini Observatory)

One suggestion is that it originated far away in the asteroid belt. It may then have been scattered outward into the Kuiper Belt region by one or more gravitational interactions with giant planets, e.g. during the migration of Jupiter during the early history of the Solar System. Quaoar might once have been an asteroid which lost most of its ice content due to rapid sublimation from solar insolation, or it may always have been a rocky object.

It is also possible that Quaoar's high density resulted from a collision that stripped away almost the entire surface layer of the parent body, leaving Quaoar as the virtually intact rocky core. This may explain the origin of Weywot.

Most of the known KBOs larger than 960 km across are now known to have companions. As already mentioned in the examples above, these are generally thought to be products of ancient collisions.

Peculiar KBOs

A handful of objects which do not fit the major groupings above have been found in recent years. One of the most unusual is Sedna, which remains far beyond the outer edge of the Kuiper Belt, never coming closer than 76 AU and reaching 975 AU at the most distant point of its 12,050 year orbit.[3] Despite its eccentric orbit, Sedna stays fairly close to the ecliptic, with an orbital inclination of 12°.

Estimated to be at least half the diameter of Pluto, Sedna has a strong reddish color. (It is the second-reddest object in the Solar System, after Mars.) Infrared spectra show its surface contains little methane ice or water ice. Instead, it seems to be covered with a coating of hydrocarbon "sludge" produced when the Sun's ultraviolet radiation and charged particles alter the chemical bonds between atoms in the ice. The absence of major collisions may explain the apparent absence of bright, icy patches.

There is some question over whether Sedna should be considered part of the Kuiper Belt, since it spends its entire life well beyond the main belt's outer edge, though it also lies well inside the postulated Oort Cloud of comets. It may be the first known member of a huge population of undiscovered objects in an "inner Oort Cloud" (see Chapter 13).

Neptune shapes the dynamics of most KBOs, but Sedna always orbits far beyond the planet's gravitational influence. Since an object as large as Sedna is unlikely to have formed by accretion

---

[2] Absolute magnitude is defined as the apparent magnitude (brightness) that an object would have if it were one astronomical unit (AU) from both the Sun and the observer, and at a phase angle of zero degrees (i.e. directly overhead).

[3] Many comets have far more eccentric orbits. Kuiper Belt object 2000 OO67 also has an extremely elongated orbit, with aphelion at 1,005 AU and perihelion at 20 AU. However, Sedna is always extremely remote.

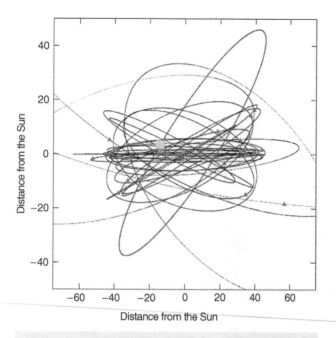

Figure 12.10   2004 XR190 has a unique orbit (red) which is highly inclined but relatively circular compared with those of other KBOs (blue triangles). (Wikimedia)

in the sparsely populated outer regions of the protoplanetary disk, it must have formed closer to the Sun or in the Kuiper Belt, and then been ejected outwards.

Since there are no known planets beyond the Kuiper Belt large enough to have reshaped Sedna's orbit, it has been suggested that passing stars could have created the Belt's outer edge, at the same time scattering some KBOs, such as Sedna, into large, eccentric orbits.

According to this hypothesis, a star passing at a distance of 150–200 AU could have captured objects from the outer Kuiper Belt, while at the same time diverting Sedna into its present orbit, without affecting Neptune or the inner planets. (Computer models also suggest that some KBOs – though not Sedna – could be captured from the passing star's system in a two-way swap of material.)

At least one other object, 2012 VP113, has an orbit similar to Sedna's. It comes no closer to the Sun than 80 AU, and moves out as far as 452 AU. This means it is the most remote object ever observed in the Solar System. Its estimated diameter is about 450 km.

The discovery of Sedna and 2012 VP113 has led astronomers to estimate that about 900 objects with similar orbits and sizes larger than 1,000 km may exist and that the total population of the inner Oort Cloud is likely bigger than that of the Kuiper Belt and main asteroid belt.

Another remote object is 2000 CR105, though a little less eccentric, with perihelion at 45 AU, aphelion at 415 AU, and an orbital period of 3,420 years. Its orbit may have resulted from the same processes that influenced Sedna.

Another object which has been puzzling theorists is 2004 XR190, which is quite bright, with a mass perhaps half that of Pluto (Figure 12.10). The KBO travels between 52 and 62 AU from the Sun, making it the first known object with a fairly circular orbit at the outer edge of the Kuiper Belt. Furthermore, its extreme orbital inclination takes it 47° above and below the ecliptic. It is currently too far from the Sun to have been affected by Neptune's gravity, and its orbit remains a mystery.

Box 12.2   Measuring the Size of a KBO

The small size and distance of KBOs mean that it is very difficult to measure their size. This calculation is further complicated by uncertainties over their surface brightness or reflectivity (albedo).

Direct optical measurements of large KBOs can be made by using the Hubble Space Telescope. Since it orbits above the turbulence of the atmosphere, the HST's angular resolution is much higher than that of most ground-based telescopes. By very carefully measuring the size of an object about 10 times over the course of an hour and comparing it to a nearby star, it is possible to calculate the diameter.

Thermal infrared measurements made by ground-based or orbiting telescopes measure the heat coming from a large KBO. This makes it possible to distinguish between a smaller, reflective object and a larger, darker object which has the same apparent brightness. For example, a dark object absorbs much more light than a white object, so it will be warmer. The diameter can be estimated by measuring the heat coming from a KBO and comparing it with the reflected optical light.

However, such estimates are difficult to make if the only information available is an object's approximate distance from the Sun and apparent brightness. Astronomers then have to estimate the object's most probable albedo, based on the likely nature of its surface, e.g. icy or covered in dark organic compounds.

It is often assumed that KBOs are like comets, with albedos of around 4%, and this figure has been used to calculate their diameters. However, observations have shown that true albedos are often much higher, so that many estimated KBO diameters have been revised downwards.

One example is 2002 AW197, which has been found to reflect 18% of its incident light, leading to an estimated diameter of about 700 km. By assuming that it reflects only 4% of its incident light, this diameter would balloon to 1500 km.

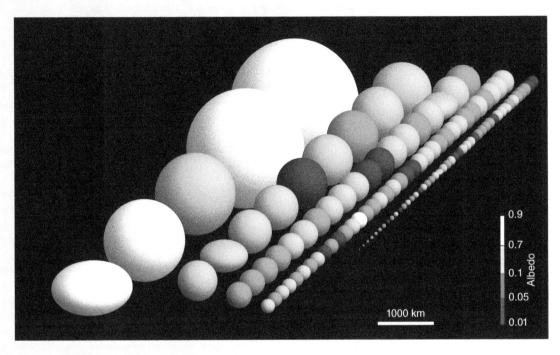

Figure 12.11 The Herschel infrared observatory determined the albedos (reflectivity) of 132 of the known 1,400 Kuiper Belt objects. The albedo implies a variety of surface compositions: low albedo (brown) is an indication of dark surface materials, such as organic material, while higher albedo (white) suggests pure ices. The KBOs shown range in size from just below 50 km to almost 2,400 km in diameter. Two of them have distinctly elongated shapes: Haumea (white) and Varuna (brown). (ESA/Herschel/PACS/SPIRE)

Binaries

Pluto and its satellites make up the only known six-body system in the Kuiper Belt, but many more KBOs are known to have at least one satellite. Indeed, some 30% of the known Cold Classical KBOs are binaries.

The first of these, 1998 WW31, was found on April 16, 2001. The two objects are thought to be approximately 150 and 130 km in diameter (Figure 12.12). By studying their orbital motions, it was calculated that the pair's combined mass is about 5,000 times smaller than Pluto–Charon. The mutual orbit of 1998 WW31 is the most eccentric ever measured for any binary Solar System object or planetary satellite, with a variation in distance between 4,000 and 40,000 km. (The eccentricity is approximately 0.82.)

Some of the largest KBOs either have companions or resemble contact binaries (see Eris and Haumea above). Their orbital periods are highly variable, ranging from 6 days (for Charon) to 876 days (2001 QT297). In cases where the pair is widely separated, the systems may be unstable over a timescale of billions of years. However, some binaries orbit only a few thousand kilometers apart and must be very stable. How did the binaries originate?

Gravitational capture of one KBO by another is considered to be impossible without some process that can dissipate some of the kinetic energy present in the bodies. On the other hand, the fact that all five of the largest objects in the Kuiper Belt have moons suggests that they may be produced by accretion of material from a disk of debris created by a cataclysmic collision. Unfortunately,

this process is unlikely with the smaller binaries, since they would seem to have insufficient mass and gravity for the creation of a sizeable accretion disk.

Another explanation is that they might have formed as the direct result of low velocity (100 m/s or less) collisions between KBOs. When the relative velocities are smaller than, or comparable to, the gravitational escape speeds from the colliding objects, peanut-shaped contact binaries may result. Alternatively, the objects may rebound with low enough energy for a binary to form (see also Chapter 13).

Even these hypotheses are not without their difficulties. Calculations indicate that collisions between 100 km-sized KBOs are currently too rare to account for the assumed number of binaries. Another problem is that the known Kuiper Belt binary objects often have comparable masses, something which would not be expected from a collisional origin.

The solution may lie in the existence of a much denser Kuiper Belt in the past – perhaps 100–1,000 times more massive than it is today. In those conditions, collisions would be more frequent. This might also make it possible for dynamical friction – the net effect of many small KBOs tugging on nearby objects – to cause a net loss of energy that would result in objects slowly spiraling together.

What happened to this massive Kuiper Belt? Some of the objects would have been removed by gravitational interactions with Neptune – either toward the Sun, or into the Oort Cloud of comets and beyond. Many others may have been ground down or shattered by

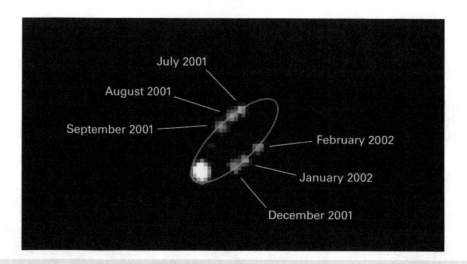

July 2001

August 2001

September 2001

February 2002

January 2002

December 2001

Figure 12.12   This HST composite shows the apparent orbit of one member of the binary KBO known as 1998 WW31. In fact, the objects revolve around a common center of gravity. The image is a composite of six exposures taken between July 2001 and February 2002. Their mutual orbital period is about 570 days. (Courtesy of NASA and C. Veillet/Canada-France-Hawaii Telescope)

Box 12.3   The Search for Planet X

With the discovery of Neptune, the catalog of major planets seemed once again to be complete. However, almost from the time of Galle's triumph (see Chapter 11), some individuals began to speculate that a ninth planet, presumably another giant, was lingering unseen in the outer Solar System. This suspicion was reinforced in the late 1800s, when observers reported that Uranus was once again straying, very slightly, from its predicted position, after all known gravitational influences were taken into account.[4]

The errors, or "residuals," were much smaller than those that prompted a search for Neptune, and might be accounted for by minute inaccuracies in the mapped positions of the background stars. However, there remained the tantalizing possibility that they might be caused by the pull of another planet.

The quest for "Planet X" was taken up by Percival Lowell, a wealthy amateur astronomer who had built his own observatory in Flagstaff, Arizona, and become a celebrity with his theories and observations concerning the existence of artificial canals on Mars. Lowell instigated a photographic survey along the ecliptic plane in 1905, the first of many attempts to find Planet X, but none of them were successful.

After a long hiatus, the Flagstaff observatory resumed the quest in 1929. The equipment included a 33 cm wide field photographic refractor and a blink comparator that would enable rapid, efficient comparisons of photographic plates of the same region of the sky taken a few days apart.

The search was entrusted to Clyde Tombaugh, a 23-year-old farmer's son and keen amateur astronomer who was appointed to the observatory's staff with the specific task of searching for Planet X. Tombaugh began his search in the constellation of Gemini, where Lowell had predicted the planet would be. He soon decided to increase the chances of a discovery by concentrating on the sky around the "opposition point," the part of the sky directly opposite the Sun, where any hidden planet would appear brighter and move a greater angular distance against the background stars over a period of a few days.

Tombaugh's routine involved taking exposures of the night sky that lasted one hour or longer. The process would be repeated a few days later, if possible, for the same area of sky, so that the plates could be exposed and developed under similar conditions. (Tombaugh took photographs around the time of New Moon.)

This was followed by the exhausting routine of manually comparing the plates, each one containing between 50,000 and 900,000 stars. Although the blink comparator helped considerably, there was always the problem of microscopic flaws in the plates and false alarms due to wandering asteroids. Fortunately, a typical asteroid trail would be about 10 times longer than the shift of a few millimeters which Tombaugh's much more remote Planet X would be expected to display.[5]

Then, on February 18, 1930, the young man's dogged persistence paid off. Having worked his way around the ecliptic, Tombaugh had returned to the region near Delta Geminorum, a bright star in the constellation of Gemini. He noticed a star-like object that shifted position against the stars. Careful study of its motion confirmed that this was no ordinary asteroid, and on March 13, 1930,

---

[4]The orbit of Neptune was less well known – it had completed only about one-third of an orbit since its discovery – so calculations of Planet X's location were based on the measured position of Uranus.

[5]In the course of his search, Tombaugh discovered numerous clusters of stars and galaxies, hundreds of asteroids, two comets, and one nova.

the discovery of the ninth planet was officially announced. (The newcomer had actually been photographed in early 1929, but it had not been recognized.)

What to call Planet X? Suggestions poured in from all over the world, but it was eventually decided to name it Pluto – a proposal sent in by Venetia Burney, a young schoolgirl from Oxford, England. The name was considered particularly appropriate for such a distant planet, since it was associated with the Roman god of the underworld. The symbol for Pluto comprises the letters P and L – the initials of Percival Lowell.

Figure 12.13   Clyde Tombaugh (1906–1997) using the blink comparator to search for Planet X among the thousands of stars on each photographic plate. The instrument enabled him to detect small, faint objects moving against the background of "fixed" stars by comparing plates taken a few days apart. (Lowell Observatory Archives)

Figure 12.14   These photos, taken on January 23 and 29, 1930, with the 33 cm refractor at Lowell Observatory, are small sections of the photographic plates on which Pluto was discovered. Pluto (arrowed) is a star-like point of light. By rapid "blinking" of the two images, Clyde Tombaugh was able to detect its movement against the "fixed" background stars. (Lowell Observatory Archives)

collisions. The fine dust that resulted would have been blown into interstellar space by the solar wind.

Puzzling Pluto – Key to the Kuiper Belt

Pluto and its satellites are, at the time of writing, the only KBOs to have been studied in detail at close range. The fly past by NASA's New Horizons spacecraft in 2015 provided new insights into the nature of these six objects and their more remote cousins, as well as numerous surprises for planetary scientists (see Box 12.5).

Although Pluto is the largest known member of the Kuiper Belt, it is much smaller than many planetary satellites. Its modest physical characteristics were not immediately recognized. At the time of its discovery in 1930, Pluto was regarded as the ninth

Box 12.4   Pluto's Eclipse Season

The most recent Pluto–Charon eclipse season took place between 1985 and 1990, when Pluto's equator was aligned with the Earth. (Fortuitously, it was also around the time of Pluto's closest approach to the Sun, so that the largest possible disk was visible.) Since Charon orbits above the equator, astronomers were able to observe it moving in front of and behind Pluto every 6.4 days.

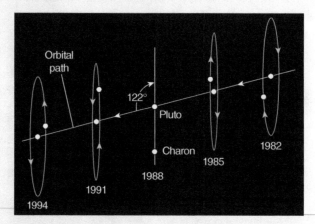

Figure 12.15   Pluto is tipped on its side with its spin axis close to the plane of its orbit. Like most satellites, Charon orbits above Pluto's equator. From 1985 to 1990, Pluto's equator and Charon's orbital plane were aligned with Earth, so that Charon passed in front of, or behind, Pluto every 3.2 days. (Peter Bond)

The location of Charon – and its shadow – varied slowly over the eclipse season. The first eclipses (also known as occultations or "mutual events") began with Charon covering more and more of Pluto's north polar region. Later eclipses blocked the equatorial region, and the final occultations took place over the south polar region.

At the beginning and end of the eclipse season, Charon blocked very little of the light reflected from Pluto, so there was little change in the light curve. Around the peak of the eclipses, in 1987, the reflected sunlight dimmed by more than 30%. The eclipses lasted up to four hours and careful timing of their beginning and end made it possible to obtain the first accurate measurements of the diameters of the two objects.

Changes in their combined brightness made it possible to produce maps of the surface reflectivity on the Charon-facing hemisphere. It was found that Pluto had a highly reflective north polar cap, a dimmer south polar cap, and a mixture of bright and dark features in the equatorial region. Pluto's geometric albedo is 0.49 to 0.66, so it is much brighter than Charon, whose albedo ranges from 0.36 to 0.39.

planet in the Solar System, but its small disk and great distance from the Sun meant that astronomers struggled to determine its mass and size.

Percival Lowell had predicted that Planet X would have more than six times the mass of Earth, but as instrumentation improved and new, more accurate, observations were obtained, it became clear that Pluto was an enigmatic lightweight whose existence was difficult to explain.

With the 1976 discovery of methane frost on its surface, Pluto was revealed to be a highly reflective object. Calculations based on its high albedo and apparent brightness showed that it must be smaller than the Moon. This was confirmed in 1978, when a large satellite, named Charon, was discovered in close orbit around Pluto.

Between 1985 and 1990, there was a rare opportunity to observe mutual occultation events as Charon and Pluto took it in turns to eclipse each other.[6] These events made it possible to constrain their sizes. (See Box 12.4).

Subsequent HST imaging and the New Horizons spacecraft's flyby in 2015 have made it possible to tie down Pluto's equatorial diameter at 2,374 km.[7] With a mass 500 times smaller than Earth's, and an orbit that causes it to travel much farther from the Sun than Neptune, Pluto's gravitational influence on Uranus is negligible. Indeed, Earth has a greater gravitational pull on Uranus than Pluto ever can.

Clearly, Pluto is not Lowell's Planet X. Instead, its discovery close to the predicted positions must rank as one of the greatest coincidences in the history of science.

---

[6] The previous series of eclipses had taken place about 120 years earlier.

[7] Pluto is 2/3 the diameter of Earth's Moon but 15,000 times farther away. Viewing surface detail is as difficult as trying to read the print on a golf ball located more than 50 km away.

Orbit

Piecing together the nature of this dim, distant world was a major challenge, although pre-discovery observations going back to 1914 soon enabled a reasonably accurate orbit to be calculated.

Although its mean distance from the Sun is 5.9 billion km (39.7 AU), well beyond Neptune, Pluto has a far more eccentric orbit than any planet in the Solar System. At perihelion, it approaches to within 4.5 billion km, while at aphelion it recedes to 7.4 billion km. Perihelion last took place in 1989 and aphelion will be reached in 2114. One entire orbit – the Plutonian year – lasts for 248 Earth years, so Pluto has completed only one-third of an orbit since its discovery. Its extreme distance from the Sun means that its orbital velocity is very low, averaging about 4.7 km/s.

The high eccentricity means that for 20 years of its 248-year orbit, Pluto is closer to the Sun than Neptune, an event that last occurred between 1979 and 1999. However, there is no possibility of a collision because Pluto's orbit is inclined 17° to the ecliptic, and because the two objects are locked in a 3:2 orbital resonance. This means that Neptune completes three orbits in the time it takes Pluto to complete two (Figure 12.16). Conjunctions always occur near Pluto's aphelion, so that they never come closer than 2.5 billion km (17 AU), whereas Pluto can approach to within 1.6 billion km (11 AU) of Uranus.

Pluto's peculiar orbit is best explained if it formed in a low eccentricity, low-inclination orbit within a primordial disk of planetesimals beyond Neptune. According to this model, outward migration of Neptune during the late phases of giant planet formation caused mean motion resonances with many of the planetesimals. The outcome was the eventual capture of Pluto (and other trans-Neptunian objects) in the 3:2 resonance, and an increase in Pluto's orbital eccentricity. The detection of many Kuiper Belt objects (Plutinos) locked in the 3:2 resonance with Neptune supports this scenario (see Classification of KBOs).

Long-term orbit integrations have uncovered other subtle resonances and near-resonances which indicate that Pluto's orbit remains stable over billions of years, i.e. most of the lifetime of the Solar System.

Seasons

For many years, Pluto's axial inclination and rotation period were unknown. The first clue came from measurements of its albedo, which showed fluctuations of more than 10% – a bigger variation than shown by all of the planets except Mars. Moreover, the fluctuations always repeated over the same period – 6 days, 9 hours, 17 minutes, plus or minus 4 minutes. It seemed clear that the observations were caused by variations in the light reflected from a solid surface rather than clouds in an atmosphere. If this was the case, then Pluto had a surprisingly long "day."

This rotation period was supported by the discovery of Charon, which orbits Pluto every 6.3872 days (see below). This is identical to Pluto's day – an example of synchronous rotation.

The next problem was to identify Pluto's axial inclination. Light curve studies suggested that the axis was tilted at least 50°, but more precise calculations were impossible.

Once again, the discovery of Charon proved to be the key. Most major satellites orbit above or very close to the equatorial plane of their primaries, and the synchronous rotation of Charon indicated that it was no exception to the rule. Since Charon's orbit is highly inclined to the plane of Pluto's orbit, it follows that – like Uranus – Pluto rotates with its poles fairly close to its orbital plane.

In fact, Pluto's rotational axis is currently tipped 120°, although this obliquity varies considerably over a period of about 3.7 million years. Like Uranus and Venus, its rotation is retrograde, so that, if you were to view Pluto from above the ecliptic, it would spin from east to west. This means that each polar region experiences prolonged "summers" and "winters," with long periods in between when the Sun is nearly overhead at the equator. Indeed, most of Pluto's surface experiences an overhead Sun at some time during the dwarf planet's year.

This is the likely explanation for the variations in brightness that have been observed since the 1950s. When Pluto was first discovered, its relatively bright south polar region – possibly coated in fresh, reflective frost – was visible from Earth. Pluto appeared to dim as the visible surface gradually shifted from nearly pole-on in 1954 to equator-on during the 1987 spring equinox. Since then,

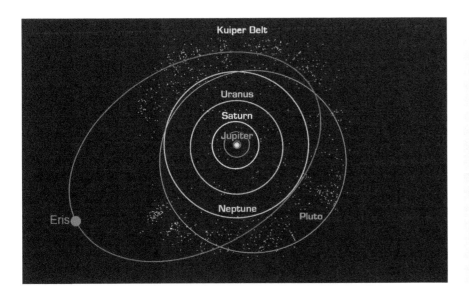

Figure 12.16 Pluto follows an eccentric orbit, with perihelion at 29.6 AU and aphelion at 50.3 AU. For 20 years of its 248-year orbital period it is closer to the Sun than Neptune, but their different orbital inclinations and 3:2 orbital resonance mean that they never come close to each other. Pluto's orbital inclination (17°) carries it well above and below the orbits of the major planets. Also shown is the orbit of Eris, another dwarf planet and the second-largest known member of the Kuiper Belt. (NASA-JPL)

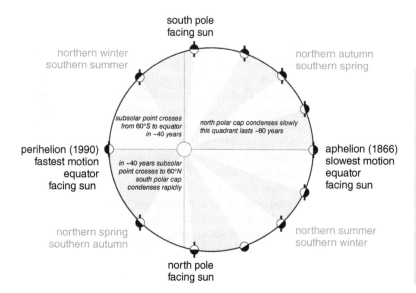

south pole
facing sun

northern winter
southern summer

northern autumn
southern spring

*subsolar point crosses
from 60°S to equator
in ~40 years*

*north polar cap condenses slowly
this quadrant lasts ~80 years*

perihelion (1990)
fastest motion
equator
facing sun

*in ~40 years subsolar
point crosses to 60°N
south polar cap
condenses rapidly*

aphelion (1866)
slowest motion
equator
facing sun

northern spring
southern autumn

northern summer
southern winter

north pole
facing sun

Figure 12.17 Pluto's axial inclination of 120° means that it appears to rotate on its side. Perihelion occurs close to the northern spring equinox, when the equator is facing the Sun. The northern and southern hemispheres then receive similar amounts of sunlight. Pluto is now moving into its northern summer, and a long period of permanent sunlight at the north pole. Note how the northern spring is much shorter than the northern fall, because Pluto is much closer to the Sun and moving faster around its orbit. (Emily Lakdawalla, after Candy Hansen)

Pluto's northern hemisphere has started to receive more sunlight than the southern hemisphere.[8]

However, the seasons on Pluto are not all the same length (see Figure 12.17). The eccentric orbit causes it to receive almost three times more sunlight at perihelion than at aphelion (the variation on Earth is a mere 5%). As it happens, perihelion almost coincides with the northern spring equinox, while aphelion is close to the northern autumnal equinox. Since planets move faster when they are closer to the Sun, Pluto's northern fall lasts much longer than the relatively short northern spring.

The ice distribution is inevitably affected as the seasons come and go during the long Plutonian year. Even on the inward leg of its orbit, Pluto (and Charon) receives 900 times less solar radiation than Earth. As a result, even the midsummer temperature registers −233°C, only 40 degrees above absolute zero. At aphelion, the incoming insolation is more than 2,400 times weaker than Earth receives, so Pluto is even colder.

Pluto's Surface Ices

Pluto is so small and far away that it is extremely difficult to see much detail on its surface with ground-based telescopes. The apparent diameter of about 0.1 arcsecond – 18,000 times smaller than the width of a full Moon – makes it impossible to resolve any features on the surface, even with the Hubble Space Telescope.

As a result, little was known about Pluto's surface for many years. The main conclusion, based on studies of the planet's light curves over many years, was that Pluto is highly reflective, with an average albedo of about 55%, similar to that of freshly fallen snow. (The Moon's average albedo is 12%.)

During the occultations of 1985–1989, when Charon moved directly in front of and behind Pluto (as seen from Earth), it was possible to map broad variations in the surface markings of both

Pluto and Charon. Several maps of Pluto's surface reflectivity were produced as a result of these efforts (see Box 12.4).

Although there were some significant differences, the maps showed general agreement on the distribution of light and dark regions. In particular, areas near the equator were generally dark, reflecting only 15% of incoming sunlight, while the north pole had an albedo of more than 70%.

Then, in 1994, the Hubble Space Telescope was able to image almost the entire surface at visible and ultraviolet wavelengths. Although each pixel (picture element) covered an area more than 160 km across, the images revealed at least 12 major regions defined by dark and light surface features. Both poles were bright and displayed a "ragged" border, but they were not identical. The northern polar region appeared to be both larger and brighter.

Subsequent observing campaigns with Hubble showed how the frosts were changing with the seasons (see Figures 12.18 and 12.19). Over time, Pluto became significantly redder, probably the result of solar ultraviolet radiation breaking up methane frost on Pluto's surface and leaving behind a dark, reddish carbon-rich residue.

The northern hemisphere was seen to be getting brighter while the southern hemisphere darkened. These seasonal changes were associated with sublimation and refreezing of ices. However, the brightening of the northern hemisphere was rather surprising, since it was experiencing a substantial increase in solar insolation, which might be expected to darken the surface. It was suggested that may be due to a change in texture and reflectivity of the sublimating nitrogen ice.

However, the reality is more complex. Infrared measurements show that the darker regions on Pluto are up to 20°C warmer than the bright regions. This can be explained if the brighter ices were not only reflecting more sunlight, but also rapidly sublimating (changing directly from ice to gas).

---

[8]Traditionally, the north pole of a planet was defined as the one at which the planet rotated to the left and the stars to the right, as on Earth. For a world with retrograde rotation, such as Pluto, this placed the north pole on the south side of the Solar System. To avoid confusion, this definition has been changed so that the north poles of retrograde rotating objects are on the north side of the Solar System, as defined by Earth's rotation.

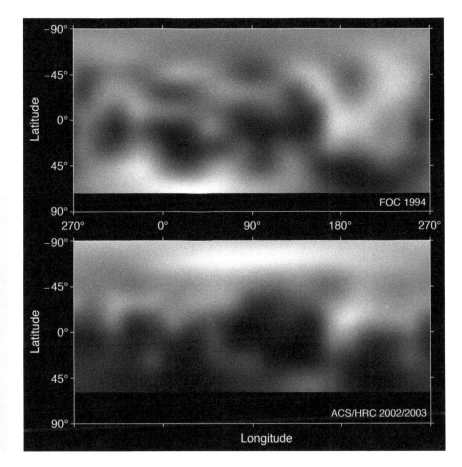

Figure 12.18 A comparison of HST maps of Pluto, obtained in 1994 (above) and 2002–2003 (below). The white areas are frosts, and the dark areas are a carbon-rich residue caused by sunlight breaking up methane. Spring is approaching in the north (top), which should cause ices to vaporize, then freeze out in the colder, shaded region at the south pole. Surprisingly, the north pole has become brighter and the visible part of the southern hemisphere is darker. More of the southern hemisphere was in darkness by 2003. (Courtesy of NASA, ESA, and M. Buie/SwRI)

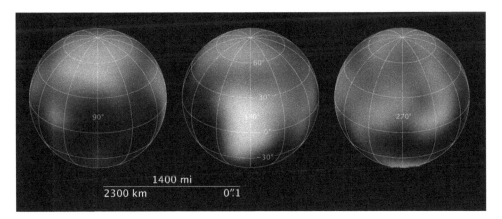

Figure 12.19 HST images of Pluto showing how the surface appeared 2002–2003. The south polar region (bottom) is not visible. The complex pattern reveals white, dark-orange, and charcoal-black terrain. Note the bright north polar region, and the mainly darker equator and southern hemisphere. (Courtesy of NASA, ESA, and M. Buie/Southwest Research Institute)

Meanwhile, spectroscopic observations at infrared wavelengths – subsequently confirmed by the New Horizons spacecraft – showed that the water ice bedrock is covered with frozen nitrogen and methane ice, as well as carbon monoxide ice.

New Horizons confirmed that the bright area on the side of Pluto facing away from Charon (longitude 180°) – now known as Sputnik Planitia – is especially rich in nitrogen ice and carbon monoxide ice, whereas the Charon-facing hemisphere is dominated by methane ice (Figure 12.20).

The older, more cratered and mountainous terrain is darker in color. Here, methane has been converted into hydrocarbons by exposure to solar ultraviolet light and high energy particles such

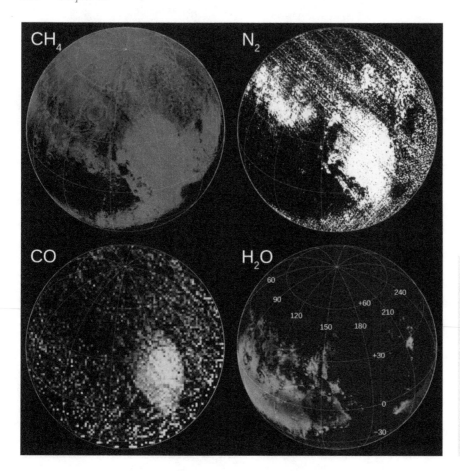

Figure 12.20 The distribution of four ices on Pluto. Brighter colors represent higher concentrations. Water (H_2O) ice makes up the "bedrock." Nitrogen (N_2), carbon monoxide (CO), and methane (CH_4) are surface deposits that sublimate and redeposit during Pluto's seasonal cycles. Sputnik Planitia is right of center. (Courtesy of NASA/Johns Hopkins University Applied Physics Laboratory/SwRI)

as cosmic rays. Since this dark material is some 13°C warmer, it prevents further deposition of bright ices.

Pluto's Icy Heart

The most noticeable feature on the anti-Charon hemisphere viewed by New Horizons was a heart-shaped depression which is now known as Tombaugh Regio (Figure 12.21). The western half of this bright basin is a smooth plain of nitrogen ice called Sputnik Planitia. This plain measures about 1,050 by 800 km in size and its center lies about 3 to 4 km below its surroundings.

Many scientists believe this depression originated in a large impact, though others argue that an ice cap would have evolved in that location anyway and that the weight of the ice would have depressed the surface (see Sputnik Planitia and Pluto's Reorientation).

An absence of impact craters suggests that the surface is young and active today. Spacecraft images show that glaciers of nitrogen ice are flowing into Sputnik Planitia from the higher terrain to the east. They move through canyons and carry dark streams of debris that resemble glacial moraines on Earth.

The nitrogen plain exhibits a pattern of interconnected polygonal cells, 10–40 km across, with shallow depressions along their margins. These probably result from convective churning of a deep layer of solid, but mobile, nitrogen ice. Some scientists have compared them to a giant, solid state lava lamp in which convection brings heat toward the surface.

Embedded in the plain's surface are "islands" of water ice up to a few kilometers across. Water ice has a lower density than nitrogen ice, and it seems that they are analogous to icebergs on Earth.

Around the western edge of Sputnik Planitia are rugged mountains of water ice coated with reddish-black organic material. The mechanism that formed these blocky mountains is still not clear.

The highlands that cover much of the terrain west of Sputnik Planitia are heavily cratered, evidence of a much older surface than Tombaugh Regio (Figure 12.22). One upland region has swarms of bright-haloed craters, each tens of kilometers across. The craters' bright walls and rims stand out from their dark floors and surrounding terrain, creating the halo effect. Exactly why the bright methane ice settles on these crater rims and walls is a mystery. The floors and terrain between craters show signs of water ice.

Two large, dome-like structures with central depressions were observed near the day–night terminator. These may be cryovolcanoes, where water mixed with ammonia to lower its melting temperature may have oozed onto the surface due to pressure from below.

A number of large, linear canyons occur to the west of Sputnik Planitia. These appear to be evidence of the ice fracturing as the result of extension, the gradual expansion of the crust – possibly due to the slow freezing of a subsurface ocean.

Near the north pole are several parallel canyons. The largest of these, which is 75 km wide, contains a shallow and winding valley. These canyons' degraded walls appear to be considerably older

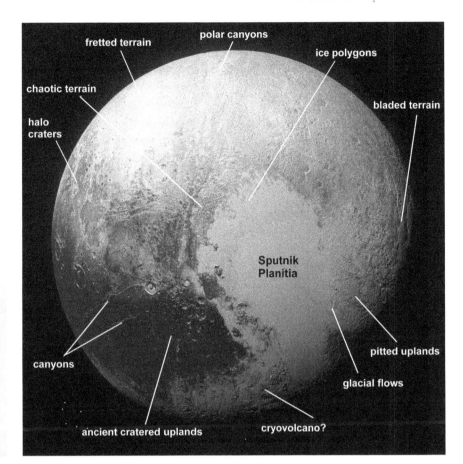

Figure 12.21 The New Horizons spacecraft captured this view of Pluto on July 14, 2015. The enhanced color image shows the main surface features – plains, craters, ridges, and troughs surrounding the icy plain of Sputnik Planitia. (Courtesy of NASA/Johns Hopkins University Applied Physics Laboratory/SwRI & Peter Bond)

Figure 12.22 This image from New Horizons shows blocks of Pluto's water ice crust that form mountains, up to several kilometers high, at the edge of the softer nitrogen ice of Sputnik Planitia. Some mountain sides are coated in dark material, while others are bright. Several sheer faces appear to show crustal layering. The plain of Sputnik Planitia displays polygonal, cellular features. These probably result from convective churning of a deep layer of solid, but mobile, nitrogen ice. Dunes made of methane sand may be related to sunlight-driven ice sublimation. This view is about 80 km wide. The top of the image is to Pluto's northwest. (Courtesy of NASA/Johns Hopkins University Applied Physics Laboratory/Southwest Research Institute)

than the more sharply defined canyon systems elsewhere on Pluto, perhaps because they are older and made of weaker material.

Other strange landforms include numerous "pits" scattered across a 100 km-wide plateau in the southeast corner of Tombaugh Regio. These may form through a combination of ice fracturing and evaporation. The pits, which are typically hundreds of meters across and tens of meters deep, appear to have formed relatively recently.

Another landform seen nowhere else in the Solar System is "bladed" terrain – also dubbed "snakeskin" terrain – to the east of Tombaugh Regio. The repetitive, blade-like hills are dozens of meters high, about 5 to 10 km apart, and are aligned from north to

Box 12.5   New Horizons

Until 2015, all of the major classes of objects in the Solar System had been explored by spacecraft, with the exception of the millions of Kuiper Belt objects. This omission came to an end on July 14, 2015, following the passage of NASA's New Horizons spacecraft through the Pluto–Charon system.

The nuclear-powered spacecraft had a launch mass of 478 kg. Most of its subsystems, including the computers and propulsion control system, are based on designs used in other spacecraft. A 2.1 m high gain antenna is available to communicate with Earth from deep space.

New Horizons carries four scientific instruments. A miniature, high-resolution camera was used to image selected regions, revealing objects down to 80 m in diameter. An infrared imaging spectrometer mapped the surface at a resolution of one kilometer and studied the composition and temperature of Pluto's ices and atmosphere. A radio science instrument probed the structure of Pluto's tenuous atmosphere and measured the surface temperatures on both day and night hemispheres. A suite of charged particle detectors sampled gases escaping from the rarefied atmosphere and determined their escape rate.

An Atlas V–Centaur rocket sent the spacecraft speeding away from Earth at approximately 58,000 km/h – the fastest human-made object ever built – on January 19, 2006.

Figure 12.23   The New Horizons spacecraft during its encounter with Pluto (foreground) and Charon. (Courtesy of NASA/JHU-APL/SwRI)

New Horizons conducted an intensive, four-month reconnaissance of the Jupiter system in early 2007, and the gravity assist put it on course for a flyby of Pluto in the summer of 2015. Observations began six months before closest approach. Once the distance was down to 100 million km – about 75 days before close encounter – New Horizons began to send back images superior to anything yet seen.

As the two worlds loomed larger, their surfaces were mapped in increasing detail, along with any atmospheric phenomena. On the day of closest approach, New Horizons swept within a few thousand kilometers of Pluto. It then turned to image the planet's night side, illuminated by dim moonshine from Charon. Measurements of the behavior of radio signals transmitted by the spacecraft revealed the vertical structure of the sparse atmosphere. Similar observations were then made of Charon. New Horizons also obtained the first close-up images of Pluto's smaller moons.

The probe continued outward toward another Kuiper Belt object, known as 2014 MU69. New Horizons made its closest approach on January 1, 2019, more than 6.5 billion km from Earth. Travelling at more than 51,000 km/h, the probe passed only 3,500 km from the KBO.

Observations showed that this "cold classical" KBO is a contact binary that measures about 31 km in length. The camera revealed details as small as 25 m across, compared to 80 m at Pluto, but because 2014 MU69 is tiny, it occupied only a few pixels in the images until the day of the flyby.

Additional flybys may be scheduled, depending on the status of the spacecraft and the availability of suitable targets and funding. New Horizons is currently funded for an extended mission until May 1, 2021. Mission planners hope to study more than two dozen other KBOs at a distance and measure the charged particle and dust environment all the way across the Kuiper Belt.

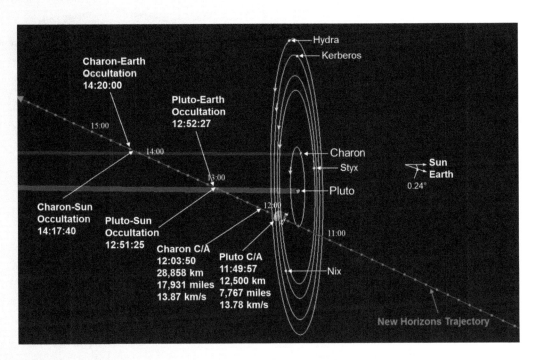

Figure 12.24 New Horizons' trajectory through the Pluto system. Times are UTC; timeline "ticks" are at 10-minute intervals. Distances are from the surfaces of Pluto and Charon; flyby speeds are relative to each body; position and lighting of each body is at the time of Pluto close approach. (NASA)

Figure 12.25 This is one of the best views New Horizons obtained of the Pluto hemisphere that faces Charon. Taken on July 11, 2015, it shows intersecting linear features, suggestive of polygonal shapes, and craters north of the dark material along the equator. The image was captured when the spacecraft was 4 million km from Pluto. The north pole is in the upper center of the image. (Courtesy of NASA/Johns Hopkins University Applied Physics Laboratory/Southwest Research Institute)

south. They are perched on a much broader set of rounded ridges that are separated by flat valley floors.

Sputnik Planitia and Pluto's Reorientation

As already mentioned, the bright, smooth terrain of Sputnik Planitia is unlike any other geological feature in the Solar System. It consists of an oval depression in Pluto's water ice shell, approximately 1,300 x 900 km in diameter, that is 3–4 km beneath the surrounding rugged uplands. The depression is filled with nitrogen, methane, and carbon monoxide ices – the same constituents that comprise the dwarf planet's sparse atmosphere.

The basin-like shape of Sputnik Planitia, a possible ejecta blanket, and a mountainous boundary to the northwest have led scientists to suggest that it originated during a major impact that occurred while the Kuiper Belt was still forming. Studies suggest that the impact would have excavated a 7 km-deep crater, with a central uplift of material reaching to a similar distance above the surface.

Liquid water welled up after the impact, creating a huge bulge of near-surface fluid This ocean upwelling resulted in a positive mass anomaly, or gravity anomaly – similar to the mass concentrations (mascons) on Earth's Moon, In other words, this region may have more mass in it than an average equivalent area of Pluto's outer shell. To maintain this anomaly, a liquid water ocean must still be present beneath the basin.

An alternative suggestion is that frost would accumulate naturally at low latitudes, even if an impact basin did not exist. A major ice deposit that is many kilometers thick could easily accumulate simply through seasonal changes in surface ices. The ices would sublimate from regions that are warmed by sunlight, and condense in cold or low-lying regions.

Modeling of surface temperatures shows that, when averaged over Pluto's 248-year orbit, the latitudes 30 degrees north and south are the coldest places on the dwarf planet, far colder than either pole. Ice would naturally form around these latitudes, including Sputnik Planitia, which is centered at 25 degrees north latitude.

As the ice deposit reflected away solar light and heat, it slowly grew in size. This positive feedback phenomenon, called the runaway albedo effect, could eventually lead to a single dominating ice cap, like the one observed on Pluto.

One interesting characteristic of Sputnik Planitia is its location in the center of the Pluto hemisphere that faces away from the large moon, Charon. Researchers believe that this alignment with Pluto's tidal axis changed the rotation of Pluto over time.

The gravity of Charon causes Pluto to be egg-shaped, so Pluto's rotation slowed gradually due to gravitational forces from Charon, just as Earth is slowly losing spin under similar forces from the Moon. As a result, the long axes of Pluto and Charon became aligned, with Pluto locking one face toward its moon in just a few million years.

The large mass of Sputnik Planitia and its resultant gravity anomaly would have resulted in a 50 percent chance of the plain either facing Charon directly or turning as far away from the moon as possible. Sputnik Planitia ended up centered at 175 degrees longitude.

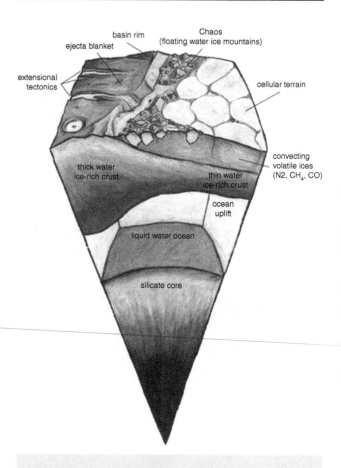

Figure 12.26 The key geological features of Sputnik Planitia and its surroundings. A global water ocean may exist beneath Pluto's surface. (James T. Keane)

The single ice cap that developed at Sputnik Planitia was an enormous weight on Pluto's surface, enough to create its own basin and shift the dwarf planet's center of mass. (A similar phenomenon happens on Earth: the massive Greenland ice sheet created a basin by pushing down the crust that it rests upon.)

As surface ice accumulated in the old crater, Sputnik Planitia grew heavy enough to reorient Pluto by around 60 degrees with respect to the rotational and tidal axes. The combination of this reorientation, loading, and global expansion due to the freezing of a possible subsurface ocean generated stresses, resulting in a global network of extensional faults that closely replicate the observed fault networks on Pluto. According to this scenario, Sputnik Planitia probably formed northwest of its present location, before Pluto reoriented.

Interior

Pluto has a bulk density of 1.86 g/cm³, or almost twice the density of water. Models of the interior indicate that rock makes up 75% of Pluto's diameter and two-thirds of its mass. The remainder is composed mostly of water ice, though more exotic ices similar to those found on the surface are also likely to be present.

Figure 12.27   One model of Pluto's interior includes a large core of hydrated rock, overlain by a thick mantle of water ice and a surface crust of water ice coated with nitrogen, methane and other ices. There may also be a subsurface ocean of liquid water, possibly mixed with ammonia or other compounds. Dark blue represents the subsurface ocean and light blue is frozen crust. (Courtesy of Pam Engebretson/NASA)

Areas with few impact craters, surface ice floes, and polygons apparently generated by subsurface heat all support the idea that convection is still active on Pluto today. Radioactive trace elements in the core warmed the interior, enabling internal differentiation to take place, melting the ice and allowing it to separate from the rock,

Pluto probably has a large core of partially hydrated rock, surrounded by a thick, low-density, mantle of water ice. Close to the surface is an icy crust, coated with volatile ices (nitrogen, methane, and carbon monoxide).

Does Pluto possess an ocean of liquid water beneath its icy shell? Observations of other icy objects, such as the Galilean moons of Jupiter, indicate that such oceans are not as rare as once believed (Figure 12.27). It is now thought to be likely that a layer of liquid water formed at the bottom of the water ice crust, and that it is still present today.

Several lines of evidence in support of this internal, global, ocean were provided by the New Horizons spacecraft.

One possibility is an internal ocean layer about 100 km thick, with roughly 30 percent salinity – comparable with the saltiness of Earth's Dead Sea. This high salinity would result from interaction between the ocean and Pluto's rock core – a similar scenario to that proposed for the salty water inside Enceladus and Europa. Another possibility is an ocean kept liquid by large amounts of ammonia, a natural antifreeze prevalent in icy bodies of the outer Solar System.

A thermal evolution model for Pluto found that if Pluto's ocean had frozen billions of years ago, it would have caused the entire planet to shrink. Overlying pressure would gradually transform the frozen water into a higher density variant known as Ice II. The result would be a contraction of Pluto's radius by up to 13 km. However, unlike on Mercury, there are no signs of a global contraction on Pluto's surface. On the contrary, images taken by New Horizons indicate that Pluto has been expanding.

New Horizons images showed extensional tectonic features, which indicate that Pluto underwent a period of global expansion. These may have been caused by the gradual freezing of the internal ocean. The "normal" ice would be less dense than water, so when it froze, it expanded. If Pluto had an ocean that was still in the process of freezing, extensional tectonics on the surface which look fairly fresh would result.

Pluto's Atmosphere

Like most of its other physical characteristics, the question of whether Pluto possessed an atmosphere remained uncertain for many years. Its low mass and density indicated that any atmosphere must be very sparse, even though the escape velocity is very low. The absence of any transient features in the light curve also suggested an absence of clouds or thick hazes.

The first evidence of a possible atmosphere came in 1976, when the spectral signature of methane ice was recorded on Pluto's surface. Since its temperature is below the freezing point of methane, there was some doubt over whether any significant sublimation of the methane would occur, thereby creating a gaseous blanket. However, much of Pluto has a high reflectivity, something that would not be expected if methane ice had been reacting with ultraviolet sunlight over millions of years. The resultant layer of hydrocarbons, such as ethane and acetylene, would result in a dark surface.

The first unambiguous detection of an atmosphere came during a stellar occultation in June 1988. Instead of suddenly disappearing behind Pluto, there was a gradual dimming of the star's light, followed by a more rapid change closer to the surface – evidence of some form of atmospheric layering.

In 1993, the signatures of frozen nitrogen and carbon monoxide were found in infrared spectra of Pluto. The following year saw the first spectral detection of methane gas in Pluto's atmosphere – but only in trace amounts. Other constituents that have been detected are hydrogen cyanide (HCN), acetylene (C_2H_2), ethylene (C_2H_4), and ethane (C_2H_6).

New Horizons confirmed that, as on Neptune's moon Triton and Saturn's moon Titan, gaseous nitrogen is the major atmospheric constituent, with a small amount of methane and carbon monoxide. These ices are volatile, even at Pluto's frigid temperature, so they sublimate to create a thin atmosphere. The gases eventually find their way to the planet's colder night side, where they condense.

Ground-based observations showed that the atmosphere is very sparse, and New Horizons confirmed a mean surface pressure of only 11.5 microbars. (One microbar is a millionth of the surface air pressure on Earth.) This is similar to the atmospheric pressure on Triton.

On both Pluto and Triton, the sparse atmosphere is capable of moving material on and above the surface. In the case of Pluto,

Figure 12.28 At least a dozen haze layers in Pluto's atmosphere are shown in this image taken by New Horizons. The haze is thought to be photochemical smog resulting from the action of sunlight on methane and other molecules in Pluto's atmosphere. (Courtesy of NASA/Johns Hopkins University Applied Physics Laboratory/Southwest Research Institute)

images show parallel, linear dunes of methane ice particles on the surface of Sputnik Planitia that have been shaped by winds that may reach 10 m/s.

Studies of stellar occultations showed that Pluto's lower atmosphere (and presumably, the surface) actually warmed by about 2°C in the decade after perihelion in 1989.

This surface warming was probably the result of "thermal lag." Just as the surplus of incoming solar heat over heat radiated to space results in the warmest time of the day on Earth occurring in mid-afternoon rather than at noon, so Pluto's highest temperature occurs after its closest approach to the Sun.

At the same time, atmospheric pressure was found to have trebled between 1988 and 2015, presumably through the sublimation of surface ices (though some cryovolcanic activity cannot be ruled out) (Figure 12.29).

There was also a reduction in the high-level hazes inferred in the 1988 observations, possibly caused by dispersion or condensation on the surface.

New Horizons returned images of numerous blue haze layers in Pluto's atmosphere that extend to altitudes of over 200 km (Figure 12.28). The haze is probably photochemical smog resulting from the action of sunlight on methane and other molecules in Pluto's atmosphere. This produces a complex mixture of hydrocarbons such as acetylene and ethylene, much like the tholins that occur above Titan (see Chapter 9).

These hydrocarbons accumulate into small particles, a fraction of a micrometer in size, which scatter sunlight to create the blue color – the same process that can make haze appear bluish on Earth. As they settle down through the atmosphere, the haze particles form numerous horizontal layers. A few low-lying cloud candidates were also imaged by New Horizons.

One of the big surprises from the New Horizons flyby was that Pluto's atmosphere is much colder than predicted at altitudes above 50 km.[9] It seems that – unlike on any other known world – the hazes have a more significant effect on atmospheric temperature than the gas molecules. Models suggest that the haze particles absorb more sunlight, and therefore generate more atmospheric heating, than the gases. At the same time, the haze particles radiate heat into space.

Pluto cannot continue to be so active for too long. It will inevitably cool as its distance from the Sun increases. Whether this will result in a gradual cooling or a dramatic freeze out and collapse of the atmosphere remains uncertain.

On the other hand, Pluto does sometimes experience major climatic upheavals. Pluto's orbital longitude of perihelion changes over millions of years, resulting in epochs of extreme seasons. The most recent of these took place 900 million years ago, when the overhead Sun at perihelion was at high northern latitude, resulting in a short, but intense, arctic summer.

Models of atmospheric pressure that take into account changes in obliquity and orbit predict possible atmospheric pressures up to hundreds of millibars. These might explain landforms that suggest possible liquids may once have flowed on the surface of Pluto, such as the dendritic, river-like terrain and lake-like features near the edge of Sputnik Planitia.

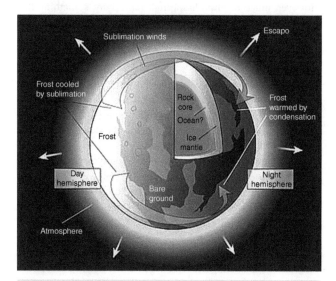

Figure 12.29 For perhaps 20% of its orbital period, Pluto is a dynamic world with active recycling of its atmosphere. Surface ices warmed by the Sun sublimate and are transported by winds to the colder night side where they condense back onto the surface as fresh white frost. With less fresh ice on its Sun-facing side, Pluto becomes slightly darker. Older surfaces, darkened by radiation and space dust absorb more heat, accelerating the warming effect. (Courtesy of NASA/JHU-APL/Southwest Research Institute)

[9] The temperature is about −200°°C instead of the anticipated −170C.

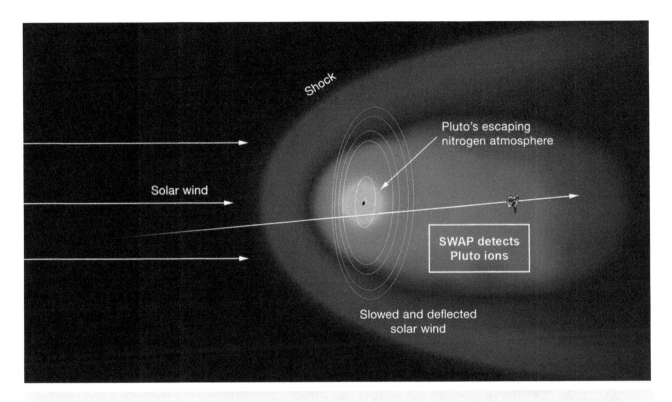

Figure 12.30   Artist's concept of the interaction of the solar wind (the supersonic outflow of electrically charged particles from the Sun) with Pluto's predominantly nitrogen atmosphere. Some of the atmospheric molecules have enough energy to overcome Pluto's weak gravity and escape into space, where they are ionized by solar ultraviolet radiation. As the solar wind encounters the obstacle formed by the ions, it is slowed and diverted (red region), possibly forming a shock wave upstream of Pluto. The ions are "picked up" by the solar wind and carried in its flow past the dwarf planet to form an ion or plasma tail (blue region). Also illustrated are the orbits of Pluto's five moons and the trajectory of the New Horizons spacecraft. (Courtesy of NASA/Johns Hopkins University Applied Physics Laboratory/Southwest Research Institute)

The current picture is of a world whose atmosphere is fed by sublimation of volatile ices for a period of perhaps 50 years around the time of perihelion. As these old, dirty surface ices vaporize, the gases are transported toward the colder, night hemisphere where they are redeposited as bright frost. The residual ices, particularly those containing frozen methane, are exposed as dark areas of hydrocarbons that have been modified by cosmic rays and ultraviolet radiation. In this way, the surface is constantly changing and most of the planet is resurfaced over thousands of years.

Since Pluto has no magnetic field, the dwarf planet has little protection from the solar wind as it flows past at roughly 400 km/s. This might be expected to strip away Pluto's atmosphere, but New Horizons found that only a small amount of methane and hardly any nitrogen is lost from the upper atmosphere.

Since the gases are concentrated so close to the surface, it is more difficult for them to escape from the grip of Pluto's gravity and fly away into space.

However, New Horizons did detect a tenuous cloud of ions and electrons, derived from gas that does escape from Pluto's upper atmosphere, and this cloud was sufficient to deflect the solar wind away from the dwarf planet (see Figure 12.30).

Charon

Pluto's largest satellite was discovered by chance in 1978. Measurements of its size, based on direct imaging with the Hubble Space Telescope and data from the New Horizons flyby, show that its diameter is 1,212 km, approximately half the diameter of Pluto (Figure 12.31). With a density of 1.6 g/cm³, Charon's mass is about one-eighth that of Pluto.

Of all the major satellites in the Solar System, Charon is the largest in comparison with its primary, so Pluto–Charon are sometimes considered as a double planet, rather than a planet with a large moon (such as the Earth–Moon system).

Furthermore, Charon orbits at a distance of only 19,596 km or 17 Pluto radii. Its relatively large mass and close proximity mean that their common center of gravity – or barycenter – actually lies outside Pluto. This situation does not occur in any planetary system.[10] Charon follows a near-circular, synchronous orbit above Pluto's equator, so it is always above the horizon on one hemisphere of Pluto, but never visible from the opposite hemisphere.

Until the New Horizons flyby, little was known about Charon's surface, although some vague light and dark markings were

[10]Typically, a planet has a much bigger mass than any of its satellites. If the objects are far apart, their center of mass is close to the center of the planet, so the planet hardly moves. In the case of the Earth and Moon, the barycenter is about 4,640 km from Earth's center (i.e. inside the planet.)

Figure 12.31   The relative sizes of Earth, Pluto, and Charon. (NASA)

Figure 12.32   NASA's New Horizons spacecraft captured this high-resolution, enhanced-color view of Charon. The image combines blue, red, and infrared images taken by the Ralph/Multispectral Visual Imaging Camera. The reddish material in the north polar region (top) is probably chemically processed methane that escaped from Pluto's atmosphere and froze onto Charon's frigid poles. Note the deep valleys near the equator – evidence of the crust pulling apart long ago. (Courtesy of NASA/Johns Hopkins University Applied Physics Laboratory/Southwest Research Institute)

mapped by studying light curves during the eclipses of 1985–1990 and by direct imaging with the Hubble Space Telescope. Its overall appearance was thought to be gray and bland, with a significantly lower albedo than Pluto (38% compared with 52%).

Even though New Horizons was only able to image one hemisphere – the side facing Pluto – in detail, it found that Charon is far from bland.

The northern part of the encounter hemisphere is extremely rugged, with craters up to 240 km in diameter and 6 km deep. Also visible is a polygonal network of broad troughs up to 10 km deep.

It displays fewer impact craters than expected, indicative of a relatively youthful surface. The southern hemisphere, in particular, has fewer craters than the north and is considerably less rugged. It is possible that a massive resurfacing event, perhaps prompted by the partial or complete freezing of an internal ocean, removed many of the earlier craters. The material that resurfaced the plains may have been a viscous fluid, perhaps an ammonia/water mixture.

Most of the craters that exist today are modest in size. A few have central peaks. The smaller craters are variously associated with bright rays, dark floors, and lobate ejecta which may be layered. At least one small crater, which seems very recent, is surrounded by ammonia-rich ejecta.

Among Charon's most notable features is Caleuche Chasma, an enormous trough at least 350 km long, and up to 14 km deep – more than seven times as deep as the Grand Canyon (Figure 12.33). There is also a series of deep valleys ("chasmata") near the equator and a reddish region around the moon's north pole. (A similar feature was observed around the dark south pole, using light reflected from Pluto.)

The enormous fracture system, which separates the moon's northern and southern plains, is at least 1,600 km long, and it probably extends onto the poorly observed "far side." In places it is more than 50 km wide and reaches depths of 7 km. It is at least four times longer than the Grand Canyon on Earth and up to twice as deep. Theorists suggest that these tectonic fractures were created long ago, when a subsurface ocean gradually froze.

Charon's outer layer is known to be primarily water ice. When the moon was young, this layer would have been warmed by the decay of radioactive elements inside Charon, as well as Charon's own internal heat of formation. Charon could have been warm enough to cause the water ice to melt deep down, creating a subsurface ocean.

However, as the moon subsequently cooled, this ocean would have slowly frozen. Since water ice expands when it freezes, the increase in volume would push the surface outward.

During this process, the satellite's surface area would have expanded around 1%, corresponding to an increase in radius of 3 km. To achieve this, Charon would have required a layer of water, about 35 km deep, which froze from the top down.

An alternative explanation put forward in 2016 suggests that strong tidal encounters may be responsible for the cracks on icy moons such as Charon, as well as Saturn's Dione and Tethys, and Uranus' Ariel. This process would affect moons such as Charon

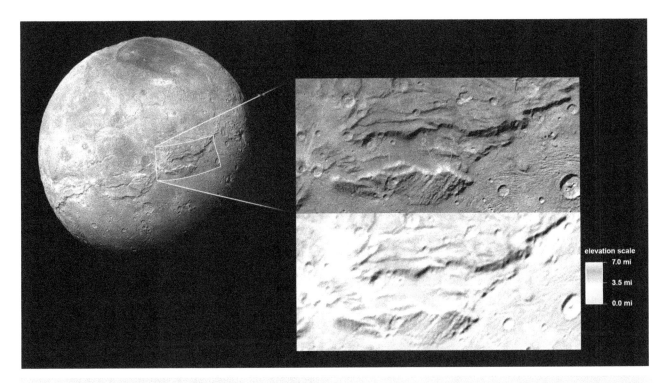

Figure 12.33 The side of Charon viewed by the New Horizons spacecraft is characterized by a huge system of ridges, scarps, and valleys which sometimes reach more than 7 km deep. These images show the widest part of the equatorial chasm system with uplands on either side. Landslides are marked by orange lines and numbers (top). The lower, color-coded, view shows the relative topography. The deep tectonic fractures may have been caused when the crust pulled apart, possibly due to the freezing and expansion of an ancient subsurface ocean. The images measure 386 x 175 km. (Courtesy of NASA/Johns Hopkins University Applied Physics Laboratory/Southwest Research Institute)

when similarly sized worlds passed close by at some undetermined time in the past.

However, none of these theories explain why the huge tectonic belt runs along the equator, as opposed to anywhere else.

Other, unusual features observed on Charon are a few isolated mountains, typically 3–4 km high, which are surrounded by depressed "moats" up to 2 km deep. The sunken mountains may be the result of flexing in Charon's lithosphere or the flow of viscous cryovolcanic materials.

A large depression at the north pole is coated with reddish material. The polar coloring is about 450 km across, with a darker inner core some 275 km across (Figure 12.33). It seems to be a thin surface coating rather than a thick layer of material that welled up from below.

The leading theory is that the material originates from methane gas that escapes from Pluto's atmosphere and is captured by the moon's gravity. The methane molecules from Pluto bounce around on Charon's surface until they either escape back into space or land on the extremely cold pole, where they freeze, forming a thin coating of methane frost that lasts until sunlight returns in the spring.

The unusual process is made possible by Pluto and Charon's 248-year orbit around the Sun. This causes extreme weather at Charon's poles, where 100 years of continuous sunlight

alternate with another century of continuous darkness. Surface temperatures during the prolonged winters dip to –257°C, cold enough to turn methane gas into solid ice.

When springtime arrives on Charon, the returning sunlight triggers conversion of the frozen methane back into gas. However, while the methane ice quickly sublimates away, the heavier hydrocarbons created from this evaporative process remain on the surface.

Chemical processing by ultraviolet light from the Sun transforms the methane into heavier hydrocarbons. Further irradiation by the Sun then changes them into reddish hydrocarbons called tholins that slowly accumulate on Charon's poles over millions of years.

New Horizons confirmed that Charon's surface is covered by water ice away from the poles. The water ice is in a crystalline (as opposed to amorphous) state.[11]

Water ice found on satellites of Jupiter, Saturn, and Uranus is always crystalline, because amorphous ice rapidly crystallizes if the temperature is above –153°C. However, on Charon, where the surface temperature is much lower, this phase transition should not take place. Instead, crystalline ice should gradually become amorphous under the constant bombardment of ultraviolet light and energetic particles. This radiation breaks hydrogen bonds in

[11] Crystalline water ice has since been found on other large Kuiper Belt objects, notably Quaoar.

the ice that subsequently reform, but not in their original crystalline positions.

The presence of crystalline ice indicates that Charon's surface is fresh or that its temperature has exceeded $-193°C$ in the recent past.

Unlike Pluto, it has no deposits of nitrogen, methane, or carbon dioxide ice. With an average temperature of $-239°C$, Charon is too small and warm to prevent these volatile materials from sublimating and escaping into space.

On the other hand, New Horizons detected ammonia in low concentrations across the visible surface, as well as exposures of ammonia hydrates (which form if the ammonia molecules lie within a water matrix), particularly near some fresh-looking, rayed impact craters. These contaminants can lower the melting temperature of water ice by around $100°C$, making water ice more ductile.

Ammonia is quite easily destroyed by radiation, so how could it survive on Charon? And why does it exist on Charon, but not on nearby Pluto? Perhaps comets have delivered interstellar ammonia ice to Charon, or maybe there is a reservoir of ammonia-rich material hidden beneath the surface.

There certainly seems to be evidence that much of Charon was resurfaced early in its history. The cratered plains that cover much of its visible surface are generally smooth, particularly south of the equator – evidence of outflows of icy material perhaps 4 billion years ago. Smaller, less cratered areas suggest that fairly viscous, nearly frozen brine may have locally erupted from the interior in several episodes of cryovolcanism. No evidence of an atmosphere has been found.

Since Charon is smaller and less dense than Pluto, its internal structure should be easier to model. If internal heating has allowed Charon to differentiate, it should be composed of about 60% rock in a large core, possibly overlain by a thin, ammonia-rich liquid layer. Above this may be a solid mantle of water ice, with an undifferentiated surface crust of rock, ice, and ammonia hydrate. However, there is also the possibility that Charon is undifferentiated, with a roughly uniform mixture of rock and ice throughout.

Pluto's Smaller Satellites

On May 15, 2005, during a search with the Hubble Space Telescope, two small satellites (later named Hydra and Nix) were found orbiting Pluto. Since then, another two satellites, Kerberos and Styx, have been discovered, also in fairly circular orbits.

They are all outside Charon's orbit: the innermost, Styx, is a little more than twice Charon's distance, while the outermost, Hydra, is a little more than three Charon distances from Pluto (Figure 12.34).

All of these satellites are in a resonant dance with Charon. Styx, Nix, Kerberos, and Hydra have orbital periods that are almost exactly 3, 4, 5, and 6 times longer than Charon's.

All of these moons orbit in the same plane as Charon, i.e. above Pluto's equator. Styx, takes about 20 days to orbit Pluto, while the outermost, Hydra, completes one orbit roughly every 38 days.

The moons spin much faster than their orbital periods. The outermost satellite, Hydra, rotates around 89 times during each 38-day orbit, and it is possible that this faster rotation may be a result of past collisions.

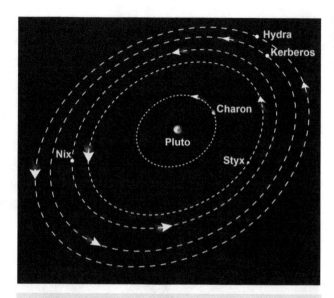

Figure 12.34   The orbits of Pluto's moons: Charon, Nix, Styx, Kerberos, and Hydra. The orbits are nearly circular and tilted to the line of sight, so they appear elongated. They orbit around the barycenter (center of mass) of the system, which lies outside Pluto. The orbits are not to scale. (Courtesy of NASA, ESA, and L. Frattare/STScI)

Despite their regular orbital distribution, the small satellites of Pluto do not keep one face pointed toward their central planet – unlike many moons in the Solar System, Whereas Charon has been slowed into synchronous rotation by tidal interaction with Pluto, the other moons tumble chaotically, so that their axial tilts and periods of rotation vary greatly over short timescales.

All four are rolling on their sides, and their rotation axes seem to be close to their orbital planes. Nix has a spin axis that seems to be precessing (wobbling) randomly. Its pole was tilted 132 degrees, meaning that it rotates "backwards" (retrograde). Furthermore, its rotation rate was found to have increased by 10% since its discovery.

Like Charon, they are neutrally colored. Since Charon's surface is known to consist primarily of crystalline water ice, the similar color of Hydra and Nix and their high albedos indicate that their surfaces are also composed of water ice (Figure 12.35). However, New Horizons did detect a sizeable region with a distinctive red tint that surrounded a large impact crater on Nix.

The Origin of the Pluto System

Ever since its discovery, astronomers have been debating the origin of Pluto. A popular theory that attempted to explain its eccentric, inclined orbit, suggested that Pluto was an escaped satellite of Neptune. However, we now know that their orbital resonance means that the two worlds can never come anywhere near each other. It is impossible for Pluto to have originated in orbit around a planet which it can never approach.

One current theory suggests that Pluto, along with the other Kuiper Belt objects, moved into its current orbit more than four

Figure 12.35   A composite view of Pluto's small moons, shown to scale, using images taken by New Horizons. At the bottom is the limb of Charon, which has a diameter of 1,212 km. Nix and Hydra are respectively 54 and 43 km across in their longest dimension. Kerberos and Styx are 12 and 10 km across in their longest dimensions. All of the small moons have highly elongated shapes, and at least two of them appear double-lobed, characteristics thought to be typical of small bodies in the Kuiper Belt. Impact craters are also visible, particularly on Nix. (Courtesy of NASA/Johns Hopkins University/APL/Southwest Research Institute)

Figure 12.36   A computer simulation showing how Pluto and Charon may have formed. After a very oblique initial impact (A), the two objects separate (B and C). During this period the smaller impactor receives a net torque from the distorted target object. After another, even more grazing encounter (D), more of the impactor is accreted onto Pluto. The remainder accretes into a moon containing 12% of the central object's mass. This is torqued once more by the ellipsoidal target (D and E), eventually forming a moon with a stable orbit (F). (Robin Canup/SwRI and AAS)

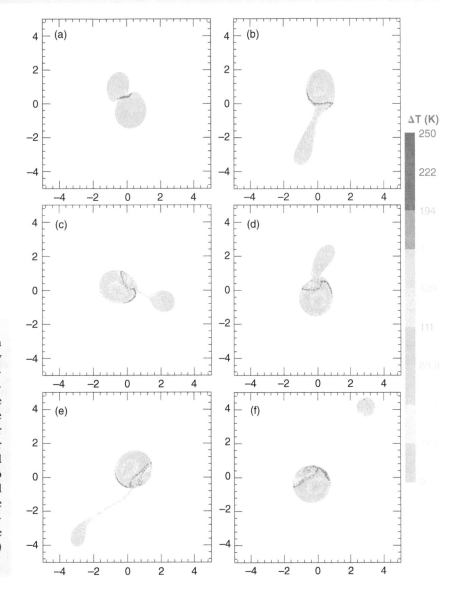

Box 12.6   Planet Nine?

In recent years, some astronomers have published papers which postulate the existence of another planet far beyond the orbit of Neptune. Using mathematical modeling and computer simulations, they suggest that the new object, dubbed Planet Nine, has a mass about 5 times that of Earth and follows a fairly elliptical, inclined orbit that may be 20–30 times farther from the Sun on average than Neptune. This new planet would take perhaps 10,000 years to make one full orbit around the Sun.

Although it has not yet been detected, the researchers suggest that its existence is indicated by a number of indirect clues, mainly associated with its gravitational footprints.

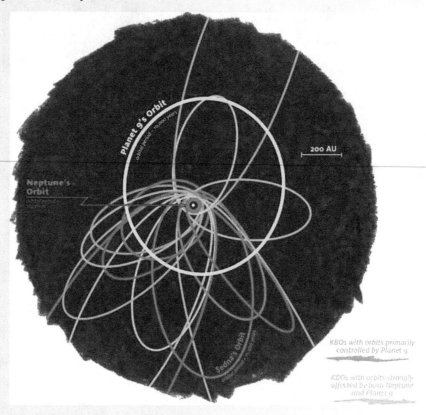

Figure 12.37   The extreme orbits of at least six trans-Neptunian objects (magenta) are aligned in one direction, a configuration which may be explained by the presence of an undiscovered Planet Nine (yellow). Some objects orbits (green) may be influenced by both Neptune and Planet Nine. (James Tuttle Keane, Caltech)

Five different lines of observational evidence pointing to the existence of Planet Nine have been put forward.

1. At least six known objects in the Kuiper Belt have been found to have extreme elliptical orbits that are pointing in the same direction.
2. These orbits also are tilted the same amount, inclined about 30 degrees away from the ecliptic plane.
3. Computer simulations of the Solar System which include Planet Nine show there should be objects in orbits tilted by around 90 degrees to the ecliptic plane. Five such objects were already known to follow such orbits.
4. Planet Nine could have tilted the orbits of the planets during the last 4.5 billion years. This could explain a longstanding mystery: why is the average plane in which the planets orbit tilted about 6 degrees compared to the Sun's equator? Over long periods of time, a massive Planet Nine could make the plane of the entire Solar System precess or wobble.
5. Planet Nine's gravitational influence could explain why some objects from the Kuiper Belt orbit in the opposite direction from everything else in the Solar System. Planet Nine could move the KBOs out of the ecliptic plane, enabling them to be scattered inward by Neptune.

If it exists, where did Planet Nine come from? One idea is that it was somehow ejected from a region closer to the Sun, or it may be a wandering planet captured from interstellar space by the Sun's gravity.

Supporters of Planet Nine's existence also note that, over recent decades, surveys of planets around other stars in our galaxy have found the most common types to be "super Earths" and "mini Neptunes" (see Chapter 14). Yet these planets are conspicuously absent from our Solar System. Weighing in at roughly 10 times Earth's mass, the proposed Planet Nine would make a good fit. Perhaps Planet Nine could be our missing super Earth.

billion years ago, during the late stages of planetary accretion. At that time, the giant planets were undergoing significant orbital migration as a result of encounters with large, residual planetesimals. As Neptune moved outwards, a small planetesimal such as Pluto, which was initially following a roughly circular orbit, could have been captured into the 3:2 resonance, causing its orbital eccentricity to rise rapidly until it reached its current Neptune-crossing orbit.

This theory has been questioned, however, because some 30% of the objects in the so-called "Cold Classical Kuiper Belt" occur in widely separated binaries which would have been disrupted during their relocation by Neptune. This suggests that they formed near their present locations and have remained largely undisturbed over the age of the Solar System (see Chapter 1).

The stable, circular orbit of Charon suggests that the two objects have been linked for a very long time. Two leading theories to explain this pairing have been put forward. One favors the idea that, like Pluto, Charon was a member of the icy debris that populated the Kuiper Belt beyond Neptune's orbit. However, computer simulations show that gravitational capture is extremely unlikely.

Today, it is widely believed that the Pluto–Charon system formed in a manner similar to the Earth and Moon – through a giant impact over four billion years ago (Figure 12.36). This scenario envisages Pluto capturing the remains of the impactor – the proto-Charon – into orbit. After the collision, tidal interaction between the pair slowed down Pluto's spin and pushed Charon's orbit outward.

The recent discovery of other, small, satellites in normal orbits that are in resonance with Charon further supports this theory. Simulations show that any remaining debris from the collision could have coalesced to form small satellites in chaotic orbits. Some of these would have been rapidly moved outward if they approached too close to Charon. Meanwhile, collisions among the small satellites would also have changed their orbits, keeping some of them away from Charon.

This scenario envisages small satellites colliding and shattering, then rebuilding, as tidal forces caused Charon to drift to its current orbit. The overall result would be that the outer satellite system we see today represents the last of many previous generations of temporary satellites.

Close-up images taken by the New Horizons spacecraft show that at least two of the satellites, Hydra and Kerberos, may be the result of mergers between still smaller moons – support for the mutual collision theory.

Questions

- Why did Kenneth Edgeworth and Gerard Kuiper believe that a large population of objects existed beyond the orbit of Neptune?
- How and when may the Kuiper Belt have been created?
- Why did it take so long to find the first members of the Kuiper Belt?
- What are the main categories of Kuiper Belt object (KBO)?
- What are the major similarities and differences between the main asteroid belt and the Kuiper Belt?
- Suggest reasons for the differences in size, density, color, and orbit between KBOs.
- How may binary KBOs have come about?
- What is a dwarf planet? Why has been Pluto been "demoted" to a dwarf planet?
- Describe and attempt to explain the differences between Pluto and Charon.
- Describe and attempt to explain the main similarities and differences between Pluto and the other recognized dwarf planets.
- Compare the atmospheres of Pluto, Triton and Titan.
- Why did mutual eclipses of Pluto and Charon take place 1985–1990? What new information was provided by these eclipses?
- What is the most likely explanation for the origin of the Pluto–Charon system?

THIRTEEN

Comets, Asteroids, and Meteorites

Once the Solar System settled down after the traumas of its birth and the final late asteroid/comet bombardment, the formation of the Sun, planets, and planetary satellites was largely complete. Yet, even today, the Sun's domain contains untold billions of pieces of planetary debris, much of it left over from the processes of accretion and collision that occurred more than four billion years ago. This debris has traditionally been classified as rocky material, which dominates the main asteroid belt and the inner Solar System, and icy material which pervades the cold, outer regions. We are not just disinterested observers of these ancient fragments. Earth is bombarded each day by interplanetary dust and meteorites. On a longer timescale, our planet periodically suffers large-scale destruction and even mass extinctions as the result of impacts by comets and asteroids.

By the late 18th century, the known population of the Solar System had risen to seven planets and 14 satellites. However, astronomers had become convinced that there must be another planet between the orbits of Mars and Jupiter. Their belief was based on a numerical relationship, known as Bode's Law, which seemed able to predict the distances of planets from the Sun.

The so-called Law was an arithmetic progression that seemed to predict the relative distances of the planets from the Sun. In the numerical series 0, 3, 6, 12, 24 ⋯ each successive number after 3 was double the previous number. After adding 4 to each number and dividing by 10, this mathematical sequence provided a good approximation to the actual distances of the known planets in astronomical units (AU).

Its apparent predictive power was reinforced in 1781, when William Herschel discovered Uranus at 19.2 AU from the Sun, very close to the Law's prediction of 19.6. The only major flaw was a missing planet in the numerical gap between Mars and Jupiter.

Among the believers in Bode's Law was a group of six astronomers, headed by Johann Schröter, who called themselves the "celestial police." In 1800, they prepared a systematic search for the missing planet along the ecliptic plane. However, before the survey could get under way, some unexpected news arrived from Sicily. On January 1, 1801, Giuseppe Piazzi, director of the Palermo Observatory, had discovered a strange object in the constellation of Taurus.

At first, Piazzi thought he had found a comet, but the absence of a gaseous coma caused him to have some doubts. Unfortunately, while Piazzi was suffering from a serious illness, the mysterious object disappeared in the Sun's glare. With limited orbital information to go on, frustrated astronomers all over Europe began to search for the newcomer.

The breakthrough came when a brilliant young mathematician, Carl Gauss, found a way of predicting planetary positions from a limited set of observations. Using his calculations, Franz Xavier von Zach, one of Schröter's group, rediscovered the faint object on December 31, only half a degree from where Gauss had predicted.

At Piazzi's suggestion, the name Ceres – after the Roman goddess who was the patron of Sicily – was soon adopted. The object was found to follow a nearly circular path between Mars and Jupiter at a distance of 2.77 AU – an almost exact fit with the

Exploring the Solar System, Second Edition. Peter Bond.
© 2020 John Wiley & Sons Ltd. Published 2020 by John Wiley & Sons Ltd.
Companion Website: www.wiley.com/go/Bond-Solar-System2e

empty 2.8 AU slot in Bode's Law. The only concern was that the 8th-magnitude object was obviously very small – less than 1,000 km across.

Then, on March 28, 1802, Wilhelm Olbers (another member of the "celestial police") found a second minor planet at roughly the same distance. Pallas, as it was named, was smaller and fainter than Ceres, and traveled in a fairly eccentric, inclined orbit, though it did approach Ceres from time to time.

It was already apparent that these objects were very different from their larger neighbors. William Herschel's suggestion that they be called "asteroids," because of their star-like appearance and absence of visible disks in even the largest telescopes, was quickly accepted.

Two more minor planets, Juno and Vesta, were discovered in 1804 and 1807. Juno was notably smaller than Pallas, while Vesta did not even approach the other known asteroids, and was sometimes bright enough to be visible with the unaided eye.

The scattered family seemed complete for nearly 40 years until Astraea was found in 1845, followed by another three objects in 1847. Since then, at least one asteroid has been discovered every year, with a marked leap in the rate of discovery after the introduction of photographic techniques in the late 19th century. The first asteroid to be discovered photographically was 323 Brucia, found by Max Wolf on December 20, 1891.

Unfortunately, unwanted asteroid tracks showed up on exposures at the most inopportune times, causing one astronomer to dub them "the vermin of the skies." Most observers lost interest in the thousands of small, seemingly unimportant, chunks of rock that occupied the broad zone between the four terrestrial planets and their giant, gaseous companions.

The vast majority of asteroids orbit the Sun in the main belt, between Mars and Jupiter (Figure 13.1). This population includes more than 200 asteroids larger than 100 km in diameter. In total, the asteroid belt is estimated to contain between 1.1 million and 1.9 million asteroids larger than 1 km in diameter, plus millions of smaller ones.

The discovery of sizeable near-Earth objects (NEOs), such as 433 Eros, caused a stir, when it was realized that asteroids sometimes paid a visit to the inner Solar System, even threatening catastrophic impacts with our planet.

The desire to track down and characterize such potentially dangerous near-Earth marauders has led to the introduction of new instruments and observing techniques. Although amateur astronomers continue to play a role, by far the majority of asteroids have been discovered since the mid-1990s using automated search programs. Most of these are located in the United States and funding is provided by NASA, the US Air Force, and various astronomical observatories. The preponderance of observing sites

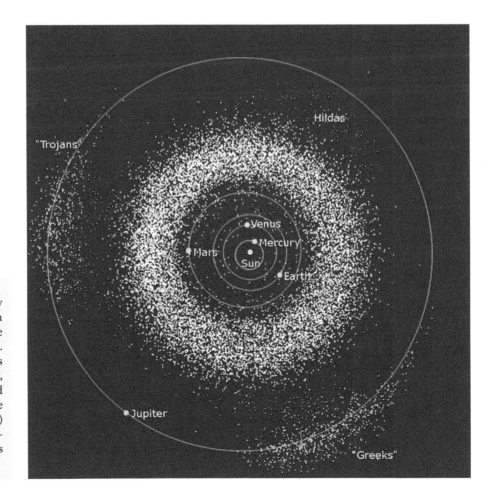

Figure 13.1 The vast majority of asteroids travel around the Sun in the "main belt," between the orbits of Mars and Jupiter (white). Two groups of Trojan asteroids (green) lie along Jupiter's orbit, centered 60° ahead and behind the giant planet. Also shown here are the Hilda asteroids (orange) and some of the near-Earth asteroids which approach or cross Earth's orbit. (Wikipedia)

Box 13.1   Naming Asteroids

After a particular asteroid has been observed on at least two nights, and it cannot immediately be matched to any known object, it is given a provisional designation by the Minor Planet Center (MPC) at the Smithsonian Astrophysical Observatory. (Note that it may take many weeks or months to determine its orbit with reasonable accuracy.) This designation is based on the date of its discovery and its place in the order of discoveries for that month and year.

The first four numbers give its year of discovery. The fifth character – a letter of the alphabet – represents the two-week period of the discovery month: A equals the first two weeks in January (1–15), B marks the third and fourth weeks in January (16–31), C the first two weeks of February, and so on. The letter I is not used and Z is not needed.

The sixth character (also a letter) tells the order of discovery during each two-week period. Hence, asteroid 1999 DA is the first asteroid discovered during the second half of February 1999. If there are more than 25 discoveries in any half-month period, the second letter is recycled and the number "1" (usually written in subscript) is added. When the number of discoveries in that fortnight exceeds 50, the number "2" is used, so 1999 DA$_1$ is the 26th asteroid found in that two-week period and 1999 DA$_2$ is the 51st.

Today, so many asteroids are being discovered that frequent recycling of the alphabet is the norm, e.g. 2008 TN$_{166}$. A lot of asteroids are reported several times in different epochs, so they may receive multiple provisional designations.

Objects with well-determined orbits are given a catalog number, and may eventually receive a name. The MPC assigns numbers in the order that observations are received and cataloged. Since the number generally gives an indication of how long ago the object was discovered, a lower number often indicates a discovery many decades ago. There are a few exceptions, such as 69230 Hermes, which was originally discovered in 1937 when it passed close to the Earth, but then lost. Only after its rediscovery in 2003 could its orbit be determined and a number assigned. Before this, it was known by its provisional designation, 1937 UB.

In most cases, the name is chosen by the astronomer who first provides enough observations to calculate the orbit with precision, NOT the first person to see the asteroid. Following the established tradition, the early discoveries were named after characters from classical mythology. Hence, the first asteroid to be discovered was designated 1 Ceres, and the second was 2 Pallas.

Today, the International Astronomical Union's Committee on Small Body Nomenclature is responsible for assigning names to asteroids. In most cases, the name proposed by the discoverer is accepted. Although controversial political, religious, or military figures such as Hitler and Stalin are not allowed, there have been some interesting – and unusual – choices. For example, asteroid 6042 is named CheshireCat and 1625 is named NORC, after the Naval Ordnance Research Calculator. Among those honored are popular singers, science fiction characters, social clubs, shipping lines, and foods.

Sometimes the IAU committee takes the initiative for special numbers. Asteroid 1000 was named Piazzia, in honor of the discoverer of Ceres. All Trans-Neptunian Objects are named after deities associated with creation myths. Thus, 50,000 Quaoar, one of the largest known members of the Kuiper Belt, is named after a Native American creation deity.

By October 2019 there were 541,132 numbered minor planets out of a total of 789,084 observed bodies, but only a small percentage of these have been given names.[1]

in the northern hemisphere means that coverage of the night sky is far from ideal.

The most successful ground-based effort has been the Lincoln Near-Earth Asteroid Research (LINEAR) program, which is funded by NASA and uses the Air Force Space Command's Space Surveillance Telescope located in Socorro, New Mexico. More than 50% of the known asteroids in our Solar System have been discovered through this program.

Another key search facility is the 1.8 m Pan-STARRS 1 telescope, located on Mt. Haleakala in Maui, which began full operation in May 2010. It has an enormous field of view – six times the width of the full Moon – and a huge 1.4 gigapixel camera that makes it the world's most powerful survey telescope.

Searches by spacecraft are a more expensive, but highly effective alternative. During a full survey of the sky in 2010, NASA's Wide-field Infrared Survey Explorer (WISE) spacecraft detected more than 150,000 main belt asteroids, including 33,000 which had never before been detected.

In 1998, NASA established a goal to discover 90% of the NEOs larger than one kilometer in diameter and in 2005, Congress extended that goal to include 90% of the NEOs larger than 140 meters. There are thought to be about 1,000 NEAs larger than one kilometer and roughly 25,000 larger than 140 meters.

The NEO surveys have been extremely successful, finding more than 90% of the near-Earth asteroids (NEAs) larger than one kilometer and numerous NEOs with diameters of more than 140 meters. In November 2017, NASA reported that almost 7,900 large NEAs had been discovered and cataloged. Each year, NASA continues to discover about 500 asteroids larger than 140 meters in size, using mostly ground-based telescopes (see Near-Earth Asteroids and The Impact Threat to Earth).

[1] http://www.minorplanetcenter.net/iau/lists/NumberedMPs.html

Asteroid Origins

Where did the asteroids come from? The realization that most of them are confined to a broad belt between Mars and Jupiter led to speculation that a small planet had once occupied that zone, then somehow been destroyed, perhaps by an explosion or a catastrophic collision with another object.

As the population of the main belt was cataloged in greater detail it became clear that the mass of the entire community – well over a million objects with diameters of 1 km or more – was only about 5% of the Moon's mass. Any single ancestor must have been far too small to be classified as a planet.

Furthermore, most of the main belt's mass is actually accounted for by the three largest asteroids: Ceres, Pallas, and Vesta. Spectral studies show that these objects are very different in composition and bulk density, so they are unlikely to have been derived from a single parent body.

Today, the consensus is that the vast majority of the bodies orbiting in the main belt are primordial debris left over from the formation of the planets, some 4.5 billion years ago.[2] They were formed through collisional accretion in the solar nebula, probably in the vicinity of the present main belt (see Chapter 1).

Figure 13.2 Most asteroids are so small and far away that they are visible as mere points of light. However, during a long-exposure photograph, an asteroid's motion results in a linear streak against the background of "fixed" stars. This Hubble Space Telescope image shows a fairly bright asteroid, with a visual magnitude of 18.7, in the constellation Centaurus. The blue trail was caused by the motion of a 2 km-diameter object in the main asteroid belt, 250 million km from the Sun. (Courtesy of R. Evans and K. Stapelfeldt, NASA-JPL)

Despite following the first steps towards planetary formation, the rocky planetesimals were prevented from accumulating into a full-grown planet because of the gravitational influence of massive Jupiter. Over the eons, the nearby giant has kept the main belt objects dynamically agitated so that collisions between them often result in their erosion and destruction, rather than the gradual agglomeration of a single large body. (Nevertheless, a number of asteroids have been shown to have very low densities, indicating that they are rubble piles with large empty spaces in their interiors. Others, such as 4769 Castalia and 4179 Toutatis, appear to be touching binaries that have come together through gentle collisions.)

Spectral studies show that the larger members of the main belt are broadly separated into zones that may be associated with different conditions in the nebula: most of the stony S-type asteroids lie in the inner regions, while the carbonacous C types are more numerous further from the Sun. This suggests that the temperature gradient played a key role, with volatile-rich objects forming in the cooler reaches of the main belt (see Box 13.2).

However, this neat pattern breaks down for smaller asteroids. Increasing numbers of "rogue" asteroids have been found, leading to the conclusion that the main belt is more compositionally diverse than previously thought. In recent years, all of the asteroid types have been found to exist throughout the main belt.

The new asteroid map suggests that the early Solar System may have undergone dramatic changes before the planets assumed their current alignment. For instance, Jupiter may have migrated closer to the Sun, dragging with it a host of asteroids that originally formed in the colder regions of the Solar System, before moving back out to its current position. At the same time, Jupiter's migration may have shaken up the orbits of the more rocky asteroids, scattering them outward.

Not all of the differences in composition and physical characteristics are accounted for by their formation in different parts of the solar nebula. Some appear to be the result of melting in the cores of newly forming asteroids. This melting was due to high-energy collisions and the decay of radioactive isotopes, such as aluminium-26 (^{26}Al), that heated the interiors of the larger rocky objects. Confirmation comes from the presence of ^{26}Al in ancient meteorites known as carbonaceous chondrites.

Many asteroids may have developed a molten core with a mantle and a crust – rather like miniature Earths. Others, such as Vesta, were probably totally molten, with short-lived surface oceans of molten rock, or magma. The variety of meteorites found on Earth shows that they originated from different regions of these parent bodies after they were partially or completely shattered by subsequent collisions.

Why do other asteroids, such as Ceres, have very low densities – which suggest that they are 25–30% water ice – and not show any evidence of melting or internal differentiation? Since aluminium-26 has a half-life of only 730,000 years, it may be that too little remained to warm any asteroids that formed later in the aggregation process.[3]

Another possibility is that Ceres, along with many smaller icy objects, was displaced from the outer Solar System when Uranus

---

[2] Some asteroids may be the nuclei of dead comets, e.g. 3200 Phaethon.

[3] The half-life is the period it takes for half of the atoms in a radioactive isotope to decay. Aluminium-26 decays into magnesium-26.

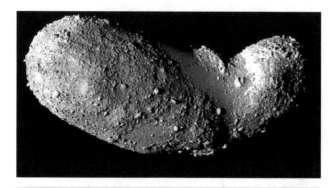

Figure 13.3 A view of near-Earth asteroid Itokawa taken by Japan's Hayabusa orbiter. Itokawa measures only 0.5 x 0.2 x 0.3 km and has an unusual shape, with a narrow waist and variable terrain that includes smooth plains, boulder-covered regions, and no obvious impact craters. (JAXA)

and Neptune were rampaging along chaotic orbits around 3.9 billion years ago (see Chapter 1). As a result, they ended up joining the rocky asteroids that were born in the main belt.

Impacts and Rubble

The most obvious characteristics of nearly all asteroids are their rocky appearance, irregular shape, and numerous craters – evidence of numerous impacts since they formed 4.5 billion years ago. The largest asteroids possess sufficient gravity to have retained roughly spherical shapes.

Only 15 asteroids have been seen at close quarters. Some of these, such as Mathilde, display large impact features that dominate their topography. These impacts must have been so energetic that they almost destroyed the asteroid. However, most of the asteroids imaged by spacecraft display numerous small and medium-sized craters rather than huge impact features.

The presence of many sizeable craters indicates ancient surfaces, the exceptions being Ryugu and Gaspra, whose craters are relatively sparse and small. Interestingly, although the average asteroid's gravity is very weak, many of these seem to be secondary craters caused by impact ejecta falling back onto the surface.

Close-up views of objects such as Bennu, Eros, and Itokawa also show numerous boulders, further evidence of impact debris being scattered over the surface (Figure 13.3). Occasionally, these have been disturbed and roll down the slopes, leaving trails in the dusty regolith.

Fine, loose material also seems widespread on many asteroids (including the Martian moon Deimos, which may be a captured main belt asteroid). Itokawa, for example, was found to be covered with unconsolidated, millimeter-sized particles and larger "gravels." This material may be deep enough to fill and cover small depressions, giving them a generally smooth, rounded appearance (see Box 13.3).

Most surprising of all is the presence of localized plains that cover 20% of Itokawa's surface and are formed from the finest granules. This was a surprise because impact ejecta on a small asteroid were expected to spread globally over its surface, resulting in continuous regolith.

One possibility is that the unconsolidated gravels have globally migrated and segregated due to fluidization caused by vibrations induced by numerous impacts of small meteoroids. When gravel is vibrated, it can be fluidized and behave as granular fluid. The most popular phenomenon related to this is the "Brazil nut effect," which causes the biggest particles to end up on the surface when granular material is shaken.

This theory is supported by close-up images which show that Itokawa's gravels are generally aligned in the same direction – exactly coinciding with the directions of local gravitational slopes. When the asteroid is shaken by an impact with a small meteoroid, the fine regolith shifts position, slowly migrating downslope and collecting in the nearest hollow. Over time, the larger boulders mixed in with the finer material are left stranded on the higher slopes, forming an area of rough terrain. Such granular sorting processes may be a major resurfacing mechanism for all small asteroids possessing regolith.

Some asteroids, including Itokawa and Mathilde, have very low bulk densities (Box 13.6). For example, Mathilde seems to be composed of material similar to that found in carbonaceous chondrite meteorites, but calculations of its volume and mass (derived from spacecraft tracking data) indicate a bulk density of only 1.3 g/cm^3, about half that of the meteorites. This suggests that Mathilde may have a rubble pile interior characterized by large pores or empty spaces.

It seems that such objects were once shattered but slowly reformed through their mutual gravitational attraction and low-velocity collisions. Today, the loosely packed rubble is barely held together by the asteroid's gravity. If an object collided with Itokawa, it would probably resemble a rock landing in a bucket of sand. The signs of impacts with small meteorites are erased as the rubble shifts after the impact, so there are very few visible craters on Itokawa.

Eros is a much larger representative of the near-Earth asteroid population, but it, too, has a surprising lack of small craters and some remarkably smooth patches (Box 13.6). Across nearly 40% of its surface, all craters up to about 0.5 km wide have been erased. Instead of the expected 400 craters per sq. km with diameters of only 20 m, it averages only about 40.

On the other hand, one of its most striking features is a fairly recent, 5.3 km-diameter impact crater. Much of the surface is covered with piles of rubble, blocks of ejecta between 30 and 100 m across. However, they are not uniformly distributed, and the NEAR spacecraft revealed some smooth, flat areas, dubbed "ponds" by mission scientists. As with Itokawa, these are attributed to the downslope movement of fine material due to seismic activity generated by meteoroid impacts.

There were few visible albedo variations, except on the inner walls of some craters where mass movement of loose material was taking place. Spectra indicate that it is probably similar in composition to ordinary chondrite meteorites. A mean density of 2.67 g/cm^3 implies an internal porosity of 10–30%, and analysis suggests that Eros has been internally fractured by large impacts. Supporting evidence is seen in a series of grooves and ridges that run across the asteroid's surface both globally and regionally.

Box 13.2   Asteroid Classes

Asteroids are classified according to their spectral and physical characteristics. Attempts to link these with the various groups of meteorites (see Meteorites) have not always been successful. The overall composition of the larger main belt asteroids also varies. Stony, S-type asteroids are more common near the inner edge, while darker, carbon-rich C-types are more numerous further out. The most common classes are C, S, and M.

- **C (carbonaceous).** The most numerous group, accounting for 75% of the main belt population. The proportion increases from about 10% at 2.2 AU to about 80% at 3 AU. They are dark gray and reflect very little sunlight (albedos 3–5 %). Their flat, generally featureless, spectra resemble those of carbonaceous meteorites, a link that was confirmed when the NEAR Shoemaker spacecraft landed on Eros.
- **S (stony or "silicaceous").** About 15% of the total population, they are most numerous in the inner main belt, accounting for some 60% of the population at 2.2 AU, decreasing to 15% at 3 AU. They are reddish and moderately reflective (albedos 15–25%). Their spectra resemble those of metal-bearing meteorites known as chondrites. Notable examples are 3 Juno and 5 Astraea. Stony ordinary chondrites are the most numerous group of meteorites arriving on Earth, comprising about 87% of all finds (see Meteorites).
- **M (metallic).** The third-most populous. They have moderate albedos and are linked to iron-nickel meteorites. They are thought to represent the remnants of the metal-rich cores of ancient asteroids that were large and hot enough to differentiate internally. 16 Psyche is thought to be almost pure iron-nickel alloy.
- **E ("enstatite").** These are fairly rare. They have high albedos (30% or more) and may contain silicates with a fairly high amount of the mineral enstatite. They are thought to be the source of aubrite meteorites. E-type asteroids include 64 Angelina, 214 Aschera, and 620 Drakonia. The first example to be observed from close range was 2867 Steins, imaged by Rosetta in October 2008.
- **V.** Very rare, they are highly reflective and were once molten. They are associated with basaltic meteorites known as eucrites. The main example is 4 Vesta, which is probably the parent body of most of the eucrites. However, several V-type asteroids do not belong to the Vesta family, including 1459 Magnya, the first basaltic object detected in the outer asteroid belt.

Many other classes and subdivisions are recognized, though they are often given different designations since a number of classification schemes based on different criteria are in use. For example, Ceres belongs to class G, a subcategory of C-class asteroids which is distinguished by a very strong ultraviolet absorption feature at wavelengths shorter than 0.4 microns due to water ice on its surface.

Double-lobed asteroids that resemble a peanut or hourglass are also quite common. In the case of Itokawa, it has been surmised that the asteroid's head-and-body structure might have arisen as the rubble pile shifted in response to impacts. Alternatively, the two rounded segments may once have been separate rubble piles that slowly collided and gradually merged.

Asteroid Families and Kirkwood Gaps

The orbits of more than half a million asteroids have been cataloged since the discovery of Ceres, and the number is rising inexorably, with thousands more reported every month (Figure 13.4). Analysis of the computed orbits shows that the vast majority lie in the main belt between 2.2 and 3.3 AU, so they are generally closer to Mars than to Jupiter.

About half of all main belt asteroids belong to families that have similar orbital characteristics. Similarities in their reflection spectra indicate that each family is often made up of fragments of a particular parent body that was shattered by a catastrophic collision billions of years ago. Certain types of meteorites have also been matched to individual large asteroids and their families (see Box 13.2).

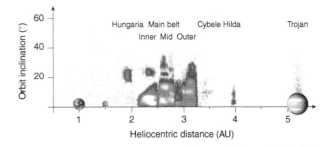

Figure 13.4   The distribution of asteroids between Earth and Jupiter, showing their orbital inclinations and number density. Yellow represents the highest number density, blue the lowest. The structure of the belt is divided by unstable regions, known as Kirkwood gaps, seen most prominently at 2.5 and 2.8 AU. (F.E. DeMeo & B. Carry)

The existence of such groups was first recognized by Japanese astronomer Kiyotsugo Hirayama (1874–1943), so they are now known as Hirayama families. Approximately 100 such families have now been identified, each being named after its largest member. The most populous examples, such as the Flora, Eos, Themis, and Koronis groups, have hundreds of members.

Jupiter is one of the major drivers behind the distribution of main belt asteroids and families. Some regions of the asteroid belt – called Kirkwood gaps, after U.S. astronomer and mathematician Daniel Kirkwood (1814–1895) – are relatively empty, due to the influence of the planet's enormous gravity.

The gaps occur where an asteroid's orbital period would be an exact multiple of Jupiter's orbital period. The most important ratios are 2:1, 3:1, and 5:2. This means that any asteroid which orbits the Sun exactly twice in the time it takes Jupiter to go around once will be removed by the repeated influence of the giant planet on the same sector of the asteroid's orbit. Jupiter's gravity simply keeps the gaps clear by changing the orbits of any asteroids within them.

One interesting exception is the region where more than 4,000 Hilda asteroids are trapped in a 3:2 mean resonance with Jupiter, at a distance of about 4 AU – well outside the main belt. Almost all Hildas have orbital eccentricities between 0.1 and 0.25. Models show that, during the early evolution of the Solar System, Jupiter probably migrated inward by perhaps 0.35–0.45 AU over a period of at least 100,000 years. During this inward movement, the planet would have shepherded many asteroids into 3:2 resonant orbits. Any family members with low eccentricity would have been ejected – hence their absence in the Hilda group today.

Not all asteroid families are ancient. In 2006, a team of U.S. scientists found a group of six objects orbiting close to the asteroid Datura, suggesting that fragments were broken off from the main member of the group during a collision some 450,000 years ago. Their continuing proximity supports the youthful age of the family, since they would be expected to drift apart and eventually mix with the general asteroid population. At least two more co-orbiting family groups with ages a little over one million years old have been identified.

On a larger scale, the collisional destruction of a 150 km-diameter main belt asteroid is thought to have created the Veritas family 8.2 million years ago. Analyses of sea floor sediments show that the collision may have been responsible for a marked, prolonged, increase in the amount of interplanetary dust falling to Earth.

Asteroid Moons

When the Galileo spacecraft flew past the 31 km-diameter main belt asteroid Ida in 1993, astronomers were surprised to see that it had a small companion (Figure 13.8). This discovery of 1 km-wide Dactyl came as a surprise because, until then, it was considered that the weak gravity of an asteroid might be insufficient to hold a satellite in orbit.

Ida is thought to be an S-type asteroid. Dactyl seems to have a slightly different composition from Ida, so it may not be a fragment that has broken away from its larger companion. However, both are thought to be remnants of the original Koronis parent body which broke apart during a collision at some undetermined time.

Although the number of asteroids encountered by spacecraft remains very small, advanced ground-based instruments have enabled dozens of asteroid satellites to be detected (Figure 13.9). In 1998, the second of these objects, 13 km-wide Petit-Prince, was found in orbit around a main belt asteroid, 215 km-diameter Eugenia.[4]

This discovery was made possible through the use of a technique called adaptive optics. When stars or asteroids are viewed through the Earth's atmosphere, turbulence makes them appear blurred. Adaptive optics remove the blurring by using an adjustable mirror, resulting in images that are comparable with those taken from by the Hubble Space Telescope.

Although this method works well for large, remote asteroids, ground-based radar is superior for observing small near-Earth asteroids. With this technique, radar waves transmitted by a large dish antenna are bounced off the asteroid and the echoes are detected. The size, surface characteristics, and orbital speed of an asteroid can be detected in this manner.

Another technique that has recently been introduced is optical interferometry, which combines the light from two or more telescopes. This method enabled two of the European Southern Observatory's 8.2 m VLT telescopes to discover that main belt asteroid 234 Barbara is composed of two bodies with diameters of 37 and 21 km. It was either shaped like a gigantic peanut or made up of two separate bodies orbiting each other.

Through these various efforts, satellites have been found orbiting asteroids of many different sizes – between 15 and 50% of all asteroids with diameters less than 10 km are estimated to have companions.

One of the strangest "binaries" is near-Earth asteroid 4179 Toutatis. Radar images indicate that it comprises two asteroids that are either loosely attached, rolling around against each other, or in close orbit (Figure 13.10). "Yam-shaped" Toutatis measures 1.92 x 2.29 x 4.6 km and has a very unusual rotation.

Whereas most Solar System objects spin around a single axis, Toutatis tumbles as the result of two different types of motion with periods of 5.4 and 7.3 Earth days. For anyone standing on its surface, the Sun and the stars would appear to cross the sky following different paths and moving at different rates during each new Toutatis "day." The asteroid follows a very eccentric orbit close to the ecliptic. It barely crosses Earth's orbit at perihelion, and retreats to the main belt at aphelion, taking 3.98 years to orbit the Sun.

Another interesting binary is 90 Antiope, which was discovered in 1866. In 2000, a team led by William Merline (Southwest Research Institute) found that it actually comprises two objects that measure about 88 km and 85 km across. Their densities are not much higher than that of water, suggesting that they are very porous. The distance between them is 171 km, with one orbit taking 16.5 hours. Their rotations are tidally locked, so they always present the same side to each other. Moreover, the asteroids rotate in the same plane as they orbit each other.

Such discoveries provide valuable insights into the nature of asteroids. Astronomers can use the orbital period of a companion to determine the mass, and then the density, of the asteroids

---

[4]**Petit-Prince was named after the son of the French Empress, Eugenie.**

Box 13.3   Sample Return Missions

Although many asteroids have been studied by ground-based remote sensing or by spacecraft, the only way to determine an asteroid's composition is to retrieve a sample of pristine material and return it to a laboratory on Earth for analysis. The first attempt to achieve such a sample return was made by Japan's Hayabusa (Falcon) spacecraft, which was launched toward near-Earth asteroid 25143 Itokawa on May 9, 2003.

Despite major technical problems, the spacecraft successfully rendezvoused with Itokawa on September 12, 2005. Remaining in a station-keeping position 20 km from the asteroid, it was able to map the surface of the tiny object at different wavelengths.

In addition to a multi-spectral camera, Hayabusa carried a laser altimeter (LIDAR) which determined Itokawa's size and shape by constantly measuring the distance between the spacecraft and the asteroid. A near-infrared spectrometer measured the mineral composition and properties of the surface. The link between asteroids and meteorites was investigated by an X-ray fluorescence spectrometer designed to detect different elements and measure their abundance.

Itokawa was found to measure only 540 x 310 x 250 m, with a surface gravity 750,000 times lower than on Earth. Other constraints to be overcome included a 12-hour rotation period and a rugged, boulder-strewn surface.

An attempt to deploy a small, hopping robot, named Minerva, on the surface failed on November 12 after the lander was jettisoned at too high an altitude. The efforts to conduct one dress rehearsal and two sampling runs also proved problematic. These involved releasing a softball-sized target marker made of highly reflective material. A flashlight-like device was to illuminate the surface marker and assist in the landing.

During a fleeting touchdown on the surface, Hayabusa would fire a "bullet" made of tantalum into the surface at a speed of 300 m/s. Tiny rock fragments ejected by the impact would enter a funnel-shaped horn and pass into a sample container which would then be sealed, ready for return to Earth.

The first dry run had to be abandoned. During a second rehearsal, the spacecraft descended to a height of 55 m but lost Minerva. The first sampling attempt on November 20–21 saw Hayabusa remain on the surface of the flat "MUSES Sea" for half an hour – much longer than planned – but the sampling gun failed to fire.

Figure 13.5   Itokawa is the smallest asteroid ever seen from close range. Despite its almost imperceptible gravity, this NEO has a very diverse surface, with numerous large boulders and several smooth areas. Note the double-lobed appearance, suggesting that it may comprise two objects which gently collided. (ISAS/JAXA)

Hayabusa made a second, brief landing on November 26. At first, it appeared that two bullets had fired and sample retrieval had been successful. However, subsequent analysis indicated that the sampling gun may not have fired, with the result that no samples were sealed inside the return capsule. The spacecraft then suffered leaks in its chemical thrusters and began to spin out of control, losing contact with the ground. The Japanese space agency delayed the return to Earth for two years in order to give engineers more time to regain control.

Figure 13.6   Hayabusa was the first mission designed to land on an asteroid and return a rock sample to Earth. The sampling horn is shown beneath the spacecraft. The spacecraft touched down twice and successfully obtained fragments of surface material. The sample return capsule successfully landed in Australia on June 14, 2010. (JAXA)

Hayabusa finally began its return journey on April 25, 2007. Overcoming all the odds, it safely returned to Earth's vicinity on June 14, 2010, three years later than originally planned.

The small re-entry capsule was ejected for a parachute descent near Woomera in Australia. The capsule was opened in Japan under controlled conditions at a new facility built for the purpose. Preliminary analysis revealed some 1,500 tiny particles of asteroid material inside one compartment of the capsule. Their mineral content seemed to match that of ordinary chondrite meteorites – an important clue to what happened during the earliest epoch of the Solar System's history.

After applying lessons learned from the Hayabusa mission, Japan designed a modified spacecraft known as Hayabusa 2. The spacecraft was launched on December 3, 2014, and it arrived at the 900 m-wide NEO Ryugu (1999 JU3) in summer 2018. If all goes according to plan, it will orbit the asteroid for a year and a half.

Hayabusa 2 revealed that the primitive C-class (carbonaceous) asteroid is pyramid-shaped with several large, shallow craters and numerous boulders. Soon after arrival, Hayabasa 2 deployed three small craft – the Minerva lander capsule with two rover/hoppers, Minerva 1a and 1b, and a German-built lander/hopper known as the Mobile Asteroid Surface Scout (MASCOT). The battery-powered MASCOT operated for 17 hours (three asteroid days), taking images and collecting data on the composition and nature of the asteroid.

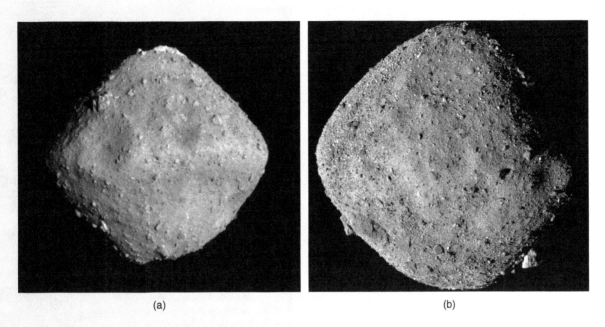

Figure 13.7   Two small near-Earth asteroids visited by sample return missions have remarkably similar shapes. (a) Ryugu is a 900 m-diameter object that is being explored by Japan's Hayabusa-2 orbiter. The carbonaceous (C-type) asteroid has a diamond shape with several large, shallow craters and numerous boulders, including one large, bright rock at the north pole (top). (b) Bennu is a B-type asteroid, which is rich in carbon. It has a diameter of about 500 m. It, too, is largely covered by boulders, with only a few smooth areas. (JAXA)

A third Minerva rover/hopper, Minerva-II2, was deployed on 2 Oct 2019, but it did not land. Hayabusa 2 is also intended to descend to the surface three times to gather samples from the surface. The first of these descents took place on February 21, 2019. During each operation, it fires a copper impactor to create a small crater, then the spacecraft briefly touches down on the asteroid to collect samples excavated by the impact. In December 2019, the spacecraft and its cargo of samples will fire its ion engines to begin a one-year return journey to Earth.

NASA has also sent a mission to return a small sample from a potentially hazardous near-Earth asteroid. The OSIRIS-REx spacecraft was launched on September 8, 2016, and it arrived at Bennu (1999 RQ36), a primitive B-type asteroid, on December 3, 2018. Bennu's composition closely resembles that of carbonaceous chondrite meteorites (see Meteorites).

The spacecraft will spend 18 months extensively surveying the asteroid. In 2020, it will briefly touch the surface to deploy its retractable arm. The sample acquisition mechanism will emit a burst of nitrogen gas that will disturb particles of loose material and blow them into the sampler head. It is hoped to collect at least 60 grams (2 ounces) of loose material. With the sample stored inside a sealed capsule, the spacecraft will then return to Earth and eject its Sample Return Capsule for landing in the Utah desert on September 24, 2023.

involved (see Chapters 1 and 8). The density gives a clue to the asteroid's composition and structure. For example, thanks to the discovery of its moon, we know that Eugenia is rather lightweight, with a lower density than its dark, rock-like appearance might suggest.[5]

Multiple asteroids are not confined to the main belt. In September 2000, NASA's Goldstone radar facility in California detected 2000 DP107, the first near-Earth binary asteroid. Subsequent observations made at Arecibo in Puerto Rico showed that its components orbit about 3 km apart, and the 300 m-wide satellite always presents the same face to its 800 m-diameter companion.

Even more complex systems are possible (Figure 13.11). In August 2005, US and French astronomers observed the first triple asteroid system. The largest member, known as 87 Sylvia, is 280 km in diameter and orbits in the outer part of the main belt, about 3.5 AU from the Sun. It spins quite rapidly, with a 5 hour, 11 minute period.

In 2001, Sylvia was found to have a much smaller satellite in a four-day orbit. Four years later, a second moon was discovered. Since Sylvia was named after Rhea Sylvia, the mythical mother of the founders of Rome, the companions were named after her sons, Romulus and Remus. The small moons follow nearly circular, prograde orbits in the same plane.

[5]Spacecraft flybys have also made it possible to calculate asteroid masses from the slight orbital changes caused by the object's gravitational tug. Previously, although the approximate size of a large, individual asteroid could be obtained through direct measurement or studies of stellar occultations, there was no accurate way to determine its mass.

Figure 13.8 Main belt asteroid 243 Ida is a member of the Koronis family. This false-color mosaic was made from images taken by the Galileo spacecraft at a distance of about 10,500 km, on August 28, 1993. Ida measures 56 x 24 x 21 km. It is probably composed of silicates and covered in impact craters, so its surface is very ancient. Brighter, bluish areas around craters (upper left, centre, and upper right) suggest a difference in the abundance or composition of iron-bearing minerals. This was the first image of an asteroid moon, now named Dactyl (far right). Both are probably fragments of a much larger asteroid. (NASA-JPL)

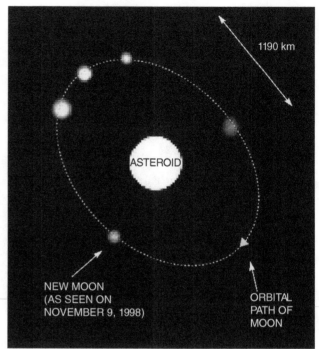

Figure 13.9 This infrared composite image, obtained with the Canada-France-Hawaii telescope in Hawaii, shows Petit-Prince, a small satellite orbiting main belt asteroid 45 Eugenia at a distance of about 1,200 km. One orbit of the moon (green dashed line) takes 4.7 days. The near-circular orbit is tilted about 45° with respect to the line-of-sight. Eugenia has a diameter of about 215 km, while the satellite's diameter is about 13 km. The moon is 300 times fainter than the asteroid. The large "cross" is an artifact caused by stray light in the telescope. (Courtesy of Laird Close/ESO and Bill Merline/SwRI)

How these gravitationally bound systems originate is not clear. At first it was thought that asteroid moons formed solely through collisions and/or close encounters with planets. Theoretical models indicate that an asteroid which resembles a pile of loose rubble can be disturbed during close encounters with the inner planets of the Solar System. By passing within 16,000 km of a planet's surface, a piece of the asteroid may break away, forming a satellite.

However, it was found that such encounters cannot account for the large number of binary asteroids that have been discovered. Some other process must have been responsible. One of the favored candidates is the YORP effect, in which sunlight spins up a small asteroid so much that it splits apart or ejects material from its equator that eventually coalesces into a satellite (See Box 13.4).

Ceres

Ceres is by far the dominant member of the main asteroid belt, accounting for about 25% of the belt's total mass. It is also the only asteroid that is deemed worthy of being classed as a "dwarf planet" (see Chapter 1).

Until the arrival of NASA's Dawn orbiter in 2015, very little was known about Ceres, although images taken by the Hubble Space Telescope revealed a slightly flattened sphere with a diameter of about 910 x 970 km (Box 13.5). By tracking features on the surface, its day was determined to last a little over nine hours.

Overall, Ceres is a dark object, reflecting only about 10% of incoming sunlight (Figure 13.13). This means that no large areas of fresh, reflective ice are visible. However, studies of its gross properties indicate that Ceres has a density of about 2.1 g/cm³,

similar to Ganymede and Callisto, suggesting that its composition is at least 25% water ice.

It seems that the ice must be largely hidden under a layer of dust or debris, only emerging in a few places where it is relatively close to the surface. This "insulation layer" probably explains why ice can survive on Ceres: because of the fairly high intensity of solar radiation in the asteroid belt, the lifetime of ice on the surface would otherwise be very short.

The presence of water ice was confirmed by the Herschel infrared space observatory, with the most marked signatures linked to areas that are slightly warmer or craters where impacts have exposed subsurface ice. Herschel also studied how the amount of water changed along the asteroid's orbit: the water signature increased near perihelion and decreased near aphelion.

Like the Moon and Mercury, Ceres possesses permanently shadowed regions around its poles, typically located on crater floors or along a crater wall facing toward the pole. Most of these areas have probably been cold enough to trap water ice for a billion years, suggesting that ice deposits could still exist there today.

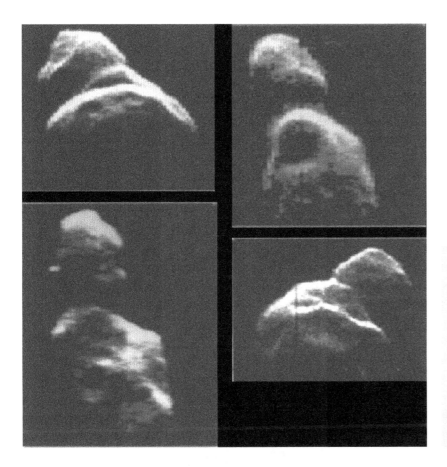

Figure 13.10 These radar images of Toutatis were obtained by the ground-based radar at Goldstone, California, in December 1992, when it passed about 2.5 million km from Earth. They reveal two irregular, cratered objects about 4 km and 2.5 km in diameter that are probably in contact with each other. The largest crater (top right image) is about 700 m across. The asteroid's tumbling rotation and cratered surface suggest a history of frequent collisions. (Steve Ostro/NASA-JPL)

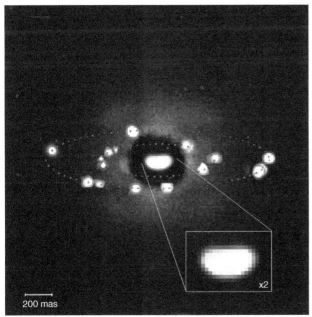

Figure 13.11 Main belt asteroid 87 Sylvia was the first triple asteroid system to be discovered. This composite image shows the positions of its two moons, Remus and Romulus over 9 nights. The inset shows the potato shape of Sylvia. The discovery was made with one of the 8.2 m telescopes of ESO's Very Large Telescope Array in Chile, using adaptive optics. (ESO)

Although Ceres is much smaller than the Moon or Mercury, and its surface gravity is much weaker, the asteroid has just enough mass to hold on to water molecules. Furthermore, the permanently shadowed regions are generally colder than most of those on the Moon or Mercury. This is because Ceres is further from the Sun, and the shaded parts of its craters receive little indirect radiation.

The ability to accumulate water ice – the trapping efficiency – is also comparable on Ceres and Mercury. Calculations suggest that about 1 out of every 1,000 water molecules generated on the surface of Ceres will end up in a cold trap during a year on the asteroid (1,682 days). That's enough to build up thin, but detectable, ice deposits over 100,000 years or so.

Dozens of sizeable permanently shadowed regions have been found across its northern hemisphere. The largest of these is inside a 16-km crater located less than 65 km from the north pole.

Altogether, Ceres' permanently shadowed regions occupy about 1,800 sq. km. This is a small fraction of the landscape – much less than 1% of the surface area of the northern hemisphere. However, they extend to lower latitudes than similar regions on Mercury or the Moon, so the shadowed regions account for roughly the same fraction of the northern hemisphere as on Mercury.

Dawn's images reveal that the dwarf planet has at least three large-scale depressions, called "planitiae," that are up to 800 km wide. These depressions may be old impact features, and they include craters that formed in more recent times.

One of them, called Vendimia Planitia, covers a broad area just north of Kerwan crater, Ceres' largest well-defined impact basin. Vendimia Planitia must have formed much earlier than Kerwan.

Box 13.4   Solar-Powered Asteroids

The rotation rates and orbits of asteroids are influenced by external gravitational forces and collisions. However, sunlight has also been shown to put asteroids in a spin.

Incoming solar radiation is absorbed on the dayside and radiated back into space as the asteroid rotates, generating a minuscule recoil effect. Since small asteroids are irregular in shape, the recoil affects one part of the asteroid more than another. The object's thermal inertia also means that the maximum surface temperature occurs on its afternoon side, rather than the morning side or sub-solar point (midday), because it has been in the Sun for longer. This is similar to Earth, where 2–3 p.m., rather than noon, is the warmest time of day.

As a result, the asteroid tends to emit more heat from its afternoon side. This unbalanced thermal radiation produces a tiny acceleration or thrust, known as the Yarkovsky effect. It also results in a twisting force or torque known as the YORP (Yarkovsky-O'Keefe-Radzievskii-Paddack) effect.[6]

The momentum generated by the Yarkovsky effect over each orbit of the Sun causes gradual drifting of the orbits of kilometer-sized asteroids, causing objects to spiral inward or outward at different rates as a function of their spin, orbit, and material properties. Since every action has an equal and opposite reaction, heat radiating from one side of the asteroid will push it out of its existing orbit in the opposite direction – towards the Sun if the asteroid spins retrograde, and away from the Sun if prograde.

Not only does it influence the dynamical spreading of asteroid families, but it also plays an important role in changing the orbits of main belt objects so that they are redirected into planet-crossing orbits, e.g. near-Earth asteroids.

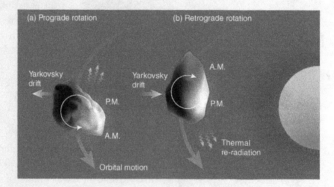

Figure 13.12   An unevenly shaped asteroid heated by solar radiation re-radiates the thermal energy at right angles to its surface. Since the afternoon side becomes hotter than other locations, that face of the asteroid re-radiates most thermal radiation, creating a recoil force (the Yarkovsky effect) and causing the asteroid to drift a little towards or away from the Sun. The direction of the drift depends on whether the asteroid is rotating in a prograde (anticlockwise) manner (a) or in a retrograde (clockwise) manner (b). It can also change the asteroid's spin rate – the YORP effect. (Nature)

The YORP effect has a subtle influence on the rotation rate of asteroids. In 2003, William F. Bottke (Southwest Research Institute) and colleagues discovered that members of the Koronis family have their spins clustered into several groups, rather than the expected random distribution. They concluded that the slow, but steady, recoil force of thermal re-radiation can be more effective than collisions in controlling the direction of asteroid spins.

The force generated by sunlight and its effect on asteroid spin rate was measured directly for the first time by a team led by Stephen Lowry (Queen's University, Belfast, UK). They monitored the spin rate of near-Earth asteroid 2000 PH5 from 2001 to 2005. The 120 m-long rock rotated once every 12 minutes or so, but its spin period was being shortened by 1 millisecond every year a rate that matched the predicted YORP effect. (It is also possible, of course, that the effect may cause asteroids to slow down.)

Similar studies have been made of a much bigger Earth-crossing asteroid, 1862 Apollo. Between 1980 and 2005 its rotation period shortened by about 4 milliseconds per year, also in line with predictions. Although it would take about 2.6 million years for YORP to double Apollo's rotation rate, the change is highly significant over the age of the Solar System.

[6]These physical processes are named after Ivan Osipovich Yarkovsky, a 19th-century Russian civil engineer and part-time scientist. Other contributors to understanding of the YORP effect were American planetary scientist John A. O'Keefe, Russian astronomer V.V. Radzievskii, and NASA aerospace engineer Stephen J. Paddack. They proposed that diurnal solar heating of a rotating object in space would cause it to experience a tiny force that could lead to large secular effects in the orbits of small bodies, especially meteoroids and small asteroids.

The YORP effect can also modify the shape of an asteroid. Many near-Earth objects, such as Ryugu, are known to have a spinning-top shape which is thought to result from a rubble-pile's response to YORP torques. Modification of such unconsolidated asteroids occurs when rotational angular momentum is added to or subtracted, causing blocks and particles to move in response to the resulting centrifugal (rotational) forces. Thus, YORP may add enough angular momentum to produce downslope movement, mass shedding, and shape changes.

The YORP effect is also important for understanding binary asteroids. Many binaries form from asteroid collisions or gravitational capture. However, a single asteroid may split in two if the YORP effect causes it to spin too fast. Lowry predicts that in about 14 million years, 2000 PH5 will spin once every 20 seconds or so, faster than any asteroid yet observed. If the object is a rubble pile, rather than solid rock, it could fly apart.

This process may sometimes create an asteroid moon. When solar energy "spins up" a rubble pile asteroid to a sufficiently fast rate, material may be thrown off from its equator. If the ejected material is able to shed a lot of its momentum through mutual collisions, then it may coalesce and form a satellite in orbit around its parent.

Figure 13.13 An enhanced-color image of Ceres made from data obtained from NASA's Dawn spacecraft. The brightest region, Cerealia Facula, lies within Occator Crater (center). Vinalia Faculae, the group of secondary bright spots in the same crater, lies to the right of Cerealia Facula. One of the darkest regions is next to Occator, and represents ejecta from the impact that formed the crater. Other craters also show a mixture of bright and dark regions. The bright areas are generally salt-rich material excavated from the crust, but the origin of the dark material is uncertain. The blueish color is generally associated with young craters. The continuous bombardment by micrometeorites alters the texture of the exposed material, leading to its reddening. (Courtesy of NASA/JPL-Caltech/UCLA/MPS/DLR/IDA)

As expected, Ceres is covered with countless impact craters. Surprisingly, almost all of the craters are small and young: only 16 are larger than 100 km across and none are more than 280 km in diameter. However, large parts of Ceres' northern hemisphere are saturated with craters 60 km in diameter or smaller. These findings are incompatible with current models of the rate of impacts with objects such as Ceres.

Somehow, Ceres has healed its largest impact scars and renewed old, cratered surfaces. One possible explanation is "relaxation" of the ice-rich surface. Because ice is less dense than rock, the topography could smooth out more quickly if ice or another lower density material, such as salt, dominates the subsurface composition.

The most notable impact feature is Occator crater, which is 92 km wide and 4 km deep. The crater contains a highly reflective central area, known as Cerealia Facula. Featuring a 400 m-high dome that sits within a broad pit, about 11 km across, it is rimmed by fractures (Figure 13.14). Cerealia Facula appears to have been created by recent cryogenic activity. Some estimates suggest that this bright feature is only about 4 million years old.

Studies indicate that the white material consists of sodium carbonate and ammonium chloride, while the secondary, smaller bright areas of Occator, called Vinalia Faculae, are comprised of a mixture of carbonates and dark material.

The carbonate deposit must be fairly thick, because the dome bears a dozen small impacts, 80 to 300 m across, and all are bright like their surroundings.

The impact that formed Occator crater probably triggered the upwelling of salty liquid. Water and dissolved gases, such as carbon dioxide and methane, rose toward the surface though a vent system. The bright material erupted through fractures, eventually forming the dome that we see today. The salts were probably left behind after the briny liquid emerged onto the surface, froze and then sublimated, so that it turned from ice into vapor.

Analysis of the center of Occator crater suggests that the salts found there could be remnants of a frozen ocean under the surface, and that liquid water could have been present in Ceres' interior.

Past hydrothermal activity, which may have influenced the salts rising to the surface at Occator, could also be linked with the erasure of craters. If Ceres had widespread cryovolcanic activity in the past – the eruption of volatiles such as water – these cryogenic materials also could have flowed across the surface, possibly burying pre-existing large craters. Smaller impacts would have then created new craters on the resurfaced area.

Another landform linked to cryovolcanism is Ahuna Mons. The only mountain on Ceres, it lies about 670 km south east of Occator crater (Figure 13.15). The dome-like feature is about 20 km across and has smooth, steep walls. On its steepest side it is about 5 km

Figure 13.14   Cerealia Facula, the brightest spot on Ceres, is about 15 km wide and lies in the center of 90 km-wide Occator crater. The landform features a dome in a smooth-walled pit. Numerous linear features and fractures crisscross the top and flanks of this dome. Like the other bright spots (faculae) scattered around Ceres, Cerealia Facula is not ice, but an exposed salty residue with a reflectivity like dirty snow. It is thought to be mostly sodium carbonate and ammonium chloride derived from a slushy brine within or below the crust. No vertical exaggeration was applied to this 3D image. (Courtesy of NASA/JPL-Caltech/UCLA/MPS/DLR /IDA/PSI)

high – about half the height of Mt. Everest – and the average overall height is around 4 km. Some of its slopes are brighter than others.

There are several dozen domes on Ceres that may be similar in nature to Ahuna Mons, although none is as tall and well-defined. It seems that they may have been tall cryovolcanoes millions or billions of years ago, but they flattened out over time.

This is likely the result of viscous relaxation, the slow flow of an apparently solid material over a long period of time – like ice in a glacier. Over time, the volcanoes would become indistinguishable from the planet's surface.

Calculations indicate that Ahuna Mons is at most 200 million years old, so it hasn't had time to deform. Viscous relaxation would occur if the dome was composed of more than 40% water ice. With this composition, it is estimated that Ahuna Mons should be flattening at a rate of 10 to 50 m per million years – a rate that would render cryovolcanoes unrecognizable after hundreds of millions to billions of years.

The dome's relatively young age suggests that cold, briny eruptions, known as cryovolcanism, emerged from a liquid reservoir trapped between a muddy icy mantle and a silicate-rich core. Once the slushy stuff breached the surface, exposing it to the cold vacuum of space, the brine would have quickly frozen and its water would have rapidly boiled or sublimated away, leaving the salts behind as a solid residue.

Whether the salts now exist as a stiff layer or as a fine fluffy powder on the surface isn't known. In late April the Dawn project planned to examine the dome with an illumination phase angle of 0° – that is, with sunlight coming from directly behind the spacecraft. Observations made at this special geometry should constrain the grain sizes in the salt deposits.

Nor is it clear how often eruptions might have occurred. "A long-lasting process appears to be prevalent," the team concludes, "whereby periodically or episodically ascending bright material from a subsurface reservoir was deposited, expelled from fractures, and extruded onto the surface, forming the present-day central dome."

An enhanced-color close-up reveals a bright dome sitting within a smooth-walled pit in the center of Occator. Numerous linear features and fractures crisscross the dome's top and flanks.

Ceres may well be one of the most important leftover relics of the early Solar System, large enough to have experienced many of the processes normally associated with planetary evolution and surviving the early accretion phase in which most of its companions were incorporated into the planets.

Models suggest that Ceres grew from a mix of silicates and water ice, possibly with considerable amounts of carbonaceous material. Early in its history, internal heating from short- and long-lived radionuclides would have caused the water ice to quickly melt, forming a mantle beneath the icy crust. The circulating water would alter the silicates. Internal thermal processing and mixing with the water could also create methane and CO_2 clathrates, in which gases are trapped within crystalline ice.

Figure 13.15 Ceres' sole mountain, Ahuna Mons (top), is 4 km high and 17 km wide. It is thought to have been formed by quite recent cryovolcanism. The elevation has been exaggerated by a factor of two. Nearby is a large impact crater, which is one of the few sites on Ceres where a significant amount of sodium carbonate has been found, shown in green and red colors in the lower right image. (Courtesy of NASA/JPL-Caltech/UCLA/MPS/DLR/IDA/ASI/INAF)

As heat was lost by conduction through the frozen surface, water began to freeze out at the base of the crust. As the crust grew in thickness, solid-state convection would become continuous, transporting more heat as well as altered materials toward the surface (Figure 13.16). After perhaps 2 billion years, Ceres' water layer eventually froze, forming a layered structure within the asteroid.

In addition, melting and freezing plus mineralization and frequent impacts would, over time, create surface topographic features. If clathrates were brought near the surface, they would suddenly vaporize, possibly producing explosive release of gases, bringing considerable quantities of altered materials to the surface, and creating surface landforms.

Ceres would also have shrunk as internal ice melted. Then, as the water froze again, the asteroid would have expanded, perhaps creating topographic features associated with extensional forces, such as cracks, faults, and dropped blocks of crust.

With so much water present and the energy to distribute it in liquid form, Ceres probably experienced complex chemistry – at least in its interior – that may have included organic materials.

It seems that Ceres may be a perfect place to study the evolution of objects at the interface between the terrestrial planets and the colder, icy objects of the outer Solar System.

Ceres has an ephemeral, transient atmosphere. The first evidence for this came in 1991, when the International Ultraviolet Explorer satellite noted a hydroxyl emission from Ceres which had not been detected a year earlier. ESA's Herschel Space Observatory subsequently made several unambiguous detections of a water vapor exosphere.

The temporary reappearances of an atmosphere do not seem to be associated with surface sublimation caused by warming as Ceres approaches perihelion. Instead, it seems that variations in solar activity are more significant than warming by the Sun.

When the Sun erupts, sending energetic particles (protons) across the Solar System, some of these strike Ceres' surface, liberating molecules of ice. These create a thin atmosphere that may last but a week or so. Such activity may also occur on other airless, water-rich bodies, including the polar regions of the Moon and some asteroids.

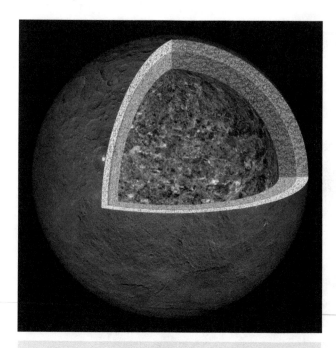

Figure 13.16   A cutaway view of Ceres shows a 40 km-thick outer crust (light blue) made largely of water ice, salts, and hydrated minerals. Beneath the crust lies a thick shell, perhaps up to 100 km deep, consisting in part of a briny liquid. The innermost layer, the "mantle," is dominated by hydrated rocks, like clays. There may be a small rocky/metallic core – not shown. (Courtesy of NASA-JPL/UCLA/MP/DLR/IDA)

Vesta

In contrast with ice-rich Ceres, Vesta, the other object on Dawn's itinerary, is much denser and drier, and seems to have experienced a very different early history. It is believed to be the only layered protoplanet remaining intact in the Solar System. The other objects like Vesta, were probably incorporated into planets or destroyed by collisions billions of years ago (see Chapter 1).

The second-most massive and second- or third-largest body in the asteroid belt after Ceres, Vesta accounts for an estimated 9% of the mass of the main belt (Figure 13.18). It is roughly spherical, with an average diameter of 525 km and a density of about 3.45 g/cm^3, corresponding to a rocky composition. Vesta takes approximately 5.34 hours to make one rotation.

Vesta's spectrum is unique among all the larger asteroids. Dawn data indicate that it has an iron core with a radius of about 110 km, suggesting that Vesta completely melted in its early history, allowing iron to sink to the center and producing a basaltic crust. The cause of the heating may have been the presence of short-lived radioisotopes in the predominantly rocky body, although significant heating would also have been produced by large impacts early in its history.

Observations with the Hubble Space Telescope show that it has one dark hemisphere and one bright hemisphere, representing two different types of solidified igneous rock. The lighter hemisphere is thought to be composed of basalt which is rich in the mineral

pyroxene, formed when magma cooled and solidified beneath the surface.

The darker hemisphere is interpreted to be composed of a type of basalt which is a mixture of pyroxene and feldspar. It may be the remains of Vesta's ancient crust, formed from lava which cooled and solidified on the surface.

Vesta also displays dark material on the rims of smaller craters or scattered around them as ejecta. This dark carbonaceous material was delivered by smaller, impacting asteroids.

The oldest, most heavily cratered terrain occurs across the northern hemisphere, whereas the southern hemisphere is dominated by terrains modified by the Veneneia and Rheasilvia impacts (Figures 13.19 and 13.20). A set of deep troughs along Vesta's equator separates these two types of terrains. These are thought to be large-scale fractures resulting from the Rheasilvia impact. The largest, named Divalia Fossa, is bigger than the Grand Canyon.

The HST discovered a giant depression near the south pole that dominates the entire hemisphere. Dawn revealed that this feature is actually composed of two overlapping impact basins which were created early in Vesta's history. Rheasilvia is 505 km across, with an escarpment along part of its perimeter which rises 4–12 km above the surrounding terrain. Its crater partially obscures an earlier impact feature, named Veneneia, that has a diameter of 395 km.

The asteroid's low gravity enabled a major crustal rebound after the impact, creating a huge central peak (see Chapter 4). The 200 km-wide mountain in the center of the crater rises 22 km from its base, more than twice the height of Everest.

Since Vesta is large enough to be internally differentiated, it was thought that the impacts may have penetrated almost all the way through the volcanic crust to expose the mantle. However, no evidence of olivine-rich mantle material was found.

Vesta is probably the source of the Vesta family and V-class asteroids. Two distinct populations of small asteroids were probably created during each of the massive cratering events, which is consistent with recent evidence for spectral color variations within the Vesta family.

The vast amount of debris that was blasted into space from the craters is also the likely source of basaltic meteorites called HEDs, for howardite-eucrite-diogenite. Approximately 6% of the meteorites that fall to Earth have a similar mineralogical signature, as indicated by their spectral characteristics. Vesta is the only major world, other than Earth, Moon, and Mars, for which there are samples of known origin.

Other Large Asteroids

Information about the other major asteroids is even more limited. Estimated to account for 7% of the total mass of the asteroid belt, Pallas is one of three objects in the main belt that can be classified as a protoplanet, rather than mere debris from the early Solar System (Figure 13.21).

Pallas is the largest of the B-type asteroids. Until, recently, the only way to determine its size with any accuracy was through stellar occultations and speckle interferometry (a technique for overcoming atmospheric turbulence by combining short photographic exposures of an object). More recent imaging with the largest ground-based telescopes and the Hubble Space Telescope

Box 13.5   Dawn

Ceres and Vesta are two of the largest members of the main asteroid belt. In order explore these contrasting objects, NASA developed the Dawn spacecraft, which was launched in September 2007. Dawn was the first spacecraft to orbit a main belt object and the first to orbit two bodies after leaving Earth. Its mission was to study two protoplanets that were never given the opportunity to grow fully. This would provide vital clues about their divergent evolutionary paths, together with the conditions and processes associated with planetary formation 4.5 billion years ago.

Figure 13.17   An artist's concept of the Dawn spacecraft with Vesta (lower left) and Ceres. Dawn was launched in September 2007. After a four-year journey using ion propulsion, it entered orbit around Vesta in August 2011. At the end of a year-long study of the large basaltic asteroid, it departed for water-rich Ceres, arriving in March 2015. (NASA/JPL)

Dawn used an innovative ion propulsion system that comprised three xenon-fueled engines powered by electricity generated by two large solar arrays. After a journey of 2.8 billion km, which included a Mars gravity assist in February 2009, it entered orbit around Vesta in August 2011. Over the next 13 months, it conducted an in-depth investigation of the basaltic asteroid from various altitudes. Dawn then departed for Ceres, with arrival in March 2015. The primary mission was completed on June 30, 2016, and Dawn then began an extended mission. Its mission ended in late 2018, but it will remain in orbit around Ceres for several decades.

Dawn carried three science instruments: a visible/near-infrared camera, a visible and infrared mapping spectrometer to map mineral distribution, and a gamma ray/neutron spectrometer to determine the elemental composition of the surface. Radiometric and optical navigation data provide information about the asteroids' gravity fields, bulk properties, and internal structure.

indicate that Pallas is a spheroid that is about 510 km in diameter, comparable in size to Vesta, though it is some 20% less massive.

Based on the differing size estimates, the mean density of Pallas is in the range 2.4–3.0 g/cm^3 – consistent with a body that formed from a rock-ice mixture. This density is significantly lower than Vesta's, but greater than Ceres, suggesting that it is largely rocky and contains less water than its larger neighbor.

The spectrum and estimated density are comparable to those of carbonaceous chondrite meteorites. However, multispectral analysis suggests that its surface is covered with hydrated minerals,

consistent with both Ceres and Pallas forming from a common reservoir of water-rich material. Both objects also appear to have differentiated interiors. The hydrated material that covers Pallas suggests that its surface and interior were thermally altered by impacts or internal heat while it contained a large amount of liquid water.

Few details have been seen on the fairly dark surface of Pallas (albedo is 12%). HST images show little variability in visible and infrared light, but significant variations at ultraviolet wavelengths suggest large surface or compositional variations. There may be a

21 Lutetia

253 Mathilde

243 Ida / 1 Dactyl

433 Eros

951 Gaspra

2867 Šteins

5535 Annefrank

25143 Itokawa

4 Vesta

Figure 13.18   A size comparison of nine asteroids that have been visited by spacecraft. (Courtesy of NASA/JPL-Caltech/UCLA/MPS/DLR/IDA)

240 km-wide impact crater within Pallas's ultraviolet-dark terrain at about 30°S, 75°E, as well as smaller impact features.

Rotation appears to be prograde, with a very large axial tilt, indicating that it is spinning on its side. This means that it experiences long summers and winters, with each hemisphere in constant sunlight or darkness for about one Earth year. The rotation period is about 7.8 hours, midway between the periods of Vesta (5.3 hours) and Ceres (9 hours).

The orbit of Pallas is inclined at an angle of 34.8° to the plane of the main asteroid belt, and the orbital eccentricity is nearly as large as that of Pluto, making Pallas a difficult target for visiting spacecraft.

Like Vesta, Pallas is linked to an asteroid family that shares its orbital and spectral parameters. The largest of these is 5222 Ioffe, with a diameter of 22 km. They may all have originated from the large crater seen in HST images.

The third asteroid to be discovered, **3 Juno**, is considerably smaller than Vesta or Pallas, but it is also more reflective. Juno has a diameter of about 240 km with an area that appears dark at near-infrared wavelengths. It seems that the asteroid has collided with another object fairly recently (in astronomical terms),

resulting in a 95 km-wide crater, or possibly a smaller crater that is surrounded by a 95-km blanket of ejecta.

Juno follows an even more eccentric orbit than Pallas, inclined about 13° to the ecliptic. It, too, is thought to be linked with a family of small asteroids. The members of the "Juno Clump" are probably derived from impact debris ejected from the parent object.

Trojan Asteroids

Although most asteroids are found between the orbits of Mars and Jupiter, a significant population can be found elsewhere in the Solar System. Many of these cross the orbits of the inner planets, including numerous near-Earth asteroids (see Near-Earth Asteroids). Others have been shepherded into groups that are co-orbital with the planets.

Known as Trojan asteroids, these latter objects are in gravitationally stable orbits around the L4 and L5 Lagrange points, 60° behind and 60° ahead of the planet in its orbit (Figure 13.1). They typically follow a path that takes them up to 30° either side of the equilibrium points over a period of 150–200 years. Many of them

Figure 13.19   Dawn mapped the distribution of hydrogen on Vesta. The hydrogen probably exists in the form of hydroxyl or water bound to minerals in Vesta's surface. The strongest signature for hydrogen occurred in regions near the equator, where water ice is not stable, rather than near the colder poles. Red indicates the greatest abundances of hydrogen and gray the least. A mountain at the south pole – more than twice the height of Mount Everest – is visible at the bottom of the image. Three impact craters nicknamed the "snowman" can be seen at top left. (Courtesy of NASA/JPL-Caltech/UCLA/PSI/ MPS/DLR/IDA)

have large orbital inclinations, up to 40° relative to the orbital plane of Jupiter.

The first Trojan, 588 Achilles, was discovered in 1906. By October 2018, 7,072 such asteroids had been discovered, nearly all of them associated with Jupiter.

Nine Trojans have also been confirmed along the orbit of Mars (eight at the L5 point and one at L4), one shares an orbit with Uranus at L4, and 22 are associated with Neptune (19 of them at L4 and three at L5). None are known to be linked with Saturn, presumably because its Lagrangian points are destabilized by Jupiter's gravity.

Many of the Trojans found in recent years are temporary. One example is 2011 QF99, the first object ever found to share the orbit of Uranus. Calculations indicate that it has traveled around the planet's L4 point for a few hundred thousand years and is likely to escape Uranus' gravitational pull in about a million years. Other examples have been found near Earth.

Despite the relatively small numbers of confirmed co-orbitals, some estimates suggest there may be in excess of one million Trojans. One model predicts that, at any given time, three percent of scattered objects located between the giant planets should co-orbit with Uranus or Neptune. Since most of the Trojans are more remote and considerably fainter than the main belt asteroids, they are more difficult to study. However, they are generally very dark (albedos of 2–5%) and classified as primitive, carbonaceous, D type asteroids which are redder than the C-type asteroids of the main belt. They have been trapped around the Lagrange points over the lifetime of the Solar System, although those in marginally stable orbits may be ejected.

Jupiter's Trojans are subdivided into the Achilles group (60° ahead) and the Patroclus group (60° behind), though their individual positions can vary considerably.[7] About two-thirds of them belong to the Achilles group. 624 Hektor is the largest known Trojan, as well as the most elongated known asteroid of its size. Measuring roughly 150 × 300 km, it seems to have a dumbbell shape, probably as a result of the partial coalescence of two spheroidal planetesimals during a relatively low-speed collision.

[7]**The names of these objects are associated with the heroes of Homer's Iliad.**

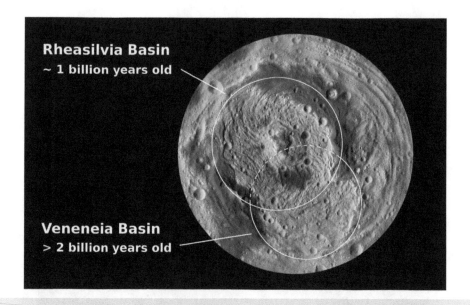

Figure 13.20 This false-color relief map of Vesta's south polar region was created from stereo images obtained by NASA's Dawn spacecraft. The map shows Rheasilvia, a 500 km-diameter circular impact structure, with a rim rising more than 15 km above the interior. The impact excavated about 1% of Vesta's entire volume. Rheasilvia is superimposed on a much older impact basin named Veneneia. The Vesta asteroid family and V-type asteroids are probable products of these impacts. (Courtesy of NASA/JPL-Caltech/UCLA/MPS/DLR/IDA)

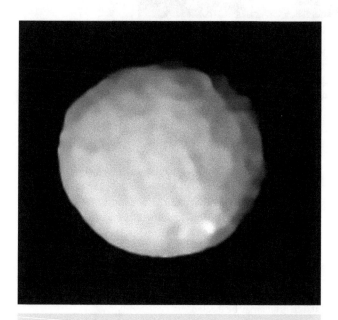

Figure 13.21 An image taken by ESO's Very Large Telescope in 2017 shows Pallas, the third-largest asteroid in the main belt. Pallas is about 510 km in diameter and roughly spherical. It contains about 7% of the mass of the entire asteroid belt. It appears to have numerous sizeable craters. (Courtesy of ESO/Vernazza et al.)

Until recently, the Trojans were generally regarded as planetesimals that formed near Jupiter during the early accretion phase of planet formation. They would have been captured into their current orbits while Jupiter was growing, possibly with the help of gas drag and/or mutual collisions which slowed their motion and modified their trajectories. This theory, however, cannot explain some properties of the Trojans, particularly the range of about 40° in orbital inclinations.

An alternative explanation envisages their formation in more distant regions of the Solar System. They were subsequently scattered and captured into co-orbital motion with Jupiter when the giant planets migrated outward to their current orbits. Such captures would have been possible during a brief period of time, just after Jupiter and Saturn crossed their mutual 1:2 resonance, when the dynamics of the Trojan region were completely chaotic.

One of the most interesting Trojans is 617 Patroclus, which is now known to comprise two objects that are nearly identical in size. The larger object has a maximum diameter of 122 km while its companion is 112 km across. They orbit their center of mass every four days, separated by a distance of about 680 km. Both of the binary components are less dense than water, suggesting that they are porous and made out of water ice.

It may be that Patroclus originated far from the Sun, about 4.5 billion years ago. Like millions of other planetesimals, it may have been redirected inward and then pulled apart when it passed close to Jupiter. Such tidal splitting has also been suggested for the formation of binary near-Earth asteroids, which often comprise objects of similar size.

Box 13.6   NEAR-Shoemaker at Eros

NASA's Near Earth Asteroid Rendezvous (NEAR) mission was the first to orbit and land on an asteroid. Launched on February 17, 1996, it was later renamed NEAR-Shoemaker in tribute to planetary geologist Eugene Shoemaker. The chosen target was the second-largest near-Earth asteroid, an S-type object known as 433 Eros.

Simplicity and low cost were the main driving factors behind NEAR's design. The science instruments, the solar panels, and the high-gain antenna were fixed and body mounted. The six science instruments included a multispectral imager and near-infrared spectrometer adapted from military remote sensing systems. Together, they mapped the surface and determined the mineralogical composition of Eros. The X-ray and gamma ray spectrometers measured and mapped elemental abundances. A laser rangefinder would measure surface topography, while a magnetometer searched for a magnetic field.

Restricted by the capability of the Delta 2 launch vehicle, the spacecraft followed a low-energy trajectory designed to last three years. En route, it flew past main belt asteroid 253 Mathilde on June 27, 1997, at a distance of 1,212 km. The Mathilde flyby was the closest spacecraft encounter carried out up to that time, and the first close-range reconnaissance of a C-class asteroid. NEAR discovered that Mathilde has at least five sizeable craters and a low density, indicative of a rubble pile. Its 17.4-day rotation is the third-longest period of any known asteroid.

On February 18, 1997, NEAR reached its most distant point from the Sun (2.18 AU), setting a new distance record for a spacecraft powered by solar cells. Unfortunately, a computer malfunction meant that the planned orbit insertion in January 1999 had to be postponed. When communications were re-established, NEAR was able to observe Eros at a distance of 3,827 km. The flyby images acquired on December 23 showed about two-thirds of its surface. The asteroid, which measured about 33 x 13 x 8 km, was shown to rotate once in just over 5 hours.

Figure 13.22   NEAR-Shoemaker investigated two asteroids, shown here at the same scale. En route to Eros (right), it imaged main belt asteroid 253 Mathilde (left), a very dark, heavily cratered object that measured 66 x 48 x 46 km. The carbon-rich asteroid reflected just 3–5% of incoming sunlight, making it twice as dark as charcoal. There were at least five craters more than 20 km across on the sunlit side alone. Mathilde displayed no color or albedo variations over the 60% of its surface that was visible to NEAR. (Courtesy of NASA/JHU-APL)

After another circuit of the Sun, NEAR was inserted into a 321 x 366 km orbit around Eros on February 14, 2000 – the first time that a spacecraft had been captured by the gravitational pull of such a small body. During its initial high-orbit phase, thousands of images of Eros' illuminated northern hemisphere were obtained. Over the next 2 ½ months, the orbit was lowered to an altitude of 50 km, close enough to resolve objects 5–10 m across and map surface composition with the X-ray/gamma ray spectrometer. Subsequent low-altitude passes resulted in images of 0.5–1 m resolution.

Figure 13.23 A mosaic of six images taken by NEAR-Shoemaker on February 29, 2000, from a distance of about 200 km. The northern hemisphere of Eros is dominated by Psyche, a 5.3 km-wide crater, with Himeros, the 10 km-wide "saddle" at the bottom. A major ridge links the two features and spans at least one-third of the asteroid's circumference. (Courtesy of NASA-JPL, JHU-APL)

NEAR showed that much of Eros is nearly saturated with craters smaller than 1 km, although its mid-section is marked by a large, saddle-shaped depression, named Himeros, which resembles a degraded crater. The largest crater, Psyche, is 5.3 km across and located on the opposite side to Himeros. Surface grooves and ridges are common, and may follow lines of weakness produced by ancient collisions. Much of the surface is covered with piles of rubble, blocks of ejecta between 30 and 100 m across. However, they are not uniformly distributed. NEAR also revealed some smooth areas, dubbed "ponds" by mission scientists.

Although NEAR was never designed to land, it was decided to end its mission by attempting a controlled descent to Eros' surface. On February 12, the spacecraft began a five-hour descent from a height of 36 km, eventually touching down close to the Himeros depression. 69 images were returned during the descent, with the last of them taken 125 m above the surface.

Since the solar panels were still pointing toward the Sun, it was possible to establish a telemetry link that sent back useful data from the gamma ray spectrometer and the magnetometer. NEAR was finally shut down on February 28.

The first Mars Trojan to be discovered was 5261 Eureka, whose infrared spectrum is typical of an A-class asteroid. It trails Mars at the L5 point at a distance varying by 0.3 AU during each revolution and has probably been in this orbit for much of the Solar System's lifetime.

Neptune's first Trojan asteroid was first imaged on August 21, 2001, and confirmed during observations over the next 16 months. Designated 2001 QR322, it is located at the L4 Lagrange point and is estimated to be about 230 km in diameter. The second such Trojan, 2004 UP10, has been given the permanent designation 385571 Otrera. Many of the Neptune family have orbital inclinations of more than 15 degrees. There may be many thousands of these objects yet to be found.

In 2011, astronomers announced the discovery of the first Earth Trojan. Roughly 300 m in diameter, 2010 TK7 has an unusual orbit that traces a complex motion near the L4 point in the plane of Earth's orbit, although the asteroid also moves above and below the ecliptic plane. The object is about 80 million km from Earth. For at least the next 100 years, it will not come closer to Earth than 24 million km.

Near-Earth Asteroids

Most asteroids spend their entire lives within the main belt. However, there are many millions of small rocky objects – most of them originating in the main asteroid belt – which follow orbits that take

Figure 13.24 There are three main groups of near-Earth asteroids: the Apollos, Atens, and Amors. The Apollos and Atens currently cross Earth's orbit. The Amors remain outside Earth's orbit, but may be diverted into Earth-crossing orbits in the future. All of these are a potential impact threat to our planet. (ESA)

them into the inner Solar System. Many of these approach or cross Earth's orbit.

Fortunately, "extinction-class" objects in Earth-crossing orbits and measuring at least 10 km across are very rare. An impact by one such object in Mexico's Yucatan peninsula is widely accepted as the cause of the dinosaurs' demise 65 million years ago (see Chapter 3).

It is estimated that there are about 1,000 near-Earth asteroids (NEAs) larger than 1 km and roughly 15,000 larger than 140 m. All of them are capable of causing large-scale devastation on our planet through the heat and shock wave generated by their immediate surface impact and the subsequent ignition of huge fires which create clouds of smoke and other pollutants that may trigger rapid climate change.[8]

As their size and explosive potential decreases, the number of NEAs increases dramatically. There are thought to be at least 100,000 objects larger than 100 m, each capable of causing regional destruction or generating a large tsunami after splashing into one of Earth's oceans.

Furthermore, it is estimated that a 100-meter-class asteroid passes closer than the Moon 50 times a year, while 100 fist-sized meteorites, fragments of NEAs, fall to Earth each day – although

most of these break up in the atmosphere, producing bright meteors or fireballs.

In 1998, NASA established a goal to discover 90% of the near-Earth objects larger than one kilometer in diameter and in 2005, Congress extended that goal to include 90% of the NEOs larger than 140 m.

NEAs typically have a survival time in deep space of about 10 million years. Since their population is continually being depleted through collisions with terrestrial planets or gravitational interactions that send them into different orbits, there must be some means of replenishing the population.

Some of the newcomers may be extinct comet nuclei, e.g. 3200 Phaethon, but many are probably removed from the Kirkwood gaps in the main belt by Jupiter's gravity and slung into highly eccentric orbits that send them into the inner Solar System. Others may be derived from high velocity impacts or drift toward the Sun as a result of the Yarkovsky effect (see Box 13.4).

The first asteroid found to cross the orbit of Mars was 433 Eros, discovered by Gustav Witt in 1898. With a maximum diameter of 33 km, it is now recognized as the second-largest NEA.[9] A great deal is now known about Eros, since it was explored by the NEAR-Shoemaker spacecraft (see Box 13.6).

Eros is not likely to collide with Earth any time soon, but there are many objects which offer a potential threat. The first object to be discovered on a path that crossed the Earth's orbit was 1862 Apollo, found in 1932 (see The Impact Threat to Earth).

NEAs are classified according to their orbital distances from the Sun at perihelion and aphelion. The three main groups are named after a representative asteroid.

The **Amors** always remain further than Earth's aphelion distance, with perihelia of 1.017–1.3 AU. Many of them cross the orbit of Mars and they may approach Earth. Close encounters with Mars or Earth can perturb them into Earth-crossing, Apollo-type orbits. They are named after 1221 Amor. Eros is the most famous example, but more than 8,000 Amors have so far been found.

The **Apollos** cross Earth's orbit. They have semi-major axes greater than 1 AU and perihelia less than 1.017 AU (Earth's aphelion distance). Since they spend most of their time beyond Earth's orbit, their orbital periods are longer than one year. The first to be found was 1862 Apollo, which is now known to have a small satellite. Other notable members include 1566 Icarus, 1866 Sisyphus (the largest, with a diameter of about 8 km), 3200 Phaethon, 1685 Toro, and 4179 Toutatis. Some Apollos have very eccentric orbits. Icarus, for example, crosses the orbit of Mercury. Approximately 9,500 Apollos have so far been found.

The **Atens** also cross Earth's orbit, but, with orbital periods of less than one year, and semi-major axes smaller than 1 AU, they spend most of their time closer to the Sun than Earth. They are difficult to find, because they are usually hidden in the Sun's glare and rarely appear in a dark sky. They are named after 2062 Aten. The current record holder for closest approach to the Sun is 1995 CR, with a perihelion distance of 0.12 AU. About 1,300 Atens had been found by late 2018.

---

[8] A 100-m rocky object has a mass of the order of 1 million tons, whereas a 1-km object has a mass of about 1 billion tonnes.

[9] 1036 Ganymed is thought to be slightly larger than Eros.

Box 13.7   The Torino and Palermo Scales

The Torino Scale is a color-coded advisory system that enables NEO researchers to place objects within a potential threat range from zero (virtually no chance of collision) to 8 and 9 (a collision is certain) and 10 (global catastrophe is certain). It was first adopted in 1999 by a working group of the International Astronomical Union (IAU) at a meeting in Torino (Turin), Italy. It is also of value in describing the impact threat to the media and the general public.

The highest Torino score yet given was a level 4, assigned briefly to asteroid 2004 MN4 (99942 Apophis) in December 2004, with a forecast 2% chance of hitting Earth in 2029. After extended tracking of the asteroid's orbit, it was reclassified to level 1, with an extremely low chance of collision.

The logarithmic Palermo Technical Impact Hazard Scale has also been developed to enable NEO specialists to categorize and prioritize potential impact risks spanning a wide range of impact dates, energies, and probabilities. There is no direct correlation between the Palermo Scale and the Torino Scale.

Figure 13.25 The extinction of the dinosaurs 65 million years ago is widely attributed to the impact of a 10 km-diameter asteroid which created the 180 km-wide Chicxulub crater in the Yucatan Peninsula, Mexico. Such impacts with Earth are now very rare, taking place on time scales of many millions of years. (NASA)

A fourth group, dubbed **Apoheles**, remains entirely within Earth's orbit.[10] Some of them cross the orbits of Mercury and Venus. They are also known as **Atiras**, after the first representative (163693) Atir, which was discovered in 2003. Apoheles are thought to account for perhaps 2% of the total NEA population, so they are rare, as well as difficult to discover because they stay in the daytime sky almost all of the time. As of October 2018, there were 30 confirmed Apoheles.

Although the Apoheles are unlikely to threaten Earth, long-term planetary perturbations can alter the orbits of the Atens and Apollos, as well as about 50% of the Amors, eventually sending them toward Earth. One consequence of this potential threat is that NEAs may one day be targets for manned space exploration, not only for scientific research but for the potential exploitation of minerals and other resources.

The Impact Threat to Earth

Fortunately, collisions of asteroids and comets (together they are known as near-Earth objects or NEOs) with our planet are rare. Estimates based on lunar crater counts and studies of the orbital dynamics of the NEO population suggest that Earth currently suffers globally catastrophic impacts with objects larger than 1 km once every 500,000 to 700,000 years on average. Regionally devastating impacts (impact energy at least 4×10^{18} J) occur on average every 41,000 to 53,000 years (Figure 13.26). Events with energies similar to that of the object that exploded over Tunguska, Siberia,

in 1908 are thought to occur once every 200 to 300 years on average (see Tunguska).

Close approaches by Potentially Hazardous Asteroids (PHAs) occur every few weeks or months. In order to establish the actual impact threat for the next century or so, NASA has been given the task of drawing up an inventory of the current NEO population. The first goal, established in 1998, was to find 90% of all NEOs larger than 1 km within 10 years. In 2005, Congress mandated NASA to extend the search to include 90% of the much more numerous objects larger than 140 m across (see Near-Earth Asteroids). There are thought to be about 1,000 NEOs larger than one kilometer and roughly 15,000 larger than 140 meters.

On October 25, 2019, there were 2,018 known PHAs, but none of them is known to be on a collision course with our planet.[11] The most notable close encounter so far predicted involves 99942 Apophis. It is now expected to skim past Earth at a distance of about 22,000 km on April 13, 2029, passing inside the orbit of

---

[10] Apohele is the Hawaiian word for orbit.

[11] Only 1% of the impact threat to Earth is thought to come from comets. The frequency of such impacts involving long period comets is about one every 32 million years. The downside is that they usually arrive at much higher speeds from the outer Solar System, providing only a few months of advance warning.

geostationary satellites. There is also a 1 in 45,000 probability that this 270-m asteroid might return and collide with Earth in April 2036. The energy release from an impact by Apophis would be in the 10,000 megatonne range (see Chapter 3).

Dense, metallic asteroids pose the greatest threat, since they generally survive all the way to the surface. Fortunately, the damage from stony asteroids falls off very rapidly for sizes smaller than Tunguska because the energy release is smaller and the explosion takes place at such a high altitude (15 km+) that there is no noticeable damage on the ground. Similarly, an ice-rich object tends to explode at very high altitude.

An example of a modest, but unexpected and damaging object was the Chelyabinsk meteoroid, which exploded in the morning sky over Russia on February 15, 2013, shattering windows and knocking people to the ground. The object, probably about 17 m across, entered the atmosphere at a shallow angle and left a vapour trail before it disintegrated at an altitude of 15–20 km. Scientists estimate the total energy of the event was equivalent to an explosion of about 500 kilotonnes of TNT, making it the largest airburst by an exploding meteorite since Tunguska in 1908. A retrieved rock sample showed it was the most common type of meteorite, an LL chondrite.

Despite their differences in size and density, the consequences of comet and asteroid impacts on Earth are roughly comparable. The key lies in the amount of energy they release. Since the kinetic energy of an object is determined by its mass and velocity (K.E. = $\frac{1}{2} mv^2$), a large icy object traveling very fast may pack at least the same punch as a much smaller rocky object traveling at a slower speed.

The average speed of asteroids colliding with Earth is about 20 km/s, whereas long period comets arrive at much greater speeds, often around 50 km/s. Even though a comet is much less dense than an asteroid, its faster speed means that impacts involving comets and asteroids of similar size release comparable amounts of energy. The energy released by the impact of a 100-m asteroid is comparable to a 50-megatonne bomb, whereas a 1 km-diameter asteroid releases 250,000 megatonnes of explosive energy – enough to wipe out most life on Earth.[12]

Tunguska

The most famous impact-related event in modern times occurred in 1908, when a stony asteroid (a comet is now generally discounted) about 60 m across exploded above Tunguska, Siberia, causing devastation over an area about 50 km in diameter.[13]

On the morning of June 30, 1908, a huge explosion occurred above the vast boreal forest near the Podkamennaya Tunguska River, north of Lake Baikal. Subsequent investigation showed that the dramatic event was caused by the destruction of an incoming cosmic object – the only entry of a large meteoroid in the modern era to be witnessed at first hand.

The explosion flattened some 80 million trees over an area of more than 2,000 sq. km. (If it had occurred over a city such as

Figure 13.26   This photo taken during the 1927 expedition shows parallel trunks of trees that were flattened by the blainst from the Tunguska event. Note how the branches have been stripped off the trees. (ESA)

London, the entire urban area and its population would have been wiped out.) Moreover, they were aligned in a radial pattern, pointing directly away from the blast's epicenter.

The scene changed at ground zero: the trees were standing upright, but their limbs and bark had been stripped away, resembling a forest of telephone poles. Such debranching requires fast-moving shock waves that cause the branches to snap off before they can transfer the impact momentum to the tree trunk. There was no sign of an impact crater and no fragments of an incoming object could be found.

The after-effects of the explosion were experienced around the world. A seismic shockwave was registered by instruments as far away as England. Massive, silvery clouds formed at high altitudes above the Tunguska region, reflecting light from the Sun after sunset. Some people saw brilliant, colored sunsets, others who lived as far away as England and Asia witnessed luminescent skies that enabled them to read newsprint at midnight. Locally, hundreds of reindeer were killed, but there seems to have been no loss of human life.

Sandia National Laboratories supercomputer simulations completed in 2007 suggest that the explosion – previously estimated to be between 10 and 20 megatonnes – was more likely to be three to five megatonnes (though still equivalent to a modern nuclear bomb).

Many explanations have been proposed over the years, but it is now generally agreed that an asteroid entered the atmosphere above Siberia and then exploded at an altitude of approximately 8 km. The rocky object may have measured 30–60 m across, depending on its speed, porosity, water content, composition, and other material characteristics.

---

[12] One megatonne is equivalent to the explosive power of one million tonnes of TNT – comparable to roughly 77 Hiroshima atomic bombs.

[13] A less well-known example is the February 1947 Sikhote-Alin strike in Siberia, about 430 km north east of Vladivostok, which created a cluster of small craters.

Box 13.8   Saving Earth

Relatively small-scale explosive events such as Tunguska may occur on average once every 200–300 years. Since larger objects are much less common, the expected frequency of Earth impactors decreases as the size of the incoming object increases (see The Impact Threat to Earth). However, this is counterbalanced by the much larger number of deaths and more widespread devastation associated with a mega-impact. Calculations show that the probability of someone dying in any one year due to a cosmic impact is of the order of 1 in one million. Surprisingly, this is comparable with the likelihood of dying in a plane crash, and more likely than death due to a tornado.

Although no one in living memory has died from an asteroid impact, the subject is now being taken quite seriously, with NASA and other agencies beginning to consider how best to deal with the potential threat. Given sufficient warning, modern technology should be able to prevent humanity from suffering a grisly extinction like the dinosaurs.

An overall Earth protection system must have three components. The first step is to identify and catalog any potential NEO impactors above a certain size, although there will always be an additional threat from long period comets that may appear without warning. This cataloging is based primarily on ground-based observations, although spacecraft, such as NASA's WISE (launched in December 2009) and ESA's Gaia (launched in December 2013), can also be used to search for small asteroids in the inner Solar System. (By the end of its mission in early 2011, WISE had discovered 134 NEOs.)

Second, a series of space missions should be launched to understand the structure, composition, rotational state, and other physical properties of potential impactors. Although more than a dozen asteroids and several comets have already been studied from close range by spacecraft, much more needs to be learned about the nature of potentially hazardous objects.

Finally, technologies must be available to prevent the intruder from impacting Earth. Several Hollywood blockbusters have shown astronauts being dispatched to save Earth from imminent destruction by a NEO, but, in reality, no practical systems to deflect or destroy such an object exist at the present time.

The most obvious solution would seem to be launching a missile with a nuclear warhead to break up a large asteroid. Unfortunately, this is likely to create a swarm of smaller objects that would continue on a similar course, resulting in multiple impacts with similar total energy release around the world – rather like the collision of comet Shoemaker-Levy 9 with Jupiter (see Breaking Up Is Easy to Do).

The less drastic alternative of deflecting an object from its collision course has been widely studied. A change in orbital speed of the order of a few cm/s will usually be sufficient. Deflection techniques generally involve the sudden introduction of a fairly large lateral force or a slow, relatively gentle, but prolonged thrust.

Once again, nuclear devices have been suggested, since they supply around a million times more energy per unit of mass than conventional explosives. A thermonuclear explosion would release highly energetic particles, X-rays, and gamma rays. The resultant vaporization of the surface would cause the object to recoil, similar to when a rifle is fired (see Newton's laws, Chapter 1).

The effect of such an explosion would be substantially increased if it took place beneath the surface. In this case, most of the explosive energy would be transferred to the target, resulting in the high-speed expulsion of a great deal of material. This would result in a much larger change in orbital velocity than a similar explosion close to the asteroid. Unfortunately, it would not be easy to place a thermonuclear charge deep inside a NEO, and the level of unpredictability and risk is very high, with the possibility that the fragile object could fragment.

A less drastic deflection method would involve the transfer of kinetic energy through an impact by a non-explosive projectile. The amount of energy exchanged would depend on the mass and relative impact speed of the projectile. For example, in order to divert a 2 km-diameter asteroid following a prograde circular orbit at 1 AU, it is necessary to deliver a 60-tonne projectile at a velocity of 10 km/s. The mass of the projectile could be greatly reduced by launching it into a retrograde orbit, so that a head-on impact speed of 60 km/s could be achieved.

A more subtle approach involves a gravity tractor, in which the tiny gravitational pull provided by a nearby thrusting spacecraft would very slowly accelerate the asteroid in the spacecraft's direction. Although the acceleration imparted to the asteroid would be very small, the technique would be able to slightly alter an asteroid's orbit. Calculations show that a 20-tonne gravity tractor could deflect a 200-m asteroid after a year of such "towing."

One of the earliest deflection proposals involved placing an electromagnetic "mass driver" on the surface of a NEA. Material excavated on the asteroid's surface would be fired into space, creating an equal and opposite reaction that gradually modifies the NEA's orbit. A similar idea involves anchoring ion engines or a reflective solar sail on an asteroid in order to apply a weak propulsive force over a long period of time.

The difficulty in anchoring such devices can be overcome if a solar sail or mirror is placed in orbit. By focusing incoming solar radiation onto the NEA, some of the surface would be strongly heated and vaporized. The resultant jet of material would modify the object's trajectory.

Whichever method is chosen, it should be borne in mind that a deflection mission may take years to develop and up to 5 years to reach its target. Furthermore, the propulsive procedure itself may require many years to have the desired effect – so the threat must be recognized well in advance.

Figure 13.27   The Hoba meteorite (also known as Hoba West) is the largest single meteorite on Earth. The 60-tonne iron-nickel slab was found by a farmer in 1920, near Grootfontein, in Namibia. It is thought to have arrived over 80,000 years ago. It landed at such a slow speed that it survived the impact and no crater was found. Flat on both major surfaces, the meteorite's unusual shape may have caused it to skip across the upper atmosphere like a flat stone on water. (Patrick Giraud/Wikimedia Commons)

Traveling at a speed of about 53,600 km/h, the asteroid would heat the surrounding air to 25,000°C before the combination of pressure and heat caused it to blow itself apart in a huge fireball. A high-temperature jet of expanding gas struck the ground beneath, causing blast waves and thermal pulses that flattened the forest beneath.

Meteorites

Meteorites are small, rocky objects that enter Earth's atmosphere at high speed and survive their fall to reach the surface. They are classified according to their chemical and physical characteristics. Some are metallic, consisting mostly of iron alloyed with nickel, but most are stony. These characteristics provide clues to their place of origin and the conditions which prevailed in their parent body. Most meteorites originated in the main asteroid belt.

There are estimated to be about 20,000 meteorite falls involving objects heavier than 100 g every year. They occur largely at random over Earth's surface, although the number is slightly reduced near the poles, since most meteorites (and asteroids) travel close to the ecliptic plane.

The majority of meteorites splash down in the oceans or arrive on land without being noticed. However, many thousands have been found on every continent. The best-known hunting ground for meteorite hunters is Antarctica, where the slowly moving ice sheets preserve the cosmic pebbles and slowly concentrate them near mountains which protrude above the ice, much as waves wash pebbles to the shore. Each summer, expeditions scoop up the dark rocks which stand out so clearly in the white wilderness.

Flat, sandy desert regions are similarly favored, although it is not always so easy to recognize a meteorite if numerous local rocks are also scattered across the surface. Like Antarctica, deserts offer ideal prospecting places, since the dry climate prevents chemical weathering. Parts of the Sahara have proved particularly productive, notably the Dar al Gani (Libya), Hammada al Hamra plateau (Libya), and Acfer (Algeria) areas.

Most meteorites are small, partly because the weaker stony objects fragment during their high speed fall to Earth, creating a meteorite shower. Iron meteorites are stronger and generally survive intact (Figure 13.27). The largest single meteorite on Earth is the 60-tonne Hoba iron-nickel meteorite which remains where it fell, near Grootfontein in Namibia, and is now a tourist attraction.

Meteorites typically enter the upper atmosphere at about 30 km/s, but their descent is slowed by friction. This causes the outer surface to heat up and melt, although the interior remains cool and unchanged. Its passage usually results in a bright fireball as molten droplets are carried away in its wake.

By the time it hits the ground, the temperature of the outer skin has dropped sufficiently for it to feel cold. The only outward signs of its fiery entry are a dark, glassy, fusion crust, possibly allied to a curved or conical shape which formed as the projectile traveled through the air.

The Origin of Meteorites

Studies of the chemical compounds that make up meteorites show that they are the oldest available samples of Solar System material, providing invaluable information about the conditions that prevailed 4,560 million years ago.

Almost all meteorites are fragments of asteroids that broke free or were ejected during collisions. These rocks were then hurled

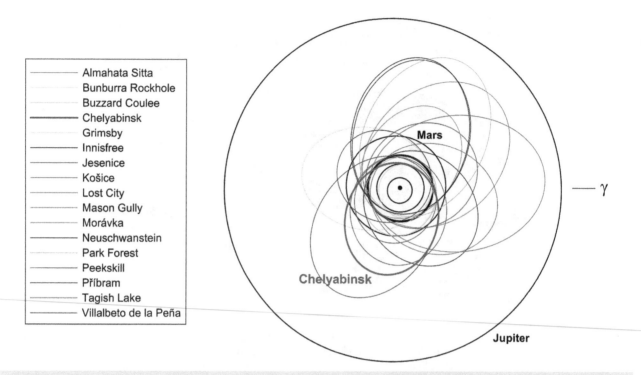

	Almahata Sitta
	Bunburra Rockhole
	Buzzard Coulee
	Chelyabinsk
	Grimsby
	Innisfree
	Jesenice
	Košice
	Lost City
	Mason Gully
	Morávka
	Neuschwanstein
	Park Forest
	Peekskill
	Příbram
	Tagish Lake
	Villalbeto de la Peña

Figure 13.28   The calculated orbits of Earth-impacting meteorites and the Chelyabinsk meteoroid, inferred from their trajectories before hitting the ground. All of them originated in the main asteroid belt between Mars and Jupiter, indicating that these meteorites are chips off asteroids. (Jiri Borovicka, Astronomical Institute of the Czech Academy of Sciences)

toward Earth, primarily by the disruptive effects of Jupiter's gravity or by the influence of sunlight (see Box 13.4).

Confirmation of the asteroid belt as the main source region has come from calculation of some meteorite orbits, based on photographic images or video records, e.g. Příbram, Lost City, Neuschwanstein, and Peekskill (Figure 13.28). Dedicated fireball detection networks, such as one established on the Nullarbor Plain in Australia, have also had some success. However, establishing a direct link between particular parent asteroids and the majority of meteorites that fall on Earth is almost impossible.[14]

The first task is to determine (or at least estimate) an asteroid's size/volume and composition, in order to calculate its mass. (This is easier for asteroids than planets or spherical moons since they are too small to compress gravitationally.) Spectroscopic data are then used in an attempt to verify differences in composition.

The next step is to compare these data with different meteorite samples in an effort to find links between them and their parent asteroid types. This is often problematic, since prolonged space weathering alters the nature of asteroid surfaces. However, there are good matches between C class asteroids and some stony meteorites. Similarly, many basaltic meteorites are thought to have originated from 4 Vesta.

Other exceptions are the relatively small number of meteorites known to have originated on Mars and the Moon. Ejected by

high-energy collisions, they were accelerated to beyond escape velocity, and traveled on circuitous routes through space until they encountered Earth. Their origins have been confirmed by chemical analysis (see Meteorites from the Moon and Mars).

Types of Meteorites

There are three main types of meteorites: stones, irons, and stony-irons (Figure 13.29). Primitive stony meteorites are known as chondrites, so-called because they contain spherical grains known as chondrules. Made of silicates, these chondrules are typically about 1 mm in diameter.[15]

Largely made of olivine and pyroxene (minerals rich in magnesium, iron, and silicon) they appear to have been small balls of dusty material in the primordial solar nebula that were suddenly heated to over 1,400°C, possibly due to shock waves in the solar nebula. They became molten before rapidly cooling and recrystallizing. During this process, water and other volatiles were vaporized, leaving behind the grains of silicates and iron.

There are three main classes of chondrites, based on their chemistry. Ordinary chondrites are the most common meteorites found on Earth, accounting for about 94% of all finds. Most of them experienced high temperatures before they recrystallized. They can

---

[14]The first definitive asteroid/meteorite link was made after a small asteroid known as 2008 TC3 was detected before it exploded over the Nubian Desert of northern Sudan on October 7, 2008. 208 fragments were subsequently recovered. 2008 TC3 had the spectral signature of an F-class asteroid, while the meteorite was a very rare and unusually fragile rock known as a polymic ureilite.

[15]The term chondrule comes from the Greek word "chondros," meaning seed or "little grain."

Figure 13.29 A 700-g piece of NWA 869, an ordinary chondrite meteorite. Numerous small, round chondrules and flecks of metal can be seen on the cut and polished face of this specimen. (Wikimedia)

Figure 13.30 A rare carbonaceous chondrite meteorite, part of the large fall at Allende in Mexico. It contains rounded chondrules and irregular patches of calcium-aluminum-rich inclusions (CAIs) which are white or light gray. CAIs are believed to pre-date chondrules by at least 2 million years. (Wikimedia)

Figure 13.31 This iron meteorite was discovered near Tamentit in the Algerian Sahara. It displays numerous rounded indentations known as regmaglypts. These features are most likely due to the selective melting and ablation of areas with different compositions during its descent though the atmosphere. (Wikimedia)

be distinguished from terrestrial rocks because their iron content makes them slightly attractive to a magnet.

Carbonaceous chondrites are made of undifferentiated silicates (including chondrules) and their chemical compositions match the chemistry of the Sun more closely than any other class of chondrites. Hence, they represent the primordial material from which planets accreted. Less than 5% of all meteorites belong to this class.

These primitive and undifferentiated meteorites formed in oxygen-rich regions of the early Solar System, so that most of the material is found as silicates, oxides, or sulfides. Most of them contain water, or minerals that have been altered in the presence of water, and some contain large amounts of carbon.

Specimens such as the Murchison meteorite contain amino acids and complex organic compounds that are important building blocks for life on Earth. They may also contain numerous white "inclusions" rich in aluminum and calcium. Dating of some inclusions has shown that they formed 4,560 million years ago – the oldest known solids in the Solar System.

Based on their distinctive compositions, there are at least seven groups that formed on different parent bodies in different regions of the solar nebula. Some groups have retained more of their volatiles in the form of carbon compounds and water because they have not undergone significant heating since their formation. It seems that their low temperature enabled water to produce clays, sulfates, and carbonates. Perhaps the most famous specimens are from the Allende meteorite shower that fell in Mexico on February 8, 1969.

Enstatite chondrites differ by being poor in volatiles and less oxidized. They contain up to 15% iron-nickel, very small chondrules and sulfides.

All other meteorite types (achondrites, stony-irons and irons) are differentiated. Since they were once completely molten, they contain no water or other volatiles. Achondrites, which account for less than 5% of all finds, are mainly igneous silicates that probably originated in the mantle or crust of differentiated asteroidal parent bodies. They rarely contain chondrules and have a low metal content.

Irons are made of nickel-iron, with some sulfur, so they are much denser than other meteorites. They probably originated in the differentiated cores of large asteroids. About 42% of all

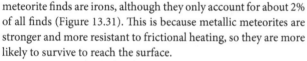

Figure 13.32   A slice through a pallasite stony-iron meteorite found near Springwater, Saskatchewan, in 1931. Pallasites are composed of about half metal and half olivine, a greenish silicate mineral. (Wikimedia)

Figure 13.33   Iron meteorites are the most common of the few meteorites that have been found by roving vehicles on Mars. These three iron meteorites – fragments of the same object – were found by NASA's Curiosity Mars rover in May 2014. The largest piece is about 2 m wide. Angular cavities on their surface may result from preferential erosion along crystalline boundaries within the metal. Another possibility is that these cavities once contained olivine crystals, which can be found in a rare type of stony-iron meteorites called pallasites. (Courtesy of NASA/JPL-Caltech/LANL/CNES/IRAP/LPGNantes/CNRS/IAS/MSSS)

meteorite finds are irons, although they only account for about 2% of all finds (Figure 13.31). This is because metallic meteorites are stronger and more resistant to frictional heating, so they are more likely to survive to reach the surface.

As their name implies, stony-irons are composed of silicates mixed with iron and nickel. Their compositions range from droplets of olivine imbedded in pure nickel-iron to droplets of nickel-iron in pure olivine. The stony-irons are rare, accounting for only about 1% of all falls.

Some (called pallasites) probably came from a core-mantle transition region inside a large asteroid, where metallic core components had not entirely separated from the silicate mantle (Figure 13.32). Others (called mesosiderites) have a jumbled texture comprising angular grains of rocky minerals and metal, suggesting they are the product of collisions between asteroids of different composition.

Meteorites from the Moon and Mars

More than 59,000 meteorites have been discovered on Earth, but fewer than 600 have been identified as originating from the Moon or Mars. All of these are pieces of debris that were ejected by a sizeable impact which excavated a large crater, then accelerated them beyond escape velocity. After a lengthy journey through space, they entered Earth's atmosphere and survived the heat of descent to reach the surface. (It is also likely that many terrestrial meteorites have made the reverse journey from Earth to the Moon or Mars during the history of the Solar System.)

Small meteoroids impact the Moon every day and the flashes of these impacts have been observed and imaged on numerous occasions. Far less frequent, particularly in the present epoch, are impacts that are sufficiently energetic to accelerate ejecta to the lunar escape velocity of 2.38 km/s – a few times the velocity of a rifle bullet – so that they can escape the Moon's gravitational influence. The lunar rocks may eventually be captured by Earth's

gravity after a journey lasting a few hundred thousand years (or less).

It is a similar story with Martian impacts. Although the escape velocity is higher (5 km/s), it is still possible for ejecta to be blasted into an interplanetary trajectory that may eventually take it close to Earth.

The history of these meteorites can often be determined from their composition and exposure to cosmic rays. By studying their textures and mineral content, it is clear that they come from many different impact events. Indeed, it has been argued that most lunar meteorites are derived from relatively small, recent, impacts that created craters only a few kilometers in diameter. Studies of Martian meteorites indicate that they originated in at least three fairly recent craters whose diameters are greater than 12 km.

How is it possible to know whether a meteorite came from the Moon? First, its abundances of major minerals and elements match those established through remote sensing and analysis of Apollo samples. Their oxygen isotopic compositions are also identical. Finally, nitrogen and noble gases implanted in them by the solar wind whilst they were on the lunar surface show that they came from a world with no atmosphere.

In the case of the SNC meteorites from Mars, all are igneous and extremely young compared to other achondrites.[16] Dating

---

[16]The SNCs are named after the first three subgroups of Martian meteorites that were identified (Shergotty, Nakhla, and Chassigny).

of their crystals shows that most of them solidified between 1.3 and 0.16 billion years ago, long after most stony meteorites. They obviously originated on a parent body that supported volcanic activity until recent times, which favors either Venus or Mars. However, analysis of gases trapped in the rocks shows that they match the composition of the Martian atmosphere, as measured by spacecraft landers (see Chapter 7).

Since impacts may occur at random over the entire Moon (or Mars), meteorites could come from anywhere on their surfaces. Although it is almost impossible to directly link an individual rock with a particular crater, some assumptions can be made about the general source region.[17] For example, meteorites that are rich in the mineral anorthite and have high concentrations of aluminum and calcium probably originated in the lunar highlands, which are also composed predominantly of the same material. Such rocks are not widely found on other Solar System bodies.

It is possible to determine how long ago a rock left the Moon or Mars by studying its isotope or nuclide content. Rocks lying on the lunar surface or traveling through space are exposed to cosmic ray hits. These are so energetic that they cause nuclear reactions in the elements, changing one nuclide into another.

Some of the isotopes produced are stable, and become more abundant over time. Hence, the duration of the meteorite's exposure to cosmic rays in space can be determined by measuring the abundance of these rare stable isotopes and elements, such as ^{3}He (helium 3), neon, argon, krypton, and xenon. This is used to calculate their time of flight to Earth and how long ago they arrived on Earth. Using this technique, it has been calculated that the Dhofar 025 meteorite left the Moon at least 13 million years ago, whereas Kalahari 008/009 took only a few hundred years to reach Earth.

Some of the nuclides produced are radioactive. However, the production of these nuclides ceases as soon as the meteorites fall to Earth, because the atmosphere absorbs nearly all cosmic rays. By measuring how much they have decayed, it is possible to determine the length of time that has passed since they arrived on Earth. Radionuclides used for dating of meteorites' arrival on Earth include ^{14}C (carbon 14), ^{10}Be (beryllium-10), ^{26}Al (aluminum 26), ^{36}Cl (chlorine 36), and ^{41}Ca (calcium 41).

As might be expected from the greater distance and higher escape velocity of Mars, the number of known meteorites from the red planet (215) is even smaller than the lunar collection (349). Some of them show evidence of having cooled rapidly from molten lava, while others are more coarse-grained, suggesting that they cooled more slowly at depth before being excavated. Some also contain minerals, such as carbonates, that were produced during exposure to water at fairly low temperatures. One famous example is ALH 84001 (see Chapter 7).

Meteors

Stare at the sky on a clear, Moonless night and you are likely to see a meteor, commonly known as a "shooting star." These brief light

Table 13.1
Major Meteor Showers

Shower	Dates	ZHR*	Parent Comet
Quadrantids	Jan 1–6	100	96P Macholz 1?
Lyrids	Apr 19–25	10–15	C/1861 G1 Thatcher
Eta Aquarids	Apr 24–May 20	50	1P Halley
Delta Aquarids	Jul 15–Aug 20	20–25	96P Machholz 1?
Perseids	Jul 25–Aug 20	80	109P Swift-Tuttle
Orionids	Oct 15–Nov 2	30	1P Halley
Leonids	Nov 15–20	100	55P Tempel-Tuttle
Geminids	Dec 7–15	100	Asteroid 3200 Phaethon

*Approximate zenithal hourly rate

shows are produced by grains of cosmic debris, perhaps 2–5 mm across, most of which have been jettisoned into space by passing comets. The meteors leave glowing trails as they are incinerated by frictional heating during their high speed (11–76 km/s) entry into Earth's upper atmosphere. The result is a short-lived streak of light caused by ionized and excited atmospheric atoms. Occasionally, larger objects may produce fireballs, or even brighter bolides, whose passage is marked by a sonic boom.

A few sporadic meteors per hour may be seen at any time during the year (Table 13.1). However, the number of visible meteors rises considerably during meteor showers. There are some 20 major showers each year, when shooting stars seem to radiate like the spokes of a wheel from one point in the sky, although the particles are actually traveling along parallel paths. The showers are named after the constellation from which they appear to originate, e.g. the Leonids, which radiate from Leo. However, they are (with one exception) all caused by Earth plowing through streamers of ice and dust that periodic comets have left trailing in their wake.[18]

The showers occur when Earth encounters a stream of comet debris around the same dates each year. However, they may vary in timing, brightness, and hourly rate of meteors. This is particularly true if the associated comet has made numerous journeys through the inner Solar System, leaving behind many different streams. Meteor showers are most prominent if the comet has passed close to Earth's orbit in the last few years. The duration of each shower ranges from a few hours to several days.

One of the most famous annual meteor showers is known as the Leonids, associated with dust particles ejected from comet 55P/Tempel-Tuttle, which visits the inner Solar System once every 33.25 years. They appear November 15–20, when Earth passes very close to the comet's orbit. However, the numbers of meteors on view vary tremendously.

---

[17] In 2014, a paper was published which claimed that 55 km-wide Mojave Crater was the ejection source for the Martian meteorites classified as shergottites, based on their ages and composition.

[18] The exception is Apollo asteroid 3200 Phaethon, which follows a similar orbit to the debris that causes the Geminid meteor shower. Phaethon's association with a meteor stream suggests that it may actually be an old comet nucleus.

Figure 13.34   This drawing shows the famous Leonid meteor storm that took place on November 12, 1833, when the skies over the United States were ablaze with shooting stars. It was, one eyewitness said, as if "a tempest of falling stars broke over the Earth . . . . The sky was scored in every direction with shining trails and illuminated with majestic fireballs." (ESA)

In most years, observers may see a peak of perhaps 5–10 per hour around November 17. But when Tempel-Tuttle approaches the Sun, it can generate a magnificent storm, with thousands of shooting stars illuminating the night sky (Figure 13.34). The most spectacular apparitions came in 1833 and 1966, with dazzling displays of over 100,000 meteors an hour.

The Leonids are also renowned for producing bright fireballs which outshine every star and planet. Their long trails are often tinged with blue and green, while their vapor trains may linger in the sky like enormous smoke rings for 5 minutes or more.

Although the incoming particles are small, ranging from specks of dust to the size of small pebbles, the Leonids glow brightly

Box 13.9   Meteors, Meteoroids, and Meteorites

Even experts are sometimes confused by the subtle differences in the meaning of these three terms. A meteoroid is a small rocky object (typically less than 100 m in diameter) traveling through space, particularly one with the potential to encounter Earth. If it enters Earth's atmosphere, its passage usually results in a meteor – a luminous trail across the sky (commonly called a "falling star" or "shooting star"). Tiny meteoroids usually vaporize completely in the atmosphere due to friction with air molecules, but larger specimens may survive to become stony or metallic remnants on Earth's surface. These rocks are known as meteorites.

Figure 13.35   A 10-micron interplanetary dust particle collected in the stratosphere with a modified U2 aircraft.[19] This particle has a similar elemental composition to primitive meteorites, but with higher abundances of carbon and volatile elements. The particle is composed of glass, carbon, and many types of silicate mineral grains. Its porosity and unusual composition suggest that it may be of cometary origin. (NASA-JPL)

It is estimated that more than 30,000 tonnes of meteoroid material enter Earth's atmosphere every year. Of this, perhaps 10% reaches the surface in the form of meteorites and micrometeorites. The latter group comprises particles of dust which are so tiny – typically less than 120 microns across – that their energy is dissipated before they burn up in the atmosphere. These particles have been collected by spacecraft and by high-flying aircraft, as well as on the ground, where they are often recognized through their magnetic properties. Some 4,000 tonnes of micrometeorite material may fall to Earth each year.

because they are the fastest of all the meteors. A typical Leonid, arriving at a speed of 71 km/s (more than 200 times faster than a rifle bullet), will start to glow at an altitude of about 155 km and leave a long trail before it is extinguished.

[19]This is a civilian version of a famous Cold War spy aircraft, adapted for use by NASA.

Long-Haired Stars

Comets are small bodies consisting of frozen volatiles and dust. They are thought to be ice-rich planetesimals left over from the formation of the planets, about 4.5 billion years ago. Some comets become splendid naked eye objects, so it is not surprising that their existence has been known since time immemorial.

At the heart of all comets is a solid nucleus, which is usually invisible through even the largest telescopes. This is partly because comets spend most of their lives far from the Sun (and Earth) and partly because their nuclei are typically only a few kilometers across. Furthermore, any comets that enter the inner Solar System are masked by a surrounding cloud of gas and dust known as a coma. The roughly spherical coma is fed by jets of material that erupt into space as the surface of the nucleus is warmed by solar radiation.

The coma is mainly composed of water vapor and carbon dioxide. Some comas display the greenish glow of cyanogen (CN) and carbon when illuminated by sunlight. This is called "resonant fluorescence." Other compounds of carbon, hydrogen, and nitrogen have been found. Ultraviolet images from spacecraft have also shown that the visible coma is surrounded by a huge, sparse cloud of hydrogen gas.

Many ancient civilizations saw them as portents of death and disaster, or omens of great social and political upheavals. Shrouded in thin, luminous veils with tails streaming behind them, they were termed "kometes" by the ancient Greeks, meaning "long-haired stars." In the pre-telescopic era this name was particularly appropriate, since naked eye comets often display one or more diaphanous tails that extend for millions of kilometers from the nucleus and surrounding coma.

Despite their size, the tails are clearly insubstantial – stars can be seen shining through them. Indeed, in 1910, news stories warning about the imminent "end of the world" proved to be unfounded when Earth passed unscathed through the tail of comet Halley. It was rather like a bowling ball speeding through a cloud of smoke.

Two tails are generally recognized (Figure 13.37). One of these is the yellowish dust tail, usually broad, stubby and curved, which is formed when tiny dust particles in the coma are pushed away by solar radiation pressure, as photons of light impact the grains. The particles ejected during a comet's passage through the inner Solar System eventually spread along its orbit. When Earth ploughs through the cloud of debris, the particles enter the upper atmosphere and burn up as meteors (see Meteors).

Meanwhile, gases released by the vaporizing nucleus are ionized by solar ultraviolet light. These electrically charged particles are influenced by the magnetic field carried by the solar wind and swept out of the coma into a long, distinctive ion tail (also called a gas or plasma tail). Since the most common ion, CO+, scatters blue light better than red, the ion tail often appears blue to the human eye. Gusts in the solar wind can cause the ion tail to swing back and forth, often developing temporary ropes, knots, and streamers that sometimes break away and then reform. These features are not seen in the dust tail. The ion tail is usually narrow and straight, sometimes streaming away from the nucleus for many millions of kilometers.[20]

Figure 13.36   Comets travel around the Sun in highly elliptical orbits. As they near perihelion, the warmth of the Sun causes the icy nucleus to vaporize, producing a bright coma and tails of gas and dust. This image of comet 96P/Machholz (top) was taken on January 8, 2002, by the LASCO C2 instrument on the SOHO spacecraft. The mask in the LASCO coronagraph blocks light from the Sun, shown by the white circle. Comet Machholz has a 63-month orbit, passing within 18.5 million km of the Sun. (Courtesy of SOHO/LASCO, ESA, NASA)

For more than 150 years, the Great March Comet of 1843 held the record for the longest ion tail, stretching 330 million km – more than two astronomical units – across space. Then, in 1998, analysis of data from the Ulysses spacecraft indicated that it had passed through the ion tail of comet Hyakutake at the remarkable distance of 570 million km from the nucleus.

One of the most interesting characteristics of the tails is a dramatic shift in their alignment as the comet moves along its orbit. As the solar wind sweeps past the comet at about 500 km/s, it shapes the tails and causes them (particularly the ion tail) to always point away from the Sun. This means that, during the outward leg of the orbit, the outward flow of the solar wind causes the tails to point ahead of the comet, rather than trail behind it.

In extreme cases, comets have been seen to lose their tails temporarily, due to strong gusts in the solar wind. In 2007, a similar disconnection was observed by NASA's STEREO spacecraft when a coronal mass ejection (CME) – a huge cloud of magnetized gas ejected into space by the Sun – collided with the tail of comet Encke and literally cut it in two. The disconnection was triggered by a process known as magnetic reconnection, when the magnetic fields around the comet and the CME were spliced together.

---

[20] A third tail, made of neutral sodium atoms, was discovered in comet Hale-Bopp in 1997.

Figure 13.37   Comet Hale-Bopp, discovered by Alan Hale and Thomas Bopp on July 23, 1995, was one of the great comets of the 20th century. As it approached the Sun, it became extremely bright and active, developing an 8° long ion tail (blue) and a yellowish, 2° long dust tail. The nucleus was estimated to be 40 km in diameter, huge compared with most comets that reach the inner Solar System. (Eckhard Slawik)

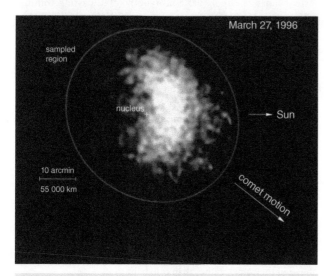

Figure 13.38   The ROSAT satellite discovered X-ray and extreme ultraviolet (EUV) emission from comet Hyakutake. Note the bright emission on the sunlit side of the nucleus. Since then, X-ray emissions have been detected from several other comets. Most of the X-rays are thought to be produced by a charge exchange caused when heavy, ionized particles carried by the solar wind collide with neutral atoms and molecules in the comet's coma. (Courtesy of NASA-GSFC/MPE)

Icy Dirtballs or Dirty Snowballs?

For hundreds of years, scientists have tried to unravel the true nature of comets. The introduction of two powerful new techniques – spectroscopy and photography – in the second half of the 19th century showed that comets are highly complex, not only in the variability of their comas, jets, and tails, but also in their composition. Bright bands in the spectrum of reflected light indicated the presence of numerous compounds made of hydrogen, carbon, oxygen, and nitrogen, as well as the silicate dust particles that produce the meteor swarms.

Then, in 1950, American astronomer Fred Whipple (1906–2004) turned cometary science on its head. Whipple knew that some periodic comets have orbited the Sun 1,000 times or more. He realized that they would have broken apart if they comprised only a large pile of sand mixed with hydrocarbons.

Whipple argued that comet cores must be "icy conglomerates" made of water ice and dust, mixed with ammonia, methane, and carbon dioxide – popularly termed "dirty snowballs." As they approach the Sun and become warmer, their outer ices begin to vaporize, creating jets that eject large amounts of dust and gas. The result is the formation of an all-enveloping coma, together with two characteristic tails if dust and gas production is sufficiently high.

Whipple's model also made it possible to understand the variations in comet orbits and predict their orbital motion by taking into account the thrust from the gaseous jets that erupt from the nucleus.

Box 13.10   Giotto

On the night of March 13–14, 1986, the Giotto spacecraft encountered comet Halley.[21] ESA's first deep-space mission was part of an international effort to investigate this famous object. Navigation to the nucleus was assisted by the Soviet Union's two Vega spacecraft, which also returned images and other data.

Giotto was launched on July 2, 1985. On March 12, 1986, after a flight of almost 150 million km, the spacecraft's instruments detected hydrogen ions at a distance of 7.8 million km from the comet. About one day later, Giotto crossed the bow shock, where a shock wave is created as the particles in the supersonic solar wind slow to subsonic speed. When Giotto entered the densest part of the coma, the camera began tracking the brightest object (the jets of the nucleus) in its field of view.

Figure 13.39   This composite image, the first detailed view of a comet's nucleus, was taken by ESA's Giotto spacecraft on March 13, 1986, from a distance of about 18,000 km. The dark nucleus measured about 8 x 8 x 16 km. The sunlit side (left) was illuminated by bright jets of gas and dust. The rugged surface revealed sizeable valleys and depressions. A "mountain" on the dark side was illuminated by sunlight. The rotation axis is approximately horizontal in this view. (ESA/MPI)

The first of 12,000 dust impacts was recorded 122 minutes before closest approach as the spacecraft closed on the nucleus at a speed of 68 km/s relative to the comet. The rate of dust impacts rose sharply when Giotto passed through a jet of material that streamed away from the nucleus.

7.6 seconds before closest approach, the spacecraft was sent spinning by an impact from a "large" (one gram) particle. Contact with Earth was temporarily lost, but, over the next 32 minutes, the spacecraft's thrusters stabilized its motion and contact was fully restored. By then, Giotto had passed within 596 km of the nucleus and was heading back into interplanetary space.

The resilient spacecraft continued to return scientific data for another 24 hours. The last dust impact was detected 49 minutes after closest approach. The historic encounter ended on 15 March, when the experiments were turned off. However, Giotto went on to encounter a second comet, Grigg-Skjellerup, on July 10, 1992.

Whipple's theories have since been tested by observing comets with orbiting observatories such as Hubble, Spitzer, and the Ultraviolet Explorer, as well as dispatching spacecraft to study them at close range.

The first close encounters came during an international campaign to study comet Halley during its return to perihelion in 1986 (see Box 13.10). Since then, spacecraft have orbited and landed on comet Churyumov-Gerasimenko, flown past several periodic comets, collected samples from the coma of Wild 2, and delivered an impactor into the nucleus of comet Tempel 1 (Box 13.13). The results have been surprising, and sometimes perplexing.

Comets are now known to display a wide range of physical characteristics, although their nuclei are generally quite small. At the top end is Hale-Bopp, with a diameter of perhaps 35 km, followed by the 16 x 8 km, potato-shaped Halley. More typical are Borrelly, which measures 8 x 4 km, the roughly spherical Tempel 1, which measures 5 x 7 km, and Wild 2 which resembles a thick hamburger with a diameter of about 5 km. The smallest nucleus yet

[21] Giotto was named after the Italian Renaissance painter, Giotto di Bondone, who included Halley's comet in one of his paintings.

Figure 13.40   Comet C/2006 P1 (McNaught) provided a spectacular sight close to the horizon in the southern hemisphere during January and February 2007. At least three jets of gas and small dust particles were seen to spiral away from the nucleus as it rotated, stretching over 13,000 km into space. The larger dust particles, ejected on the sunlit side of the nucleus, followed a different pattern. They produced a bright fan, which was then blown back by the pressure of sunlight. (Courtesy of ESO/Sebastian Deiries)

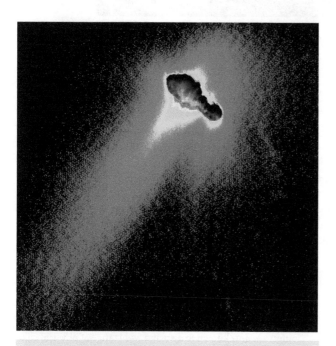

Figure 13.41   A false-color, composite view of comet Borrelly taken by NASA's Deep Space 1 spacecraft on September 22, 2001. Apart from the dark nucleus, it shows details in the dust jets (purple and blue) and the coma of dust and gas. The nucleus was the blackest object in the Solar System, in places reflecting less than 1% of the sunlight that it receives. It is about 8 km long and 4 km wide. (NASA-JPL)

observed belongs to Hartley 2, which is about 2 km long, narrowing to 0.4 km in its mid-section.

Some nuclei appear to double-lobed, most notably Churyumov-Gerasimenko, which was likened to a duck with a head, a narrow neck, and a wide body (Figure 13.45). The larger lobe of the comet measured 4.1 x 3.2 x 1.3 km, while the smaller lobe was 2.5 x 2.5 x 2.0 km.

Of course, the size of the nucleus shrinks each time it approaches the Sun and becomes active. During peak activity near the Sun, Halley's comet was releasing about 20 tonnes of gas and 10 tonnes of dust per second, though the amount of dust ejected varied considerably (Figure 13.39). Calculations suggested that the surface of its nucleus is lowered by about 10 m during each apparition. However, since the mass of the nucleus was estimated at 10^{14} kg, it is likely to remain active for many millennia. On the other hand, smaller comets trapped in shorter periodic orbits will shrink much more rapidly.

Measuring the density of a nucleus is not easy, but estimates for various comets indicate that they are typically only 0.3–0.5 g/cm^3, well below the density of water. This may be due to a porous, fluffy texture, or to a rubble pile structure with large empty spaces, in combination with a largely icy composition.

Their surfaces are blacker than coal with albedos of 2–4%. Some spots on comet Borrelly were so unreflective that their albedos were only 0.8%, making it the darkest object known in the Solar System (Figure 13.41). Despite their considerable ice content, comets all appear to be coated with a "crust" of carbon-rich organic material.

The prevalence of fine dust in comet nuclei was demonstrated when the Impactor probe from NASA's Deep Impact spacecraft hit comet Tempel 1. Observations showed only weak emissions

Figure 13.42 A composite image of comet Wild 2 taken during Stardust's flyby on January 2, 2004. The unusual surface displayed several large crater-like features, depressed regions, cliffs, and pinnacles. Shaped like a thick hamburger, Wild 2 is about 5 km in diameter. A short exposure image that shows surface detail is overlain on a long exposure image taken 10 seconds later that reveals at least five jets of dust and gas. (Courtesy of NASA/JPL-Caltech)

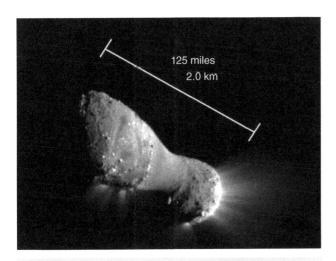

Figure 13.43 Comet Hartley 2 was imaged by the Deep Impact orbiter from a distance of about 700 km. The nucleus is approximately 2 km long, but only 0.4 km wide at its smooth "waist." Jets of carbon dioxide gas and ice particles can be seen streaming from the rough parts of the nucleus. The Sun is to the right. (Courtesy of NASA/JPL-Caltech/UMD)

from water vapor and all the other gases that were expected to erupt from the impact site. The most conspicuous feature of the blast was brightening due to sunlight scattered by the extremely fine, ejected dust, which was described as more like talcum powder than beach sand. It seems that comets should be regarded more as "icy dirtballs" than "dirty snowballs."

The main Deep Impact spacecraft went on to complete a second comet flyby on November 4, 2010, as part of its extended EPOXI mission. Although 103P/Hartley 2 had the smallest nucleus yet observed from close range, it was emitting numerous jets of gas and ice particles.

As anticipated, water and carbon dioxide dominated the outgassing, with organics, including methanol, present at lower levels. However, there was a lot more carbon dioxide escaping than expected. This was associated with numerous bright jets that were erupting from the rough surfaces at either end of the nucleus. Most of the water vapor was derived from beneath the smooth central "waist," as water ice vaporized before escaping through the overlying porous material. The smooth surface in the middle was lower than the rest of the comet and may be coated by fine dust.

Previously, it was thought that water vapor from sublimating water ice was the main propulsive force behind jets of material coming from comet nuclei, but the data from the Deep Impact/EPOXI flyby provided unambiguous evidence that solar heating causes sublimation of subsurface frozen carbon dioxide. It is this primordial gas, rather than water vapor, which powers the numerous jets of material that erupt from comet nuclei.

Comet 67P Churyumov-Gerasimenko

Further, dramatic insights into the nature of comet nuclei and the processes that transform them as they near perihelion came from ESA's Rosetta mission to periodic comet 67P/Churyumov-Gerasimenko (Box 13.11).

The Rosetta orbiter accompanied the comet on the inward leg of its orbit, until it passed perihelion, whilst a small lander, named Philae, made the first soft touchdown on a comet. For two years, its 11 experiments investigated every aspect of the comet as it was transformed from an inert chunk of ice to a warm, active mini-world surrounded by a coma that was fed by jets of gas and dust.

Rosetta revealed a two-lobed object which was popularly likened to a duck with a lengthy neck. During its 4.5 billion-year lifetime, 67P has probably been involved in numerous collisions which the narrow "neck" would not survive, indicating that its shape is not primordial, but has evolved over billions of years.

Its current shape is thought to be the result of its most recent collision. This resulted in the fracturing of a sizeable nucleus into two parts, which, due to the effects of their mutual gravitational force, later merged into a structure with two lobes.

The first jet activity, which was detected months before the rendezvous, was limited to the "neck" region (Figure 13.45). Subsequent images showed that dust was being emitted along almost the whole body of the comet.

Box 13.11   Rosetta and Philae

Launched in 2004, ESA's Rosetta spacecraft embarked on a 10-year trek to periodic comet 67P/Churyumov-Gerasimenko, which it encountered more than 5.2 AU from the Sun. In 2014, it became the first spacecraft to enter orbit around a comet. After gradually closing to within 25 km of the nucleus, it released a small lander, named Philae, which made the first soft touchdown on a comet.

The OSIRIS cameras obtained Rosetta's first images of 67P at a distance of about 5 million km from the comet. A critical series of manoeuvres, beginning on May 6, gradually reduced Rosetta's velocity relative to the comet.

The first ever spacecraft rendezvous with a comet was achieved on August 6, 2014, at a distance of 405 million km from Earth. Rosetta was now just 100 km from the comet's surface. Over the next six weeks, its instruments began to obtain detailed scientific data, scrutinizing the surface for potential target sites for the Philae lander.

As it moved closer to the nucleus, Rosetta obtained close-up images of cliffs, small and large boulders, dusty areas, and cracks. On October 10, Rosetta began its Close Observation Phase, moving to only 10 km from the comet's surface.

The historic landing by Philae took place on November 12, 2014. During the seven-hour descent, which was made without propulsion or guidance, Philae took images and recorded information about the comet's environment.

The selected landing site was on the smaller of the comet's two lobes (the "head"), a region that offered unique scientific potential, with hints of activity nearby. However, the landing did not go to plan.

Touchdown was planned to take place at a speed of around 1 m/s, with the three-legged landing gear absorbing the impact to prevent a rebound, and an ice screw in each foot driving into the surface. At the same time, two harpoons were to fire, fixing the probe to the surface, with a small thruster on top counteracting the recoil of the harpoons and pushing the lander down onto the surface.

Prior to its deployment, the probe's cold gas thruster system was deemed inoperable. Then neither of the harpoons fired upon touchdown. Subsequent analysis showed that the lander bounced, touching the surface three times. The final location was undetermined, although images confirmed that it was close to a steep cliff, which meant that it could not receive enough sunlight to recharge its batteries.

The lander completed its primary science mission after nearly 57 h. In that time, the lander returned all of its housekeeping data, as well as science data from the targeted instruments. Less than a month before the end of the mission, Rosetta's high-resolution camera revealed Philae wedged into a dark crack, at a location later named Abydos, on the comet's smaller lobe.

Figure 13.44   Rosetta imaged a plume of dust erupting from the surface of 67P/Churyumov-Gerasimenko on July 3, 2016. It lasted for about 1 hour, producing around 18 kg of dust every second and numerous particles of water ice. The plume erupted from within a circular pit several hundred kilometers across. It was probably ejected by pressurized gas, possibly caused by sudden vaporization of subsurface ices. (ESA/Rosetta/MPS for OSIRIS Team MPS/UPD/LAM/IAA/SSO/INTA/UPM/DASP/IDA)

Meanwhile, the orbiter moved back into a higher orbit around the comet. It continued to study 67P as the comet became more active, en route to perihelion on August 13, 2015. Rosetta began a series of more distant, unbound orbits, while performing a series of close flybys, some within just 8 km of the comet's centre. These allowed scientists to watch the short- and long-term changes that took place on the comet.

ESA extended the mission to enable observations as the comet moved away from the Sun. From August 9, 2016, Rosetta began flying elliptical orbits that eventually brought it within 1 km of the surface.

The remarkable mission ended with a touchdown on the nucleus on September 29. The landing site was a region of active pits on the comet's "head" – a region known as Ma'at, where some of the comet's dust jets originated. The final image received on Earth was a slightly blurred view of a patch of ground about 2.4 m across, obtained when the spacecraft was about 51 m above the surface. Rosetta operated in the harsh environment of the comet for 786 days.

Figure 13.45 This Rosetta navigation camera image of comet 67P shows its two-lobed shape and numerous dust jets. The image was taken on July 7, 2015, as 67P approached perihelion, at a distance of 154 km from the comet's center. The image has a resolution of 13.1 m/pixel and measures 13.4 km across. (ESA/Rosetta/NAVCAM – CC BY-SA IGO 3.0)

Rosetta studies confirmed that the numerous jets of dust that the comet emitted were not driven solely by the sublimation of frozen water. This process involves frozen gases on a comet's surface passing directly into the gaseous state; the gas streaming into space entrains dust particles with it, producing the visible jets. Often these occur shortly after sunrise.

Rosetta demonstrated that other scenarios are often involved, including the release of pressurized gas stored below the surface or the conversion of one kind of frozen water into an energetically more favorable one (Figure 13.44).

Particularly valuable insights were provided on July 3, 2016, when Rosetta's trajectory took it right through a jet erupting from the comet's Imhotep region High-resolution images of the surface showed the starting point of the jet to be a circular depression, about 10 m in diameter, which contained water ice (Figure 13.44).

With a dust production rate of approximately 18 kg per second, the jet was a lot "dustier" than conventional models predict. An additional process must release energy from beneath the surface to support the plume.

One possibility is that subsurface cavities are filled with compressed gas; after sunrise radiation begins to warm the overlying surface, cracks develop and the gas escapes.

Another theory suggests that deposits of amorphous ice beneath the surface play a decisive role. In this type of frozen water, the individual molecules are not aligned in a lattice-like structure, as is customary in the case of crystalline ice, but arranged in a far more disorderly fashion. Under solar warming, energy is released during the transition from amorphous to crystalline ice, boosting the force of the jet. Exactly which process took place on July 3 is unclear.

In the course of the Rosetta mission, more than 35,000 dust grains were collected and analysed. The smallest of them measured only 0.01 mm in diameter, the largest about 1 mm. Organic molecules accounted for about 45% of the weight of the solid cometary material – clear evidence that the comet is among the most carbon-rich bodies in the Solar System.

The other 55% was provided by mineral substances, mainly silicates. They were almost exclusively non-hydrated minerals, i.e. missing water compounds. This was an indication that 67P contains very pristine material, since it has spent most of its life in the icy outer regions of the Solar System, where its ices have remained largely frozen and unable to react chemically with the minerals.

In June 2014, Rosetta first detected water vapor from the comet's coma, when it was 350,000 km from the nucleus. Although water ice was predominant, Rosetta also made the first detection of solid CO_2 on a comet's surface.

CO_2 is a volatile compound that quickly sublimates when exposed to sunlight, so it was only expected to be present below the surface. It seems that, when 67P's dark "winter" begins, a drastic drop in surface illumination and temperature occurs, freezing some gaseous CO_2 onto the surface as it continues moving outwards from the still-warm interior. The patch of CO_2 was seen when sunlight returned to the area, but, as expected, it disappeared completely after three weeks.

Images returned from Rosetta showed that comet 67P was very active, with numerous gaseous jets, growing fractures, collapsing

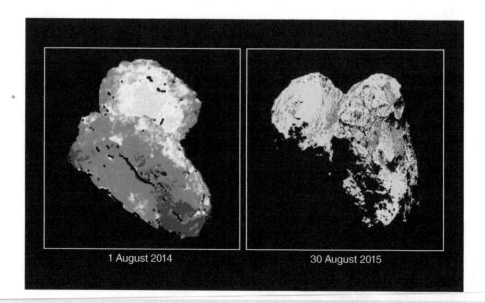

Figure 13.46   The color of visible light reflected by comet 67P on August 1, 2014 (left), shortly before Rosetta's arrival, and on August 30, 2015 (right), soon after perihelion. The entire surface became increasingly blue as it approached the Sun, and gradually turned redder again as it moved away. Bluer colours indicate areas that are richer in water ice. As the comet's activity increased, the outgassing of water vapor and other gases lifted large amounts of dust, exposing more of the ice-rich terrain underneath. (ESA/Rosetta/MPS for OSIRIS Team MPS/UPD/LAM/IAA/SSO/INTA/UPM/DASP/IDA)

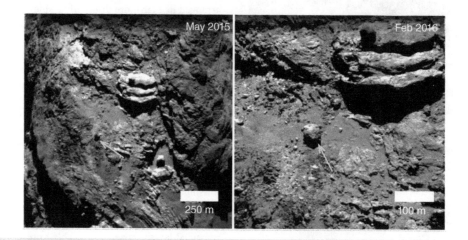

Figure 13.47   Images taken by Rosetta show that a 30 m-wide boulder moved 140 m across the surface of comet 67P as it neared perihelion, when the comet's activity was at its highest. In both images, an arrow points to the boulder; in the right-hand image, the dotted circle outlines the original location of the boulder. The movement could have been triggered if the material on which it was sitting eroded, allowing it to roll downslope, or a forceful gaseous outburst could have directly lifted it to the new location. Several outbursts were detected close to the original position of the boulder during perihelion. (ESA/Rosetta/MPS for OSIRIS Team)

cliffs and massive rolling boulders. Moving material buried some features on the comet's surface while exhuming others.

One 30 m-wide boulder was shown to have moved 140 m from its original position (Figure 13.47). The massive rock probably moved as a result of several gaseous outbursts that were detected close to its original position.

The warming of 67P and its increased jet activity caused the comet's rotation rate to speed up. The increasing spin rate prior to

perihelion is thought to be responsible for a 500 m-long fracture that ran through the comet's neck. The fracture was found to have increased in width by about 30 m by December 2014. Subsequent images taken in June 2016 revealed a new 150- to 300-m fracture parallel to the original fracture. These observations led scientists to suggest that the comet may split up one day.

Rosetta also observed the collapse of cliffs at two locations on the comet, where pre-existing fractures gave way, causing sections of material tens of meters long to crumble. Sequences of images taken at different times reveal that, as the comet approached perihelion, steep slopes at one location retreated as fast as 5.4 m per day. Other areas revealed circular ripple-like features on the surface that reached a diameter of about 100 m in less than three months, before fading away and being replaced by a new set of ripples.

Another surprising discovery was the widespread presence of dune-like forms approximately 10 m apart. Comparison of two images of the same spot taken 16 months apart provided evidence that the dunes had moved and were therefore active.

Despite the absence of an atmosphere, 67P was being modified by a wind blowing across its surface. This breeze was caused by the pressure difference between the sunlit side, where the surface ice sublimated, and the night side. Although this transient atmosphere is extremely tenuous, with a maximum pressure at perihelion 100,000 times lower than on Earth, the comet's gravity is also very weak. Analysis of the forces exerted on the surface grains shows that these thermal winds can transport centimeter-scale grains, whose presence was confirmed by spacecraft images. These conditions allow the formation of the dunes.

Breaking Up Is Easy To Do

How many passages through the inner Solar System can comets complete before they cease to illuminate the sky? What happens to them when their activity ceases?

There is no doubt that some comets break apart and disintegrate as they approach the Sun. Indeed, catastrophic breakups may be the ultimate fate of most comets. One of the first objects to reveal the fragility of nuclei was 3P/Biela, which orbited the Sun roughly once every six years. During its apparition of 1846, observers were surprised to discover two comets on almost identical paths, each with its own tail. For a while, the two nuclei were enveloped in a common coma.

Six years later, the pair was still flying in tandem, but some two million kilometers apart. They were never seen again. However, starting in 1872, thousands of meteors illuminated the November sky at six-yearly intervals. They became known as the Andromedid shower. When the orbits of the meteors were traced back, they matched the path followed by comet Biela. Clearly, the comet(s) had disintegrated, leaving only a swarm of debris. After the turn of the 20th century, the meteor shower largely disappeared.

Several comets have been seen to break into multiple fragments in modern times (Figure 13.48). Comet 73P/Schwassman-Wachmann 3 began to splinter in 1995 during one of its many passages through the inner Solar System. Shortly after experiencing a huge outburst of activity, four separate nuclei were identified. Since then, the comet has disintegrated into dozens of fragments.

Images taken by the Hubble Space Telescope showed that a hierarchical destruction process was taking place as large fragments continued to break into smaller chunks. Several dozen "mini-fragments" were found trailing behind each main object. The images suggested that the pieces were propelled down the tail by outgassing from their sunward-facing surfaces. Since the smaller chunks had the lowest mass, they were accelerated away

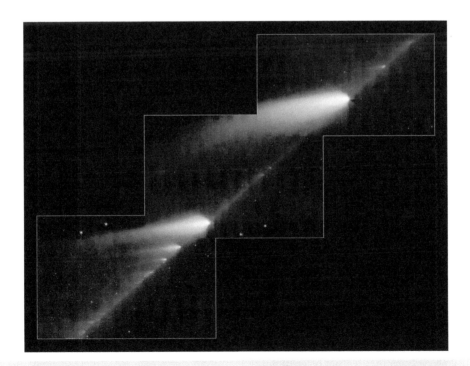

Figure 13.48 This Spitzer Space Telescope infrared image, taken May 4–6, 2006, shows at least 36 sizeable fragments of comet 73P/Schwassman-Wachmann 3 amidst a stream of dust and smaller debris. The flame-like tails were warmed by sunlight so that they glowed at infrared wavelengths. The debris ranged in size from pebbles to large boulders. (Courtesy of NASA/JPL-Caltech)

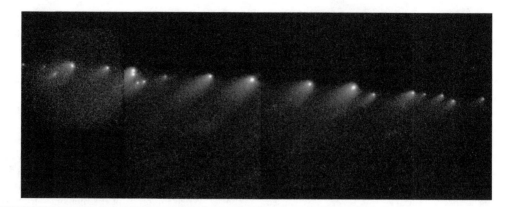

Figure 13.49   The Hubble Space Telescope imaged 22 pieces of comet Shoemaker-Levy 9, each with its own coma, after the nucleus broke apart during a close approach to Jupiter in 1992. The remaining "string of pearls" was made up of one remnant around 3 km across and other fragments measuring 0.5–2 km. (Courtesy of NASA/STScI)

from the parent nucleus faster than the larger fragments. Some of the smaller pieces seemed to dissipate completely over the course of several days.

Occasionally, close encounters with planets can have a similar effect. Nuclei are probably porous, fragile mixes of dust and ices and they can be broken apart by gravitational tidal forces when they pass near large bodies (Figure 13.49). This is what happened to comet Shoemaker-Levy 9 when it passed near Jupiter in 1992, prior to plunging into the planet's atmosphere two years later (see Chapter 7).

An even more catastrophic example was comet LINEAR (C/1999 S4), which suddenly disintegrated on July 25, 2000, during its first passage through the inner Solar System. Having previously appeared completely normal, the comet was seen to undergo a rapid change as it evolved into a fuzzy, extended, and much fainter object. By early August, all that could be seen was a cloud of debris, with no sign of the nucleus or any active sub-nuclei larger than a few meters across.

Analysis of the sequence of events led to the conclusion that rapid vaporization near perihelion caused the small comet to run out of ice. When all of its ice was exhausted, nothing remained to bond the solid material together. The nucleus began to fall apart, leaving behind a loose conglomerate of particles that dispersed into space.

Comet nuclei can also be broken up by rapid rotation, thermal stresses as they pass near the Sun, or explosive disruption – like corks from champagne bottles – due to a sudden outburst of trapped volatile gases.

One of the most surprising examples is comet 17P/Holmes, which has exploded on two separate occasions (November 1892 and October 2007) as it approached the asteroid belt. During its 2007 return, Holmes unexpectedly changed from magnitude 17 to 2.8, becoming almost a million times brighter in only 42 hours so that it was visible to the naked eye. Its coma gradually expanded outward, eventually growing larger than the Sun before it finally dissipated.

Observations made with Spitzer in November 2007 revealed a lot of fine silicate dust, or crystallized grains smaller than sand. The dust was apparently created by the violent explosion that destroyed

larger particles in the interior of the nucleus. This was confirmed when observations made in March 2008 found only large dust particles, about one millimeter in size, in the coma. Calculations indicated that the energy of the blast was about 10^{14} joules – equivalent to 24 kilotonnes of TNT – and the total mass of ejected material was some 10 million tonnes.

The presence of jets and a spherical cloud, not to mention two similar events more than a century apart, indicated that a collision with a meteoroid was an unlikely explanation. Thermal stresses are also unlikely to be the cause since, in both cases, the comet was well past perihelion and heading away from the Sun. Rather, it seems more likely that the nucleus behaved like a pressure cooker, in which trapped gases suddenly erupted through weaknesses in the overlying surface.

Some comets have undergone violent eruptions far from the warmth of the Sun. Perhaps the most impressive was the discovery on February 15, 1991, that comet Halley was surrounded by a huge coma, at least 300,000 km in diameter – evidence of a tremendous outburst. At the time of the observation, the comet was at a heliocentric distance of 14.3 AU. This renewed activity was attributed to sublimation of a more volatile compound than water, for example CO or CO_2, which had built up sufficient pressure beneath the dust mantle to initiate a major outburst.

Centaurs

The term Centaurs is applied to objects that have been found orbiting among the giant planets of the outer Solar System. The first Centaur, Chiron, was discovered in 1977 by Charles Kowal. It was found that Chiron follows a highly elliptical (e=0.383) orbit with a perihelion of 8.46 AU and an aphelion of about 19 AU, i.e. between the orbits of Saturn and Uranus. Its orbital period is 50.7 years and the orbit is inclined about 7° to the ecliptic.

Although it followed such an unusual orbit, Chiron was originally classified as an asteroid, with an estimated diameter of 148–208 km. Light curve studies gave a rotational period of 5.9 hrs. This classification came into question in 1988 when Chiron unexpectedly brightened. The following year, a faint, coma-like, shroud of dust and gas was discovered, suggesting that it was more likely to be a very large comet nucleus. Today, it has a classification both as an asteroid (2060 Chiron) and as a periodic comet (95 P/Chiron).

The diameter of the coma has occasionally expanded to almost 2 million km and spectroscopic analysis has revealed the presence of carbon monoxide (CO) and cyanogen (CN) radicals. The existence of surface activity and coma production at the low temperatures beyond Saturn suggests that sublimation of super-volatile substances, such as methane, carbon monoxide, and molecular nitrogen, is the source of the coma. Curiously, observations show that Chiron's activity was lower near perihelion (February 1996) than near aphelion.

The largest Centaur is Chariklo, with a diameter of 250 km. It is the first known asteroid-like object to have rings. A 7 km-wide ring is located at a distance of 396 km and a second ring, which is 3 km wide, lies at a distance of 405 km. The rings are no more than a few hundred meters thick.

The Centaurs are now generally regarded as icy objects with diameters typically greater than 100 km – transitional in size between typical short-period comets and the even larger Kuiper Belt objects that orbit beyond Neptune.

They have reddish surfaces – though not as dark or red as the Kuiper Belt objects. Near-infrared spectroscopy has revealed water ice on the surfaces of a number of Centaurs, while methanol (CH_3OH) ice (or another light hydrocarbon or oxidized derivative) has been seen in the near-infrared spectrum of Pholus. (Pholus' orbit takes it from inside the orbit of Saturn to beyond the orbit of Neptune. Like many Centaurs, no comet-like activity has been recorded for Pholus.)

One estimate puts the number of Centaurs with diameters larger than 1 km at about 44,000, assuming an inward flux of one new short-period comet every 200 years. Although their orbital inclinations vary considerably, they are all thought to originate in the Kuiper Belt. Their chaotic orbits are influenced by the giant planets, so that objects such as Chiron will ultimately either collide with a planet or be ejected from the Solar System.[*]

Comet or Asteroid?

Calculations related to the rate of capture of long-period comets from the Oort Cloud imply that there should be about 1,000 times more Halley-type objects than are actually observed. If, on the other hand, they disintegrate to dust, the debris would create a bright, near-spherical zodiacal cloud[22] and some 15–30 strong annual meteor showers, also contrary to observation.

One possibility is that many active comets rapidly become dormant when they lose most of their volatiles and/or develop a thick, dark surface deposit of organic material that insulates the inner nucleus and prevents all further solar heating of the interior. Such inactive comets may resemble cometary meteoroids, with extremely low reflectivities in visible light. This implies a large population of fast-moving, multi-kilometer-sized bodies that are too dark to be seen with current near-Earth object surveys.

A number of extremely dark objects that may belong to this postulated population of dead comet nuclei have been identified, including some transitional objects that were originally classified as asteroids and then observed to undergo comet-like outbursts of activity.

Figure 13.50 The zodiacal light is a cone of faint light which is visible above the horizon in the tropics. It lies along the ecliptic and is caused by scattering of sunlight due to dusty debris from asteroid collisions and comets. (Courtesy of ESO/Yuri Beletsky)

[22] The zodiacal cloud is a disk of dust in the inner Solar System. It is believed to be derived from comet dust and the debris of asteroid collisions, and causes a band of light (the zodiacal light) that extends along the ecliptic and is visible in the west after sunset or in the east before sunrise.

The first indications that the classical distinction between rocky asteroids and icy comets was inadequate came in the 1970s with the discovery that many comets (including Halley) have black surfaces, often as dark as carbonaceous asteroids. Since a comet loses its coma and tails as it retreats from the Sun, it becomes very difficult to distinguish between an asteroid and a bare comet nucleus – particularly if the latter is in an asteroid-like orbit.

One famous example is asteroid 4015, which was discovered in 1979. Subsequent checks revealed that it was actually the periodic comet 107P/Wilson-Harrington, which had displayed a diffuse coma and tail when it was first found in 1949. Since then, all activity had subsided, enabling the object to take on the external mask of an asteroid.

More indirect evidence has pointed to the likelihood that Earth-crossing asteroid 3200 Phaethon is also a dead (or dormant) comet. Phaethon follows an unusual orbit that takes it to within 0.14 AU of the Sun and it is linked with the Geminid meteor stream that follows the same orbit around the Sun. Although it is currently a faint, dark object, Phaethon must have been active in the not-too-distant past in order to have generated such a major trail of debris.

Other comets that may be masquerading as Apollo asteroids include 2101 Adonis, which may be associated with the fairly unspectacular Capricornid-Sagittariid meteor shower, and 2201 Oljato, which has been linked with minor outgassing as well as a minor meteor shower. Modeling of Oljato's chaotic orbital evolution also suggests a cometary origin.

Another group of objects that fits this category is the Damocloids, named after 5335 Damocles, a peculiar asteroid that follows a highly inclined (61.95°) elliptical orbit that takes it from just outside Earth's orbit almost to the orbit of Uranus.

Well over 100 Damocloids are known. Although they are often classified as asteroids, they follow eccentric orbits typical of longer period comets such as Halley. About 25% of their orbits are retrograde – unlike any other asteroids. They also seem to be similar in size to Halley's nucleus, with an average radius of 8 km. The albedos of some Damocloids have been measured, and they are among the darkest objects known in the Solar System. Damocloids are reddish in color, though not as red as many Kuiper Belt objects or Centaurs.

Damocloids are believed to be the nuclei of Halley-type comets that have lost most of their volatile materials during many approaches to the Sun over thousands or millions of years. This hypothesis is strengthened by the fact that some objects thought to be Damocloids subsequently developed a coma and were confirmed to be comets, and some follow retrograde orbits.

Such cases have led to the suggestion that perhaps half of the near-Earth asteroid population consists of rocky objects, while the remainder are dead comets.

Main Belt Comets

The population of active comets that we see today consists almost entirely of objects from the Kuiper Belt and Oort Cloud that have been scattered into Jupiter-crossing orbits by gravitational interactions with the giant planets. Even the dynamically peculiar comet 2P/Encke is believed to have originated in the Kuiper Belt, though its orbital evolution has been strongly influenced by non-gravitational forces induced by cometary outgassing.

However, while the dominant cometary reservoirs are located beyond the orbit of Neptune, their main volatile constituent, water ice, is stable at much smaller heliocentric distances and it has long been suspected that dormant comets might exist closer to the Sun (e.g. the Hilda asteroids at 4 AU and the Jovian Trojans at 5 AU).

There is clear evidence that the definitions of a comet and an asteroid are not as clear-cut as once thought. As mentioned, a number of asteroid-like objects are almost certainly dead or dormant comets. It has also been discovered that comets can inhabit the main asteroid belt between Mars and Jupiter.

Twelve objects that sometimes display comet-like behavior by ejecting clouds of dust and sprouting tails are currently known to occupy a typical orbit in the main belt (Figure 13.51). The first to be recognized, in 1996, was asteroid (7968) Elst-Pizarro (also known as comet 133P/Elst-Pizarro). Since then, other objects have exhibited similar activity while following stable orbits within the asteroid belt. Together, these objects form a new class known as main belt comets (MBCs).

In some cases, their activity persists for several weeks or months, consistent with ejection caused by sublimation of volatiles. Others are more likely to be activated by collisions that release carbon-rich dust and ice.

In October 2013, observations of main belt comet P/2013 R3 showed that it was breaking apart. Hubble Space Telescope images showed that the object had split into 10 pieces, each with comet-like dust tails, embedded within the asteroid's dusty envelope.

Observers suggested that the object was disintegrating due the YORP effect (see Box 13.4) in which sunlight caused the rotation rate to slowly increase over time. If the object was a loosely bound rubble pile, its component pieces could pull apart gently due to centrifugal force.

Similar activity was found in main belt comet P/2013 P5, which developed a unique system of six dust tails in 2013. Once again, the protracted period of dust release appeared to be inconsistent with an impact origin, but may have been caused by a body that was losing mass through rotational instability.

The discovery of comets which are in stable orbits in the main asteroid belt sheds some doubt on the idea that these primitive members of the Solar System only come from the Kuiper Belt and Oort Cloud. Instead, they appear to have formed within the main belt and presumably have remained there all their lives. This would suggest that they formed in a warmer environment than other comets, and that they may differ in composition from the Kuiper Belt and Oort Cloud comets. They may even have been a major source of water for the early Earth.

The existence of these objects for billions of years can be explained if a thin, insulating layer of dusty material has, until relatively recently, prevented solar heating from triggering the vaporization of their ices and the creation of gaseous comas and tails.

Sungrazing Comets

Although the nuclei of most comets that enter the inner Solar System measure only a few kilometers in diameter, there is clear

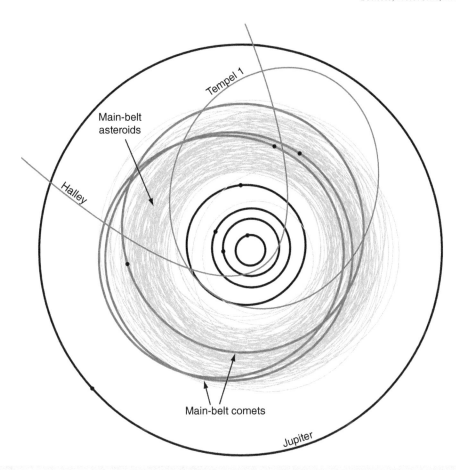

Figure 13.51 Orbits of the first three known main belt comets (red), the five innermost planets (black), a sample of 100 main belt asteroid orbits (orange), and two "typical" comets – Halley and Tempel 1 (blue). Positions of the main belt comets and planets on March 1, 2006, are plotted with black dots. (Henry Hsieh and David Jewitt/IfA, University of Hawaii)

evidence that much larger objects exist. Direct measurements of Kuiper Belt Objects such as Eris and Pluto show that icy planetesimals can have diameters exceeding several thousand kilometers. Furthermore, analysis of comet orbits shows that many thousands of icy fragments may be derived from one large parent body.

The most notable example of this is the Kreutz comet family, first recognized by German astronomer Heinrich Kreutz (1854–1907). They were all found to have similar retrograde orbits with orbital inclinations of about 142°, longitudes of perihelion at 280–282°, and orbital periods of more than 500 years. They are often dubbed "sungrazers" because (like several other comet families) their perihelion distances lie within 0.02 AU (Figure 13.52).

The remarkable similarities in their orbits have led to the belief that they are all derived from a single, large parent comet which fragmented in the remote past, probably more than 1,000 years ago. Detailed analysis indicates that there are two principal subgroups, possibly the result of separate fragmentation events at different perihelia. Since then, there has been progressive fragmentation of the remnants, creating smaller and smaller pieces of debris. Most of these tiny objects, only a few meters across, plunge into the Sun or vaporize completely during their passage through the inner Solar System.

Kreutz-group comets have been observed for many hundreds of years. The number of known Kreutz family comets has soared since 1995, following the launch of ESA's SOHO spacecraft, which takes images of the Sun's neighborhood, using a coronagraph to blank out the brilliant solar disk. By September 2017, the number of SOHO comets had reached 3,400, with more being discovered (mainly by amateur web watchers) on an almost daily basis. This meant that SOHO had been indirectly responsible for the discovery of more comets than all other observers throughout history. Kreutz comets have also been discovered in images taken by NASA's twin Stereo spacecraft, launched in October 2006.

Not all of them are invisible to the unaided eye. Larger fragments have produced some of the brightest comets in history, notably comet C/1965 S1 Ikeya-Seki.

Some unusual features have been observed as Kreutz objects sweep towards the Sun. For example, on May 24, 2003, two comets swept to within 0.1 solar radii (70,000 km) of the Sun's visible surface, deep within its searing multimillion-degree atmosphere, the corona. As usual, one of the intruders was vaporized by the intense heat and radiation pressure. However, SOHO-614 left a faint, tail-like feature that moved away from the Sun, apparently emanating from a point in the orbit beyond the comet's closest

Figure 13.52 Two "sungrazing" comets with extended tails of dust and ice are seen heading towards the Sun's corona – the tenuous outer atmosphere of our nearest star. The comets followed similar, but not identical, orbits, and did not reappear on the other side. The shaded disk in the center is a mask in the coronagraph instrument that blots out direct sunlight. The white circle added within the disk shows the size and position of the visible Sun. (ESA, NASA)

approach. Although the intense solar heat and radiation vaporized its nucleus and coma, the dust tail survived the encounter and was pushed outwards by the radiation pressure. Such headless comets are very rare.

The Riddle of Stardust

One of the most ambitious space missions to explore a comet was NASA's Stardust. Launched in 1999, the spacecraft encountered comet P81/Wild 2 in January 2004 (Figures 13.53 and 13.54). Passing through the coma, it captured an estimated 1,000 particles, ranging in size range from 5 to 300 microns, in an incredibly low-density material known as aerogel. The comet dust and samples of interstellar dust were successfully brought back to Earth in January 2006, when the return canister parachuted onto the Utah desert.

Early analysis of the comet samples led mission scientists to the surprising conclusion that Wild 2's coma contained minerals that formed at very high temperatures – despite a likely birthplace in the frigid outer regions beyond the orbit of Neptune.

Until now, the consensus opinion has been that comets are made of pristine, pre-solar materials that have never been altered by thermal processes. In addition to the icy material that forms comets at a temperature of around –243°C, their dust component was thought to be primarily interstellar material, tiny (submicron)

Box 13.12   Naming Comets

Comets have traditionally been named after their discoverers. If a comet suddenly brightened so that its discovery was made by several people, independently or jointly, the comet's name could include up to three of them. If the same team made more than one discovery, a number was added after their names, e.g. comet Shoemaker-Levy 9. Sometimes, as in the case of comets Halley and Encke, they were named after the astronomer who carried out special studies of them, since the actual discovery could not be attributed to any particular individual.

In the case of a periodic comet (with a revolution period of less than 200 years), the letter P is placed before the name, e.g. 160P/LINEAR. Hence, 1P/Halley was the first periodic comet to be discovered and confirmed. Other prefixes are C/ for a comet that is not periodic, X/ for a comet for which a meaningful orbit can not be computed, D/ for a periodic comet that no longer exists or is believed to have disappeared, and A/ for an object first thought to be a comet but later reclassified as an asteroid.

Some comets have never been named, including most comets from the 18th century or earlier, comets discovered years after they were first visible (e.g. from photographs), and others whose orbits cannot be determined due to a lack of observations.

In March 2003, the procedure for naming comets was made more formal by the International Astronomical Union, and now resembles the system applied to asteroids. A comet is immediately given a provisional designation, comprising the year of its discovery, followed by a capital letter and a number. The letter (in alphabetical order) identifies successive periods of 15 consecutive days or half months, while the number indicates its order of discovery during that time. Hence, comet 1996 B1 was the first to be discovered in the second half of January 1996.

After its perihelion passage, each comet receives a definitive designation. This includes the year of perihelion passage, followed by a Roman numeral specifying the order of such passage in that year. Obviously, the year of discovery does not always coincide with the year of perihelion passage. The full designation of Halley's comet is 1P/1682 Q1 (Halley), since Edmond Halley predicted its return to perihelion on September 15, 1682 – the first time such a prediction had been successfully fulfilled.

With the arrival of the Space Age and dedicated sky surveys, the number of discoveries made by a single spacecraft or ground-based telescope has often led to large numbers of comets with the same name, e.g. SOHO or LINEAR, so, in addition to the normal designation, numbers are often added to clarify their order of discovery, e.g. LINEAR 43.

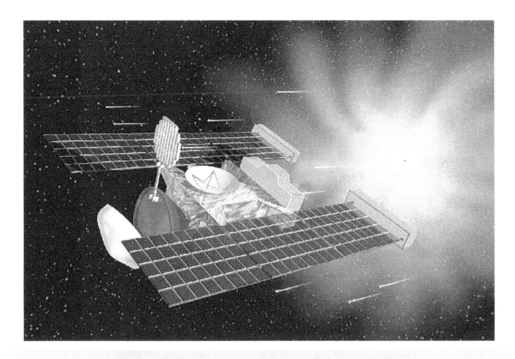

Figure 13.53   Stardust spacecraft brought back samples of comet (and interstellar) dust after passing through the coma of P81/Wild 2 in January 2004. During closest approach, the spacecraft was protected from incoming particles by bumper shields. (NASA)

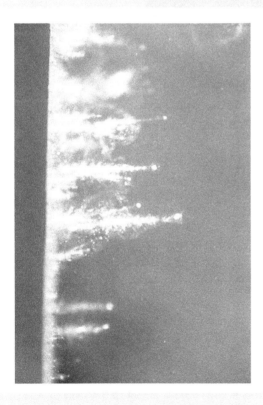

Figure 13.54   In order to prevent damage and preserve them during and after the collection process, the dust particles from comet Wild 2 were captured in a low-density material known as aerogel. Despite their high velocities, the particles were gently slowed, leaving trails in the aerogel from which they could be extracted back on Earth. (NASA-JPL)

grains that formed around other stars. However, Stardust showed that, although it does contain some interstellar dust, Wild 2's dirt content is composed mainly of relatively large silicate particles. Their composition indicates that some of these formed at temperatures of 1,227°C, in the innermost part of the Solar System, very close to the Sun.

The analysis of comet dust also identified a few grains that contain calcium/aluminum-rich inclusions (CAIs) similar to those found in chondrite meteorites. Thought to be the oldest objects in the Solar System, chondrites have always been a puzzle because they contain minerals that formed at low temperatures, as well as minerals, like CAIs, that must have formed at high temperatures.

Some high-temperature, crystalline, minerals – rich in calcium, aluminum, and titanium – apparently formed near the Sun during the early history of the Solar System. They were subsequently transported as far out as the Kuiper Belt and then incorporated into the object which eventually became comet Wild 2. The majority of the dust particles appear to have formed over a very broad range of solar distances and perhaps over an extended time range.

The process which enabled this transportation is not clear. Large-scale turbulence in the solar nebula has been suggested, although other evidence, such as the way in which planetary and asteroidal composition changes with solar distance, suggests that the cloud was not well mixed.

Another theory favors ballistic transport above the main disk of the nebula by a strong solar wind outflow, called the X-wind. This theory suggests that some of the dust falling into the young Sun was melted and ejected outward in a fiery spray. This condensed into small chondrules, or beads of melted rock, which were pushed out of the nebular disk by the X-wind. When they eventually reached the distant reaches of the Solar System, they combined

Box 13.13   Deep Impact

Comet nuclei have been observed by several spacecraft, but there is still considerable uncertainty about the nature of their interiors. In an attempt to address this knowledge gap, NASA decided to send a spacecraft that would send a high-speed projectile into periodic comet 9P/Tempel 1 as it neared perihelion.

The Deep Impact spacecraft, which would observe the impact from a safe distance, carried two imagers. The High-Resolution Instrument was attached to a multispectral camera and an infrared spectrometer. This system would take detailed pictures of the comet and the impact crater. The Medium-Resolution Instrument took visible light images with a wide field-of-view, enabling it to observe the cloud of ejected material as well as the crater created by the impact event. A duplicate of the Medium Resolution Instrument on the impactor recorded the vehicle's final moments before it collided with Tempel 1.

On July 4, 2005, the Deep Impact spacecraft released the 1 m-wide, copper-fortified impactor. Over the next 22 hours, the mother craft and 364-kg impactor headed toward the comet using an auto-navigation system. During most of the encounter phase, the instruments on the flyby spacecraft were aimed at the comet in order to capture the impact. Protected by dust shields, the spacecraft was oriented so that its cameras could continue operating until it came to within about 700 km of the nucleus. At this point, the spacecraft stopped taking pictures and turned so that its dust shields could provide as much protection as possible during the passage through the inner coma.

Meanwhile, the impactor continued to send back images until three seconds before its head-on collision. The final image was taken from a distance of about 30 km above the comet's surface, resolving features on the surface less than 4 m across. It then smashed into the nucleus – with kinetic energy comparable to detonating 4.5 tonnes of TNT – and vaporized.

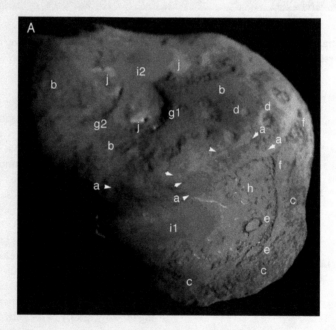

Figure 13.55   A composite image of comet Tempel 1 taken by the high-resolution camera on Deep Impact's flyby craft, before the arrival of the Impactor. The nucleus has ridges, scalloped edges, and possible impact craters that formed long ago. Examples of different features are marked: (a) linear outcrops; (b) intermediate roughness pitted surface; (c) pitted and rough surface; (d) isolated, rimless depressions; (e) rim remnants; (f) close-packed depressions; (g) large arcuate depressions; (h) stripped surface; (i) smooth units; (j) exposures of ice. (Courtesy of NASA/UMD/Cornell/Peter Thomas)

The Deep Impact cameras and other telescopes on the ground and in Earth orbit observed a cone-shaped cloud of ejecta rising from the nucleus at an average speed of about 110 m/s. The cone was seen to rise from the impact site for more than an hour, evidence that the comet's surface must consist of weak, easily pulverized material, and continued to expand outward over the following days.

The impact blew out so much fine, powdery material that the crater remained totally hidden. However, a subsequent visit by the Stardust-NExT mission revealed a crater about 150 m in diameter, from which perhaps 1,000 tonnes of material had been excavated. The amount of dust compared with the water vapor content indicated that Tempel 1 is best described as an "icy dirtball" rather than a "dirty snowball." Based on the rate at which the ejecta cone expanded near its base, the science team calculated that the comet has an extremely low bulk density, roughly 0.6 g/cc. 50–70% of Tempel 1 must, therefore, consist of empty space.

Spitzer Space Telescope observations after the impact returned detailed spectra of the ejecta, showing that the nucleus was composed of primordial material formed in the solar nebula. Emission signatures due to amorphous silicates and carbon, and crystalline silicates, carbonates, phyllosilicates, polyaromatic hydrocarbons (PAHs), water gas/ice, and sulfides were found. The presence of crystalline silicates indicates that they were heated to temperatures of more than 700°C before somehow being transported beyond the orbit of Neptune, the formation region of the comet.

Three small patches of water ice were observed on the surface of the nucleus, surrounded by dark, organic material. This was the first definitive observation of water ice on the surface of any comet (although it had been spectrally identified in near-infrared observations of Kuiper Belt objects and Centaurs). The fact that the nucleus did not exhibit any enhanced activity in the days after the impact suggested that, in general, meteoroid impacts are not the cause of comets' outbursts.

with colder material to form chondrites, the building blocks for asteroids and planets. Some particles were also incorporated into comets. Alternatively, there may even have been ejection of material by bipolar jets shooting out of the young Sun.

Whichever theory is closest to the truth, the unexpected origin of the dust in comet Wild 2 adds up to more evidence for the "icy dirtball" theory of comet composition. In recent years, astronomers have come to think of comets not as snowballs coated in dust, but as dirtballs crusted with ice.

Periodic Comets

By the beginning of the 18th century, it was understood that comets were celestial objects that occasionally appeared without warning, illuminated the skies for a few weeks or months as they moved closer to the Sun, and then disappeared whence they came – presumably never to be seen again. However, their composition and origin remained a mystery.

The breakthrough came in 1705, when British astronomer Edmond Halley (1656–1742) began to calculate the orbits of 24 comets. He noticed that the path followed by a bright comet observed in 1682 was very similar to the orbits of other comets recorded in 1607 and 1531. Halley concluded that the only reasonable explanation was that the same comet had reappeared over a period of 75–76 years. The slight variations in the timing of each return were accounted for by small tugs on the comet by the giant planets.

Working forward in time, Halley predicted that the comet should return in December 1758. Although he did not live to see the event, his theory was proved to be correct when the comet duly reappeared on schedule. The first periodic comet to be recognized was named 1P/Halley in his honor.

Figure 13.56 An image of comet 1P/Halley taken on March 8, 1986 by W. Liller, on Easter Island, as part of the International Halley Watch. Note the large dust tail and ion tail. (Courtesy of NASA/W. Liller)

Trawls through ancient records have shown that the famous comet was recorded by the Chinese as long ago as 240 BC. Halley is currently heading outward, and will reach its furthest point from the Sun (aphelion) in December 2023. It will then begin its long return journey, arriving in the inner Solar System in 2062.

Since Edmond Halley successfully predicted the return of a comet in 1682, more than 230 periodic comets have been discovered. As time went by and more orbits were calculated, it became clear that comets were hard to classify. At one extreme are the Jupiter family comets, which typically take no more than 20 years to circle the Sun. Most of these have orbits that have been drastically altered by the gravitational pull of the largest planet. The shortest period belongs to comet Encke, which scoots around the Sun every 3.3 years.

Fewer than 100 known comets, including Halley, follow a more leisurely route, traveling beyond the orbit of Neptune before returning to the inner Solar System. Although these short-period comets have also been influenced by encounters with the giant planets, their orbits are more random, often tilted steeply to the ecliptic. Many of these – including Halley's comet – travel in a retrograde (backward) direction.

New comets arrive from the Oort Cloud, far beyond the orbit of Pluto, and, after making a brief appearance in our skies, disappear from view for thousands of years. The orbits of others are modified by close encounters with the planets, particularly Jupiter. These periodic comets may reappear every five or six years.

Comets with orbits of less than 200 years are thought to originate in the Kuiper Belt, a doughnut-shaped region that begins at the orbit of Neptune and extends to at least 50 Sun-Earth distances. They were probably ejected to their present location billions of years ago by gravitational interactions with Uranus and Neptune. The first Kuiper Belt object was discovered in 1992 and many hundreds are now known (see Chapter 12).

The Oort Cloud

Approximately a dozen "new" comets are discovered every year. In contrast to the relatively predictable, well-behaved periodic comets, these intruders arrive without warning. They can appear in any part of the sky, heading sunwards from all directions. Known as long period comets, they follow highly elliptical or near-parabolic orbits which carry them many billions of kilometers from the Sun, and may take many thousands or millions of years to return. Some may never return.

By analyzing the paths of 19 such comets, Dutch astronomer Jan Oort (1900–1992) calculated that they must originate in a huge spherical reservoir located at a distance of 100,000 to 300,000 AU. Based on the frequency of their appearance in our skies, he estimated that this cloud may contain around 190 billion unseen, dormant comets. Oort suggested that every 100,000 to 200,000 years, the gravity from a star passing within 200,000 AU of the Sun would send some of the comets toward the inner Solar System.

In the last half century, the availability of more detailed information and new analysis of comet orbits has led to a modified version of Oort's cloud (Figure 13.57). It is now thought that the region can be subdivided into a doughnut-shaped inner cloud (the Hills cloud), at 2,000–20,000 AU and a spherical outer cloud at 20,000–200,000 AU (Figure 1.9 and Figure 13.60). Models predict that the inner cloud should have tens or hundreds of times as many cometary nuclei as the outer halo, so it may be a source of new comets that resupply the relatively sparsely populated outer cloud as the latter's numbers are gradually depleted.

It is now thought that the average aphelion distance of a new long period comet lies at about 44,000 AU, markedly closer than Oort suggested. The outer edge of the cloud is now thought to reside at about 200,000 AU (3 light years) from the Sun, since comets beyond this distance are so loosely bound by the Sun's gravity that they are easily nudged into interstellar space.

Meanwhile, analysis of stellar orbits shows that it is quite common for a star to pass right through the Oort Cloud. On average, a star will approach within 10,000 AU of the Sun once every 35 million years, with an even closer passage (about 3,000 AU) once every 400 million years.

Tidal forces related to the motion of the Solar System through the galactic plane and encounters with molecular clouds also influence the objects within the Oort Cloud. A giant molecular cloud made of cold hydrogen gas is likely to be encountered about every 300–500 million years, but its massive gravity is capable of altering the orbits of the Oort population, sending comets toward the Sun or outward into interstellar space.

The most widely accepted theory of the Oort Cloud's formation is that the objects initially formed much closer to the Sun as part of the same accretion process that formed the planets and asteroids, but that gravitational interaction with the young gas giants ejected them into extremely elliptical or parabolic orbits. This process also scattered the objects out of the ecliptic plane, explaining the cloud's spherical distribution.

The ejection of comets may have been a violent process involving numerous collisions, so that many of the objects which survived to reach the Oort Cloud were quite small. As time went by, gravitational interactions with nearby stars further modified their orbits to make them more circular.

It has also been suggested that the Sun captured comets from other stars that were close by in the original stellar cluster that formed within a large dusty-rich nebula. In this scenario, a substantial fraction of the Oort cloud comets originate from the protoplanetary discs of other stars.

Although there are probably billions of objects in the cloud, the area they occupy is so vast that they rarely approach to within tens of millions of kilometers of each other. Collisions must be extremely rare. Since the majority of the objects are expected to be small (most cometary nuclei have diameters of ~20 km or less), the amount of material contained in the Oort Cloud is thought to be fairly modest – perhaps five times Earth's mass. However, even that amount of material is more than would be expected from current theories about the Solar System's formation.

No members of the main Oort Cloud have yet been observed or identified, but a number of unusual objects have been been found in extremely elliptical orbits which carry them far beyond the Kuiper Belt.

The best known of these is Sedna, which was discovered in 2003 (See Chapter 1). Its closest approach to the Sun is 76 AU, far beyond the Kuiper Belt's outer edge, but its 10,500 year orbit takes it out to around 990 AU at aphelion, equal to 33 times Neptune's solar distance. However, this is still some 10 times closer than the

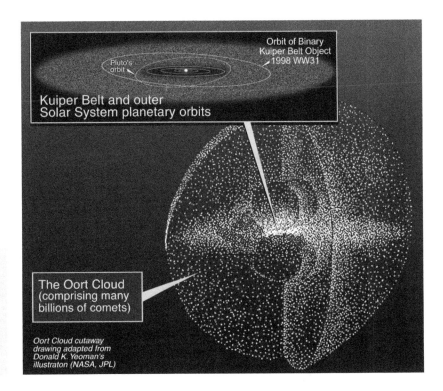

Figure 13.57 The Oort Cloud is a hypothetical, spherical swarm of icy bodies located 10,000 to 200,000 AU from the Sun. This diagram shows its presumed size and shape in relation to the Kuiper Belt and the region inside Pluto's orbit. (Courtesy of NASA-JPL/Yeomans)

predicted distance of the Oort Cloud. This has led to the suggestion that it may belong to an inner cloud, made up of bodies trapped between the Kuiper Belt and the classical Oort Cloud.

Recent deep sky surveys, undertaken partly in response to suggestions that an undiscovered planet (Planet Nine) may exist far beyond Neptune, have revealed TNOs in even more extreme orbits that never approach to within hundreds of AU from the Sun. These are also thought to be members of an inner Oort Cloud.

One of these extreme objects is 2015 TG387, which has the third-most distant perihelion so far known (about 65 AU) and an aphelion that reaches all the way out to about 2,300 AU. It takes some 40,000 years to complete one of its very elongated orbits. At its furthest point, 2015 TG387 is one of the few known objects that never comes close enough to the solar system's giant planets, like Neptune and Jupiter, to have significant gravitational interactions with them.

Even more extreme is 2014 FE72, which has the most elliptical orbit of any Solar System body that is not a long-period comet (eccentricity = 0.98). Its orbital parameters are not well known, but it appears to have a perihelion of 36.3 AU, an aphelion of ∼3,200 AU and an orbital period of ∼66,000 years. It takes roughly five times longer than Sedna to orbit the Sun.

Objects such as Sedna, 2015 TG387, 2014 FE72, and 2013 SY99 are believed to be members of the inner Oort Cloud. How these "extreme trans-Neptunian objects" were placed on their orbits is uncertain. Their closest approach to the Sun is so far beyond Neptune that they are thought to be "detached" from the giant planet's gravitational influence. On the other hand, at their furthest points, they are still too close to be nudged by neighbouring stars or the slow tides of the galaxy itself.

One explanation is a gradual gravitational nudging over tens or hundreds of millions of years. This so-called "diffusion" can cause the orbits of minor planets to become more eccentric, then less eccentric over time. In this process, the size of each orbit would vary by a random amount.

For example, when 2013 SY99 comes to its closest approach every 20,000 years, Neptune is often on the opposite side of the Solar System. However, at other encounters, when both SY99 and Neptune are "close," the ice giant's gravity will subtly nudge SY99, minutely changing its velocity and the shape of ts orbit.

During such encounters, the long axis of SY99's ellipse will alter, becoming either larger or smaller. Although each adjustment would be fairly modest, over the 4.5 billion-year lifetime of the Solar System the long axis of SY99's orbit would change by hundreds of astronomical units. Several other extreme trans-Neptunian objects with smaller orbits also show diffusion, on a smaller scale. It is possible that the gradual effects of diffusion act in a similar way on the tens of millions of objects orbiting in the inner region of the Oort Cloud.

However, diffusion cannot explain the orbit of Sedna, whose perihelion is too far from Neptune for the planet to change its orbital shape. One possibility is that, early in the history of the Solar System, many small icy bodies such as Sedna were orbiting the Sun and ejected outward by close encounters with planets. Far from the Sun, the orbits of these bodies were affected by distant stars, causing them to slow down and stay gravitationally attached to the Sun. This inner Oort cloud would have been formed in the same manner as the previously postulated Oort Cloud.

Sedna probably suffered a similar fate, although any star which affected it must have been extremely close in astronomical terms. This may be evidence that the Sun formed in a tight-knit cluster, together with many other stars.

A second explanation for Sedna's orbit is that a larger body, perhaps Mars-sized or larger, could exist in a near-circular orbit at around 70 AU. This object could have caused objects such as Sedna to be thrown into such extreme orbits (Box 12.6).

Box 13.14   'Oumuamua, An Interstellar Intruder

Astronomers have long believed that objects can be scattered from one star system to another. The first proof of this theory came on October 19, 2017, when the Pan-STARRS telescope detected a faint point of fast-moving point of light. Subsequent observations over the following days and weeks allowed its orbit to be computed accurately.

These calculations revealed beyond any doubt that this body was arriving from interstellar space. The unique object had been traveling through space for millions of years before its chance encounter with our Solar System.

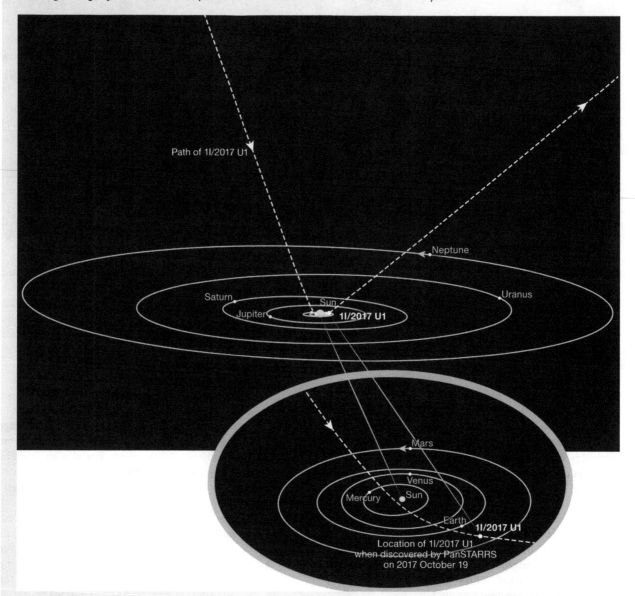

Figure 13.58   On September 19, 2017, interstellar object 1I/2017 U1 ('Oumuamua) sped past the Sun at about 315,400 km/h, so rapidly that it could not be captured by the Sun's gravity. Analysis of its trajectory indicates that it is more likely to be a comet than an asteroid. (Courtesy of ESO/K. Meech et al.)

Although originally classified as a comet, visual observations revealed no signs of a coma or jets after it passed closest to the Sun in September 2017. The object was reclassified as an interstellar asteroid and named 1I/2017 U1 ('Oumuamua).

Figure 13.59 'Oumuamua, the first interstellar object discovered in the Solar System, is estimated to be at least 400 m long and about 40 m wide. (Courtesy of ESA/Hubble, NASA, ESO, M. Kornmesser)

However, subsequent analysis of its trajectory suggested that some outgassing may have taken place, just as it would from a comet, causing it to move away from the Sun faster than expected. Perhaps it was a comet located far from its star, and was kicked out of the star system, something which would be more likely than ejection of an asteroid.

'Oumuamua seems to resemble a cosmic baguette, with an estimated length of at least 400 m and width of about 40 m. Its brightness varied by a factor of 10 as it turned on its axis every 7.3 hours. This unusually large variation in albedo means that the object must be highly elongated – about 10 times as long as it is wide. Its surface is dark and reddened due to the effects of irradiation from cosmic rays over millions of years. It lacks significant amounts of water or ice and is either rocky or has a high metallic content.

Traveling at a speed of about 95,000 kmh, it may have been wandering through the Milky Way, unattached to any star system, for hundreds of millions of years before its chance encounter with the Solar System. Astronomers estimate that an interstellar asteroid similar to 'Oumuamua passes through the inner Solar System about once per year, but they are faint and hard to spot, causing them to be missed until now. It is only recently that survey telescopes, such as Pan-STARRS, have become powerful enough to have a chance to discover them.

Questions

- What are the links between (a) asteroids and meteorites, (b) comets and meteors?
- Asteroids have traditionally been described as small, rocky objects in the inner Solar System, while comets are small icy objects in the outer Solar System. Explain why these descriptions have been challenged in recent years.
- Which are the two most common meteorites found on Earth? Describe their main characteristics.
- Why are Ceres, Vesta, and Pallas of particular interest to planetary scientists?
- Describe and account for the main characteristics of (a) Ceres and (b) Vesta.

- Why are Ceres and Vesta so different?
- Describe the main characteristics of a comet.
- Describe the major scientific discoveries made by: (a) Rosetta, (b) Stardust, (c) Deep Impact.
- What are the main differences between the Oort Cloud and the Kuiper Belt?
- What evidence is there for the existence of the inner and outer Oort Cloud?
- What are the unique characteristics of interstellar object 'Oumuamua?
- If a comet or asteroid was predicted to impact Earth in 10 years' time, what scientific advice would you give your government and the United Nations?

FOURTEEN
Exoplanets

Since ancient times, astronomers and philosophers have debated and speculated about the existence of exoplanets (also known as extrasolar planets) – planets in orbit around distant stars. Is our Solar System unique, or is it just one of many such systems scattered throughout the Galaxy? Are there worlds out there that support alien life forms, and are there other advanced civilizations in the Galaxy?

The distances between the stars are so immense that, for many centuries, there seemed little likelihood of answering any of these questions. Then, as measurements of stellar positions became more accurate, suspicions arose that planets might be the cause of the irregular proper motions of some stars. Barnard's Star, one of our closest stellar neighbors, was a favorite candidate, but its postulated planetary companions proved to be elusive until evidence for a nearby exoplanet was uncovered in 2018.

Then came the breakthrough (Box 14.1 and Figure 14.1). In 1991, Aleksander Wolszcan (Pennsylvania State University) and Dale Andrew Frail (National Radio Astronomy Observatory) were using the giant radio telescope at Arecibo Observatory in Puerto Rico to study a rapidly rotating neutron star known as PSR B1257+12. Tiny variations in the arrival time of its radio signals led them to the surprising conclusion that the dense, dead star was orbited by at least three fairly small planets.

Four years later, the first undisputed planet orbiting another main sequence star (51 Pegasi) was found by Michel Mayor and Didier Queloz of Geneva Observatory, Switzerland. Detected indirectly by observing tiny changes in the star's radial velocity, their planet was at least half the mass of Jupiter. The rapidly orbiting world was very close to 51 Pegasi and blisteringly hot.

Three months later, a team led by Geoffrey W. Marcy and Paul Butler of San Francisco State University and the University of California at Berkeley confirmed the Swiss discovery and reported the presence of two more planets.

These announcements marked the beginning of a flood of discoveries. Since then, the number of confirmed exoplanets has exceeded 3,900 – mostly detected by indirect observations – with more than 3,000 planet candidates awaiting confirmation. The total number is likely to rise rapidly as new planet-hunting techniques are introduced. This has led to estimates that there may be up to 140 billion alien worlds in the Milky Way galaxy – more than one planet for each star.

Many of the confirmed objects are like 51 Pegasi, giant planets with short periods (up to about 10 days) and orbits that bring them extremely close to their star. However, improved detection techniques are now discovering planets that more closely resemble those in our Solar System, with circular orbits and longer orbital periods.

Exploring the Solar System, Second Edition. Peter Bond.
© 2020 John Wiley & Sons Ltd. Published 2020 by John Wiley & Sons Ltd.
Companion Website: www.wiley.com/go/Bond-Solar-System2e

Box 14.1   Pulsar Planets

The first exoplanets to be discovered were detected in 1991, during a search for pulsars conducted with the 305 m Arecibo radio telescope in Puerto Rico. The find was most unexpected, since pulsars are rapidly spinning neutron stars that emit regular, pulsing radio emissions, rather like a rotating lighthouse beam. These extremely small, dense objects are the remnants of supernova explosions that mark the deaths of massive stars. The collapsed core measures only about 15 km across, but a teaspoonful of their matter would weigh a billion tonnes.

The discovery was made by a team led by Polish astronomer Aleksander Wolszczan. They were studying an old neutron star, known as PSR B1257+12, which is spinning once every 6.22 milliseconds. He recognized tiny variations in the time of arrival of the pulses from the star, instead of their usual precise regularity. Wolszczan analyzed these variations and came to the remarkable conclusion that the small, but unexpected, departures from the regular beat of the radio signals were caused by three small planets that were tugging on the pulsar and causing slight wobbles as it rotated.

Figure 14.1   An artist's impression of a planetary system around pulsar PSR B1257+12. The planet nearest the pulsar is very lightweight, while the other two (foreground) are more massive than Earth. Radiation from charged pulsar particles probably rains down on the planets, causing their night skies to light up with auroras similar to our Northern Lights. One such aurora is illustrated on the planet at the bottom of the picture. (Courtesy of NASA/JPL-Caltech/R. Hurt, SSC)

Skepticism about the reality of these unexpected worlds was overcome when predictions of how they should gravitationally interact were subsequently confirmed. The perturbations between the two larger planets were used to measure their masses and orbital inclinations.

The calculations indicate that the three pulsar planets have masses of 0.02, 4.3, and 3.9 Earth masses, and their orbital periods are 25, 66, and 98 days. Compared to our Solar System, they would all fit within the orbit of Mercury, and the spacing of their orbits is similar in proportion to those of Mercury, Venus, and the Earth. There is also some evidence that the pulsar may have an asteroid belt similar to the one that occurs beyond the orbit of Mars.

The fact that the planets have almost coplanar orbits indicates that they all evolved from a protoplanetary disk, very much like the planets around our Sun. However, it seems unlikely that could have existed before the supernova occurred and then survived the stellar holocaust. Indeed, they may well represent a second generation of planets, replacing an earlier family that was destroyed by the cataclysmic explosion.

One possibility is that metal-rich debris from the supernova formed an orbiting disk around the neutron star. Small, but massive planets with large iron cores could have coalesced from the disk. Powerful dynamos in the cores would generate very strong magnetic fields, so, although the dead star would be a poor source of illumination, dense streams of energetic particles emitted by the pulsar would fill their night skies with magnificent auroras. However, the highly energetic particles would also gradually erode each planet's surface, generating a low-level haze.

Evidence for possible exoplanets and planet-forming disks has since been found around other pulsars.

Beta Pictoris

Until recently, exoplanets had always been too small and distant to be observed directly. However, since planets were thought to be born inside huge, dusty nebulas surrounding young stars, the obvious starting place was to look for circumstellar disks that might contain fledgling planetary nurseries.

Prospects for finding new worlds around other stars brightened in 1984 when infrared observations by Bradford A. Smith (University of Arizona in Tucson) and Richard J. Terrile, (Jet Propulsion Laboratory) showed a disk of dust surrounding the young star Beta Pictoris, located 63 light years from Earth.

Their discovery provided the first unambiguous proof that flattened disks of matter exist around stars other than the Sun. The Beta Pictoris disk appeared to be a young planetary system in the making, and thus supported the standard model of Solar System birth, which theorizes that planets accrete from a disk of dust and gas surrounding a young star.

Since then, Beta Pictoris has been the subject of intense study. In 1995, Hubble Space Telescope observations revealed an apparent warp in the disk, which was later confirmed. Observers suggested that the warp might be associated with a secondary disk tilted about 4° from the main disk, possible due to a planet in an inclined orbit.

Further Hubble observations made in 2003 confirmed the existence of two dust disks encircling the star. To see the faint disks, astronomers used the Advanced Camera for Surveys' coronagraph, which blocked the light from Beta Pictoris. The secondary disk was visible out to roughly 250 AU from the star, and probably extended even farther.

Since dust grains are only expected to survive for a few hundred thousand years, the continuing existence of these disks around a star no more than 20 million years old means that the dust is probably being replenished by collisions between larger planetesimals.

The observers theorized that Beta Pictoris may have two planet-forming regions, one associated with each disk. An unseen planet, up to 20 times the mass of Jupiter and orbiting within the secondary disk, was sweeping up planetesimals from the primary disk. When these collide, they create the secondary dust disk. The pale red color of the disk may be due to the presence of tiny grains of graphite and silicates, a theory supported by data from NASA's Far Ultraviolet Spectroscopic Explorer (FUSE) spacecraft, which discovered abundant amounts of carbon gas in the dust disk.

To complicate matters still further, infrared images taken in 2002 by the Keck II telescope in Hawaii showed that another, smaller, inner disk may exist around the star in a region the size of our Solar System. The possible inner disk was tilted in the opposite direction from the secondary disk seen in the Hubble images.

In November 2008, a team of French astronomers using ESO's Very Large Telescope reported the discovery of a postulated giant planet, about 8 times as massive as Jupiter (Figure 14.3). With a projected distance from the star of 8–15 AU, comparable to Saturn's distance from the Sun, it orbits inside the main dust disk of Beta Pictoris. Further observations have confirmed the existence of a giant planet, known as Beta Pictoris b, which takes 21 years to complete one orbit of its star at an orbital distance of 9.1 AU. Its mass is unknown,

Dusty Disks

Since the discovery of the complex disk surrounding Beta Pictoris, similar structures have been found around numerous young stars. Many of these embryonic planetary systems – also known as protoplanetary disks (proplyds) – are buried deep within dense nebulas where large-scale star formation is taking place. Although

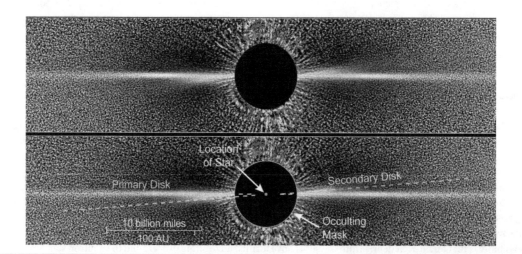

Figure 14.2   Images of Beta Pictoris, taken by the Advanced Camera for Surveys' coronagraph on the Hubble Space Telescope, confirmed the existence of two dust disks encircling the star and suggested the presence of at least one giant planet. The visible-light image shows a distinct secondary disk that is tilted by about 4° to the main disk. The secondary disk may be created by a giant planet that is removing material from the primary disk. (Courtesy of NASA, ESA, D. Golimowski/Johns Hopkins University, D. Ardila/IPAC, J. Krist/JPL, M. Clampin/GSFC, H. Ford/JHU, G. Illingworth/UCO/Lick and the ACS Science Team)

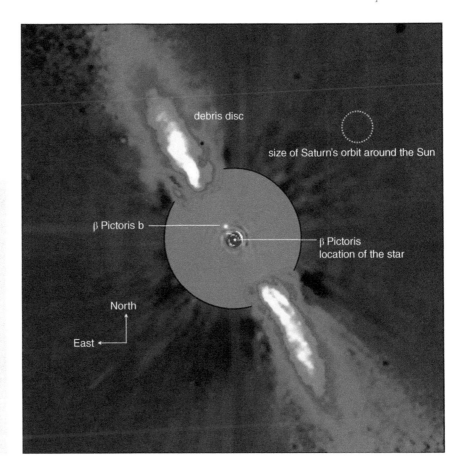

Figure 14.3 This composite infrared image shows a giant exoplanet inside the dust disk of Beta Pictoris. This very faint planet was revealed after subtraction of the much brighter stellar halo. The outer part of the image shows light reflected from the dust disk, as observed in 1996 by ESO's 3.6 m telescope. The inner part of the system was imaged at 3.6 microns with the NACO instrument on the Very Large Telescope. The newly detected source is more than 1,000 times fainter than Beta Pictoris, and aligned with the disk, at a distance of about 8 AU. (Courtesy of ESO/A.-M. Lagrange et al.)

they appear as fairly indistinct clouds surrounding newborn stars, the proplyds are helping to clarify the mechanisms behind planet formation.

The most famous examples of proplyds are found in the Orion Nebula, a huge star-forming region about 1,500 light years away, where dozens of examples have been found with the Hubble Space Telescope (Figures 1.17 and 14.4). Estimates indicate that at least half of the stars in the nebula are surrounded by such disks.

Two different types of disk have been identified. Those close to the brightest star in the Trapezium cluster, Theta 1 Orionis C, shine brightly as a result of the energetic radiation they receive. These hot, glowing disks – some seen edge-on and some face-on – are sculpted into different shapes by the stellar wind and radiation, a process known as photoevaporation. In some cases, emerging jets and curved shock waves are apparent. The shock waves occur when the stellar wind collides with the gaseous disks. Proplyds further away from Theta Orionis C are not illuminated and appear as dark smudges silhouetted against the bright nebula in the background.

The dark proplyds are composed of very cold gas and dust with a temperature of –250°C. Their masses vary considerably, but those that are in the range 0.01–0.025 solar masses are thought to have sufficient disk material bound to their central stars to form planetary systems on the scale of our Solar System. However, those that are being blasted by intense radiation from Theta 1 Orionis C are likely to have much of their material dissipated and vaporized, possibly destroying their potential for planets to form.

It is interesting to note that dusty disks have also been found around brown dwarfs, often known as "failed stars." Although the disks are often less massive than those around ordinary stars, the material is less likely to be blown away by the weak radiation from the small, cool brown dwarfs. Some crystallization and clumping of dust particles has been detected, showing that the disks are undergoing at least the same initial steps of the planet-building process as circumstellar clouds.

Disk Evolution and Exoplanet Formation

If the disks survive, significant changes take place over time, as the central stars and their disks evolve, and accretion processes begin (see Chapter 1). Structure within the disks appears, often in the form of belts of asteroidal or cometary debris. "Empty" zones which have been largely cleared of material, probably by one or more large protoplanets, become apparent. Although they are extremely difficult to detect, exoplanets orbiting within the disks have occasionally been found.

The structure and composition of protoplanetary disks can be studied with infrared telescopes, such as the Spitzer Space Observatory (Figure 14.5). Infrared observations detect dust that is created during collisions between planetesimals left over from planetary formation. In the majority of cases, the warm dust in the inner regions of the disk is made of common terrestrial minerals such as olivine, pyroxene, and silica. Further out, the colder material is dominated by ices.

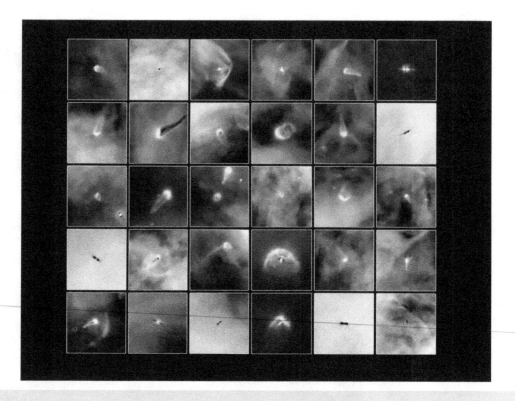

Figure 14.4   Thirty protoplanetary disks in the Orion Nebula, imaged by the Hubble Space Telescope's Advanced Camera for Surveys. The images show two different types of disk around young and newly forming stars: those that lie close to the brightest star in the cluster (Theta 1 Orionis C) and those further away from it. The star heats up the gas in the nearby disks, causing them to shine brightly. The cooler, more remote, disks can only be detected as dark silhouettes against the background of the bright nebula. Some are viewed edge-on, some face-on. (Courtesy of NASA, ESA and L. Ricci/ESO)

The formation of dust disks around stars seems to be almost universal. However, studies of intermediate mass stars, similar or slightly larger than the Sun, show that dust within the terrestrial planet zone is soon removed. Only stars with ages in the range 10–30 million years display dust in their inner disks, created during the epoch of final catastrophic mass accretion for rocky, terrestrial planets.

Observations of star-forming regions have shown that about 10% of type A and B stars, about two to 15 times as massive as the Sun, appear to possess dusty disks. However, there is a fairly small chance that the disks will survive long enough for planets to evolve.

Planet formation around such massive stars is a battle between opposing forces. On one hand, the circumstellar disks tend to be larger and contain more of the raw materials to build planets. On the other hand, fierce UV radiation and stellar winds tend to destroy the disks rapidly. Within two to five million years of their birth, most of the massive stars have already lost the raw materials needed to form planets. This indicates that, at least for type A and B stars, planets must form quickly or not at all.

Although direct imaging of exoplanets around dwarf stars is very difficult with current instrumentation, the coalescence of planets from the less massive nebulas surrounding smaller, cooler stars seems to be a much easier proposition. Observations indicate that at least 20%, and possibly as many as 60%, of stars similar to the Sun are candidates for forming rocky planets. One possibility is that the most massive disks form their planets quite rapidly, whereas the low mass disks take 10 to 100 times longer.

The composition of the disk is also an important factor. When a star forms by collapsing from a dense cloud into a luminous sphere, the new arrival and its circumstellar disk reflect the composition of the original molecular cloud. Some nebulas are much richer in "metals" than others.[1] Elements such as iron and oxygen are the raw materials for rocks and ice, so it is not surprising that the stars with more heavy elements tend to have more planets. In contrast, there are few opportunities for planets to coalesce and grow from gaseous disks that contain very little solid material.

The dust particles, which are the building blocks of planets, accumulate into comets and small asteroid-size bodies, and then clump together to form planetary embryos, and finally full-fledged planets. In the process, some rocky objects collide and grow into planets, while others shatter into dust.

Observations of fledgling systems show bands of material separated by regions that have been cleared by planets. In the case of Epsilon Eridani, a younger, slightly cooler and fainter version of the Sun, infrared observations indicate the presence of two

---

[1] Astronomers refer to all the elements heavier than helium as "metals."

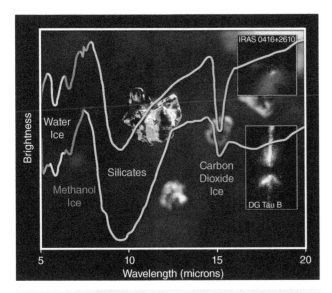

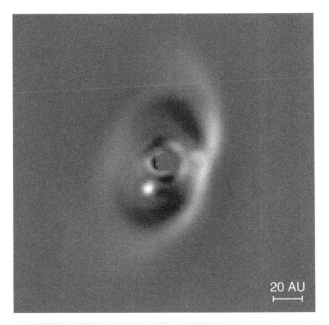

Figure 14.5 Spectra obtained by the Spitzer infrared space observatory reveal the chemical composition of protoplanetary disks around two young stars. The broad depression in the center of each spectrum signifies the presence of crystalline silicates. Tiny olivine crystals are thought to make up at least some of the dust grains, becoming coated with ice deep within the disk. The depth of the silicate absorption feature indicates that the dusty cocoon around the embedded protostar is extremely thick. Other dips due to absorption are produced by water ice (blue), methanol ice (red), and carbon dioxide ice (green). Their existence in solid form suggests that the material close to the protostar is cold. (Courtesy of NASA/JPL-Caltech/D. Watson – University of Rochester)

Figure 14.6 This is the first clear image of a planet caught in the act of forming. Known as PDS 70b, it is visible as the bright spot to the right of the masked-out star. The young (10 million-year-old) dwarf star PDS 70, located 370 light years from Earth, is surrounded by a disk with a distinctive gap. The newborn planet is orbiting within the gap. The infrared image was taken by the SPHERE instrument on ESO's Very Large Telescope. (Courtesy of ESO/A. Müller et al.)

asteroid belts and a broad, outer ring of icy comets similar to our own Kuiper Belt, in addition to at least two possible planets.

By the time a Sun-like star reaches an age of around 300 million years, the dusty disk has almost completely dissipated, either because the dust particles are blown away or dragged onto the star, or the particles clump together to form much larger objects.

It is interesting to note that systems containing asteroids, comets, and, possibly, planets are at least as abundant in binary star systems as they are in those with only one star. Since more than half of all stars are binaries, this suggests that planets with two Suns may be quite common across the Universe.

Disks have been found not only around widely spaced pairs, but also around very tight binaries separated by only a few astronomical units. Far fewer disks have been found in intermediately spaced binary systems, three to 50 AU apart. In other words, if two stars are as far from each other as the Sun is from Jupiter (5 AU) or Pluto (40 AU), they would be less likely to host a family of planetary bodies.

Spitzer data have also revealed that debris disks circle all the way around both members of a close-knit binary – these are called circumbinary disks – but only a single member of a wide duo

(Figure 14.8). This could explain why the intermediately spaced systems are inhospitable to planetary disks: they are too far apart to support one large disk around both stars, and they are too close together to have enough room for a disk around a single star.

Another curiosity is that at least some systems may contain planets that orbit the central star in opposing directions (Figure 14.9). The first evidence for such a system was provided by radio observations of a disk around a young protostar some 500 light years from Earth in the constellation Ophiuchus. Doppler shifts in the frequencies emitted by molecules within the cloud show the direction in which the gas is moving relative to Earth. The shifts show that the inner and outer parts of the disk are rotating in opposite directions. This is likely to result in the formation of a system where the inner planets orbit in one direction and the outer planets go around their star in the opposite direction, unlike our own Solar System, where all the planets travel in the same direction.[2]

The Sun and its planets are thought to have formed during the collapse of a single molecular cloud, so they all rotate in the same direction. However, the system in Ophiuchus lies in a large, star-forming region where chaotic motions and eddies in the gas and dust result in small clouds with independent rotations. The protoplanetary disk may, therefore, contain substantial amounts

[2]It is not too surprising to find counter-rotation in a protostellar disk, since the phenomenon has been previously reported in the disks of galaxies.

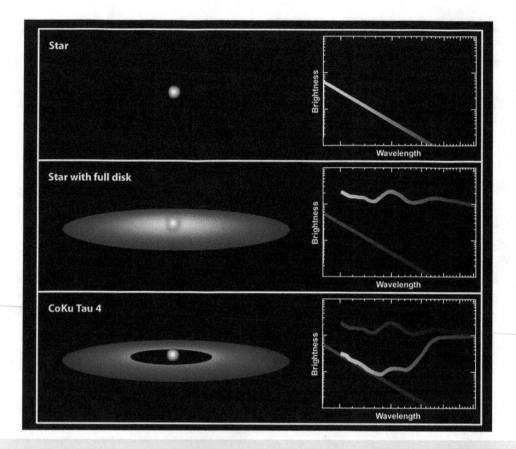

Figure 14.7 Spectroscopy enables scientists to deduce the temperature and chemical composition of material around a star, even if the disk itself cannot be seen. In the case of a star with no circumstellar disk (top), most of the light is produced at shorter wavelengths (left side of graph), due to the high temperature of the star's surface. If a star has a disk around it (center), the warm dust and gas produce their own infrared light, which changes the shape of the spectrum. The circumstellar material is cooler than the star, so it emits most at longer, infrared wavelengths. If the inner part of the disk has been cleared, perhaps by the formation of a planet, there is a normal spectrum near the star, with an excess of infrared light further away. (Courtesy of NASA/JPL-Caltech, D. Watson/University of Rochester)

of material from two such clouds, each rotating in a different direction.

Another unusual arrangement involves a planet in a triple star group. Fewer than 40 of such systems have so far been found. One example involves three gravitationally bound stars known as KELT-A, KELT-B, and KELT-C. The first of these is orbited by KELT-4Ab, a gas giant planet, similar in size to Jupiter, which takes approximately three days to complete one circuit. The other two stars are much farther away and orbit one another over a period of about 30 years. It takes the pair approximately 4,000 years to orbit KELT-A.

KELT-4Ab must be a sweltering world, where KELT-A would appear roughly 40 times as big as our Sun does to us, due to its close proximity. The other two orbiting stars would appear much dimmer due to their great distance, shining no brighter than our Moon.

Another planet, HD 131399Ab, appears to orbit much further out in a triple star system. The exoplanet seems to orbit the main star in the group, which is roughly 10 times brighter than the Sun. It lies about 82 AU from the star HD 131399A and completes one orbit in 400 to 700 years. It is located more than one-quarter of the distance to the pair of much smaller, stars that whirl around the primary star at a distance of about 300 AU.

Brown Dwarfs or Exoplanets?

Many exoplanets are much larger and more massive than Jupiter. This has led to some confusion over whether these gas giants should be regarded as planets or as sub-stellar brown dwarfs (Figure 14.10). The latter are objects that are not undergoing nuclear fusion in their cores, but display some stellar characteristics. With surface temperatures of between 200 and 2,500°C, they appear as faint, dark red objects – just like gas giants.

One of the most extreme exoplanets yet found is CoRoT-3b, which was found during a stellar transit (Figure 14.11). This strange object is about the size of Jupiter, but more than 20 times as massive. It takes 4 days and 6 hours to orbit its star, which is slightly larger than the Sun.

Whether CoRoT-3b is a giant planet or a brown dwarf depends on how this oddity originally formed. If it was born

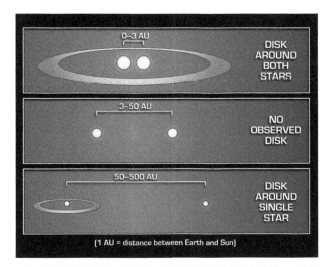

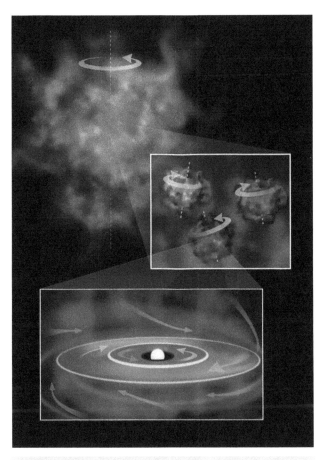

Figure 14.8 Infrared observations with NASA's Spitzer Space Telescope show that debris disks are common around binary stars. In particular, they are more abundant around pairs that are 0–3 AU apart (top panel) and 50–500 AU apart (bottom) than binary stars with orbital separations of 3–50 AU. (Courtesy of NASA/JPL-Caltech/T. Pyle-SSC)

Figure 14.9 The stages in the formation of a counter-rotating protostellar disk. (Top) A huge star-forming nebula is rotating as it gives birth to multiple stellar systems. (Middle) A detailed view inside the nebula shows three protostars forming as the cloud collapses. The chaotic collapse process creates eddies, allowing newly forming stars to rotate in different directions and at different speeds. (Bottom) One protostellar cloud collapses further into a disk-like structure that rotates counter-clockwise about the newly formed star. The protostar also draws material from a second, passing, cloud which is rotating in the opposite direction. As a result, the outer part of the disk rotates clockwise, so any planets forming in this region will orbit the star in the opposite direction from the inner planets. (Bill Saxton, NRAO/AUI/NSF)

from a disk-shaped nebula that circled the central star, then it must be classified as a planet. However, if it formed simultaneously with its neighboring star, due to the gravitational collapse of a molecular cloud, then it is a brown dwarf. Unfortunately, this question is extremely difficult to resolve, since there are few tell-tale signs in the object's orbit or physical characteristics.

What is clear is that COROT-3b is one of the most massive exoplanets yet found, and, with a density more than twice that of lead, the densest exoplanet so far discovered. The surface gravity is over 50 times that of Earth. This density is so high because of the extreme compression of matter in the object's interior.

It is hoped that further studies may throw some light on how this heavyweight object formed and ended up so close to its star. One intriguing possibility is that it might just be the first member of a family of very massive planets that encircle stars more massive than our Sun.

Further confusion can arise because isolated planetary mass objects (sometimes dubbed "planemos"), only a few times more massive than Jupiter, have been found far from any stellar companion. Some astronomers believe that these fairly cool objects were planets that were ejected from orbit around their stars. Others believe that they may form on their own through gas cloud collapse, similar to star formation.

If they are tiny brown dwarfs (or sub-brown dwarfs), they probably originated as failed stars that formed during the collapse of small molecular clouds. However, like *bona fide* stars, some sub-brown dwarfs and planemos are surrounded by disks of gas and dust. It seems possible that small planets or asteroids could form in these disks.

Detecting Exoplanets: Radial Velocity

Until recently, it was thought that current technology was not sufficiently advanced to achieve direct imaging of an exoplanet. The only alternative was to search for indirect evidence for the existence of such objects. The first exoplanets were detected by measuring the tiny wobbles that they induced in the motion of their parent stars. Since then, the vast majority of ground-based planetary detections has been achieved with telescopes that use the so-called radial velocity technique.

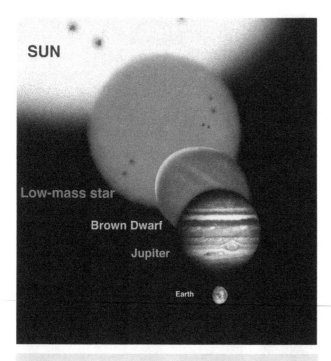

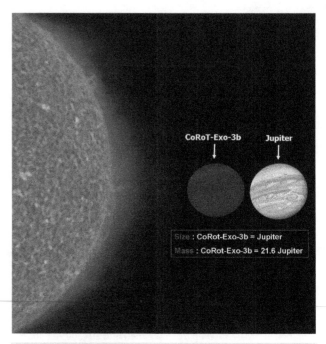

Figure 14.10 Brown dwarfs are classified as gaseous, sub-stellar objects with masses of between 13 and 80 Jupiter masses. They are similar in size to Jupiter, but their temperatures vary from 200°C to more than 2,000°C. Their atmospheres are rich in methane, also a major constituent of Jupiter's atmosphere. With so many characteristics in common, it is very difficult to distinguish a brown dwarf in a binary system from a giant gaseous planet. (Gemini Observatory)

Figure 14.11 The exoplanet CoRoT-3b compared with the Sun and Jupiter. The exoplanet is similar in size to Jupiter, but much more massive, so there is some uncertainty over whether it can be classified as a giant planet or a brown dwarf. (Courtesy of ESO/OAMP)

The method requires the light from a star to be split into a spectrum, similar to the way water droplets in the atmosphere separate sunlight into a colorful rainbow. When the spectrum is magnified, straight black lines can be seen superimposed on the colors. These spectral lines correspond to the wavelengths of light that have been absorbed by chemicals in the atmosphere of the star from which the light originated.

By studying a sequence of tiny Doppler shifts (frequency shifts) in these lines, it is possible to show that a star is orbited by one or more planets (Figure 14.13). When a star is moving toward Earth, the wavelengths of incoming light "bunch up" and shorten. Astronomers say they are "blue-shifted." When the star is moving away from Earth, the wavelengths stretch out, so they are "red-shifted." A sequence of abnormal, but predictable, shifts in the spectrum can reveal minor changes in a star's motion in response to the gravitational pull of a nearby planet. The star appears to wobble, or pirouette, around a point in space.

This technique is now able to detect Earth-sized worlds if they have very short periods and orbit low-mass stars, e.g. GJ 273c. At present, the best spectroscopes can confidently detect motions of about 0.5 m/s. Although Earth's modest gravity only causes the

Sun to wobble at 0.1 m/s, new instruments under development will soon be able to detect such tiny motions. Techniques are also being introduced to extract the effects of the bubbling activity of the star's gaseous surface, which tends to hide the Doppler effect of such a planet.

Since the discovery of 51 Pegasi in 1995, almost 700 planets have been discovered using this radial velocity (RV) technique. Many of them occur in multiplanet systems that occasionally lie in resonant orbits with one another. Given the period of the orbit and the maximum radial velocity observed for a system, it is possible to deduce the mass of the exoplanet (assuming the star's mass is known) as well as its orbital distance.

Unfortunately, the inclination of the planet's orbit is often unknown, since the only data available relate to the observed radial velocity changes of the star. As already mentioned, the other disadvantage of the RV technique is that it is most sensitive to large, Jupiter-like, exoplanets that are orbiting close to their star. Small, rocky worlds located 1 AU or more from the star cause such small RV shifts that they are very difficult to detect.

The most successful ground-based exoplanet finder to date has been the High Accuracy Radial Velocity Planet Searcher (HARPS) spectrograph on ESO's 3.6 m telescope at La Silla, Chile.[3] HARPS is capable of detecting back-and-forward motions in a star's radial velocity as small as 3.5 km/h – a slow walking pace.

---

[3] On April 23, 2012, the almost identical HARPS–North (HARPS-N) was inaugurated at the Italian 3.6 m Telescopio Nazionale Galileo, located on the island of San Miguel de La Palma in the Canary Islands.

Box 14.2   How Normal is Our Solar System?

Knowing whether our Solar System is unique among planetary systems can help scientists to better understand how planets form and evolve. Furthermore, it sheds light on the possibility of life existing elsewhere in the Universe.

To date, some 3,900 exoplanets have been confirmed, as well as thousands of additional candidates. Amidst this vast swarm of solar systems, how special is our own?

Planet Masses and Densities: Those of the gas giants in our Solar System are fairly typical. Although the terrestrial planets are on the low side for mass, that may be a selection effect, because it is very difficult to detect low-mass planets.

Age of the Solar System: Roughly half the stars in the disk of our galaxy are younger than the Sun, and half are older. There is nothing special about the age of our Solar System.

Orbital Locations of the Planets: Our Solar System has no planets very close to its star. All of them orbit at a distance that is larger than the mean distance observed in known exoplanetary systems. However, this might be a selection effect, because it is easier to detect large planets orbiting very close to their stars.

Eccentricities of the Planets' Orbits: Planets in our Solar System follow very circular orbits, something which is less likely compared to typical exoplanet systems. One possible explanation is that the eccentricity of orbits tends to decrease with more planets in the system. The presence of eight planets may help to explain why their eccentricities are so low.

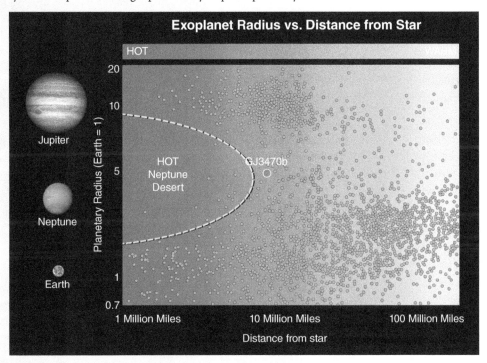

Figure 14.12   This graph plots exoplanets based on their size and distance from their star. Each dot represents an exoplanet. Planets the size of Jupiter (top) and planets the size of Earth and so-called super-Earths (bottom) are found both close to and far from their star. Neptune-sized planets (middle) are scarce close to their star. This "desert" of hot Neptunes shows that such alien worlds are rare, or they were once plentiful but have since disappeared. GJ 3470b, a warm Neptune at the border of the "desert," is rapidly losing its hydrogen atmosphere. Over time, hotter Neptunes may be eroded so that they become smaller, rocky super-Earths. (Courtesy of NASA, ESA, and A. Feild/STScI)

Super-Earths: A super-Earth is a planet with a mass between 1 and 10 times that of Earth. The super-Earth classification refers only to the mass of the planet, and does not imply anything about its surface conditions or habitability. There are no planets in this mass range in our Solar System, which is quite unusual – super-Earths are common in other planetary systems (see Earth-Like Planets and Super-Earths).

Super-Neptunes: Another mass-class of planets that is found only outside our Solar System consists of planets more massive than Neptune at 0.05 Jupiter masses, but smaller than Saturn at 0.3 Jupiter masses. These low-mass gas planets are at the very transition between planets that have bulk compositions dominated by hydrogen/helium and those with lower hydrogen/helium content. These objects have a wide variation in densities, e.g. HATS-8b has a mean density of just 0.259 g/cm$_3$, whereas Kepler-101b has a bulk density of 1.45g/cm$_3$.

In summary, our Solar System has many characteristics in common with exoplanetary systems. Its most unusual features are the lack of a super-Earth, the lack of any close-in planets, and the low eccentricities of the planetary orbits. In terms of habitability, there is probably nothing special about planetary family, so life elsewhere is a clear possibility.

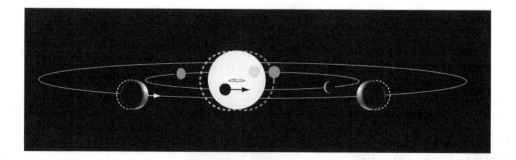

Figure 14.13   An illustration of how an Earth-sized planet orbiting close to a star may be discovered and studied. The planet lies inside the orbit of a giant planet and both are shown at different positions in their orbits. The presence of the unseen planets can be inferred from the periodic Doppler shift of starlight (blue and red dashed circles) as the star is pulled toward and away from the observer. The smaller planet's diameter can be estimated when it transits the star and blocks a tiny fraction of the starlight. At the opposite phase, the planet is eclipsed by the star, and its infrared flux can be measured by comparing the total flux before and during eclipse. The inner planet perturbs the outer giant planet on its orbit (dashed lines), causing times of transit of the latter to deviate from simple periodicity by a measurable amount. (Courtesy of Eric Gaidos et al./Science, 12 Oct. 2007/AAAS)

The instrument has led to the discovery of more than 150 exoplanets. Some of these orbit low-mass stars known as M dwarfs, or red dwarfs. Its extremely precise measurements also make it possible to search for smaller planets. Up until 2011, about two-thirds of all the known exoplanets with masses less than that of Neptune were discovered by HARPS. Since then, the transit method has become the most important method of detecting exoplanets (see Transits).

Exoplanet-hunting surveys that rely on the Radial Velocity Method are expected to benefit greatly from the deployment of the James Webb Space Telescope (JWST), which is scheduled for launch in 2021. Once operational, this infrared space observatory will obtain Doppler measurements of stars to determine the presence of exoplanet candidates.

Transits

During the last decade, the most important method for detecting small worlds has been to look for slight dips in a star's brightness caused by a planet passing in front of its star (Figure 14.14). Such a celestial alignment is known as a transit. From Earth, both Mercury and Venus occasionally pass across the front of the Sun, appearing as tiny black dots passing across the brilliant photosphere (see Chapters 5 and 6).

Unfortunately, this possibility only opens up for star–planet systems that happen to be viewed nearly edge-on from Earth. Furthermore, detection is not easy, since transits only block a tiny fraction of the star's light. Even if a distant star is transited by a giant, Jupiter-like planet, only 1% or less of the starlight will be blocked. Detection is usually limited to planets that orbit close to their primary. However, if one or more exoplanets can be observed in transit across the star's disk, then a much clearer understanding of the system can be obtained.

In some cases, a discovery made using the radial velocity method may be confirmed by subsequently detecting a transit. One example was the star HD 209458, which was being affected by the gravitational pull of a hot Jupiter in a tight circular orbit.

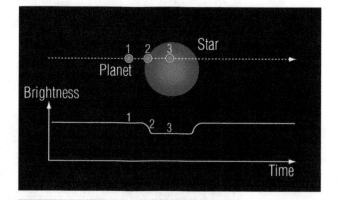

Figure 14.14   An exoplanet in transit across a star's disk causes a tiny drop in the apparent brightness of the star. Such transits are only possible if the planet and star are in our line of sight. However, they enable the diameter of the planet to be calculated, from which the mean density of the planet can be derived. It is also possible to study the exoplanet's atmosphere by comparing the combined spectrum of the star and planet with a spectrum taken when the planet passes behind the star. (Courtesy of NASA)

Its existence was later confirmed when the planet passed in front of the star, exactly as predicted.

The key is to measure the time each transit takes and the relative radii of the planet and the star. The radius of a planet can be derived from the star's light curve during the transit. Since radial velocity data enable the minimum mass to be calculated, by combining the minimum mass and the orbital inclination, it is possible to obtain the true mass. Then, by combining the latter with the radius, the mean density can be calculated.

By studying transits, it is also possible to learn about the exoplanet's atmospheric composition and weather, by comparing the combined spectrum of the star and planet with a spectrum taken when the planet passes behind the star. By subtracting the second spectrum from the first, it is possible to derive the spectrum of the planet itself. [4]

By November 2018, more than 3,000 transiting exoplanets had been detected. The first of these was HD 209458b. Originally discovered by using the indirect radial-velocity method, its existence was confirmed in 1999, when it crossed in front of its star.

Nearly all of the transiting planets so far observed are quite close to their central star; the furthest examples include KIC 5010054 b, a Neptune-class planet with a 904-Earth day orbit and KIC 5522786 b, which is a little larger than Earth and has a 757-day orbit. Both of these would be located beyond Mars in our Solar System.

The majority of known transiting exoplanets have been discovered by the Kepler spacecraft, which was designed to detect tiny dips in luminosity among 150,000 stars in the Milky Way (see below) (Box 14.3). Kepler's mission ended in 2018, but it has been replaced by another NASA observatory known as the Transiting Exoplanet Survey Satellite (TESS), whose mission is to survey the entire sky for two years. TESS is expected to detect hundreds of small exoplanets around fairly bright, nearby stars.

Direct Imaging

Predictions based on radial velocity measurements are, of course, very useful precursors to follow-up observations by the most powerful telescopes. If images of exoplanets can be obtained, it is then possible to measure their orbits and extrapolate various physical characteristics. Infrared imaging will also give an idea of an object's temperature, as well as its chemical composition and physical state.

Although direct imaging would seem to be an obvious method to use, it is extremely difficult to carry out, mainly due to the overwhelming brightness of the central star. It is rather like looking for the light from a glow worm flying next to a searchlight. (Of course, the problem diminishes if the central star is a small, faint object, such as a red dwarf or a white dwarf.)

At visible wavelengths, a star like the Sun will outshine a planet like the Earth by a thousand million times. This is because planets simply reflect some of the star's light. If, however, observations are made at longer wavelengths, such as the mid-infrared, the contrast in luminosity between the star and the planet drops to a million, because the amount of infrared light given out by the star is relatively small, while the planet itself begins to glow.

With the advent of adaptive optics, high-resolution observations in near-infrared light can be undertaken from large ground-based telescopes, whereas mid-infrared radiation is most easily viewed from space (Figure 14.6). However, in the former case, it is essential to use a coronagraph to block out the light from the star. In recent years, more than 40 direct images of exoplanets have been obtained, either through the detection of their infrared emission or from their reflected visible light. By November 2018, about 1% of all confirmed exoplanets had been detected through direct imaging.

The first visible-light images of an exoplanet were taken by the Hubble Space Telescope. In 2004, the HST's High Resolution Camera produced the first-ever resolved visible-light image of a large dust belt surrounding Fomalhaut, a naked eye star located 25 light years away in the southern constellation Piscis Australis (Figure 14.15). It showed a ring of protoplanetary debris approximately 34 billion km across with a sharp inner edge.

This large debris disk is similar to the Kuiper Belt in our Solar System. A team led by Paul Kalas, of the University of California at Berkeley, proposed in 2005 that the ring was being gravitationally modified by a planet orbiting between the star and the ring's inner edge.

Circumstantial supporting evidence came from HST's confirmation that the ring is offset from the center of the star. The sharp inner edge of the ring was also consistent with the presence of a planet that gravitationally "shepherds" ring particles.

In 2008, Hubble photographed a point source of light lying 2.9 billion km inside the ring's inner edge. This was the first extrasolar planet to be imaged directly by an optical telescope, demonstrating the advantages of observing from above the turbulent atmosphere. The detection was also made easier by the relatively large angular separation (more than 12 arc seconds) between the planet and its star.

The mass of Fomalhaut b is uncertain, but it is unlikely to be more than twice the mass of Jupiter. Observations show that it follows a highly elliptical orbit that crosses a wide belt of debris encircling Fomalhaut. Its orbit ranges between 7.4 billion km (50 AU) and 43 billion km (300 AU) from the star. The orbital period is approximately 2,000 years.

Such discoveries can cause scientists to question the current theories of planet formation. The extreme faintness of Fomalhaut b and the fact that it could not be seen in infrared light seem to remove any doubts over whether it is a brown dwarf. Since the planet must have cooled to its current temperature since the system formed, 100–300 million years ago, it is likely to contain up to two Jupiter masses. Nevertheless, it is hard to explain how such a gas giant could form so far from the star.

The ultimate goal of obtaining images and spectra of terrestrial planets around nearby stars will require new instruments and techniques developed for a future generation of extremely large ground-based facilities, such as the European Extremely Large Telescope, as well as sophisticated space telescopes.

One example is a technique known as nulling interferometry. This method relies on the wave nature of light. A wave has peaks and troughs. Usually when combining light in an interferometer, the peaks are lined up with one another, boosting the signal. In nulling interferometry, however, the peaks are lined up with the troughs so that they cancel each other out and the star disappears. Planets in orbit around the star show up, however, because they are offset from the central star and their light takes different paths through the telescope system. Several ground-based nulling interferometers are being developed around the world to perfect the technique.

[4] The first direct spectrum of a giant exoplanet was announced in January 2010. The planet orbits the bright, very young star HR 8799. Unfortunately, the spectrum was difficult to interpret, possibly because of large amounts of dust in the upper atmosphere.

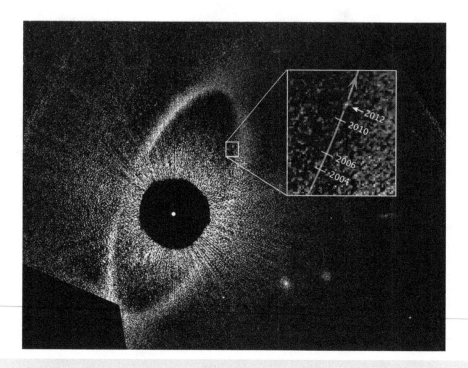

Figure 14.15   The first visible-light images of an exoplanet were taken by the Hubble Space Telescope. They revealed a planet in orbit around the star Fomalhaut. This subsequent, false-color composite view combines images taken with HST's Space Telescope Imaging Spectrograph in 2010 and 2012. They reveal the orbital motion of the planet Fomalhaut b. The observations indicate that the planet is in a 2000 year-long, highly elliptical orbit. The planet may enter a vast belt of debris around the star roughly 20 years into the future. The black circle at the center of the image blocks out light from the bright star, allowing reflected light from the belt and planet to be photographed. (Courtesy of NASA, ESA and P. Kalas/University of California, Berkeley and SETI Institute)

Astrometry

Astrometry involves precise measurement of a star's position in relation to background stars in the sky and detection of any changes in its position over time. The technique is related to the radial-velocity detection method and is used to detect any wobbling of a star due to the gravitational influence of one or more planets. Such observations are restricted by Earth's atmosphere, so they have to be made from space.

Various mistaken claims over the years show how difficult it is to use this technique to detect exoplanets. For instance, in 2009, it was announced that a world named VB 10b was the first distant planet to be discovered through astrometry, but this was later retracted after radial velocity measurements. However, astrometric measurements with the HST have been used to characterize a previously discovered planet around the star Gliese 876.

ESA's Gaia spacecraft, launched in December 2013, is the most precise astrometric satellite ever launched. During its detailed survey of a billion stars, Gaia is expected to find between 10,000 and 50,000 giant exoplanets, although even it is not sufficiently sensitive to detect the tiny wobbling motion caused by an Earth-sized planet.

Gravitational Microlensing

Gravitational microlensing also has potential to find exoplanets, though the number of objects found is always likely to be mod-est. The effect comes into play when two stars are almost exactly aligned to an observer on Earth (Figure 14.16). When one of the stars moves in front of the other, its gravitational field acts like a lens, magnifying the light of the more distant star.

If the foreground lensing star is orbited by a planet, then that world's gravity can make a detectable contribution to the lensing effect, but only if it happens to be in the right place at the right time. Since such alignments are highly improbable, it is necessary to monitor continuously millions of stars in order to detect a handful of planetary microlensing events. The chances of detection are improved considerably by looking towards the galactic centre, where there is a huge population of background stars.

Such lensing events may last a matter of days or weeks. However, the duration of an event decreases when the mass of the lens decreases. In the case of low-mass planets, the duration may be only a few hours, so they are extremely difficult to detect unless continuous observation is possible through networks of telescopes.

By November 2018, 70 exoplanets had been detected through microlensing. The first detection of an exoplanet by this method was announced in 2004 by two international research teams – Microlensing Observations in Astrophysics (MOA) of New Zealand/Japan, and the Optical Gravitational Lensing Experiment (OGLE). The Jupiter-sized object (OGLE-2003-BLG-235/MOA-2003-BLG-53Lb) and its red dwarf star are three times farther apart than Earth and the Sun. Together,

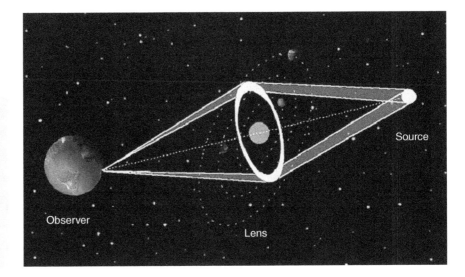

Figure 14.16 Gravitational micro-lensing occurs if the light rays from a star pass sufficiently close to a massive foreground object, e.g. another star, so that their path is bent, or lensed to create multiple images. If the foreground star has a planetary companion, additional lensing may occur, producing another small change in the light curve. (STScI)

they magnify a distant, background star some 24,000 light years away, near the Milky Way's center.

Weird Worlds

Although the study of exoplanets is still a new discipline, it is already clear that these alien worlds are far from uniform. One-third of the exoplanets so far discovered are at least as massive as Jupiter and most of these are located within 5 AU of their star (Figure 14.12). The proportion rises considerably if the search is extended to include planets the size of Saturn or Neptune within a radius of 10 AU.

At the time of writing, 42% of those so far discovered are comparable in size to Neptune, about one-quarter are classified as super-Earths – more massive than Earth but lighter than Neptune and Uranus – and 4% are terrestrial planets, comparable to Earth, Venus, and Mars. The difficulty in detecting small, low-mass planets means that the vast majority of those so far detected lie within 1 AU.

Although lower-mass planets are currently in the minority, it is almost certain that they far outnumber the gas giants. An extrapolation of the number of low-mass planets found by radial velocity shows that at least 30% of stars have at least one super-Earth companion. Furthermore, 80% of circumstellar disks contain enough material to form Earth-like rocky planets.

What can we say so far about the physical characteristics of exoplanets? All giant planets seem to be made of a gaseous envelope of hydrogen and helium – the same elements that make up our Sun – surrounding a central dense core that is probably made of compressed water (ices) and rock. This core is at the heart of the formation of giant exoplanets.

Standard planet formation theories predict that the core equals about 10 Earth masses. This works well for Uranus and Neptune, but not so well for Jupiter and Saturn. Jupiter appears to have a rather small core of a few Earth masses, while Saturn seems to have a larger core, around 10 to 25 times the mass of the Earth. What about exoplanets? Although they are more difficult to model, astronomers have confirmed that the largest planets are made mostly of hydrogen and helium. However, according to recent studies, some of them possess surprisingly large cores, up to 100 times the mass of the Earth.

Planets only a few times bigger than Earth are apparently enveloped in thick atmospheres, which suggests that they formed very quickly after the birth of their star, while there was still a gaseous accretionary disk. By contrast, Earth is thought to have formed much later, after the gas disk disappeared.

Furthermore, there seems to be a correlation between how rich the star is in "heavy" elements, such as iron, silicate, and oxygen, and the mass of the planetary cores. Higher metallicity in stars increases the probability of forming large, Jupiter-like planets. This supports the theory that such cores grow by the sticking together of grain particles and subsequent accretion, but the process may be more effective than previously thought.

The range of densities observed to date is also startling, varying by more than a factor of 20. This suggests that planetary systems may form, as well as evolve, in different ways. Our experience is already broadening far beyond the three broad categories of planet that occur in the Solar System: those made mostly of hydrogen and helium; those composed largely of iron, and those made up mainly of silicates.

The menagerie of exoplanets already contains everything from gigantic balls of gas that would float in water, like TrES 4, to carbon-rich planets that may contain vast deposits of graphite or diamonds, and dense, rocky worlds that may have huge, iron-rich cores, e.g. CoRoT-7b (see Earth-Like Planets and Super-Earths).

Modeling of all the possible exoplanets results in some very weird worlds. The models take into account how the size of a planet with a given mass varies according to its composition. This relationship is not as straightforward as might be expected because some substances can be compressed more easily than others.

One exotic possibility is a planet composed mainly of carbon monoxide. This could form in a disk around a star that is stealing material from a nearby white dwarf, the remnant of a Sun-like star that is made mostly of carbon and oxygen.

Another extreme environment associated with a super-dense neutron star could result in a helium planet, which started life as a binary companion but lost all of its outer layers to the overwhelming gravitational attraction of the nearby stellar cinder.

Box 14.3 Exoplanet Surveys

Exoplanet science is advancing rapidly with the launch of space-based missions designed to search for such objects. The first of these was the European CoRoT (Convection, Rotation and planetary Transits) spacecraft, which was launched on December 27, 2006, and operated until November 2012. CoRoT carried a 27 cm-diameter telescope designed to detect tiny changes in the brightness of nearby stars. Its camera was designed to look for planets and subtle variations in a star's light caused by sound waves rippling across the surface. To date, 29 exoplanets have been discovered from the CoRoT data and confirmed by ground-based follow-up campaigns.

One of the most notable discoveries was CoRoT-7b, the first rocky exoplanet to be confirmed. With about 5 Earth masses and a high density (about 5.6 g/cm^3), CoRoT-7b is an example of a super-Earth. A second planet has also been found in the CoRoT-7 system. Known as CoRoT-7c, it is another super-Earth of about 8 Earth masses.

Many more discoveries came from NASA's Kepler mission, launched in March 2009 into an Earth-trailing heliocentric orbit. In this orbit, it slowly drifted away from our planet, reaching a distance of up to 0.5 AU after four years. Kepler was specially designed to search for exoplanets that transit their primary stars. It used a special 1.4 m telescope, called a photometer, to measure small changes in brightness caused by exoplanet transits.

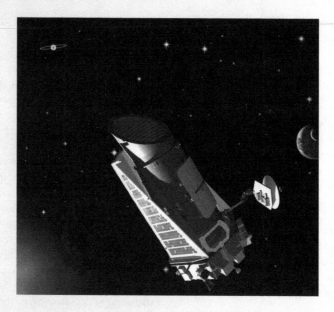

Figure 14.17 Kepler was designed to survey a single star field in the Cygnus-Lyra region of the Milky Way. The single instrument, a photometer, had a 0.95 m aperture and a 1.4 m primary mirror. It had an array of 42 CCDs and a very large field of view (105 sq. deg.). It stared at the same star field, continuously monitoring the brightness of more than 100,000 stars over many years, in a search for transiting planets. (NASA)

Equipped with a 95-megapixel camera, the largest ever launched into space, Kepler was able to detect changes in a star's brightness of only 20 parts per million. Over a period of almost 5 years, it simultaneously measured every 30 minutes tiny variations in the brightness of more than 100,000 stars. The large sample of stars helped to determine the percentage of exoplanets in or near the habitable zones around many different types of stars and made it possible to compare the sizes and shapes of the orbits of these planets.

During its four-year primary mission and a 5.5-year extension, Kepler discovered 2,662 exoplanets and a similar number of candidate planets. The mission showed that there are more planets than stars in our galaxy. Many of these may be similar to Earth in size and distance from their parent stars. Analysis of Kepler's data concludes that 20 to 50% of the stars in the sky are likely to have small, possibly rocky planets that are in the habitable zones of their stars where liquid water could pool on the surface. Kepler also discovered a remarkable diversity of planet types. The most common size of planet found by Kepler is super-Earths – worlds between the size of Earth and Neptune – that do not exist in our Solar System.

A successor telescope, NASA's Transiting Exoplanet Survey Satellite (TESS), is continuing Kepler's work. TESS was launched in April 2018 on a two-year mission to search for Earth-like exoplanets around the closest stars, 30–300 light years from Earth. (Kepler targeted stars 300–3,000 light years away.)

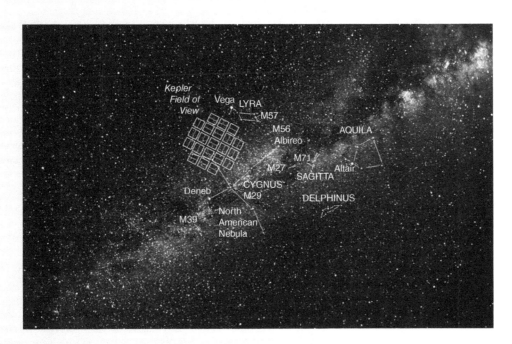

Figure 14.18   Since exoplanet transits are very brief, Kepler had to monitor the brightness of all the stars in its field of view (FOV) at least once every few hours. This continuous monitoring of the targeted stars meant that the FOV had to be clearly visible throughout the year. To avoid the Sun, the telescope pointed well away from the ecliptic plane. In order to include the largest possible number of stars in the FOV, a region along the Cygnus arm of our Galaxy was selected. The squares show the FOV of each of Kepler's 21 CCD camera modules. Each covers 5 sq. deg. (Courtesy of NASA – Carter Roberts)

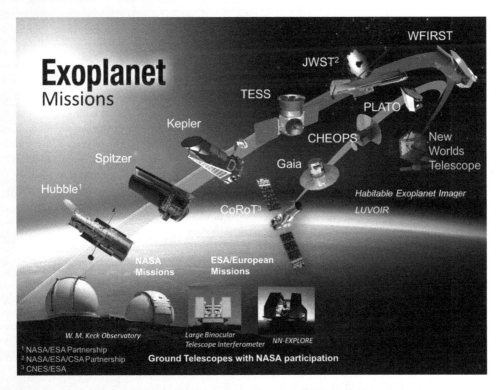

Figure 14.19   A graphic showing exoplanet missions flown or planned by NASA, ESA, and the Canadian Space Agency. (Courtesy of NASA, ESA, CSA)

The European Space Agency also intends to play a leading role in exoplanet studies. Its CHEOPS (Characterising Exoplanet Satellite) mission involves the launch of a small satellite that will define the properties of planets orbiting nearby stars.

CHEOPS was launched on 18 December 2019, CHEOPS will observe individual bright stars that are known to host exoplanets, in particular those in the Earth-to-Neptune size range. By targeting known planets, CHEOPS will know exactly when and where to point in order to observe an exoplanet as it transits across the disk of its host star. Its ability to observe multiple transits of each planet will provide the high-precision transit signatures that are needed to measure the sizes of small planets.

The combination of the precise sizes determined by CHEOPS with masses determined from other measurements will be used to establish the bulk density of the planets, placing constraints on their composition. The data, together with information on the host stars and the planets' orbits, will provide key insights into the formation and evolutionary history of planets in the super-Earth to Neptune size range.

ESA is also developing the PLATO (Planetary Transits and Oscillations of stars) mission, with a launch date targeted for 2026. PLATO will be a follow-on to missions such as CoRoT, Kepler, and TESS. During its six-year mission, PLATO will scan and observe about half the sky, including the brightest and nearest stars. Its 26 cameras will search for evidence of planetary transits as it observes one million stars, leading to the likely discovery and characterization of thousands of new exoplanets. PLATO will operate from the L2 Lagrange point in space, 1.5 million km beyond Earth as seen from the Sun.

NASA's much-delayed James Webb Space Telescope will also operate from L2. The largest infrared space observatory ever launched, JWST will be in demand for many different astronomical studies, including investigations of dust clouds where stars and planetary systems are born, and characterisation of the atmospheres of extrasolar planets. It is currently scheduled to be launched no earlier than March 2021.

Looking further ahead, NASA is planning to launch a Hubble-class observatory known as the Wide Field Infrared Survey Telescope (WFIRST). Using its 2.4 m primary mirror and wide field camera, WFIRST will be able to search for exoplanets that show up as a result of microlensing – when a foreground star in our galaxy acts as a lens to reveal a more distant planet (see Gravitational Microlensing). WFIRST's microlensing survey will monitor 100 million stars for hundreds of days and is expected to find about 2,500 planets, with significant numbers of small, rocky planets in and beyond the region where liquid water may exist.

In parallel to the space missions, studies of exoplanets will be advanced by attaching more sensitive and capable instruments to existing ground-based observatories, together with the introduction of powerful new observatories, such as the European Extremely Large Telescope, the Giant Magellan telescope and the proposed Thirty Meter Telescope. These facilities will be used to study planet-forming regions and to characterise exoplanet systems, including the discovery of Earth-like worlds and their atmospheres.

Carbon planets, with more than 50% of their mass in the form of carbon compounds such as graphite, could also potentially form under certain circumstances, for example a circumstellar disk that is rich in carbon grains and methane, like the disk around Beta Pictoris. This has even led to speculation that planets formed in such environments might be covered by tar, with mountains made of diamonds.

Exoplanets can differ drastically from the worlds of the Solar System in other unexpected ways. Although many comets, plus Triton and many smaller, captured satellites, follow retrograde orbits, all of the major planets orbit in the same direction as the rotation of the Sun. However, at least nine exoplanets have now been found to orbit the "wrong way" around their star.

To account for these retrograde hot Jupiters orbiting close to their stars, it has been suggested that their orbits moved slowly inward, over hundreds of millions of years, due to a gravitational tug-of-war with more distant planetary or stellar companions. These disturbances would move a planet into a tilted, elongated orbit that would cause it to experience tidal friction, losing energy every time it swung close to the star. It would eventually enter a near circular, but randomly tilted, orbit close to the star. This theory is supported by the fact that more than half of all known hot Jupiters have orbits that are misaligned with the rotation axis of their stars.

Although comets, together with some asteroids and small satellites, have highly eccentric orbits, the planets in our Solar System follow near-circular orbits, and all of them orbit close to the plane of the ecliptic. This is a rare occurrence in the exoplanet community, where most planets have a high orbital eccentricity. Some orbits are extremely elongated, e.g. Jupiter-sized HD 20782 b which follows a comet-like path (eccentricity 0.97). Over a period of 592 days it travels from very close to its star to the equivalent of the asteroid belt in our Solar System.

Hot Jupiters

Partly due to the limits of current technology, most of the exoplanets so far discovered lie close to their central star, so that they are visible in transit across the star's disk or sufficiently massive to cause a noticeable wobble in that star's motion (Figure 14.12). This combination of factors has inevitably led to a bias in the population of known exoplanets. Many of them are gas giants, orbiting closer to their star than Mercury's orbit around the Sun.

Apart from the facts that they are very large and massive, sometimes dwarfing Jupiter, and many of them zip around their orbits in a matter of days, little is known about the majority of hot Jupiters. However, it is clear that those orbiting very close to their star are

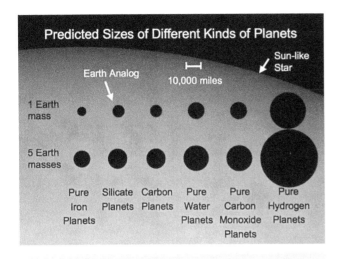

Figure 14.20 If a planet transits the face of its star, it is possible to calculate its diameter, mass, and density. However, the composition of planets with similar masses may vary considerably. Hot gas giants may also become bloated if they orbit close to their star. This diagram shows the predicted sizes of transiting planets according to their composition. (Courtesy of Marc Kuchner/NASA-GSFC)

tidally locked, so they always keep the same face toward the star. Their proximity to their star also means that they are blasted by thousands of times more energy per second than Jupiter, so their star-facing surface temperatures soar to hundreds, or even thousands of degrees.

Some hot Jupiters follow surprisingly eccentric orbits, very different from the well behaved, almost circular orbits of the Sun's retinue. HD 20782b has already been mentioned above. Another extreme example is HD 80606b, a planet four times the mass of Jupiter which follows a path that is almost as elongated as that of Halley's comet and markedly inclined to the star's equator (Figure 14.22). Astronomers believe the giant planet may be migrating slowly inward, a process that would eventually relocate it in a much tighter orbit typical of hot Jupiters.

During its 111-day orbit, HD 80606b travels as far out as 0.85 AU (almost equal to Earth's distance from the Sun) before swinging back to a mere 0.03 AU. An observer sitting on its cloud tops would see the star's apparent size increase almost 1,000 times in only a few weeks. At the same time, its temperature rockets from about 100°C to 1,700°C as the planet rushes towards its star. The extreme temperature swing indicates that its outer atmosphere must absorb and lose heat very quickly. It may even sprout a giant, comet-like tail.

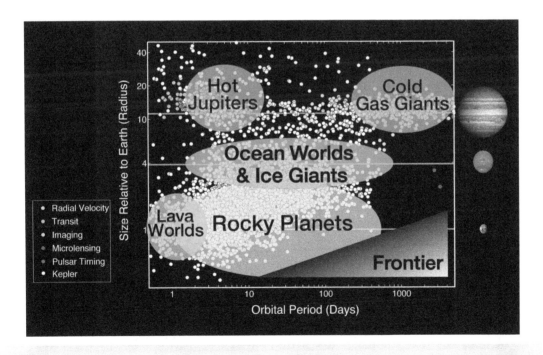

Figure 14.21 The population of exoplanets detected by the Kepler mission (yellow dots) compared to those detected by other surveys using various methods: radial velocity (light blue dots), transit (pink dots), imaging (green dots), microlensing (dark blue dots), and pulsar timing (red dots). The horizontal lines mark the sizes of Jupiter, Neptune, and Earth, which are shown on the right side of the diagram. The colored ovals denote different types of planets: hot Jupiters (pink), cold gas giants (purple), ocean worlds and ice giants (blue), rocky planets (yellow), and lava worlds (green). The shaded gray triangle (lower right) will be explored by future exoplanet surveys. (Courtesy of NASA-Ames Research Center/Natalie Batalha/Wendy Stenzel)

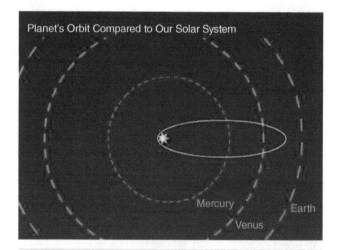

Planet's Orbit Compared to Our Solar System

Figure 14.22   HD 80606b has the second most eccentric orbit yet found for an exoplanet. During its 111-day orbit, it travels along a comet-like path that takes it from 0.85 AU to 0.03 AU from its star. (NASA-JPL)

The driving force behind this unusual arrangement may be a binary companion star, known as HD 80607. It seems likely that the nearby star caused long-term gravitational perturbations to the planet, not only stretching its orbit but also causing it to be tilted.

With such a massive bully sweeping through the star's inner system, any smaller planets would be thrown into chaotic orbits, and possibly even ejected from the system altogether. Under such conditions, it seems unlikely that any habitable worlds exist around HD 80606.

Another consequence of the elliptical orbit is that, unlike hot Jupiters which remain in tight orbits around their stars, HD 80606b does not always keep the same face towards its star. Instead, it rotates around its axis roughly every 34 hours. This means that it experiences the most extreme temperatures swings on any known planet. Observations with the Spitzer Space Observatory, made when HD 80606b disappeared behind its star around the time of periastron (closest approach), show that the planet's temperature rose from about 230°C to 1,230°C in just six hours (Figure 14.23).

Although simple weather patterns have been detected on HD 80606b, little is known about the atmospheres and appearance of the hot Jupiters. However, some interesting – and unusual – discoveries have been made by space observatories. For example, Spitzer has observed HD 189733b, which is located 63 light years away in the constellation Vulpecula. The gas giant crosses in front of and behind its star every 2.2 days.

Observations of longitudinal surface temperature variations on HD 189733b have shown that there is a fairly modest difference between the day and night hemispheres: the temperatures range from 650°C on the dark side to 930°C on the sunlit side. Since the planet's overall temperature variation is much less than predicted

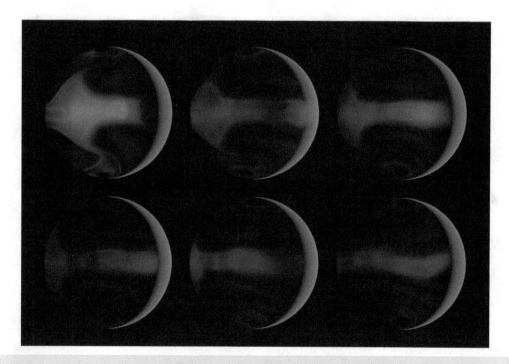

Figure 14.23   These computer-generated images chart the development of thermally driven storm patterns on the nightside of the highly eccentric exoplanet HD 80606b. The simulations were based on measurements of heat radiating from the planet obtained by the Spitzer Space Telescope. The six frames are evenly spaced in time, between 4.4 days and 8.9 days after closest approach. The blue glow of the crescent is starlight that has been scattered and reflected by the planet. The blue color is mainly caused by absorption of red light by sodium and potassium in its atmosphere. The nightside appears reddish orange as it glows with its own internal heat. (Courtesy of NASA/JPL-Caltech, J. Langton/UC Santa Cruz)

by numerical models, it is thought that winds reaching speeds of 9,600 km/h or more must be spreading heat from its permanently sunlit side around to its nightside.

HD 189733b has a warm spot 30° east of the substellar point, the point directly facing the star (Figure 14.24). Assuming the planet is tidally locked with regard to its star, this implies that fierce winds are blowing eastward. Direct heating from the star combined with the rapid delivery of hot gas by the hurricane force wind results in a particularly warm region.

The redistribution of heat observed on HD 189733b does not apply to all giant exoplanets. Another very hot exoplanet, HD 149026b, has been studied in some detail. Its temperature of over 2,000°C was calculated by observing the drop in infrared light that occurs as it moves behind its star. The planet's extreme temperature suggests that its heat is not being spread around, so that the dayside is very hot, and the nightside is much cooler.

One explanation is that the planet's surface is blacker than charcoal, which is unprecedented for planets. In this case, it would reflect almost no starlight, instead absorbing all of the radiation that reaches it. HD 149026b would glow a dull red, like an ember in space.

Another "ultra-hot" Jupiter is KELT-9b, which orbits a star whose surface temperature is over 10,000°C, almost twice as hot as the Sun. This giant gas planet orbits 30 times closer than Earth's distance from the Sun, completing one circuit in 36 hours. It is heated to a temperature of over 4,000°C, hotter than many stars. Iron and titanium vapors have been detected in its atmosphere.

It is rivalled for the title of hottest exoplanet by Kepler 70b, one of two planets in orbit around a subdwarf star. Kepler-70b completes one orbit around its star in only 5.76 hours, one of the shortest orbital periods yet discovered. Its surface temperature is 6,900°C – comparable to the Sun. Its density is 5,500 kg/m³, similar to Earth. It survived the expansion of its star into a red giant (see also V 391 Pegasi).

The spectra of a handful of exoplanets have been measured, but determination of their atmospheric characteristics is very difficult. The first determination of the true colour of a planet orbiting another star was announced in 2013, when the Hubble Space Telescope was able to measure the brightness (albedo) and visible color of a hot Jupiter known as HD 189733b. If seen up close, this planet would be a deep cobalt blue, reminiscent of Earth's colour as seen from space.

HD 189733b orbits very close to its star, so its scorching atmosphere has a temperature of over 1000° C. Glass rains down sideways from its silicate clouds, in howling 7000 m/h winds. The planet's azure blue colour does not come from the reflection of a tropical ocean, but is due to a hazy, turbulent atmosphere thought to contain a scattering of silicate particles, which scatter blue light.

Studies of the system revealed a drop in the blue part of the spectrum when the planet passed behind its star, whereas the

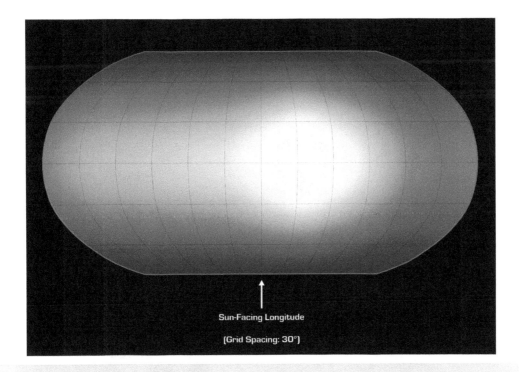

Figure 14.24 Jupiter-sized HD 189733b orbits close to a star 63 light years from Earth. Like other hot Jupiters, its rotation is tidally locked, so one side always faces the star during each 2.2 day orbit. Infrared data from the Spitzer Space Telescope enabled its surface temperature variations to be mapped – the first map ever made of an exoplanet. The brighter colors show higher temperatures. The hottest region is about 30° to the east (right) of the substellar point, evidence that fierce, planet-circling winds transfer heat toward the east. The temperatures vary from about 930 to 650°C. (Courtesy of Heather Knutson/Harvard-Smithsonian CfA et al., NASA/JPL-Caltech)

Figure 14.25 Spectral observations of hot Jupiter HD 209458b, depicted in this artist's concept, have revealed molecules of carbon dioxide, methane, and a tiny amount of water vapor in its atmosphere. Larger than Jupiter, HD 209458b, occupies a tight, 3.5-day orbit around a Sun-like star about 150 light years away. (NASA, ESA, G. Bacon/STScI, and N. Madhusudhan/UC)

signal remained constant at the other colours measured – clear evidence that HD 189733b is blue.

One of the most observed hot Jupiters is HD 209458b, the first extrasolar world discovered to transit a star, and the first to have its atmosphere analyzed (Figure 14.25). The planet is located 150 light years from Earth in the constellation Pegasus. It lies about 7.5 million km from its star and completes one orbit every 3.5 days.

The spectrum shows an unexpected peak at 9.65 microns, which has been attributed to tiny grains of silicate dust. Silicates form rocks on Earth, but on scorching hot worlds they may exist as sand-like grains that occur in high level clouds.

Observations with the Hubble Space Telescope have also revealed individual elements, such as sodium, oxygen, carbon, and hydrogen, in the upper atmosphere, where molecules like water and methane break apart. This was achieved by measuring changes in the light from the star as the planet passed in front of it. The observations indicated less sodium than predicted, again supporting the presence of high clouds. It seems that some hot Jupiters are cloaked with black silicate clouds, making them darker than any planet in our Solar System.

In 2010, astronomers announced that had measured a super-storm, with wind speeds of up to 7,000 km/h, for the first time in the atmosphere of HD 209458 b. The observations showed that it is sweeping from the extremely hot dayside to the cooler nightside of the planet.

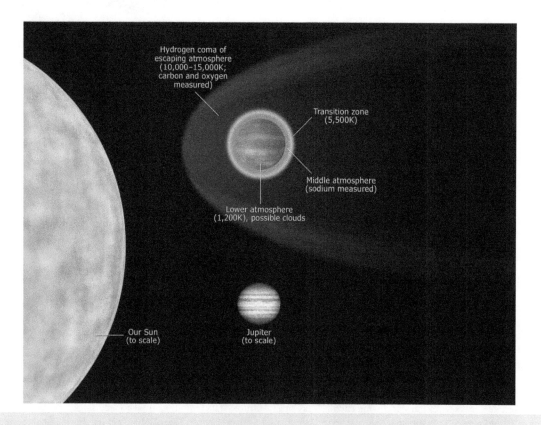

Figure 14.26 Hot exoplanet HD 209458b may be surrounded by an extended envelope of escaping gas. In 2003, the Hubble Space Telescope showed that its hydrogen-rich outer atmosphere appears to be evaporating in the extreme heat from its nearby star. Hubble later observed the first signs of oxygen and carbon in an exoplanet's atmosphere. (Courtesy of NASA, ESA, and A. Field/STScI)

However, the most striking characteristic of HD 209458b is the presence of an elongated envelope of escaping gas, the first ever detected (Figure 14.26). HST observations made in 2003 revealed evidence of a cloud of evaporating hydrogen, resembling a comet's tail. Later studies indicated that oxygen and carbon atoms are also being swept up from the lower atmosphere with the flow of the escaping atomic hydrogen, a process called atmospheric blow off.

This "blow-off" effect is explained by the fact that the planet orbits so close to its Sun-like star. Its atmosphere is heated to about 1,000°C by intense ultraviolet radiation, inflating it like a balloon. Under these scorching conditions, the gas is streaming into space at a speed of more than 35,000 km/h. The amount of hydrogen gas escaping HD 209458b is estimated to be at least 10,000 tonnes per second, and the hydrogen tail is 200,000 km long.[5]

Despite this rapid erosion, HD 209458b is so massive that is expected to survive for more than 5 billion years. However, this remarkably powerful evaporation mechanism may, in some cases, give rise to a strange new class of exoplanets, unlike anything seen in our Solar System: solid, remnant cores of evaporated gas giants, orbiting even closer to their parent star than HD 209458b.

Close-in planets may also be disturbed and disrupted in other ways. WASP-12b orbits a Sun-like star at extremely close range – about two million km or 75 times closer than Earth is to the Sun – once every 26 hours. It is also larger than expected. Its mass is estimated to be almost 50% larger than Jupiter's and its diameter 80% larger, giving it six times Jupiter's volume. It is also unusually hot, with a daytime temperature of 2,330°C.

Wasp-12b is also unusual because it reflects almost no visible light, making it appear essentially pitch black (Figure 14.27). It is so hot that clouds cannot form and alkali metals are ionized. It is even hot enough to break up hydrogen molecules into atomic hydrogen, which causes the atmosphere to act more like the atmosphere of a low-mass star. A blanket of atomic hydrogen and helium would not reflect light at any wavelength, explaining why it is so dark, although WASP-12b does emit light because of its high temperature, giving it a red hue similar to a hot, glowing metal.

Why is it so big? Part of the reason is that the planet is being so strongly heated by its star. However, powerful tidal forces acting on the planet also play a part. These cause the planet's shape to be distorted into something resembling a rugby ball or American football. These tides not only mold the shape of WASP-12b, but they also create friction in its interior. The friction produces heat, which causes the planet to expand. There are also hints that a second planet may be orbiting nearby. This close neighbor could increase the eccentricity of WASP-12b's orbit, thus enhancing the star's tidal effects.

Observations also show that the planet has ballooned outward so much that it can no longer withstand the star's gravity. Material is being stripped off WASP-12b at a rate of six billion metric tons per second. At this rate, the planet's atmosphere will be completely destroyed by its host star in about 10 million years. Instead of falling directly onto the star, the lost material forms a disk around it and slowly spirals inwards.

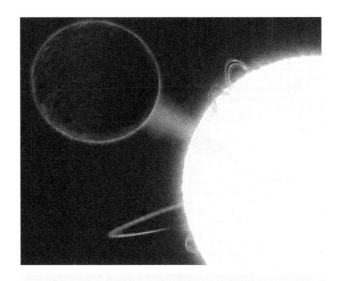

Figure 14.27 Hot Jupiter WASP-12b is being destroyed by strong heating and tidal forces which are causing the planet to balloon outward. Material is being stripped away from the outer atmosphere by the star's gravity, creating a disk of material that is spiraling inward toward the star. This rapid rate of erosion will eventually destroy the planet's atmosphere. The planet is as black as fresh asphalt. (Courtesy of NASA, ESA, and G. Bacon/STScI)

Another unusual, puffed up planet is TrES-4b, a Jupiter mass world that was discovered by the Trans-atlantic Exoplanet Survey. TrES-4 is one of a family of planets which orbits a star about the same age as our Sun. However, the star's greater mass means that it has evolved much faster. Having exhausted all of the hydrogen fuel in its core, the star has grown into a subgiant and is on its way to becoming a huge, cool, red giant.

TrES-4b completes one orbit in 3.5 days at a distance of about 7.2 million km, so its temperature is about 1,300°C. It is about 70% bigger than Jupiter, but considerably less massive. Its mean density is only about 0.24 grams per cubic centimeter, or about the density of balsa wood. This makes it one of the lowest-density planets yet found. The planet's relatively weak pull on its upper atmosphere may allow some of the atmosphere to escape.

The reason why TrES-4b is so bloated is unclear. The planet's proximity to its star cannot fully account for the inflation, since it is not as close or as hot as WASP-12b. One possibility is that internal heat may play a part. If their hydrogen and helium separate, with the heavier helium sinking into the core, this contraction could release gravitational energy as heat. However, it is not clear why this would happen to some giant planets and not others. Planets such as TrES-4 are testing current theories of planet formation, since they are larger, relative to their mass, than current models of superheated giant planets can presently explain.

---

[5] The spectral evidence is open to another interpretation. If the exoplanet has a strong magnetic field, it may be able to deflect the stellar wind and trap enough ionized gas to produce the observed spectral signature. In such a case, the planet would exhibit spectacular auroras, where particles spiral down towards its magnetic poles.

The future of TrES-4b is uncertain. In less than a billion years, the swelling star will engulf the planet. However, despite the odds, there is a chance that it will survive the fiery furnace. The planet V 391 Pegasi b provides a precedent by being the first planet to be discovered around a star that has gone through its red giant phase. This escape act occurred despite an orbital distance of only 1.7 AU. With an age of about 10 billion years, it is one of the oldest planets ever discovered.

Another exotic hot Jupiter is the transiting planet WASP-18b, which is about 10 times the mass of Jupiter. It orbits its star in 23 hours, one of a small number of hot Jupiters confirmed to have an orbital period of less than a day. Its large size and close proximity (three stellar radii) should result in powerful tidal interactions between the planet and star. As they rotate, the resulting bulges and torques should cause WASP-18b to spiral inwards, causing it to be engulfed and destroyed by its star in less than a million years.

At first, its modest amount of magnetic activity was thought to indicate that WASP-18 was around one billion years old, which would suggest that the planet was a similar age (Figure 14.28). However, subsequent observations have caused researchers to believe that tidal forces caused by the gravitational pull of the massive planet may be responsible for disrupting the star's magnetic field and making it behave as if it is much older than it really is. This tidal interaction will likely cause WASP-18b to merge with the star within the next million years.

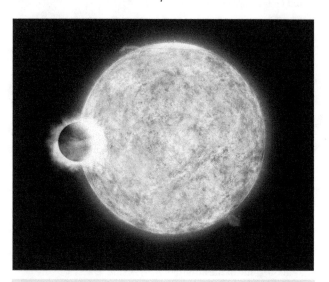

Figure 14.28 WASP-18b orbits a star about 330 light years from Earth. The planet has a mass about 10 times that of Jupiter and completes one orbit around its star in less than 23 hours. This "hot Jupiter" is the first known example of an orbiting planet that has apparently caused its star, which is roughly the mass of our Sun, to display traits of an older star. Its upper atmosphere contains a large amount of carbon monoxide, but no water has been detected. (Courtesy of NASA/CXC/M. Weiss)

The strength of the magnetic field in a star depends on the amount of convection, the process by which hot gas moves around the stellar interior. The planet's gravity may cause motions inside the star that weaken the convection. Because WASP-18 has a thinner convection zone than most stars, it is more vulnerable to the impact of tidal forces from such a close neighbor.

WASP-18b also seems to have a highly unusual atmospheric composition, with a great deal of carbon monoxide but no sign of water vapor. Compared to other hot Jupiters, this planet's atmosphere likely would contain 300 times more "metals," or elements heavier than hydrogen and helium. This extremely high metallicity would indicate WASP-18b might have accumulated greater amounts of solid ices during its formation than Jupiter, suggesting it may not have formed the way other hot Jupiters did.

Like other hot gas giants, WASP-18b may have formed further away from the star and then migrated inward. This seems quite feasible, since few, if any, of the hot Jupiters are likely to have been born in their current locations, and inward migration of many planets almost certainly occurred during the evolution of the young solar systems (including ours).

Yet another weird exoplanet is WASP-17b, which travels the "wrong" way around its host star.[6*] The first sign that WASP-17b was unusual was its large size. Although it has only half the mass of Jupiter, it is bloated to nearly twice Jupiter's size, making it the largest planet known and the planet with the lowest known density. Although it weighs in at 1.6 Saturn masses, its density is only between 6% and 14% that of Jupiter. In 2013, the Hubble Space Telescope detected water in its atmosphere, although the infrared signal appeared to be dimmed by a layer of haze or dust.

Subsequent observations have shown that it has a retrograde orbit. Since planets form out of the same swirling gas cloud that creates a star, they are expected to orbit in the same direction that the star spins. The likely explanation is that WASP-17b was involved in a near collision with another planet early in its history. Ejected into a highly elliptical, retrograde orbit, it would have been subjected to intense tides. Tidal compression and stretching would have heated the gas giant planet so that it became hugely bloated. As a consequence, its density is extremely low, comparable to expanded polystyrene.

From the examples mentioned above, it should also be abundantly clear that many hot Jupiters are undergoing rapid changes, involving factors such extreme heating, stellar winds, mass loss, tidal forces, and orbital evolution. There is still a great deal to be learned about these mysterious and exotic worlds.

Earth-Like Planets and Super-Earths

As instruments and observational techniques have improved, so the sizes and masses of the newly found exoplanets have been falling. The list now includes several dozen that are known to weigh in at three Earth masses or less, and almost 400 which are similar in size to the Earth.[7] Indeed, as the inventory of alien worlds grows, the smaller members of the family are expected to

---

[6]WASP-17b and WASP-18b were found by the Wide Area Search for Planets (WASP) consortium of UK universities. The planets were detected using an array of cameras that monitor hundreds of thousands of stars, searching for small dips in their light when a planet transits in front of them.

[7]The radii and masses of many exoplanets have yet to be accurately determined.

Figure 14.29   CoRoT-7b was the first confirmed rocky exoplanet. It is larger than Earth but five times more massive, so its density is similar to Earth's. It orbits 2.5 million km from its star, or 23 times closer than Mercury is to the Sun. (ESO)

far outnumber their larger brethren. Analysis of Kepler's discoveries shows that 20 to 50% of the stars in the sky are likely to have small, potentially rocky planets in their habitable zones.

In recent years, Kepler and other observatories have begun to discover numerous examples of a class of planet that does not exist in our Solar System – more massive, rocky versions of Earth, known as super-Earths (Box 14.2). (They are sometimes also known as mini-Neptunes if they are thought to possess dense, hydrogen-rich atmospheres.[8])

For the sake of convenience, super-Earths can be defined as planets with 1.5 to 10 Earth masses. This approximate upper limit is based on the idea that more massive rocky planets would pull in large amounts of gas from the protoplanetary disk in which they were born, so becoming inflated gas giants like Jupiter and Saturn.

There is, however, no reason to believe that all of the exoplanets in this category are alike. At the heavier end, many of the planets may be mini versions of Uranus and Neptune, with small rocky cores and mantles of pressurized ices topped by deep atmospheres of hydrogen and helium. Moving down the scales, a world of five Earth masses could have a rock-iron core of three Earth masses surrounded by a deep atmosphere, or it could be a solid sphere with no appreciable atmosphere at all.

One of the best known examples of a super-Earth is CoRoT-7b, discovered by the French CoRoT spacecraft in 2009 (Figure 14.29). Although it has about five Earth masses, its diameter is about 70% greater than that of Earth, so its density is quite similar to our planet's, suggesting a solid, rocky world. It travels around a star

which is slightly smaller and cooler than our Sun, with an age of about 1.5 billion years.

CoRoT-7b orbits only 2.5 million km from its star, so it completes one orbit every 20.4 hours, zipping around at a speed of more than 750,000 km/h. CoRoT-7b is so close that the temperature on its day hemisphere may soar above 2,000°C, plunging to −200°C on its nightside. Theoretical models suggest that it may have oceans of molten lava on its surface.

With such a high dayside temperature, any rocky surface facing the star must be molten, and the planet would be unable to retain anything more than a tenuous atmosphere, even one of vaporized rock. One possibility is that CoRot-7b weighed in at 100 Earth masses – about the same as Saturn – when it first formed, some 50% further from its star than it is now. After a history of mass loss and migration closer to its star, it could now represent a new class of planet – an evaporated remnant core.

Observations with the High Accuracy Radial velocity Planet Searcher (HARPS) spectrograph attached to the 3.6 m telescope at the La Silla Observatory in Chile, indicate that a second super-Earth orbits slightly further away than CoRoT-7b. Designated CoRoT-7c, it completes one circuit in 3 days and 17 hours, and has a mass about eight times that of Earth. Unlike CoRoT-7b, it does not pass in front of its star as seen from Earth, so astronomers cannot measure its radius and determine its density.

At the time of writing, the smallest planet yet detected is Kepler-37b, which is slightly larger than our Moon. Although the Kepler-37 star is similar to the Sun and Kepler-37b is probably

---

[8]Super-Earths and mini-Neptunes have similar sizes and masses, but a super-Earth would be expected to be rocky, perhaps with an atmosphere containing nitrogen, water vapor, and carbon dioxide; a mini-Neptune would have an atmosphere dominated by hydrogen and helium.

rocky in composition, the planet orbits every 13 days at less than one-third Mercury's distance from the Sun. The planet's estimated surface temperature is around 470°C.

The lowest mass exoplanet yet discovered is Kepler 138b, an object with a mass similar to that of Mars. It is the innermost of three known planets that orbit a red dwarf star about 200 light years away. Its composition is uncertain, though it is likely to be mostly rock. All three planets orbit close to their star, so they are probably too hot to be habitable.

Most Earth-like planets and super-Earths are located too close or too far away from their star to be likely candidates for life to evolve (see Alien Life). The first identified Earth-sized exoplanet to orbit a star in the habitable zone is Kepler-186f. The planet's primary is a red dwarf star, which is much smaller and less luminous than the Sun, so the habitable zone is much closer to the star than in our Solar System. The planet orbits at a similar distance to Mercury (0.4 AU) and its orbital period is 129.9 days.

Although Kepler-186f lies within the zone where liquid water can exist, other factors may affect its habitability. Red dwarfs can emit high fluxes of harmful UV radiation, especially when young. Other unknown factors include the composition and density of the planet's atmosphere, its axial inclination and rotation rate – it may be tidally locked.

Another candidate for habitability is Kepler-452b, a relatively small super-Earth that orbits in the habitable zone of a G2-type star, like our Sun (Figure 14.30). Kepler-452b is 60% larger in diameter than Earth. While its mass and composition are not yet determined, previous research suggests that planets the size of Kepler-452b are likely to be rocky.

Although Kepler-452b is larger than Earth, its 385-day orbit is only 5 percent longer. The planet is 5% farther from its primary star than Earth is from the Sun. Kepler-452 is 6 billion years old, 1.5 billion years older than our Sun, has the same temperature, and is 20% brighter.

Multi-Planet Systems

Individual planets alone cannot give an accurate picture of the planetary systems that are spread throughout the Galaxy. More significant, in many ways, are the multi-planet systems, and the distribution and types of planet within those systems. Only by studying such alien systems is it possible to compare and contrast their characteristics with those of our Solar System, and to discover whether the Sun's family is rare, or even unique.

At present, the limitations of instrumentation and the time it takes for remote exoplanets to orbit their stars make it impossible to state with any certainty whether our Solar System is unusual, since it is still very difficult to detect planets far from a star. However, the data are improving all the time, so that some preliminary conclusions may be drawn.

So far, surveys have discovered only a handful of planetary systems like our own, with two gas giants in their outer reaches. This lack of discoveries leads to the conclusion that only about 15% of stars in our Galaxy host systems of planets that contain several gas giant planets in fairly remote orbits. Despite this apparently pessimistic outcome, the sheer number of stars in the Milky Way would mean that it could contain a few hundred million systems like our own.

At the time of writing there were over 500 systems with more than one known exoplanet (Figure 14.32). Most of these contain two or three planets. Among the exceptions are Kepler-90, which is tied with our Solar System for the most number of planets around a single star (eight); TRAPPIST-1 with seven known exoplanets; and six systems with six planets, e.g. Kepler-11, HD 10180, and GJ 667 C.

Kepler-90 is a Sun-like star 2,545 light years from Earth. Its planetary system is like a mini version of our Solar System: the six inner planets are either super-Earths or mini-Neptunes while two gas giants orbit farther out (Figure 14.33). The five innermost exoplanets, Kepler-90b, c, i, d, and e, may be tidally locked,

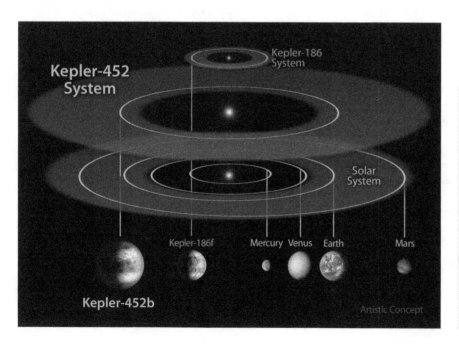

Figure 14.30 A comparison between three systems with planets in habitable zones: Kepler-452, Kepler-186, and the Solar System. When its discovery was announced in 2015, Kepler-452b was the smallest planet known to orbit in the habitable zone of a Sun-like star. The orbits of the five Kepler-186 planets would fit entirely inside the orbit of Mercury, but at least one planet is inside the habitable zone, because the central star is a low luminosity red dwarf. (Courtesy of NASA/JPL-Caltech/R. Hurt)

Box 14.4   The TRAPPIST-1 System

The TRAPPIST-1 system contains a total of seven planets, more than almost every other known exoplanetary system. All of the planets, labelled TRAPPIST-1b, c, d, e, f, g, and h in order of increasing distance from their star, are comparable in size to Earth.

TRAPPIST-1 is an ultracool red dwarf star, with only 8% the mass of our Sun and much lower luminosity. The star is estimated to be between 3 billion and 8 billion years old. The system is located about 40 light years away in the constellation of Aquarius.

All of the planets in the TRAPPIST-1 system transit their star, and they were discovered from the regular and repeated shadows that they cast during each passage. The transit data enabled observers to measure their orbital periods and calculate the sizes of the planets. The exact time at which the planets transited also provided a means to measure their masses, densities, and bulk properties. The planets are consistent with a rocky composition.

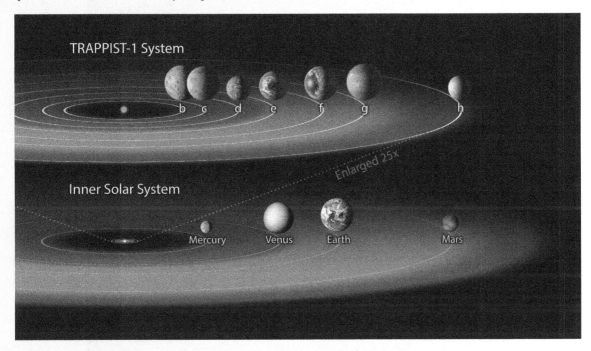

Figure 14.31   The TRAPPIST-1 system contains a total of seven known planets, all similar in size to Earth. Three of them – TRAPPIST-1e, f, and g – orbit in their star's habitable zone (shown here in green) where temperatures are right for liquid water to occur on their surfaces. (NASA-JPL)

The planetary orbits are crammed into a region of space not much larger than that of Jupiter's Galilean moon system, and much smaller than the orbit of Mercury. However, TRAPPIST-1's small size and low temperature mean that the energy input to its planets is similar to that received by the inner planets in our Solar System. TRAPPIST-1c, d, and f receive similar amounts of light and heat to Venus, Earth, and Mars, respectively.

Three of them – TRAPPIST-1e, f, and g – are thought to orbit in their star's habitable zone where the temperatures are just right for liquid water to occur on their surfaces. However, the close orbits of the planets make it likely that most, if not all of them, perpetually show the same face to the star, just as the Moon shows the same face to Earth. This would result in an extreme range of temperatures from the day to night sides, which would modify the definition of the habitable zone.

Stellar flares, as well as X-ray and extreme UV radiation from the star, would also affect the planets' atmospheres and their potential habitability. Little is known about their atmospheres at present.

The orbital periods of TRAPPIST-1's seven planets are all linked in a series of gravitational resonances, so the system is extremely stable. Hence, the second planet completes five orbits in almost exactly the time the first planet makes eight. The third planet completes three orbits for every five orbits of the second planet, and the fourth planet makes two orbits for every three orbits of the third. The previous record holders for the number of resonances were the Kepler-80 and Kepler-223 systems, each with four resonant planets.

The resonances are thought to have resulted from interactions between the planets as they migrated inward within the residual protoplanetary disk after forming at greater distances from the star.

The system was discovered through observations from NASA's Spitzer Space Telescope and the ground-based TRAPPIST (TRAnsiting Planets and PlanetesImals Small Telescope) telescope, as well as other ground-based observatories. It was named after the TRAPPIST instrument.

Systems such as TRAPPIST-1 are important because they are centered on a red dwarf, the most common class of star in the Galaxy, and they are, therefore, located in the most likely environment in which exoplanets occur.

Finding such planets with conditions, architectures, and properties different from those of the Solar system allows scientists to place Earth in context and improve their understanding of how Earth-like planets form in different environments.

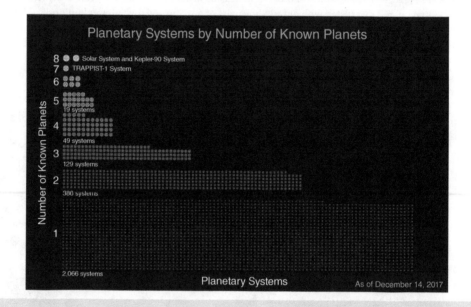

Figure 14.32 The number of known exoplanets in each planetary system, as of December 14, 2017. Each dot represents one planetary system. The graph shows more than 2,000 one-planet systems, and progressively fewer systems with many planets. The discovery of Kepler-90, the first known exoplanet system with eight planets, may indicate that more highly populated systems will be found in future. (Courtesy of NASA/Ames Research Center/Wendy Stenzel and The University of Texas at Austin/Andrew Vanderburg)

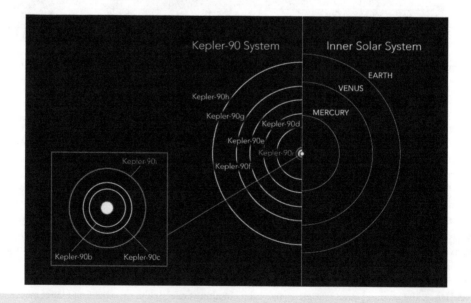

Figure 14.33 Kepler-90 is a Sun-like star that is orbited by at least 8 exoplanets. All of these orbit closer than Earth is to the Sun. The inner planets have extremely tight orbits with a "year" on Kepler-90i lasting only 14.4 days. In comparison, Mercury's orbital period is 88 days. Consequently, Kepler-90i has an average surface temperature of about 430°C. The planets around Kepler-90 may have formed more spread out, like the planets in our own Solar System, and then somehow migrated to their current orbits. (Courtesy of NASA-Ames Research Center/Wendy Stenzel)

so that one hemisphere permanently faces the star. The planets are designated by their order of discovery, so the most recent addition is known as Kepler-90i.

All of the Kepler-90 planets are close to being in orbital resonance, suggesting a very stable system, similar to the seven planets in the TRAPPIST-1 system (see Box 14.4).

The most interesting member of the family is Kepler-90h, a Jupiter-sized world which orbits at a distance of about 1 AU, within the habitable zone (Figure 14.34). However, its likely gas giant status means that it is not considered to be a good candidate for life to exist. The smaller, rocky planets are all too close to the star to be hospitable to life. For example, Kepler-90i is about 30% larger than Earth but probably has an average surface temperature of around 430°C, comparable with Mercury.

Kepler-11, located approximately 2,000 light years from Earth, is a Sun-like star which possesses the most tightly packed planetary system yet discovered. All six of the planets orbit in more or less the same plane. Five of them have orbits smaller than Mercury's, with orbital periods of less than 50 days. They have 2.3 to 13.5 times the mass of the Earth. The sixth planet is larger and farther out, with an orbital period of 118 days and an undetermined mass. All of them are larger than Earth, with the largest of them being comparable in size to Uranus and Neptune.

Radial velocity measurements reveal that the planetary system around the Sun-like star HD 10180 contains at least six planets in almost circular orbits. Five of these are between 13 and 25 Earth masses, orbiting at distances between 0.06 and 1.4 AU (closer than Mars to their star), with periods ranging from about 6 to 600 days. There is also a Saturn-like planet (minimum 65 Earth masses)

orbiting in 2,200 days at a distance of 3.38 AU – which would place it in the main asteroid belt of our Solar System.

55 Cancri is a yellow dwarf star located 41 light years away in the constellation Cancer, with nearly the same mass and age as our Sun (Figure 14.35). Although this five-planet system has a dominant gas giant in an orbit similar to Jupiter's, and most of the planets follow nearly circular orbits, it has some significant differences compared with the Sun's family.

The family includes a super-Earth (55 Cancri e) which orbits very close to its star. The star-facing side of the planet apparently has an average temperature of 2,300°C, while the "cold" side averages 1,300 to 1,400°C. The temperatures are moderated somewhat by the presence of an atmosphere, which may be rather Earthlike, containing nitrogen, water vapor, and even oxygen. The density of the planet is also similar to Earth, suggesting that it, too, is rocky, perhaps with widespread lava flows.

55 Cancri b and c orbit closer to the star than Mercury does to the Sun. 55 Cancri d is four times the mass of Jupiter and completes one orbit every 14 years at a distance of approximately 867.6 million km. It is one of the few gas giants so far known to orbit as far away from its star as Jupiter does from the Sun. 55 Cancri f has about 45 times the mass of Earth and may be similar to Saturn in composition and appearance. It completes one orbit every 260 days, at a distance of about 116.7 million km (slightly closer than Earth to our Sun). This places the planet inside the habitable zone, where the temperature would permit liquid oceans to exist.

Characterization of many exoplanet systems is still uncertain, so it is difficult to make accurate comparisons. However, there is

Figure 14.34 The sizes of the Kepler-90 planets compared to the planets of our Solar System. (Courtesy of NASA-Ames Research Center/Wendy Stenzel)

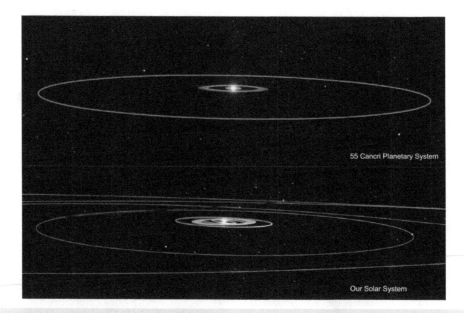

Figure 14.35   A comparison between 55 Cancri (top) and our Solar System. The stars are similar in mass and age, and all of their planets have nearly circular orbits. In addition, both systems have giant planets in their outer regions. The furthest exoplanet is 55 Cancri d, four times the mass of Jupiter, which completes one orbit every 14 years. It is the only gas giant so far known to have an orbital distance comparable to Jupiter's. (Courtesy of NASA/JPL-Caltech)

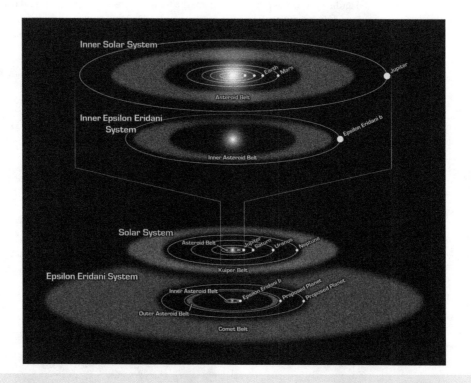

Figure 14.36   Epsilon Eridani is one of the closest known planetary systems, about 10 light years from Earth. Its central star is a younger, fainter version of the Sun. The system is quite similar to our own, containing asteroids (brown), comets (blue), and planets (white dots). The system hosts two asteroid belts, one at about 3 AU (similar to our asteroid belt) and another at about 20 AU. There is also a dense comet ring at 35–90 AU. Three planets have been suggested as shepherds that shape and confine the ring material. (Courtesy of NASA/JPL-Caltech)

evidence of systems similar to ours that contain debris belts as well as fully grown planets.

One example is the Epsilon Eridani system. Located only 10.5 light years away, it is one of the closest yet found to Earth (Figure 14.36). The main sequence star is much younger than the Sun, with an estimated age of about 850 million years. It is also smaller and less massive than the Sun, with a lower metallicity. However, the star has a higher level of magnetic activity than the Sun, with a stellar wind 30 times as strong.

Its close proximity has encouraged various searches for exoplanets, but radial velocity observations over the past 20 years have only yielded tantalizing hints of a gas giant orbiting once every 7 years at a distance of about 3.4 AU. The problem of confirmation arises because the star's magnetic activity and surface storms create considerable background noise, from which any potential wobbles caused by nearby planets are difficult to extract.

What is certain is that the system also includes two rocky asteroid belts and an outer icy ring, similar to the Sun's Kuiper Belt. The inner asteroid belt is a virtual twin of the belt in our Solar System, located at about 3 AU, while the outer asteroid belt holds 20 times more material and lies at a distance of 20 AU, comparable to the position of Uranus. The outer icy ring extends about 35 to 100 AU from Epsilon Eridani, and contains about 100 times more material than our Kuiper Belt.

This arrangement bears a close resemblance to the way our Solar System must have appeared in its youth. Moreover, the presence of the three rings of debris implies that unseen planets confine and shape them. This indicates the presence of at least three planets with masses between those of Neptune and Jupiter.

The possible planet near the innermost ring remains a problem. Some radial velocity studies suggested that it follows a highly elliptical path around Epsilon Eridani, but such an orbit would have cleared out the inner asteroid belt long ago through gravitational disruption. If Epsilon Eridani b exists, it must follow a much more circular orbit that keeps it just beyond the outer edge of the first asteroid belt. A second planet may lurk near the second asteroid belt, at about 20 AU, while a third could be hiding at about 35 AU, near the inner edge of Epsilon Eridani's Kuiper Belt. Future studies may detect these currently unseen worlds, as well as any terrestrial planets that may orbit inside the innermost asteroid belt.

There is much greater certainty about the multiple planet system around a young, massive star called HR 8799, which lies about 129 light years from Earth (Figure 14.37). The star has about 1.5 times the mass of the Sun and is 5 times more luminous, though significantly younger. Infrared observations have shown evidence for a massive disk of cold dust orbiting the star.

Four planets have also been directly imaged by the Gemini North and Keck II ground-based telescopes. Based on the age of the star and their observed luminosities, the objects formed about 60 million years ago, and they are still glowing from heat released as they contracted.

The system is highly unusual, since the planets orbit far from the star, at 14, 24, 38, and 68 AU, and their masses are estimated to be 9, 10, 10, and 7 Jupiter masses, respectively. The outermost planet, HR 8799b, takes 460 Earth years to orbit the star. It orbits just inside a disk of dusty debris, similar to that produced by the comets of our Kuiper Belt, so, in some ways, this planetary system seems to be a scaled-up version of our Solar System in orbit around a larger, brighter star.

However, the fact that such large planets exist so far from the star is hard to explain using the most popular theory of planet formation – steady gravitational coalescence within a disk which leads to the creation of a rocky core capable of drawing in large amounts of gas. Perhaps they came about through a different process, such as the gravitational collapse of gaseous clumps. Scattering due to gravitational interactions with other massive objects is another possibility.

Exoplanet Moons

Most of the planets in our Solar System have natural satellites. Some of these moons are as large as terrestrial planets and one of them, Titan, has a dense, nitrogen-rich atmosphere.

The larger satellites, such as the Galilean moons of Jupiter, are thought to have formed from a disk of material surrounding their planet, like a mini-Solar System. A number of giant exoplanets, such as Fomalhaut b, seem to be embedded in such a disk, supporting speculation that they may also be in the process of gaining a family of moons. Other satellites, such as the Moon, are more likely to have coalesced from the debris cloud ejected during a collision between two large planetesimals.

There is also evidence that some of the satellites in the Solar System have liquid water beneath their icy surfaces. It would, therefore, be logical to assume that a large proportion of the exoplanet

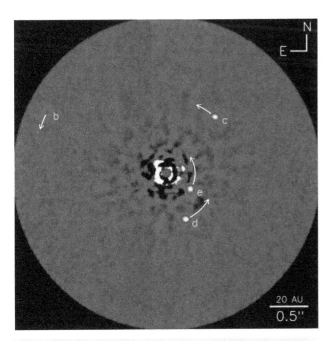

Figure 14.37 HR 8799 was the first exoplanet system to be directly imaged. In 2008, Keck Observatory near-infrared imagery revealed three planets (labeled "b," "c," and "d") orbiting a dust-shrouded young star (center). In 2010, a fourth planet was detected (labeled "e"). HR 8799 is located 129 light years from Earth. (NRC-HIA/C. Marois/W.M. Keck Observatory)

Figure 14.38 An artist's impression of a Neptune-sized exomoon which Hubble Space Telescope observations indicate may orbit the gas giant Kepler-1625b. (Courtesy of NASA/ESA/L. Hustak)

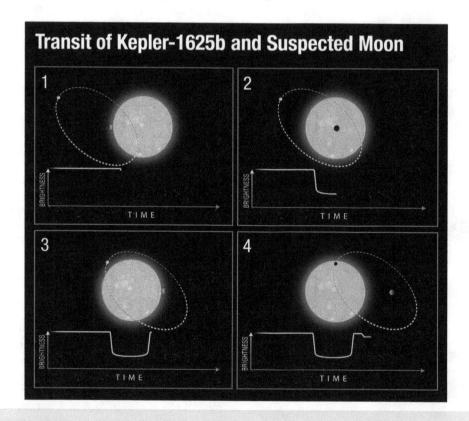

Figure 14.39 This diagram represents Hubble Space Telescope photometric observations of a transit by exoplanet Kepler-1625b. The planet blocks a small fraction of the star's light, shown in the light curve (green line) as a slight dip in the star's brightness. After the 19-hour-long transit was completed, a second, smaller dip in the light curve occurred about three and a half hours later (panel 4). This may be the signature of a moon trailing the planet. The exomoon may be as big as planet Neptune. The inclination of the candidate moon's orbit is uncertain. If confirmed, this would be the first exomoon to be discovered. (Courtesy of NASA, ESA, D. Kipping/Columbia University, and A. Feild/STScI)

Box 14.5   The Drake Equation

How can we estimate the number of advanced civilizations that might exist in our Galaxy? In 1961, radio astronomer Frank Drake produced an equation that included all of the terms required to estimate the number of technological civilizations that may exist. The Drake Equation identifies specific factors thought to play a role in the development of such civilizations.

The equation is usually written: $N = R^* \bullet fp \bullet ne \bullet fl \bullet fi \bullet fc \bullet L$, where,

N = The number of civilizations in the Milky Way galaxy whose electromagnetic emissions are detectable.

R^* = The rate of formation of stars suitable for the development of intelligent life.

fp = The fraction of those stars with planetary systems.

ne = The number of planets in each system with an environment suitable for life.

fl = The fraction of suitable planets on which life actually appears.

fi = The fraction of life bearing planets on which intelligent life emerges.

fc = The fraction of civilizations that develop a technology that releases detectable signs of their existence into space.

L = The length of time such civilizations release detectable signals into space.

Although there is no unique solution to this equation, since many of the numbers are unknown, it is a generally accepted scientific tool that is used to examine the place of humanity in the natural evolution of the Milky Way.

community has satellites, and that a sizeable proportion of these are small and rocky, perhaps with a coating of ice.

Dozens of exoplanets have already been found in the habitable zone of their host star, where liquid water can exist, but many of them are inhospitable giants. However, there is a strong possibility that they may be orbited by large moons, which, of course, will also be in the habitable zone and, therefore, have the potential to harbor life.

Until 2018, no such moons had been found because of the difficulty in detecting such small, dim objects. However, one possible Neptune-sized exomoon, dubbed Kepler-1625b-i, has been reported in orbit around a gas giant known as Kepler-1625b (Figures 14.38 and 14.39). Although it is so large, the moon candidate is estimated to be only 1.5% the mass of its companion planet.

Evidence for the existence of the exomoon was found while the exoplanet was in transit in front of its star, causing a dimming of the starlight. The researchers first reported small deviations and wobbles in the light curve that caught their attention.

Subsequent observations with the Hubble Space Telescope of a subsequent transit by Kepler 1625b revealed a second, much smaller, decrease in the star's brightness approximately 3.5 hours later. This small decrease was consistent with a gravitationally bound moon trailing the planet.

The exoplanet, which is several times more massive than Jupiter, orbits its star at a distance similar to the distance between the Sun and Earth. Hence, both the planet and its candidate moon lie at the inner edge of the habitable zone of their star system. Precise measurements of variations in the duration of multiple transits can also provide information about an exomoon's mass and its orbital separation from the planet. A moon's gravity would tug on the planet and either speed or slow its transit, depending on whether the moon leads or trails the planet. Once it is possible to calculate a satellite's mass and distance from its host planet, the likely habitability of a moon may be determined.

There is also the potential, as measurement techniques are refined, to seek them out by looking for wobbles in the velocity of the planets they orbit.

Looking further ahead, the James Webb Space Telescope may be able to detect an atmosphere of gases such as water, oxygen, carbon dioxide, and methane on a transiting, Earth-like moon.

Small, dim, red dwarf stars are considered to be better targets in the hunt for habitable planets or moons, since the habitable zone is closer to the star and the probability of a transit is markedly higher. In such systems, a planet close enough to be in the habitable zone would also be close enough for the star's gravity to slow it until one side always faces the star. One side of the planet would then be baked in constant sunlight, while the other side would freeze in constant darkness.

However, an exomoon circling such a planet would be much more hospitable. The satellite would be tidally locked to its planet, not to the star, so it would have regular – if lengthy – day–night cycles, just like Earth's Moon. Its atmosphere would moderate temperatures, and plant life would have a source of energy moon-wide. Alien moons orbiting gas giant planets may be more likely to be habitable than tidally locked Earth-sized planets or super-Earths.

Alien Life?

The only place in the Universe known to have life is planet Earth, though radio and optical telescopes are searching for signals from intelligent aliens. No one knows how life started on our world. One theory is that comets added "organic" (carbon-based) molecules to the planet's early oceans. Another idea favors chemical reactions near hot springs or submarine black smokers.

Some scientists believe simple life forms may have evolved long ago on Mars. There is even a faint possibility that life exists in a subsurface ocean on Jupiter's moon Europa. In our Galaxy, there may be billions of planets and moons capable of supporting life.

In the standard view of what makes life possible, the presence of liquid water is the key factor. In the Sun's vicinity this restricts the habitable zone to between 0.8 AU and 2 AU. Hence, our Earth is the "Goldilocks" planet in our Solar System – located where the

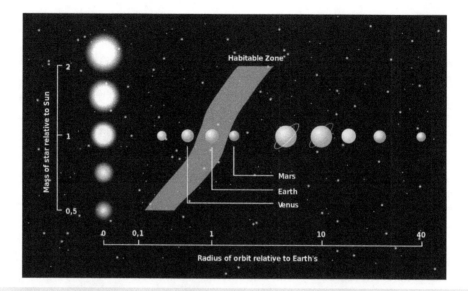

Figure 14.40 The location of the habitable zone varies according to the mass and luminosity of main sequence stars. Regions that could theoretically support life are much closer to cool, red dwarf stars than hot, massive stars. However, the size of the habitable zone around a red dwarf is correspondingly small. On the other hand, the planetary systems around small stars are also scaled down, so rocky planets may exist within their narrow habitable zones. The planets of our Solar System are shown for reference (not to scale) with Venus and Mars just outside the habitable zone around the Sun. (Wikimedia, based on data from James Kasting)

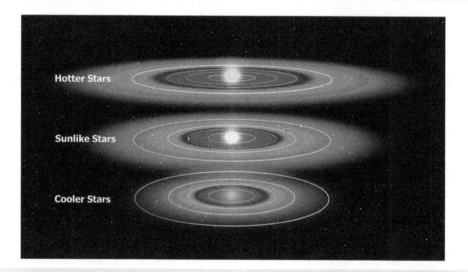

Figure 14.41 The width of the habitable zone (green) and its distance from the central star increases with the size and luminosity of the star. Small, rocky planets may be able to form in the habitable zone around cool red dwarfs, despite its limited dimensions. The red area is too hot for life and the blue area is too cold. (NASA)

temperature is (currently) just right – while Venus is too hot and Mars is too cold.

Water-rich worlds may be fairly common across the Galaxy. One of the best examples so far of such an exoplanet is GJ 1214b, which circles a nearby red dwarf that is much smaller than our Sun. It orbits its star once every 1.6 days at a distance of about 15 stellar radii. Earth, by contrast, is located at about 215 solar radii. The planet is about 2.5 times the diameter of Earth and about 6.5 times its mass.

Together, these parameters suggest that the density of GJ 1214b is about twice that of water. Although its surface temperature is high, between 120°C and 280°C, the higher gravitational field, and hence atmospheric pressure, would ensure that any water is in liquid form. It would be shrouded by a thick, steamy atmosphere. It seems likely that the exoplanet has a small rocky core, surrounded by a massive water ocean.

The size and mass of an exoplanet are also key factors. Planets such as Mars, which form with less than about 0.5 Earth masses, are unlikely to have enough surface gravity to hold onto a life-sustaining atmosphere for very long. On the other hand, planets with at least 10 Earth masses can hold onto even the lightest and most abundant elements, hydrogen and helium, so they grow into gas giants – highly unsuitable for life as we know it.

Based on these criteria, the search for habitable planets should preferentially concentrate on finding planets with masses between about 0.5 and 10 Earth masses, or, assuming a similar density for terrestrial planets, about 0.8 to 2.2 Earth radii.

Of course, the habitable zone is not static and unchanging. It depends on the radiation output of the star and on the nature of the planet's atmosphere, e.g. the inventory of greenhouse gases. Stars tend to become more active and luminous as they grow older, so the zone moves outward as a star ages. Some 3.8 billion years ago, when life was first evolving on Earth, the planet would have been too cold without the plentiful presence of greenhouse gases to ameliorate the surface conditions.

Looking into the future, the Sun will continue to grow and Earth's oceans will boil away, leaving our planet a lifeless, arid desert. We have perhaps 500 million to one billion years to move house and find another habitable world. About 7 billion years from now, the Sun will swell to become a red giant and Earth may even be swallowed by the bloated star. For a short time, the habitable zone will shift outward, first to Mars and then to the satellites of Jupiter and Saturn. However, this balmy period in the outer Solar System will be short-lived, and as the Sun collapses to a white dwarf, the remaining rocky planets and moons will turn into balls of ice.

The nature of the central star is a key determinant of a planet's potential habitability. Although the habitable zones around cool stars, such as red dwarfs, are very narrow, there is plenty of evidence that they can host rocky planets (Figures 14.40 and 14.41). According to one estimate, a large fraction of red dwarfs, which make up at least three quarters of the stars in the Universe, is likely to be orbited by low-mass planets.

On the other hand, although the habitable zones are wider around stars significantly more massive than the Sun, such environments are not very hospitable for life to develop. The stars evolve very quickly (over millions rather than billions of years), so that the habitable zone migrates quickly outward. The intense UV flux may also be a problem.

The habitability of a world may also be influenced by the density and composition of its atmosphere, which, in turn, influences the surface temperature and provides protection against ultraviolet light, X-rays, meteorites, and cosmic rays.

Massive neighbors may influence the shape of a rocky planet's orbit, just as Earth's orbit periodically evolves from purely circular to slightly eccentric, giving rise to alternating warm and glacial periods. Giant planets in the outer reaches of the planetary system may also provide protection from comet impacts that can sterilize planetary surfaces or cause mass extinctions. However, worlds such as Jupiter may also lead to the formation of asteroid belts, which turn young planetary systems into shooting galleries. Giant planets in the inner regions may also scatter smaller planets or even eject them altogether from the system.

Despite the devastation caused by asteroid and comet impacts, a planetary system free of these bodies may not be the best thing for the evolution of life. Indeed, ice-rich comets and asteroids may be necessary for life to emerge on terrestrial planets, since they deliver the organic material and water that are necessary for the formation of oceans and the origin and development of life. In addition, although such impacts cause mass extinctions which wipe out the dominant organisms, they clear the way for other life forms to fill the recently vacated ecological niches.

The transfer of organisms from one world to another may actually be enabled by large impacts. A comet slamming into a terrestrial planet blasts billions of tons of rock into space. A few per cent of this ejected material will escape from the planet and eventually fall onto a neighboring world, possibly seeding a new generation of life forms.

It seems inevitable that, if there were once abundant microbes on Mars, some of them would have reached Earth by hitching a ride on Martian meteorites (and vice versa). Cocooned inside a large rock, they would be screened from the worst effects of radiation and protected from burning up as they plunged into the Earth's atmosphere.

It is interesting to note that the absence of gas giants may open the door to the formation of more small, rocky planets. Hence, if Jupiter and Saturn had never formed in our Solar System, three Earth-like planets might have developed instead of just one. Lacking the gas giants, icy planets like Uranus and Neptune may have formed at 5–10 AU. Such a system would be fairly asteroid-free, since most of the rocky debris would have coalesced into planets. The presence of life is further complicated by factors such as tidal locking and internal heating. Tidal locking causes a planet to keep the same face towards the star, so that one hemisphere is boiling hot and the other freezing cold (unless a thick atmosphere and strong winds enable transfer of heat from the dayside to the dark side). Furthermore, the presence of large moons may stabilize the axis of rotation and create tidal effects in the crust and any large, surface water masses.

Internal heating within a rocky planet or satellite, either due to the decay of radioactive elements or tidal flexing, can enable the presence of subsurface water, or even oceans, where alien life could evolve without sunlight. The amino acids and other organic molecules thought to be the building blocks of life are plentiful throughout the Galaxy, as are chemical energy sources such as hydrogen and iron oxides.

It is now clear that life forms on Earth are remarkably resilient, taking hold in extreme environments, e.g. where there is no sunlight, temperatures are above the boiling point of water or below freezing, and high acidity or alkalinity occur. Some organisms can even survive exposure to ultraviolet light and high-energy particles, or eke out a bare existence inside "solid" rock, deep below the surface.

This apparent hardiness has led to speculation that primitive organisms may exist on worlds where surface conditions are much harsher than those on Earth. Exoplanets with subsurface oceans may also offer potential habitable environments, even though they may orbit well outside their star's traditional habitable zone.

SETI

A Search for Extraterrestrial Intelligence (SETI) has been taking place for half a century, since Frank Drake began Project OZMA. The most common method is to use large radio telescopes and

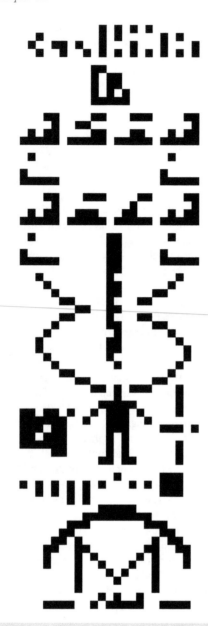

Figure 14.42 The Arecibo message was sent from the giant radio telescope in Puerto Rico on November 16, 1974. It was aimed at the globular star cluster M13, some 25,000 light years away. The message forms an image when translated into graphic characters and spaces. Each of the seven sections provides basic information that would inform an alien civilization about human life and the origin of the signal. (Cornell University)

high-speed computers to seek out unusual radio signals or messages from the blizzard of electromagnetic signals that reach our planet from space (and from sources much closer to home!).

NASA began a SETI program in 1992, but it was cancelled soon after, owing to lack of political support and funding. The program was revived in 1995 as a privately funded initiative known as Project Phoenix. Directed by the non-profit SETI Institute in

California, the search used some of the most powerful instruments in the world, including the Parkes radio telescope in Australia, the Green Bank telescope in West Virginia, and Arecibo in Puerto Rico. In all, Phoenix observed about 800 stars.

Another effort, called SERENDIP (Search for Extraterrestrial Radio Emissions from Nearby Developed Intelligent Populations), is based at UC Berkeley. The data used in SERENDIP are currently obtained using the giant Arecibo radio telescope. A similar system is being installed for the Southern SETI program in Argentina. The latest spectrometers can track 128 million channels simultaneously and listen instantaneously to a radio band 80 MHz wide.

Most of the SETI programs include powerful computers that analyze data from a large radio telescope in real time. However, a lack of computing power means that none of the programs can scrutinize the data very closely for weak signals or specific signal types. In order to overcome this drawback, the UC Berkeley SETI team came up with the idea of using thousands of home computers to analyze the data streaming from Arecibo. The SETI@home initiative takes advantage of spare computing power whenever a machine switches to screen saver, and then send the results back to the project center.

Private funding plays a major role in SETI. One example is the Allen Telescope Array (ATA) in northern California, which comprises 42 small dishes. It can operate as a "snapshot" instrument, rapidly surveying large swaths of the sky. It operates over multiple frequency bands concurrently. The ATA can also operate as a single large dish (by pointing all dishes in the same direction) or as 42 independent dishes, searching in different areas of the sky simultaneously.

The ATA searches a large area near the galactic center, which contains perhaps 40 billion stars, at frequencies from 1,420 to 1,720 MHz. This "waterhole" frequency range is defined by the spectral lines for the dissociation products of water, the key to life as we know it. It is also a radio quiet zone, since it features the minimum background noise caused by emission from the Galaxy and by cosmic microwave background radiation. As such, it is where we can best listen for a faint whisper from across the vast expanse of interstellar space.

Breakthrough Listen has been conducting SETI searches using giant radio telescopes such as the Green Bank Observatory and Parkes Observatory, but in 2018 the group signed up to use the new South Africa-based MeerKAT array of 64, 13.5 m-wide dishes to scan up to one million stars over the next 5 years.

Radio observations are not the only technique available. In recent years, optical SETI has been introduced. This involves a dedicated search for nanosecond flashes of intense light, using instruments capable of detecting laser beacons from civilizations many light years distant. Unlike its radio counterpart, optical SETI requires an extraterrestrial civilization to be deliberately signaling in the direction of our Solar System.

The Planetary Society's All-Sky Optical SETI (OSETI) program uses a 1.8 m optical telescope in Massachusetts to observe the entire sky over periods of 200 clear nights. One thousand detectors are used to measure the incoming light nearly a billion times a second.

A number of efforts have been made to establish contact with extraterrestrials over the years. The most famous of these took place in October 1974, when the Arecibo radio telescope beamed

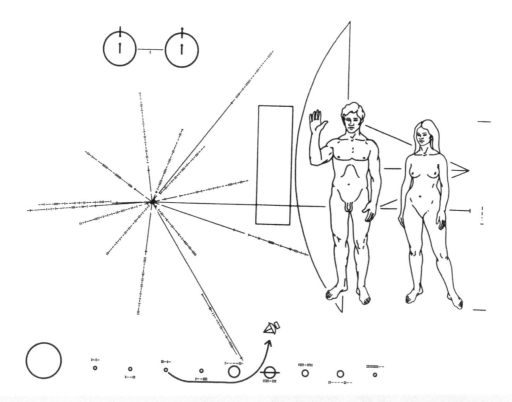

Figure 14.43   Pioneer 10 and 11 carry a 15 × 23 cm, gold-anodized aluminum plaque which carries information about their origin. The radiating lines (left) represent the positions of 14 pulsars, arranged to indicate the Sun's location. The "1-" symbols at the ends of the lines are binary numbers that represent the frequencies of these pulsars relative of that to the hydrogen atom (upper left). The human figures show Pioneer's creators. The man's hand is raised in a gesture of good will. At the bottom are the Sun and planets with the spacecraft's trajectory. (NASA)

a simple, coded message towards 500,000 mature stars in the globular cluster M13 (Figure 14.42). There was no expectation of a quick response: M13 is so far away that we can't expect an answer for another 50,000 years.

The message was devised by a group led by Frank Drake of Cornell University, creator of the famous Drake equation. It consisted of seven sections, encoded into 1,679 on-off pulses – a multiple of the prime numbers 23 and 73. The message is arranged in a rectangle as 73 rows with 23 columns. The sections were as follows:

1. The numbers 1 to 10 in binary code.
2. The atomic numbers of the five elements that make up deoxyribonucleic acid (DNA): hydrogen, carbon, nitrogen, oxygen, and phosphorus.
3. The chemical formulas for some key biological molecules in DNA, sugar, phosphate, and the nucleotides.
4. A graphic of the twisted, double-helix structure of DNA.
5. An outline of a human, with the physical height of an average man (equal to 14 wavelengths of signal), flanked by the number of humans on Earth.
6. A graphic of the Solar System, with the main bodies approximately to scale. Earth is displaced to highlight it.
7. A graphic of the Arecibo radio telescope and how it works.

Some more recent attempts at interstellar communication have been more commercial than scientific. For example, on October 9, 2008, a complex four and a half hour transmission containing 501 photos, drawings, and text messages was beamed toward the Gliese 581 planetary system by the RT-70 radar telescope in Evpatoria, Ukraine. It was part of a competition sponsored by a social networking site.

Meanwhile, our TV and radio broadcasts have been spreading out into space for many decades, and, weak as they are, they may one day be picked up by alien civilizations tuning in on a nearby planet.

The chances of ET finding a tiny spacecraft sent from Earth as it travels through the Galaxy are even more remote, but this has not prevented simple attempts at communication involving the four spacecraft that are heading out of the Solar System.

NASA's Pioneer 10 and 11 spacecraft both carry a plaque which includes a man and a woman in front of the spacecraft, and Earth's position in the Solar System (Figure 14.43). A ray diagram shows the location of the Sun.

The two Voyager spacecraft carry a record engraved with Earth's location and playing instructions (Figure 14.44). The 30.5 cm, gold-plated, copper disks contain greetings in 60 languages, natural sounds, 90 minutes of music, and 118 pictures of Earth and the other planets. They also contain electronic information that an advanced technological civilization could convert into diagrams and photographs.

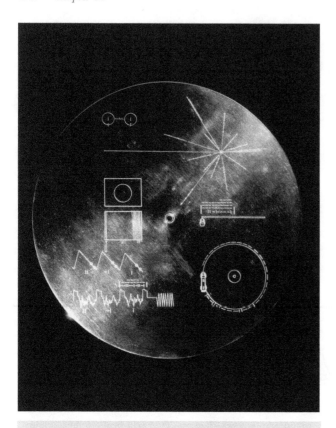

Figure 14.44 This gold-anodized aluminum cover protects the Voyager 1 and 2 "Sounds of Earth" records from micrometeorite bombardment, and includes a guide to playing the record. The explanatory diagram appears on the inner and outer surfaces, as the outside will be eroded in time. (NASA-JPL)

A number of important questions arise from SETI research. Despite decades of searching, why have no signs of ET ever been found? So far, only false alarms have been detected, the most famous example being a solitary "wow!" signal detected in 1977, the most powerful unidentified transmission ever detected. Is this because such civilizations are very rare in the Galaxy, or is no one using the same technology and listening to the Universe in the same way?[9]

There is also the question of how to react if an alien signal, or message, is ever detected and decoded. How would we know whether the senders were hostile or friendly? Although the vast distances between star systems mean that interstellar travel is highly unlikely, is it wise to advertise our presence, particularly if the receivers may be more technologically advanced?

Questions

- Why were no exoplanets discovered until the 1990s?
- Summarize the various methods currently used to find evidence of exoplanets.
- How are exoplanet detection methods likely to improve in the future?
- What are the characteristics of (a) hot Jupiters, (b) super-Earths, (c) free-floating planets?
- Select three exoplanets that are very different from any found in our Solar System and describe their main characteristics.
- Explain the meaning of the term "habitable zone." Describe the factors that may influence the habitability of a planetary system.
- What is the probability of finding habitable, Earth-like planets in the future? Explain your answer.
- Explain why the Solar System may not be a typical example of a planetary system in our Galaxy.
- Select one of the known multi-planet systems and compare it with our Solar System.
- What are pulsar planets? Why was their discovery a surprise?
- "SETI programs are not a worthwhile use of government funds." Discuss.

---

[9] Radio and TV transmissions, radar, and laser beams have only been introduced in the last 100 years. For more than 99.9% of Earth's existence, any advanced aliens would have been unable to detect the presence of intelligent life on Earth. Since technologies may also change rapidly, there is no reason to believe that ET would be at the same stage of technological innovation as ourselves.

Appendices

Appendix 1: Planetary Data

(a) Planets

	Mercury	Venus	Earth	Mars	Jupiter	Saturn	Uranus	Neptune
Equatorial diam. (km)	4,879	12,104	12,756	6,792	142,984	120,536	51,118	49,528
Rotation period (equat.)	58 d 15 h 30 m	243 d 0 h 36 m (R)	23 h 56 m	24 h 37 m	9 h 50 m	10 h 14 m	17 h 14 m (R)	16 h 6 m
Density (water = 1)	5.43	5.24	5.52	3.91	1.33	0.69	1.29	1.64
Mass (Earth = 1)	0.055	0.814	1	0.11	317.8	95.2	14.53	17.14
Surface gravity (Earth = 1)	0.378	0.903	1	0.38	2.69	1.07	0.91	1.14
Equatorial inclination (deg)	0.034	177.3	23.44	25.19	3.13	26.73	97.77	28.32
Orbital period	87.97 d	224.7 d	365.25 d	687 d	11.86 y	29.46 y	84.01 y	164.79 y
Av. distance from Sun (million km)	57.9	108.2	149.6	227.9	778.3	1,427	2,871	4,497.1
Av. orbital velocity (km/s)	47.88	35.02	29.8	24.1	13.06	9.65	6.81	5.43
Orbital eccentricity	0.2	0.007	0.0167	0.093	0.048	0.056	0.046	0.0097
Orbital inclination (deg)	7.0	3.4	0.0	1.9	1.3	2.5	0.8	1.8
Average temp. (°C)	350 (day), −170 (nt)	467	7	−63	−148	−178	−218	−220
Atmosphere	Potassium, sodium	Carbon dioxide	Nitrogen, oxygen	Carbon dioxide	Hydrogen, helium	Hydrogen, helium	Hydrogen, helium	Hydrogen, helium
Satellites	0	0	1	2	79	82	27	14

R = retrograde

Exploring the Solar System, Second Edition. Peter Bond.
© 2020 John Wiley & Sons Ltd. Published 2020 by John Wiley & Sons Ltd.
Companion Website: www.wiley.com/go/Bond-Solar-System2e

(b) Dwarf Planets

	Pluto	Eris	Makemake	Haumea	Ceres
Equatorial diameter (km)	2,324	Approx. 2,300	Approx. 1,500	Approx. 1,960 × 1,520 × 1,000	909 × 975
Rotation period	6d 9h 17m (R)	3h 55 m	7 h 46 m	3 h 55 m	9 h 5 m
Density (water = 1)	2.05	2.6–3.3	Approx. 2?	Approx. 2?	2.08
Mass (Earth = 1)	0.0021	0.0027?	0.00067?	0.00067?	0.00015
Surface gravity (Earth = 1)	0.06	?	?	?	0.027
Equatorial inclination (deg)	122	?	?	?	Approx. 3
Orbital period (years)	248.54	557.8	309.88	285.4	4.6
Av. distance from Sun (million km)	5,913.5	10,132	6,850.3	6,484	414.1
Orbital eccentricity	0.249	0.44	0.159	0.189	0.08
Orbital inclination (deg.)	17.14	44.19	28.96	28.19	10.58
Av. orbital velocity (km/s)	4.72	3.44	4.419	4.484	17.882
Average temp. (°C)	−228	−235	−240	−241	−106
Atmosphere	Nitrogen, methane?	Nitrogen, methane?	Nitrogen, methane, ethane?	?	–
Satellites	5	1	0	2	0

Appendix 2: Satellite Data

Earth

Satellite	Discoverer	Year of discovery	Diameter (km)	Visual magnitude	Distance (km)	Orbital period (d)	Orbital ecc.	Orbital inc. (degrees)
Moon	–	–	3,475	−12.7	384,400	27.3	0.055	5.2

Mars

Satellite	Discoverer	Year of discovery	Diameter (km)	Visual magnitude	Distance (km)	Orbital period (d)	Orbital ecc.	Orbital inc. (degrees)
Phobos	A. Hall	1877	13.4 × 11.2 × 9.2	11.4	9,380	0.3	0.015	1.1
Deimos	A. Hall	1877	7.5 × 6.1 × 5.2	12.5	23,460	1.3	0.0005	1.8

Jupiter

Satellite	Discoverer	Year of discovery	Diameter (km)	Visual magnitude	Distance (km)	Orbital period (d)	Orbital ecc.	Orbital inc. (degrees)
Metis	S. Synott/Voyager 2	1979	44	17.5	128,100	0.294	0.001	0.0
Adrastea	D. Jewitt, E. Danielson	1979	13 × 10 × 8	18.7	128,900	0.298	0.002	0.0
Amalthea	E. Barnard	1892	270 × 165 × 150	14.1	181,400	0.498	0.003	0.4
Thebe	S. Synott/Voyager 1	1980	116 × 98 × 84	16.0	221,900	0.68	0.018	1.1
Io	Galileo	1610	3,643	5.0	421,800	1.77	0.004	0.0
Europa	Galileo	1610	3,122	5.3	671,100	3.6	0.009	0.5
Ganymede	Galileo	1610	5,262	4.6	1,070,400	7.2	0.001	0.2
Callisto	Galileo	1610	4,821	5.7	1,882,700	16.7	0.007	0.3
Themisto	C. Kowal, E. Roemer	1975	9	21.0	7,507,000	130.0	0.242	43.1
Leda	C. Kowal	1974	18	19.5	11,165,000	240.9	0.164	27.5
Himalia	C. Perrine	1904	160	14.6	11,461,000	250.6	0.162	27.5
Ersa	S. Sheppard & others	2018	3	22.9	11,483,000	252	0.094	30.61
Pandia	S. Sheppard & others	2017	3	23.0	11,525,000	252.1	0.18	28.15
Lysithea	S. Nicholson	1938	38	18.3	11,717,000	259.2	0.112	28.3
Elara	C. Perrine	1905	78	16.3	11,741,000	259.6	0.217	26.6
Dia	S. Sheppard & others	2000	4	22.4	12,118,000	287.0	0.211	28.23
Carpo	S. Sheppard & others	2003	3	23.2	16,989,000	456.1	0.430	51.4
Valetudo	S. Sheppard & others	2017	1	24.0	18,980,000	533.3	0.222	34
S/2003 J12	S. Sheppard & others	2003	1	23.9	19,002,480	533.3R	0.376	145.8
Euporie	S. Sheppard & others	2001	4	23.1	19,302,000	550.7R	0.144	145.8
S/2011 J1	S. Sheppard & others	2011	1	23.6	20,155,290	580.7R	0.296	162.8
S/2003 J18	B. Gladman & others	2003	2	23.4	20,274,000	588R	0.105	146.4
S/2010 J2	C. Veillet	2010	1	23.9	20,307,150	588.1R	0.307	150.4
S/2017 J7	S. Sheppard & others	2017	2	23.6	20,627,000	602.6R	0.215	143.4
S/2016 J1	S. Sheppard & others	2016	1	24.0	20,650,845	602.7R	0.141	139.8
S/2017 J3	S. Sheppard & others	2017	2	23.4	20,694,000	606.3R	0.148	147.9
Orthosie	S. Sheppard & others	2001	4	23.1	20,721,000	622.6R	0.281	145.9
Euanthe	S. Sheppard & others	2001	6	22.8	20,799,000	620.6R	0.232	148.9
Thyone	S. Sheppard & others	2001	8	22.3	20,940,000	627.3R	0.229	148.5
S/2003 J16	B. Gladman & others	2003	2	23.3	21,000,000	595.4R	0.27	148.6
Mneme	S. Sheppard & others	2003	2	23.3	21,069,000	620.0R	0.227	148.6
Harpalyke	S. Sheppard & others	2000	4	22.2	21,105,000	623.3R	0.226	148.6

Satellite	Discoverer	Year of discovery	Diameter (km)	Visual magnitude	Distance (km)	Orbital period (d)	Orbital ecc.	Orbital inc. (degrees)
Hermippe	S. Sheppard & others	2001	8	22.1	21,131,000	633.9R	0.210	150.7
Praxidike	S. Sheppard & others	2000	7	21.2	21,147,000	625.3R	0.230	149.0
Thelxinoe	S. Sheppard & others	2004	2	23.5	21,162,000	628.1R	0.221	151.4
Eupheme	S. Sheppard & others	2003	2	23.4	21,199,710	631.5R	0.216	149.4
Helike	S. Sheppard & others	2003	4	22.6	21,263,000	634.8R	0.156	154.8
Iocaste	S. Sheppard & others	2000	5	21.8	21,269,000	631.5R	0.216	149.4
Ananke	S. Nicholson	1951	28	18.8	21,276,000	610.5R	0.244	148.9
S/2003 J9	S. Sheppard & others	2003	1	23.7	22,441,680	683.0R	0.269	164.5
S/2017 J6	S. Sheppard & others	2017	2	23.5	22,455,000	683R	0.557	155.2
S/2017 J9	S. Sheppard & others	2017	3	23.7	21,487,000	639.2R	0.229	152.7
S/2003 J19	B. Gladman & others	2003	2	23.7	22,757,000	697.6R	0.257	166.7
Philophrosyne	S. Sheppard & others	2003	2	23.5	22,819,950	668.4R	0.194	143.6
Eurydome	S. Sheppard & others	2001	6	22.7	22,865,000	717.3R	0.276	150.3
Arche	S. Sheppard & others	2002	3	22.8	22,931,000	723.9R	0.259	165.0
Autonoe	S. Sheppard & others	2001	8	22.0	23,039,000	762.7R	0.334	152.9
Pasithee	S. Sheppard & others	2001	4	23.2	23,096,000	719.5R	0.267	165.1
Herse	B. Gladman & others	2003	2	23.4	23,097,000	715.4R	0.200	164.2
Chaldene	S. Sheppard & others	2000	4	22.5	23,179,000	723.8R	0.251	165.2
Isonoe	S. Sheppard & others	2000	4	22.5	23,217,000	725.5R	0.246	165.2
Kale	S. Sheppard & others	2001	4	23.0	23,217,000	729.5R	0.264	165.1
Aitne	S. Sheppard & others	2001	6	22.7	23,231,000	730.2R	0.264	165.1
S/2017 J5	S. Sheppard & others	2017	2	23.5	23,232,000	719.5R	0.284	164.3
S/2017 J8	S. Sheppard & others	2017	1	24.0	23,232,700	719.6R	0.312	164.7
S/2003 J4	S. Sheppard & others	2003	2	23.0	23,257,920	723.2R	0.204	144.9
Erinome	S. Sheppard & others	2000	3	22.8	23,279,000	728.3R	0.266	164.9
S/2017 J2	S. Sheppard & others	2017	2	23.5	23,303,000	723.1R	0.236	166.4
S/2010 J1	R. Jacobson & others	2010	2	23.3	23,314,335	723.2R	0.320	163.2
Taygete	S. Sheppard & others	2000	5	21.9	23,360,000	732.2R	0.252	165.2
Carme	S. Nicholson	1951	46	17.9	23,404,000	702.3R	0.253	164.9
S/2011 J2	S. Sheppard & others	2011	1	23.6	23,463,885	730.5R	0.332	148.8
Sponde	S. Sheppard & others	2001	4	23.0	23,487,000	748.3R	0.312	151.0
S/2017 J1	S. Sheppard & others	2017	2	23.8	23,547,105	734.2R	0.397	149.2
Kalyke	S. Sheppard & others	2000	5	21.8	23,583,000	743.0R	0.245	165.2
Pasiphae	P. Melotte	1908	60	17.0	23,624,000	708.0R	0.409	151.4
Eukelade	S. Sheppard & others	2003	4	22.6	23,661,000	746.4R	0.272	165.5
Eirene	S. Sheppard & others	2003	4	22.5	23,731,770	759.7R	0.220	163.1
Megaclite	S. Sheppard & others	2000	5	21.7	23,806,000	752.8R	0.421	152.8
Sinope	S. Nicholson	1914	38	18.1	23,939,000	724.5R	0.250	158.1
Hegemone	S. Sheppard & others	2003	3	23.2	23,947,000	739.6R	0.328	155.2
Aoede	S. Sheppard & others	2003	4	22.5	23,981,000	761.5R	0.432	158.3
Kallichore	S. Sheppard & others	2003	2	23.7	24,043,000	764.7R	0.264	165.5
S/2003 J23	S. Sheppard & others	2003	2	23.6	24,055,500	759.7R	0.309	149.2
Callirrhoe	Spacewatch	1999	9	20.7	24,102,000	758.8R	0.283	147.1
S/2003 J10	S. Sheppard & others	2003	2	23.6	24,249,600	767.0R	0.214	164.1
Cyllene	S. Sheppard & others	2003	2	23.2	24,349,000	737.8R	0.319	149.3
Kore	S. Sheppard & others	2003	2	23.6	24,543,000	779.2R	0.325	145.0
S/2003 J2	S. Sheppard & others	2003	2	23.2	28,570,410	982.5R	0.380	151.8

Saturn

Satellite	Discoverer	Year of discovery	Diameter (km)	Visual magnitude	Distance (km)	Orbital period (d)	Orbital ecc.	Orbital inc. (degrees)
S/2009 S1	C. Porco & others/Cassini	2009	0.3	28	117,000	?	0.000	0.0
Pan	M. Showalter/Voyager 2	1990	20	19.4	133,600	0.575	0.000	0.0
Daphnis	C. Porco & others/Cassini	2005	7	24	136,500	0.59	0.000	0.0
Atlas	R. Terrile/Voyager 1	1980	37 × 34.4 × 27	19.0	137,700	0.60	0.001	0.0
Prometheus	S. Collins/Voyager 1	1980	148 × 100 × 68	15.8	139,400	0.61	0.002	0.0
Pandora	S. Collins/Voyager 1	1980	110 × 88 × 62	16.4	141,700	0.63	0.004	0.1
Epimetheus	J. Fountain & others/ Voyager 1	1980	138 × 110 × 110	15.6	151,400	0.69	0.021	0.4
Janus	A. Dollfus	1966	194 × 190 × 154	14.4	151,500	0.7	0.007	0.2
Aegaeon	C. Porco & others/Cassini	2008	0.5	27	167,500	0.808	0.000	0.0
Mimas	W. Herschel	1789	418 × 392 × 383	12.8	185,600	0.9	0.021	1.6
Methone	C. Porco & others/Cassini	2004	3	23	194,300	1.0	0.001	0.0
Anthe	C. Porco & others/Cassini	2007	1	26	197,700	1.04	0.001	0.1
Pallene	C. Porco & others/Cassini	2004	4	23	212,300	1.1	0.004	0.2
Enceladus	W. Herschel	1789	512 × 494 × 489	11.8	238,100	1.4	0.000	0.0
Calypso	D. Pascu & others	1980	30 × 16 × 16	18.7	294,700	1.9	0.001	1.5
Telesto	B. Smith & others/ Voyager 1	1980	30 × 25 × 15	18.5	294,700	1.9	0.001	1.2
Tethys	G. Cassini	1684	1,060	10.2	294,700	1.9	0.000	0.17
Dione	G. Cassini	1684	1,118	10.4	377,400	2.7	0.000	0.0
Helene	P. Laques, J. Lacacheux	1980	36 × 32 × 30	18.4	377,400	2.7	0.000	0.2
Polydeuces	C. Porco & others/Cassini	2004	13 × 4	23.0	377,400	2.7	0.018	0.2
Rhea	G. Cassini	1672	1,528	9.6	527,100	4.5	0.001	0.3
Titan	C. Huygens	1655	5,150	8.4	1,221,900	15.95	0.029	1.6
Hyperion	W. Bond, W. Lassell	1848	360 × 280 × 225	14.4	1,464,100	21.3	0.018	0.6
Iapetus	G. Cassini	1671	1,436	11.0	3,560,800	79.3	0.028	7.6
Kiviuq	B. Gladman & others	2000	14	22.0	11,111,000	449.2	0.334	46.1
Ijiraq	J. Kavelaars & others	2000	10	22.6	11,124,000	451.4	0.322	46.44
Phoebe	W. Pickering	1898	230 × 220 × 210	16.4	12,944,300	550.2R	0.164	174.8
Paaliaq	B. Gladman & others	2000	19	21.3	15,200,000	686.9	0.363	45.1
Skathi	J. Kavelaars & others	2000	6	23.6	15,541,000	728.2R	0.269	152.6
Albiorix	M. Holman & others	2000	26	20.5	16,182,000	783.5	0.479	34.0
S/2007 S2	S. Sheppard & others	2007	6	24.4	16,560,000	800R	0.218	176.7
Bebhionn	D. Jewitt & others	2004	6	24.1	17,119,000	834.8	0.469	35.01
Erriapus	J. Kavelaars & others	2000	9	23.0	17,343,000	871.2	0.474	34.62
Siarnaq	B. Gladman & others	2000	32	20.1	17,531,000	895.6	0.295	45.56
Skoll	D. Jewitt & others	2006	6	24.5	17,665,000	878.3R	0.464	161.2
Tarvos	J. Kavelaars & others	2000	13	22.1	17,983,000	926.2	0.531	33.82
Tarqeq	S. Sheppard & others	2007	7	23.9	18,009,000	887.5	0.16	46.09
Greip	D. Jewitt & others	2006	6	24.4	18,206,000	921.2R	0.326	179.8
Hyrrokkin	D. Jewitt & others	2004	8	23.5	18,437,000	931.8R	0.333	151.4
S/2004 S13	D. Jewitt & others	2004	6	24.5	18,450,000	906R	0.273	167.4
S/2004 S17	D. Jewitt & others	2004	4	25.2	18,600,000	986R	0.259	166.6
Mundilfari	B. Gladman & others	2000	6	23.8	18,685,000	952.6R	0.21	167.3
Jarnsaxa	D. Jewitt & others	2006	6	24.7	18,811,000	964.7R	0.216	163.3
S/2006 S1	D. Jewitt & others	2006	6	24.6	18,981,135	970R	0.13	154.2

Satellite	Discoverer	Year of discovery	Diameter (km)	Visual magnitude	Distance (km)	Orbital period (d)	Orbital ecc.	Orbital inc. (degrees)
Narvi	S. Sheppard & others	2003	8	24.0	19,007,000	1003.9R	0.431	145.8
Bergelmir	D. Jewitt & others	2004	6	24.2	19,338,000	1005.9R	0.142	158.5
Suttungr	B. Gladman & others	2000	6	23.9	19,459,000	1,016.7R	0.114	175.8
S/2004 S12	D. Jewitt & others	2004	5	24.8	19,650,000	1,048R	0.401	164.0
S/2004 S7	D. Jewitt & others	2004	6	24.8	19,800,000	1,103R	0.58	165.1
Hati	D. Jewitt & others	2004	6	24.4	19,856,000	1,038.7R	0.372	165.8
Bestla	D. Jewitt & others	2004	7	23.8	20,129,000	1,083.6R	0.521	145.2
Farbauti	D. Jewitt & others	2004	5	24.7	20,390,000	1,086.1R	0.206	156.4
Thrymr	B. Gladman & others	2000	6	23.9	20,474,000	1,094.3R	0.47	176.0
S/2007 S3	S. Sheppard & others	2007	5	24.9	20,518,500	1,100R	0.13	177.2
Aegir	D. Jewitt & others	2004	6	24.4	20,735,000	1116.5R	0.252	166.7
S/2006 S3	D. Jewitt & others	2006	6	24.6	21,132,000	1,142R	0.471	150.8
Kari	D. Jewitt & others	2006	7	23.9	22,118,000	1233.6R	0.478	156.3
Fenrir	D. Jewitt & others	2004	4	25.0	22,453,000	1260.3R	0.136	164.9
Surtur	D. Jewitt & others	2006	6	24.8	22,707,000	1,297.7R	0.451	177.5
Ymir	B. Gladman & others	2000	16	21.7	23,040,000	1,315.4R	0.334	173.1
Loge	D. Jewitt & others	2006	6	24.6	23,065,000	1,312R	0.187	167.9
Fornjot	D. Jewitt & others	2004	6	24.6	25,108,000	1,491R	0.206	170.4

Note: This list does not include 20 small outer satellites of Saturn whose discovery by Scott Sheppard et al. was announced in October 2019.

Uranus

Satellite	Discoverer	Year of discovery	Diameter (km)	Visual magnitude	Distance (km)	Orbital period (d)	Orbital ecc.	Orbital inc. (degrees)
Cordelia	R. Terrile/Voyager 2	1986	40	23.6	49,800	0.4	0.000	0.1
Ophelia	R. Terrile/Voyager 2	1986	42	23.3	53,800	0.4	0.01	0.1
Bianca	B. A. Smith/Voyager 2	1986	51	22.5	59,200	0.4	0.001	0.2
Cressida	S. Synott/Voyager 2	1986	80	21.6	61,800	0.5	0.000	0.0
Desdemona	S. Synott/Voyager 2	1986	64	22.0	62,700	0.5	0.000	0.1
Juliet	S. Synott/Voyager 2	1986	94	21.1	64,400	0.5	0.001	0.1
Portia	S. Synott/Voyager 2	1986	135	20.4	66,100	0.5	0.000	0.1
Rosalind	S. Synott/Voyager 2	1986	72	21.8	69,900	0.6	0.000	0.3
Cupid	M. Showalter, J. Lissauer	2003	24	26.0	74,800	0.6	0.000	0.0
Belinda	S. Synott/Voyager 2	1986	81	21.5	75,300	0.6	0.000	0.0
Perdita	E. Karkoschka/Voyager 2	1999	20	23.6	76,400	0.6	0.000	0.0
Puck	S. Synott/Voyager 2	1985	162	19.8	86,000	0.8	0.000	0.3
Mab	M. Showalter, J. Lissauer	2003	32	26.0	97,734	0.9	0.000	0.0
Miranda	G. Kuiper	1948	472	15.8	129,900	1.4	0.001	4.3
Ariel	W. Lassell	1851	1,158	13.7	190,900	2.5	0.001	0.0
Umbriel	W. Lassell	1851	1,169	14.5	266,000	4.1	0.004	0.1
Titania	W. Herschel	1787	1,578	13.5	436,300	8.7	0.001	0.1
Oberon	W. Herschel	1787	1,523	13.7	583,500	13.5	0.001	0.1

Satellite	Discoverer	Year of discovery	Diameter (km)	Visual magnitude	Distance (km)	Orbital period (d)	Orbital ecc.	Orbital inc. (degrees)
Francisco	M. Holman, J. Kavelaars	2001	22	25.0	4,276,000	266.6R	0.146	145.2
Caliban	B. Gladman & others	1997	72	22.4	7,231,000	579.7R	0.159	140.9
Stephano	B. Gladman & others	1999	32	24.1	8,004,000	677.4R	0.23	144.1
Trinculo	M. Holman & others	2001	18	25.4	8,504,000	759.0R	0.208	167.0
Sycorax	B. Gladman & others	1997	150	20.8	12,179,000	1,288.3R	0.522	159.4
Margaret	D. Jewitt, S. Sheppard	2001	20	25.2	14,345,000	1,694.8	0.783	50.7
Prospero	M. Holman & others	1999	50	23.2	16,256,000	1,977.3R	0.445	152.0
Setebos	J. Kavelaars & others	1999	47	23.3	17,418,000	2,234.8R	0.591	158.2
Ferdinand	M. Holman, B. Gladman	2001	21	25.1	20,901,000	2,823.4R	0.368	169.8

Neptune

Satellite	Discoverer	Year of discovery	Diameter (km)	Visual magnitude	Distance (km)	Orbital period (d)	Orbital ecc.	Orbital inc. (degrees)
Naiad	R. Terrile/Voyager 2	1989	58	24.1	48,227	0.29	0.000	4.7
Thalassa	R. Terrile/Voyager 2	1989	80	23.4	50,075	0.31	0.000	0.2
Despina	S. Synott/Voyager 2	1989	148	22.0	52,526	0.33	0.000	0.1
Galatea	S. Synott/Voyager 2	1989	158	22.0	61,953	0.43	0.000	0.1
Larissa	H. Reitsema & others	1989	192	21.5	73,548	0.55	0.001	0.2
Hippocamp	M. Showalter & others	2013	34	26.5	105,300	0.95	0.000	0.0
Proteus	S. Synott/Voyager 2	1989	416	20.0	117,647	1.12	0.000	0.0
Triton	W. Lassell	1846	2,707	13.0	354,760	5.87R	0.000	156.8
Nereid	G. Kuiper	1949	340	19.2	5,513,400	360.1	0.751	7.2
Halimede	M. Holman & others	2002	61	24.5	15,728,000	1,879.7R	0.571	134.1
Sao	M. Holman & others	2002	40	25.4	22,422,000	2,914.1	0.293	48.5
Laomedeia	M. Holman & others	2002	40	25.4	23,571,000	3,167.9	0.424	34.7
Psamathe	S. Sheppard & others	2003	38	25.6	46,695,000	9,115.9R	0.45	137.4
Neso	M. Holman & others	2002	60	24.6	48,387,000	9,374.0R	0.495	132.6

Pluto

Satellite	Discoverer	Year of discovery	Diameter (km)	Visual magnitude	Distance (km)	Orbital period (d)	Orbital ecc.	Orbital inc. (degrees)
Charon	J. Christy	1978	1,212	16.8	19,596	6.39R	0.002	0.08
Styx	M. Showalter et al.	2012	16 × 9 × 8	27.0	42,410	20.16	0.006	0.80
Nix	M. Showalter et al.	2005	50 × 35 × 33	23.7	57.750	24.85	0.000	0.00
Kerberos	M. Showalter et al.	2011	19 × 10 × 9	26.0	57.750	32.17	0.000	0.426
Hydra	H. Weaver, S. Alan Stern & others	2005	65 × 45 × 25	23.3	64,720	38.20	0.005	0.304

Eris

Satellite	Discoverer	Year of discovery	Diameter (km)	Visual magnitude	Distance (km)	Orbital period (d)	Orbital ecc.	Orbital inc. (degrees)
Dysnomia	M. Brown & others	2005	150?	18.8	37,350 ± 140	15.774	<0.013	142±3°

Haumea

Satellite	Discoverer	Year of discovery	Diameter (km)	Visual magnitude	Distance (km)	Orbital period (d)	Orbital ecc.	Orbital inc. (degrees)
Namaka	M. Brown & others	2005	170?	Approx. 24	25,657 ± 91	18.28?	0.249 (variable)	113?
Hi'iaka	M. Brown & others	2005	310?	Approx. 24	49,880 ± 198	49.4?	0.25	126.3?

Note: R = retrograde

Appendix 3: Planetary Rings

Rings of Jupiter

Name	Distance from Center of Planet (planet radii)	Distance from Center of Planet (km)	Width (km)	Thickness (km)
Halo	1.40–1.71	100,000–122,000	22,000	20,000
Main	1.71–1.81	122,000–129,000	7,000	30–300
Gossamer (inner)	1.81–2.55	129,200–182,000	52,800	2,600?
Gossamer (outer)	2.55–3.15	182,000–224,900	42,900	8,800?

Rings of Saturn

Name	Distance from Center of Planet (planet radii)	Distance from Center of Planet (km)	Width (km)	Thickness (km)
D	1.11–1.236	66,900–74,510	7,610	?
C	1.239–1.527	74,658–92,000	17,342	5 m
B	1.527–1.951	92,000–117,580	25,580	5–10 m
A	2.027–2.269	122,170–136,775	14,605	10–30 m
F	2.326	140,180	30–500	1 km
Janus-Epimetheus ring	2.47–2.55	149,000–154,000	5,000	?
G	2.82–2.90	170,000–175,000	5,000	100 km

Name	Distance from Center of Planet (planet radii)	Distance from Center of Planet (km)	Width (km)	Thickness (km)
Methone ring arc	3.22	194,230	?	?
Anthe ring	3.28	197,665		
Pallene ring	3.5–3.54	211,000–213,500	2,500	
E	3–8	181,000–483,000	302,000	10,000 km
Phoebe ring	59–300?	~4,000,00–>18,000,000	> 14,000,000?	2,400,000 km?

Rings of Uranus

Name	Distance from Center of Planet (planet radii)	Distance from Center of Planet (km)	Width (km)	Thickness (km)
zeta (ζ)	1.448–1.545	37,000–39,500	Approx. 2,500	(100)
6	1.637	41,837	Approx. 1.5	(100)
5	1.652	42,235	Approx. 2	(100)
4	1.666	42,571	Approx. 2.5	(100)
alpha (α)	1.75	44,718	4–10	(100)
beta (β)	1.786	45,661	5-11	(100)
eta (η)	1.834	47,176	1.6	(100)
gamma (γ)	1.863	47,626	1–4	(100)
delta (δ)	1.900	48,303	3–7	(100)
lambda (λ)	1.958	50,024	Approx. 2	(100)
epsilon (ε)	2.00	51,149	20–96	150
nu (ν)	2.586–2.735	66,100–69,900	3,800	?
mu (μ)	3.265–4.03	86,000–103,000	17,000	?

Rings of Neptune

Name	Distance from Center of Planet (planet radii)	Distance from Center of Planet (km)	Width (km)	Thickness (km)
Galle	1.692	41,900–43,900	Approx. 2,000	?
Le Verrier	2.148	~53,200	Approx. 110	?
Lassell	2.148–2.31	~53,200–57,200	Approx. 4,000	?
Arago	2.31	52,200	Less than 100	?
Unnamed (v. faint)	2.501	61,950	?	?
Adams	2.541	62,933	15-50	?

Arcs in Adams Ring are called Courage, Liberté, Egalité 1, Egalité 2, and Fraternité.

Appendix 4: The Largest Known Trans-Neptunian Objects

N.B. The sizes of most of these objects are very uncertain.

Object	Discoverer	Year of discovery	Diameter (km)	Mean distance (AU)	Orbital period (y)	Orbital ecc.	Orbital inc. (degrees)	Satellites
Eris	M. Brown, C. Trujillo, D. Rabinowitz	2003	Approx. 2,400	67.668	557.8	0.442	44.19	1
Pluto	C. Tombaugh	1930	2,374	39.482	247.94	0.249	17.16	5
Makemake	M. Brown, C. Trujillo, D. Rabinowitz	2005	Approx. 1,500?	45.791	309.88	0.159	28.96	1
Sedna	M. Brown, C. Trujillo, D. Rabinowitz	2003	Approx. 1,400?	518.57	11,809	0.853	11.93	–
Haumea	M. Brown et al/J. Ortiz et al	2004	Approx. 1,960 × 1,520 × 1,000?	43.132	285.4	0.195	28.22	2
Charon	J. Christy	1978	1,207		Satellite of Pluto			–
2007 OR10	M. E. Schwamb, M. Brown, D. Rabinowitz	2007	Approx. 1,200?	67.21	550.98	0.5	30.7	1
84522 (2002 TC302)	NEAT	2002	Approx. 1,150?	55.23	409.16	0.293	35.1	–
Orcus	M. Brown, C. Trujillo, D. Rabinowitz	2004	Approx. 950?	39.187	247.94	0.226	20.56	1
Quaoar	C. Trujillo & M. Brown	2002	Approx. 890?	43.574	285.97	0.039	7.98	1
Varuna	R. McMillan (Spacewatch)	2000	Approx. 800?	43.129	283.2	0.051	17.2	–

Appendix 5: Lunar and Planetary Missions

Moon

Name	Country	Launch Date	Purpose	Results
Pioneer 0	U.S.	17 Aug. 1958	Orbiter	Launch failure
[Unnamed]	USSR	23 Sep. 1958	Impact	Launch failure
Pioneer 1	U.S.	11 Oct. 1958	Orbiter	Failed to reach escape velocity
[Unnamed]	USSR	11 Oct. 1958	Impact	Failed to reach escape velocity
Pioneer 2	U.S.	8 Nov. 1958	Orbiter	Failed to reach escape velocity
[Unnamed]	USSR	4 Dec. 1958	Impact	Launch failure
Pioneer 3	U.S.	6 Dec. 1958	Flyby	Failed to reach escape velocity
Luna 1	USSR	2 Jan. 1959	Impact	Lunar flyby at distance of 6,400 km on 4 Jan. 1959

Name	Country	Launch Date	Purpose	Results
Pioneer 4	U.S.	3 Mar. 1959	Flyby	Lunar flyby 4 Mar. 1959 at distance of 59,545 km
[Unnamed]	USSR	18 Jun. 1959	Impact	Launch failure
Luna 2	USSR	12 Sep. 1959	Impact	Lunar impact
Luna 3	USSR	4 Oct. 1959	Circumlunar	Flyby at distance of 7,900 km on 6 Oct.
Pioneer P-3	U.S.	26 Nov. 1959	Orbiter	Launch failure
[Unnamed]	USSR	15 Apr. 1960	Circumlunar	Failed to reach escape velocity
[Unnamed]	USSR	19 Apr. 1960	Circumlunar	Launch failure
Pioneer P-30	U.S.	25 Sep. 1960	Orbiter	Launch failure
Pioneer P-31	U.S.	15 Dec. 1960	Orbiter	Launch failure
Ranger 3	U.S.	26 Jan. 1962	Impact	Flyby on 28 Jan. at distance of 36,793 km
Ranger 4	U.S.	23 Apr. 1962	Impact	Impact on lunar farside 26 Apr., no data
Ranger 5	U.S.	18 Oct. 1962	Impact	Flyby on 21 Oct. at distance of 724 km
[Unnamed]	USSR	4 Jan. 1963	Lander	Failed to leave Earth orbit
[Unnamed]	USSR	3 Feb. 1963	Lander	Launch failure
Luna 4	USSR	2 Apr. 1963	Lander	Flyby on 6 Apr. at distance of 8,500 km
Kosmos 21	USSR	11 Nov. 1963	Flyby	Failed to leave Earth orbit
Ranger 6	U.S.	30 Jan. 1964	Impact	Impact 2 Feb., but no data returned
[Unnamed]	USSR	21 Mar. 1964	Lander	Failed to reach Earth orbit
[Unnamed]	USSR	20 Apr. 1964	Lander	Launch failure
Ranger 7	U.S.	28 Jul. 1964	Impact	Impact 31 Jul. Returned 4,316 pictures
Ranger 8	U.S.	17 Feb. 1965	Impact	Impact 20 Feb. Returned 7,137 pictures
Ranger 9	U.S.	21 Mar. 1965	Impact	Impact 24 Mar. Returned 5,814 pictures
[Unnamed]	USSR	10 Apr. 1965	Lander	Failed to reach Earth orbit
Luna 5	USSR	9 May 1965	Lander	Crashed on lunar surface
Luna 6	USSR	8 Jun. 1965	Lander	Flyby on 11 Jun. at distance of 161,000 km
Zond 3	USSR	18 Jul. 1965	Flyby	Flyby 20 Jul. at distance of 9,220 km
Luna 7	USSR	4 Oct. 1965	Lander	Crashed on lunar surface
Luna 8	USSR	3 Dec. 1965	Lander	Crashed on lunar surface
Luna 9	USSR	31 Jan. 1966	Lander	Landed on Ocean of Storms 3 Feb. Operated until 6 Feb.
Kosmos 111	USSR	1 Mar. 1966	Orbiter	Failed to leave Earth orbit
Luna 10	USSR	31 Mar. 1966	Orbiter	Arrived 3 Apr., operated until 30 May
Surveyor 1	U.S.	30 May 1966	Lander	Landed on Ocean of Storms 2 Jun. Operated until 7 Jan. 1967
Explorer 33	U.S.	1 Jul. 1966	Orbiter	Entered eccentric Earth orbit

Name	Country	Launch Date	Purpose	Results
Lunar Orbiter 1	U.S.	10 Aug. 1966	Orbiter	Arrived 14 Aug., deliberately crashed on Moon on 29 Oct.
Luna 11	USSR	24 Aug. 1966	Orbiter	Arrived 27 Aug., operated until 1 Oct.
Surveyor 2	U.S.	20 Sep. 1966	Lander	Crashed on lunar surface
Luna 12	USSR	22 Oct. 1966	Orbiter	Arrived 25 Oct., operated until 19 Jan. 1967
Lunar Orbiter 2	U.S.	6 Nov. 1966	Orbiter	Arrived 10 Nov., deliberately crashed on Moon on 11 Oct. 1967
Luna 13	USSR	21 Dec. 1966	Lander	Landed on Ocean of Storms 24 Dec. Operated until 28 Dec.
Lunar Orbiter 3	U.S.	5 Feb. 1967	Orbiter	Arrived 7 Feb., deliberately crashed on lunar surface 9 Oct.
Surveyor 3	U.S.	17 Apr. 1967	Lander	Landed on Ocean of Storms 20 Apr. Operated until 4 May.
Lunar Orbiter 4	U.S.	4 May 1967	Orbiter	Arrived 8 May, operated until 17 Jul.
Surveyor 4	U.S.	14 Jul. 1967	Lander	Contact lost during descent
Lunar Orbiter 5	U.S.	1 Aug. 1967	Orbiter	Arrived 5 Aug., deliberately crashed on Moon 31 Jan. 1968
Surveyor 5	U.S.	8 Sep. 1967	Lander	Landed on Sea of Tranquillity 11 Sep. Operated until 16 Dec. 1967
Surveyor 6	U.S.	7 Nov. 1967	Lander	Landed on Central Bay 10 Nov. Operated until 24 Nov.
Surveyor 7	U.S.	7 Jan. 1968	Lander	Landed near Tycho 10 Jan. Operated until 21 Feb.
Luna 14	USSR	7 Apr. 1968	Orbiter	Arrived 10 Apr. Operated until Jun/Jul.
Zond 5	USSR	14 Sep. 1968	Circumlunar	Flyby 18 Sep. at distance of 1,950 km
Zond 6	USSR	10 Nov. 1968	Circumlunar	Flyby 14 Nov. at distance of 2,420 km
Apollo 8	U.S.	21 Dec. 1968	Orbiter	Arrived 24 Dec., returned to Earth 25 Dec. Crew: Borman, Lovell, Anders
[Unnamed]	USSR	19 Feb. 1969	Lander/rover	Launch failure
[Unnamed]	USSR	14 Jun. 1969	Sample return	Failed to reach Earth orbit
Apollo 10	U.S.	18 May 1969	Orbiter	Arrived 21 May, returned to Earth 24 May. Crew: Stafford, Young, Cernan
Luna 15	USSR	13 Jul. 1969	Sample return	Crashed on lunar surface
Apollo 11	U.S.	16 Jul. 1969	Orbiter/lander/ sample return	Arrived 19 Jul., returned to Earth 21 Jul. Landed on Sea of Tranquillity 20 Jul. Crew: Armstrong, Aldrin, Collins
Zond 7	USSR	7 Aug. 1969	Circumlunar	Flyby 10 Aug. at distance of 1,200 km

Name	Country	Launch Date	Purpose	Results
Kosmos 300	USSR	23 Sep. 1969	Sample return	Failed to leave Earth orbit
Kosmos 305	USSR	22 Oct. 1969	Sample return	Failed to reach Earth orbit
Apollo 12	U.S.	14 Nov. 1969	Orbiter/lander/ sample return	Arrived 18 Nov., returned to Earth 21 Nov. Landed on Ocean of Storms 19 Nov. Crew: Conrad, Bean, Gordon
[Unnamed]	USSR	6 Feb. 1970	Sample return	Failed to reach Earth orbit
Apollo 13	U.S.	11 Apr. 1970	Orbiter/lander/ sample return	Flyby 14 Apr. Crew: Lovell, Haise, Swigert
Luna 16	USSR	12 Sep. 1970	Sample return	Landed on Sea of Fertility 20 Sep., returned to Earth 21 Sep. with 101 g (3.5 oz) of soil
Zond 8	USSR	20 Oct. 1970	Circumlunar	Flyby 24 Oct. at distance of 1,200 km
Luna 17 / Lunokhod 1	USSR	10 Nov. 1970	Lander/rover	Landed 17 Nov. Rover operated until 14 Sep. 1971, travelling 10.54 km
Apollo 14	U.S.	31 Jan. 1971	Orbiter/lander/ sample return	Arrived 4 Feb., returned to Earth 6 Feb. Landed at Fra Mauro 5 Feb. Crew: Shepard, Roosa, Mitchell
Apollo 15	U.S.	26 Jul. 1971	Orbiter/ subsatellite/ lander/rover/ sample return	Arrived 29 Jul., returned to Earth 4 Aug. Landed at Hadley-Apennine 30 Jul. Released subsatellite 4 Aug. Crew: Scott, Irwin, Worden
Luna 18	USSR	2 Sep. 1971	Sample return	Contact lost during descent
Luna 19	USSR	28 Sep. 1971	Orbiter	Arrived 2 Oct., operated until Oct. 1972
Luna 20	USSR	14 Feb. 1972	Sample return	Landed on Sea of Fertility 21 Feb., returned to Earth 22 Feb. with 55 g (2 oz) of soil
Apollo 16	U.S.	16 Apr. 1972	Orbiter/ subsatellite/ lander/rover/ sample return	Arrived 19 Apr., returned to Earth 24 Apr. Landed at Descartes 20 Apr. Released subsatellite 24 Apr. Crew: Young, Duke, Mattingley
Apollo 17	U.S.	7 Dec. 1972	Orbiter/lander/ rover/sample return	Arrived 10 Dec., returned to Earth 14 Dec. Landed at Taurus-Littrow 11 Dec. Crew: Cernan, Schmitt, Evans
Luna 21/ Lunokhod 2	USSR	8 Jan. 1973	Lander / rover	Landed 15 Jan. Rover operated until 9 May, travelling 37 km (23 ml)
Explorer 49	U.S.	10 Jun. 1973	Orbiter	Arrived 15 Jun. No lunar data – conducted studies of radio emissions from distant sources. Operated until Jun. 1975
Luna 22	USSR	29 May 1974	Orbiter	Arrived 2 Jun., operated until 2 Sep. 1975
Luna 23	USSR	28 Oct. 1974	Sample return	Damaged during landing on 6 Nov. Contact lost 9 Nov.

Name	Country	Launch Date	Purpose	Results
[Unnamed]	USSR	16 Oct. 1975	Sample return	Failed to reach Earth orbit
Luna 24	USSR	9 Aug. 1976	Sample return	Landed on Sea of Crises 18 Aug., returned to Earth 19 Aug. with 170 g of soil
Galileo	U.S.	18 Oct. 1989	Venus, Earth and asteroid flybys, Jupiter orbiter/probe	Moon flyby 8 Dec. 1990 at distance of 565,000 km; Moon flyby 8 Dec. 1992 at distance of 110,000 km
Hiten/Hagomoro	Japan	24 Jan. 1990	Circumlunar/ orbiter	First flyby 19 Mar. at distance of 16,742 km; orbiter released but contact lost
Clementine	U.S.	25 Jan. 1994	Lunar orbiter/ asteroid flyby	Arrived at Moon 19 Feb., operated until 3 May
Cassini-Huygens	U.S./ESA	15 Oct. 1997	Venus, Earth and asteroid flybys, Saturn orbiter, Titan probe/ lander	Moon flyby 18 Aug. 1999 at distance of 377,000 km
Lunar Prospector	U.S.	7 Jan. 1998	Orbiter	Arrived 11 Jan. Deliberately crashed on Moon 31 Jul. 1999
Nozomi	Japan	3 Jul. 1998	Earth–Moon flybys, Mars orbiter	Moon flyby 24 Sep. 1998 at distance of 2,809 km; 2nd flyby 18 Dec. 1998 at distance of 1,003 km
SMART-1	ESA	27 Sep. 2003	Orbiter	Arrived 15 Nov. 2004; Deliberately crashed on Moon 3 Sep. 2006
Kaguya (SELENE)	Japan	14 Sept. 2007	Orbiter/2 subsatellites	Arrived 3 Oct. Released subsatellites on 9 and 12 Oct. Deliberately crashed on Moon 11 Jun. 2009.
Chandrayaan-1	India	22 Oct. 2008	Orbiter / Moon impactor	Arrived 8 Nov. Moon Impact Probe hit surface on 14 Nov. Radio contact lost on 28 Aug. 2009.
Chang'e-1	China	24 Oct. 2007	Orbiter	Arrived 5 Nov. Deliberately crashed on Moon 1 Mar. 2009.
Lunar Reconnaissance Orbiter	U.S.	18 June 2009	Orbiter	Arrived 23 Jun. 2009.
LCROSS	U.S.	18 June 2009	Impactor	Impacted Cabeus crater on 9 Oct.
Chang'e-2	China	1 October 2010	Orbiter	Arrived 9 Oct. Left lunar orbit 9 June 2011.
GRAIL (Gravity Recovery and Interior Laboratory)	U.S.	10 Sep. 2011	Two orbiters	Arrived 31 Dec. 2011 and 1 Jan. 2012. Mission ended 14 Dec. 2012.
LADEE (Lunar Atmosphere and Dust Environment Explorer)	U.S.	7 Sep. 2013	Orbiter	Arrived 20 Nov. Crashed into the Moon on 18 Apr. 2014.
Chang'e-3/Yutu	China	1 Dec. 2013	Lander/rover	Entered lunar orbit on 6 Dec. Landed on Sea of Showers on 13 Dec. Operations ended 3 Aug. 2016.

Name	Country	Launch Date	Purpose	Results
Chang'e-4/Yutu-2	China	7 Dec. 2018	Lander/rover	First landing on lunar farside, in Von Karman crater, on 3 Jan. 2019; data returned via Queqiao data relay satellite.
Beresheet	Israel	21 Feb. 2019	Lander	First private mission. Crash landed on Mare Serenitatis on 11 Apr. 2019.
Chandrayaan-2	India	22 July 2019	Orbiter, lander/ rover	Entered lunar orbit 20 Aug. Vikram lander crashed on surface on 6 Sep. 2019.
Chang'e-5	China	2020?	Sample return	
Korean Pathfinder Lunar Orbiter	S. Korea	2020?	Orbiter	

Mercury

Name	Country	Launch Date	Purpose	Results
Mariner 10	USA	3 Nov. 1973	Venus/Mercury flybys	Mercury flybys 29 Mar. 1974, 21 Sep. 1974, 16 Mar. 1975
MESSENGER	USA	3 Aug. 2004	Venus flyby, Mercury flybys/ orbiter	Mercury flybys 14 Jan. 2008, 6 Oct. 2008 and 29 Sept. 2009. Entered orbit 17 Mar. 2011. Deliberately crashed into planet 30 Apr. 2015.
BepiColombo	ESA/Japan	20 Oct. 2018	2 orbiters (ESA and Japan)	Arrival planned in 2025.

Venus

Name	Country	Launch Date	Purpose	Results
Sputnik 7	USSR	4 Feb. 1961	Impact	Did not reach Earth orbit
Venera 1	USSR	12 Feb. 1961	Impact	Contact lost 22 Feb. 1961
Mariner 1	U.S.	22 Jul. 1962	Flyby	Did not reach Earth orbit
Sputnik 19	USSR	25 Aug. 1962	Impact	Failed to leave Earth orbit
Mariner 2	U.S.	27 Aug. 1962	Flyby	Flyby 14 Dec. 1962.
Sputnik 20	USSR	1 Sep. 1962	Flyby	Failed to leave Earth orbit
Sputnik 21	USSR	12 Sep. 1962	Flyby	Destroyed in Earth orbit
[Unnamed]	USSR	19 Feb. 1964	Flyby	Did not reach Earth orbit
[Unnamed]	USSR	1 Mar. 1964	Flyby	Did not reach Earth orbit
Kosmos 27	USSR	27 Mar. 1964	Atmospheric probe	Failed to leave Earth orbit

Name	Country	Launch Date	Purpose	Results
Zond 1	USSR	2 Apr. 1964	Atmospheric probe	Contact lost 25 May
Venera 2	USSR	12 Nov. 1965	Flyby	Flyby 27 Feb. 1966, no return of data
Venera 3	USSR	16 Nov. 1965	Atmospheric probe	Arrived 1 Mar. 1966, no return of data
Kosmos 96	USSR	23 Nov. 1965	Flyby	Failed to leave Earth orbit
Venera 4	USSR	12 Jun. 1967	Atmospheric probe	Arrived 18 Oct. 1967, sent back data for 93 min. during descent
Mariner 5	U.S.	14 Jun. 1967	Flyby	Flyby 19 Oct.
Kosmos 167	USSR	17 Jun. 1967	Atmospheric probe	Failed to leave Earth orbit
Venera 5	USSR	5 January 1969	Atmospheric probe	Arrived 16 May 1969, sent back data for 53 min. during descent
Venera 6	USSR	10 Jan. 1969	Atmospheric probe	Arrived 17 May 1969, sent back data for 51 min. during descent
Venera 7	USSR	17 Aug. 1970	Atmospheric probe/lander	Landed on 15 Dec. 1970, sent back data from surface for 23 min.
Kosmos 359	USSR	22 Aug. 1970	Atmospheric probe/lander	Failed to leave Earth orbit
Venera 8	USSR	27 Mar. 1972	Atmospheric probe/lander	Landed on 22 Jul. 1972, sent back data from surface for 50 min.
Kosmos 482	USSR	31 Mar. 1972	Atmospheric probe/lander	Failed to leave Earth orbit
Mariner 10	U.S.	4 Nov. 1973	Venus/Mercury flybys	Venus flyby 5 Feb. 1974
Venera 9	USSR	8 Jun. 1975	Orbiter/lander	Arrived 22 Oct. 1975. Lander sent back data from surface for 53 min, including first photo
Venera 10	USSR	14 Jun. 1975	Orbiter/lander	Arrived 25 Oct. 1975. Lander sent back data from surface for 65 min., including a photo
Pioneer Venus 1	U.S.	20 May 1978	Orbiter	Arrived 4 Dec. 1978, operated until 8 Oct. 1992
Pioneer Venus 2	U.S.	8 Aug. 1978	Main bus/4 atmospheric probes	Arrived 9 Dec. 1978. Day probe sent back data from surface for 67 min.
Venera 11	USSR	9 Sep. 1978	Orbiter/lander	Arrived 25 Dec. 1978. Lander sent back data from surface for 95 min.
Venera 12	USSR	14 Sep. 1978	Orbiter/lander	Arrived 9 Dec. 1978. Lander sent back data from surface for 110 min.
Venera 13	USSR	30 Oct. 1981	Flyby/lander	Arrived 1 Mar. 1982. Lander sent back data from surface for 127 min., including 8 colour photos

Name	Country	Launch Date	Purpose	Results
Venera 14	USSR	4 Nov. 1981	Flyby/lander	Arrived 3 Mar. 1982. Lander sent back data from surface for 57 min., including 8 colour photos
Venera 15	USSR	2 Jun. 1983	Orbiter	Arrived 10 Oct. 1983. Mapped N. hemisphere with radar until 10 Jul. 1984
Venera 16	USSR	7 Jun. 1983	Orbiter	Arrived 14 Oct. 1983. Mapped N. hemisphere with radar until 10 Jul. 1984
Vega 1	USSR	15 Dec. 1984	Flyby/lander/ balloon; comet Halley flyby	Venus flyby and landing 11 Jun. 1985. Lander sent back 56 min. of data. Balloon operated until 13 Jun.
Vega 2	USSR	21 Dec. 1984	Flyby/lander/ balloon; comet Halley flyby	Venus flyby and landing 15 Jun. 1985. Lander sent back 57 min. of data. Balloon operated until 17 Jun.
Magellan	U.S.	4 May 1989	Orbiter	Arrived 10 Aug. 1990. Mapped planet with radar, operational until 12 Oct. 1994.
Galileo	U.S.	18 Oct. 1989	Venus, Earth and asteroid flybys, Jupiter orbiter/ probe	Venus flyby 10 Feb. 1990.
Cassini-Huygens	U.S./ESA	15 Oct. 1997	Venus, Earth and asteroid flybys, Saturn orbiter, Titan probe/ lander	Venus flybys 26 Apr. 1998 and 24 Jun. 1999.
MESSENGER	U.S.	3 Aug. 2004	Venus flyby/ Mercury orbiter	Venus flybys 24 Oct. 2006 and 5 Jun. 2007.
Venus Express	ESA	9 Nov. 2005	Orbiter	Arrived 11 Apr. 2006. Mission ended 16 Dec. 2014, followed by atmospheric entry.
Akatsuki	Japan	21 May 2010	Orbiter	Arrived in orbit 7 Dec. 2015.

Mars

Name	Country	Launch Date	Purpose	Results
[Unnamed]	USSR	10 Oct. 1960	Flyby	Did not reach Earth orbit
[Unnamed]	USSR	14 Oct. 1960	Flyby	Did not reach Earth orbit
[Unnamed]	USSR	24 Oct. 1962	Flyby	Achieved Earth orbit only
Mars 1	USSR	1 Nov. 1962	Flyby	Radio failed at 106 million km (65.9 million ml)
[Unnamed]	USSR	4 Nov. 1962	Flyby	Achieved Earth orbit only

Name	Country	Launch Date	Purpose	Results
Mariner 3	U.S.	5 Nov. 1964	Flyby	Launcher shroud failed to jettison
Mariner 4	U.S.	28 Nov. 1964	Flyby	First successful Mars flyby 14 Jul. 1965, returned 21 photos
Zond 2	USSR	30 Nov. 1964	Flyby	Passed Mars but radio failed, returned no planetary data
Mariner 6	U.S.	24 Feb. 1969	Flyby	Mars flyby 31 Jul. 1969, returned 75 photos
Mariner 7	U.S.	27 Mar. 1969	Flyby	Mars flyby 5 Aug. 1969, returned 126 photos
[Unnamed]	USSR	27 Mar. 1969	Orbiter	Did not reach Earth orbit
[Unnamed]	USSR	2 Apr. 1969	Orbiter	Did not reach Earth orbit
Mariner 8	U.S.	8 May 1971	Orbiter	Failed during launch
Kosmos 419	USSR	10 May 1971	Orbiter	Achieved Earth orbit only
Mars 2	USSR	19 May 1971	Orbiter/lander	Arrived 27 Nov. 1971, no useful data, lander burned up due to steep entry
Mars 3	USSR	28 May 1971	Orbiter/lander	Arrived 3 Dec. 1971, lander operated on surface for 20 seconds before failing
Mariner 9	U.S.	30 May 1971	Orbiter	Arrived 13 Nov. 1971. End of mission 27 Oct. 1972. Returned 7,329 photos
Mars 4	USSR	21 July 1973	Orbiter	Flew past Mars 10 Feb. 1974
Mars 5	USSR	25 July 1973	Orbiter	Arrived 12 Feb. 1974, lasted a few days. Returned 43 photos
Mars 6	USSR	5 Aug. 1973	Flyby module/lander	Arrived 12 Mar. 1974, lander failed due to fast impact
Mars 7	USSR	9 Aug. 1973	Flyby module/lander	Arrived 9 Mar. 1974, lander missed the planet
Viking 1	U.S.	20 Aug. 1975	Orbiter/lander	Arrived 19 Jun. 1976. Orbiter operated until 7 Aug. 1980. Lander operated on surface 20 Jul. 1976–13 Nov. 1982
Viking 2	U.S.	9 Sep. 1975	Orbiter/lander	Arrived 7 Aug. 1976. Orbiter operated until 24 Jul. 1978. Lander operated on surface 3 Sep. 1976–11 Apr. 1980. Viking orbiters and landers returned 50,000+ photos
Phobos 1	USSR	7 Jul. 1988	Mars/Phobos orbiter/lander	Contact lost 29 Aug. 1988 en route to Mars
Phobos 2	USSR	12 Jul. 1988	Mars/Phobos orbiter/lander	Contact lost 27 Mar. 1989 near Phobos. Returned photos of Mars and Phobos
Mars Observer	U.S.	25 Sep. 1992	Orbiter	Contact lost just before Mars arrival 21 Aug. 1993
Mars Global Surveyor	U.S.	7 Nov. 1996	Orbiter	Arrived 12 Sep. 1997. Contact lost 2 Nov. 2006

Name	Country	Launch Date	Purpose	Results
Mars 96	Russia	16 Nov. 1996	Orbiter and landers	Did not reach Earth orbit
Mars Pathfinder	U.S.	14 Dec. 1996	Lander/rover	Landed 4 Jul. 1997, operated until 27 Sep. 1997
Nozomi	Japan	4 Jul. 1998	Earth–Moon flybys, Mars orbiter	Mars flyby 14 Dec. 2003 at distance of 1,000 km. No Mars data returned
Mars Climate Orbiter	U.S.	11 Dec. 1998	Orbiter	Contact lost on arrival, 23 Sep. 1999
Mars Polar Lander/ Deep Space 2	U.S.	3 Jan. 1999	Lander/ penetrators	Contact lost on arrival 3 Dec. 1999
Mars Odyssey	U.S.	7 Mar. 2001	Orbiter	Arrived 24 Oct. 2001, still operational
Mars Express/ Beagle 2	ESA	2 Jun. 2003	Orbiter/lander	Orbiter arrived 25 Dec. 2003, still operational. Lander lost during entry 25 Dec. 2003
Mars Exploration Rover A (Spirit)	U.S.	10 Jun. 2003	Lander/rover	Landed 4 Jan. 2004, lost contact 22 Mar. 2010
Mars Exploration Rover B (Opportunity)	U.S.	7 Jul. 2003	Lander/rover	Landed 25 Jan. 2004, lost contact 10 June 2018
Mars Reconnaissance Orbiter	U.S.	12 Aug. 2005	Orbiter	Arrived 10 Mar. 2006, still operational.
Rosetta/Philae	ESA	24 Feb. 2004	Earth, Mars and asteroid flybys; comet orbiter/ lander	Flyby 25 Feb. 2007
Phoenix	U.S.	4 Aug. 2007	Lander	Arrived 25 May 2008, contact lost 2 Nov. 2008
Dawn	U.S.	27 Sep. 2007	Dual asteroid orbiter	Flyby 17 Feb. 2009
Phobos-Grunt/ Yinghuo-1	Russia/China	9 Nov. 2011	Phobos lander and sample return/orbiter	Failed to leave Earth orbit
Mars Science Laboratory/ Curiosity	U.S.	26 Nov. 2011	Lander / rover	Landed in Gale crater 6 Aug. 2012
Mars Orbiter Mission (Mangalyaan)	India	5 Nov. 2013	Orbiter	Arrived 24 Sep. 2014.
MAVEN (Mars Atmosphere and Volatile Evolution)	U.S.	18 Nov. 2013	Orbiter	Arrived 22 Sep. 2014.
InSight	U.S.	5 May 2018	Lander	Landed on Elysium Planitia 26 Nov. 2018
Hope	U.A.E.	July 2020	Orbiter	Arrival planned in 2021.
Mars 2020	U.S.	July 2020	Lander/rover	Landing planned in Jezero crater on 18 Feb. 2021

Asteroids

Name	Country	Launch Date	Purpose	Results
Galileo	U.S.	18 Oct. 1989	Venus, Earth and asteroid flybys, Jupiter orbiter/ probe	Flyby of Gaspra 29 Oct. 1991; flyby of Ida and Dactyl 28 Aug. 1993
Clementine	U.S.	25 Jan. 1994	Lunar orbiter/ asteroid flyby	Mission to Geographos cancelled 3 May 1994 after thruster malfunction.
NEAR-Shoemaker	U.S.	17 Feb. 1996	Earth and asteroid flybys, Eros orbiter	Mathilde flyby 27 Jun. 1997; Eros flyby 23 Dec. 1998. Orbit around Eros 14 Feb. 2000. Landed on Eros 12 Feb. 2001.
Cassini-Huygens	U.S./ESA	15 Oct. 1997	Venus, Earth and asteroid flybys, Saturn orbiter, Titan probe/ lander	Flyby of Masursky 23 Jan. 2000.
Deep Space 1	U.S.	24 Oct. 1998	Asteroid and comet flybys	Flyby of Braille 29 Jul. 1999.
Stardust	U.S.	7 Feb. 1999	Earth, asteroid and comet flybys/ comet sample return	Flyby of Anne Frank 2 Nov. 2002.
Hayabusa	Japan	9 May 2003	Orbiter/mobile landers/sample return	Arrival at Itokowa 12 Sep. 2005; sample return to Earth 13 Jun. 2010.
Rosetta	ESA	24 Feb. 2004	Earth, Mars and asteroid flybys; comet orbiter/ lander	Flyby of Steins 5 Sep. 2008; flyby of Lutetia 10 Jul. 2010.
Dawn	U.S.	27 Sep. 2007	Dual asteroid orbiter	Arrival at Vesta 15 Jul. 2011. Departed 5 Sep. 2012. Arrival at Ceres 6 Mar. 2015. Mission ended 31 Oct. 2018.
Hayabusa 2	Japan	3 Dec. 2014	Orbiter/four mobile landers/ sample return	Arrival at 1999 JU3 (Ryugu) 27 Jun. 2018. Deployed Minerva lander capsule with two rover/hoppers, Minerva 1a and 1b on 21 Sep. 2018, MASCOT lander/hopper on 3 Oct. 2018. Three sample collection touchdowns planned. Sample return to Earth in 2020.
OSIRIS-REx	U.S.	8 Sep. 2016	Earth flyby, asteroid orbiter/ sample return	Arrival at Bennu 3 Dec. 2018.

Jupiter

Name	Country	Launch Date	Purpose	Results
Pioneer 10	U.S.	2 Mar. 1972	Flyby	Jupiter flyby 4 Dec. 1973
Pioneer 11	U.S.	6 Apr. 1973	Jupiter and Saturn flybys	Jupiter flyby 1 Sep. 1979
Voyager 2	U.S.	20 Aug. 1977	Jupiter, Saturn, Uranus, and Neptune flybys	Jupiter flyby 9 July 1979
Voyager 1	U.S.	5 Sep. 1977	Jupiter and Saturn flybys	Jupiter flyby 5 Mar. 1979
Galileo	U.S.	18 Oct. 1989	Venus, Earth and asteroid flybys, Jupiter orbiter/ atmospheric probe	Orbiter arrived 8 Dec. 1995, operational until atmospheric entry 21 Sep. 2003. Probe entry 7 Dec. 1995.
Ulysses	ESA/U.S.	6 Oct. 1990	Solar orbiter/ Jupiter flybys	Jupiter flyby 8 Feb. 1992 at distance of 378,400 km; distant flyby 5 Feb. 2004
Cassini-Huygens	U.S./ESA	15 Oct. 1997	Venus, Earth and asteroid flybys, Saturn orbiter, Titan probe/lander	Jupiter flyby 30 Dec. 2000 at distance of 9.7 million km
New Horizons	U.S.	19 Jan. 2006	Jupiter, Pluto and KBO flybys	Jupiter flyby 28 Feb. 2007
Juno	U.S.	5 Aug. 2011	Jupiter orbiter	Arrived 4 Jul. 2016.
JUICE (Jupiter Icy Moons Explorer)	ESA	June 2022 (planned)	Jupiter orbiter	Arrival planned 2030

Saturn

Name	Country	Launch Date	Purpose	Results
Pioneer 11	U.S.	6 Apr. 1973	Jupiter and Saturn flybys	Saturn flyby 1 Sep. 1979
Voyager 2	U.S.	20 Aug. 1977	Jupiter, Saturn, Uranus, and Neptune flybys	Saturn flyby 22 Aug. 1981
Voyager 1	U.S.	5 Sep. 1977	Jupiter and Saturn flybys	Saturn flyby 12 Nov. 1980
Cassini-Huygens	U.S./ESA	15 Oct. 1997	Venus, Earth and asteroid flybys, Saturn orbiter, Titan probe/lander	Orbiter arrived 1 Jul. 2004. Huygens landing on Titan 14 Jan. 2005. Orbiter's atmospheric entry 15 Sep. 2017.

Uranus

Name	Country	Launch Date	Purpose	Results
Voyager 2	U.S.	20 Aug. 1977	Jupiter, Saturn, Uranus, and Neptune flybys	Uranus flyby 24 Jan. 1986

Neptune

Name	Country	Launch Date	Purpose	Results
Voyager 2	U.S.	20 Aug. 1977	Jupiter, Saturn, Uranus, and Neptune flybys	Neptune flyby 25 Aug. 1989

Pluto/Kuiper Belt

Name	Country	Launch Date	Purpose	Results
New Horizons	U.S.	19 Jan. 2006	Jupiter, Pluto, Kuiper Belt object flybys	Pluto–Charon flyby 14 Jul. 2015. Flyby of KBO 2014-MU69 on 1 Jan. 2019.

Comets

Name	Country	Launch Date	Purpose	Results
International Cometary Explorer	U.S.	12 Aug. 1978	Study of solar wind etc. from L1; comet flyby	Flyby of Giacobini-Zinner 11 Sep. 1985; distant flyby of Halley 28 Mar. 1986
Vega 1	USSR	15 Dec. 1984	Flyby/lander/ balloon; comet Halley flyby	Halley flyby 6 Mar. 1986
Vega 2	USSR	21 Dec. 1984	Flyby/lander/ balloon; comet Halley flyby	Halley flyby 9 Mar. 1986
Sakigake	Japan	7 Jan. 1985	Comet and Earth flybys	Halley flyby 11 Mar. 1986
Giotto	ESA	2 Jul. 1985	Comet and Earth flybys	Halley flyby 14 Mar. 1986; Grigg-Skjellerup flyby 10 Jul. 1992

Name	Country	Launch Date	Purpose	Results
Suisei	Japan	18 Aug. 1985	Comet flyby	Halley flyby 8 Mar. 1986
Deep Space 1	U.S.	24 Oct. 1998	Asteroid and comet flybys	Flyby of Borrelly 22 Sep. 2001
Stardust/NExT	U.S.	7 Feb. 1999	Earth, asteroid, and comet flybys/ comet sample return	Flyby of Wild 2 on 2 Jan. 2004. Returned sample capsule to Earth 15 Jan. 2006. Flyby of Tempel 1 on 14 Feb. 2011
Rosetta/Philae	ESA	24 Feb. 2004	Earth, Mars and asteroid flybys; comet orbiter/ lander	Arrival at Churyumov-Gerasimenko 6 Aug. 2014; Philae landing 12 Nov. 2014. Mission ended 29 Sept. 2016.
Deep Impact/EPOXI	U.S.	12 Jan. 2005	Flyby/impact	Flyby/impact at Tempel 1 on 4 Jul. 2005. Flyby of Hartley 2 on 4 Nov. 2010

Appendix 6: Lunar and Planetary Firsts

Mercury

First flyby:
Mariner 10 (U.S.), 29 March 1974

First orbiter:
MESSENGER (U.S.), 17 March 2011

Venus

First successful flyby:
Mariner 2 (U.S.), 14 December 1962

First impact:
Venera 3 (USSR), 1 March 1966

First successful atmospheric entry:
Venera 4 (USSR), 18 October 1967

First successful soft-landing:
Venera 7 (USSR), 15 December 1970

First surface photos:
Venera 9 (USSR), 22 October 1975

First orbiter:
Venera 9 (USSR), 22 October 1975

First surface colour photos:
Venera 13 (USSR), 1 March 1982

First soil analysis:
Venera 13 (USSR), 1 March 1982

First multi-probe atmospheric entry:
Pioneer Venus 2 (U.S.), 9 December 1978

First balloon flight:
Vega 1 (USSR/France), 11 June 1985

Moon

First impact:
Luna 2 (USSR), 14 September 1959

First flyby:
Luna 3 (USSR), 6 October 1959

First photos of the far side:
Luna 3 (USSR), 6 October 1959

First survivable landing:
Luna 9 (USSR), 3 February 1966

First orbiter:
Luna 10 (USSR), 3 April 1966

First lift-off from the Moon:
Surveyor 6 (U.S.), 17 November 1967

First robotic sample return:
Luna 16 (USSR), 12–21 September 1970

First automated rover:
Lunokhod 1 (USSR), 17 November 1970

First landing on far side:
Chang'e-4 (China), 3 January 2019

First crewed orbital mission:
Apollo 8 (U.S.), 21–27 December 1968

First human landing:
Apollo 11 (U.S.), 20 July 1969

Mars

First successful flyby:
Mariner 4 (U.S.), 15 July 1965

First orbiter:
Mariner 9 (U.S.), 14 November 1971

First impact:
Mars 2 (USSR), 27 November 1971

First surface photos:
Viking 1 (U.S.), 20 July 1976

First successful rover:
Sojourner (U.S.), 5 July 1997

Asteroids

First flyby (Gaspra):
Galileo (U.S.), 29 October 1991

First orbiter (Eros):
NEAR-Shoemaker (U.S.), 14 February 2000

First soft landing (Eros):
NEAR-Shoemaker (U.S.), 12 February 2001

First sample return (Itokawa):
Hayabusa (Japan), 13 June 2010

Jupiter

First flyby:
Pioneer 10 (U.S.), 4 December 1973

First atmospheric entry:
Galileo Probe (U.S.), 7 December 1995

First orbiter:
Galileo Orbiter (U.S.), 8 December 1995

Saturn

First flyby:
Pioneer 11 (U.S.), 1 September 1979

First orbiter:
Cassini (U.S.), 30 June 2004

First landing on Titan:
Huygens (ESA), 14 January 2005

Uranus

First flyby:
Voyager 2 (U.S.), 24 January 1986

Neptune

First flyby:
Voyager 2, (U.S.), 25 August 1989

Kuiper Belt

First flyby (Pluto–Charon):
New Horizons (U.S.), 14 July 2015

Comets

First flyby (Giacobini-Zinner):
ICE / ISEE-3 (U.S.), 11 September 1985

First impact (Tempel 1):
Deep Impact (U.S.), 4 July 2005

First orbiter (Churyumov-Gerasimenko)
Rosetta (ESA), 6 August 2014

First landing (Churyumov-Gerasimenko)
Philae (ESA), 12 November 2014

Glossary

absolute magnitude the brightness a Solar System object would have if placed 1 AU from both the observer and the Sun, and seen with a phase angle of 0° (fully illuminated).

accretion the collection of gas and dust to form large bodies such as stars, planets, and moons.

achondrite a stony meteorite that lacks chondrules.

active region a disturbed area on the Sun where strong magnetic fields emerge from the interior, resulting in prominences, sunspots, flares, and other phenomena.

albedo the proportion of light falling on a body that is reflected.

Alfvén waves magnetic waves that can travel through an electrically conducting fluid or plasma in a magnetic field.

Amor a class of asteroids with perihelion distances between 1.017 and 1.3 AU.

anorthosite a granular, igneous rock composed almost wholly of anorthite, a calcium-rich silicate.

anticyclone a region in a planetary atmosphere where pressure increases toward the center.

aphelion: the point in the orbit of an object at which it is furthest from the Sun.

apogee the point in the orbit of an Earth-orbiting object at which it is furthest from Earth.

Apollo a class of asteroids that cross Earth's orbit, with semimajor axes greater than 1.0 AU and perihelion distances less than 1.017 AU.

apparent magnitude a measure of the apparent brightness of a planet or star. Very bright objects, such as Venus and the Sun have negative magnitudes.

asthenosphere the least rigid region of a planet's mantle, located immediately below the lithosphere.

astronomical unit (AU) 149,597,870 km, which is approximately the average Sun–Earth distance.

Aten a class of asteroids with semimajor axes less than 1.0 AU and aphelion distances greater than 0.983 AU.

aurora a colorful, variable glow in a planetary ionosphere, produced by atoms and ions excited by collisions with energetic particles.

bar a unit of pressure equal to 0.987 atmospheres, the average air pressure on Earth's surface.

barchan a crescent-shaped sand dune.

barycenter the center of mass of a system of celestial bodies, such as the Earth–Moon system.

basalt a volcanic igneous rock which consists primarily of silicon, oxygen, iron, aluminum, and magnesium.

biosphere the zone occupied by living organisms on a planet.

black dwarf a dead star that is no longer hot or luminous.

bow shock a sharp, curved boundary in the solar wind, upstream of a planetary magnetosphere or a cometary/planetary ionosphere, where a shock wave is created when the solar wind suddenly slows and is diverted; marked by heating, compression, and deflection of the solar wind.

breccia a rock composed of broken rock fragments that were fused together by finer-grained material during an impact.

brown dwarf a gaseous object with a mass between 13 and 80 times the mass of Jupiter, which is too low for fusion to occur at its core.

caldera a large volcanic crater, often resulting from the partial collapse of the summit of a shield volcano or an explosive eruption.

Centaur a small body in the outer Solar System that has been scattered inward from the Kuiper belt by gravitational interactions.

chaotic terrain areas of the Martian surface where the ground has collapsed, possibly due to the removal of subsurface groundwater or ice, to form a surface of jumbled blocks and isolated hills.

chasmata deep, steep-sided trenches on Venus which may be examples of giant rift valleys.

Exploring the Solar System, Second Edition. Peter Bond.
© 2020 John Wiley & Sons Ltd. Published 2020 by John Wiley & Sons Ltd.
Companion Website: www.wiley.com/go/Bond-Solar-System2e

chondrite a very common class of stony meteorites that usually contain chondrules; carbonaceous chondrites also include carbon compounds.

chondrules small spherical grains, usually composed of iron, aluminum or magnesium silicates, found in abundance in primitive stony meteorites.

chromosphere the region of the solar atmosphere lying between the photosphere and corona.

clathrate a solid in which molecules of one compound are trapped in cavities in the crystalline lattice of another compound.

coma the spherical, gaseous envelope that surrounds the nucleus of a comet.

composite volcano a volcano composed of both solidified lava and ash.

conjunction an alignment of two objects so that they appear to be in approximately the same position in the sky, as seen from Earth. Mercury and Venus can move directly between Earth and the Sun (**inferior conjunction**) or directly opposite Earth on the far side of the Sun (**superior conjunction**). Planets further than Earth can only come to superior conjunction. This term is also used when planets and the Moon come within a few degrees of each other in the sky.

convection transport of heat energy by vertical atmospheric or fluid motions.

convective zone a region inside the Sun where convection is the dominant process of energy transfer.

core the center of a massive object such as the Sun or a planet.

Coriolis effect the modification of wind direction due to a planet's Coriolis force.

Coriolis force the force that appears to deflect anything moving freely between the pole of a rotating body and its equator (e.g. winds); produces cyclonic and anticyclonic weather patterns on Earth.

corona (1) the Sun's hot, highly ionized atmosphere; (2) a circular formation of ridges enclosing a central area of jumbled relief on Venus.

coronal hole an extended region of low density and temperature in the solar corona.

coronal mass ejection (CME) an outward eruption of billions of tons of plasma from the Sun's corona.

cosmic rays high-energy atomic nuclei (mostly protons) that enter the Solar System from interstellar space.

cosmogony the study of how the universe and objects within it (e.g. the Solar System) were formed.

crater density a measure of the closeness of impact crater distribution on a planetary surface, usually given as the number of craters of a given size per unit area.

crater ray a streak of ejected material extending radially beyond a crater's rim.

crust the outer solid layer of a planet.

cryovolcanism volcanism that occurs in icy materials rather than rock.

current sheet the two-dimensional surface within a magnetosphere that separates magnetic fields of opposite polarities.

cyclone a large atmospheric system characterized by the rapid inward circulation of air around a low-pressure center.

deflation hollow a surface depression caused by removal of loose material by the wind.

depression a mid-latitude storm on Earth associated with a region of low atmospheric pressure.

deuterium a heavy isotope of hydrogen; its nucleus contains one proton and one neutron.

dune an elevated region of sand, usually in a desert or coastal region.

eccentricity (e): One of the six Keplerian orbital elements, it describes the shape of an orbit, in particular, how much an elliptical orbit deviates from a circle. When e = 0, the orbit is a circle. When e is close to 1, it is a very elongated and narrow ellipse. Eccentricity is calculated by dividing the distance from the center to the focus of an ellipse (c) with the semimajor axis of the ellipse (a). The eccentricity is then given by the ratio c/a. The eccentricity of an ellipse must always be less than unity (1).

eclipse an event in which the light from a celestial object is partly or totally cut off by the passage of another object in front of it, e.g. a solar eclipse, when the Moon passes directly in front of the Sun. An eclipse of a star by another object is called an **occultation**.

ecliptic the mean plane of Earth's orbit around the Sun; the Sun's apparent path across the celestial sphere.

ejecta material that was blasted out of an impact crater during its formation.

ellipse a closed, symmetrical curve that has two perpendicular axes, one of which is longer than the other.

equinox one of two dates on which the Sun crosses the celestial equator. At the vernal (or spring) equinox, the Sun crosses the equator from south to north on or near 21 March each year. At the autumnal equinox, the Sun crosses the equator from north to south on or near 23 September. On these dates, the duration of day and night is approximately equal all over Earth (equinox is Latin for "equal nights").

exoplanet/extrasolar planet a planet orbiting a distant star.

exosphere the outermost, tenuous layer of a planet's atmosphere where it merges into interplanetary space, or the entire gaseous envelope of a planetary object where the atmosphere has similar characteristics to an exosphere.

extremophiles microbes that exist in environmentally extreme regions, such as sea ice, deep-sea vents, hot springs, and salt or soda lakes.

facula (plural faculae) (1) a bright region of the Sun's photosphere.

facula (2) – a bright spot on a planet's surface, particularly Ganymede.

feldspar an aluminum-rich silicate mineral found in meteorites and other rocks.

Ferrel cell a mid-latitude circulation in Earth's atmosphere in which air near the surface moves poleward, then rises and returns toward lower latitudes.

filament a solar prominence projected against the bright disk of the Sun (i.e. seen from above) which appears as a dark streak across the photosphere.

flare see **solar flare**.

flux the amount of radiation or number of particles arriving over a given period of time.

geocentric theory: a theory in which the Earth lies at the centre of the Universe so that the Sun, Moon, stars and all the other planets move around it.

graben a linear surface depression bounded on its long sides by normal faults and caused by stretching of the crust; commonly known as a rift valley.

granulation a mottled effect in the Sun's photosphere caused by columns of gas rising as a result of convection.

greenhouse effect the trapping of infrared radiation by gases in a planet's atmosphere, resulting in an elevated surface temperature.

habitable zone a region around a star where the temperature is thought to be suitable for the evolution of Earth-like life on a planetary body.

Hadley cell circulation in a planet's atmosphere in which gas heated by the Sun rises and moves poleward, then sinks and returns toward the equator.

heliocentric theory a theory in which the Sun lies at the centre of the Solar System, so that all of the planets move around it.

heliopause the outer edge of the heliosphere, where the pressure of the solar wind equals that of the interstellar medium.

helioseismology the study of the Sun's interior by analysis of oscillations in the solar convective zone.

heliosphere the region of space dominated by the Sun's magnetic field and the solar wind.

horst a block of crust that has been uplifted between parallel faults, sometimes known as a block mountain.

hot spot site above a plume of hot, rising material in the mantle which may result in the formation of volcanoes at the surface.

hydrocarbon a chemical compound made up of hydrogen and carbon atoms.

inclination the angle between an object's orbital plane and a reference plane, usually the ecliptic (for heliocentric orbits) or a planet's equator (for satellites).

inferior conjunction when a planet lies directly between Earth and the Sun.

insolation incoming solar radiation.

ion an atom that has gained or lost one or more electrons so that it has become electrically charged.

ionopause the lower boundary of the ionosphere.

ionosphere the upper region of a planet's atmosphere in which many atoms are ionized.

irregular satellite a satellite whose orbit is highly inclined and/or highly eccentric.

isostatic uplift a very slow local uplift of crust due to the removal of surface material, e.g. ice sheets.

jet stream a narrow ribbon of high-speed wind circulating around a planet at high altitude.

Kirkwood gaps empty regions in the asteroid belt where the orbital periods of the asteroids are simple fractions of the orbital period of Jupiter.

KREEP acronym for lunar basaltic material rich in radioactive elements; K for potassium, REE for rare-earth elements, P for phosphorus.

Kuiper belt a disk-shaped region beyond the orbit of Neptune that contains millions of icy objects; the source region for short period comets.

Lagrangian points five points in the orbital plane of two massive bodies where a third body can remain in equilibrium, so that all three bodies have a fixed geometrical configuration. Two Lagrangian points (L4 and L5) form equilateral triangles with the two primaries and are stable; the other three are unstable and lie on the line connecting the two primaries.

latent heat heat released or absorbed as the result of a phase change or change of physical state, e.g. from water vapor (a gas) to liquid water.

libration a small oscillation, or apparent rocking motion, exhibited by a synchronously rotating satellite in a slightly eccentric orbit, e.g. the Moon.

lithosphere rigid outer layer of a planet, including the crust and part of the upper mantle.

magma hot, fluid rock material beneath the crust from which igneous rock is formed.

magnetic reconnection the breaking and reconnection of magnetic field lines which results in the release of magnetic energy.

magnetic substorm a sudden instability of the plasma in Earth's magnetic field which results from a build up of energy deposited by the solar wind. Such events cause a dramatic increase in the intensity and activity of the aurora.

magnetopause the outer boundary of a planet's magnetosphere.

magnetosheath the region between a planetary bow shock and magnetopause in which the solar wind plasma flows around the magnetosphere.

magnetosphere the region of space surrounding a planet in which the planet's magnetic field dominates that of the solar wind.

magnetotail the portion of a planetary magnetosphere pulled downstream by the solar wind.

mantle the layer inside a planet between its crust and core.

mare an area on the Moon or Mars that appears darker and smoother than its surroundings; plural "maria."

mascon subsurface mass concentration that causes large-scale gravity anomalies on the Moon and other bodies.

mesosphere an atmospheric region between the stratosphere and the thermosphere.

meteor a streak of light in the sky caused when a particle of interplanetary dust burns up in the upper atmosphere; commonly known as a "shooting star."

meteorite a natural rocky/metallic object that survives its fall to Earth from space.

meteoroid a small natural body in orbit around the Sun; may be cometary or asteroidal in origin.

meteor shower: a period of enhanced meteor activity caused by a planet's passage (usually at the same time each year) through the particles distributed along a comet's orbit.

micrometeorite a very small meteorite or meteoritic particle less than a millimeter in diameter.

nanoflares small explosions on the Sun.

near-Earth object (NEO) an asteroid or comet whose orbit sometimes brings it close to Earth.

nebula a large region of interstellar gas and dust. (The word "nebula" is Latin for "cloud.")

nebular hypothesis the theory that the Solar System formed from a nebula around the Sun.

neutrino a neutral relativistic particle of very small rest mass.

node one of two points on the celestial sphere where an orbit crosses a reference plane, e.g. the equatorial plane.

nucleus the solid core of a comet, consisting of a mixture of ices and solid silicate and carbonaceous grains.

nuée ardente a hot, glowing cloud of gas and pulverized rock which spreads sideways from an erupting volcano.

oblateness the equatorial distortion of an otherwise spherical planet, caused by its rotation; defined as the difference between the equatorial and polar radii, divided by the equatorial radius.

obliquity angle between a planet's equatorial plane and orbital plane; the tilt of the rotational axis. It causes the planet to have seasons.

occultation:
the passage of an object in front of another object with a smaller apparent angular size, e.g. when an asteroid or planet temporarily blocks a star from view.

olivine a green, metal-rich silicate mineral found in meteorites and other rocks.

Oort cloud a roughly spherical region, extending more than 100,000 AU from the Sun, occupied by up to a trillion small icy bodies; the source region for long-period comets.

opposition the time when a planet orbiting beyond Earth is directly opposite the Sun in the sky, i.e. when its elongation is 180 degrees.

orbit the path followed by a small object located within the gravitational field of a larger object.

organic a complex compound containing carbon, though not necessarily associated with life.

outflow channels large channels that have few if any tributaries and are believed to have been formed by huge floods.

ozone a form of oxygen that contains three atoms instead of the normal two.

ozone hole a region in the ozone layer of Earth's upper atmosphere where the amount of ozone gas has been depleted by pollution.

palimpsest a roughly circular spot on icy satellites thought to identify a former crater and rim.

parallax the angular distance by which a celestial object appears to be displaced with respect to more distant objects when observed from two widely separated locations. The value of its parallax can be used to calculate an object's distance.

penumbra (1) the lighter, outer area around the dark central region (umbra) of a sunspot.

penumbra (2) the outer part of the shadow cast by a celestial body which is illuminated by an extended light source such as the Sun. An observer in the penumbral region of a shadow sees only part of the source obscured, e.g. in a partial solar eclipse.

perigee the point in the orbit of an Earth-orbiting object at which it is closest to Earth.

perihelion the point in the orbit of an object at which it is nearest to the Sun.

permafrost near-surface zone within which temperatures are always below 0°C. It may or may not contain ground ice.

phase the extent to which the surface of an illuminated object, e.g. the Moon, is visible.

phase angle the angle between the direction of illumination (usually the Sun) and a planet or other celestial object as seen by an observer.

photodissociation the breakdown of molecules due to the absorption of light, especially ultraviolet sunlight.

photometry the branch of astronomy concerned with measuring the brightness of a celestial object at various wavelengths.

photosphere the visible surface of the Sun.

photosynthesis a chemical process in organisms, such as plants and some bacteria, that converts carbon dioxide into organic compounds, especially sugars, using the energy from sunlight.

plagioclase a common rock-forming silicate mineral.

planetesimal a primordial body of intermediate size, up to perhaps 1 km across, which accreted into planets or asteroids.

plasma completely ionized gas, consisting of free electrons and atomic nuclei, in which the temperature is too high for neutral atoms to exist.

plasma torus a donut-shaped ring of plasma.

plate see **tectonic plate**.

plate tectonics the movement of segments of the lithosphere under the influence of convection in the mantle.

polarization the process of affecting radiation, especially light, so that its electromagnetic vibrations are not randomly oriented.

polar layered deposits layers of dust and ice which surround and lie beneath the permanent ice caps on Mars.

precession (1) a slow, periodic conical motion of the rotation axis of a spinning body due to external gravitational influences. (2) a gradual rotation of a planet's elliptical orbit, shown by a shift in the alignment of the line joining aphelion and perihelion.

primordial existing at or near the very beginning of the Solar System, about 4.6 billion years ago.

prograde anticlockwise direction of rotation and orbital motion, as seen from above Earth's north pole. The Sun and most of the planets rotate in this direction, and all of the planets in the Solar System have prograde orbits.

prominence name given to large, flame-like structures in the Sun's chromosphere and corona.

protoplanetary disk a disk of gas and dust around a young star where a planetary system may form.

protoplanets the largest planetary embryos which grew from material in the solar nebula and eventually collided to form planets.

radiation pressure the pressure exerted by photons of electromagnetic radiation when they transfer momentum to matter.

radiative zone a region inside the Sun where high-energy photons of radiation collide with electrons and ions and are then reradiated in the form of light and heat.

red giant a large, cool, red star that has grown greatly in size after consuming all of the hydrogen fuel needed for nuclear fusion in its core.

refractory an element that vaporizes at high temperatures, e.g. aluminum, calcium, and uranium.

regolith fragmented rocky debris, produced by numerous meteorite impacts, which forms the upper surface on planets, satellites, and asteroids.

regular satellite a prograde-moving satellite whose orbit has low eccentricity and inclination.

remote sensing any technique for measuring an object's characteristics from a distance.

resonance a state in which one orbiting object is subject to periodic gravitational perturbations by another.

retrograde clockwise direction of rotation and orbital motion, as seen from above Earth's north pole. Retrograde motion is unusual in the planets of the Solar System, but common among comets and smaller, irregular planetary satellites.

rille a trenchlike valley, up to several hundred km long and 1 to 2 km wide, on the surface of the Moon or other satellites.

Roche limit the imaginary boundary around a planet within which a "rubble pile" satellite will be torn apart by gravitational forces. Particles in this region cannot gravitationally coalesce to form a satellite because of the disruptive effect of tidal forces. Located at 2.46 planetary radii for objects with little internal strength, and 1.44 radii for small, rocky satellites.

rock pedestal a rock pinnacle whose base has been mostly eroded, leaving it perched on a narrow support.

saltation a process of particle transport by wind or water which involves a series of hops or leaps. When a particle completes each "hop" it may dislodge other particles and set them in motion.

salt flat a deposit of salt left behind when an inland lake has disappeared due to evaporation.

scarp a cliff produced by faulting or erosion.

secondary crater an impact crater produced by ejecta from a primary impact.

seif dune a type of longitudinal dune, generally oriented in a direction parallel to the prevailing wind.

seismology the science of the study of earthquakes.

semimajor axis (a): the mean orbital distance of an object from its primary; half of the longest axis (diameter) of an ellipse.

shepherd satellite a satellite that sustains the structure of a planetary ring through its close gravitational influence.

shield volcano a broad volcano with a large summit pit (caldera) formed by collapse and gently sloping flanks, built mainly from numerous overlapping flows of fluid, basaltic lava.

sidereal day Earth's rotation period with respect to the background of "fixed" stars (23 hours, 56 minutes, and 4 seconds).

sidereal month the time taken for one revolution of the Moon around Earth, relative to a fixed star.

sidereal period the orbital period of a planet or other celestial body with respect to a particular background star; the true orbital period.

SNC meteorites group of meteorites (Sher-gotty-Nakhla-Chassigny) believed to have come from Mars because of their young ages, basaltic composition, and inclusion of gases with the same composition as the Martian atmosphere.

solar constant the total amount of solar energy irradiating a given surface area in a given interval of time; at 1 AU, this value is 1,367 watts per square meter.

solar day the period of time Earth takes to rotate once with respect to the Sun (24 hours).

solar flare a sudden release of energy in or near the Sun's photosphere that accelerates charged particles into space.

solar nebula the disk-shaped cloud of gas and dust where the Solar System formed (also known as the protosolar nebula or protoplanetary nebula).

solar wind the high-speed outflow of energetic charged particles (mainly protons and electrons) and entrained magnetic field lines from the solar corona.

solstice time when the Sun reaches the most northerly or southerly position in its yearly path along the ecliptic plane. At the solstices the Sun reaches a declination of 23.5 deg N or 23.5 deg S. The summer solstice occurs around 21 June and the winter solstice occurs on or about 22 December.

spectrum the distribution of wavelengths of electromagnetic radiation emitted from or reflected by a body.

spectroscopy the study of the light emitted from or reflected by a body (its spectrum).

spicules spike-like structures in the Sun's chromosphere which may be observed above the solar limb.

spin-orbit coupling a mathematical relationship between an object's orbital period and its period of rotation. The most common example is a 1:1 coupling, which results in the same hemisphere of the Moon always facing toward Earth.

stratosphere the layer of a planet's atmosphere above the troposphere and below the ionosphere; on Earth, its temperature increases with altitude.

stratovolcano a volcano composed of layers of solidified lava and ash.

subduction the process by which by one crustal plate is forced under another.

sunspot a relatively cool, dark area of the solar photosphere.

super-rotation the characteristic of an atmosphere that rotates faster than its planet, e.g. Venus.

superior conjunction the point in the orbit of an inferior planet (Mercury or Venus) when it lies on the opposite side of the Sun to Earth.

synchronous rotation when the axial rotation of a celestial body has the same duration as the orbital period of that body, e.g. the Moon.

synodic month the period between two identical phases of the Moon.

tectonic plate a large slab of a planet's lithosphere with independent motion.

terminator the boundary between night and day.

tesserae areas of very rough, elevated terrain on Venus which show evidence of a complex deformational history – ridges, fractures, and rift valley structures.

tholin a complex hydrocarbon created by the action of ultraviolet radiation on a mixture of volatile compounds containing carbon, hydrogen, nitrogen, and oxygen.

tidal heating the heating of a satellite's interior due to the tidal friction induced by the strong gravitational field of its planet.

thermosphere the atmospheric region where temperature rises due to ionospheric heating.

transform fault a type of crustal fault where slabs of crust or tectonic plates slide sideways past each other, e.g. the San Andreas fault in California.

transit the passage of Mercury or Venus across the visible disk of the Sun; the passage of an exoplanet across the visible disk of a distant star; the passage of a satellite across the disk of its planet; the passage of a star or celestial object across the observer's meridian as the result of the daily apparent motion of the celestial sphere (see **occultation**).

transverse dune a large, asymmetrical, elongated dune lying at right angles to the prevailing wind direction.

tropopause the boundary between the troposphere and the stratosphere.

troposphere the lowest layer of Earth's atmosphere, where most weather takes place, and the convection-dominated region of other planetary atmospheres.

umbra (1) the dark central area of a sunspot.

umbra (2) the darkest, inner part of the shadow cast by a celestial body which is illuminated by an extended light source such as the Sun. An observer in the umbra sees the light source totally obscured, e.g. in a total solar eclipse.

Van Allen radiation belts ring-shaped regions around Earth, discovered by James van Allen in 1958, where high-energy charged particles have been trapped by the planet's magnetic field.

volatiles elements or molecules with low melting temperatures, e.g. water and ammonia.

weathering the *in situ* breakdown of rock by elements of weather – rain, ice, temperature change, etc.

white dwarf a very small, hot stellar remnant created when a star collapses after consuming all of its possible sources of fuel for thermonuclear fusion.

yardang a large area of soft rock which has been eroded by wind action into alternating ridges and furrows, mostly parallel to the dominant wind direction.

zenith the overhead point.

zodiac a belt of 12 constellations through which the Sun's path across the sky – the ecliptic – passes.

zodiacal light a faint glow in the night sky caused by sunlight scattering off interplanetary dust near the plane of the ecliptic.

Further Reading

General

Beatty, J. Kelly, et al., *The New Solar System, 4th Edition*, Sky Publishing Corp./Cambridge Univ. Press, 1999.

Bond, Peter, *Distant Worlds: Milestones in Planetary Exploration*, Copernicus/Praxis, 2007.

Greeley, Ronald, & Batson, Raymond, *The Compact NASA Atlas of the Solar System*, Cambridge Univ. Press, 2001.

Hartmann, William K., *Moons & Planets, 5th Edition*, Thomson, 2005.

Jones, Barrie W., Life in the Solar System and Beyond, Springer-Praxis, 2004.

Jones, Barrie W., *Discovering the Solar System, 2nd Edition*, Wiley, 2007.

Kluger, Jeffrey, *Journey Beyond Selene*, Simon & Schuster, 1999.

Kraemer, Robert S., *Beyond the Moon: Golden Age of Planetary Exploration 1971-1978*, Smithsonian, 2000.

Laughlin, Greg, *Hanging in the Balance*, Sky & Telescope, April 2010.

Logsdon, John M. ed., Exploring the Unknown: Selected Documents in the History of the U.S. Civil Space Program, *Volume V, Exploring the Cosmos*, (NASA SP-4407), 2001.

Morrison, David & Owen, Tobias, *The Planetary System, 3rd Edition*, Addison Wesley, 2003.

Murray, Bruce W., *Journey into Space: The First Thirty Years of Space Exploration*, W. Norton & Co., 1989.

NASA Science Mission Directorate: https://science.nasa.gov/.

NASA Solar System Exploration: http://solarsystem.nasa.gov/.

Nicks, Oran W., Far Travelers: *The Exploring Machines, (NASA SP-480)*, 1985.

Planetary Society: www.planetary.org/.

Sagan, Carl, *Cosmos*, MacDonald & Co., 1980.

Sagan, Carl, *Pale Blue Dot*, Headline, 1995.

Siddiqi, Asif A., *Deep Space Chronicle: A Chronology of Deep Space and Planetary Probes, 1958-2000 (NASA Monograph in Aerospace History #24)*, 2002.

Solar System Exploration: http://solarsystem.nasa.gov/index.cfm.

Stephenson, David J., *Planetary Oceans*, Sky & Telescope, November 2002.

The 8 Planets: http://nineplanets.org/.

Tyson, Peter, *Written in the Star*, Sky & Telescope, October 2017.

Views of the Solar System: http://www.solarviews.com/eng/homepage .htm.

Wilson, Andrew, *Solar System Log*, Jane's, 1987.

Chapter 1: Beginnings

Alibert, Y., et al., *Origin and Formation of Planetary Systems*, Astrobiology, vol. 10, no. 1, p. 19–32, January 2010.

Doody, Dave, *Basics of Interplanetary Flight*: http://www.jpl.nasa.gov/ basics.

Gomes, R., Levison, R.H.F., Tsiganis, K., & Morbidelli A., *Origin of the cataclysmic Late Heavy Bombardment period of the terrestrial planets*, Nature, May 26, 2005a.

Gomes, R., Levison, R.H.F., Tsiganis, K., & Morbidelli A., *Origin of the orbital architecture of the giant planets of the Solar System*, Nature, May 26, 2005b.

Grand Tack Hypothesis: https://en.wikipedia.org/wiki/Grand_tack_ hypothesis.

Jayawardhana, Ray, *Planets in Production: Making New Worlds*, Sky & Telescope, April 2003.

Lakdawalla, Emily, *Pummeling the Planets*, Sky & Telescope, August 2011.

Laughlin, Gregory P., *From Here to Eternity: The Fate of the Sun and the Earth*, Sky & Telescope, June 2007.

Laughlin, Greg, *Hanging in the Balance*, Sky & Telescope, April 2010.

Malhotra, Renu, & Minton David A., A record of planet migration in the main asteroid belt,

Nature, vol. 457, p. 1109–1111, February 26, 2009.

Re-Imagining the Heliosphere special issue: Science, vol. 326, no. 5951, p. 959–971, November 13, 2009.

Schilling, Govert, *From a Swirl of Dust, a Planet Is Born*, Science, vol. 286, p. 66–68, October 1, 1999.

Scott, Edward R.D., *Meteoritical and dynamical constraints on the growth mechanisms and formation times of asteroids and Jupiter*, Icarus, June 2006: http://arxiv.org/ftp/astro-ph/papers/0607/ 0607317.pdf.

Soter, Steven, *What Is a Planet?* Scientific American, January 2007.

Stevenson, David J., *Giant Planets and their Satellites: What are the Relationships Between Their Properties and How They Formed?*

Exploring the Solar System, Second Edition. Peter Bond.
© 2020 John Wiley & Sons Ltd. Published 2020 by John Wiley & Sons Ltd.
Companion Website: www.wiley.com/go/Bond-Solar-System2e

National Academies Press, 1991: http://www.nap.edu/books/0309043336/html/163.html.

Strom, Robert G., et al., *The Origin of Planetary Impactors in the Inner Solar System*, Science, vol. 309, p. 1847–1850, September 6, 2005.

Taylor, G. Jeffrey, *Hit-and-Run as Planets Formed*, PSR Discoveries, November 27, 2006a: http://www.psrd.hawaii.edu/Nov06/hit-and-run.html.

Taylor, G. Jeffrey, *Cosmochemistry from Nanometers to Light-Years*, PSR Discoveries, January 31, 2006b: http://www.psrd.hawaii.edu/Jan06/protoplanetary.html.

Taylor, G. Jeffrey, *Triggering the Formation of the Solar System*, PSR Discoveries, May 21, 2003: http://www.psrd.hawaii.edu/May03/SolarSystemTrigger.html.

Taylor, G. Jeffrey, *Wandering Gas Giants and Lunar Bombardment*, PSR Discoveries, August 24, 2006c: http://www.psrd.hawaii.edu/Aug06/cataclysmDynamics.html.

Thommes, Edward W., Matsumura, Soko, & Rasio, Frederic A., *Gas Disks to Gas Giants: Simulating the Birth of Planetary Systems*, Science, vol. 321, p. 814–817, August 8, 2008.

Torino Impact Hazard Scale: https://cneos.jpl.nasa.gov/sentry/torino_scale.html

Tsiganis, K., et al., *Origin of the orbital architecture of the giant planets of the Solar System*,
Nature, vol. 435, p. 459–461, May 26, 2005.

Tyson, Peter, *Written in the Star*, Sky & Telescope, October 2017.

U.S. National Near-Earth Object Preparedness Strategy and Action Plan: https://www.whitehouse.gov/wp-content/uploads/2018/06/National-Near-Earth-Object-Preparedness-Strategy-and-Action-Plan-23-pages-1MB.pdf.

Vinković, Dejan, *Radiation-pressure mixing of large dust grains in protoplanetary disks*, Nature, vol. 459, p. 227–229, May 14, 2009.

Voyager – The Interstellar Mission: http://voyager.jpl.nasa.gov/.

Walsh, Kevin J., *Asteroids: when planets migrate*, Nature, vol. 457, 1091–1093, February 26, 2009.

Wetherill, George W., *Formation of the Terrestrial Planets from Planetesimals*, National Academies Press, 1991: http://www.nap.edu/books/0309043336/html/98.html.

Wood, John A., *Forging the Planets*, Sky & Telescope, January 1999.

Chapter 2: The Sun

A Super Solar *Flare*, Science@NASA, May 6, 2008: http://science.nasa.gov/headlines/y2008/06may_carringtonflare.htm.

Are Sunspots Disappearing? Science@NASA, September 3, 2009: http://science.nasa.gov/headlines/y2009/03sep_sunspots.htm.

Bahcall, John N., *Solving the Mystery of the Missing Neutrinos*, 28 April 2004. http://nobelprize.org/nobel_prizes/physics/articles/bahcall/.

Baker, Daniel N., & Green, James L., *The Perfect Solar Superstorm*, Sky & Telescope, February 2011.

Big Bear Solar Observatory: http://www.bbso.njit.edu/.

Bobra, Monica, *New Scrutiny of the Sun's Secrets*, Sky & Telescope, February 2011.

Bobra, Monica, *Superflares*, Sky & Telescope, November 2015.

Ehrenstein, David, *SOHO Traces the Sun's Hot Currents*, Science, vol. 277, p. 1438, September 5, 1997.

Flack, Bernhard, et al., *10 Years of SOHO*, ESA Bulletin 126, p. 24–32, May 2006: http://sohowww.nascom.nasa.gov/publications/ESA_Bull126.pdf.

Foukal, P., et al., *Variations in solar luminosity and their effect on the Earth's climate*, Nature, vol. 443, p. 161–166, September 14, 2006.

Global Oscillation Network Group (GONG): http://gong.nso.edu/.

Golub, Leon, and Pasachoff, Jay M., *Nearest Star: The Surprising Science of Our Sun*,
Harvard University Press (November 2002).

Gough, Douglas, News from the Solar Interior, Science, vol. 287, p. 2434–2435, March 31, 2000. Gough, D.O., et al, *The Seismic Structure of the Sun*, Science, vol. 272, p. 1296–1300, May 31, 1996.

Haisch, Bernhard, *and Schmitt, Jürgen, The Solar-Stellar Connection*, Sky & Telescope, October 1999.

Hall, Jeffrey, *Lessons from Solar Twins*, Sky & Telescope, July 2010.

Hassler, Donald M., et al., *Solar Wind Outflow and the Chromospheric Magnetic Network*, Science, vol. 283, p. 810–813, February 5, 1999.

Hill, Steele, and Carlowicz, Michael, *The Sun*, Harry N. Abrams, *Inc.* (May 2006).

Hinode (Solar-B) mission (NASA): http://solarb.msfc.nasa.gov/.

Hinode (Solar-B) mission (JAXA): http://www.isas.jaxa.jp/e/enterp/missions/solar-b/index.shtml.

Hudson, H.S. and Kosugi, T., *How the Sun's Corona Gets Hot*, Science, vol. 285, p. 849, August 6, 1999.

James, C. *Renée, Solar Forecast: Storm Ahead*, Sky & Telescope, July 2007.

Jess, David B., et al., *Alfven Waves in the Lower Solar Atmosphere*, Science, vol. 323, p. 1582, March 20, 2009.

Lang, Kenneth R., *SOHO Reveals the Secrets of the Sun*, Magnificent Cosmos – Scientific American Special, Spring 1998.

Lang, Kenneth R., *The Cambridge Encyclopedia of the Sun*, Cambridge University Press (September 2001).

Lang, Kenneth R., *Sun, Earth and Sky*, Springer, 2nd edition (September 2006).

Laughlin, Gregory P., *From Here to Eternity: The Fate of the Sun and the Earth*, Sky & Telescope, June 2007.

Lang, Kenneth R., *The Sun from Space*, Springer; 2nd edition (December 2008).

Layton, Laura, & Pesnell, Dean, *NASA Sets Its Sights on the Sun*, Sky & Telescope, January 2010.

Lindsey C., and Braun, D. C., *Seismic Images of the Far Side of the Sun*, Science, vol. 287, p. 1799–1801, March 10, 2000.

Lockheed Martin Solar and Astrophysics Laboratory: http://www.lmsal.com/.

McComas, D.J., et al., *Ulysses observations of very different heliospheric structure during the declining phase of solar activity cycle 23*, Geophysical Research Letters, vol. 33, L09102, May 9, 2006.

Moran, Thomas G., and Davila, Joseph M., *Three-Dimensional Polarimetric Imaging of Coronal Mass Ejections*, Science, vol. 305, p. 66–70, July 2, 2004.

Mystery of the Missing Sunspots Solved? Science@NASA, June 17, 2009: http://science.nasa.gov/headlines/y2009/01apr_deepsolarminimum.htm.

National Solar Observatory: http://www.nso.edu/.

NOAA Space Environment Center: http://www.sec.noaa.gov/.

Polar Crown Prominences, Science@NASA September 17, 2008: https://science.nasa.gov/science-news/science-at-nasa/2008/17sep_polarcrown/.

Pontieu, Bart de, et al., *Solar chromospheric spicules from the leakage of photospheric oscillations and flows*, Nature, vol. 430, p. 536–539, July 29, 2004.

Rempel, M, et al., *Penumbral Structure and Outflows in Simulated Sunspots*, Science, vol. 325, p. 171–174, July 10, 2009.

Results from Hinode, Science special issue, vol. 318, p. 1572–1599, December 7, 2007.

Schriver, Carolus J., *The Science Behind the Solar Corona*, Sky & Telescope, April 2006.

Science from Hinode, Astronomy and Astrophysics special issue, vol. 481, April 1, 2008: http://www.aanda.org/index.php?option=toc&url=/articles/aa/abs/2008/13/contents/contents.html.

Solar Physics Group, NASA Marshall Space Flight Center: http://solarscience.msfc.nasa.gov/index.html.

Shepherd, Debra, Stellar Origins: *From the Cold Depths of Space*, Sky & Telescope, June 2007.

Smith, E.J., et al, *The Sun and Heliosphere at Solar Maximum*, Science, vol. 302, p. 1,165–1,169, 14 November 14, 2003.

Smith, J. Kelly, & Smith, David L., Discovering the Radio Sun, Sky & Telescope, October 2014.

SOHO mission: http://soho.nascom.nasa.gov/.

Solar Dynamics Observatory: https://sdo.gsfc.nasa.gov/.

Solar Wind Loses Power, *Hits 50-year Low*, Science@NASA, September 23, 2008: http://science.nasa.gov/headlines/y2008/23sep_solarwind.htm.

STEREO mission: http://stereo.jhuapl.edu/.

Sudbury Neutrino Observatory: http://www.sno.phy.queensu.ca/.

Swedish 1-meter Solar Telescope: http://www.solarphysics.kva.se.

Taylor, G. Jeffrey, *The Sun's Crowded Delivery Room*, PSR Discoveries, July 6, 2007: http://www.psrd.hawaii.edu/July07/iron-60.html.

Tyson, Peter, *Written in the Star*, Sky & Telescope, October 2017.

Ulysses above the sun's south pole, Science, vol. 268, p. 1005–1033, May 19, 1995.

Ulysses mission: http://helio.esa.int/ulysses/.

Yohkoh mission: http://www.lmsal.com/YPOP/homepage.html.

Young, C. Alex, *Shadow Science*, Sky & Telescope, August 2018.

Zimmerman, Robert, *What's Wrong With Our Sun?* Sky & Telescope, August 2009.

Zimmerman, Robert, *Finding the Sun's Lost Nursery*, Sky & Telescope, March 2012.

Chapter 3: Earth

Barry, T.L., et al, *Whole-mantle convection with tectonic plates preserves long-term global patterns of upper mantle geochemistry*, Nature Scientific Reports 7, published online May 12, 2017: http://www.nature.com/articles/s41598-017-01816-y.

Bintanja, R., & van de Wal, R.S.W., *North American ice-sheet dynamics and the onset of 100,000-year glacial cycles*, Nature, vol. 454, p. 869–872, August 14, 2008.

Destination Earth (NASA): http://www.earth.nasa.gov/.

Donnadieu, Yannick, Cracking continents caused 'snowball Earth', Nature, vol. 428, p. 303–306, March 18, 2004.

Dorminey, Bruce, *Is Plate Tectonics Necessary for Sentient Life?* Sky & Telescope, July 2013.

Durda, Daniel D., *The Most Dangerous Asteroid Ever Found*, Sky & Telescope, November 2006.

Earth (NASA): https://www.nasa.gov/connect/ebooks/earth&uscore;detail.html and https://earthobservatory.nasa.gov/features/earthbook-2019.

Earth Impact Database, University of New Brunswick: http://www.passc.net/EarthImpactDatabase/index.html.

Earth Impact Effects Program: http://impact.ese.ic.ac.uk/.

Earth pages (Blackwell): http://www.earth-pages.com/default.asp.

Earth Science World: http://www.earthscienceworld.org.

Earth's Fidgeting Climate: http://science.nasa.gov/headlines/y2000/ast20oct_1.htm?list153136.

Encyclopedia of the Atmospheric Environment: http://www.ace.mmu.ac.uk/eae/english.html.

ESA Earth Observation: http://www.esa.int/esaEO/.

Grifantini, Kristina, *Where DID Earth's Water Come From?*, Sky & Telescope, January 2011.

Hays, J.D., Imbrie, J., & Shackleton, N.J., *Variation in the Earth's orbit: Pacemaker of the ice ages*, Science, vol. 194, p. 1121–1132, 1976.

Hecht, Jeff, *Glimpses of an Evolving Planet*, Sky & Telescope, August 2010.

Intergovernmental Panel on Climate Change:

James, C. Renée, *Solar Forecast: Storm Ahead*, Sky & Telescope, July 2007.

Kasting, J.F., *Earth's early atmosphere*, Science, vol. 259, p. 920–926, 1993.

Kious, W. Jacquelyne & Tilling, Robert I., *This Dynamic Earth: The Story of Plate Tectonics*, USGS, 1996: http://pubs.usgs.gov/gip/dynamic/dynamic.html.

Kring, David A., *Calamity at Meteor Crater*, Sky & Telescope, November 1999.

McPhaden, Michael J., et al., *ENSO as an Integrating Concept in Earth Science*, Science, vol. 314, p. 1740–1745, December 15, 2006.

Milutin Milankovitch and Ice Ages: https://earthobservatory.nasa.gov/Features/Milankovitch/.

Naeye, Robert, *Earth's Changing Magnetic Field*, Sky & Telescope, March 2018.

NASA Earth Observatory: http://earthobservatory.nasa.gov/.

NASA EOS IDS Volcanology: http://eos.higp.hawaii.edu/.

NASA Global Climate Change: https://climate.nasa.gov/.

NASA Near Earth Object Program: http://neo.jpl.nasa.gov/.

NASA Magnetospheric Multiscale Mission Near Earth Object Program: https://www.nasa.gov/mission_pages/mms/index.html.

NASA Ozone Hole Watch: http://ozonewatch.gsfc.nasa.gov/index.html.

NASA Sun-Earth environment: https://www.nasa.gov/mission_pages/sunearth/index.html.

Ocean Surface Topography from Space (NASA): http://sealevel.jpl.nasa.gov/.

Ocean Surface Currents (Univ. of Miami): http://oceancurrents.rsmas.miami.edu/index.html.

Pacchioli, David, *Reflections From a Warm Little Pond*, Astrobiology magazine, April 27, 2001: http://www.astrobio.net/news/article5.html.

Palaeos: The Trace of Life on Earth: http://www.palaeos.com/Default.htm.

Paleomap Project (Christopher R. Scotese): http://www.scotese.com/Default.htm.

Salomonsen, Ole C., *Secrets of the Northern Lights*, Sky & Telescope, February 2013.

Solar Views – Terrestrial Volcanoes: http://www.solarviews.com/eng/tervolc.htm.

Space Weather: http://www.spaceweather.com/.

Tarduno, John A., et al., *The Emperor Seamounts: Southward Motion of the Hawaiian Hotspot Plume in Earth's Mantle*, Science, vol. 301, p. 1064–1069, August 22, 2003.

USGS Earthquake Hazard Program: http://earthquake.usgs.gov/.

USGS Volcano Hazards Program: http://volcanoes.usgs.gov/.

Volcano World: http://www.volcanoworld.org/.

Walker, A.S., *Deserts: Geology and Resources*, USGS: http://pubs.usgs.gov/gip/deserts/contents/.

Chapter 4: The Moon

Andrews-Hanna, Jeffrey C., *Structure and evolution of the lunar Procellarum region as revealed by GRAIL gravity data*, Nature, vol. 514, p. 68–71, October 2, 2014.

Apollo Lunar Surface Journal: http://www.hq.nasa.gov/alsj/.

Apollo Image Atlas: http://www.lpi.usra.edu/resources/apollo/.

Barbuzano, Javier, *The Moon Mess*, Sky & Telescope, August 2018.

Beatty, J. Kelly, *NASA Slams the Moon*, Sky & Telescope, February 2010.

Bell, Jim, *Seeing the Moon Like Never Before*, Sky & Telescope, June 2012.

Brandon, Alan, A Younger Moon, Nature, vol. 450, p. 1169–1170, December 20/27, 2007.

Cameron, A.G.W., The Origin of the Moon and the Single Impact Hypothesis: http://www.xtec.es/recursos/astronom/moon/camerone.htm.

Campbell, B.A., et al, *Radar imaging of the lunar poles*, Nature, vol. 426, p. 137–138, 2003.

Change-3 mission: https://directory.eoportal.org/web/eoportal/satellite-missions/c-missions/chang-e-3#mission-status.

Cohen, Barbara A., *Lunar Meteorites and the Lunar Cataclysm*, PSR Discoveries, January 2001: http://www.psrd.hawaii.edu/Jan01/lunarCataclysm.html.

Consolidated Lunar Atlas: http://www.lpi.usra.edu/resources/cla/.

Digital Lunar Orbiter Photographic Atlas of the Moon: http://www.lpi.usra.edu/resources/lunar_orbiter/?info1083.

Foust, Jeffrey A., *NASA's New Moon*, Sky & Telescope, September 1998.

Geologic History of the Moon: cps.earth.northwestern.edu/GHM/ghm_hplates.html.

Google Moon: https://www.google.com/moon/.

Harland, David M., *Exploring the Moon: The Apollo Expeditions*, Springer Praxis, 1999.

Introduction to Cratering Studies, Planetary Science Institute, Tucson, Arizona: http://www.psi.edu/projects/mgs/cratering.html.

Investigating the Moon's Far Side (3 articles), Science, vol. 323, p. 897–912, February 13, 2009.

LCROSS and Lunar Reconnaissance Orbiter special issue, Science, October 22, 2010.

Lunar meteorites (Randy Korotev): http://epsc.wustl.edu/admin/resources/moon_meteorites.html.

Lunar photo of the day: http://www.lpod.org/.

Lunar Prospector Results, Science, vol. 281, p. 1475–1500, September 4, 1998.

Lunar Reconnaissance Orbiter mission: http://lunar.gsfc.nasa.gov/.

Martel, Linda M.V., *The Moon at its Core*, PSR Discoveries, September 24, 1999: http://www.psrd.hawaii.edu/Sept99/MoonCore.html.

Martel, Linda M.V., The Moon's Dark, Icy Poles, PSR Discoveries, June 4, 2003: http://www.psrd.hawaii.edu/June03/lunarShadows.html.

Martel, Linda M.V., *Lunar Crater Rays Point to a New Lunar Time Scale*, PSR Discoveries, September 28, 2004: http://www.psrd.hawaii.edu/Sept04/LunarRays.html.

Martel, Linda M.V., *Composition of the Moon's Crust*, PSR Discoveries, December 10, 2004: http://www.psrd.hawaii.edu/Dec04/LunarCrust.html.

Martel, Linda M. V., *Celebrated Moon Rocks*, PSR Discoveries, December 21, 2009: http://www.psrd.hawaii.edu/Dec09/Apollo-lunar-samples.html.

Masursky, Harold, Colton, G.W., El-Baz, Farouk, eds., *Apollo Over The Moon: A View From Orbit* (NASA SP-362), 1978: http://www.hq.nasa.gov/office/pao/History/SP-362/contents.htm.

Melosh, H.J., et al, *The Origin of Lunar Mascon Basins*, Science, June 28, 2013.

NASA Astromaterials Acquisition and Curation Office: https://www-curator.jsc.nasa.gov/.

Nozette, S., et al., *The Clementine Bistatic Radar Experiment*, Science, vol. 274, p. 1495–1498, 1996.

Selene (Kaguya) mission science reports, Science, vol. 323, p. 897–912, February 13, 2009.

Shirao, Motomaro, *Kaguya's High-Def Highlights*, Sky & Telescope, February 2010.

Smart-1 mission: http://sci.esa.int/science-e/www/area/index.cfm?fareaid=10.

Spudis, Paul D., *The Once and Future Moon*, Smithsonian Institution Press, 1996.

Spudis, Paul D., *Ice on the Bone Dry Moon*, PSR Discoveries, December 21, 1996: http://www.psrd.hawaii.edu/Dec96/IceonMoon.html.

Spudis, Paul D., *The New Moon*, Scientific American, January 12, 2003.

Spudis, Paul D., *Our Two-Faced Moon*, Sky & Telescope, April 2016.

Taylor, G. Jeffrey, The Moon Beyond 2002, PSR Discoveries, October 8, 2002: http://www.psrd.hawaii.edu/Oct02/moon.html.

Taylor, G. Jeffrey, *Moonbeams and Elements*, PSR Discoveries, October 20, 1997: http://www.psrd.hawaii.edu/Oct97/MoonFeO.html.

Taylor, G. Jeffrey, *Origin of the Earth and Moon*, PSR Discoveries, December 31, 1998: http://www.psrd.hawaii.edu/Dec98/OriginEarthMoon.html.

Taylor, G. Jeffrey, *The Surprising Lunar Maria*, PSR Discoveries, June 23, 2000: http://www.psrd.hawaii.edu/June00/lunarMaria.html.

Taylor, G. Jeffrey, A *New Moon for the Twenty-First Century*, PSR Discoveries, August 31, 2000: http://www.psrd.hawaii.edu/Aug00/newMoon.html.

Taylor, G. Jeffrey, *Recipe for High-Titanium Lunar Magmas*, PSR Discoveries, December 5, 2000: http://www.psrd.hawaii.edu/Dec00/highTi.html.

Taylor, G. Jeffrey, *New Lunar Meteorite Provides its Lunar Address and Some Clues about Early Bombardment of the Moon*, PSR Discoveries, 31 October 2004: http://www.psrd.hawaii.edu/Oct04/SaU169.html.

Taylor, G. Jeffrey, *Finding Basalt Chips from Distant Maria*, PSR Discoveries, April 30, 2006: http://www.psrd.hawaii.edu/April06/basaltFragments.html.

Taylor, G. Jeffrey, Gamma Rays, *Meteorites, Lunar Samples, and the Composition of the Moon*, PSR Discoveries, November 22, 2005: http://www.psrd.hawaii.edu/Nov05/MoonComposition.html.

Taylor, G. Jeffrey, *Two Views of the Moon's Composition*, PSR Discoveries, April 3, 2007: http://www.psrd.hawaii.edu/April07/Moon2Views.html.

Terada, Kentaro, et al., *Cryptomare magmatism 4.35 Gyr ago recorded in lunar meteorite Kalahari 009*, Nature vol. 450, p. 849–852, December 6, 2007.

Weitz, Catherine M., *Explosive Volcanic Eruptions on the Moon*, PSR Discoveries, February 12, 1997: http://www.psrd.hawaii.edu/Feb97/MoonVolcanics.html.

Wilhelms, Don E., *To a Rocky Moon, A Geologist's History of Lunar Exploration*, University of Arizona Press, Tucson, 1993. http://www.lpi.usra.edu/publications/books/rockyMoon/.

Wilhelms, Don, *Geologic History of the Moon*, U.S. Geological Survey Professional Paper 1348 (1987): http://ser.sese.asu.edu/GHM/.

Wood, Charles A., *The Moon's Far Side: Nearly a New World*, Sky & Telescope, January 2007.

Chapter 5: Mercury

Beatty, J. Kelly, *Mercury's Marvels*, Sky & Telescope, April 2012.

BepiColombo mission: http://www.sci.esa.int/bepicolombo.

Blewett, David T., et al., *Hollows on Mercury: MESSENGER Evidence for Geologically Recent Volatile-Related Activity*, Science, September 30, 2011.

Byrne, Paul K., et al., *Mercury's global contraction much greater than earlier estimates*, Nature Geoscience, published online March 16, 2014.

Correia, Alexandre C.M., and Laskar, Jacques, Mercury's capture into the 3/2 spin-orbit resonance as a result of its chaotic dynamics, Nature, vol. 429, p. 848–850, June 24, 2004.

Dunne, James, and Burgess, Eric, *The Voyage of Mariner 10*, NASA SP-424, 1978.

Harmon, J.K., *Mercury Radar Imaging at Arecibo, LPI abstract* 2001: www.lpi.usra.edu/meetings/mercury01/pdf/8001.pdf.

Ice on Mercury, NSSDC: http://nssdc.gsfc.nasa.gov/planetary/ice/ice_mercury.html.

Johnson, Catherine L., et al., *Low-altitude magnetic field measurements by MESSENGER reveal Mercury's ancient crustal field*, Science, May 22, 2015.

Lucey, Paul G., et al., *A Wet and Volatile Mercury*, Science, January 18, 2013.

Mariner 10 Archive Project (Mark Robinson, Northwestern University): http://cps.earth.northwestern.edu/merc.html.

Mariner 10 Preliminary Science Report, Science, vol. 185, p. 141–180, 1974.

Mercury's Lively Core, Science Now, January 14, 2005.

MESSENGER mission: http://messenger.jhuapl.edu.

MESSENGER's First Mercury Flyby (special issue), Science, vol. 321, p. 59–94, July 4, 2008.

MESSENGER's Second Mercury Flyby, Science, vol. 324, p. 613–621, May 1, 2009.

MESSENGER's Third Mercury Flyby, Science, vol. 329, p. 665–675, August 6, 2010.

MESSENGER Orbits Mercury, Science Special Issue, September 30, 2011.

Namur, Olivier, & Charlier, Bernard, *Silicate mineralogy at the surface of Mercury*, Nature Geoscience, December 19, 2016.

Neumann, Gregory A., et al., *Bright and Dark Polar Deposits on Mercury: Evidence for Surface Volatiles*, Science, January 18, 2013.

Paige, David A., et al., *Thermal Stability of Volatiles in the North Polar Region of Mercury*, Science, January 18, 2013.

Peplowski, Patrick N. et al., Remote sensing evidence for an ancient carbon-bearing crust on Mercury, Nature Geoscience, published online March 7, 2016.

Rothery, David A., et al., *Mercury's Caloris basin: Continuity between the interior and exterior plains*, Journal of Geophysical Research: Planets, March 20, 2017.

Smith, David E., et al., *Gravity Field and Internal Structure of Mercury from MESSENGER*, Science, April 13, 2012.

Smith, David E., et al., *Topography of the Northern Hemisphere of Mercury from MESSENGER Laser Altimetry,* Science, April 13, 2012,

Strom, Robert, *Mercury,* The Elusive Planet, Smithsonian Institution Press, 1987.

Strom, Robert G. and Sprague, Ann L., *Exploring Mercury: The Iron Planet,* Springer-Praxis, 2003.

Taylor, G. Jeffrey, New Data, New Ideas, and Lively Debate About Mercury, PSR Discoveries, October 22, 2001: http://www.psrd.hawaii.edu/Oct01/MercuryMtg.html.

Taylor, G. Jeffrey, Mercury Unveiled, PSR Discoveries, January 23, 1997: http://www.psrd.hawaii.edu/Jan97/MercuryUnveiled.html.

The Atlas of Mercury, NASA SP-423 (NASA History website): http://history.nasa.gov/SP-423/mariner.htm.

Vilas, F., Chapman, C.R., and Matthews, M.S. (editors), *Mercury*, The University of Arizona Press, 1988.

Watters, Thomas R., et al., *Thrust faults and the global contraction of Mercury*, Geophysical Research Letters, vol. 31, 2004: http://www.earth.northwestern.edu/people/craig/publish/pdf/grl04.pdf.

Watters, Thomas R., et al., *The mechanical and thermal structure of Mercury's early lithosphere*, Geophysical Research Letters, vol. 29, June 14, 2002: http://www.mines.unr.edu/geo-eng/schultz/pdf%20reprints/Watters,+Merc.GRL02.pdf.

Watters, Thomas R., et al., *Recent tectonic activity on Mercury revealed by small thrust fault scarps*, Nature Geoscience, published online September 26, 2016.

Chapter 6: Venus

Bullock, Mark A., & Grinspoon, David H., *Global Climate Change on Venus*, Scientific American Special, New Light on the Solar System, 2003.

Cattermole, Peter, *Venus: The Geological Story*, UCL Press, London, 1994.

Dunne, James, and Burgess, Eric, *The Voyage of Mariner 10*, NASA SP-424, 1978.

Fimmel Richard, Colin, Lawrence, and Burgess, Eric, Pioneer Venus, NASA SP-461, 1983.

Fimmel Richard, Colin, Lawrence, and Burgess, Eric, Pioneering Venus, NASA SP-518, 1995: http://ntrs.nasa.gov/archive/nasa/casi.ntrs.nasa.gov/19960026995_1996055001.pdf.

Grinspoon, David, *Venus Revealed*, Perseus Publishing, New York, 1998.

Hansen, Vicki L., & Phillips, Roger J., *Tectonics and Volcanism of Eastern Aphrodite Terra, Venus: No Subduction, No Spreading*, Science, vol. 260, p. 526–530, April 23, 1993.

Hunten, D.M., Colin, L., Donahue, T.M., and Moroz, V. I. (eds), *Venus*, University of Arizona Press, 1993.

Kaula, William M., *Venus: A Contrast in Evolution to Earth*, Science, vol. 247, p. 1191–1196, March 9, 1990.

Lopes, Rosaly, *Where the Hot Stuff Is*, Sky & Telescope, July 2011.

Magellan at Venus, Journal of Geophysical Research, vol. 97, p. 13,062–13,689, August 25, 1992.

Magellan mission: http://www2.jpl.nasa.gov/magellan/.

Malin, Michael C., *Mass Movements on Venus: Preliminary Results from Magellan Cycle I Observations*, Journal of Geophysical Research, vol. 97, No. E10, p. 16,337–16,352, 1992: http://www.msss.com/venus/landslides/venus_paper.html.

Pettengill, Gordon H., et al., *The Surface of Venus*, Scientific American, August 1980.

Prinn, Ronald, *The Volcanoes and Clouds of Venus*; Scientific American Special, Exploring Space, 1990.

Sagdeev, R.Z. et al, *Overview of VEGA Venus balloon in situ meteorological measurements*, Science, vol. 231, p.1,411–1,414, 1986.

Sánchez-Lavega, A. et al., *Variable winds on Venus mapped in three dimensions*, Geophysical. Research Letters, vol. 35, p. L13204, July 10, 2008.

Saunders, R. et al., *Magellan: A First Overview of Venus Geology*, Science, vol. 252, p. 249–252, 1991.

Schubert, Gerald, & Covey, Curt, *The Atmosphere of Venus*, Scientific American, July 1981.

Shalygin E.V., et al., *Active volcanism on Venus in the Ganiki Chasma rift zone*, Geophysical Research Letters, June 17, 2015: http://onlinelibrary.wiley.com/doi/10.1002/2015GL064088/full.

Smrekar, Suzanne E., et al., *Recent hot-spot volcanism on Venus from VIRTIS emissivity data*, Science, vol. 328, p. 605–608, April 8, 2010.

Solomon, Sean C., et al., *Venus Tectonics: Initial Analysis from Magellan*, Science, vol. 252, p. 297–312, April 12, 1991.

Solomon, S, et al., *Climate Change as a Regulator of Tectonics on Venus*, Science, vol. 286, p. 87–90, October 1999.

Taylor, F.W., Read, P.L., and Lewis S.R., The Venusian Environment. http://www.atm.ox.ac.uk/user/fwt/WebPage/Venus%20Review%204.htm.

The Soviet Exploration of Venus (Don Mitchell): http://www.mentallandscape.com/V_Venus.htm.

Transits of Venus and black drop effect: http://www.transitofvenus.info.

Venus Climate Orbiter (Akatsuki): http://www.jaxa.jp/projects/sat/planet_c/index_e.html.

Venus Express mission: http://sci.esa.int/venusexpress/

Venus Express special – The Long-Lost Twin: Nature, vol. 450, p. 629–662, November 29, 2007.

Zuber, M.T. and Parmentier, E.M., *Formation of fold-and-thrust belts on Venus by thick-skinned deformation*, Nature, vol. 377, p. 704–707, October 26, 2002.

Chapter 7: Mars

Andrews-Hanna, Jeffrey C., Phillips, Roger J., & Zuber, Maria T., *Meridiani Planum and the global hydrology of Mars*, Nature, vol. 446, p. 163–166, March 8, 2007.

Andrews-Hanna, Jeffrey C., Zuber, Maria T., & Banerdt, W. Bruce, *The Borealis basin and the origin of the martian crustal dichotomy*, Nature, vol. 453, p. 1212–1215, June 26, 2008.

Bandfield, Joshua L., Hamilton, Victoria E., & Christensen, Philip R., *A Global View of Martian Surface Compositions from MGS-TES*, Science, vol. 287, p. 1626–1630, March 3, 2000.

Barabash, Stas, et al., *Martian Atmospheric Erosion Rates*, Science, vol. 315, p. 501–503, January 26, 2007.

Beatty, J. Kelly, *Curiosity Hits the Road*, Sky & Telescope, January 2013.

Beatty, J. Kelly, *In Search of Martian Seas*, Sky & Telescope, November 1999.

Beatty, J. Kelly, *Mars Under the Microscope*, Sky & Telescope, September 2010.

Bell, Jim, *Uncovering Mars's Secret Past*, Sky & Telescope, July 2009.

Bell, Jim, *Mineral Mysteries and Planetary Paradoxes*, Sky & Telescope, December 2003.

Bibring, Jean-Pierre, et al., *Perennial water ice identified in the south polar cap of Mars*, Nature, vol. 428, p. 627–630, April 8, 2004.

Bibring, J.-P., *Coupled Ferric Oxides and Sulfates on the Martian Surface*, Science Express, August 2, 2007.

Bishop, J.L., et al., *Phyllosilicate Diversity and Past Aqueous Activity Revealed at Mawrth Vallis, Mars*, Science, vol. 321, p. 830–833, August 8, 2008.

Brasch, Klaus, *Canal Mania*, Sky & Telescope, July 2018.

Canup, Robin, & Salmon, Julien, *Origin of Phobos and Deimos by the impact of a Vesta-to-Ceres sized body with Mars*, Science Advances, April 18, 2018, vol. 4, no. 4, doi: 10.1126/sciadv.aar6887.

Carlisle, Camille M., *Deciphering Mars*, Sky & Telescope, September 2014.

Carr, M.H., *Water on Mars*, Oxford University Press, New York, 1996.

Christensen, P.R., *Formation of recent martian gullies through melting of extensive water-rich snow deposits*, Nature, vol. 422, p. 45–48, published online February 19, 2003.

Citron, Robert I., et al., *Timing of oceans on Mars from shoreline deformation*, Nature, vol. 555, p 643–646, March 29, 2018.

Clifford, Stephen M., *The Iceball Next Door*, Sky & Telescope, August 2003.

Committee on an Astrobiology Strategy for the Exploration of Mars, *An Astrobiology Strategy for the Exploration Of Mars*, The National Academies Press, 2007: http://www.nap.edu/catalog.php?record_id=11937.

Connerney, J.E.P., et al., *Tectonic implications of Mars crustal magnetism*, PNAS vol. 102 no. 42, p. 14970–14975, October 18, 2005: http://www.pnas.org/content/102/42/14970.full.

Corrigan, Catherine M., *Carbonates in ALH84001: Part of the Story of Water on Mars*, PSR Discoveries, July 1, 2004: http://www.psrd.hawaii.edu/July04/carbonatesALH84001.html.

Cull, Selby, Reverse Panspermia: *Seeding Life in the Solar System*, Sky & Telescope, January 2007.

DiGregorio, Barry E., *Life on Mars?*, Sky & Telescope, February 2004.

Dundas, Colin M., et al., *Granular Flows at Recurring Slope Lineae on Mars Indicate a Limited Role for Liquid Water*, Nature Geoscience,

November 20, 2017: https://www.nature.com/articles/s41561-017-0012-5.

Ezell, Edward Clinton, and Ezell, Linda Neuman, *On Mars - Exploration of the Red Planet 1958-1978*, NASA History Office, NASA SP-4212, 1984: http://www.hq.nasa.gov/office/pao/History/SP-4212/on-mars.html.

Fenton, Lori K., et al., *Global warming and climate forcing by recent albedo changes on Mars*, Nature, vol. 446, p. 646–649, April 5, 2007.

Forget, F, et al., *Formation of Glaciers on Mars by Atmospheric Precipitation at High Obliquity*, Science vol. 311, p. 368–371, January 20, 2006.

Google Mars: http://www.google.com/mars/.

Gurnett, D.A., et al., *Radar Soundings of the Ionosphere of Mars*, Science, volume 310, p. 1929-1933, December 23, 2005.

Halevy, Itay, Zuber, Maria T., Schrag, Daniel P., A Sulfur Dioxide Climate Feedback on Early Mars, Science, vol. 318, p. 1903–1907, December 21, 2007.

Hartmann, William K., *Mysteries of Mars*, Sky & Telescope, July 2003.

Hartmann, William K., *A Traveler's Guide to Mars*, Workman Publishing Company, 2003.

Herkenhoff, Kenneth E., & Keszthelyi, Laszlo P., *MRO/HiRISE Studies of Mars*, Icarus, vol. 205, issue 1, p. 1–320, January 2010.

Holt, J.W., et al., *Radar Sounding Evidence for Buried Glaciers in the Southern Mid-Latitudes of Mars*, Science, vol. 322, p. 1235–1238, November 21, 2008.

Holt, John W., & Smith, Isaac B., *Onset and migration of spiral troughs on Mars revealed by orbital radar*, Nature, vol. 465, p. 450–453, May 27, 2010.

Holt, J.W., et al., *The construction of Chasma Boreale on Mars*, Nature vol. 465, p. 446–449, May 27, 2010.

Jaeger, W.L., et al., Athabasca Valles, Mars: A Lava-Draped Channel System, Science, vol. 317, p. 1709–1711, September 21, 2007.

Jakosky, Bruce, *Mars's Lost Atmosphere*, Sky & Telescope, July 2018.

Jeffrey, *The Multifarious Martian Mantle*, PSR Discoveries, June 23, 2004: http://www.psrd.hawaii.edu/June04/martianMantle.html.

Kraal, Erin R., et al., Martian stepped-delta formation by rapid water release, Nature, vol. 451, p. 973–977, February 21, 2008.

Lakdawalla, Emily, *Face to Face with a Giant*, Sky & Telescope, December 2011.

Lakdawalla, Emily, *Touchdown on the Red Planet*, Sky & Telescope, November 2012.

Lakdawalla, Emily, *The History of Water on Mars*, Sky & Telescope, September 2013.

Lakdawalla, Emily, *Curiosity's Discoveries on Mars*, Sky & Telescope, April 2017.

Lakdawalla, Emily, *Mars: The INSIDE Story*, Sky & Telescope, December 2018, p. 34–38.

Lefevre, Franck, et al., *Heterogeneous chemistry in the atmosphere of Mars*, Nature, vol. 454, p. 971–975, August 21, 2008.

Lowell, Percival, *Mars*, 1895: http://www.wanderer.org/references/lowell/Mars/.

Lundin R., et al., *Plasma Acceleration Above Martian Magnetic Anomalies*, Science, vol. 311, p. 980–983, February 17, 2006.

Malin, M.C., and Edgett, K.S., *Oceans or Seas in the Martian Northern Lowlands: High Resolution Imaging Tests of Proposed Coastlines*, Geophys. Res. Letters, vol. 26, no. 19, p. 3049–3052, 1999.

Malin, Michael C., & Edgett, Kenneth S., Evidence for Recent Groundwater Seepage and Surface Runoff on Mars, Science, vol. 288, p. 2330–2335, June 30, 2000.

Malin, Michael C., & Edgett, Kenneth S., Evidence for Persistent Flow and Aqueous Sedimentation on Early Mars, Science, vol. 302, p. 1931–1934, December 12, 2003.

Malin Space Science Systems – Images of Mars and Other Features From Mars Spacecraft: http://www.msss.com/mars_images/index.html.

Mangold N., et al., *Evidence for Precipitation on Mars from Dendritic Valleys in the Valles Marineris Area*, Science, vol. 305, p. 78–81, July 2, 2004.

Mars Exploration Rovers mission: http://marsrovers.jpl.nasa.gov/home/index.html.

Mars Express mission: http://www.esa.int/SPECIALS/Mars_Express/index.html.

Mars Global Surveyor mission: http://mars.jpl.nasa.gov/mgs/.

Mars Odyssey mission: http://mars.jpl.nasa.gov/odyssey/.

Mars Odyssey Themis instrument web site: http://themis.asu.edu/.

Mars Pathfinder mission: http://mars.jpl.nasa.gov/MPF/index1.html.

Mars Phoenix mission: http://phoenix.lpl.arizona.edu.

Mars Reconnaissance Orbiter HiRise instrument website: http://hirise.lpl.arizona.edu.

Mars Reconnaissance Orbiter mission: http://mars.jpl.nasa.gov/mro/.

Mars Reconnaissance Orbiter special section, Science, vol. 317, p. 1705–1719, September 21, 2007.

Martel, Linda M.V., Outflow Channels May Make a Case for a Bygone Ocean on Mars, PSR Discoveries, June 14, 2001: http://www.psrd.hawaii.edu/June01/MarsChryse.html.

Martel, Linda M.V., Ancient Floodwaters and Seas on Mars, PSR Discoveries, July 16, 2003: http://www.psrd.hawaii.edu/July03/MartianSea.html.

Martel, Linda M.V., Gullied Slopes on Mars, PSR Discoveries, August 29, 2003: http://www.psrd.hawaii.edu/Aug03/MartianGullies.html.

Martel, Linda M. V., Did Martian Meteorites Come from These Sources? PSR Discoveries, January 29, 2007: http://www.psrd.hawaii.edu/Jan07/MarsRayedCraters.html.

McKay, D. S., et al., *Search for Past Life on Mars: Possible Relic Biogenic Activity in Martian Meteorite ALH84001*, Science, vol. 273, p. 924–930, August 16, 1996.

Morton, Oliver, *Mapping Mars*, Picador, 2002.

Mumma, Michael J., et al., *Strong Release of Methane on Mars in Northern* Summer 2003, Science, vol. 323, p. 1041–1045, February 20, 2009.

Murray, John B., et al., Evidence from the Mars Express High Resolution Stereo Camera for a frozen sea close to Mars' equator, Nature, vol. 434, p. 352–356, March 17, 2005.

Mustard, John F., et al., *Hydrated silicate minerals on Mars observed by the Mars Reconnaissance Orbiter CRISM instrument*, Nature, vol. p. 305–309, July 17, 2008.

Naeye, Robert, *Red-Letter Days*, Sky & Telescope, May 2004.

Neukum, G., et al., *Recent and episodic volcanic and glacial activity on Mars revealed by the High Resolution Stereo Camera*, Nature, vol. 432, p. 971–979, December 23, 2004.

Okubo, Chris H. and McEwen, Alfred S., Fracture-Controlled Paleo-Fluid Flow in Candor Chasma, *Mars*, Science, vol. 315, p. 983–985, February 16, 2007.

Perron, J. Taylor, et al, Evidence for an ancient martian ocean in the topography of deformed shorelines, Nature, vol. 447, p. 840–843, June 14, 2007.

Safe on Mars: Precursor Measurements Necessary to Support Human Operations on the Martian Surface, National Research Council, 2002: http://www.nap.edu/catalog/10360.html.

Phillips, R.J., et al., *Mars North Polar Deposits: Stratigraphy, Age, and Geodynamical Response*, Science, vol. 320, p. 1182–1185, May 30, 2008.

Plaut, Jeffrey J., et al., *Subsurface Radar Sounding of the South Polar Layered Deposits of Mars*, Science, vol. 316, p. 92–95, 6 April 6, 2007.

Raeburn, Paul, *Mars: Uncovering the Secrets of the Red Planet*, National Geographic Society, 1998.

Schofield, J.T., et al., *The Mars Pathfinder Atmospheric Structure Investigation/Meteorology (ASI/MET) Experiment*, Science, vol. 278, p. 1752–1758, December 5, 1997.

Schorghofer, Norbert, *Dynamics of ice ages on Mars*, Nature, vol. 449, p. 192–195, September 13, 2007.

Solomon, Sean C., et al., *New Perspectives on Ancient Mars*, Science, vol. 37, p. 1214–1219, February 25, 2005.

Space Studies Board/Board on Life Sciences, *An Astrobiology Strategy for the Exploration of Mars*, The National Academies Press, 2007: http://www.nap.edu/openbook.php?record_id=11937&page=23.

Squyres, Steve, Roving Mars: *Spirit, Opportunity, and the Exploration of the Red Planet*, Hyperion, 2005.

Squyres, S.W., et al., *Two Years at Meridiani Planum: Results from the Opportunity Rover*, Science, vol. 313, p. 1403–1407, September 7, 2006.

Taylor, G. Jeffrey, Liquid Water on Mars: The Story from Meteorites, PSR Discoveries, May 24, 2000: http://www.psrd.hawaii.edu/May00/wetMars.html.

Taylor, G. Jeffrey, Hafnium, Tungsten, and the Differentiation of the Moon and Mars, PSR Discoveries, November 28, 2003: http://www.psrd.hawaii.edu/Nov03/Hf-W.html.

Taylor, G. Jeffrey, *Life on Mars?* PSR Discoveries, October 18, 1996: http://www.psrd.hawaii.edu/Oct96/LifeonMars.html.

Taylor, G. Jeffrey, Magma and Water on Mars, PSR Discoveries, December 27, 2005: http://www.psrd.hawaii.edu/Dec05/Magma-WaterOnMars.html.

Taylor, G. Jeffrey, Mars Crust: Made of Basalt, PSR Discoveries, May 8, 2009: http://www.psrd.hawaii.edu/May09/Mars.Basaltic.Crust.html.

Taylor, G. Jeffrey, Recent Activity on Mars: Fire and Ice, PSR Discoveries, January 31, 2005: http://www.psrd.hawaii.edu/Jan05/MarsRecently.html.

The Mars Exploration Rover Mission, Nature, vol. 436, p. 42–69, July 7, 2005.

The Mars Journal: http://themarsjournal.org/.

The Quarantine and Certification of Martian Samples, Committee on Planetary and Lunar Exploration, National Research Council, 2002: http://www.nap.edu/catalog/10138.html.

Viking Lander Imaging Team, *The Martian Landscape*, NASA History Office, NASA SP-425, 1978: http://history.nasa.gov/SP-425/cover.htm.

Viking mission: http://nssdc.gsfc.nasa.gov/planetary/viking.html.

Viking Orbiter Imaging Team, *Viking Orbiter Views of Mars*, NASA History Office, NASA SP-441, 1980: http://history.nasa.gov/SP-441/cover.htm.

Villanueva, G., et al., *Strong water isotopic anomalies in the Martian atmosphere: probing current and ancient reservoirs*, Science, March 5, 2015.

Withers, Paul, & Smith, Michael D., *Atmospheric entry profiles from the Mars Exploration Rovers Spirit and Opportunity*, Icarus, vol. 185, p. 133–142. Published online August 23, 2006: http://sirius.bu.edu/aeronomy/withersmericarus2006.pdf.

Zuber, Maria T., *Mars: The Inside Story*, Sky & Telescope, December 2003.

Zuber, Maria T., et al., *Density of Mars' South Polar Layered Deposits*, Science, vol. 317, p. 1718–1719, September 21, 2007.

Chapter 8: Jupiter

Adriani, A., et al., Clusters of cyclones encircling Jupiter's poles, Nature, vol. 555, p 216–219, March 8, 2018.

Anderson, J.D., et al., *Europa's Differentiated Internal Structure: Inferences from Two Galileo Encounters*, Science, vol. 276, p. 1236–1239, 1998.

Anderson, J.D., et al., *Distribution of Rock, Metals, and Ices in Callisto*, Science, vol. 280, p. 1573–1576, June 5, 1998.

Anderson, John D., *Amalthea's Density Is Less Than That of Water*, Science, vol. 308, p. 1291–1293, May 27, 2005.

Bagenal, Fran, *Revealing Jupiter's Inner Secrets*, Sky & Telescope, July 2016.

Baker, David, & Ratcliff, Todd, *Nature's Wrath*, Sky & Telescope, September 2012.

Bolton, S.J., Adriani, A., Adumitroaie, V., et al., *Jupiter's interior and deep atmosphere: the first pole-to-pole passes with the Juno spacecraft*, Science, May 26, 2017: DOI:10.1126/science.aal2108.

Burns, Joseph A., et al., *The Formation of Jupiter's Faint Rings*, Science, vol. 284, p. 1146–1150, May 14, 1999.

Cassini Jupiter Millennium Flyby: http://www.jpl.nasa.gov/jupiterflyby/.

Connerney, J.E.P., et al., Jupiter's magnetosphere and aurorae observed by the Juno spacecraft during its first polar orbits, Science, May 26, 2017: DOI: 10.1126/science.aam5928.

Galileo mission (NASA): http://www2.jpl.nasa.gov/galileo/.

Greeley, Ronald, *The Partially Watery World of Europa, One of Jupiter's Moons*, Earth in Space, vol. 10, December 1997, p. 5–8: http://www.agu.org/sci_soc/greeley.html.

Guillot, T., et al., *A suppression of differential rotation in Jupiter's deep interior*, Nature, vol. 555, p. 227–230, March 8, 2018.

Harland, David M., Jupiter Odyssey: *The Story of NASA's Galileo Mission*, Springer-Praxis, 2000.

Iess, Luciano, et al., Measurement of Jupiter's asymmetric gravity field, Nature, Vol. 555, p 220–222, March 8, 2018.

Ingersoll, Andrew P., *Jupiter and Saturn, Scientific American Special 'Exploring Space'*, 1990.

Jewitt, David, Sheppard, Scott S., & Kleyna, Jan, The Strangest Satellites in the Solar System, Scientific American, August 2006: http://www.ifa.hawaii.edu/~jewitt/papers/2006/JSK06.pdf.

Johnson, Torrence V., The Galileo Mission to Jupiter and Its Moons, Scientific American Special 'New Light on the Solar System', p. 54–63, 2003.

Kaspi, Yohai, *Jupiter's atmospheric jet streams extend thousands of kilometres deep*, Nature, Vol. 555, March 8, 2018.

Kivelson, Margaret G., et al, *Galileo Magnetometer Measurements: A Stronger Case for a Subsurface Ocean at Europa*, Science, vol. 289, p. 1340–1343, August 25, 2000.

Malhotra R., *Migrating Planets*, Scientific American, September 1999.

Marcus, Philip S., *Prediction of a global climate change on Jupiter*, Nature, vol. 428, p. 828–831, April 22, 2004.

Martel, Linda M.V., Big Mountain, Big Landslide on Jupiter's Moon, Io, April 27, 1998: http://www.psrd.hawaii.edu/April98/io.html.

Martel, Linda M.V., The Europa Scene in the Voyager-Galileo Era, PSR Discoveries, February 26, 2001: http://www.psrd.hawaii.edu/Feb01/EuropaGeology.html.

Martel, Linda M.V., Bands on Europa, PSR Discoveries, November 25, 2002: http://www.psrd.hawaii.edu/Nov02/EuropanBands.html.

McCord, Thomas B., et al., *Hydrated Salt Minerals on Ganymede's Surface: Evidence of an Ocean Below*, Science, vol. 292, p. 1523–1525, May 25, 2001.

McEwen A.S., et al., High-Temperature Silicate Volcanism on Jupiter's Moon Io, Science, vol. 281 p. 87–90, July 3, 1998.

McEwen, A.S., et al., *Galileo: Io Reports*, Science, vol. 288, p. 1193–1220, May 19, 2000.

Moore, Kimberly M., et al., *A complex dynamo inferred from the hemispheric dichotomy of Jupiter's magnetic field*, Nature, vol. 561, p 76–78, September 5, 2018. DOI: 10.1038/s41586-018-0468-5.

Morrison, David, & Samz, Jane, *Voyage to Jupiter*, NASA SP-439, 1980.

New Horizons at Jupiter special section, Science, vol. 318, p. 215–243, October 12, 2007.

Nimmo, F., & Pappalardo, R.T., *Ocean worlds in the outer solar system*, Journal of Geophysical Research Planets, August 8, 2016.

Pappalardo, Robert T., et al., *The Hidden Ocean of Europa*, Scientific American Special 'New Light on the Solar System', p. 64–73, 2003.

Planet Hunters: https://www.planethunters.org.

Porco, Carolyn C., et al., *Cassini Imaging of Jupiter's Atmosphere, Satellites and Rings*, Science, vol. 299, p. 1541–1547, March 7, 2003.

Regas, Dean, *Jupiter Since Galileo*, Sky & Telescope, January 2014.

Reports on Comet Shoemaker-Levy 9 Collision with Jupiter, Science, vol. 267, p. 1277–1323, March 3, 1995.

Sánchez-Lavega, A., et al., *Depth of a strong jovian jet from a planetary-scale disturbance driven by storms*, Nature, vol. 451, p. 437–440, January 24, 2008.

Saur, Joachim, et al., *The search for a subsurface ocean in Ganymede with Hubble Space Telescope observations of its auroral ovals*, Journal of Geophysical Research: Space Physics, February 3, 2015.

Scharf, Caleb, *A Universe of Dark Oceans*, Sky & Telescope, December 2014.

Schenk, Paul M., et al., Flooding of Ganymede's bright terrains by low-viscosity water-ice lavas, Nature, vol. 410, p. 57–60, March 1, 2001.

Sheppard, Scott S. and Jewitt, David C., An abundant population of small irregular satellites around Jupiter, Nature, vol. 423, p. 261–263, May 15, 2003.

Sheppard, Scott S., *Dancing with Planets*, Sky & Telescope, June 2016.

Showman, Adam P. and Malhotra, Renu, The Galilean Satellites, Science, vol. 286 p. 77–84, October 1, 1999.

Simon, Amy A., et al., Historical and Contemporary Trends in the Size, Drift, and Color of Jupiter's Great Red Spot, The Astronomical Journal, Vol. 155, No. 4, March 13, 2018.

Simon, Amy A., *The Not-So-Great Red Spot*, Sky & Telescope, March 2016.

Soderblom, Laurence A., *The Galilean Moons of Jupiter*, Scientific American, January 1980.

Stevenson, David, *Europa's Ocean – the Case Strengthens*, Science, vol. 289, p. 1305–1307, August 25, 2000.

Taylor, G. Jeffrey, Europa's Salty Surface, PSR Discoveries, September 24, 1998: http://www.psrd.hawaii.edu/Sept98/EuropaSalts.html.

Taylor, G. Jeffrey, Jupiter's Hot, Mushy Moon, PSR Discoveries, February 15, 2000: http://www.psrd.hawaii.edu/Feb00/IoMagmaOcean.html.

Voyager missions (NASA): http://voyager.jpl.nasa.gov/.

Young, Richard E., et al., *Galileo Probes Jupiter's Atmosphere*, Science, vol. 272, p. 837–860, May 10, 1996.

Zander, Jon, *Peering Beneath Jupiter's Clouds*, Sky & Telescope, September 2011.

Chapter 9: Saturn

30-day reports from the Voyager 1 and Voyager 2 instrument teams, Science, April 10, 1981 and January 29, 1982.

Adamkovics, Mate, et al., *Widespread Morning Drizzle on Titan*, Science Express, October 11, 2007.

Atreya, Sushil K., *The Mystery of Methane on Mars and Titan*, Scientific American, January 15, 2009.

Bagenal, Fran, *Saturn's mixed magnetosphere*, Nature, vol. 433, p. 695–696, February 17, 2005.

Barnes, Jason, *Titan: Earth in Deep Freeze*, Sky & Telescope, p. 26–32, December 2008.

Batson, Raymond M., *Voyager 1 and 2 Atlas of Six Saturnian Satellites*, NASA SP-474, 1984: http://ntrs.nasa.gov/archive/nasa/casi.ntrs.nasa.gov/19840027171_1984027171.pdf.

Beatty, J. Kelly, *Saturn's Amazing Rings*, Sky & Telescope, May 2013.

Bosh, A.S., and Rivkin, A.S. *Observations of Saturn's inner satellites during the May 1995 ring-plane crossing*, Science, vol. 272, p. 518–521, 1996.

Brown, R.H., et al., *The identification of liquid ethane in Titan's Ontario Lacus*, Nature, vol. 454, p. 607–610, July 31, 2008.

Brown, Robert, *Lebreton*, Jean-Pierre and Waite, Hunter, Titan from Cassini-Huygens, Springer, 2009.

Cassini at Enceladus, Science, vol. 311, p. 1388–1428, March 10, 2006.

Cassini's Final Year: Science Highlights and Discoveries (October 4, 2018): https://agupubs.onlinelibrary.wiley.com/doi/toc/10.1002/(ISSN)1944-8007.CASSINI_FINALE1.

Cassini mission (NASA): http://saturn.jpl.nasa.gov.

Charnoz, S., et al., Cassini Discovers a Kinematic Spiral Ring Around Saturn, Science, vol. 310, p. 1300–1304, November 25, 2005.

Closing with Saturn (special issue), Science, vol. 362, Issue 6410, October 5, 2018.

Coustenis, Athéna and Taylor, F.W., *Titan:* exploring an earthlike world, World Scientific, 2008.

Crida, A., and Charnoz, S., Formation of Regular Satellites from Ancient Massive Rings in the Solar System, Science, vol. 338, p. 1196–1199, November 30, 2012. DOI: 10.1126/science.1226477.

Cruikshank, D.P., et al., *Surface composition of Hyperion*, Nature, vol. 448, p. 54–56, 5 July 2007.

Cuzzi, J.N. et al., *An Evolving View of Saturn's Dynamic Rings*, Science, vol. 327, p. 1470–1475, March 19, 2010.

Dyudina, U.A., et al., *Dynamics of Saturn's South Polar Vortex*, Science, vol. 319, p. 1801, March 28, 2008.

Elachi, C., et al., *Titan Radar Mapper observations from Cassini's T3 fly-by*, Nature, vol. 441, p. 709–713, June 8, 2006.

Fischer, G., et al., A giant thunderstorm on Saturn, Nature, vol. 475, p 75–77, July 7, 2011.

Gehrels, T., and Matthews, M.S. (Eds.), *Saturn*, University of Arizona Press.

Gombosi, T.I. and Ingersoll, A.P., *Saturn: Atmosphere, Ionosphere, and Magnetosphere*, Science, vol. 327, p. 1476–1479, March 19, 2010.

Greenberg, R. and Brahic, A. (Eds.), *Planetary Rings*, University of Arizona Press, 1984.

Griffith, C.A., Penteado, P., and Brown R., *Evidence for a Polar Ethane Cloud on Titan*, Science, September 2006.

Hansen, C.J., et al., *Water vapour jets inside the plume of gas leaving Enceladus*, Nature, vol. 456, p. 477–479, November 27, 2008.

Hedman, Matthew M., *The Source of Saturn's G Ring*, Science, vol. 317, p. 653–656, August 3, 2007.

Horst, Sarah, *Titan's Veil*, Sky & Telescope, February 2019, p. 22–29.

Hörst, S.M., *Titan's atmosphere and climate*, Journal of Geophysical Research: Planets, Volume 122, Issue 3, March 2017.

Hueso, R., and Sánchez-Lavega, A., *Methane storms on Saturn's moon Titan*, Nature, vol. 442, p. 428–431, July 27, 2006.

Huygens mission (ESA): http://sci.esa.int/huygens.

Ingersoll, Andrew P., *Jupiter and Saturn*, Scientific American Special 'Exploring Space', 1990.

Initial Cassini Results, Science, vol. 307, p. 1226–1251, February 25, 2005.

Jewitt, David, Sheppard, Scott S., and Kleyna, Jan, The Strangest Satellites in the Solar System, Scientific American, August 2006: http://www.ifa.hawaii.edu/~jewitt/papers/2006/JSK06.pdf.

Jones, G.H. et al., *The Dust Halo of Saturn's Largest Icy Moon, Rhea*, Science, vol. 319, p. 1380–1384, March 7, 2008.

Kerr, Richard A., *How Saturn's Icy Moons Get a (Geologic) Life*, Science, vol. 311, p. 29, January 6, 2006.

Kieffer, Susan, & Jakosky, Bruce M., Enceladus–Oasis or Ice Ball?, Science, vol. 320, p. 1432–1433, June 13, 2008.

Krimigis, S.M., et al, A dynamic, rotating ring current around Saturn, Nature, vol. 450, p. 1050–1053, December 13, 2007.

Lakdawalla, Emily, *Ice Worlds of the Ringed Planet*, Sky & Telescope, June 2009.

Li, Cheng, & Ingersoll, Andrew P., Moist convection in hydrogen atmospheres and the frequency of Saturn's giant storms, Nature Geoscience, vol. 8, p 398–403, April 13, 2015.

Lorenz, Ralph, *The Weather on Titan*, Science, vol. 290, p. 467–468, October 20, 2000.

Lorenz, Ralph, et al., *The Sand Seas of Titan: Cassini RADAR Observations of Longitudinal Dunes*, Science, vol. 312, p. 724–727, May 5, 2006.

Lorenz, Ralph, and Mitton, Jacqueline: Lifting Titan's Veil: Exploring the Giant Moon of Saturn, Cambridge University Press, New York, 2002.

Murray, Carl D., et al., The determination of the structure of Saturn's F ring by nearby moonlets, Nature, vol. 453, p. 739–744, June 5, 2008.

New views of Titan (special issue), Nature, vol. 438, p. 538–539 and 765–802, December 1, 2005.

Nicholson, P.D., et al., *Observations of Saturn's ring-plane crossings in August and November 1995*, Science, vol. 272, p. 509–515, 1996.

Nimmo, F. & Pappalardo, R.T., *Ocean worlds in the outer solar system*, Journal of Geophysical Research Planets, August 8, 2016.

Porco, Carolyn, Cassini at Saturn, Scientific American, October 2017, vol. 317, p. 78–85. Published online September 19, 2017. DOI:10.1038/scientificamerican1017-78.

Rannou, P., Montmessin, F., Hourdin, F., Lebonnois, S., The Latitudinal Distribution of Clouds on Titan, Science, vol. 311 p. 201–205, January 13, 2006.

Robrtson, Donald F., *Where Goes the Rain?*, Sky & Telescope, March 2013.

Sanchez-Lavega, Agustin, et al., Deep winds beneath Saturn's upper clouds from a seasonal long-lived planetary-scale storm, Nature, vol. 475, p. 71–74, July 7, 2011.

Sanchez-Lavega, Agustin, *Saturn's Raging Superstorm*, Sky & Telescope, May 2012.

Schaller, E.L., et al., Storms in the tropics of Titan, Nature, vol. 460, p. 873–875, August 13, 2009.

Schmidt, Jürgen, Slow dust in Enceladus' plume from condensation and wall collisions in tiger stripe fractures, Nature, vol. 451, p. 685–688, February 7, 2008.

Showalter, Mark R., Saturn's Strangest Ring Becomes Curiouser and Curiouser, Science, vol. 310, p. 1287–1288, November 25, 2005.

Soderblom, Laurence A., and Johnson, Torrence V., *The Moons of Saturn*, Scientific American Special 'Exploring Space', 1990.

Sohl, Frank, Revealing Titan's Interior, Science, vol. 327, p. 1338–1339, March 12, 2010.

Spitale, Joseph N., and Porco, Carolyn C., Association of the jets of Enceladus with the warmest regions on its south-polar fractures, Nature, vol. 449, p. 695–697, October 11, 2007.

Sremcevic, Miodrag, et al., A belt of moonlets in Saturn's A ring, Nature, vol. 449, p. 1019–1021, October 25, 2007.

The Saturn System through the Eyes of Cassini, NASA and the Lunar and Planetary Institute (LPI): https://saturn.jpl.nasa.gov/resources/7777.

Thomas, P.C., et al., *Hyperion's sponge-like appearance*, Nature, vol. 448, p. 50–53, 5 July 2007.

Tiscareno, Matthew S., *Ringworld Revelations*, Sky & Telescope, February 2007.

Tiscareno, Matthew S., et al., 100-metre-diameter moonlets in Saturn's A ring from observations of 'propeller' structures, Nature, vol. 440, p. 648–650, March 30, 2006.

Tobie, Gabriel, Lunine, Jonathan I., and Sotin, Christophe, Episodic outgassing as the origin of atmospheric methane on Titan, Nature, vol. 440, p. 61–64, March 2, 2006.

Tokano, Tetsuya, et al., Methane drizzle on Titan, Nature, vol. 442, p. 432–435, July 27, 2006.

Verbiscer, Anne, et al., *Enceladus: Cosmic Graffiti Artist Caught in the Act*, Science, vol. 315, p. 815, February 9, 2007.

Verbiscer, Anne J., et al., Saturn's largest ring, Nature, vol. 461, p. 1098–1100, October 22, 2009.

Zarka, Philippe, et al., Modulation of Saturn's radio clock by solar wind speed, Nature, vol. 450, p. 265–267, November 8, 2007.

Chapter 10: Uranus

30-day reports from the Voyager 2 instrument teams, Science, vol. 233, July 4, 1986.

Benedetti, Laura Robin, Nguyen, Jeffrey H., Caldwell, Wendell A., Liu Hongjian, Kruger, Michael and Jeanloz, Raymond, Dissociation of CH4 at High Pressures and Temperatures: Diamond Formation in Giant Planet Interiors? Science, vol. 286 p. 100–102, October 1, 1999.

Bergstralh J.T., Miner E.D., and Matthews M.S., (Eds.), *Uranus*, University of Arizona Press, 1991.

Cavazzoni, C., et al., *Superionic and Metallic States of Water and Ammonia at Giant Planet Conditions*, Science, vol. 283, p. 44–46, January 1, 1999.

Crida, A., and Charnoz, S., Formation of Regular Satellites from Ancient Massive Rings in the Solar System, Science, vol. 338, p. 1196–1199, November 30, 2012. DOI: 10.1126/science.1226477.

de Pater, Imke, et al., New Dust Belts of Uranus: One Ring, Two Ring, Red Ring, *Blue Ring*, Science vol. 312, p. 92–94, April 7, 2006.

de Pater, Imke, et al., *The Dark Side of the Rings of Uranus*, Science, vol. 317, p. 1888–1890, September 28, 2007.

Desch, S., Mass Distribution and Planet Formation in the Solar Nebula, Astrophysical Journal, vol. 671, p. 878–893, December 10, 2007.

Guillot, Tristan, *Interiors of Giant Planets Inside and Outside the Solar System*, Science, vol. 286, p. 72–77, October 1, 1999.

Hubbard, W.B., et al., Interior Structure of Neptune: Comparison with Uranus, Science vol. 253, p. 648–651, August 9, 1991.

Infrared observations of Uranus and its rings (Imke de Pater): http://astron.berkeley.edu/~newstar/Infrared/UranusAo/UranusAO.htm.

Ingersoll, Andrew, *Uranus*, Scientific American, January 1987.

Johnson T.V., Brown R.H. and Soderblom L. A., *The Moons of Uranus*, Scientific American, April 1987.

Lamy, L., et al., *Uranus' aurorae past equinox*, Journal of Geophysical Research: Space Physics, published online March 9, 2017: http://onlinelibrary.wiley.com/doi/10.1002/2017JA023918/abstract.

Laeser, Richard, *McLaughlin, William, & Wolff, Donna, Engineering Voyager 2's Encounter with Uranus*, Scientific American Special, Exploring Space, 1990.

Littmann, Mark, Planets Beyond: *Discovering the Outer Solar System*, John Wiley, Revised edition 1990.

Miner, Ellis D., *Uranus, Wiley-Praxis*, 2nd Edition 1998.

O'Donel, Alexander, *The Planet Uranus: A History of Observation, Theory and Discovery*, Faber & Faber, 1965.

Showalter, Mark R., and Lissauer, Jack J., The Second Ring-Moon System of Uranus: Discovery and Dynamics, Science, vol. 311, p. 973–977, February 17, 2006.

Srmovsky, L.A., et al., Episodic bright and dark spots on Uranus, Icarus, 220, p. 6–22, 21 April 2012.

Sromovsky, Lawrence and Fry, Patrick, *Most detailed views ever of weather on Uranus*, 18 December 2012, SPIE Newsroom. DOI: 10.1117/2.1201212.004620.

Stanley, Sabine, and Bloxham, Jeremy, Convective-region geometry as the cause of Uranus' and Neptune's unusual magnetic fields, Nature, vol. 428, p. 151–153, March 11, 2004.

Taylor, G. Jeffrey, Uranus, Neptune, and the Mountains of the Moon, PSR Discoveries, August 21, 2001: http://www.psrd.hawaii.edu/Aug01/bombardment.html.

Thommes, E.W., Duncan, M.J., and Levison H.F., The formation of Uranus and Neptune in the Jupiter-Saturn region of the Solar System, Nature, vol. 402, p. 635–638, December 9, 1999.

Thommes, E.W., Duncan, M.J., and Levison H.F., The formation of Uranus and Neptune Among Jupiter And Saturn, The Astronomical Journal, vol. 123. p. 2862–2883, May 2002.

Voyager missions (NASA): http://voyager.jpl.nasa.gov/.

Chapter 11: Neptune

30-day reports from the Voyager 2 instrument teams, Science, vol. 246, December 15, 1989.

Agnor, Craig B., and Hamilton, Douglas P., *Neptune's capture of its moon Triton in a binary–planet gravitational encounter*, Nature, vol. 441, p. 192–194, May 11, 2006.

Cruikshank, Dale P., et al., *Ices on the Surface of Triton*, Science, vol. 261, p. 742–745, August 6, 1993.

Cruikshank D.P., (Ed.), *Neptune and Triton*, University of Arizona Press, 1994.

de Pater, Imke, et al., The Dynamic Neptunian Ring Arcs: Evidence for a Gradual Disappearance of Liberte and Resonant Jump of Courage, Icarus, September 27, 2004.

Desch, S., Mass Distribution and Planet Formation in the Solar Nebula, Astrophysical Journal, vol. 671, p. 878–893, December 10, 2007.

Dumas, C., Terrile, R.J., Smith, B.A., Schneider, G., and Becklin, E.E., Stability of Neptune's ring arcs in question, Nature vol. 400, p. 733–735, August 19, 1999.

Dumas, C., Terrile, R., Smith, B.A., and Schneider, G., *Astrometry and Near-Infrared Photometry of Neptune's Inner Satellites and Ring Arcs*, The Astronomical Journal, vol. 123, p. 1776–1783, March 2002.

Elliot, J.L., et al., *Global Warming on Triton*, Nature, vol. 293, p. 785–787, June 25, 1998.

Holman, Matthew J., et al., Discovery of five irregular moons of Neptune, Nature, vol. 430, p. 865–867, August 19, 2004.

Hubbard, W.B., Brahic, A., Sicardy, B., Elicer, L.R, Roques, F., and Vilas, F., *Occultation detection of a Neptunian ring-like arc*, Nature, vol. 319, p. 636–640, February 20, 1986.

Hubbard, W.B., et al, *Interior Structure of Neptune: Comparison with Uranus*, Science vol. 253, p. 648–651, August 9, 1991.

Hubbard, W.B., *Neptune's Deep Chemistry*, Science, vol. 275, p. 1279–1280, February 28, 1997.

Kaspi, Yohai, et al., Atmospheric confinement of jet streams on Uranus and Neptune, Nature, vol. 497, p. 344–347, May 16, 2013.

Kinoshita, June, *Neptune, Scientific American Special 'Exploring Space'*, 1990.

Lellouch, E., et al., *Detection of CO in Triton's atmosphere and the nature of surface-atmosphere interactions*, Astronomy & Astrophysics, vol. 512, L8, March–April 2010.

Miner, Ellis, *and Wessen, Randii R., Neptune:* The Planet, Rings and Satellites, Springer-Praxis, 2002.

Moore, Patrick, *The Planet Neptune: An Historical Survey Before Voyager*, 2nd *Edition*, Wiley-Praxis, 1996.

Porco, Carolyn C., An explanation for Neptune's ring arcs, Science, vol. 253, p. 995–1001, August 30, 1991.

Roddier F., Roddier C., Graves J.E., Northcott M.J., and Owen T., Neptune's Cloud Structure and Activity: Ground-based Monitoring with Adaptive Optics, Icarus, vol. 168–172, 1998.

Sheehan, William, *Kollerstrom, Nicholas, and Waff, Craig B.*, The Case of the Pilfered Planet, Scientific American, December 2004.

Sicardy B., Roddier F., Roddier C., Perozzi E., Graves J.E., Guyon O., and Northcott M.J., Images of Neptune's ring arcs obtained by a ground-based telescope, Nature, vol. 400, p. 731–733, August 19, 1999.

Stanley, Sabine, and Bloxham, Jeremy, Convective-region geometry as the cause of Uranus' and Neptune's unusual magnetic fields, Nature vol. 428, p. 151–153, March 11, 2004.

Thommes, E.W., Duncan, M.J., and Levison H.F., *The formation of Uranus and Neptune in the Jupiter-Saturn region of the Solar System*, Nature, vol. 402, p. 635–638, December 9, 1999.

Thommes, E.W., Duncan, M.J., and Levison H.F., *The formation of Uranus and Neptune Among Jupiter And Saturn*, The Astronomical Journal, vol. 123. p. 2862–2883, May 2002.

Triton – Reports from Voyager instrument teams, Science, vol. 250, p. 410–443, October 19, 1990.

Wong, Michael H., et al., *A New Dark Vortex on Neptune*, The Astronomical Journal, 155:117, March 2018. https://doi.org/10.3847/1538-3881/aaa6d6.

Chapter 12: Pluto & Kuiper Belt

Beatty, J. *Kelly, Pluto & Charon: The Odd Couple*, Sky & Telescope, November 2015.

Beatty, J. *Kelly, Pluto's Amazing Story*, Sky & Telescope, October 2016.

Beatty, J. *Kelly, Pluto's Perplexing Atmosphere*, Sky & Telescope, November 2016.

Beatty, J. *Kelly, Charon & Company*, Sky & Telescope, December 2016.

Bertoldi, F., et al., *The trans-neptunian object UB313 is larger than Pluto*, Nature, vol. 439, p. 563–564, February 2, 2006.

Bertoldi, F., et al., *Comment on the recent Hubble Space Telescope size measurement of 2003 UB313 by Brown et al.*, April 13, 2006: http://www.astro.uni-bonn.de/~bertoldi/ub313/.

Binzel, R., et al., *The Detection of Eclipses in the Pluto-Charon System*, Science, vol. 228, p. 1193–1195, June 7, 1985.

Bown, M.E., *Dwarf Planets* http://web.gps.caltech.edu/~mbrown/dwarfplanets/.

Brown M.E. & Schaller E.L., *The Mass of Dwarf Planet Eris*, Science, vol. 316, p. 1585, June 15, 2007.

Brown, M.E., et al., *Direct measurement of the size of 2003 UB313 from the Hubble Space Telescope* (preprint): http://www.gps.caltech.edu/%7Embrown/papers/ps/xsize.pdf:

Brown, M.E., et al., *Satellites of The Largest Kuiper Belt Objects*, The Astrophysical Journal, vol. 639, p. L43–L46, March 1, 2006.

Brown, Michael E. and Calvin, Wendy M., Evidence for Crystalline Water and Ammonia Ices on Pluto's Satellite Charon, Science, vol. 287, p. 107–109, January 7, 2000.

Brown, Michael E., *Discovery of a Candidate inner Oort Cloud planetoid*, ApJ Letters, August 10, 2004: http://www.gps.caltech.edu/%7Embrown/papers/ps/sedna.pdf.

Brown, M.E., Trujillo, C.A., and Rabinowitz, D.L., *Discovery of A Planetary-Sized Object In The Scattered Kuiper Belt*, The Astrophysical Journal, December 10, 2005: http://web.gps.caltech.edu/˜mbrown/papers/ps/xena.pdf.

Buie, M.W., et al., *Pluto and Charon with the Hubble Space Telescope: I. Monitoring global change and improved surface properties from light curves*, The Astronomical Journal, vol. 139, p. 1117–1127, 2010: http://www.boulder.swri.edu/~buie/biblio/pub072.html.

Buie, M.W., et al., *Pluto and Charon with the Hubble Space Telescope: II. Resolving changes on Pluto's surface and a map for Charon*, The Astronomical Journal, vol. 139, p. 1128–1143, 2010: http://www.boulder.swri.edu/~buie/biblio/pub073.html.

Buie, Marc W., et al., Orbits and photometry of Pluto's satellites: Charon, *S/2005 P1 and S/2005 P2*: http://arxiv.org/PS_cache/astro-ph/pdf/0512/0512491.pdf.

Canup, Robin M., *A Giant Impact Origin of Pluto-Charon*, Science, vol. 307, p. 546–550, January 28, 2005.

Clyde Tombaugh: https://www.kshs.org/kansapedia/clyde-tombaugh/12222.

Cook, Jason C., et al., Near-Infrared Spectroscopy of Charon: Possible Evidence for Cryovolcanism on Kuiper Belt Objects, Astrophysical Journal, vol. 663, p. 1406–1419, July 10, 2007.

Cruikshank, D.P., & Sheehan, W., *Discovering Pluto*, University of Arizona Press, 2018.

Davies, John, Beyond Pluto: *Exploring the Outer Limits of the Solar System*, Cambridge University Press, 2001.

Distant EKOs – The Kuiper Belt Electronic Newsletter: http://www.boulder.swri.edu/ekonews.

Elliot, J.L., et al., The recent expansion of Pluto's atmosphere, Nature, vol. 424, p. 165–168, July 10, 2003.

Exploring the Trans-Neptunian Solar System, Committee on Planetary and Lunar Exploration, National Research Council (1998): http://www.nap.edu/catalog/6080.html.

Gingerich, Owen, *Losing It in Prague: The Inside Story of Pluto's Demotion*, Sky & Telescope, November 2006.

Gladman, Brett, *The Kuiper Belt and the Solar System's Comet Disk*, Science, vol. 307, p. 71–75, January 7, 2005.

Goldreich, P., Lithwick, Y., & Sari, R., Formation of Kuiper-belt Binaries by Dynamical Friction and Three-body Encounters, Nature, vol. 420, p. 643–646, August 28, 2002.

Jewitt, David C., and Luu, Jane, *Crystalline water ice on the Kuiper Belt object (50000) Quaoar*, Nature, vol. 432, p. 731–733, December 9, 2004.

Keane, James T., et al., Reorientation and faulting of Pluto die to volatile loading within Sputnik Planitia, Nature, vol. 540, p. 90–93, December 1, 2017.

Kenyon, Scott J., and Bromley, Benjamin C., *Stellar encounters as the origin of distant Solar System objects in highly eccentric orbits*, Nature, vol. 432, December 2, 2004.

Kuiper Belt (David Jewitt): http://www2.ess.ucla.edu/~jewitt/kb.html.

Lakdawalla, Emily, *Pluto and the Kuiper Belt*, Sky & Telescope, February 2014.

Lakdawalla, Emily, *Pluto at Last*, Sky & Telescope, July 2015.

Levison, H.F., Morbidelli, A., *The formation of the Kuiper belt by the outward transport of bodies during Neptune's migration*, Nature, vol. 426, p. 419–421, November 27, 2003.

Levy, David H., Clyde Tombaugh: *Discoverer of Planet Pluto*, University of Arizona Press, 1992.

Littmann, Mark, *From Chaos to the Kuiper Belt*, Sky & Telescope, September 2007.

Littmann, Mark, *Dark Beasts of the Trans-Neptunian Zoo*, Sky & Telescope, November 2007.

Luu, Jane, and Jewitt, David C., *Kuiper Belt Objects: Relics from the Accretion Disk of the Sun*, Ann. Rev. Astron. Astrophys., vol. 40, p. 63–101, 2002.

Minor Planet Center: http://www.minorplanetcenter.org/iau/mpc.html.

New Horizons mission: http://pluto.jhuapl.edu/.

Nimmo, F., et al., *Reorientation of Sputnik Planitia implies a subsurface ocean on Pluto*, Nature, vol. 540, p 94–96, December 1, 2016, doi:10.1038/nature20148.

Owen, Tobias, et al., *Surface Ices and the Atmospheric Composition of Pluto*, Science, vol. 261, p. 745–748, August 6, 1993.

Petit, J.-M., et al., *The Extreme Kuiper Belt Binary* 2001 QW322, Science, vol. 322, p. 432–434, October 17, 2008.

Rabinowitz, David L., et al., *Photometric Observations Constraining the Size, Shape, and Albedo of 2003 EL61, a Rapidly Rotating, Pluto-sized Object in the Kuiper Belt*, The Astrophysical Journal, vol. 639, p. 1238–1251, March 10, 2006.

Schaller, E.L., & Brown, M.E., *Volatile Loss and Retention on Kuiper Belt Objects*, The Astrophysical Journal, vol 659, p. L61–L64, April 2007.

Sheppard, Scott, *Beyond the Kuiper Belt*, Sky & Telescope, March 2015.

Sheppard, Scott, *The Hunt for Planet X*, Sky & Telescope, October 2017.

Sicardy B., et al., *Large changes in Pluto's atmosphere as revealed by recent stellar occultations*, Nature, vol. 424, p. 168–170, July 10, 2003.

Stern, Alan, and Mitton, Jacqueline, *Pluto and Charon*, 2nd edition, Wiley-VCH, 2005.

Stern, S. Alan, *Journey to the Farthest Planet*, Scientific American Special Edition "New Light on the Solar System", September 2003.

Stern, S. Alan, *The 3rd Zone: Exploring the Kuiper Belt*, Sky & Telescope, November 2003.

Stern, S. Alan, *The evolution of comets in the Oort cloud and Kuiper belt*, Nature, vol. 424, p. 639–642, August 7, 2003.

Stern, S.A., et al., *A giant impact origin for Pluto's small moons and satellite multiplicity in the Kuiper belt*, Nature, vol. 439, p. 946–948, February 23, 2006.

The Discovery of Eris / 2003 UB313 (Mike Brown website): http://www.gps.caltech.edu/~mbrown/planetlila/.

Tombaugh, Clyde, and Moore, Patrick, *Out of the Darkness: the Planet Pluto*, New American Library, 1981.

Two New Satellites of Pluto: http://www.boulder.swri.edu/plutomoons/

Ward, William R. and Hahn, Joseph M., *Neptune's Eccentricity and the Nature of the Kuiper Belt*, Science vol. 280, p. 2104–2106, June 26, 1998.

Ward, William R., and Canup, Robin M., *Forced Resonant Migration of Pluto's Outer Satellites by Charon*, Science Express web site, June 30, 2006.

Wetterer, Margaret K., *Clyde Tombaugh and the Search for Planet X*, Carolrhoda Books, 1996.

Young, Eliot, *Charon's First Detailed Spectra Hold Many Surprises*, Science, vol. 287, p. 53–54, January 7, 2000.

Young, Eliot F., Binzel, Richard P., Crane, Keenan, *A Two-Color Map Of Pluto's Sub-Charon Hemisphere*, The Astronomical Journal, vol. 121, p. 552–561, January 2001 http://www.boulder.swri.edu/recent/pluto_map.pdf.

Zhang, Xi, Strobel, Darrell F., & Imanaka, Hiroshi, *Haze heats Pluto's atmosphere yet explains its cold temperature*, Nature, November 2017.

Chapter 13: Comets, Asteroids and Meteorites

Arctic Asteroid! (Tagish Lake meteorite): http://science.nasa.gov/headlines/y2000/ast01jun_1m.htm.

Asteroid and Comet Impact Hazards: http://impact.arc.nasa.gov/.

Beatty, J. Kelly, *The Falcon's Wild Ride*, Sky & Telescope, September 2006.

Bell, Jim, *Dawn's Early Light: A Vesta Fiesta!*, Sky & Telescope, November 2012.

Bell, Jim, *Protoplanet Close-up*, Sky & Telescope, September 2012.

Bland, Philip A., et al., An Anomalous Basaltic Meteorite from the Innermost Main Belt, Science, vol. 325, p. 1525–1527, September 18, 2009.

Boehnhardt, Hermann, *The Death of a Comet and the Birth of Our Solar System*, Science, vol. 292, p. 1307–1309, May 18, 2001.

Borovicka, Jiří, et al., The trajectory, structure and origin of the Chelyabinsk asteroidal impactor, Nature, vol. 503, p 235–237, November 14, 2013.

Bottke, William F., & Martel, Linda M.V., *Iron Meteorites as the Not So Distant Cousins of Earth*, PSR Discoveries, July 21, 2006: http://www.psrd.hawaii.edu/July06/asteroidGatecrashers.html.

Bottke, William F., *Spun in the Sun*, Nature, vol. 446, p. 382–383, March 22, 2007.

Bottke, William F., et al., An asteroid breakup 160 Myr ago as the probable source of the K/T impactor, Nature, vol. 449, p. 48–53, September 6, 2007.

Comets (David Jewitt): http://www.ifa.hawaii.edu/faculty/jewitt/comet.html.

Comet Shoemaker-Levy Collision with Jupiter (NASA-JPL website): http://www2.jpl.nasa.gov/sl9/.

Comet Shoemaker-Levy 9 Reports, Science, vol. 267, p. 1277–1323, March 3, 1995.

Chiron: http://nssdc.gsfc.nasa.gov/planetary/chiron.html.

Cruikshank, Dale P., & Sheehan, William, Discovering Pluto: *Exploration at the Edge of the Solar System*, University of Arizona Press, 2018.

Dawn mission: http://dawn.jpl.nasa.gov/.

Deep Impact mission: http://deepimpact.jpl.nasa.gov/home/index.html.

Deep Impact special section, Science, vol. 310, p. 258–283, October 14, 2005.

Deep Space 1 mission: http://nmp.jpl.nasa.gov/ds1/.

DeMeo, F.E., & Carry, B., *Solar System evolution from compositional mapping of the asteroid belt*, Nature, vol. 505, p. 629–634, January 30, 2014.

Dorminey, Bruce, *Earth's Come and Go Moons*, Sky & Telescope, September 2015.

Durda, Daniel D., *The Chelyabinsk Super-Meteor*, Sky & Telescope, June 2013.

Durda, Daniel, *Space Rock Rendezvous*, Sky & Telescope, June 2018.

Festou, M.C., Keller, H.U., and Weaver, H.A, eds., *Comets II*, University of Arizona Press, 2005.

Floss, Christine, *QUE 93148: A Part of the Mantle of Asteroid 4 Vesta?* PSRD Discoveries, January 23, 2003: http://www.psrd.hawaii.edu/Jan03/QUE93148.html.

Giotto mission: http://sci.esa.int/giotto/.

Hayabusa special section, Science, vol. 312, p. 1330–1353, June 2, 2006.

Hayabusa - dust from Itokawa, Science special issue, vol. 333, p. 1113–1131, August 26, 2011.

Jenniskens, P., et al., *The impact and recovery of asteroid 2008 TC3*, Nature, vol. 458, p. 485–488, March 26, 2009.

Jewitt, David, & Hsieh, Henry, *Physical Observations of 2005 UD: A Mini-Phaethon*, The Astronomical Journal, Vol. 132, October 2006: http://www.ifa.hawaii.edu/faculty/jewitt/papers/2006/JH06.pdf.

Jewitt, David, *Mysterious Travelers*, Sky & Telescope, December 2013.

Jewitt, David, *The Active Asteroids*: http://www2.ess.ucla.edu/~jewitt/mbc.html.

Kring, D.A., *Meteorites and Their Properties*, Lunar and Planetary Laboratory, Department of Planetary Sciences, University of Arizona, 1998: http://meteorites.lpl.arizona.edu/index.html.

Kuiper Belt (David Jewitt): http://www.ifa.hawaii.edu/faculty/jewitt/kb.html.

Küppers, Michael, et al., A large dust/ice ratio in the nucleus of comet 9P/Tempel 1, Nature, vol. 437, p. 987–990, October 13, 2005.

Lakdawalla, Emily, *Dawn Arrives at Ceres*, Sky & Telescope, April 2015.

Li, Aigen, *Cosmic crystals caught in the act*, Nature, vol. 459, p. 173–176, May 14, 2009.

Laughlin, Greg, *'Oumuamua's Dramatic Visit*, Sky & Telescope, October 2018.

LINEAR NEO program: http://www.ll.mit.edu/mission/space/linear/.

Lisse, C.M., et al., *Spitzer Spectral Observations of the Deep Impact Ejecta*, Science, vol. 313, p. 635–640, August 4, 2006.

Main belt comets: http://www.ifa.hawaii.edu/~hsieh/mbcs.html.

Marchi, S., et al., *The missing large impact craters on Ceres*, Nature Communications. Published online: 26 July 2016: https://www.nature.com/articles/ncomms12257.

Marchis, Franck, A low density of 0.8 g cm^{-3} for the Trojan binary asteroid 617 Patroclus, Nature, vol. 439, p. 565–567, February 2, 2006.

Marcus Robert, Melosh, H. Jay, & Collins, Gareth, Earth Impact Effects Program: http://www.lpl.arizona.edu/impacteffects/.

Martel, Linda M.V., Meteorites on Ice, PSR Discoveries, November 2001: http://www.psrd.hawaii.edu/Nov01/metsOnIce.html.

Martel, Linda M.V., Searching Antarctic Ice for Meteorites, PSR Discoveries, February 28, 2002: http://www.psrd.hawaii.edu/Feb02/meteoriteSearch.html.

Martel, Linda M.V., *Searching Antarctic Ice for Meteorites*, PSR Discoveries, February 28, 2002: http://www.psrd.hawaii.edu/Feb02/meteoriteSearch.html.

Martel, Linda M.V., Using Chondrites to Understand the Inside of Asteroid 433 Eros, PSR Discoveries, June 28, 2002: http://www.psrd.hawaii.edu/June02/ErosPorosity.html.

Martel, Linda M.V., Getting to Know Vesta, PSR Discoveries, November 27, 2007: http://www.psrd.hawaii.edu/Nov07/HEDs-Vesta.html.

Marzari, Francesco, *Puzzling Neptune Trojans*, Science Express web site, June 15, 2006.

Meech, Karen J., 1997 *Apparition of Comet Hale-Bopp*, PSR Discoveries, February 14, 1997: http://www.psrd.hawaii.edu/Feb97/Hale-Bopp.html.

Meteor showers (American Meteor Society): http://www.amsmeteors.org/showers.html.

Meteor Showers Online (Gary W. Kronk): http://meteorshowersonline.com/.

Morbidelli, A., et al., *Chaotic capture of Jupiter's Trojan asteroids in the early Solar System*, Nature, vol. 435, p. 462–465, May 26, 2005.

Nakamura, T., et al., Chondrule-like Objects in Short-Period Comet 81P/Wild 2, Science, vol. 321, p. 1664–1667, September 19, 2008.

NASA 2006 Near-Earth Object Survey and Deflection Study - Final Report: http://www.hq.nasa.gov/office/pao/FOIA/NEO_Analysis_Doc.pdf.

NASA 2017 Report of the Near-Earth Object Science Definition Team: https://www.nasa.gov/sites/default/files/atoms/files/2017_neo_sdt_final_e-version.pdf.

NASA Center for Near Earth Object Studies: https://cneos.jpl.nasa.gov/.

NASA-JPL Asteroid Watch: https://www.jpl.nasa.gov/asteroidwatch/.

NASA Planetary Defense Coordination Office: https://www.nasa.gov/planetarydefense.

Near-Earth Object Survey and Definition - Analysis of Alternatives, Report to Congress, NASA, March 2007: http://www.nasa.gov/pdf/171331main_NEO_report_march07.pdf.

NEAR-Shoemaker mission: http://near.jhuapl.edu/.

NEAR-Shoemaker special edition, Science, vol. 289, p. 2085–2105, September 22, 2000.

Paolicchi, Paolo, *Solar-powered asteroids*, Nature, vol. 428, p. 400–401, March 25, 2004.

Parker, Joel, *The Comet Chaser*, Sky & Telescope, August 2014.

Radar observations of near-Earth asteroids: http://echo.jpl.nasa.gov/asteroids/.

Rayman, Marc D., *Dawn of Discovery at Ceres*, Sky & Telescope, December 2016.

Rosetta mission: http://sci.esa.int/rosetta/.

Russo, N. Dello, et al., *Compositional homogeneity in the fragmented comet 73P/ Schwassmann–Wachmann 3*, Nature, vol. 448, p. 172–175, July 12, 2007.

Ruzicka, Alex, & Hutson, Melinda, *Portales Valley: Not Just Another Ordinary Chondrite*, PSR Discoveries, September 30, 2005: http://www.psrd.hawaii.edu/Sept05/PortalesValley.html.

Schmidt, B.E., et al., Hubble Takes a Look at Pallas: Shape, *Size, and Surface*, 39th Lunar and Planetary Science Conference (XXXIX, 2008): http://www.lpi.usra.edu/meetings/lpsc2008/pdf/2502.pdf.

Schmidt, B.E., et al., *The Shape and Surface Variation of 2 Pallas from the Hubble Space Telescope*, Science, vol. 326, p. 275–278, October 9, 2009.

Scott, Edward R.D., Meteoritical and dynamical constraints on the growth mechanisms and formation times of asteroids and Jupiter, Icarus, June 2006: http://arxiv.org/ftp/astro-ph/papers/0607/0607317.pdf.

Shepard, Michael, *Why Do Asteroids Come In Pairs?*, Sky & Telescope, December 2012.

Sheppard, Scott S., and Trujillo, Chadwick A., *A Thick Cloud of Neptune Trojans and Their Colors*, Science Express web site, June 15, 2006.

Soderblom L.A., et al, *Observations of Comet 19P/Borrelly by the Miniature Integrated Camera and Spectrometer Aboard Deep Space 1*, Science, vol. 296, p. 1087–1091, May 10, 2002.

Srinivasan, G., *The Crystallization Age of Eucrite Zircon*, Science, vol. 317, p. 345–347, July 20, 2007.

Stardust mission: http://stardust.jpl.nasa.gov/.

Stardust special section, Science, vol. 314, p. 1707–1739, December 15, 2006.

Stern, S. Alan, *The evolution of comets in the Oort cloud and Kuiper belt*, Nature, vol. 424, p. 639–642, August 7, 2003.

Stern, Alan, & Campins, Humberto, *Chiron and the Centaurs: escapees from the Kuiper belt*, Nature, vol. 382, p. 507–510, August 8, 1996.

Sungrazing comets: http://sungrazer.nrl.navy.mil/.

Sunshine, J.M., et al., *Exposed Water Ice Deposits on the Surface of Comet Tempel 1*, Science, vol. 311, p. 1453–1455, March 10, 2006.

Sunshine, J.M., et al., *Ancient Asteroids Enriched in Refractory Inclusions*, Science, vol. 320, p. 514–517, April 25, 2008.

Taylor, G. Jeffrey, Interstellar Organic Matter in Meteorites, PSR Discoveries, May 26, 2006: http://www.psrd.hawaii.edu/May06/meteoriteOrganics.html.

Taylor, G. Jeffrey, Organic Globules from the Cold Far Reaches of the Proto-Solar Disk,

PSR Discoveries, *January 25,* 2007: http://www.psrd.hawaii.edu/Jan07/organicGlobules.html.

Taylor, G. Jeffrey, Making Sense of Droplets Inside Droplets, PSR Discoveries, May 31, 2005: http://www.psrd.hawaii.edu/May05/chondrulesCAIs.html.

Taylor, G. Jeffrey, Asteroidal Lava Flows, PSR Discoveries, April 28, 2003.

Taylor, G. Jeffrey, Honeycombed Asteroids, PSR Discoveries, August 24, 1999: http://www.psrd.hawaii.edu/Aug99/asteroidDensity.html.

Taylor, G. Jeffrey, The Complicated Geologic History of Asteroid 4 Vesta, PSR Discoveries, June 25, 2009: http://www.psrd.hawaii.edu/June09/Vesta.granite-like.html.

Taylor, G. Jeffrey, *The Composition of Asteroid 433 Eros*, PSR Discoveries, February 26, 2002: http://www.psrd.hawaii.edu/Feb02/eros.html.

Taylor, G. Jeffrey, and Martel, Linda M.V., *Wee Rocky Droplets in Comet Dust*, PSR Discoveries, December 14, 2008: http://www.psrd.hawaii.edu/Dec08/cometDust.html.

The Meteoritical Society: http://www.meteoriticalsociety.org/.

Thomas, Peter C., et al., *Impact Excavation on Asteroid 4 Vesta: Hubble Space Telescope Results*, Science, vol. 277, p. 1492–1495, September 5, 1997.

Thomas P.C., et al., *Differentiation of the asteroid Ceres as revealed by its shape*, Nature, vol. 437, p. 224–226, September 8, 2005.

Tyson, Peter, *Space Invaders*, Sky & Telescope, June 2018.

Vernazza, P., et al., *Compositional differences between meteorites and near-Earth asteroids*, Nature, vol. 454, p. 858–860, August 14, 2008.

Vernazza, P., et al., *Solar wind as the origin of rapid reddening of asteroid surfaces*, Nature, Vol 458, p. 993–995, April 23, 2009.

Vinković, Dejan, *Radiation-pressure mixing of large dust grains in protoplanetary disks*, Nature, vol. 459, p. 227–229, May 14, 2009.

Walsh, Kevin J., Richardson, Derek C., & Michel, Patrick, Rotational breakup as the origin of small binary asteroids, Nature, vol. 454, p. 188–191, July 10, 2008.

Weissman, Paul, *The Oort Cloud*, Scientific American Special Edition, 2003.

Weissman, Paul, *A Comet Tale*, Sky & Telescope, February 2006.

Werner, Stephanie, et al, The Source Crater of Martian Shergottite Meteorites, Science. March 21, 2014: vol. 343, issue 6177, p. 1343–1346. DOI:10.1126/science.1247282.

Chapter 14: Exoplanets

An Astrobiology Strategy for the Search for Life in the Universe, The National Academies Press, 2018: https://www.nap.edu/catalog/25252/an-astrobiology-strategy-for-the-search-for-life-in-the-universe.

Belikov, Rusklan & Bendek, Eduardo, The Next Blue Dot, Sky & Telescope, October 2015.

Boss, Alan, *Brave new worlds*, Physics World Online, March 2, 2009: http://physicsworld.com/cws/article/print/37980.

California & Carnegie Planet Search: http://exoplanets.org/.

Carlisle, Camille M., *7 Earth-Size Planets Orbit Dim Star*, Sky & Telescope, June 2017.

Carlisle, Camille M., *The Race to Find Alien Earths*, Sky & Telescope, January 2009.

Charbonneau, David, et al., A super-Earth transiting a nearby low-mass star, Nature, vol. 462, p. 891–894, December 17, 2009.

Corot mission: http://smsc.cnes.fr/COROT/index.htm.

Currie, Thayne M., & Grady, Carol A., *Pictures of a Baby Solar System*, Sky & Telescope, August 2012.

Davies, Paul, *New Hope for Life Beyond Earth*, Sky & Telescope, June 2004.

Deming, Drake & Seager, Sara, Light and shadow from distant worlds, Nature, vol. 462, p. 301–306, November 19, 2009.

DiGregorio, Barry E., *Life on Mars?* Sky & Telescope, February 2004.

ESO Exoplanet media kit: http://www.eso.org/public/products/presskits/exoplanets/.

Exoplanet Education Articles, special issue, Astrobiology, vol. 10, no. 1, p. 5–126, January 2010.

Exoplanet Science Strategy; The National Academies Press, 2018: https://www.nap.edu/catalog/25187/exoplanet-science-strategy.

Extrasolar Planets Encyclopedia: http://exoplanet.eu.

Fortney, Jonathan, *Weird Weather on Alien Worlds*, Sky & Telescope, May 2014.

Gaidos, Eric, et al., *New Worlds on the Horizon: Earth-Sized Planets Close to Other Stars*, Science, vol. 318, p. 210–213, October 12, 2007.

Gaudi, B.S., et al., *Discovery of Jupiter/Saturn Analog with Gravitational Microlensing*, Science, vol. 319, p. 927–930, February 15, 2008.

Gillon, Michael et al., Seven temperate terrestrial planets around the nearby ultracool dwarf star TRAPPIST-1, Nature, published online February 22, 2017.

Greaves, Jane S., *Disks Around Stars and the Growth of Planetary Systems*, Science, vol. 307, p. 68–71, January 7, 2005.

Hall, Shannon, *The Secrets of Super-Earths*, Sky & Telescope, March 2017.

Impey, Chris, *The New Habitable Zones*, Sky & Telescope, October 2009.

Jakosky, Bruce M., *Searching for Life in Our Solar System*, Magnificent Cosmos – Scientific American Special, Spring 1998.

Johnson, John A., *The Stars That Host Planets*, Sky & Telescope, April 2011.

JPL PlanetQuest: http://planetquest.jpl.nasa.gov/.

Kepler mission: http://www.kepler.arc.nasa.gov/.

Laughlin, Gregory, et al., *Rapid heating of the atmosphere of an extrasolar planet*, Nature, vol. 457, p. 562–564, January 29, 2009.

Laughlin, Greg, *How Worlds Get Out of Whack*, Sky & Telescope, May 2013.

Lissauer, Jack, et al., *A closely packed system of low-mass, low-density planets transiting Kepler-11*, Nature, vol. 470, p. 53–58, February 3, 2011.

Lovis, C., et al., *The HARPS search for southern extra-solar planets. XXVII. Up to seven planets orbiting HD 10180: probing the architecture of low-mass planetary systems*. Astronomy & Astrophysics manuscript no. HD10180, August 13, 2010: http://www.eso.org/public/archives/releases/sciencepapers/eso1035/eso1035.pdf.

Malhotra R., *Migrating Planets*, Scientific American, September 1999.

Marcy, Geoffrey W., and Butler, R. Paul, *Giant Planets Orbiting Faraway Stars*, Magnificent Cosmos – Scientific American Special, Spring 1998.

Masses and Orbital Characteristics of Extrasolar Planets: http://exoplanets.org/almanacframe.html.

Naeye, Robert, *Planetary Harmony*, Sky & Telescope, January 2005.

Naeye, Robert, *Exoplanets Imaged At Last*, Sky & Telescope, March 2009.

Naeye, Robert, *Amateur Exoplanets*, Sky & Telescope, December 2009.

NASA Astrobiology website: https://astrobiology.nasa.gov/.

NASA Astrobiology Institute: http://nai.arc.nasa.gov/.

NASA Astrobiology magazine: http://www.astrobio.net/.

NASA Exoplanet Archive: https://exoplanetarchive.ipac.caltech.edu/index.html.

NASA Exoplanet Exploration Program: https://exoplanets.nasa.gov/exep/.

O'Donoghue, J., et al, *Heating of Jupiter's upper atmosphere above the Great Red Spot*, Nature, vol. 536, p 190–192, August 11, 2016.

Planetary Society Catalog of Exoplanets: http://www.planetary.org/exoplanets/.

Planets Around Pulsars: http://www.astro.psu.edu/users/alex/pulsar_planets_text.htm.

Raymond, S.N., et al., *Exotic Earths: Forming Habitable Worlds with Giant Planet Migration*, Science, vol. 313, p. 1413–1416, September 7, 2006.

Redd, Nola Taylor, *Phoenix Planets*, Sky & Telescope, March 2015.

Rivera, E., et al., *A 7.5 Earth-Mass Planet Orbiting the Nearby Star, GJ 876: Technical Figures and Table*, Astrophysical Journal, vol. 634, p. 625–640, 2005.

Schilling, Govert, *From a Swirl of Dust, a Planet Is Born*, Science, vol. 286, p. 66–68, October 1, 1999.

Seager, Sara, *Unveiling Distant Worlds*, Sky & Telescope, February 2006.

Seager, Sara, *Alien Earths from A to Z*, Sky & Telescope, January 2008.

Seager, Sara, *The Hunt for Super-Earths*, Sky & Telescope, October 2010.

Seager, Sara, *Exoplanets Everywhere*, Sky & Telescope, August 2013.

Seager, Sara, *TESS The Transiting Exoplanet Hunter*, Sky & Telescope, March 2018.

SETI Institute: https://seti.org/.

Setiawan, J, et al., *A young massive planet in a star–disk system*, Nature, vol. 451, p. 38–41, January 3, 2008.

Shiga, David, *Imaging Exoplanets*, Sky & Telescope, April 2004.

Silvotti, R., et al., *A giant planet orbiting the 'extreme horizontal branch' star V391 Pegasi*, Nature, vol. 449, p. 189–191, September 13, 2007.

Snellen, Ignas A.G., *The changing phases of extrasolar planet CoRoT-1b*, Nature, vol. 459, p. 543–545, May 28, 2009.

The Search for Life, Astronomy & Geophysics special issue, Wiley-Blackwell, February 2011.

Teachey, Alex, & Kipping, David M., *Evidence for a Large Exomoon Orbiting Kepler-1625b*, Science Advances, October 3, 2018.

Thommes, Edward W., Matsumura, Soko, & Rasio, Frederic A., *Gas Disks to Gas Giants: Simulating the Birth of Planetary Systems*, Science, vol. 321, p. 814–817, August 8, 2008.

Tinetti, Giovanna, et al., Water Vapour in the Atmosphere of a Transiting Extrasolar Planet, Nature, vol. 448 p. 169–171, July 12, 2007.

Tingley, Brandon, *The First Earth-Size Exoplanet*, Sky & Telescope, May 2009.

Transiting Exoplanet Survey Satellite (TESS): http://tess.gsfc.nasa.gov/.

Trappist-1: http://www.trappist.one/#.

Triaud, Amaury, *Migration of giants*, Nature, vol 537, p 496–497, September 22, 2016.

Vogel, Gretchen, *Expanding the Habitable Zone*, Science, vol. 286 p. 70–71, October 1, 1999.

Vogt, Steven S., et al., *The Lick-Carnegie Exoplanet Survey: A 3.1M_ Planet in the Habitable Zone of the Nearby M3V Star Gliese 581*, Astrophysical Journal: http://www.ucolick.org/%7Evogt/ms_press-1.pdf and http://arxiv.org/abs/1009.5733.

Wagner K., et al., *Direct imaging discovery of a Jovian exoplanet within a triple-star system*, Science, vol. 353, issue 6300, p 673–678, August 12, 2016 DOI: 10.1126/science.aaf9671. http://science.sciencemag.org/content/353/6300/673.

Weinberger, Alycia J., *Building Planets in Disks of Chaos*, Sky & Telescope, November 2008.

Wolszczan, Alexander, *Confirmation of Earth-Mass Planets Orbiting the Millisecond Pulsar PSR B1257 + 12*, Science, vol. 264, p. 538–542, April 22, 1994.

Index

Abulfeda crater chain (Moon), 106
Acheron Fossae (Mars), 182
Achilles asteroids, 403
achondrite meteorites, 413, 414
Acidalia Planitia (Mars), 180, 199
Adams, John Couch, 336
Adonis, 428
Adrastea, 264, 265, 268
Aegaeon, 292
Africa, 68, 75–77, 79, 87, 157
Akatsuki spacecraft *see* Venus Climate Orbiter (Akatsuki)
Akna Montes (Venus), 157
Alabama, 67
Alba Patera (Mars), 182, 183
Aldrin, Edwin (Buzz), 103
Alexandria, 76
Alfvén, Hannes, 38
Alfvén waves, 37, 38, 46, 53
Algeria, 411
ALH 84001, 220, 223, 225
Allan Hills (Antarctica), 116, 220
Allende meteorite, 413
Allen Telescope Array, 472
Alpha Regio (Venus), 155, 158, 160, 163, 166
Alphonsus crater (Moon), 108, 111
Alps mountains (Moon), 101
Amalthea, 263–265, 267, 268
Amazonis Planitia (Mars), 179, 207
Ames Research Center, 455, 464, 465
Amirani (Io), 251
Amor asteroids, 407, 408
Ananke, 267
Andes mountains (Earth), 75, 80
Andromeda galaxy, 24
Andromedid meteors, 425
Angelina (asteroid), 389

Antarctica, 63–64, 68, 70, 71, 76, 79, 82, 116, 154, 173, 220, 221, 223, 411
Anthe, 284
Antiope, 390
Apennine mountains (Moon), 101
Aphrodite Terra (Venus), 154, 156–162, 165
Apohele asteroids, 408
Apollinaris Patera (Mars), 183
Apollo 8, 105
Apollo 11, 103, 112, 119
Apollo 12, 105, 119
Apollo 14, 101, 117, 119
Apollo 15, 104, 109, 110, 112, 113, 119
Apollo 16, 65, 104, 112, 113, 119
Apollo 17, 100, 104, 108, 119
Apollo asteroids, 396, 407, 408, 415, 428
Apollodorus crater (Mercury), 136
Apollo Lunar Surface Experiments Package (ALSEP), 119
Apollo program, 92, 96, 99, 100, 104, 105, 110–112, 116, 119, 414
Apophis (asteroid), 408–409
Appalachian mountains (Earth), 75, 80, 302
Aqua (satellite), 62
Arabia Terra (Mars), 194, 201, 208, 219
arachnoids *see* Venus
Archimedes crater (Moon), 101, 117
Arctic (Earth), 63, 73, 88, 187, 202, 254
Arden Corona (Miranda), 328, 329
Arecibo Observatory, 124, 140, 160, 438, 439, 472, 473
Ares Vallis (Mars), 195, 196, 212
Argentina, 472
Argyre Basin (Mars), 177, 179, 180, 195, 201, 209, 215
Ariadaeus Rille (Moon), 110
Ariel, 21, 326–328, 334, 379
Aristarchus, 1, 2
Aristarchus crater (Moon), 108, 109, 111

Exploring the Solar System, Second Edition. Peter Bond.
© 2020 John Wiley & Sons Ltd. Published 2020 by John Wiley & Sons Ltd.
Companion Website: www.wiley.com/go/Bond-Solar-System2e

Arizona,   84, 186, 440
Armstrong, Neil,   112
Arram Chaos (Mars),   195
Arrhenius, Svante,   148
Arsia Mons (Mars),   183–185
Artemis Chasma (Venus),   158, 159, 165
Artemis Corona (Venus),   158, 165
Aschera (asteroid),   389
Ascraeus Mons (Mars),   183, 184, 194
Asgard crater (Callisto),   262
Asia,   68, 180, 409
Assyrians,   269
asteroids,   4, 6–8, 10, 16, 18, 20, 21, 201, 225, 229, 231, 252, 264,
     266, 384–411, 427–428, 433, 434, 439, 441–443, 445, 454,
     465–467, 471
    albedo,   388, 389, 401, 403, 405, 428
    composition,   387–391, 393, 394, 397–400, 403
    density,   387, 388, 390, 393, 394, 397, 400, 401, 405, 409
    designations,   386
    distances,   384–385, 389, 407–408
    families,   389–390, 396
    gaps in main belt,   389–390, 407
    impact prevention,   410
    magnetic field,   405
    main belt,   9, 355, 385–390, 393–396, 400, 401, 403–407, 428,
     429
    near-Earth,   390, 391, 393, 396, 397, 402, 404–408
    orbits,   384–387, 389, 390, 396, 402–403
    origin,   20, 387–388, 404
    rotation,   390, 391, 395–397, 400, 402, 405
    sample return,   391–393
    satellites,   390, 393, 394, 397, 407
    search programs,   385–386, 410
    shape,   388–393, 395–397, 403
    size,   385–387, 390, 393–395, 400–402, 404, 406–409
    surfaces,   387, 388, 391, 393–395, 397–406
    types,   389
Aston, Francis,   31
Astraea,   385, 389
astrometry,   450
astronomical unit (AU),   4, 14, 232, 417, 435
Aswan,   76
Aten asteroids,   407
Athabasca Valles (Mars),   195–196
Atir (asteroid),   408
Atlantic Ocean,   57, 66, 68, 71, 74, 75, 77, 79, 84
Atla Regio (Venus),   158, 159, 168
Atlas,   287, 288, 312, 313
Atlas V–Centaur rocket,   372
A-Train,   62
auroras,   53, 90, 169, 218, 239, 245, 247, 258, 261, 276, 279–281,
     316, 322–326, 439, 459
Australia,   68, 74, 76, 79, 90, 157, 180, 392, 412
2002 AW197,   362

Babbar Patera volcano (Io),   250
Babylon,   121
Bach crater (Mercury),   135

Bacolor crater (Mars),   193
Baghdad Sulcus (Enceladus),   306
Baikal lake,   409
Baltis Vallis (Venus),   163
Barbara (asteroid),   390
Barnacle Bill rock (Mars),   196
Barnard, Edward Emerson,   264
Barnard's Star,   438
Barringer crater *see* Meteor crater (Earth)
Beagle 2,   224
Beethoven Basin (Mercury),   134, 135
Belinda,   332–334
Bennu,   388, 393
BepiColombo mission,   143–144
Berlin Observatory,   336
Beta Pictoris,   440, 441, 454
Beta Regio (Venus),   155, 160, 161, 167, 168
Betelgeuse,   26
Bianca,   332, 334
Bode's Law,   384
Bond, William,   293
Bondone, Giotto di,   419
Bonneville crater (Mars),   228
Bopp, Thomas,   418
Bottke, William,   396
Brahe, Tycho,   4, 13
Brown, Mike,   6, 358–360
brown dwarf,   7, 441, 444–446, 449
Brucia,   385
Burney, Venetia,   365
Butler, Paul,   438

Cabeus crater (Moon),   116
Caliban,   334, 335
California,   68, 71, 76, 80, 358, 393, 395
California Institute of Technology,   358
Callisto,   14, 20, 21, 247, 248, 252, 253, 257, 261–264, 267, 293,
     294, 328, 394
Caloris Basin (Mercury),   125–127, 131–137
Calypso,   313
Cambridge University (UK),   336
Canada-France-Hawaii Telescope,   335, 394
Canary Islands,   84, 446
55 Cancri,   465, 466
Candor Chasma (Mars),   187
Canup, Robin,   21
Capricornid-Sagittariid meteors,   428
carbonaceous chondrite meteorites,   225, 326, 387–389, 393,
     401, 413
Carme,   267
Carpathians (Moon),   101
Carrington, Richard,   53
Cascade mountains (Earth),   75
Caspian Sea,   300
Cassini, Giovanni Domenico,   4, 281, 293
Cassini Division,   281, 284, 285
Cassini mission,   22, 267, 270–274, 276–279, 281, 284–292,
     297–306, 308, 310, 312

Cassini Regio (Iapetus), 310, 311
Castalia, 387
Cat's Eye nebula, 55
Caucasus mountains (Moon), 101
Cavendish Laboratory, 31
Centaurs, 426–428, 433
Ceraunius Tholus (Mars), 182
Cerberus Fossae (Mars), 195–196, 200
Cerberus Palus (Mars), 196
Ceres, 6, 7, 384–387, 389, 394–395, 397–402
Challis, James, 336
Chandra X-ray Observatory, 113
Chandrayaan spacecraft, 100, 116
Chang'e spacecraft, 99, 100
Chariklo, 427
Charon, 4, 17, 20, 96, 366–369, 372–374, 377–381, 383
Chasma Boreale (Mars), 188, 190
Chelyabinsk, 87, 409, 412
CHEOPS (Characterising Exoplanet Satellite) mission, 454
CheshireCat (asteroid), 386
Chicxulub crater (Earth), 88
Chile, 68, 395
China, 68, 74, 84, 99, 100
Chiron, 426–427
chondrites 16, 326, 389, 412–413, 431, 433 *see also* carbonaceous chondrites
chromosphere *see* Sun
Chryse Planitia (Mars), 180, 195–196, 213, 214
circumstellar disks, 440–445, 451, 454
Clementine spacecraft, 99, 102, 114
Cleopatra crater (Venus), 157, 169
CloudSat, 62, 70
Colette Patera (Venus), 157
Colombo, Giuseppe (Bepi), 124
Columbia Hills (Mars), 202, 203, 228
Columbus, Christopher, 76
coma *see* comets
comet(s), 2, 6, 7, 17, 19, 23, 133, 213, 266, 354, 355, 361–364, 380, 408–410, 415, 417–431, 433–435, 441–443, 454, 466, 467, 469, 471
    albedo, 420
    Biela, 425
    Borrelly, 419, 420
    Churyumov-Gerasimenko, 419–425
    coma, 384, 417–421, 423, 425–428, 430–432, 436
    composition, 418, 420, 423, 424, 428, 431, 433
    density, 420, 433
    disintegration, 425–428
    distances, 426, 428, 434, 435
    early observations, 433, 434
    Encke, 417, 428, 430, 434
    extinct, 387, 407, 427–428
    formation, 417, 421
    Grigg-Skjellerup, 419
    Hale-Bopp, 417–419
    Halley, 157, 158, 419, 420, 426–430, 433–434
    Hartley 2, 420, 421
    Holmes, 426

Hyakutake, 417, 418
Ikeya-Seki, 429
Jupiter family, 434
Kreutz family, 429–430
Linear (C/1999 S4), 426, 430
long period, 10, 408–410, 427, 434, 435
Machholz, 415
main belt, 428, 429
McNaught, 420
names, 430
nucleus, 417–428, 430, 432, 433
orbits, 417, 428–430, 433, 434
P/2013 P5, 428
P/2013 R3, 428
Schwassman-Wachmann 3, 425
Shoemaker-Levy 9, 23, 266, 410, 426, 430
short period, 7, 355, 427, 434
Siding Spring, 213
size, 419–421, 423
SOHO-614, 429–430
sungrazing, 46, 428–430
tails, 417–418, 425, 428–430, 433
Tempel 1, 419, 420, 429, 432–433
Tempel-Tuttle, 415, 416
Wild 2, 421, 430–431, 433
Wilson-Harrington, 428
Conamara Chaos (Europa), 254, 255
Convection, Rotation and planetary Transits spacecraft *see* CoRoT mission
Cook, Capt. James, 4
Copernicus, Nicolaus, 4
Copernicus crater (Moon), 106, 117
Cordelia, 330, 331, 334
Cordillera mountains (Moon), 101
Coriolis force, 66, 212, 233, 277
Cornell University, 473
corona *see* Sun
coronae *see* Venus; Miranda
coronal mass ejection *see* Sun
CoRoT-3b, 444–446
CoRoT mission, 452, 454
CoRoT-7 system, 451, 452, 461
cosmic rays, 48, 143, 238, 279, 280, 357, 370, 377, 471
1995 CR, 407
2000 CR105, 362
CryoSat, 69
cryovolcanism, 252–254, 260, 294, 295, 302–303, 308, 397–399
Cupid, 332–334
Curiosity rover, 414
cyclones, 66–67, 217, 233, 240–242, 341

Dactyl, 23, 390, 394
Dali Chasma (Venus), 157
Damascus Sulcus (Enceladus), 306
Damocles (asteroid), 428
Damocloids, 428
Danu Montes (Venus), 157
Daphnis, 285, 286, 312

Dark Spot 2 (D2) (Neptune), 341, 342
D'Arrest, Heinrich, 6, 336
Datura, 390
Davy crater chain (Moon), 106
Dawes crater (Moon), 108
Dawn mission, 394, 395, 397–401, 403, 404
Debussy crater (Mercury), 134
Deccan, 79, 88
Deep Impact mission, 420, 421, 432–433
Deep Space 1, 420
Deep Space Network, 300
Deimos, 21, 223–225, 229–230, 388
Descartes plateau (Moon), 99
Despina, 349, 351
Devana Chasma (Venus), 161
Diana Chasma (Venus), 156
dinosaurs, 86, 87, 407, 408, 410
Dione, 21, 282, 289, 293, 307–310, 313, 378
Dione Regio (Venus), 163
Discovery Rupes (Mercury), 139
Doppler shift, 34, 35, 124, 147, 305, 443, 446, 448
2000 DP107 (asteroid), 393
Drake, Frank, 469, 471, 473
Drake equation, 469, 473
Drakonia (asteroid), 389
dunes, 83–84, 159, 179, 180, 187–189, 195, 197, 199,
    207–209, 213, 214, 229, 301–303
dwarf planets, 6, 7, 354, 359, 360, 365–371, 374–377, 385–387,
    389, 394–395, 397–402
Dyce, Rolf, 124
Dynamics Explorer, 91
Dysnomia, 358

Eagle crater (Mars), 204, 227, 229
Earth, 2, 4, 10, 12, 14, 16, 17, 20, 23, 47, 49, 57–97, 101, 102,
    104–107, 111–114, 116–122, 124, 126, 128–130, 133, 137,
    142, 173–177, 180, 181, 184, 187, 191, 193–195, 197, 199,
    200, 202, 205, 211, 212, 215, 216, 218, 220, 221, 223, 233,
    235–238, 241–245, 249–252, 254, 255, 258, 260–262, 265,
    266, 301, 302, 305, 310, 315, 337, 338, 340, 343, 348,
    389–391, 396, 400, 403, 406–411, 451
  albedo, 60
  atmosphere, 10, 60–67, 211, 212, 294, 295, 345, 411
  auroras, 53, 90, 91, 218, 279
  axial inclination, 11, 57–59, 70, 174, 175
  biosphere, 84–87, 221, 223
  clouds, 60, 61, 83, 298
  density, 8, 16, 73, 126, 174
  deserts, 82–84, 205, 393, 411, 412, 430
  distance from Sun, 4, 14, 57
  earthquakes, 69, 73, 75, 76
  El Niño, 68, 72
  equinoxes, 59
  exosphere, 64
  extinctions, 87–88, 384, 407, 408, 410
  geocorona, 65
  global warming, 72–73
  Great Oxidation Event, 86
  ice ages, 68, 70, 72, 84, 193, 194
  impacts, 87–88, 385, 407–411
  interior, 74, 77, 78
  ionosphere, 47, 63, 64, 90, 216
  La Niña, 68, 72
  magnetic field, 47, 49, 88, 90–91, 129, 180, 215, 281, 343, 346
  monsoons, 68
  mountains, 68, 74, 75, 77, 79, 80, 82, 86, 101, 154, 168
  oceans, 23, 60, 62, 64, 66–68, 80, 82, 176, 252–253, 255–256,
      258, 260–262, 294, 295, 300, 305, 307, 308, 370, 374, 375,
      378, 379
  orbit, 14, 23, 57–59, 70, 73, 368, 406–408
  ozone hole, 63–64
  plate tectonics, 74–75, 77, 87, 180, 181
  radiation belts, 90, 91
  rotation, 57–58, 66, 89, 152, 174
  seasons, 58–59, 174
  shape, 2, 73, 89
  size, 2, 7, 21, 74, 76, 176, 231, 233, 269, 270, 314, 315, 359,
      378
  solstices, 59, 76
  South Atlantic Anomaly, 91
  storms, 66–67, 212
  stratosphere, 63–64, 80, 81, 87
  temperature, 12, 58–64, 66–68, 73, 82, 86, 173
  tides, 89, 93
  Trojans, 406
  troposphere, 62–63
  vegetation, 60, 82–87
  volcanoes, 77, 79–81, 159, 161, 184
  water cycle, 83, 85
  winds, 63–67, 190
Earth Observing System, 62
Eberswalde crater (Mars), 198
eccentricity of orbits, 6, 13
Echus Chasma (Mars), 187
eclipses (lunar or solar), 2, 27–29, 32, 35, 37
ecliptic plane, 1–3, 6, 8, 16, 121–123, 147, 173, 269, 281, 315,
    338, 354, 355, 361, 362, 364, 367, 382, 384, 390, 402, 406,
    411, 426, 427, 434
Ecuador, 68
Eddington, Arthur, 31
Edgeworth, Kenneth, 354, 355
Edgeworth-Kuiper Belt *see* Kuiper Belt
Einstein, Albert, 31, 122, 123
Eistla Regio (Venus), 164
Elliot, James, 330
El Niño *see* Earth
Elsinore Corona (Miranda), 328
Elst-Pizarro, 428
Elysium (Mars), 182, 183, 195, 196, 200, 202, 207, 219, 220
Elysium Mons (Mars), 182
Enceladus, 21, 273, 279, 282, 283, 289, 291, 293, 303, 305–308
Encke Division, 281, 285, 288, 312
Encke, Johann, 336
Endeavour crater (Mars), 229
Endurance crater (Mars), 228, 229
England, 409

enstatite chondrites, 413

Eos asteroids, 390

Epimetheus, 284, 287, 291, 292, 313

Epsilon Eridani system, 442, 466, 467

equinoxes, 2, 59, 176, 274, 287, 297–299, 302, 367, 368

Eratosthenes, 2, 76

Eratosthenes crater (Moon), 117

Erebor Mons (Titan), 303

Erebus crater (Mars), 229

Erech Sulcus (Ganymede), 260

Eris, 6–8, 356, 358, 367–359–361

Eros asteroid, 4, 385, 388, 389, 406, 407

ERS-2 satellite, 70

escape velocity, 96

Euboea Montes (Io), 251

Eugenia, 390, 394

Eureka, 406

Europa, 14, 20, 21, 247, 248, 252–256, 258–262, 264, 267, 375

Europe, 66, 68, 70–72, 75–77, 79, 80, 84, 86, 224, 384

European Extremely Large Telescope, 440

European Southern Observatory (ESO), 390, 440

European Space Agency (ESA), 48, 69, 70, 90, 99, 113, 154, 165, 225, 399, 410, 419, 421422–424, 429

Eve Corona (Venus), 158

Evpatoria, 473

ExoMars, 225

exoplanets, 7, 15, 19, 438–474
  atmospheres, 447, 448, 451, 454–461, 463, 465, 469–471
  composition, 447–449, 451, 454–456, 460–462, 465
  densities, 445, 447, 448, 451, 452, 455, 457, 459–463, 465, 470, 471
  detection, 438, 445–446, 448–451
  distances, 440, 441, 443, 449, 451, 452, 456–458, 460, 462, 463, 465–467
  formation, 440–444, 459, 460, 467
  hot Jupiters, 448, 454–460
  masses, 438–440, 444–452, 454, 455, 459–463, 465–467, 469–471
  missions, 449, 452–453
  moons, 467–469
  multiple systems, 451, 457, 461–467
  orbits, 444–450, 454–467, 470, 471
  sizes, 448, 450–452, 454, 455, 457, 459, 461, 462, 470
  super-Earths, 447, 451, 452, 460–462, 465, 469
  temperatures, 457, 459, 461–465, 470
  weather, 456–457

Explorer 1 satellite, 91

extrasolar planets *see* exoplanets

Extremely Large Telescope, 449

faculae *see* Sun

Far Ultraviolet Spectroscopic Explorer *see* FUSE spacecraft

2014 FE72, 435

Ferdinand, 335

Ferrel cells, 65

Flagstaff, 364

Flamsteed, John, 314

Flora asteroids, 390

Florida, 67, 84

Fomalhaut b, 449, 450, 467

Fortuna Regio (Venus), 165

Frail, Dale, 438

Fra Mauro (Moon), 101, 111, 117

Francisco, 335

Fraunhofer lines, 30, 35

Freyja Montes (Venus), 157

FUSE spacecraft, 440

Gaea crater (Amalthea), 264

Gaia mission, 410, 450, 453

Galatea, 349–352

Galaxy *see* Milky Way

Galilean satellites, 24, 247–264, 326, 344, 375, 463, 467

Galilei, Galileo, 5, 247, 281, 283

Galileo Regio (Ganymede), 259–260

Galileo spacecraft, 97, 151, 236, 237, 248–251, 254, 255, 258–263, 265–267, 390

Galle, Johann, 6, 336

gamma rays, 29–31, 119, 143, 167, 191, 201, 202, 224, 401, 405, 406, 410

Ganges Chasma (Mars), 202, 204

Ganges river, 82

Ganymed, 407

Ganymede, 14, 20, 21, 126, 247, 248, 252, 253, 259–261, 264, 293, 294, 344, 394

Gaspra, 388, 402

Gauss, Carl, 384

Geminid meteors, 415, 428

Geneva Observatory, 438

geocentric theory, 2, 5

geomagnetic storms, 45, 54, 91

George III, 314

Geostationary Operational Environmental Satellites *see* GOES (satellites)

geysers, 191, 205, 250, 252–258, 347

Gilgamesh Basin (Ganymede), 260

Giotto spacecraft, 419

GJ 1214b, 470

Gliese 581 system, 473

Gliese 876 system, 450

Global Oscillation Network Group, 34, 46

globular cluster, 472

GOES (satellites), 49, 51, 62

Goldstone radar, 393, 395

Gondwana supercontinent, 76–77, 86

GONG *see* Global Oscillation Network Group

GRAIL mission, 96, 101, 102

Grand Canyon (Earth), 84, 186, 378

Grand Tack Model, 18–20

Gravitational Microlensing, 450–451, 454, 455

gravity, 5, 96, 101, 149, 168, 180, 205, 223, 230, 231, 250–252, 257, 258, 264, 337, 338, 348, 356, 360, 364, 374, 377, 379, 395, 400, 401, 403, 407, 410, 412, 414, 425, 434–436, 445, 446, 455, 459, 460, 469, 471

gravity tractor, 410

Great Dark Spot *see* Neptune

Great March Comet,　417
Great Red Spot *see* Jupiter
Great Rift Valley (Earth),　75, 168
Greek astronomy,　1, 2, 76
Green Bank Telescope,　472
Greenland,　70, 76, 82, 188
Grimaldi crater (Moon),　111
Gruithuisen domes (Moon),　105
Guinevere Planitia (Venus),　155
Gulf of Mexico,　67, 71
Gulf Stream,　68, 71
Gusev crater (Mars),　202, 203, 205, 211, 227

habitable zone,　12, 60, 452, 456, 461–463, 465, 469–471
Hadley cells,　64–66, 152, 153, 212
Hadley Rille (Moon),　99, 103, 104, 109, 110
Hadriaca Patera (Mars),　183
Hale, Alan,　418
Halimede,　349, 350
Hall, Asaph,　223
Halley, Edmond,　430, 433, 434
Hamlet crater (Oberon),　327
HARPS,　446, 448, 461
HATS-8b,　447
Haumea,　359–360, 363
Hawaii,　75, 77, 80, 161, 183, 184, 248, 251, 355, 394
Hayabusa,　388, 391–393
HD 10180 system,　460, 465
HD 80606b,　455, 456
HD 80607,　456
HD 20782 b,　454, 455
HD 131399A,　444
HD 149026b,　457
HD 189733b,　456–457
HD 209458 system,　448, 449, 458–459
Hektor,　403
Helene,　313
heliocentric theory,　4, 173
heliopause,　22, 23
helioseismology,　34, 43, 45
heliosheath,　22, 23
heliosphere,　8, 22, 23, 45, 48, 53
Hellas Basin (Mars),　177–179, 182, 188, 194, 195, 201, 205, 207–210, 212, 215, 217, 230
Herbig Haro objects,　15
Hermes (asteroid),　386
Hermes (planet),　1
Hermite crater (Moon),　114
Hero Rupes (Mercury),　139
Herschel, Caroline,　314
Herschel, Sir William,　6, 27, 293, 314, 384
Herschel crater (Mimas),　308, 309
Herschel Space Observatory,　399
Hesperia Planum (Mars),　179, 208
Hesperus,　1
Hey, James,　27
High Accuracy Radial Velocity Planet Searcher *see* HARPS
Hi'iaka,　359, 360

Hi'iaka Patera (Io),　251
Hilda asteroids,　385, 389, 390, 428
Himalaya mountains (Earth),　68, 80, 82
Himalia,　267
Himeros crater (Eros),　406
Hinode (spacecraft),　37, 51, 52
Hipparchus,　2
Hippocamp,　349
Hirayama families,　390
Hirayama, Kiyotsugo,　390
Hiroshima,　409
Hoba meteorite,　411
Hokusai crater (Mercury),　126, 131
Home Plate (Mars),　228
Homer,　403
Hooke, Robert,　238
hot Jupiters *see* exoplanets,　19
HR 8799 system,　467
Hubble Space Telescope (HST),　15, 189, 210, 217, 240, 241, 245–247, 256, 261, 271, 280, 300, 316, 318, 319, 322, 324, 326, 329, 330, 332, 336, 341, 344, 345, 352, 358, 359, 362, 364, 366, 368, 369, 377, 378, 387, 390, 394, 400–402, 419, 425, 426, 428, 449, 450, 459
Hungaria asteroids,　389
hurricanes *see* Earth, storms
Husband Hill (Mars),　228
Huygens, Christiaan,　281, 283, 284, 293
Huygens (probe),　289, 300, 304–305
Hydra,　380, 381, 383
Hyperion,　21, 283, 289, 292, 293, 310, 311

Iapetus,　21, 292, 293, 308, 310–312
IBEX mission,　22
Icarus,　407
Iceland,　74, 77, 79
Ida,　21, 266, 390, 394
Idunn Mons (Venus),　163, 164
Iliad,　403
Imbrium Basin (Moon),　100, 101, 103, 105, 108, 109, 111, 115–117
Imdr Regio (Venus),　164
impact basins and craters,　97, 98, 101–103, 105, 106, 108–111, 115–117, 126, 127, 131–141, 157, 159, 166, 178–181, 202, 213, 215, 225, 227, 248, 252–255, 259–265, 302, 305, 308, 309, 311, 348–350
India,　68, 74, 76, 77, 79, 82, 88, 100, 114, 116
Indian Ocean,　66, 68, 71, 75
Indonesia,　68, 72, 75, 80
interferometry,　390, 400, 449
International Astronomical Union (IAU),　6, 7, 334, 360, 386, 408, 430
International Ultraviolet Explorer (IUE),　399
interplanetary dust,　390
Inverness Corona (Miranda),　328, 329
Io,　14, 20, 21, 245–253, 258, 264
Ioffe,　402
ion propulsion,　144, 401
IRIS spacecraft,　37

iron meteorites, 414
Irwin, James, 110
Ishtar Terra (Venus), 156–158, 160, 168
Isidis Basin (Mars), 182, 198, 215
Italy, 70, 79, 408
Ithaca Chasma (Tethys), 308
Itokawa, 388, 389, 391–392, 402
Ius Chasma (Mars), 187
Ixion, 356

James Webb Space Telescope, 448, 454, 469
Janus, 284, 286, 287, 291, 292, 313
Japan, 97, 100, 143, 144, 150, 151, 154, 156, 338, 388, 389, 391–393, 450
Jason crater (Phoebe), 312
Jet Propulsion Laboratory, 440
jet streams, 63, 66, 151, 153, 207, 208, 212, 233, 237, 242, 270–273, 275–277, 283, 290, 320, 321, 340–342
Jewitt, David, 313, 355, 356
Jezero crater (Mars), 198
Juno (asteroid), 6, 385, 389, 402
Juno (spacecraft), 233, 237, 239, 241, 242, 245
Jupiter, 1, 3, 5, 7, 8, 11, 14, 17–21, 26, 48, 58, 111, 173, 231–275, 277–279, 282, 289, 293, 314, 316, 319, 320, 326, 337, 339–342, 361, 372, 379, 390, 407, 410, 412, 426, 428, 429, 434, 435
    atmosphere, 232, 236–243, 272, 273
    auroras, 245, 247, 258, 261
    axial inclination, 11, 58
    clouds, 232–235, 237–242, 272, 273
    composition, 242–243
    density, 231, 232, 320
    distance, 231, 232
    early observations, 231
    formation, 11, 15, 17, 451
    gravity, 231, 235, 412, 426
    Great Red Spot, 234, 236–240
    interior, 17, 239, 242–243, 277, 278
    ionosphere, 246, 247
    lightning, 235, 237, 239, 258, 266
    magnetic field, 231, 233, 242–247, 278
    migration, 17–20, 361, 387
    orbit, 8, 14, 17–20, 231, 232
    radiation belts, 243
    radio emissions, 231, 237, 243, 247, 248
    Red Spot Junior (Oval BA), 241
    rings, 264, 267–268
    rotation, 231–233, 269
    satellites, 5, 20–21, 231, 232, 245–256, 258–265, 267
    shape, 231
    size, 7, 231–233, 444, 446, 447, 455
    storms, 232, 237–242, 340, 341
    temperature, 12, 236
    Trojans, 231, 403–404
    Voyager 1 mission, 238, 239, 248, 249, 257–258, 267
    Voyager 2 mission, 248, 257–259, 262
    winds, 232–233, 236–238, 240–242, 271

Kaguya mission, 97, 98, 100
Kasei Valles (Mars), 187, 194, 195
Katrina (hurricane), 67
Kawelu Planitia (Venus), 155
Keck telescopes, 251, 276, 317–321, 331, 358, 440, 467
Keeler Gap, 285, 286, 313
KELT-9b, 457
KELT star system, 444
Kepler, Johannes, 4–5, 13–14, 121, 281
Kepler-70b, 457
Kepler 138b, 462
Kepler crater (Moon), 117
Kepler mission, 449, 452–455, 461
Kepler 11 system, 462, 465
Kepler 37 system, 461
Kepler 80 system, 463
Kepler-90 system, 462, 464, 465
Kepler-101 system, 447
Kepler-186 system, 462
Kepler 223 system, 463
Kepler-452 system, 462
Kepler-1625 system, 468, 469
Kerberos, 380, 381, 383
Kerwan crater (Ceres), 395
KIC 5010054, 449
KIC 5522786, 449
Kirkwood, Daniel, 390
Kirkwood gaps, 389–390, 407
Kleyna, Jan, 313
Koronis asteroids, 390
Kowal, Charles, 426
Kozyrev, Nikolai, 111
Kraken Mare (Titan), 300–301
Kreutz, Heinrich, 429
Kuiper Airborne Observatory, 330
Kuiper, Gerard, 287, 354, 355
Kuiper Belt, 4, 7–9, 18–20, 96, 283, 312, 334, 348, 354–383, 449
    atmospheres, 375–377
    binaries, 363–365
    brightness, 356, 358, 361, 362, 366, 367
    classification, 356
    colors, 356–359, 361, 369, 376, 378–380
    compositions, 356, 357, 363, 368–370, 374–375, 378
    distance, 8, 9, 356, 358, 359, 367
    orbits, 362–364, 366–368, 374, 376, 377, 379–382
    origin, 19–20, 361, 367
    satellites, 354, 358–360, 377–381, 383
    sizes, 355, 357–359, 361, 362, 366, 368, 377, 378, 381
    temperatures, 356, 359, 368, 370, 374–376, 379, 380
Kuiper crater (Mercury), 134

Lagrangian points, 100, 313, 402, 403, 406
Lake Tanganyika (Earth), 75
Lakshmi Planum (Venus), 157, 160, 165, 166
Landsat, 62, 84
La Niña *see* Earth
Laomedeia, 350
Larissa, 349, 350

La Silla Observatory, 446, 461
Lassell, William, 326, 344
late heavy bombardment, 18, 20, 105, 132, 134, 139, 177, 261
latent heat, 67
Latona Chasma (Venus), 157
Latona Corona (Venus), 168
Laurasia, 77
LCROSS mission, 100, 116
Lei-Kung Fluctus volcano (Io), 249
Leonid meteors, 415, 416
Le Verrier, Urbain Jean Joseph, 122, 336
Libya, 411
Libya Linea (Europa), 254
life in other planetary systems, 462, 465, 469–472, 474
life in Solar System, 10, 12, 24, 57, 60, 62, 63, 68, 86–88, 198, 206, 207, 213, 220–224, 227, 258, 295, 305, 413
Lincoln Laboratory, 386
Lincoln Near-Earth Asteroid Research program *see* LINEAR program
LINEAR program, 386
Lissauer, Jack, 333
Loki volcano (Io), 249
London, 409
Lost City meteorite, 412
Louisiana, 67
Louros Valles (Mars), 187
Lowell, Percival, 364–366
Lowell Observatory, 111, 354, 364, 365
Lowry, Stephen, 396
Lucifer (planet), 1
Lunae Planum (Mars), 182
Lunar Orbiter spacecraft, 99
Lunar Prospector, 105, 115
Lunar Reconnaissance Orbiter, 102, 106, 111, 112, 116
Lunar Roving Vehicle, 100, 110, 112, 119
Lunokhod rovers, 119
Lutetia, 402
Luu, Jane, 355, 356

Ma'adim Vallis (Mars), 227
Maat Mons (Venus), 158, 160, 162
Mab, 332–333
Magellan mission, 155, 158–161, 163, 164, 169, 171–172
magnetic fields, 47, 49, 88, 90–91, 113, 417
Magnya, 389
Mahilani Plume (Triton), 346–347
Makemake, 359, 360
Marcy, Geoffrey, 438
Marduk volcano (Io), 249
Mare Crisium, 103
Mare Fecunditatis, 102, 106
Mare Humorum, 102, 103, 115
Mare Imbrium, 100, 101, 103, 105, 108, 109, 111, 115–117
Mare Marginis, 93
Mare Moscoviense, 101
Mare Muscoviense, 101
Mare Nectaris, 103, 115, 117
Mare Nubium, 102, 103, 115

Mare Orientale, 101, 103, 115, 117
Mare Serenitatis, 103, 108, 115
Mare Smythii, 93, 103, 115
Mare Tranquillitatis, 102, 103, 108, 110
Margaret, 335
Marianas Trench (Earth), 74
Mariner 4, 224
Mariner 9, 186, 208, 224
Mariner 10, 126, 127, 129, 132, 133, 135–137, 142
Mars, 1, 3, 7, 10, 12, 13, 17, 24, 126, 133, 134, 139, 173–230, 252, 302, 337, 361, 364, 367, 389, 400, 403, 406, 407, 412, 414–415, 449, 451, 462, 463, 465, 469, 470
atmosphere, 10, 177, 179–181, 184, 188, 190, 191, 193, 194, 197, 200–203, 205, 207–223, 225, 415
auroras, 218
axial inclination, 10, 11, 174, 175
canals, 186, 364
canyons, 177, 180, 182, 184, 186–190, 195, 202, 204, 207, 211, 212, 224
channels, 177, 179, 183, 184, 186, 187, 191, 194–198, 200, 201, 204
clouds, 183, 189, 210, 212, 214, 217, 218
composition, 175, 177, 181, 205, 209
density, 174, 175
deserts, 175, 195, 204–205
distance, 4, 173, 174, 176, 211
dunes, 179, 180, 187–189, 195, 197, 199, 203, 205, 207, 208, 213, 214, 229
dust devils, 188, 205, 209, 211, 212
dust storms, 188, 205, 207–212, 216, 224, 229
early observations, 1, 3, 173, 205
geological history, 178–182, 198, 200–202
glaciation, 177, 178, 188, 193–194
gravity, 174, 175, 180, 181, 205
gullies, 183, 194, 199, 200
highlands, 177–181, 183, 188, 196, 198–202, 215, 222, 227
ice ages, 175, 193
impact craters, 111, 133, 175, 177–182, 188, 189, 191–199, 202, 213, 302, 308, 309, 311, 312
interior, 181
ionosphere, 213, 215, 216, 218, 223
life on, 173, 198, 206, 207, 220–223, 227, 469
lowlands, 177, 178, 181, 187, 188, 194, 195, 200–202, 215
magnetic field, 130, 177, 180, 214–216, 218, 220, 222, 223
meteorites, 175, 182, 202, 218, 220, 222, 223, 414–415, 471
mountains, 177, 178, 180–182, 184, 194
ocean, 179, 199–202, 212, 218
orbit, 13, 173, 174, 176
permafrost, 179, 187, 191–193, 195, 198, 214, 220
polar ice caps, 139, 175, 177, 188–193, 199, 200, 207, 208, 210–212, 218, 221
polygons, 187, 188, 192, 198–199
precipitation, 188, 193, 196, 201, 212
radiation, 173, 188, 207, 209, 214–216, 218, 220, 221, 223
rotation, 174, 175
satellites, 17, 21, 223–225, 229–230, 388
seasons, 10, 173–176, 188, 202, 211, 219
shape, 175

size, 7, 126, 174–176
tectonics, 180–182, 184
temperature, 12, 188–193, 195, 200, 204, 207–209, 211–212, 215, 216, 218, 220
topography, 178, 179, 190, 195, 201, 202, 212, 224
Trojans, 403, 406
volcanic activity, 177–186, 190, 194, 195, 201–203, 208, 220, 222, 224, 228, 229
water, 175, 177–180, 183, 184, 187–189, 191–207, 209–225, 227–229
winds, 177, 187, 188, 190, 197, 198, 203–205, 207–209, 211, 212
Mars (Soviet spacecraft), 208
Mars Atmosphere and Volatile EvolutionMission (MAVEN), 218, 225
Mars Exploration Rovers 205, 211, 227–229 *see also* Spirit rover and Opportunity rover
Mars Express, 182, 186, 207–209, 216, 221, 224
Mars Global Surveyor, 181, 193, 199–202, 210, 217, 219, 222–224
Mars 2020 mission, 198
Mars Observer, 224
Mars Odyssey, 183, 186, 192, 193, 201, 202, 215, 224
Mars Pathfinder, 195, 196, 202, 212
Mars Phoenix *see* Phoenix mission
Mars Reconnaissance Orbiter, 180, 190, 192, 194, 202, 225, 230
Mars Science Laboratory *see* Curiosity rover
Marte Vallis (Mars), 195
Maskelyne crater (Moon), 103
Massachusetts, 472
mass driver, 410
Mathilde, 388, 402, 405
Mauna Loa, 77, 162, 183, 184
Maunder Minimum, 42
Maxwell, James Clerk, 160, 281
Maxwell Gap, 288
Maxwell Montes (Venus), 149, 155–157, 160, 166, 168, 169
Mayor, Michel, 438
Mead crater (Venus), 159, 161
Mediterranean Sea, 77
Medusae Fossae (Mars), 183
Menrva crater (Titan), 302
Mercury, 1, 3–5, 7, 8, 11, 12, 16, 55, 58, 77, 88, 121–145, 187, 247, 248, 394, 395, 407, 408, 454, 461, 464, 465
   atmosphere, 8, 121, 124, 126, 130, 133, 140, 142, 145
   axial inclination, 11, 121
   composition, 126, 128–130, 139
   craters, 124, 125, 130–134, 139, 140
   density, 16, 126, 128
   distance, 121, 231
   early observations, 121
   formation, 128–129
   impact basins, 132, 134–137
   interior, 126, 128, 139, 145
   magnetic field, 88, 126–131, 142, 144
   mountains, 131, 135, 137, 138
   orbit, 4, 5, 121–123, 464, 465
   phases, 122, 126

plains, 131–139, 187
regolith, 132
rotation, 123–125, 127, 129, 139
scarps, 131, 133, 135, 137–139
size, 7, 126, 128, 248
temperature, 12, 58, 121, 124–126, 140–142, 465
transits, 122–124, 448
volcanic activity, 131–138, 187
water ice, 138–142
Meridiani Planum (Mars), 180, 202, 227
Merline, William, 390
MESSENGER mission, 123, 126, 127, 129–132, 134, 135, 138–141, 143
Messier craters (Moon), 101, 106
Meteor crater (Earth), 87
meteorites, 7, 10, 16, 96, 116, 126, 140, 145, 175, 182, 202, 218, 220, 222, 223, 225, 326, 387–389, 400, 401, 407, 411–416
meteors, 111, 213, 415–418, 425, 427, 428
Meteosats, 62
methane cycle, 219–220
Methone, 284
Metis, 264, 268
Mexico, 407, 408, 413
M13 globular cluster, 472
Microlensing Observations in Astrophysics *see* MOA
micrometeorites, 132, 397, 416
migration of planetary orbits, 17–20, 264, 308, 355, 361, 367, 381, 383, 387, 404
Milankovitch, Milutin, 70, 72
Milky Way, 24–26, 438, 449, 451, 452, 461, 469
Millis, Robert, 330
Mimas, 21, 23, 281, 282, 285–287, 291–293, 308, 309
Minor Planet Center, 386
Miranda, 21, 325–329, 332, 334
Mississippi, 67
Mississippi river, 195
MOA, 450
MOA-2003-BLG-53Lb, 450
Moon, 1, 2, 4, 5, 10, 17, 20, 21, 28–29, 76, 87, 89, 92–120, 126, 128–137, 139, 173, 177, 178, 187, 223, 224, 270, 284, 337, 344, 368, 374, 377, 383, 387, 394, 395
   age, 111, 115, 117
   albedo, 94, 96, 102, 262, 366, 368
   atmosphere, 116
   axial inclination, 92
   density, 96, 117, 120
   distance, 4, 28, 93, 119
   early observations, 1, 2, 96
   eclipses, 2, 76, 94–96
   formation, 17, 20, 105, 117–118, 120, 383
   gravity, 96–98, 100–102, 106, 126, 133, 235
   highlands, 96, 97, 101, 104–105, 111, 112, 116, 415
   ice, 96, 99, 100, 114–116, 139, 394
   impact features, 101–102, 106–107, 111, 115, 131, 134–136
   interior, 101, 102, 104, 112–113, 129
   KREEP (rock), 105, 117
   libration, 92, 93
   magnetic field, 113

Moon (*continued*)
  maria, 96–98, 101–104, 106, 108, 110, 111, 114, 117, 134, 136
  mascons, 97, 101, 374
  meteorites, 96, 116, 412, 414–415
  moonquakes, 105, 111, 112, 119
  mountains, 100, 101, 103, 109
  orbit, 5, 28–29, 92–93, 95, 117, 120, 377
  phases, 28, 94
  regolith, 100, 102, 104, 111–116, 119
  rilles, 103, 106, 108–110
  rock types, 103–105, 108, 113–114, 116–117
  rotation, 92–93, 96
  shape, 92
  size, 2, 21, 28, 76, 96, 126, 344
  tectonics, 113
  temperature, 96, 112, 114
  topography, 97
  transient lunar phenomena, 111
  volcanic activity, 102–106, 108–111, 115, 117, 132, 187
  water, 96, 99, 100, 114–116
Moon Mineralogy Mapper, 115
Morabito, Linda, 248
mountains, 68, 74, 75, 77, 79, 80, 82, 86, 96, 100, 101, 103, 108,
    109, 119, 135, 137, 138, 165, 166, 168, 169, 177, 178, 181,
    194, 249, 251, 297, 302, 303
Mount Everest (Earth), 63, 74, 149, 184, 251, 311
Mount Fuji (Earth), 79
Mount Haleakala (Earth), 386
Mount Pinatubo (Earth), 80, 81
Mount St. Helens (Earth), 79
Mozamba crater (Triton), 348
2014 MU69 (Ultima Thule), 356, 357
Murchison meteorite, 413
MUSES Sea (Itokawa), 391

Naiad, 349, 351
Namaka, 359, 360
Namibia, 84, 411
Naples, 70
NASA, 38, 53, 62, 91, 98–100, 102, 106, 111, 115, 116, 119, 123,
    129, 143, 151, 155, 160, 170, 171, 183, 187, 198, 202, 206,
    208, 209, 213, 218, 219, 223–225, 227–229, 265, 359, 365,
    372, 385, 386, 393, 394, 396, 397, 401, 404, 405, 407, 408,
    410, 414, 416, 417, 420, 429–432
National Radio Astronomy Observatory, 438
National Solar Observatory, 43
near-Earth asteroids *see* asteroids
near-Earth objects (NEOs), 87, 385, 397, 407, 408 *see also*
    asteroids, near-Earth
NEAR Shoemaker mission, 388, 389, 405–407
nebular hypothesis, 10
Nectaris Basin (Moon), 103, 115, 117
Neptune, 4–12, 18–21, 122, 126, 232, 243, 270, 273, 276, 308,
    318, 319, 321, 329, 336–353, 355–357, 361–364, 366, 367,
    375, 388, 403, 406, 427, 428, 430, 433–435
  atmosphere, 270, 339–342
  auroras, 344
  axial inclination, 11, 21, 337, 340, 344

  brightness, 341, 342
  clouds, 318, 340–342
  composition, 337, 339, 340, 342
  density, 337
  discovery, 5, 6, 122, 336–337
  distance, 8, 9, 337
  formation, 320
  Great Dark Spot, 318, 319, 340–342, 345
  interior, 337, 342–343
  magnetic field, 273, 337, 338, 343–344, 346
  migration, 19–20, 367, 383
  orbit, 10, 14, 18, 232, 336, 337, 364
  rings, 4, 329, 338, 344, 349–352
  rotation, 337
  satellites, 21–22, 337, 344–352
  seasons, 337, 341, 342
  shape, 337
  size, 337
  spots, 318, 319, 338, 340–343, 345
  temperature, 12, 276, 339–340, 342–343
  Trojans, 403
  Voyager 2 mission, 336–339, 341–345, 347–351
  winds, 321, 337, 340–342
Nereid, 21, 348, 349
Neso, 349, 350
Neuschwanstein meteorite, 412
neutron star, 438, 439, 451
Newfoundland, 68
New Horizons mission, 356, 357, 365, 366, 369–373,
    375–379380, 381, 383
New Mexico, 386
Newton, Sir Isaac, 5, 122, 123, 235, 410
Newton Basin (Mars), 199
New Zealand, 450
Nice model, 17, 19–20
Nile river (Earth), 82, 163
Nili Fossae (Mars), 202
Nilosyrtis (Mars), 208
Nix, 380, 381
NOAA, 49
Noachis Terra (Mars), 179, 199, 209
Noctis Labyrinthus (Mars), 184, 186
NORC asteroid, 386
North America, 68, 71, 72, 75, 79, 84, 86
North Atlantic Drift, 71
North Massif (Moon), 119
North Sea, 200
Norway, 68
Nubian desert, 412
nuclear power, 48, 372
Nullarbor Plain, 412

Oberon, 21, 315, 326, 327, 334, 335
occultations, 1, 4, 247, 285, 290, 294, 316, 329, 330, 338, 345,
    350, 351, 360, 366, 368, 375, 376, 393, 400
oceans, 23, 60–62, 64–69, 71–75, 77–80, 82–89, 91, 149, 150,
    152, 156, 168, 176, 179, 180, 199–179, 199, 202, 212,
    218–202, 212, 218, 252–253, 255–256, 258, 260–262, 266,

267, 293–295, 300, 305, 307, 308, 370, 374–376, 378, 379, 397, 407, 411

Oceanus Procellarum, 102, 105, 110, 117

Odysseus crater (Tethys), 308, 309

OGLE, 450

OGLE-2003-BLG-235, 450

O'Keefe, John A., 396

Olbers, Wilhelm, 385

Old Faithful, 250

Oljato, 428

Olympus Mons (Mars), 179, 182–185, 192, 209, 222

2000 OO67, 361

Oort, Jan, 434

Oort Cloud, 8, 9, 18, 361–363, 427, 428, 434–435

Ophelia, 330, 331

Ophiuchus, 443

Opportunity rover, 202, 204, 209, 211, 214, 227–229

Optical Gravitational Lensing Experiment *see* OGLE

optical SETI, 472

2007 OR10, 359

orbital elements, 6, 11

Orcus, 356, 359, 360

Orientale Basin (Moon), 101–103, 115, 117

Orion Nebula, 10, 15, 441, 442

OSIRIS-REx, 393

'Oumuamua, 436–437

Oval BA *see* Jupiter, Red Spot Junior(Oval BA)

Ovda Regio (Venus), 157, 166

Pacific Ocean, 66, 68, 71, 72, 74, 75, 80, 86

Paddack, Stephen J., 396

Palermo Observatory, 384

Palermo Scale, 408

Pallas, 6, 385–387, 400–402, 404

pallasite meteorites, 414

Pallene, 291, 292

Pan (satellite), 285, 287, 288, 312, 313

Pan crater (Amalthea), 264

Pandora, 287, 290, 291, 313

Pangaea supercontinent, 77, 79

Pan-STARRS telescope, 386, 436, 437

Pantheon Fossae (Mercury), 136

parallax, 3–4

Parga Chasma (Venus), 161

Parker Solar Probe, 53

Parkes Telescope, 338, 472

Pasiphae, 267

Patroclus asteroids, 403

Pavonis Mons (Mars), 183–185

Peary crater (Moon), 114

Peekskill meteorite, 412

51 Pegasi, 438, 446

Pele volcano (Io), 248–250

Pennsylvania State University, 438

Perdita, 332–334

Perth Observatory, 330

Peru, 68

Petit-Prince, 390, 394

Pettengill, Gordon, 124

2000 PH5, 396

Phaethon, 387, 407, 415, 428

Pharos crater (Proteus), 349

Philae lander, 421–423

Philippines, 75, 80, 81

Phobos, 21, 223–225, 229–230

Phoebe, 21, 283, 284, 289, 292, 293, 310–313

Phoebe Regio (Venus), 168

Phoenix mission, 187, 188, 212

Pholus, 427

Phosphorus, 1

photosphere *see* Sun

photosynthesis, 62

Piazzi, Giuseppe, 6, 384

Piazzia, 386

Picard, Jean, 4

Pickering, William, 293

Pillan Patera volcano (Io), 249, 250

Pinatubo volcano (Earth), 80, 81

Pioneer 10, 243, 248, 473

Pioneer 11, 290, 473

Pioneer 12 *see* Pioneer Venus

Pioneer 13 *see* Pioneer Venus

Pioneer Venus, 151, 152, 155, 160, 165, 170–171

plage *see* Sun

planetary nebula, 23, 55, 355

planetesimals, 7, 11, 15–20, 105, 114, 118, 129, 132, 243, 348, 354, 355, 367, 383, 440, 441, 467

Planet Nine, 382

planets
  definition of, 6–7
  formation of, 10–11, 15–17, 387, 434, 441–445, 449, 451, 459, 460, 467, 471
  migration of *see* migration of planetary orbits

Planet X, 364–365

Plato crater (Moon), 101, 117

PLATO (Planetary Transits and Oscillations of stars) mission, 454

Pleiades, 3

Plutinos, 356, 367

Pluto, 6, 7, 11, 12, 14, 18, 20, 344, 345, 348, 354–383
  albedo, 366–368, 374
  atmosphere, 354, 359, 367, 372, 374–377
  axial inclination, 11, 367, 368
  composition, 368–371, 374–375
  density, 359, 374, 375
  discovery, 364–365
  distance, 366, 367
  eclipses, 366, 378
  interior, 374–375
  ocean, 374, 375
  orbit, 14, 18, 366–368, 376
  origin, 380–381, 383
  rotation, 368, 374
  satellites, 20, 96, 354, 355, 358, 360, 363, 366–370, 374, 380–381, 383
  seasons, 377–381

Pluto (*continued*)
  size, 7, 358, 359, 368, 374, 377
  temperature, 12, 359, 368, 374–376
Plutoids, 360
Polydeuces, 313
Pope, Alexander, 334
Portia, 332, 334
precession of orbits, 58, 122, 123
Pribram meteorite, 412
Project OZMA, 471
Project Phoenix, 472
Promethei Terra (Mars), 194, 212
Prometheus (moon), 287, 288, 290, 291, 313
Prometheus volcano (Io), 250, 251
prominences *see* Sun
proplyds *see* protoplanetary disks
Prospero, 335
Proteus, 349, 350
protoplanetary disks, 10, 11, 15, 20, 434
protoplanets, 11, 15, 16, 20, 400, 401
PSR B1257+12., 438, 439
Psyche (asteroid), 389
Psyche crater (Eros), 406
Puck, 21, 332–334
Puerto Rico, 393
pulsars, 438, 439, 455, 473
Pwyll crater (Europa), 254, 255

1992 QB1, 355, 356
2001 QT297, 363
Quaoar, 356, 360, 361, 379
Queen's University, Belfast, 396
Queloz, Didier, 438

Rachmaninov crater (Mercury), 135
radar, 4, 69–70, 124, 126, 131, 134, 138–140, 142, 147, 148,
    155–161, 163–165, 169–172, 188, 190, 194, 220–221, 224,
    225, 289, 300–303, 390, 393, 395
Raditladi impact basin (Mercury), 138
Radzievskii, V. V., 396
Ramsey, W., 320
Ranger spacecraft, 99
red dwarf, 448–450, 462–464, 469–471
red giant, 457, 459, 460, 471
Red Sea (Earth), 75
Relativity, General Theory of, 122, 123
Rembrandt Basin (Mercury), 134
Remus (asteroid), 393, 395
resonances (orbital), 18–20, 124, 127, 139, 264, 285, 286, 288,
    307, 308, 310, 328, 330, 355–357, 367, 390, 404
Rhea, 21, 282, 292, 293, 308–310
Rhea Mons (Venus), 161
Rheasilvia impact basin (Ceres), 400, 404
Richer, Jean, 4
rings, 269, 270, 273, 275, 277–281, 283–288, 290–293,
    313, 360

Ritter, Johann, 27
Roche Division, 288, 290, 291
Roche, Edouard, 281
Roche limit, 281, 283, 360
Roman Catholic Church, 5
Romulus (asteroid), 393, 395
Rook mountains (Moon), 101
Rosalind, 332, 334
ROSAT, 418
Rosetta mission, 389, 421–425
Royal Greenwich Observatory, 44
Rozhdestvensky crater (Moon), 114
1996 RQ20, 356
R/2004 S1, 288
R/2004 S2, 288
RT-70 telescope, 473
Ruach Planitia (Triton), 348
Rupes Recti *see* Straight Wall (Moon)
Rutherford, Ernest, 31
1997 RX9, 356
Ryugu (asteroid), 388, 392, 393, 397

Sacajawea Patera (Venus), 158
Sagan, Carl, 213
Sahara desert, 65, 82, 84, 411
saltation, 83, 152, 205
San Andreas fault (Earth), 76
Sandia National Laboratories, 409
San Francisco State University, 438
Santa Maria Rupes (Mercury), 133
Sao, 350
Saturn, 1, 3, 7, 10–12, 14, 16–21, 58, 231, 232, 243, 269–313,
    337, 339, 340, 342, 350, 351, 403, 404, 427
  atmosphere, 270–278, 281, 316
  auroras, 279–281
  axial inclination, 11, 270, 273
  clouds, 270–278
  composition, 269, 272–274, 277–278, 342
  density, 16, 269, 270, 277
  distance, 3, 4, 269, 270
  early observations, 269
  interior, 17, 277–278, 342
  lightning, 274, 277
  magnetic field, 273, 277–281, 288, 293
  migration, 19–20, 404
  orbit, 14, 18, 269, 270
  radiation belts, 279–280
  radio signals, 273, 274, 276, 278, 279
  rings, 269, 270, 273, 275, 277–281, 283–288, 290–293, 313,
    350, 351
  ring spokes, 282, 287–288
  rotation, 269–271, 273, 279
  satellites, 21, 270, 279, 281, 282, 286, 288–291, 293–313
  seasons, 297–298
  shape, 269
  size, 7, 269, 270, 320

spots, 270, 271, 273–277
storms, 272–276, 290, 342
temperature, 12, 272–275,-276–278, 286
Voyager 1, 273, 274, 282–283, 293
Voyager 2, 273, 282–283, 288, 294
winds, 270–276, 319, 340
Schröter, Johann, 384
Schwabe, Heinrich, 42
Scottish Highlands, 75, 80
Sea of Clouds *see* Mare Nubium
Sea of Crises *see* Mare Crisium
Sea of Fertility *see* Mare Fecunditatis
Sea of Muscovy *see* Mare Moscoviense
Sea of Rains *see* Mare Imbrium
Sea of Serenity *see* Mare Serenitatis
Sea of Smyth *see* Mare Smythii
Sea of the Margin *see* Mare Marginis
Sea of Tranquillity *see* Mare Tranquillitatis
Search for Extraterrestrial Intelligence *see* SETI
Sedna, 9, 356, 359–362, 434–435
Selk crater (Titan), 302
semi-major axis, 6, 14
Sentinel satellites, 62
SERENDIP, 472
Setebos, 335
SETI, 471–474
SETI Institute, 472
Shackleton crater (Moon), 114
Shakespeare, William, 334
shepherd moons, 282, 290, 291, 313, 326, 330, 331, 333, 334,
349, 351
Sheppard, Scott, 313
Shoemaker, Eugene, 405
Shorty crater (Moon), 104
Showalter, Mark, 333
Siberia, 87, 88, 408
sidereal month, 94
sidereal rotation, 10
Sierra Nevada (Earth), 80
Sikhote-Alin, 409
Sinus Aestuum (Moon), 108
Sippar Sulcus (Ganymede), 260
Sirenum Terra (Mars), 199
Sirius, 121
Sirsalis Rille (Moon), 110
Sisyphus, 407
SMART-1 spacecraft, 99, 100, 113
Smith, Bradford A., 440
Smithsonian Astrophysical Observatory, 386
SNC meteorites, 201, 414–415
Socorro (New Mexico), 386
SOHO spacecraft, 3, 27, 33–35, 37–39, 43, 46, 47, 49, 50, 53, 54,
417, 429, 430
Sojourner rover, 196
Solar and Heliospheric Observatory *see* SOHO spacecraft
solar cycle *see* Sun, sunspot cycle

Solar Dynamics Observatory, 37, 38, 40, 43, 53
solar flares *see* Sun
solar nebula, 8, 10, 11, 16–18, 20, 117, 120, 126, 128, 132, 155,
175, 277, 342, 345, 354, 355, 357, 387, 412, 413, 431
Solar Orbiter, 53
solar sail, 410
Solar Terrestrial Relations Observatory *see* STEREO (spacecraft)
solar wind, 1, 22, 23, 32, 37, 40, 45–49, 51, 53, 55, 91, 106,
111–114, 116, 119, 129–130, 142, 144, 145, 169, 171, 215,
216, 218, 223, 244, 245, 338, 343, 357, 365, 377, 417, 419, 431
solstices, 59, 76, 176, 212, 275, 277, 298, 299, 344
Sotra Patera (Titan), 303
South America, 68, 72, 75–77, 79, 80
South Pole–Aitken Basin (Moon), 97, 100, 102, 105, 114, 115,
177
Southwest Research Institute, 390, 396
Soviet space program, 92, 96, 111, 116, 119, 151, 152, 157, 159,
160, 162, 167–168, 208, 419
Space Shuttle, 48, 81, 266
spicules *see* Sun
spin-orbit coupling, 92–93, 124
Spirit rover, 202, 203, 205, 209, 211, 213, 216, 227–228
Spitzer Space Telescope, 292, 293, 419, 425, 426, 433
Sputnik Planitia (Pluto), 369–371, 374, 376
S/2004 S6, 291
Stanford University, 223
Stardust (mission), 421, 430–431, 433
Steins, 389
Stephano, 335
STEREO (spacecraft), 46, 53, 417, 419
Stickney crater (Phobos), 225, 230
stony-iron meteorites, 412–414
Straight Wall (Moon), 103
Styx, 380, 381
Sudan, 412
Sumeria, 121
Sun, 2–5, 7, 10, 23–61, 73, 89, 93, 121–124, 218, 235, 295, 297,
387, 390, 393, 396, 399, 407, 430, 431, 433–437
active regions, 27, 34, 37–40, 42–44, 46, 48, 51–53
axial inclination, 26
black dwarf stage, 56
chromosphere, 32, 35–36, 38, 41, 43, 46, 53
composition, 26, 27, 29,-31, 33, 34, 278
corona, 25–29, 32, 35–39, 44–47, 49, 51–54, 430
coronal heating, 27, 36–38, 46
coronal holes, 45, 47
coronal mass ejections, 37, 39, 42, 45–47, 49, 51, 53–54, 417
density, 26
eclipses of, 28–29, 32, 35, 37
faculae, 35, 41, 42
flares, 25, 27, 34, 37–39, 42, 46–47, 49–51, 53
formation of, 25
gamma rays, 29–32, 47, 52
granulation, 33, 35, 36
interior, 32–34, 38, 39, 41–43, 45, 47
magnetic field, 32, 33, 35–49, 51–54, 417

mass, 5, 10
nuclear fusion, 25, 31–32, 55
photosphere, 26–28, 32–42, 46, 47, 50
plage, 36, 41
planetary nebula stage, 24, 55
prominences, 27–29, 35, 36, 38, 39
red giant stage, 24, 54–55, 471
rotation, 10, 26, 273
size, 2, 7, 23, 54–55
solar cycle, 40, 42–45, 48, 341
spectrum, 27, 29–30
spicules, 36–38, 46
sunspots, 26, 28, 35–37, 39–47, 49, 51, 53, 73
temperature, 25–27, 29–35, 37–39, 41, 42, 45, 47, 52, 55
white dwarf stage, 24, 55–56
X-rays, 27, 30, 32, 51, 52
supernova, 10, 439
Surveyor program, 99
Swedish Solar Telescope, 42
Swift crater (Deimos), 225
Switzerland, 438
2013 SY99, 435
Sycorax, 334, 335
Syene, 76
Sylvia (asteroid), 393, 395
synchronous rotation, 92, 367, 377, 380, 390
Syrtis Major (Mars), 202

Taurus-Littrow valley (Moon), 99, 100, 104, 108, 119
tectonics, 74–75, 77, 87, 88, 113, 132–134, 137, 139, 155, 159,
     165–166, 169, 180–182, 184, 258, 302, 303, 305, 308, 329,
     358, 374, 375, 378, 379
Telescopio Nazionale Galileo, 446
Telesto, 313
Terra Sirenum (Mars), 200
Terrile, Richard, 440
TESS (Transiting Exoplanet Survey Satellite), 449, 452, 454
Tethys, 21, 279, 280, 282, 293, 307–310, 313, 378
Texas, 188
Thalassa, 349, 351
Tharsis (Mars), 178, 179, 181–186, 194, 201, 208, 217, 219, 220,
     222
Thebe, 264, 268
Theia Mons (Venus), 161
Themis asteroids, 390
Themis Regio (Venus), 163
Themisto, 267
Thera Macula (Europa), 254
Theta Orionis C, 441, 442
Thetis Regio (Venus), 157, 166
Thrace Macula (Europa), 254
Tibet, 77, 79
tides, 57, 89, 93, 111, 124, 147, 201, 225, 247, 251, 253, 261, 262,
     302, 328, 329, 374, 378, 380, 383, 435
Tirawa crater (Rhea), 309
Titan, 21, 24, 126, 252, 269, 279, 281, 282, 289, 292–303, 310,
     375, 376, 467

Titania, 21, 315, 326–328, 334
Tolstoj Basin (Mercury), 132–134
Tombaugh, Clyde, 6, 354, 364–365
Tombaugh Regio (Pluto), 370, 371
Torino Scale, 408
Toro (asteroid), 407
Toutatis (asteroid), 387, 390, 395, 407
TRACE spacecraft, 37
Tractus Catena (Mars), 182
Tractus Fossae (Mars), 182
Trans-atlantic Exoplanet Survey (TrES), 459–460
Transiting Exoplanet Survey Satellite *see* TESS
transits, 4, 122–124, 444, 448–449, 452–455, 458, 460, 463, 468
TRAPPIST-1 system, 462–465
TrES-4, 459–460
Trinculo, 335
Triton, 21, 338, 343–348, 354, 375
TRMM spacecraft, 70
Trojans (asteroids and satellites), 313, 402–404, 428
Tropical Rainfall Measuring Mission *see* TRMM spacecraft
Tsiolkovsky crater (Moon), 105
T Tauri phase, 11, 21, 129
Tunguska, 408–411
Tvashtar Catena (Io), 249, 250
Twin Peaks (Mars), 196
Tycho crater (Moon), 97, 106, 117
Tyrrhena Patera (Mars), 183

U2 aircraft, 416
Ukraine, 473
Ultraviolet Explorer, 399, 419
Ulysses spacecraft, 48–49, 417
Umbriel, 21, 326–328, 334
United States of America, 385, 416
University of Arizona in Tucson, 440
University of California at Berkeley, 438, 449, 472
University of Hawaii Telescope, 355, 356
Uranius Tholus (Mars), 182
Uranus, 4–6, 11, 12, 14, 18–21, 126, 232, 235, 243, 273, 308,
     314–340, 342, 343, 346, 350, 351, 355, 360, 364, 366, 367,
     403, 451, 461, 465, 467, 471
     atmosphere, 314–320, 322–324, 326, 329, 331, 339, 340, 343
     auroras, 323, 324, 326
     axial inclination, 11, 314–315, 323
     clouds, 315–321, 323, 340
     composition, 317, 318, 320, 322, 323
     density, 314, 315, 342
     discovery, 6, 314, 336
     distance, 314, 316
     formation, 326
     interior, 243, 320, 322, 323, 342, 346
     magnetic field, 273, 315, 322–326, 343
     migration, 19–20
     orbit, 14, 18, 314–316
     radiation belts, 326
     radio signals, 315, 322, 325
     rings, 315, 317, 323, 324, 326, 329–332, 350, 351

rotation, 323
satellites, 21, 326–335
seasons, 318, 319, 323
shape, 314, 329
size, 7, 314, 315
storms, 318–320, 322
temperature, 12, 320, 322, 339, 342
Trojans, 403
Voyager 2 flyby, 315, 316, 318, 319, 322, 324–326, 329–332
winds, 318, 319, 324, 340
Uruk Sulcus (Ganymede), 259, 260
US Air Force, 385, 386
US Congress, 386, 407, 408
Utah, 393
Utopia Planitia (Mars), 193, 213–216, 230

Valhalla (Callisto), 261–263
Valles Marineris (Mars), 178–181, 184, 186–187, 192, 194–196, 199, 202, 204
Van Allen belts *see* Earth, radiation belts
Van de Graaff crater (Moon), 101
VB 10b, 450
VEGA spacecraft, 157, 168, 419
Vendimia Planitia (Ceres), 395
Venera program, 151, 152, 159, 160, 162, 167–168
Venus, 1, 3–5, 7, 8, 23, 55, 122, 127, 128, 130, 146–173, 175, 266, 305, 329, 367, 408, 439, 448, 451
arachnoids, 165, 171
atmosphere, 8, 12, 148–155, 165, 168, 169, 171
auroras, 169
axial inclination, 10, 11, 147, 149
black drop effect, 148
brightness, 146
channels, 162–163
chasmata, 165, 168
clouds, 147, 149–155, 157
composition, 146, 157, 163, 167, 168
coronae, 158, 159, 163–165, 168, 171, 329
density, 146, 149, 168
distance, 8, 146
dunes, 159
early observations, 147
formation, 147
greenhouse effect, 149–150
impact craters, 157–159, 161, 166, 169, 171
interior, 155, 157
ionosphere, 152, 169, 171
lightning, 157, 167
magnetic field, 130, 169, 171
mountains, 154, 156, 157, 165, 166, 168, 169
novae, 165
ocean, 150, 152
orbit, 14, 23, 121, 146–148
phases, 5, 122, 146–147
rotation, 10, 11, 147–149, 151, 152, 367
shape, 146

size, 7, 128, 146, 147
tectonics, 155, 159, 165–166, 168–169
temperature, 12, 148–152, 154, 155, 157, 162–164, 167, 168
tesserae, 155, 166
topography, 155, 156, 168
transits, 4, 147, 148, 448
volcanic activity, 151, 155, 157–159, 162–169
winds, 147, 151–154, 156, 157, 159
Venus Climate Orbiter (Akatsuki), 150, 151, 154, 156
Venus Express, 153–156, 163–165
Veritas asteroids, 390
Verona Rupes (Miranda), 328, 329
Very Large Telescope (VLT), 390, 395, 440
Vesta, 6, 385, 387, 389, 400–404, 412
Vesuvius, 70, 79
Victoria crater (Mars), 180, 229
Viking program, 180, 182, 201, 202, 209, 211–214, 216, 220–224, 226
Vladivostok, 409
volcanoes, 77, 79–81, 102–106, 108, 151, 155, 157–159, 162–169, 177–186, 190, 194, 195, 201–203, 208, 211–214, 217, 219, 220, 222, 228, 229, 246–252, 264, 301–303
Voltaire crater (Deimos), 225
Voyager program, 21, 22, 234, 235, 238, 239, 248, 249, 257–259, 262, 267, 271, 273, 315, 316, 318, 319, 322, 324–326, 329–332, 473, 474
2012 VP113, 362
V 391 Pegasi system, 457, 460
Vredefort impact structure (Earth), 87

Ward, William, 21
WASP-12b, 459
WASP-17b, 460
WASP-18b, 460
Wegener, Alfred, 75
West Virginia, 472
Weywot, 360, 361
Whipple, Fred, 418–419
white dwarf, 24, 55, 471
Wide Area Search for Planets (WASP), 459, 460
Wide Field Infrared Survey Explorer *see* WISE spacecraft
Wildt, Rupert, 243, 277
WISE spacecraft, 386, 410
Witt, Gustav, 407
Wolf, Max, 385
Wolszcan, Aleksander, 438, 439
Woomera, 392
Wunda crater (Umbriel), 327
1998 WW31, 363, 364

Xanadu (Titan), 299, 302, 303
2004 XR190, 362
X-rays, 27, 30, 32, 37, 38, 44, 46, 47, 49–52, 129, 130, 143, 168, 169, 196, 206, 213, 391, 405, 410, 418

Yale University,   277
yardangs,   83, 207, 209
Yarkovsky, Ivan,   396
Yarkovsky effect,   396, 407
Yellowknife Bay (Mars),   207
Yellowstone National Park,   250
Yogi rock (Mars),   196
Yohkoh spacecraft,   44, 52–53
YORP effect,   394, 396–397, 428

Young, John,   65, 112
Yucatan (Earth),   88, 407, 408

Zach, Franz Xavier von,   384
Zeeman effect,   39
zodiac,   1, 3
zodiacal cloud,   16, 427
zodiacal light,   427

www.ingramcontent.com/pod-product-compliance
Lightning Source LLC
Chambersburg PA
CBHW080552270125
20834CB00020B/246